B. G. Teubners Sammlung von Lehrbüchern dem Gebiete der Mathematischen Wissenschaf mit Einschluß ihrer Anwendungen.

gr. 8. Geb.

Diese Sammlung bietet in einzelnen in sich abgeschlossenen We zusammenfassende Darstellungen der wichtigsten Abschnitte der ma matischen Wissenschaften und deren Anwendungen. Im einzelnen wollen diese Werke in ihrer ausführlichen, neben der rein wissenschaftlichen auc pädagogische Momente berücksichtigenden Darstellung die Möglichkeit z selbständigem und von umfangreichen Quellenstudien unabhängigem dringen in die verschiedenen Disziplinen geben; in ihrer Gesamtheit sollen sie durch ihre eingehenden literarischen und historischen Nach ein genaues Bild von der modernen Entwicklung und von dem geg wärtigen Stande der mathematischen Wissenschaften und ihrer Anwendungen darbieten.

Bisher erschienen folgende, mit Figuren, Tafeln usw. versehene, in Leinwand geb. Bände:

P. Bachmann, niedere Zahlentheorie. In 2 Bänden. [Bd. X, 1 u. 2.] I. Band. X, 402 S. 1902. n. *M* 14.—. II. Band. X, 480 S. 1910. n. *M* 17.—

E. Blaschke, Vorlesungen über mathematische Statistik. Die Lehre von den statistischen Maßzahlen. VIII, 268 S. 1906. n. *M* 7.40. [Bd. XXIII.]

H. Bruns, Wahrscheinlichkeitsrechnung und Kollektivmaßlehre. VIII, 310 S. und Anhang 18 S. 1906. n. *M* 8.40. [Bd. XVII.]

G. H. Bryan, Thermodynamics. An introductory Treatise dealing mainly with first Principles and their direct Applications. XIV, 204 S. 1907. n. *M* 7.—. (Englisch.) [Bd. XXI.]

E. Czuber, Wahrscheinlichkeitsrechnung und ihre Anwendung auf Fehlerausgleichung, Statistik und Lebensversicherung. 2. Aufl. in 2 Bänden. [Bd. IX, 1 u. 2.]
I. Band: Wahrscheinlichkeitstheorie. Fehlerausgleichung. Kollektivmaßlehre. X, 410 S. 1908. n. *M* 12.—
II. — Mathematische Statistik. Mathematische Grundlagen der Lebensversicherung. X, 470 S. 1910. n. *M* 14.—

L. E. Dickson, linear Groups with an Exposition of the Galois Field theory. X, 312 S. 1901. n. *M* 12.—. (Englisch.) [Bd. VI.]

F. Dingeldey, Sammlung von Aufgaben zur Anwendung der Differential- und Integralrechnung. In 2 Teilen. I. Teil. IV, 202 S. 1910. n. *M* 6.—. [Bd. XXXII, 1.]

O. Fischer, theoretische Grundlagen für eine Mechanik der lebenden Körper mit speziellen Anwendungen auf den Menschen sowie auf einige Bewegungsvorgänge an Maschinen. In möglichst elementarer und anschaulicher Weise dargestellt. X, 372 S. 1906. n. *M* 14.—. [Bd. XXII.]

A. Gleichen, Lehrb. d. geometrisch. Optik. XIV, 511 S. 1902. n. *M* 20.—. [Bd. VIII.]

L. Henneberg, graphische Statik der starren Systeme. XV, 782 S. 1911. n. *M* 24.—. [Bd. XXXI.]

A. Krazer, Lehrbuch der Thetafunktionen. XXIV, 509 S. 1903. n. *M* 24.—. [Bd. XII.]

H. Lamb, Lehrbuch der Hydrodynamik. Deutsch von Joh. Friedel. XVI, 787 S. 1907. n. *M* 20.—. [Bd. XXVI.]

R. von Lilienthal, Vorlesungen über Differentialgeometrie. In 2 Bänden. I. Band: Kurventheorie. VI, 368 S. 1908. n. *M* 12.—. [Bd. XXVIII, 1.]

H. A. Lorentz, on the Theory of Electrons and its Applications to the Phenomena of Light and Radiant Heat. IV, 332 S. 1909. n. *M* 9.—. (Englisch.) [Bd. XXIX.]

B. G. TEUBNERS SAMMLUNG VON LEHRBÜCHERN
AUF DEM GEBIETE DER
MATHEMATISCHEN WISSENSCHAFTEN
MIT EINSCHLUSS IHRER ANWENDUNGEN
BAND XI

THE DYNAMICS

OF PARTICLES

AND OF RIGID, ELASTIC, AND FLUID BODIES

BEING LECTURES ON MATHEMATICAL PHYSICS

BY

ARTHUR GORDON WEBSTER, A.B. (HARV.), PH. D. (BEROL.)
PROFESSOR OF PHYSICS, DIRECTOR OF THE PHYSICAL LABORATORY,
CLARK UNIVERSITY, WORCESTER, MASSACHUSETTS

SECOND EDITION

LEIPZIG: B. G. TEUBNER.
POSTSTRASSE 3.
1912

NEW YORK
G. E. STECHERT & Co.
151–155 WEST
25 th STREET

LONDON
WILLIAMS & NORGATE
14 HENRIETTA STREET
COVENT GARDEN W. C.

 PRINTED BY B. G. TEUBNER, DRESDEN.

PREFACE.

The science of Dynamics may be variously classified. In by far the greater number of universities in both Europe and America it is dealt with by professed mathematicians, and is properly considered an essential part of mathematical discipline. Nevertheless it is but an application of mathematics to the most fundamental laws of Nature, and as such is of the highest importance to the physicist. The whole of modern Physics experiences the attempt to "explain" or describe phenomena in terms of motion, with conspicuous success in the departments of Light, Electricity, and the Kinetic Theory of Gases. It is therefore evident that no one can except to materially advance our knowledge of Physics who is ignorant of the principles of Dynamics. It is nevertheless to be feared that this subject is often slighted by the physical student, partly on account of its difficulty, and partly because of the fact that the many excellent treatises on Dynamics existing in English address themselves chiefly to the mathematician, and often seem to lay more stress on examples in analysis or trigonometry than on the elucidation of physical laws. The aim of this book is to give in compact form a treatment of so much of this fundamental science of Dynamics as should be familiar to every serious student of physics (and in my opinion no less should suffice for the student of mathematics). The classical English treatises usually fill one or even two large volumes with one of the subdivisions of the subject, such as Dynamics of a Particle, Rigid Dynamics, Hydrodynamics or Elasticity. The student confronted with the five volumes of Routh, the three of Love, and the large work of Lamb is likely to be appalled at the size of the task before him. It is practically impossible for the physical student, while spending the necessary amount of time in the laboratory, to read through all these or similar works, and thus his knowledge of the whole subject generally remains fragmentary. The great work of Lord Kelvin and Tait, while treating the whole subject, is far too difficult for most students, though it must ever remain a mine of information for those sufficiently advanced.

This book has grown out of the lectures which I have given at Clark University during the last fourteen years primarily to my own students of Physics. It is obvious that it leads to no particular examinations, from which we in America are to a large extent fortunately free. The text is not interrupted by examples for the student to work, which are found in great numbers in the usual treatises, and to which I could hardly add. The attempt has been made to treat what is essential to the understanding of physical phenomena, leaving out what is chiefly of mathematical interest. Thus the subject of Kinematics is not treated as a subject by itself, but is introduced in connection with each subdivision of Dynamics as it comes up. The student is supposed to have a fair knowledge of the Calculus, but not of Differential Equations or the Higher Analysis. Many explanations are therefore necessary, some of which are given in the form of notes.

Two opposing tendencies have at various times made themselves manifest in the treatment of Dynamics, both of which have been very fruitful. Lagrange, in the advertisement to his great work, the "Mécanique Analytique", proudly says, "On ne trouvera point de Figures dans cet Ouvrage. Les méthodes que j'y expose ne demandent ni constructions, ni raisonnements géometriques ou mécaniques, mais seulement des opérations algébriques, assujetties a une marche régulière et uniforme. Ceux qui aiment l'Analyse verront avec plaisir la Mécanique en devenir une nouvelle branche, et me sauront gré d'en avoir étendu ainsi la domaine." Lagrange's boast of having made Mechanics a branch of Analysis has been amply justified by the results obtained by means of his general method for solving mechanical problems, and his pleasure would have been greatly enhanced could be have foreseen the results of extending it to wider fields in the hands of Maxwell, of Helmholtz, and of J. J. Thomson. Nevertheless in attempting to do without figures or mental images we may rob ourselves of a precious aid. Thus Maxwell, speaking of the motion of the top, says that "Poinsot has brought the subject under the power of a more searching analysis than that of the calculus, in which ideas take the place of symbols, and intelligible propositions supersede equations". There is certainly no doubt of the advantage, particularly to the physicist, of having ideas take the place of symbols. The introduction by Hamilton of the notion of vector quantities was a great step in this direction, which has assumed very great value to the physicist, and it was to a particular case of this that Maxwell alluded, namely to the idea of the moment of momentum, or impulsive couple, as it was termed by Poinsot. The importance of this physical or geometrical conception may be seen from the use made of it, under the name of the Impulse, by Klein and Sommerfeld in their very interesting work on the Top. On the other hand this notion of impulse, while in this particular case a vector, is but one case of the general notion of the momentum in Lagrange's generalized coordinates. Will it not then be an additional advantage if, keeping both the analytical and the geometrical modes of expression, we attempt to introduce into Lagrange's analytical method geometrical analogies and terminology? This it is perfectly possible to do, for it turns out, as was shown by Beltrami, and beautifully worked out in detail by Hertz, that the properties of Lagrange's equations have to do with a quadratic form, of exactly the sort that represents the arc of a curve in geometry. Analytically it is of no importance whether the number of variables is more or less than three — how natural it is accordingly to employ the terminology of geometry, which must result in giving a more definite image of the quantities involved. For this reason I hope that no physicist will accuse me of having dragged in the subject of hyperspace into a physical treatise. I have insisted that what is involved is merely a mode of speaking, and has the advantage of logical consistency with the results of geometry, which is to most of us a physical subject. At all events this matter has been so introduced that it may be completely passed over by those to whom such analogies are repugnant. The advantage of a good terminology, as well as of clear physical conceptions, must be plain to all and every physicist will acknowledge the indebtedness which our science owes in this respect to Kelvin and Tait.

he work divides itself naturally into three parts, the first of which considers aws of Motion in general and those methods which are applicable to s of all sorts. Although not addressed to students who are beginning anics, it seemed necessary to begin at the beginning, and to explain hibition of Newton's Laws of Motion in mathematical form. For this e the Principle of Hamilton is of so universal application that it has ntroduced near the beginning, and considerable attention devoted to it. consider this principle, together with the equations of Lagrange, a ractical subject, of the highest importance for the physical student. ame may be said of the subject of Energy, upon which it has even attempted to found the laws of Physics. Although such attempts seem d to fail, for the reason that the principle of Energy, though afford- integral, is insufficient to deduce the differential equations, the notion ergy must remain one of the most important in Dynamics, and is therefore considered in every problem. The subject of Oscillations, of very great physical interest, with its accompanying phenomena of Resonance, is next taken up. After this follows a treatment of the so-called Cyclic Systems, from which, since the labors of Helmholtz and Hertz, it seems that Physics has so much to expect. In fact the first steps have been taken to explain the nature of Potential Energy by means of Motion, perhaps the chief desideratum of Physics. In this connection we way again point to the epoch-making work of Lord Kelvin, both in Mechanics and in the Theory of Light.

The second part is devoted to the Motion of Rigid Bodies, particularly to their rotation, a matter of the greatest importance practically, especially to the engineer, but one which is often avoided by the physical student. To this subject Maxwell again called the attention of physicists, and created a charming instrumental demonstration in his celebrated Dynamical Top. To this the writer has ventured to add a small detail, which permits of a number of interesting additional verifications. A number of practical illustrations, of interest to the physicist and engineer, are also included.

The third part divides itself from the other two from the fact that in ifferential equations are partial, while in the others they are ordinary. preparation for this subject is introduced the theory of the Potential n, which introduces the most important mathematical theorems, and res for the subsequent chapters. Most of this chapter has already d in the author's treatise on the Theory of Electricity and Magnetism, several matters have been added, especially on applications to Geodesy. follows the subject of Stress and Strain, with applications to the simpler ms of Elasticity, including the problem of de St. Venant on the flexion rsion of prisms. Finally in Hydrodynamics the main questions of nd vortex motion are taken up, with a brief account of the pheno- of the tides and of viscous fluids. Thus the student is prepared for dy of Sound, Light, and Electricity. The only work in English of know having the same purpose is Professor Tait's admirable treatise amics. While this book has been in preparation, there has appeared st volume of Professor Gray's Treatise on Physics, the scope of which h broader, but the aim of which is not greatly different from that book. I have however attempted to provide a treatise which would in not over a year's time offer to the student an amount of knowledge of Dynamics sufficient te prepare him for the study of Mathematical Physics in general.

My obligations to previous authors are obvious, and where possible explicit mention is made. A list of works which have been of service to me is appended, but I wish particularly to acknowledge my indebtedness to Thomson and Tait, to Kirchhoff and to Appell. I am under great obligations to Dr. Margaret E. Maltby for valuable assistance in the preparation of the manuscript, and for frequent suggestions, and to Messrs. J. G. Coffin and J. C. Hubbard, Fellows of Clark University, for efficient aid in the preparation of the drawings. I take this opportunity of expressing my thanks to my colleague Professor William E. Story for his continual willingness during fourteen years to aid me by putting at my disposition his unusual knowledge in matters connected with Algebra. My thanks are due to the publisher for the fine mechanical execution of the work in the style for which the house of Teubner is noted.

As the proof has been read only by myself, it is hoped that errors will be dealt with lightly. In conclusion I venture to hope that my attempt to make Dynamics more of an experimental science by subjecting some of its conclusions to quantitative experimental verification may deserve notice.

Worcester, Mass., July 22, 1904.

A. G. WEBSTER.

PREFACE TO THE SECOND EDITION.

The unexpected success of the first edition of this work, as shown by its early exhaustion, and by the kind reception that it has met with on both sides of the Atlantic, has necessitated the preparation of a second edition. This edition is substantially identical with the first, except that a few errors kindly pointed out by friends have been corrected. Unfortunately the time has failed me to rewrite or add to certain sections, nor has it been possible to adequately treat certain new applications of the gyroscope, which have been merely indicated. Again it is hoped that errors of proof reading will be pardoned.

Worcester, Mass., Feb. 27, 1912.

A. G. WEBSTER.

CONTENTS.

PART I.

GENERAL PRINCIPLES.

CHAPTER I.

Kinematics of a Point. Laws of Motion.

CHAPTER II.

Important Particular Motions of a Material Point.

CHAPTER III.

General Principles. Work and Energy.

CHAPTER IV.

Principle of Least Action. Generalized Equations of Motion.

CHAPTER V.

Oscillations and Cyclic Motions.

PART II.

DYNAMICS OF RIGID BODIES.

CHAPTER VI.

Systems of Vectors. Distribution of Mass. Instantaneous Motion.

CHAPTER VII.

Dynamics of Rotating Bodies.

PART III.

THEORY OF THE POTENTIAL, DYNAMICS OF DEFORMABLE BODIES.

CHAPTER VIII.

Newtonian Potential Function.

CHAPTER IX.

Dynamics of Deformable Bodies.

CHAPTER X.

Statics of Deformable Bodies.

CHAPTER XI.

Hydrodynamics.

NOTES.

LIST OF WORKS CONSULTED BY THE AUTHOR.

Appel, Traité de Mécanique Rationelle.
Boltzmann, Vorlesungen über die Prinzipe der Mechanik.
Budde, Allgemeine Mechanik.
Christiansen, Elemente der Theoretischen Physik.
Clifford, Elements of Dynamic.
Despeyrous, Cours de Mécanique.
Föppl, Vorlesungen über technische Mechanik.
Gray, Treatise on Physics.
Helmholtz, Dynamik diskreter Massenpunkte.
Hertz, Die Prinzipien der Mechanik.
Jacobi, Vorlesungen über Mechanik.
Kirchhoff, Vorlesungen über mathematische Physik-Mechanik.
v. Lang, Theoretische Physik.
Klein und Sommerfeld, Theorie des Kreisels.
Love, Theoretical Mechanics.
Loudon, Rigid Dynamics.
Mach, Die Mechanik in ihrer Entwickelung.
Mathieu, Cours de Physique Mathématique.
Maxwell, Treatise on Electricity and Magnetism.
Poisson, Traité de Mécanique.
Rausenberger, Lehrbuch der analytischen Mechanik.
Résal, Traité de Physique Mathématique.
Ritter, Lehrbuch d. Ingenieurmechanik.
Routh, Dynamics of a Particle.
Routh, Dynamics of a System of Rigid Bodies.
Schell, Theorie der Bewegung und der Kräfte.
Somoff, Theoretische Mechanik.
Sturm, Cours de Mécanique.
Thomson and Tait, Treatise on Natural Philosophy.
Voigt, Elementare Mechanik.
Voigt, Kompendium der theoretischen Physik.
Williamson and Tarleton, Elementary Treatise on Dynamics.
Ziwet, Elementary Treatise on Theoretical Mechanics.
Boussinesq, Applications des Potentiels.
Betti, Teorica delle Forze Newtoniane.
Basset, Treatise on Hydrodynamics.
Bjerknes, Vorlesungen über hydrodynamische Fernkräfte.
Clebsch, St. Venant, Théorie de l'Elasticité des Corps Solides.
Greenhill, Treatise on Hydrostatics.
Helmholtz, Dynamik kontinuierlich verbreiteter Massen.
Ibbetson, Mathematical Theory of Elasticity.
Lamé, Leçons sur la Théorie mathématique de l'Élasticité des Corps solides.
Lamb, Hydrodynamics.
Lévy, Théorie des Marées.
Love, Treatise on the Mathematical Theory of Elasticity.
Mathieu, Theorie de l'Élasticité des Corps solides.
Minchin, Treatise on Statics.
Neumann, Das Logarithmische und Newtonsche Potential.
Peirce, Elements of the Theory of the Newtonian Potential Function.
Poincaré, Leçons sur la Théorie de l'Élasticité.
Poincaré, Théorie des Tourbillons.
Poincaré, Théorie du Potentiel Newtonien.
Poincaré, Figures d'Equilibre d'une Masse Fluide.
Rayleigh, Theory of Sound.
Riemann, Schwere, Elektrizität und Magnetismus.
Routh, Treatise on Analytical Statics.
Tarleton, Introduction to the Mathematical Theory of Attraction.
Todhunter and Pearson, History of the Theory of Elasticity.
Wien, Lehrbuch der Hydrodynamik.
Williamson, Mathematical Theory of Stress and Strain.

PART I

GENERAL PRINCIPLES

CHAPTER I.

KINEMATICS OF A POINT. LAWS OF MOTION.

1. Dynamics. Dynamics or Mechanics is the science of motion. It is the fundamental subject of Physics, since it is the aim of scientists to reduce the characterization of all physical phenomena to description of states of motion. The problem of dynamics according to Kirchhoff[1]), is to describe all motions occurring in nature in an unambiguous and the simplest manner. In addition it is our object to classify them and to arrange them on the basis of the simplest possible laws. The success which has attended the efforts of physicists, mathematicians, and astronomers in achieving this object, from the time of Galileo and Newton through that of Lagrange and Laplace to that of Helmholtz and Kelvin, constitutes one of the greatest triumphs of the human intellect.

2. Kinematics. That which moves is *matter*. The properties of matter may be left for later consideration. We may, however, describe motions without considering the nature of that which is moved, — this forms a special branch of our subject known as *Kinematics*.

Kinematics is merely an extension of geometry and may be called geometry of motion, for while in geometry we consider the properties of space, in Kinematics we consider also the idea of time, giving us another variable. Since the position of a point in space is known when its three rectangular Cartesian coordinates with respect to a definite system of axes are given, its motion is completely described if its coordinates are given for all instants of time, or are known functions of the time. Analytically

$$1)\qquad x = f_1(t), \quad y = f_2(t), \quad z = f_3(t).$$

The functions f_1, f_2, f_3 must be continuous, since in no actual motion does a point considered disappear in one position to reappear after

1) Kirchhoff, Vorlesungen über mathematische Physik. Mechanik, p 1.

1*

a vanishingly small interval of time in a new position at a finite distance from the old. The functions are also supposed to have definite derivatives for every value of t.

Since the motion of a point involves four variables, Kinematics was called by Lagrange Geometry of four dimensions. We shall not here discuss the nature of time, nor the mode of measuring it, reserving the latter until we have considered motions that actually occur in nature, upon which all methods for measuring time are based.

We may accept the fact that the idea of time, like that of space, is the intuitive possession of us all. Its exact definition must depend on the science of dynamics.

3. Scalars and Vectors. In mathematics we have to consider two sorts of quantities, those which do not involve the idea of direction, called by Hamilton *scalars* (because they may be specified by numbers marked off on a scale), and those which do, called *steps* or *vectors.* The distance between two points $x_1, y_1, z_1, x_2, y_2, z_2$

2) $$s = \sqrt{(x_2 - x_1)^2 + (y_2 - y_1)^2 + (z_2 - z_1)^2}$$

is a scalar, whereas the geometrical difference in position of the two points is known only when we specify not merely the length, but also the direction of the line joining them. This is usually done by giving its length s and the cosines of the angles made by the line with the three rectangular axes,

$$\cos\lambda, \quad \cos\mu, \quad \cos\nu,$$

which in virtue of the identical relation

3) $$\cos^2\lambda + \cos^2\mu + \cos^2\nu = 1,$$

leaves three independent data. We may otherwise make the specification by giving the three projections of the line upon the coordinate axes,

4) $$\begin{aligned} s_x &= s\cos\lambda = x_2 - x_1, \\ s_y &= s\cos\mu = y_2 - y_1, \\ s_z &= s\cos\nu = z_2 - z_1. \end{aligned}$$

Squaring and adding we have in virtue of relation 3)

5) $$s_x^2 + s_y^2 + s_z^2 = s^2.$$

By the vector AB we mean the line in the direction *from* A to B, and its projections have the sign of the coordinates of B minus those of A, the vector being defined as that which *carries* us from A to B. We may write symbolically

$$pt \cdot A + \overline{AB} = pt\ B$$

$$\overline{AB} = pt \cdot B - pt \cdot A.$$

$\overline{AB}$ is to be understood, *vector* AB. Similarly when we wish to specify that s is to be regarded as a vector (i. e. its *direction* is to be considered as well as length), we shall write $\bar{s}$.

We have from 4) and 5)

$$\cos\lambda = \frac{s_x}{s} = \frac{s_x}{\sqrt{s_x^2 + s_y^2 + s_z^2}},$$

6)
$$\cos\mu = \frac{s_y}{s} = \frac{s_y}{\sqrt{s_x^2 + s_y^2 + s_z^2}},$$

$$\cos\nu = \frac{s_z}{s} = \frac{s_z}{\sqrt{s_x^2 + s_y^2 + s_z^2}}.$$

Also multiplying the equations 4) respectively by $\cos\lambda$, $\cos\mu$, $\cos\nu$, and adding,

7)
$$s_x \cos\lambda + s_y \cos\mu + s_z \cos\nu = s.$$

Whatever quantities are needed to completely specify a quantity are called its coordinates. A point has three, and we have seen that a vector also has three, which may be taken as s_x, s_y, s_z. In this sense all vectors are to be considered as equal whose lengths are equal and directions parallel irrespective of the absolute positions of their ends. It is, however, sometimes necessary to distinguish vectors equal in this sense, but whose ends do not respectively coincide. To determine such a vector we must know not only its length and direction, but also the position of one end. It will therefore be specified by six coordinates, which may be the three coordinates of one end, x_1, y_1, z_1, with the projections, s_x, s_y, s_z, of the coordinates of both ends, $x_1, y_1, z_1, x_2, y_2, z_2$. In any case there will be six coordinates. Such a vector may be called a fixed vector to distinguish it from the ordinary or free vector.

4. Addition of Vectors. To add two vectors means to take successively the steps denoted by them, their sum being a single step equivalent thereto. For example, (Fig. 1)

$$\overline{AB} + \overline{BC} = \overline{AC}.$$

The vectors $\overline{AB}$ and $\overline{BC}$ are called the *components* of $\overline{AC}$, which is called their *resultant*, or geometrical sum.

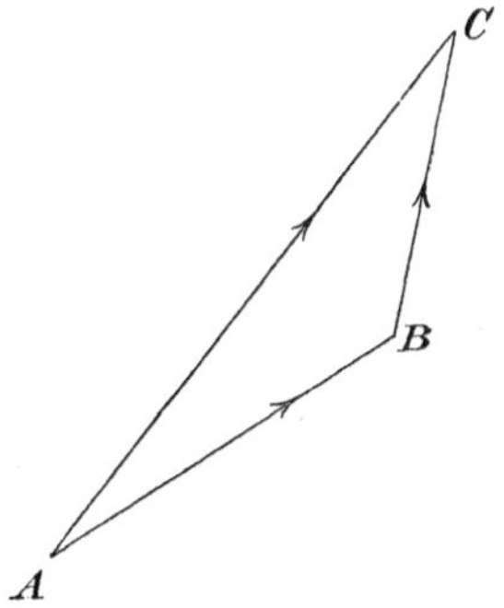

We may state the rule: Place the initial point of the second vector at the terminal point of the first, the resultant or geometrical sum is the vector from the initial point of the first component to the terminal point of the second. This construction gives us the so-called *triangle* of vectors. By continuing the process any number of vectors may be added, giving us the *polygon* of vectors.

C
B
A
Fig. 1.

The nature of the construction shows that the resultant is independent of the order of taking the components.

Since a negative quantity is defined as that which added to a given positive quantity produces zero, the negative of $\overline{AB}$ must be $\overline{BA}$, for by the above rule,

$$A + \overline{AB} = B,$$

$$B + \overline{BA} = A,$$

therefore

$$A + \overline{AB} + \overline{BA} = A,$$

$$\overline{AB} + \overline{BA} = 0,$$

$$\overline{BA} = -\overline{AB}.$$

The coordinates of $\overline{BA}$ are also the negatives of $\overline{AB}$. The scalar length of a vector is called by Hamilton its *tensor*, so that the tensor of the negative of a vector is the same as that of the vector itself.

It is evident from the definition of a vector that the projection of the sum of two vectors on any direction is the algebraic sum of the projections of the components. Projecting on the three directions of the coordinate-axes, and distinguishing the projections of the components by suffixes, we have for the projections of the resultant,

$$s_x = s_{1x} + s_{2x},$$

$$s_y = s_{1y} + s_{2y},$$

$$s_z = s_{1z} + s_{2z},$$

$$s^2 = (s_{1x} + s_{2x})^2 + (s_{1y} + s_{2y})^2 + (s_{1z} + s_{2z})^2,$$

and for the sum of any number of vectors,

8) $$s^2 = (\Sigma s_x)^2 + (\Sigma s_y)^2 + (\Sigma s_z)^2.$$

We may easily find an expression for the projection of any vector $\bar{s}$ upon any direction, which is given by its direction cosines, $\cos\lambda$, $\cos\mu$, $\cos\nu$. We have for the angle ϑ between two lines whose direction cosines are $\cos\lambda$, $\cos\mu$. $\cos\nu$, $\cos\lambda'$, $\cos\mu'$, $\cos\nu'$,

$$\cos\vartheta = \cos\lambda\cos\lambda' + \cos\mu\cos\mu' + \cos\nu\cos\nu',$$

but by 6), we have for $\bar{s}$,

$$\cos\lambda' = \frac{s_x}{s}, \quad \cos\mu' = \frac{s_y}{s}, \quad \cos\nu' = \frac{s_z}{s},$$

so that

9) $$s\cos\vartheta = s_x\cos\lambda + s_y\cos\mu + s_z\cos\nu,$$

which is the expression for the projection. Taking for the direction of projection the direction of the vector itself, this becomes equation 7).

If $\cos\lambda$, $\cos\mu$, $\cos\nu$ are the direction cosines of a second vector $\overline{s_2}$

$$\cos\lambda = \frac{s_{2x}}{s_2}, \quad \cos\mu = \frac{s_{2y}}{s_2}, \quad \cos\nu = \frac{s_{2z}}{s_2};$$

multiplying by s_2 we have the expression symmetrical with respect to both vectors

10) $$s_1 s_2 \cos\vartheta = s_{1x} s_{2x} + s_{1y} s_{2y} + s_{1z} s_{2z}.$$

This expression, which may be defined either as the product of the tensors of the vectors and the cosine of their included angle, or as the tensor of either multiplied by the value of the projection on its direction of the other, is so important that it has received a special name, and will be called the *geometric product* of the two vectors. It is not a vector, but is essentially a scalar quantity, and its negative was called by Hamilton the scalar product of the vectors.

The condition of perpendicularity of two vectors is that their geometric product vanishes.

11) $$s_{1x} s_{2x} + s_{1y} s_{2y} + s_{1z} s_{2z} = 0.$$

5. Moments. Consider a *fixed* vector $\overline{AB}$, Fig. 2. — The product of the length AB and the perpendicular distance of O from AB is called the *moment* of $\overline{AB}$ about O. It is arithmetically equal to twice the area of the triangle OAB. The sign of the moment will change with the direction of $\overline{AB}$. If we draw a line through O whose length is equal to the magnitude of the moment and whose direction is perpendicular to the plane OAB, this line is called the *axis* of the moment, and in a certain way represents the latter. We shall draw it in such a direction that a person standing on O with his back against the axis would see motion from A to B as from right to left.

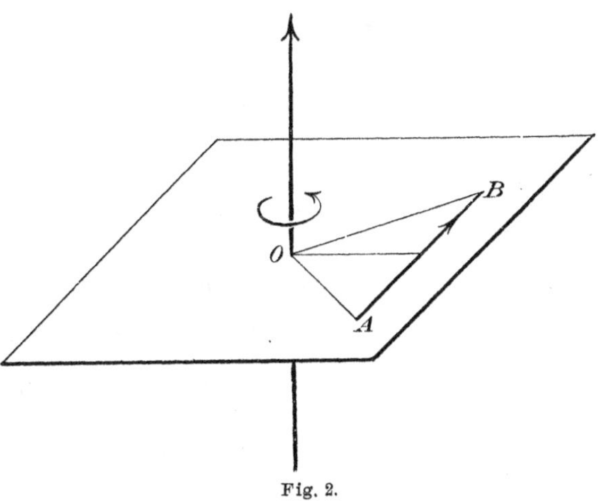

Fig. 2.

The coordinates of the axis may be found from those of the vector $\overline{AB}$ and of O. If we choose O for origin, OAB for the plane of XY, (Fig. 3) and let the coordinates of A be x, y, the projections of $\overline{AB}$, s_x, s_y, we have for the area of the triangle OAB

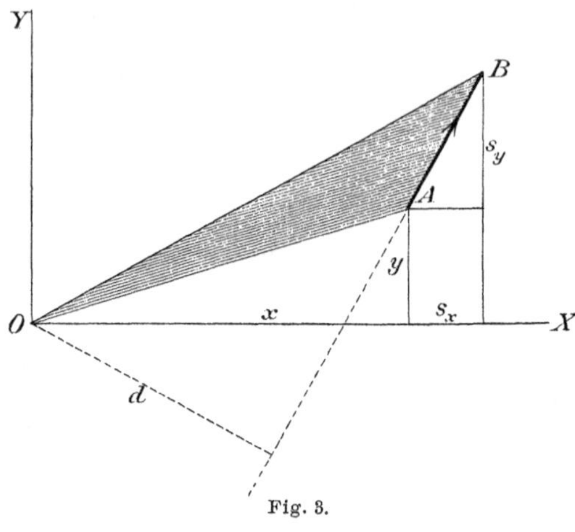

Fig. 3.

$$\frac{1}{2}\{(x+s_x)(y+s_y)-xy-s_x s_y\} - y s_x = \frac{1}{2}(x s_y - y s_x).$$

Accordingly we have for m, the moment about O of a vector whose projections are s_x, s_y and whose initial point has the coordinates x, y,

$$m = x s_y - y s_x.$$

To find the moment of the resultant of two vectors drawn from the same initial point, whose plane contains O, their projections being s_x', s_y', s_z', s_x'', s_y'', s_z'', we have

$$m = x(s_y' + s_y'') - y(s_x' + s_x'')$$
$$= x s_y' - y s_x' + x s_y'' - y s_x'' = m' + m'',$$

thus the moment of the resultant is equal to the sum of the moments. If the plane of OAB is not one of the coordinate-planes, we may project the triangle OAB upon the three coordinate-planes, and obtain three moments m_x, m_y, m_z. If the direction cosines of the axis of m are $\cos\alpha$, $\cos\beta$, $\cos\gamma$, we have by the rule for the projection of areas,

$$m_x = m\cos\alpha, \quad m_y = m\cos\beta, \quad m_z = \cos\gamma,$$
$$m^2 = m_x^2 + m_y^2 + m_z^2.$$

Therefore the moment m has three coordinates, m_x, m_y, m_z, and may itself be considered a vector $\overline{m}$. Since the coordinates of the projections of A and $\overline{AB}$ on the YZ plane are y, z, s_y, s_z, we have by the preceding formula

$$12) \qquad m_x = y s_z - z s_y, \quad m_y = z s_x - x s_z, \quad m_z = x s_y - y s_x.$$

In the language of Hamilton m is the *vector product* of the vector $\overline{OA}$ into the vector $\overline{AB}$. We have evidently

$$xm_x + ym_y + zm_z = 0$$

$$s_x m_x + s_y m_y + s_z m_z = 0,$$

that is, the vector product of the two vectors is perpendicular to their plane. From the definition of moment, or by reference to Fig. 3, its magnitude or tensor is equal to the product of their tensors times the sine of the angle included between them. (It is to be noted that the projections of the first factor in the vector product follow each other in cyclic order in equations 12), those of the second factor in reverse order.) It is at once evident that the moment of the resultant of two vectors with the same initial point is the resultant of their individual moments. Thus moments are to be considered in all respects like vectors. It is evident that the moment of a vector, s_x, s_y, s_z, with initial point x, y, z, about a point ξ, η, ζ, has the projections:

13)
$$\begin{aligned} m_x &= (y-\eta)\,s_z - (z-\zeta)\,s_y \\ m_y &= (z-\zeta)\,s_x - (x-\xi)\,s_z \\ m_z &= (x-\xi)\,s_y - (y-\eta)\,s_x. \end{aligned}$$

6. Velocity. As a second means of description of the motion of a point we may give the geometrical locus of the positions that it occupies at different instants. This is called the *path* of the point, and if it is straight, the motion is said to be rectilinear. This alone does not suffice to describe the motion, for the same path may be described with different speeds. We must therefore give something which shall determine what positions are reached at various instants. If we call s the distance the point has traversed in its path, counting from a fixed point, and give the value of s for every value of t $s = \varphi(t)$, this together with the equations of the path, which may be

14)
$$F_1(x, y, z) = 0, \quad F_2(x, y, z) = 0,$$

completely specifies the motion, making as before three equations. The *velocity* of the point is defined as the limit of the ratio of the length of the path Δs described in an interval of time Δt to the time Δt when both decrease without limit, that is,

15)
$$v = \lim_{\Delta t = 0} \frac{\Delta s}{\Delta t} = \frac{ds}{dt}.$$

Velocities of the same numerical magnitude may however have different directions, accordingly to completely specify a velocity we must give not only its magnitude, but also its direction. It is therefore a vector quantity. Its direction is that of the tangent to the path at the point in question, and its direction cosines are

$$\frac{dx}{ds}, \quad \frac{dy}{ds}, \quad \frac{dz}{ds}.$$

A velocity, like any other vector, may be resolved into components,

$$
\begin{aligned}
v_x &= v\frac{dx}{ds} = \frac{ds}{dt}\frac{dx}{ds} = \frac{dx}{dt},\\
v_y &= v\frac{dy}{ds} = \frac{ds}{dt}\frac{dy}{ds} = \frac{dy}{dt},\\
v_z &= v\frac{dz}{ds} = \frac{ds}{dt}\frac{dz}{dt} = \frac{dz}{dt},
\end{aligned}
\tag{16}
$$

or the projection of the velocity on any direction is the velocity of the projection of the point on that direction.[1] We have therefore

$$v^2 = \left(\frac{ds}{dt}\right)^2 = \left(\frac{dx}{dt}\right)^2 + \left(\frac{dy}{dt}\right)^2 + \left(\frac{dz}{dt}\right)^2. \tag{17}$$

A third method of description of a motion would be to give as before the equations of the path and to give the velocity as a function of the time,

$$v = \frac{ds}{dt} = \psi(t).$$

An integration of this differential equation would give us

$$s = \text{const} + \int \psi(t)\,dt = \varphi(t),$$

and we should have the same form as before.

Fourthly we might have

$$v_x = \frac{dx}{dt} = F_1(t), \quad v_y = \frac{dy}{dt} = F_2(t), \quad v_z = \frac{dz}{dt} = F_3(t) \tag{18}$$

together with the initial conditions

$$x = x_0, \quad y = y_0, \quad z = z_0, \text{ when } t = t_0.$$

An integration of these three simultaneous equations would give us a description equivalent to 1).

In equations 1), if t is any parameter, not necessarily the time, we have what is called the parametric representation of a curve. By the elimination of t, we may obtain two coordinates as functions of the third. If, on the contrary, we have only the path given, whereas the geometry of the motion is known, kinematically the description is incomplete, as the specification of the time is lacking. To remedy this defect of the geometrical representation, Hamilton introduced the *Hodograph*, which is a curve, the locus of a point related to the moving point on the path by having its position

---

1) It is to be noticed that in stating that velocity is a vector we assume the mode of composition of velocities as a matter of definition.

vector with respect to a point taken as origin equal to the vector *velocity* of the moving point. Thus the radius vector of any point on the hodograph is parallel to the tangent at the corresponding point of the path. If X, Y, Z are the coordinates of a point on the hodograph, we have for the relation between the two curves,

19) $$X = \frac{dx}{dt}, \quad Y = \frac{dy}{dt}, \quad Z = \frac{dz}{dt},$$

so that having established the correspondence of point to point, we obtain the time from

20) $$t = \int \frac{ds}{\sqrt{X^2 + Y^2 + Z^2}}.$$

We shall call any vector which is related to another vector as the vector X, Y, Z is to the vector x, y, z, the *velocity* of the vector, and by a natural extension, shall call the locus of the end of the second vector drawn from a fixed origin the hodograph of the first vector. Thus we call Hamilton's hodograph the hodograph of the position of the first point.

7. Polar Coordinates. If a point moves in a plane it may be convenient to specify its position by means of polar coordinates Let r be the distance of the point from the origin O, φ the angle that the radius vector makes with a fixed line through the origin. If now the point moves from A to B (Fig. 4) in the time Δt, describing the space Δs, so that r turns through the angle $\Delta\varphi$, at the same time increasing by Δr, we may resolve the velocity into two components, one proportional to AC, where AC is perpendicular to OB, the other proportional to CB. We have then the following vector equation

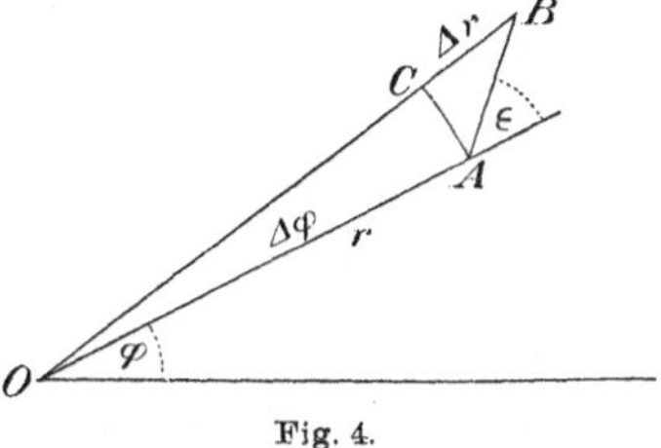

Fig. 4.

$$\bar{v} = \lim_{\Delta t = 0} \left(\frac{\overline{AC}}{\Delta t} + \frac{\overline{CB}}{\Delta t} \right),$$

or, passing to the limit,

21) $$\bar{v} = \overline{r\frac{d\varphi}{dt}} + \overline{\frac{dr}{dt}}.$$

The two components of v may be called the *radial* velocity, $v_r = \frac{dr}{dt}$, and the *transverse* velocity, $v_\varphi = r\frac{d\varphi}{dt}$. The rate of increase of the angle φ is called the *angular* velocity $\frac{d\varphi}{dt}$. The vector equation 21) gives rise to the scalar equation

that is,
$$v^2 = v_r^2 + v_\varphi^2,$$

22)
$$\left(\frac{ds}{dt}\right)^2 = r^2\left(\frac{d\varphi}{dt}\right)^2 + \left(\frac{dr}{dt}\right)^2,$$

which might have been obtained from the expression for the length of the arc in polar coordinates.

8. Sector Velocity. Let the polar coordinates of a point at the time t be r, φ, and let the area of the sector enclosed between the path, the fixed line of reference, and the radius vector be denoted by S. If ε denote the angle made by the tangent to the path, in the direction of motion, with the direction of the radius vector *from* the origin, we have (Fig. 5)

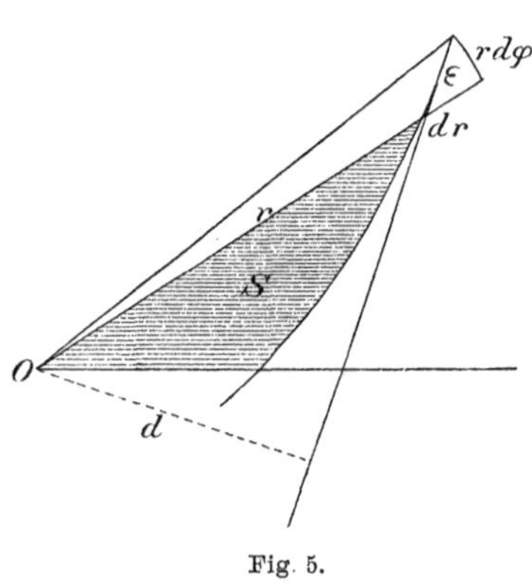

Fig. 5.

$$dr = ds\cos\varepsilon,$$
$$rd\varphi = ds\sin\varepsilon,$$

and if in the time dt the area of the sector increases by dS, we have

$$dS = \frac{1}{2}rds\sin\varepsilon = \frac{1}{2}r^2 d\varphi.$$

The rate of increase of the area of the sector

23)
$$\frac{dS}{dt} = \frac{1}{2}r^2\frac{d\varphi}{dt},$$

may be called the sector velocity, and making use of the value $\frac{1}{2}\frac{ds}{dt}r\sin\varepsilon$, we see that it is equal to one-half the product of the magnitude of the velocity $\frac{ds}{dt}$ and the perpendicular distance, $d = r\sin\varepsilon$ from the origin to the line of direction of the velocity, that is, to one-half the *moment* of the velocity. Therefore the sector velocity may be represented by a vector perpendicular to the plane OAB, the components of which will be

24)
$$\frac{dS_x}{dt} = \frac{1}{2}(yv_z - zv_y),$$
$$\frac{dS_y}{dt} = \frac{1}{2}(zv_x - xv_z),$$
$$\frac{dS_z}{dt} = \frac{1}{2}(xv_y - yv_x),$$

and we also have

25)
$$\left(\frac{dS}{dt}\right)^2 = \left(\frac{dS_x}{dt}\right)^2 + \left(\frac{dS_y}{dt}\right)^2 + \left(\frac{dS_z}{dt}\right)^2.$$

9. Acceleration. If the velocity of a point is variable with the time we define the acceleration of the point as the limit of the ratio of the increment of velocity Δv to the increment of time Δt, as both approach zero. We may consider either the numerical change

$$\lim_{\Delta t=0} \frac{\Delta v}{\Delta t} = \frac{dv}{dt} = \frac{d^2 s}{dt^2},$$

or the geometrical change. If we draw a vector $\overline{AB}$ (Fig. 6) to represent the velocity at the time t and the vector $\overline{AC}$ to represent the velocity at the time $t + \Delta t$, and draw the arc of a circle BD, DC will represent the numerical change of velocity, Δv, not considering its direction, while $\overline{BC}$ represents its geometrical, or vector change, $\Delta \bar{v}$, for

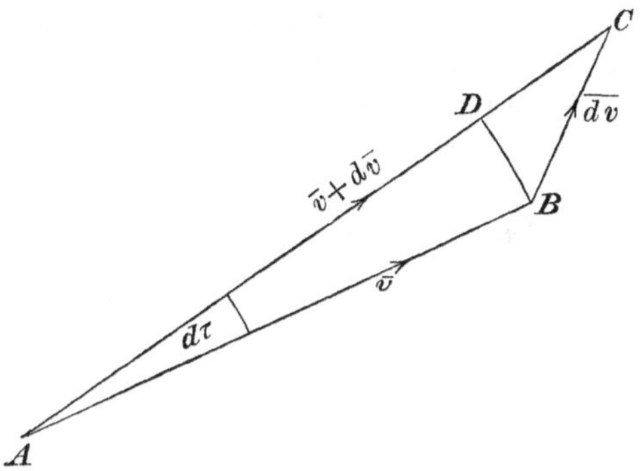

Fig. 6.

$$\overline{AB} + \overline{BC} = \overline{AC}$$

$$\overline{BC} = \overline{AC} - \overline{AB} = \Delta \bar{v}.$$

Accordingly $\lim_{\Delta t=0} \frac{\Delta v}{\Delta t} = \lim_{\Delta t=0} \frac{\overline{BC}}{\Delta t}$ is the vector acceleration a.

Since the projections of the geometrical difference of two vectors are the differences of the projections, the components of a in any direction will be proportional to the changes of the corresponding components of the velocities, that is

$$
\begin{aligned}
a_x &= \frac{dv_x}{dt} = \frac{d^2 x}{dt^2}, \\
a_y &= \frac{dv_y}{dt} = \frac{d^2 y}{dt^2}, \qquad (26)\\
a_z &= \frac{dv_z}{dt} = \frac{d^2 z}{dt}.
\end{aligned}
$$

In the language of § 6, the acceleration is the velocity of the velocity-vector.

The vector acceleration $\bar{a}$ being the resultant of the components a_x, a_y, a_z, has the numerical value or tensor

$$a = \sqrt{\left(\frac{d^2 x}{dt^2}\right)^2 + \left(\frac{d^2 y}{dt^2}\right)^2 + \left(\frac{d^2 z}{dt^2}\right)^2}.$$

This is not in general equal to $\frac{d^2 s}{dt^2}$, which is the acceleration of the scalar velocity. The direction of $\bar{a}$ is given by its direction cosines,

$$27)\qquad \cos(ax) = \frac{d^2x}{dt^2}\Big/a, \quad \cos(ay) = \frac{d^2y}{dt^2}\Big/a, \quad \cos(az) = \frac{d^2z}{dt^2}\Big/a.$$

10. Acceleration Components. We may now find the component of the acceleration in the direction of the tangent to the path. The direction cosines of the tangent being, by § 6,

$$\frac{v_x}{v}, \quad \frac{v_y}{v}, \quad \frac{v_z}{v},$$

we have for the tangential component by 9)

$$28)\qquad \begin{aligned} a_\tau &= a_x\frac{v_x}{v} + a_y\frac{v_y}{v} + a_z\frac{v_z}{v} \\ &= \frac{1}{v}\left\{\frac{dx}{dt}\frac{d^2x}{dt^2} + \frac{dy}{dt}\frac{d^2y}{dt} + \frac{dz}{dt}\frac{d^2z}{dt^2}\right\}. \end{aligned}$$

But differentiating equation 17)

$$v\frac{dv}{dt} = \frac{ds}{dt}\frac{d^2s}{dt^2} = \frac{dx}{dt}\frac{d^2x}{dt^2} + \frac{dy}{dt}\frac{d^2y}{dt^2} + \frac{dz}{dt}\frac{d^2z}{dt^2}$$

and dividing by x and comparing with 28) we find

$$a_\tau = \frac{d^2s}{dt^2},$$

that is the acceleration of the *scalar* velocity is the projection of the vector acceleration on the tangent. This is called the *tangential acceleration*, or acceleration in the path.

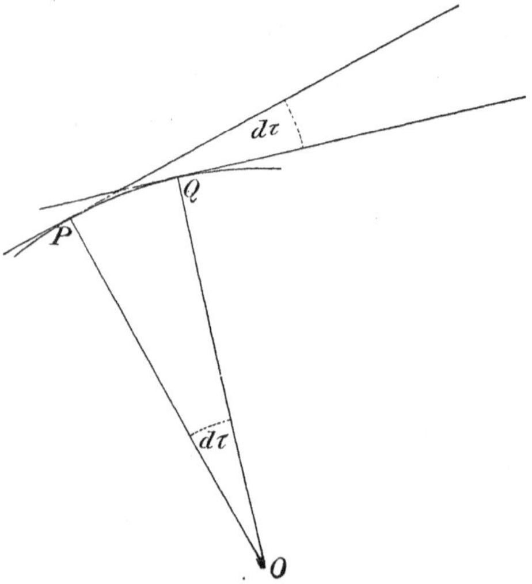

Fig. 7.

We may obtain a convenient expression for the remaining component of the acceleration. If P and Q (Fig. 7) be two "consecutive" points of the path, the plane containing the tangents at P and Q is called the osculating plane, or plane of principal curvature Normals drawn in this plane are called principal normals, and the point O where they intersect, the center of curvature. The radius $OP = \varrho$ is called the radius of curvature. If the angle between the consecutive tangents is $\Delta\tau$ and the distance between the points P, Q is Δs, the curvature is defined as

$$\lim_{\Delta s = 0} \frac{\Delta \tau}{\Delta s} = \frac{d\tau}{ds} = \varkappa.$$

Since the angle between the tangents is equal to that between the normals

$$ds = \varrho\, d\tau$$

29) $$\varkappa = \frac{d\tau}{\varrho\, d\tau} = \frac{1}{\varrho}.$$

If as before we draw lines AB, AC (Fig. 6) representing the velocities at P and Q, the acceleration, $\lim \frac{\overline{BC}}{\Delta t}$, is in the plane of AB and AC, that is the vector acceleration is in the osculating plane. As we have already found the component parallel to the tangent, there remains only the component parallel to the principal normal. Since BC is proportional to the acceleration, DC is proportional to the tangential acceleration a_τ, BD to the normal acceleration a_ν, and since the angle at A is $d\tau$ and the side AB is v.

$$BD = v\, d\tau$$

$$a_\nu = \frac{v\, d\tau}{dt}.$$

Also since $$ds = \varrho\, d\tau$$

30) $$a_\tau = \frac{v}{\varrho} \frac{ds}{dt} = \frac{v^2}{\varrho}.$$

This normal acceleration is always directed *toward* the center of curvature, and is otherwise called the *centripetal* acceleration.

Inserting the above values in the equation

$$a^2 = a_\tau^2 + a_\nu^2,$$

we may obtain an analytical expression for the radius of curvature,

31) $$\left(\frac{d^2 x}{dt^2}\right)^2 + \left(\frac{d^2 y}{dt^2}\right)^2 + \left(\frac{d^2 z}{dt^2}\right)^2 = \left(\frac{d^2 s}{dt^2}\right)^2 + \frac{1}{\varrho^2}\left(\frac{ds}{dt}\right)^4.$$

Let us change the independent variable from t to s. We have

$$\frac{dx}{dt} = \frac{dx}{ds}\frac{ds}{dt},$$

and differentiating again by t,

$$\frac{d^2 x}{dt^2} = \frac{dx}{ds}\frac{d^2 s}{dt^2} + \frac{d^2 x}{ds^2}\left(\frac{ds}{dt}\right)^2,$$

and similarly

$$\frac{d^2 y}{dt^2} = \frac{dy}{ds}\frac{d^2 s}{dt^2} + \frac{d^2 y}{ds^2}\left(\frac{ds}{dt}\right)^2,$$

$$\frac{d^2 z}{dt^2} = \frac{dz}{dz}\frac{d^2 s}{dt^2} + \frac{d^2 z}{ds^2}\left(\frac{ds}{dt}\right)^2.$$

Squaring and adding, 31) becomes

$$32)\quad \left(\frac{d^2s}{dt^2}\right)^2\left\{\left(\frac{dx}{ds}\right)^2+\left(\frac{dy}{ds}\right)^2+\left(\frac{dz}{ds}\right)^2\right\}+2\frac{d^2s}{dt^2}\left(\frac{ds}{dt}\right)^2\left\{\frac{dx}{ds}\frac{d^2x}{ds^2}\right.$$
$$\left.+\frac{dy}{ds}\frac{d^2y}{ds^2}+\frac{dz}{ds}\frac{d^2z}{ds^2}\right\}+\left(\frac{ds}{dt}\right)^4\left\{\left(\frac{d^2x}{ds^2}\right)^2+\left(\frac{d^2y}{ds^2}\right)^2+\left(\frac{d^2z}{ds^2}\right)^2\right\}$$
$$=\left(\frac{d^2s}{dt^2}\right)^2+\frac{1}{\varrho^2}\left(\frac{ds}{dt}\right)^4.$$

Now we have $\left(\frac{dx}{ds}\right)^2+\left(\frac{dy}{ds}\right)^2+\left(\frac{dz}{ds}\right)^2=1.$

Differentiating this by s gives

$$\frac{d}{ds}\left\{\left(\frac{dx}{ds}\right)^2+\left(\frac{dy}{ds}\right)^2+\left(\frac{dz}{ds}\right)^2\right\}=2\left\{\frac{dx}{ds}\frac{d^2x}{ds^2}+\frac{dy}{ds}\frac{d^2y}{ds}+\frac{dz}{ds}\frac{d^2z}{ds}\right\}=0.$$

Therefore equation 32) reduces to

$$\left(\frac{d^2s}{dt^2}\right)^2+\left(\frac{ds}{dt}\right)^4\left\{\left(\frac{d^2x}{ds^2}\right)^2+\left(\frac{d^2y}{ds^2}\right)^2+\left(\frac{d^2z}{ds^2}\right)^2\right\}=\left(\frac{d^2s}{dt^2}\right)^2+\frac{1}{\varrho^2}\left(\frac{ds}{dt}\right)^4,$$

or

$$33)\qquad \frac{1}{\varrho^2}=\left(\frac{d^2x}{ds^2}\right)^2+\left(\frac{d^2y}{ds^2}\right)^2+\left(\frac{d^2z}{ds^2}\right)^2.$$

If $\frac{ds}{dt}=v=1$, $ds=dt$ and the right hand member of 33) becomes the square of the acceleration. We thus have a kinematical definition of curvature, viz., the acceleration of a point traversing the curve with unit velocity. This agrees with the original expression 30), $a_\nu=\frac{v^2}{\varrho}$, for if $v=1$, $a_\tau=0$, the acceleration is entirely normal and $a_\nu=\frac{1}{\varrho}=\varkappa$.

We may in like manner resolve the acceleration into components along the radius vector and at right angles to it. Let us consider the case of motion in a plane, that of XY. We will call the radial component of the acceleration, or the *radial acceleration*, a_r, and we shall find that it is not equal to $\frac{d^2r}{dt^2}$, which is the scalar acceleration of the radial velocity. We will denote the component perpendicular to the radius or the *transverse acceleration* by a_φ which is not equal to the angular acceleration $\frac{d^2\varphi}{dt^2}$, nor to the acceleration of the transverse velocity, $\frac{dv_\varphi}{dt}$.

Differentiating the formulae for the change of coordinates

$$x=r\cos\varphi,\quad y=r\sin\varphi,$$

gives

$$\frac{dx}{dt}=\frac{dr}{dt}\cos\varphi-r\sin\varphi\frac{d\varphi}{dt}$$

$$\frac{dy}{dt}=\frac{dr}{dt}\sin\varphi+r\cos\varphi\frac{d\varphi}{dt}.$$

Differentiating again

$$\frac{d^2x}{dt^2} = \frac{d^2r}{dt^2}\cos\varphi - 2\sin\varphi\frac{d\varphi}{dt}\frac{dr}{dt} - r\sin\varphi\frac{d^2\varphi}{dt^2} - r\cos\varphi\left(\frac{d\varphi}{dt}\right)^2$$
$$\frac{d^2y}{dt^2} = \frac{d^2r}{dt^2}\sin\varphi + 2\cos\varphi\frac{d\varphi}{dt}\frac{dr}{dt} + r\cos\varphi\frac{d^2\varphi}{dt^2} - r\sin\varphi\left(\frac{d\varphi}{dt}\right)^2.$$

The direction cosines of the radius vector are:

$$\cos(rx) = \cos\varphi, \quad \cos(ry) = \sin\varphi,$$

so that we obtain by resolution,

$$34)\qquad a_r = \frac{d^2x}{dt^2}\cos(rx) + \frac{d^2y}{dt^2}\cos(ry) = \frac{d^2r}{dt^2} - r\left(\frac{d\varphi}{dt}\right)^2,$$

being less than the scalar acceleration of the radial velocity by the product of the radius vector and the square of the angular velocity. (If $\frac{dr}{dt} = 0$, the motion is circular, and a_r is the *normal* acceleration.)

The direction cosines of a line perpendicular to r and in the direction of increasing φ are, $-\sin\varphi$, $\cos\varphi$, so that for the transverse acceleration we obtain,

$$a_\varphi = -\frac{d^2x}{dt^2}\sin\varphi + \frac{d^2y}{dt^2}\cos\varphi = r\frac{d^2\varphi}{dt^2} + 2\frac{d\varphi}{dt}\frac{dr}{dt},$$

which may be written

$$35)\qquad a_\varphi = \frac{1}{r}\frac{d}{dt}\left(r^2\frac{d\varphi}{dt}\right).$$

Of course we have

$$a^2 = a_r^2 + a_\varphi^2.$$

11. Moment of Acceleration. The expression in the parenthesis of 35) is by 23) equal to twice the sector velocity $\frac{dS}{dt}$. Let us call $\frac{d^2S}{dt^2}$ the sector acceleration. Thus in plane motion

$$36)\qquad a_\varphi = \frac{2}{r}\frac{d^2S}{dt^2}, \qquad r a_\varphi = 2\frac{d^2S}{dt^2}.$$

Suppose (Fig. 8) AB represents the acceleration a, then AC perpendicular to r represents a_φ, therefore $r a_\varphi$ is twice the area of the triangle OAB. But that is the *moment* of the acceleration about the point O. Accordingly twice the sector acceleration is equal to the moment of the acceleration about the origin, or

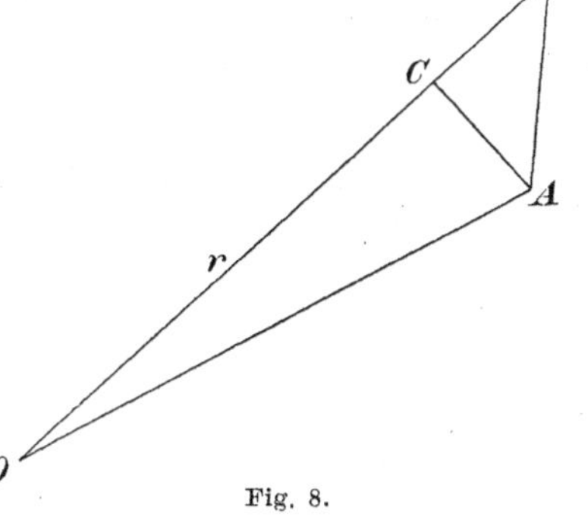

Fig. 8.

$$2\frac{d^2S}{dt^2} = x a_y - y a_x = x\frac{d^2y}{dt^2} - y\frac{d^2x}{dt^2}.$$

If the motion is not in one plane, we have, differentiating the sector velocity components 24)

37)
$$2\frac{d^2 S_x}{dt^2} = y\frac{d^2 z}{dt^2} - z\frac{d^2 y}{dt^2},$$
$$2\frac{d^2 S_y}{dt^2} = z\frac{d^2 x}{dt^2} - x\frac{d^2 z}{dt^2},$$
$$2\frac{d^2 S_z}{dt^2} = x\frac{d^2 y}{dt^2} - y\frac{d^2 x}{dt^2}.$$

The resultant of these is the moment of the acceleration. The fact that the moment of the acceleration is the exact time derivative of the moment of velocity leads to an important general principle of mechanics, the so-called Law of Areas.

12. Kepler's Laws. We may now obtain Newton's conclusions from Kepler's three laws of planetary motion, which were purely kinematical and based on a great amount of observational material collected by Tycho Brahe. The first law states that the areas swept over by the radius vector drawn from the sun to a planet in equal times are equal. (The motion is in one plane.) That is

$$\frac{dS}{dt} = \text{const.},$$
$$\frac{d^2 S}{dt^2} = 0,$$

therefore from 37) the moment of the acceleration with respect to the sun is zero. Consequently the line of direction of the acceleration passes through the sun, or the acceleration is central.

The second law states that the planets describe ellipses about the sun as a focus. The ellipse being always concave toward the focus, the acceleration is directed toward the sun. In order to deduce the quantitative meaning of the second law, we will use the polar equation of a conic section referred to the focus,

$$r = \frac{p = a(1-e^2)}{1+e\cos\varphi},$$ [1])

1) If d is the distance from focus to directrix, e the eccentricity, by the definition of a conic section,

$$\frac{r}{d - r\cos\varphi} = e, \quad r = \frac{ed = p}{1+e\cos\varphi}.$$

When $\cos\varphi = 1$,

$$r_1 = \frac{p}{1+e},$$

$\cos\varphi = -1$,

$$r_2 = \frac{p}{1-e}, \quad r_1 + r_2 = \frac{2p}{1-e^2} = 2a, \quad p = a(1-e^2)$$

and will find the value of the central acceleration. We have 34)

$$a_r = \frac{d^2 r}{dt^2} - r\left(\frac{d\varphi}{dt}\right)^2,$$

but from Kepler's first law,

$$2\frac{dS}{dt} = r^2\frac{d\varphi}{dt} = \text{const.} = h, \text{ say},$$

$$\frac{d\varphi}{dt} = \frac{h}{r^2}.$$

Now changing the variable from t to φ,

$$\frac{dr}{dt} = \frac{dr}{d\varphi}\frac{d\varphi}{dt} = \frac{h}{r^2}\frac{dr}{d\varphi} = -h\frac{d}{d\varphi}\left(\frac{1}{r}\right).$$

Differentiating by t,

$$\frac{d^2 r}{dt^2} = -h\frac{d^2}{d\varphi^2}\left(\frac{1}{r}\right)\frac{d\varphi}{dt} = -\frac{h^2}{r^2}\frac{d^2}{d\varphi^2}\left(\frac{1}{r}\right).$$

From the equation of the path we obtain

$$\frac{1}{r} = \frac{1}{p} + \frac{e}{p}\cos\varphi,$$

$$\frac{d}{d\varphi}\left(\frac{1}{r}\right) = -\frac{e}{p}\sin\varphi,$$

$$\frac{d^2}{d\varphi^2}\left(\frac{1}{r}\right) = -\frac{e}{p}\cos\varphi = \frac{1}{p} - \frac{1}{r}.$$

Inserting this value above gives

$$\frac{d^2 r}{dt^2} = -\frac{h^2}{pr^2} + \frac{h^2}{r^3},$$

and finally,

$$a_r = \frac{d^2 r}{dt^2} - r\left(\frac{d\varphi}{dt}\right)^2 = -\frac{h^2}{pr^2}.$$

Thus the fact that the path is a conic section shows that the central acceleration varies inversely as the square of the length of the radius vector. The negative sign shows that the acceleration is *toward* the sun.

The third law states that for different planets the squares of the times of describing the orbits are proportional to the cubes of the major axes.

Since

$$2\frac{dS}{dt} = h,$$

if T is the time of a complete period hT is twice the area of the orbit.

$$hT = 2\pi ab = 2\pi a^2\sqrt{1-e^2}.$$

2*

From which

$$h^2 = \frac{4\pi^2 a^4}{T^2}(1-e^2),$$

$$a_r = -\frac{h^2}{p r^2} = -\frac{h^2}{a(1-e^2)} \cdot \frac{1}{r^2} = \frac{4\pi^2 a^3}{T^2 r^2}.$$

Now since $\frac{a^3}{T^2}$ is by the third law constant for all the planets, the factor by which the inverse square of the radius vector is multiplied in order to obtain the central acceleration is *the same for all the planets* and depends only on the sun. We have thus obtained a complete kinematical statement of the law of gravitation for the planets.

Newton tested the law of the inverse square by applying it to the motion of the moon about the earth, and comparing its acceleration with that of a body at the surface of the earth as directly observed. Supposing the moon's orbit to be circular, of radius a, with period T, since the tangential acceleration is zero, its velocity is constant, and equal to $\frac{2\pi a}{T}$. Its acceleration, which is entirely normal, will accordingly be by 30)

$$a_m = \frac{v^2}{a} = \frac{4\pi^2 a}{T^2}.$$

If the acceleration varies inversely as the square of the distance, the acceleration experienced by a body at the earth's surface a_s will be given by

$$\frac{a_s}{a_m} = \frac{a^2}{R^2},$$

where R is the earth's radius. Therefore

$$a_s = a_m \frac{a^2}{R^2} = \frac{4\pi^2 a^3}{R^2 T^2}.$$

Now we have $T = 27\,\text{d.}\ 7\,\text{h.}\ 43\,\text{m.} = 39,343\,\text{m.}$, $2\pi R = 4\cdot 10^7$ meters, $a = 60R$, from which

$$a_s = \frac{2\pi\cdot 60^3\cdot 4\cdot 10^7\ \text{meters}}{(39,343\cdot 60\ \text{sec.})^2} = 9.74\,\frac{\text{meters}}{\text{sec.}^2}.$$

Now terrestrial observations give for the mean acceleration of bodies at the earth's surface $9.82\,\frac{\text{meters}}{\text{sec.}^2}$, which by a more exact calculation is in agreement with the predicted result.

13. Physical Axioms. Laws of Motion. It is necessary in order to pass from the kinematical specification of motion to the dynamical one to make use of knowledge drawn from a consideration of terrestrial phenomena. This knowledge is summed up by Newton in his three *Axiomata sive Leges Motus.* An axiom is defined by

Thomson and Tait[1]) as a proposition, the truth of which must be admitted as soon as the terms in which it is expressed are clearly understood. These physical axioms rest not on intuitive perception, but on convictions drawn from observation and experiment.

The manner of summing up the results of our experience is to a great extent unimportant, provided that it is sufficiently all-embracing. We are not concerned with the metaphysical question of the causes of motions, but merely with the physical question of stating what is actually found to take place in nature. The statement may be made by means of a single analytical formula, as was done in different ways by Lagrange, Hamilton and Hertz, or we may consider the various assumptions upon which such formulae are founded, making detailed statements, employing conceptions with which we are familiar.

This is what was done by Newton, and although his laws have received considerable criticism, they have, when properly understood, been generally admitted to be better than anything that has been proposed in their place.

Lex I. Corpus omne perseverare in statu suo quiescendi vel movendi uniformiter in directum, nisi quatenus illud a viribus impressis cogitur statum suum mutare.

Every body persists in its state of rest or of uniform motion in a straight line, except in so far as it may be compelled by force to change that state.

The property of persistence thus defined is called *Inertia.*

This gives a criterion for finding whether a force is acting on a body or not, or in other words a negative definition of force. Force is acting on a body when its motion is not uniform. By uniform we mean such motion that the vector velocity is constant. If the body be a *material point*, that is a body so small that the distances between its different parts may be neglected, the motion is uniform if

$$\frac{dx}{dt} = c_1, \quad \frac{dy}{dt} = c_2, \quad \frac{dz}{dt} = c_3,$$

38) that is

$$\frac{d^2x}{dt^2} = \frac{d^2y}{dt^2} = \frac{d^2z}{dt^2} = 0.$$

Accordingly we see that the force and acceleration vanish together. Integrating the equations 38),

$$x = c_1 t + d_1, \quad y = c_2 t + d_2, \quad z = c_3 t + d_3,$$

39)

$$\frac{x - d_1}{c_1} = \frac{y - d_2}{c_2} = \frac{z - d_3}{c_3},$$

1) Thomson and Tait, Natural Philosophy, § 243.

the path is a straight line, and since

$$v = \sqrt{\left(\frac{dx}{dt}\right)^2 + \left(\frac{dy}{dt}\right)^2 + \left(\frac{dz}{dt}\right)^2} = \sqrt{c_1^2 + c_2^2 + c_3^2},$$

it is traversed with constant velocity. We may on the other hand interpret the statement as giving us a means of measuring time. *Intervals of time are proportional to the corresponding distances traversed by a material point not acted on by forces.*

Obviously this statement gives us an absolute definition neither of time nor of force, but only a relation between them. It is difficult or impossible for us to realize experimental conditions in wich a body shall be withdrawn from the influence of all force. However we may approximate toward this condition, which must at any rate give us the ideal measurement of time. Fortunately we find in nature angular motions which, by an application of the first law, give us a practical means for the measurement of time.

The second law gives us in a more positive manner than the first a measure of a force.

Lex II. Mutationem motus proportionalem esse vi motrici impressae, et fieri secundum lineam rectam qua vis illa imprimitur.

Change of motion is proportional to force applied, and takes place in the direction of the straight line in which the force acts.

By change of motion is meant acceleration. If all our experiments were made with a single body, there would be no advantage in the introduction of the term force over that of acceleration, the multiplication of names being useless when no new ideas are thereby introduced. The convenience of the term force arises from the consideration of the third law. In the case of more than one body the factor of proportionality mentioned above requires separate definition for the different bodies.

Lex III. Actioni contrariam semper et aequalem esse reactionem: sive corporum duorum actiones in se mutuo semper esse aequales et in partes contrarias dirigi.

To every action there is always an equal and contrary reaction: or, the mutual actions of any two bodies are always equal and oppositely directed.

If we have a certain action between two bodies 1 and 2, and if the *actio* were proportional only to the accelerations, we should have

$$\frac{d^2x_1}{dt^2} = -\frac{d^2x_2}{dt^2}, \quad \frac{d^2y_1}{dt^2} = -\frac{d^2y_2}{dt^2}, \quad \frac{d^2z_1}{dt^2} = -\frac{d^2z_2}{dt^2},$$

which is not found to be the case. We must accordingly introduce a factor of proportionality, or (for symmetry) two factors, so that we write

$$m_1 \frac{d^2 x_1}{dt^2} = - m_2 \frac{d^2 x_2}{dt^2},$$

40)
$$m_1 \frac{d^2 y_1}{dt^2} = - m_2 \frac{d^2 y_2}{dt^2},$$

$$m_1 \frac{d^2 z_1}{dt^2} = - m_2 \frac{d^2 z_2}{dt^2}.$$

Experiment shows that the factors m_1 and m_2 are constant for a given body that undergoes no changes other than those of position. These factors are called the *masses* of the bodies. The nature of the actions between the two bodies may be of any sort, and may be transmitted by the help of any number of intervening bodies. For instance, the actions of two heavenly bodies on each other, transmitted we know not how, or the actions of two bodies kept at a fixed distance by means of a rod or string or connected by an elastic spring, or attracting or repelling each other by magnetic or electric agencies, are all illustrations of the third law. It is obvious that if we could observe the motions so as to obtain the coordinates of both bodies as functions of the time, equations 40) would enable us to determine the ratio of the masses. For example, consider the toy consisting of two horse-chestnuts or bullets connected by a string, and suppose this to be whirled about and projected into the air so that the two bodies describe complicated paths, the whole apparatus describing in general a parabolic path. If we take a series of photographs of it in rapid succession, by means of a kinetoscope or similar device, we may by measurement obtain the coordinates of the two bodies as functions of the time. This illustrates perfectly the dynamical measurement of mass and the means of obtaining the relative masses of the heavenly bodies. We have no means of defining the *absolute* mass of a body. As a further example of the third law, let us suppose the action is transmitted from one body to the other by means of a flexible string passing over frictionless pulleys, as in the case of Atwood's machine. The assumption here made is that the tension of the string is unchanged by passing over the pulleys.

A more practical means of realizing the dynamical comparison of masses would be by experimentally establishing the equality of both sides of equations 40) with the same quantity. For example let the body be made to describe a horizontal circular path, say by means of a whirling machine. It will be found that it must be retained in this path by external means such as the tension of a string. Let this be passed over a pulley at the center of the path and exactly balance its pull against that of a weight suspended from it. The resultant of the components $m\frac{d^2x}{dt^2}$, $m\frac{d^2y}{dt^2}$, is by 30) equal to $\frac{mv^2}{r}$, where r is the radius of the circular path. The resultant is directed toward the center, and measures the effect of the tension of the string on the motion. If we repeat the experiment with another body for which the corresponding quantities are denoted by accents, making use of the same counterbalancing weight, the tensions of the string in the two cases are obviously equal and consequently we have,

$$\frac{m v^2}{r} = \frac{m' v'^2}{r'}.$$

Measuring the velocities and radii therefore enables us to compare the masses.

The vector defined by the product of the scalar quantity mass by the vector quantity acceleration, whose components are

$$41) \qquad X = m\frac{d^2 x}{dt^2}, \quad Y = m\frac{d^2 y}{dt^2}, \quad Z = m\frac{d^2 z}{dt^2}$$

is called the *force* acting upon the body, and is the *vis impressa* of the second law. The second and third laws taken together accordingly give us a complete definition and mode of measurement of force. The introduction of the new term is justified by the third law. For we find that force is capable of representing the dual nature of the interaction between two bodies, while the acceleration is not, there being two different accelerations for the two different bodies.

The two sided nature of the action between two bodies is often expressed by calling it a *stress.*

The equations 41) are called the differential equations of motion of the body. This statement needs some explanation. The introduction of the term force has given us no explanation of the cause of motion, for whereas the second law tells us that the change of motion is proportional to the force applied, and we are accustomed to say that the force is the *cause* of the change, no additional knowledge of the motion is given us by this statement. When we say that a body moves because we push it, all we mean is that the motion and the push exist simultaneously. Were we accustomed to a different point of view, we might be as much struck with the fact that the body pushes back when it moves as that we push it. This is what the third law calls to our attention.

It is undoubtedly true that our fundamental notions of dynamics are derived through what may be called the muscular sense, which is affected when we make ourselves one of the bodies of a system. We then perceive the reactions, and we have learned to correlate our perceptions to the motions of the other bodies of the system. Nevertheless, had we not possessed this extremely important sense, we might have elaborated the same system of dynamics that we now have merely by the sense of sight, as illustrated by the example of the two particles, fixing our attention on the facts embodied in equations 40). Had we been merely astronomers this is what we should have been obliged to do. We may perhaps doubt whether we should have in this way arrived at the conceptions of force which we possess with the aid of both senses. At any rate no one can doubt that an individual newly arrived in this world learns its properties as much through the muscular sense as through the more generally appreciated sense of sight.

Let us now reverse the mode of looking at equations 41). Suppose that we find that under given conditions a certain agency will produce a certain force, as shown by the motion of some body, and suppose that as the circumstances are changed we can always measure the force. If then it is possible to submit a second body to the action of the same agent under similarly varying circumstances, we shall be able to find the motion of the second body. The equations 41) under these circumstances furnish merely another means of describing motions. We might go on obtaining still further descriptions by means of higher derivatives of the coordinates, but experience shows us that nothing is gained thereby, for, in the great majority of cases with which we have to deal, it is found that the components, X, Y, Z, are expressible as functions of only the coordinates of the bodies involved, or at most of the coordinates and their first time derivatives.

There is a further advantage in the introduction of the notion of force, in that if a body be submitted to the action of two agencies at different times, so as to move under the influence of definite forces, and then be submitted to the action of both simultaneously, the force now found to be acting will be the resultant of the two original forces. This statement, that forces are compounded as vectors, being the equivalent of the so-called statement of the parallelogram of forces, is implicity contained in Newton's second law of motion.

Under certain circumstances, an agent which would under other conditions cause motion, may cause no motion. We then say that its effect is counteracted by that of some other agent, or otherwise, that the two forces are in equilibrium. According to the third law, the two forces are equal and opposite, either being the reaction with respect to the other. Such reactions are called *static* reactions, as opposed to the *kinetic* reactions exerted by bodies undergoing acceleration.

As has been stated above, most of the forces which occur in nature depend only on the positions of the bodies upon which they act, or at most upon their positions and velocities, but not upon the higher derivatives of the coordinates. Forces of the former sort are called *positional* forces, those of the latter *motional* forces. As an example of the latter, we know that a body moving through the air experiences a negative acceleration which is greater the greater the velocity of the body, and we say that the motion is retarded by a force, which we call the *resistance* of the air.

Supposing now X, Y, Z to be given functions of the coordinates and velocities, the integration of the differential equations 41) constitutes the problem of the mechanics of a single particle. It is in this sense that the problems of mechanics in general are to be considered. (See Note I.)

Returning to the "change of motion" mentioned in the second law, it is customary to characterize the product of the mass by the vector velocity as the *momentum* of the body, a vector whose components are

$$42)\qquad M_x = m\frac{dx}{dt}, \quad M_y = m\frac{dy}{dt}, \quad M_z = -\frac{dz}{dt}.$$

This is the *motus* whose rate of change measures the force, so that equations 41) may be written

$$43)\qquad \frac{dM_x}{dt} = X, \quad \frac{dM_y}{dt} = Y, \quad \frac{dM_z}{dt} = Z.$$

These equations are in fact more general than those of 41) for they apply even when m is changing, as in the case of a rocket being burned.

14. Units. The specification of any quantity, scalar or vector, involves two factors, first a numerical quantity or *numeric*, and secondly a concrete quantity in terms of which all quantities of that kind are numerically expressed, called a *unit*. The simplest unit is that of the geometrical quantity, length. We shall adopt as the unit of length the *centimeter*, defined as the one-hundredth part of the distance at temperature zero degrees Centigrade, and pressure 760 millimeters of mercury, between two parallel lines engraved on a certain bar of platinum-iridium alloy, deposited in a vault in the laboratory of the *"Comité International des Poids et Mesures"*, at Sèvres, near Paris. This bar is known as the *"Mètre Prototype"*, and serves as the basis of length measurements for the civilized world[1]) (except the British Empire and Russia[2]).

It was proposed by Maxwell to use a natural unit of length, namely the length of a wave of light corresponding to some well defined line in the spectrum of some element, at a definite temperature and pressure, as it is highly probable that such a wave-length is extremely constant. Measurements were carried out at Sèvres by Michelson, with this end in view, which established the ratio between

1) See Guillaume, La Convention du Mètre.

2) The United States yard is defined as 3600/3937 meters.

the above meter and the wave-length in air of a red cadmium ray as 1,553,163.5.[1])

The unit of mass will be assumed to be the *gram*, defined as the one-thousandth part of a piece of platinum-iridium, deposited at the place above mentioned and known as the "*Kilogramme Prototype*".

As the unit of time we shall take the mean solar second, obtained from astronomical observations on the rotation of the earth. The unit of time cannot be preserved and compared as in the case of the units of length and mass, but is fortunately preserved for us by nature, in the nearly constant rotation of the earth. As the earth is gradually rotating more slowly, however, this unit is not absolutely constant, and it has been proposed to take for the unit of time the period of vibration of a molecule of the substance giving off light of the standard wave-length. To obtain such a unit would involve a measurement of the velocity of light, which cannot at present be made with the accuracy with which the mean solar second is known.

15. Derived Units and Dimensions. It can be shown that the measurements of all physical quantities with which we are acquainted may be made in terms of three independent units. These are known as fundamental units, and are most conveniently taken as those of length, mass, and time. Other units, which depend on these, are known as derived units. If the same quantity is expressed in terms of two different units of the same kind, the numerics are inversely proportional to the size of the units. Thus six feet is otherwise expressed as two yards, the numerics 6 and 2 being in the ratio 3, that of a yard to a foot. If we change the magnitude of one of the fundamental units in any ratio r, the numeric of a quantity expressed in derived units will vary proportionately to a certain power of r, r^{-n}, the derived unit is then said to be of *dimensions*[2]) n in the fundamental unit in question. For instance, if we change the fundamental unit of length from the foot to the yard, $r = 3$, an area of 27 sq. ft. becomes 3 sq. yds., the numeric has changed in the ratio $3 : 27 = 1 : 3^2 = r^{-2}$, and the unit of area is of dimensions 2 in the unit of length. We may express this by writing

$$[\text{Area}] = [L^2].$$

The derived unit increases in the same ratio that the numeric of the quantity decreases. In our system the unit of area is the square

1) Travaux et Mémoires du Bureau International des Poids et Mesures. Tome 11, p. 85.

2) The idea of dimensions of units originated with Fourier: Théorie analytique de la Chaleur, Section IX.

centimeter, written 1 cm^2. In like manner the unit of volume is of the dimensions $[L^3]$ and the unit is 1 cm^3. The dimensions of velocity are $\left[\frac{L}{T}\right]$, or as we write for convenience,

$$\text{Velocity} = \text{Length/Time}.$$

Two quantities of different sorts do not have a ratio in the ordinary arithmetical sense, but such equations as the above are of great use in physics, and give rise to an extended meaning of the terms ratio and product.

The above equation is to be interpreted as follows. If any velocity be specified in terms of units of length and time the numerical factor is greater in proportion directly as the unit of length is smaller, and as the unit of time is greater. For instance we may write the equation expressing the fact that a velocity of 30 feet per second is the same as a velocity of 10 yards per second or 1800 feet per minute.

$$30\,\frac{\text{ft.}}{\text{sec.}} = 10\,\frac{\text{yd.}}{\text{sec.}} = 1800\,\frac{\text{ft.}}{\text{min.}}\cdot$$

We may operate an such equations precisely as if the units were ordinary arithmetical quantities, for the ratio of two quantities of the same kind is always a number. For instance

$$\frac{30}{10} = \frac{\text{yd. sec.}}{\text{ft. sec.}}\cdot$$

The ratio $\frac{\text{yd.}}{\text{ft.}}$ is the number 3, while $\frac{\text{sec.}}{\text{sec.}} = 1$. Also

$$\frac{1800}{10} = \frac{\text{yd.}}{\text{ft.}}\,\frac{\text{min.}}{\text{sec.}} = 3 \cdot 60.$$

Such an expression as $\frac{\text{ft.}}{\text{sec.}}$ is read *feet per second.*

This unit of velocity is one centimeter-per-second, written,

$$\frac{\text{cm.}}{\text{sec.}} = \text{cm. sec.}^{-1}.$$

Since acceleration is defined as a ratio of increment of velocity to increment of time, we have

$$[\text{Acceleration}] = \frac{[\text{Velocity}]}{[\text{Time}]} = \frac{[\text{Length}]}{[\text{Time}^2]} = \left[\frac{L}{T^2}\right],$$

or the numeric of a certain acceleration varies inversely as the magnitude of the unit of length, and directly as the square of the unit of time. For instance, an acceleration in which a velocity of 10 feet per second is gained in 2 seconds is equal to one in which a velocity of 9000 feet per minute is gained in a minute,

$$\frac{10\text{ ft.}}{(2\text{ sec.})^2} = \frac{10}{4}\,\frac{\text{ft.}}{\text{sec.}^2} = 9000\,\frac{\text{ft.}}{\text{min.}^2}\cdot$$

The unit of acceleration is one centimeter-per-second per second, written $\frac{\text{cm.}}{\text{sec.}^2} = \text{cm. sec.}^{-2}$. (It is to be noted that in a derivative such as $\frac{d^2 s}{dt^2}$, the numerator being a differential of no matter what order is of the same dimensions as s, while the denominator being the *square* of a differential is of dimensions $[T^2]$).

Since momentum = mass · velocity, we have

$$[\text{Momentum}] = \frac{[\text{Mass}] \cdot [\text{Length}]}{[\text{Time}]} = \left[\frac{ML}{T}\right].$$

Since force = mass · acceleration,

$$[\text{Force}] = \frac{[\text{Mass}] \cdot [\text{Length}]}{[\text{Time}^2]} = \left[\frac{ML}{T^2}\right].$$

The unit of force is one gram-centimeter-per-second per second. It is called a *dyne.*

Moment of a force being force · length is of dimensions

$$\left[\frac{ML^2}{T^2}\right].$$

The dimensions of an angular magnitude, being those of the ratio of two quantities of the same kind, arc and radius, are zero. Angular velocity being defined as $\frac{\text{angle}}{\text{time}}$ is of dimensions $\left[\frac{1}{T}\right]$.

All physical equations must be homogeneous in the various units, that is, the dimensions of every term must be the same. This gives us a valuable check on the correctness of our equations.

For an excellent account of the theory of dimensions the reader may consult Everett, The C. G. S. System of Units.

16. Universal Gravitation. We may now convert the kinematical statement of *§ 12* regarding the planetary motion into the dynamical one, that the sun attracts the different planets with forces proportional directly to their masses and inversely to the square of their distances from itself. By observations on other celestial bodies that possess satellites, it is found that the attractions are proportional to the masses of the attracting bodies (as determined by their own motions). From this we may pass to Newton's great generalization: *Every particle of matter in the universe attracts every other particle, with a force whose direction is that of the line joining the two, and whose magnitude is directly as the product of their masses, and inversely as the square of their distance from each other*[1]),

$$F_{12} = \gamma \frac{m_1 m_2}{r_{12}^2},$$

the factor of proportionality γ being the same for all bodies. This is the law of *Universal Gravitation.*

1) Thomson and Tait, Treatise on Natural Philosophy, Part II, p. 9.

The numerical value of γ, the Newtonian constant of gravitation, depends upon the system of units used. Its dimensions are those of

$$\frac{[\text{Force}]\cdot[\text{Length}^2]}{[\text{Mass}^2]} = \left[\frac{L^3}{MT^2}\right].$$

It is possible, and in astronomy is convenient to choose the units in such a manner as to make γ equal to unity. If this were done, we should get a relation between the dimensions of mass, length and time, for by supposing that γ has no dimensions, we should have

$$[M] = [L^3\,T^{-2}].$$

Thus we should need only two fundamental units instead of three. This is an example of the somewhat arbitrary nature of the dimensions of physical quantities. What is not arbitrary however is the statement that every physical equation must be dimensionally homogeneous. For the purposes of physics it is customary to retain the three fundamental units, giving γ the dimensions specified above. It is not possible to determine γ by astronomical observations, since the knowledge of the absolute magnitude of the force is lacking. In the laboratory we may use two known masses, and determine the force between them Determinations undertaken to ascertain the numerical value of γ by terrestrial observations have been made in great numbers from the time of Cavendish to the present. One of the most accurate, that of Boys[1]), gives in the units which we have adopted,

$$\gamma = 6.6576\cdot 10^{-8}\frac{\text{cm.}^3}{\text{gm. sec.}^2},$$

that is, two spherical masses each of mass one gram with centers one centimeter apart attract each other with the force of γ dynes.[2])

If two particles have coordinates $x_1, y_1, z_1, x_2, y_2, z_2$ and distance apart r_{12}, the direction cosines of the line drawn from 1 to 2 are

$$\frac{x_2-x_1}{r_{12}},\quad \frac{y_2-y_1}{r_{12}},\quad \frac{z_2-z_1}{r_{12}},$$

and, since the force exerted by 2 on 1 has the direction of this line, the equations of motion for 1 are

$$m_1\frac{d^2x_1}{dt^2} = \gamma\frac{m_1 m_2}{r_{12}^2}\cdot\frac{x_2-x_1}{r_{12}},$$

$$m_1\frac{d^2y_1}{dt^2} = \gamma\frac{m_1 m_2}{r_{12}^2}\cdot\frac{y_2-y_1}{r_{12}},$$

$$m_1\frac{d^2z_1}{dt^2} = \gamma\frac{m_1 m_2}{r_{12}^2}\cdot\frac{z_2-z_1}{r_{12}},$$

---

1) Boys, Phil. Trans. 1895, I.

2) It will be shown later that homogeneous spheres attract each other as if their masses were all concentrated at their centers.

and for 2 are

$$m_2 \frac{d^2 x_2}{dt^2} = \gamma \frac{m_1 m_2}{r_{12}{}^2} \frac{x_1 - x_2}{r_{12}},$$

$$m_2 \frac{d^2 y_2}{dt^2} = \gamma \frac{m_1 m_2}{r_{12}{}^2} \frac{y_1 - y_2}{r_{12}},$$

$$m_2 \frac{d^2 z_2}{dt^2} = \gamma \frac{m_1 m_2}{r_{12}{}^2} \frac{z_1 - z_2}{r_{12}}.$$

The integration of these six equations is easily carried out (see § 102), and gives us for the case of the sun and a planet a slight modification of Kepler's laws, for the sun does not remain absolutely at rest. If there are three bodies their equations of motion are similarly,

$$\frac{d^2 x_1}{dt^2} = \gamma \left(m_2 \frac{x_2 - x_1}{r_{12}{}^3} + m_3 \frac{x_3 - x_1}{r_{13}{}^3} \right),$$

$$\frac{d^2 y_1}{dt^2} = \gamma \left(m_2 \frac{y_2 - y_1}{r_{12}{}^3} + m_3 \frac{y_3 - y_1}{r_{13}{}^3} \right),$$

$$\frac{d^2 z_1}{dt^2} = \gamma \left(m_2 \frac{z_2 - z_1}{r_{12}{}^3} + m_3 \frac{z_3 - z_1}{r_{13}{}^3} \right),$$

$$\frac{d^2 x_2}{dt^2} = \gamma \left(m_3 \frac{x_3 - x_2}{r_{23}{}^3} + m_1 \frac{x_1 - x_2}{r_{21}{}^3} \right),$$

$$\frac{d^2 y_2}{dt^2} = \gamma \left(m_3 \frac{y_3 - y_2}{r_{23}{}^3} + m_1 \frac{y_1 - y_2}{r_{21}{}^3} \right),$$

$$\frac{d^2 z_2}{dt^2} = \gamma \left(m_3 \frac{z_3 - z_2}{r_{23}{}^3} + m_1 \frac{z_1 - z_2}{r_{21}{}^3} \right),$$

$$\frac{d^2 x_3}{dt^2} = \gamma \left(m_1 \frac{x_1 - x_3}{r_{31}{}^3} + m_2 \frac{x_2 - x_3}{r_{32}{}^3} \right),$$

$$\frac{d^2 y_3}{dt^2} = \gamma \left(m_1 \frac{y_1 - y_3}{r_{31}{}^2} + m_2 \frac{y_2 - y_3}{r_{32}{}^3} \right),$$

$$\frac{d^2 z_3}{dt^2} = \gamma \left(m_1 \frac{z_1 - z_3}{r_{31}{}^2} + m_2 \frac{z_2 - z_3}{r_{32}{}^3} \right).$$

The problem of integrating these equations is known as "the problem of three bodies" and has not been completely solved. The problem of the solar system is still more complicated, but by means of approximations, the perturbations of the different planets upon each other, causing slight variations from Kepler's laws, have been calculated. It is in this manner that the observations of astronomers from the time of Newton until the present have furnished the most brilliant verification of Newton's great discovery.

17. Absolute Systems. The above system of units, which has for its fundamental units the centimeter, gram, and second, is called the C. G. S. system, and was recommended by a committee of the British Association for the Advancement of Science in 1861. It is sometimes incorrectly spoken of as the *absolute system* of units.

An absolute system is any system, irrespective of the magnitudes of the units, by which physical quantities can be specified in terms of the least number of fundamental units, which shall be independent of time or place, and reproducible by copying from standards. A system based on the foot, pound, and minute is just as much an absolute system as the C. G. S. system. The idea of an absolute system is due to Gauss.[1])

The ordinary method of measuring force, used by non-scientific persons and engineers, though very convenient, does not belong to the absolute system of measurements. The unit of force is taken as the weight of, or downward force exerted by the earth upon, the mass of a standard piece of metal, such as the standard pound or kilogram. To measure the force in absolute units, we must know what acceleration the earth's pull would cause this mass to receive, if allowed to fall. As stated above, the attraction according to the Newtonian law exercised by the earth is the same as it would be if the whole mass were concentrated in a very small region at its center. Consequently the more remote a body is from the center the less will be the earth's pull upon it, or its weight. If however we consider a region so small that its dimensions may be neglected in comparison with those of the earth, the force exerted[2]) upon a given body at any point of the region may be considered as constant, and exerted in a constant direction, called the vertical of the place. Dividing the weight, which is proportional to the mass of the body, by the mass, we find that the acceleration experienced by all bodies at a given place is the same. This was proved experimentally by Galileo, to the great astonishment and scandal of the philosophers of the time. (On account of the disturbing action of the air, this statement is exactly true only for bodies falling *in vacuo.*) The value of this acceleration is denoted by g, and its value at the sea-level in latitude 45^0 is

$$g = 980.62 \frac{\text{cm.}}{\text{sec.}^2}.$$

Accordingly the force exerted by the earth on a mass of m grams is mg dynes, or the

weight of a kilogram in latitude $45^0 = 980{,}620$ *dynes.*

Now the value of the acceleration g is not constant, but varies as we go from place to place on the earth's surface, ascend mountains or descend into mines. Accordingly, the weight of a kilogram is

1) Gauss, *Intensitas vis magneticae terrestris ad mensuram absolutam revocata.* Göttingen, 1832. Ges. Werke, V. p. 80.

2) For the effect of the earth's rotation, see § 104.

not an invariable, or *absolute* standard of force. At the center of the earth, a kilogram would weigh nothing. Its mass is, however, invariable.

The ordinary method of comparing masses by means of the balance is in reality a comparison of two forces, the weights of the bodies. As these are proportional to the masses, the method becomes one for the comparison of masses, being a statical one, as distinguished from the kinetic method of § 13. If, however, we should make use of a balance with arms so long that the two masses compared were situated in regions for which the values of g were different, equality of weights would not connote equality of masses. An instrument which shows the variable weight of a body as it changes locality is found in the spring-balance, another in the pendulum.

The value of g at points on the earth in latitude λ and h centimeters above the sea-level, is given by the formula, originally given by Clairaut[1]),

$$g = 980.62 - 2.6 \cos 2\lambda - 0.000003\,h.$$

For further information with regard to units, the reader may consult Everett's *The C. G. S. System of Units.*

CHAPTER II.

IMPORTANT PARTICULAR MOTIONS OF A MATERIAL POINT.

18. Constant Accelerations. Let us examine the motion of a particle experiencing a constant vertical downward acceleration g. If the axis of Z be taken vertically upward, we have for the equations of motion,

$$1)\qquad \frac{d^2x}{dt^2} = 0, \quad \frac{d^2y}{dt^2} = 0, \quad \frac{d^2z}{dt^2} = -g.$$

Integrating with respect to t we have

$$2)\qquad \frac{dx}{dt} = V_x, \quad \frac{dy}{dt} = V_y, \quad \frac{dz}{dt} = -gt + V_z,$$

where V_x, V_y, V_z are constants representing the component velocities at the time $t = 0$.

Integrating again,

$$3)\qquad x - x_0 = V_x t, \quad y - y_0 = V_y t, \quad z - z_0 = -\frac{1}{2}gt^2 + V_z t,$$

1) Everett, *The C. G. S. System of Units,* Chap. VI. The above constants are adopted by Helmert.

where x_0, y_0, z_0 are the coordinates of the point at the time $t = 0$. Eliminating t between the first two of equations 3), we obtain

$$4)\qquad \frac{x - x_0}{V_x} = \frac{y - y_0}{V_y}$$

which shows that the motion is in a vertical plane. (The twisted curves sometimes described by a base-ball, golf or tennis-ball or rifled shot are the results of actions due to the air and the rotation of the ball and not here contemplated.) If we choose this vertical plane for the plane of XZ, we shall have $y = 0$, $V_y = 0$, and the equation of the path is found by eliminating t between the first and third of equations 3) giving

$$5)\qquad z - z_0 = \frac{V_z}{V_x}(x - x_0) - \frac{1}{2} g \frac{(x - x_0)^2}{V_x^2},$$

the equation of a parabola with axis vertical. If V_z is positive, the projectile will rise until $\frac{dz}{dx} = 0$, or $\frac{dz}{dt} = 0$, that is $x - x_0 = \frac{V_x V_z}{g}$. The height reached at this point is

$$6)\qquad z - z_0 = \frac{V_z^2}{2g} = h.$$

It will be observed that this is independent of the horizontal component of the velocity, V_x, and is therefore the height that would be reached by a projectile thrown vertically upward, or in other words

$$7)\qquad V_z = \sqrt{2gh}$$

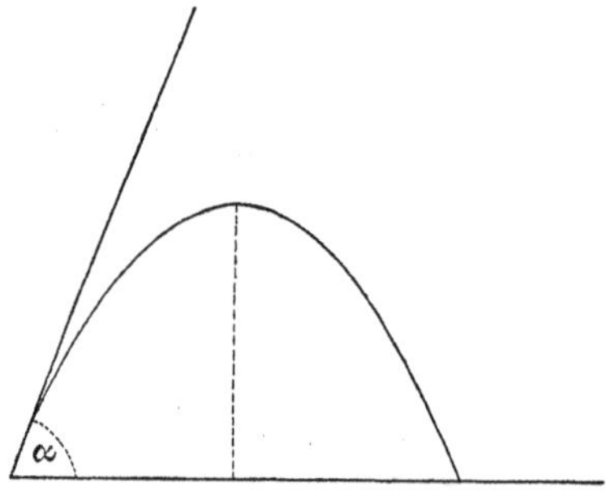

is the velocity that would be attained by a body falling from rest vertically through the height h.

If α be the angle of elevation of the path at the start, V the velocity of projecton, we have,

$$V_x = V \cos \alpha,$$

$$V_z = V \sin \alpha,$$

Fig. 9.

and the range or horizontal distance traversed by the projectile until it has fallen to the original level is twice the value of $(x - x_0)$ calculated for the highest point, or

$$\frac{V^2 \sin 2\alpha}{g}.$$

As we vary the elevation the range is accordingly greatest when $\alpha = 45^\circ$.

These formulae due to Galileo, are of little practical value in gunnery, because beside the attraction of the earth a projectile is acted on by retarding forces due to its motion relative to the air, and depending on the velocity of the projectile.

19. Harmonic Motions. Next in simplicity to motions under constant forces are those in which the force is directed toward a fixed point, and depends upon the distance of the particle from it. The simplest way in which it can depend upon the distance is by being proportional to it. If the particle moves in a straight line with an acceleration toward the origin proportional to its distance x from it, we have

8) $$\frac{d^2x}{dt^2} = -n^2x, \text{ where } n \text{ is a constant.}$$

The integral of this differential equation is

9) $$x = A\cos nt + B\sin nt,$$

where A and B are arbitrary constants. If we put $A = a\cos\alpha$, $B = a\sin\alpha$, this may be written

10) $$x = a\cos(nt - \alpha),$$

which as before contains two arbitrary constants, a and α.

Obviously by giving α a value differing by $\frac{\pi}{2}$ we may use the sine instead of cosine. If we increase nt by 2π the value of the sine and cosine is unchanged, consequently the motion is *periodic*, or the point is found in a given position at times separated by an interval T, called the *period*, given by $nT = 2\pi$, so that we may write

11) $$x = a\cos\left(\frac{2\pi t}{T} - \alpha\right).$$

The maximum excursion of the point on either side of the origin is called the *amplitude* a, and it is to be noticed that it does not occur in the differential equation 8). Since x takes on positive and negative values in symmetrical succession, the motion is an oscillation with period T, and *frequency*, that is the number of oscillations in unit time,

$$\frac{1}{T} = \frac{n}{2\pi}.$$

An oscillation expressed as above, 10), by a single sine or cosine function of a linear function of the time is called a *simple harmonic motion*, the name arising from the occurrence of such motions in musical sounds. The frequency of harmonic motions in nature is due to the fact that in any system which is disturbed from a position of rest forces are called into play which depend in general on the magnitude of the displacement. Suppose such a displacement depends upon a single variable x, then as the force F developed will usually be a uniform, continuous function of x, we may develop by Taylor's Theorem,

$$F = F_0 + x\left(\frac{dF}{dx}\right)_0 + \frac{x^2}{2}\left(\frac{d^2F}{dx^2}\right)_0 + \cdots$$

If now x be small enough, the term in the first power of x is more important than those that follow, which may therefore be neglected. If we measure x from the configuration of equilibrium, when $x = 0$, $F = 0$, so that we have

$$F = x\left(\frac{dF}{dx}\right)_0 = cx.$$

If the coefficient of x is negative, the force tends to restore the system to the configuration of equilibrium, and being proportional to the displacement, the system will execute harmonic vibrations about this configuration. Thus *small* vibrations are harmonic, which explains the extreme frequency of such motions in nature. A common method of realizing such vibrations is by the use of a tuning-fork.

If a point moves so as to describe the resultant of two simple harmonic motions of the *same* frequency in lines intersecting at right angles, its equations of motion are

$$12)\qquad \frac{d^2x}{dt^2} + n^2x = 0, \quad \frac{d^2y}{dt^2} + n^2y = 0.$$

The resultant acceleration is directed toward the origin and is directly proportional to the radius vector. The path is obtained by the elimination of t between the integrals

$$13)\qquad \begin{aligned} x &= a\sin(nt - \alpha) = a(\sin nt\cos\alpha - \cos nt\sin\alpha) \\ y &= b\sin(nt - \beta) = b(\sin nt\cos\beta - \cos nt\sin\beta), \end{aligned}$$

where a, b, α, β are constants of integration. Solving for $\sin nt$ and $\cos nt$,

$$\sin nt = \frac{\frac{x}{a}\sin\beta - \frac{y}{b}\sin\alpha}{\sin(\beta - \alpha)},$$

$$\cos nt = \frac{\frac{x}{a}\cos\beta - \frac{y}{b}\cos\alpha}{\sin(\beta - \alpha)}.$$

Squaring and adding we have the equation of the path

$$14)\qquad \frac{\frac{x^2}{a^2} + \frac{y^2}{b^2} - 2\frac{x}{a}\frac{y}{b}\cos(\alpha - \beta)}{\sin^2(\alpha - \beta)} = 1,$$

which represents an ellipse. The motion is called *elliptic harmonic motion*. If $\alpha = \beta$, that is if both components vanish together, the denominator above vanishes, therefore the numerator must also, and the path is

$$\frac{x^2}{a^2} + \frac{y^2}{b^2} - 2\frac{x}{a}\frac{y}{b} = 0,$$

which represents a pair of coincident straight lines

$$\frac{x}{a} - \frac{y}{b} = 0,$$

so that the path of the point is rectilinear and the motion is simple harmonic. Similarly if $\alpha - \beta = \pi$, the motion is rectilinear.

The angle $(\alpha - \beta)$ is called the *phase difference* of the two vibrations. If this is a right angle one component reaches its maximum when the other vanishes, we then have

$$\frac{x^2}{a^2} + \frac{y^2}{b^2} = 1,$$

and the coordinate axes are the principal axes of the ellipse. The amplitudes, a and b, of the component vibrations are in this case the semi-axes of the ellipse.

It is obvious from 13) that whatever the value of $\alpha - \beta$ the maximum values attained by x and y are a and b respectively, so that the ellipse is always inscribed in a rectangle of sides $2a$ and $2b$ (Fig. 10). If we allow the phase difference, $\alpha - \beta$, to change its value, the points of tangency will run along the sides of the rectangle, the axes of the ellipse will turn, and it will flatten out, in two positions degenerating into the straight lines forming the diagonals of the rectangle, as above stated.

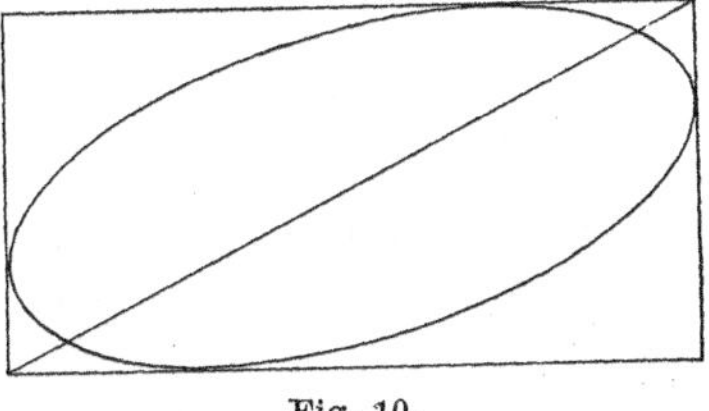

Fig. 10.

If when the phase difference is a right angle the two amplitudes are equal, the ellipse becomes a circle and the acceleration being toward the center and constant in magnitude the motion must be uniform circular motion. A harmonic motion is often defined as the projection of uniform circular motion on a line in its plane. From the value of the central acceleration in a circle we may by projection obtain the properties of simple harmonic motion.

The composition of two simple harmonic motions in intersecting perpendicular lines when their frequencies are *different* gives a class of curves of great interest in acoustics known by the name of Lissajous.

If the ratio of the frequencies is a rational number the least common multiple of the periods of the component vibrations will be a period for both and the curves are reentrant and algebraic. In the simplest case, where one frequency is twice the other, and the phase difference is ε, we have

$$15)\qquad \begin{aligned} x &= a \sin (nt - \varepsilon), \\ y &= b \sin 2nt. \end{aligned}$$

Expressing $\sin 2nt$ in terms of $\cos nt$ and eliminating the functions of t we obtain

$$16)\qquad \frac{x^2}{a^2} \pm \frac{y}{2b} \sin 2\varepsilon = \frac{1}{2} + \frac{1}{2}\sqrt{1 - \frac{y^2}{b^2}} \cos 2\varepsilon.$$

Rationalizing this we shall obtain a curve of the fourth order having one double point, shown in Fig. 11, for $\varepsilon = 0$. If $\varepsilon = \frac{\pi}{4}$, 16) becomes

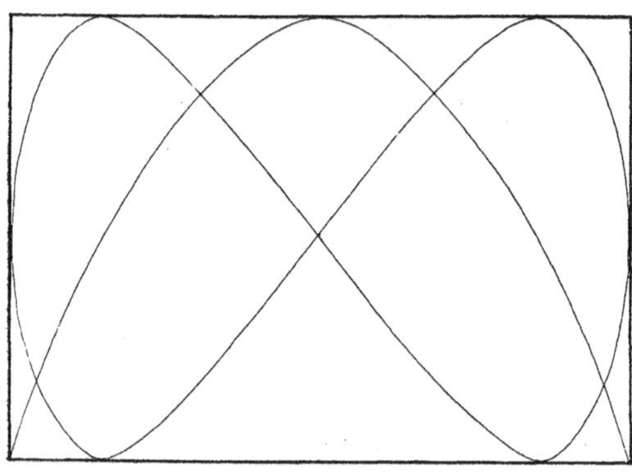

Fig. 11.

$$17)\qquad \frac{x^2}{a^2} - \frac{1}{2}\frac{y}{b} = \frac{1}{2},$$

a parabola (Fig. 11). Since we may always express $\sin mx$ rationally in terms of $\sin x$, $\cos x$, when m is an integer, the elimination may always be performed and the curves will be algebraic.

20. Central Forces. Having now dealt with two cases in which the acceleration passes through a fixed point, — that of the motion of the planets and harmonic motions, it will be convenient to treat the general case. In § 12 we found the nature and magnitude of the acceleration by the differentiation of the equations expressing the motions. We will now consider the inverse problem, that of obtaining the equations describing the motion by integration of the differential equations of motion when the force is given.

We have by § 10, 34) and 35) for the radial acceleration in the direction away from the center,

$$18)\qquad a_r = \frac{d^2 r}{dt^2} - r\left(\frac{d\varphi}{dt}\right)^2$$

and for the transverse acceleration,

$$19)\qquad a_\varphi = \frac{1}{r}\frac{d}{dt}\left(r^2 \frac{d\varphi}{dt}\right).$$

If the acceleration is central $a_\varphi = 0$ and we have by integration

$$20)\qquad r^2 \frac{d\varphi}{dt} = h,$$

Kepler's law of areas.

It will now be convenient to change the independent variable from t to φ and at the same time to introduce the reciprocal of r, $u = \frac{1}{r}$,

$$\frac{dr}{dt} = \frac{dr}{d\varphi}\frac{d\varphi}{dt},$$

and introducing the value of $\frac{d\varphi}{dt}$ from 20) gives

$$\frac{dr}{dt} = \frac{h}{r^2}\frac{dr}{d\varphi} = -h\frac{du}{d\varphi}.$$

Differentiating again and proceeding in like manner,

$$\frac{d^2 r}{dt^2} = -h\frac{d^2 u}{d\varphi^2}\frac{d\varphi}{dt} = -h^2 u^2 \frac{d^2 u}{d\varphi^2},$$

so that finally,

21) $$-a_r = h^2 u^2 \left(\frac{d^2 u}{d\varphi^2} + u\right).$$

If a_r is given as a function of the distance, this is the differential equation of the path. As an example let us consider attractions varying according to the Newtonian law. We have then

$$-a = \frac{\gamma}{r^2} = \gamma u^2,$$

and the differential equation becomes

22) $$\frac{d^2 u}{d\varphi^2} + u = \frac{\gamma}{h^2},$$

or as we may write it,

$$\frac{d^2}{d\varphi^2}\left(u - \frac{\gamma}{h^2}\right) + u - \frac{\gamma}{h^2} = 0.$$

Thus $u - \frac{\gamma}{h^2}$ is given in terms of φ by an equation like equation 8), whose integral is

$$u - \frac{\gamma}{h^2} = a\cos(\varphi - \alpha)$$

or putting $\frac{ah^2}{\gamma} = e$,

23) $$u = \frac{1}{r} = \frac{\gamma}{h^2}(1 + e\cos[\varphi - \alpha]).$$

This is the equation of a conic section with which we started the investigation of § 12. In order to find the eccentricity e let us consider the initial circumstances, or the magnitude and direction of the velocity for a given position of the body. Let the body be projected from a point $\varphi = 0$, $r = R$ with a velocity V, making an angle ε with the radius vector. Now we have

24) $$\tan\varepsilon = r\frac{d\varphi}{dr}, \quad \cot\varepsilon = \frac{1}{r}\frac{dr}{d\varphi} = -\frac{1}{u}\frac{du}{d\varphi}.$$

Putting $\varphi = 0$ in equation 23) gives

$$\frac{1}{R} = \frac{\gamma}{h^2}(1 + e\cos\alpha) \quad \text{or}$$

25) $$\frac{h^2}{R\gamma} - 1 = e\cos\alpha.$$

Differentiating equation 23)

$$\frac{du}{d\varphi} = -\frac{\gamma}{h^2} e \sin(\varphi - \alpha).$$

Introducing this into 24) and putting $\varphi = 0$,

$$\cot\varepsilon = -\frac{R\gamma e}{h^2}\sin\alpha, \quad \text{or}$$

26) $$e\sin\alpha = -\frac{h^2\cot\varepsilon}{R\gamma}.$$

Squaring and adding 25) and 26)

27) $$e^2 = \frac{h^4}{R^2\gamma^2}\operatorname{cosec}^2\varepsilon - \frac{2h^2}{R\gamma} + 1.$$

Also dividing 26) by 25)

28) $$\tan\alpha = \frac{h^2\cot\varepsilon}{R\gamma - h^2}.$$

Now h being the constant moment of velocity (§ 8), is equal to the value when $\varphi = 0$,

29) $$h = VR\sin\varepsilon.$$

Inserting this in 27) and 28) gives

30) $$1 - e^2 = \frac{V^2R^2\sin^2\varepsilon}{\gamma^2}\left(\frac{2\gamma}{R} - V^2\right),$$

31) $$\tan\alpha = \frac{V^2R\sin\varepsilon\cos\varepsilon}{\gamma - V^2R\sin^2\varepsilon}.$$

According as V^2 is less than, equal to, or greater than $\frac{2\gamma}{R}$, e will be less than, equal to, or greater than 1, and the orbit will be respectively an ellipse, parabola, or hyperbola.

The critical velocity, V, has a simple physical significance. Suppose we consider a particle falling from infinity straight toward the center of attraction. Its equation of motion is

$$\frac{d^2r}{dt^2} = -\frac{\gamma}{r^2}.$$

Myltiply by $\frac{dr}{dt}$, both sides become exact derivatives and we may integrate, obtaining

32) $$\frac{1}{2}\left(\frac{dr}{dt}\right)^2 = \frac{\gamma}{r} + \text{const.}$$

If it starts with no velocity the constant is zero, consequently V the velocity at a distance R is given by

$$V^2 = \frac{2\gamma}{R}.$$

Therefore we may state the result by saying that the path will be an ellipse, parabola, or hyperbola according as the body is projected in any direction with a velocity less than, equal to, or greater than the velocity that it would acquire in falling from an infinite distance to the point of projection.

21. Constrained Motion. We have so far considered the moving particle as free to move in any direction. This is however by no means usually the case, since in the majority of cases with which we have to deal the particle forms part of a body which is possibly itself a part of a machine, and is guided by contact with other bodies to travel in certain definite paths, although the velocities with which it travels may be left undetermined. Such limitations to the freedom of movement of a body are known as *constraints*, and they are specified by certain equations having a geometrical significance. In the case of a single particle, the simplest constraint is that in which the particle is constrained to move upon a certain surface. For instance, if the surface is a material one, the particle may, during the whole motion, press against its inner, or concave side, the material preventing the particle from passing across the geometrical surface. The surface may itself be in motion, in this case the constraint is said to be varying, and the equation of the surface will contain the time. Let the equation expressing the constraint be

33) $$\varphi(x, y, z, t) = 0.$$

It is evident that a particle cannot move subject to constraints without calling into play certain reactions due to the constraints. In other words the acceleration experienced by the particle under the influence of given forces will no longer be the same as if the particle were free, but there will be a certain action and reaction between the surface and particle which may be represented by an extra force whose components are X_1, Y_1, Z_1, applied to the particle. The equations of motion may then be written

34) $$m\frac{d^2x}{dt^2} = X + X_1, \quad m\frac{d^2y}{dt^2} = Y + Y_1, \quad m\frac{d^2z}{dt^2} = Z + Z_1,$$

where X, Y, Z are the components of the given forces and X_1, Y_1, Z_1 are the components of the force exercised by the surface upon the particle, that is the reaction of the surface. These are to be found by means of the equation of condition, $\varphi = 0$, which holds for all values of t. Differentiating by t,

$$35)\qquad 0=\frac{d\varphi}{dt}=\frac{\partial\varphi}{\partial x}\frac{dx}{dt}+\frac{\partial\varphi}{\partial y}\frac{dy}{dt}+\frac{\partial\varphi}{\partial z}\frac{dz}{dt}+\frac{\partial\varphi}{\partial t},$$

$$\begin{aligned}36)\qquad 0=\frac{d^2\varphi}{dt^2}&=\frac{\partial\varphi}{\partial x}\frac{d^2x}{dt^2}+\frac{\partial\varphi}{\partial y}\frac{d^2y}{dt^2}+\frac{\partial\varphi}{\partial z}\frac{d^2z}{dt^2}\\&+\frac{\partial^2\varphi}{\partial x_2}\left(\frac{dx}{dt}\right)^2+\frac{\partial^2\varphi}{\partial y^2}\left(\frac{dy}{dt}\right)^2+\frac{\partial^2\varphi}{\partial z_2}\left(\frac{dz}{dt}\right)^2+\frac{\partial^2\varphi}{\partial t^2}\\&+2\frac{\partial^2\varphi}{\partial x\,\partial y}\frac{dx}{dt}\frac{dy}{dt}+2\frac{\partial^2\varphi}{\partial y\,\partial z}\frac{dy}{dt}\frac{dz}{dt}+2\frac{\partial^2\varphi}{\partial x\,\partial z}\frac{dx}{dt}\frac{dz}{dt}\\&+2\frac{\partial^2\varphi}{\partial x\,\partial t}\frac{dx}{dt}+2\frac{\partial^2\varphi}{\partial y\,\partial t}\frac{dy}{dt}+2\frac{\partial^2\varphi}{\partial z\,\partial t}\frac{dz}{dt}.\end{aligned}$$

If we put the unknown forces, X_1, Y_1, Z_1, equal to an unknown function λ multiplied by certain known functions, by inserting the values of $\frac{d^2x}{dt^2}$, $\frac{d^2y}{dt^2}$, $\frac{d^2z}{dt^2}$ from 34) in 36) we obtain an equation, linear in λ, permitting us to find its value in terms of x, y, z, t, $\frac{dx}{dt}$, $\frac{dy}{dt}$, $\frac{dz}{dt}$.

If the surface is smooth, it is evident that it cannot affect a motion of the particle which would naturally take place on the surface. Consequently the reaction has no component tangential to the surface, but is in the direction of the normal. This is otherwise a *definition* of a smooth or frictionless surface. The components of the reaction X_1, Y_1, Z_1 are accordingly proportional to the direction cosines of the normal to the surface $\varphi=0$, so that we may write

$$37)\qquad X_1=\lambda\frac{\partial\varphi}{\partial x},\quad Y_1=\lambda\frac{\partial\varphi}{\partial y},\quad Z_1=\lambda\frac{\partial\varphi}{\partial z}.$$

When λ has been determined as above we have for the magnitude of the reaction,

$$38)\qquad R=\lambda\sqrt{\left(\frac{\partial\varphi}{\partial x}\right)^2+\left(\frac{\partial\varphi}{\partial y}\right)^2+\left(\frac{\partial\varphi}{\partial z}\right)^2}.$$

As an example let us consider the motion of a particle acted upon by gravity and constrained to move on the surface of a fixed sphere of radius l. If the constraint is caused by attaching the particle to a fixed point by means af an inextensible string whose mass is negligible, we have the so-called ideal pendulum. The equation of constraint is

$$39)\qquad \varphi\equiv\frac{1}{2}(x^2+y^2+z^2-l^2)=0,$$

and does not contain t, so that $\frac{\partial\varphi}{\partial t}=0$. If the Z-axis be taken vertically downward the equations of motion are

$$40) \quad \begin{aligned} \frac{d^2x}{dt^2} &= \lambda\frac{\partial\varphi}{\partial x} = \lambda x, \\ \frac{d^2y}{dt^2} &= \lambda\frac{\partial\varphi}{\partial y} = \lambda y, \\ \frac{d^2z}{dt^2} &= \lambda\frac{\partial\varphi}{\partial z} + g = \lambda z + g. \end{aligned}$$

Now inserting these in 36) we have to determine λ,

$$\lambda(x^2 + y^2 + z^2) + gz + \left(\frac{dx}{dt}\right)^2 + \left(\frac{dy}{dt}\right)^2 + \left(\frac{dz}{dt}\right)^2 = 0$$

and since $x^2 + y^2 + z^2 = l^2$,

$$41) \quad \lambda = \frac{1}{l^2}\left[-gz - \left\{\left(\frac{dx}{dt}\right)^2 + \left(\frac{dy}{dt}\right)^2 + \left(\frac{dz}{dt}\right)^2\right\}\right].$$

Using this value of λ in the differential equations 40) we have to integrate

$$42) \quad \begin{aligned} \frac{d^2x}{dt^2} &= -\frac{gxz}{l^2} - \frac{x}{l^2}\left\{\left(\frac{dx}{dt}\right)^2 + \left(\frac{dy}{dt}\right)^2 + \left(\frac{dz}{dt}\right)^2\right\}, \\ \frac{d^2y}{dt^2} &= -\frac{gyz}{l^2} - \frac{y}{l^2}\left\{\left(\frac{dx}{dt}\right)^2 + \left(\frac{dy}{dt}\right)^2 + \left(\frac{dz}{dt}\right)^2\right\}, \\ \frac{d^2z}{dt^2} &= -\frac{gz^2}{l^2} - \frac{z}{l^2}\left\{\left(\frac{dx}{dt}\right)^2 + \left(\frac{dy}{dt}\right)^2 + \left(\frac{dz}{dt}\right)^2\right\} + g. \end{aligned}$$

Now differentiating the equation of constraint 39) by t gives

$$43) \quad x\frac{dx}{dt} + y\frac{dy}{dt} + z\frac{dz}{dt} = 0.$$

Multiplying the equations 42) respectively by $\frac{dx}{dt}, \frac{dy}{dt}, \frac{dz}{dt}$, adding and making use of 43) we may integrate at once and obtain

$$44) \quad \left(\frac{dx}{dt}\right)^2 + \left(\frac{dy}{dt}\right)^2 + \left(\frac{dz}{dt}\right)^2 = 2(gz + h),$$

where h is an arbitrary constant of integration. This integral gives us the square of the velocity and shows that it depends only upon the initial velocity and the height through which the particle has fallen, for if it has a velocity v_0 when $z = z_0$, we have

$$v_0^2 = 2(gz_0 + h),$$

to determine h.

Making use of 44) in 41) we have

$$45) \quad \lambda = -\frac{3gz + 2h}{l^2}$$

and from 38)

$$46) \quad R = -\frac{3gz + 2h}{l}.$$

Multiplying the second of equations 42) by x and subtracting from it the first multiplied by y we obtain

$$x\frac{d^2 y}{dt^2} - y\frac{d^2 x}{dt^2} = 0,$$

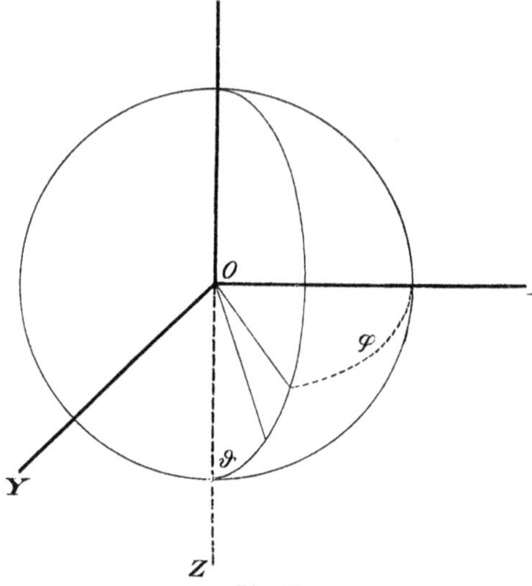

Fig. 12.

which expresses the fact that the horizontal component of the acceleration has no moment about the origin, as in § 12. We may therefore integrate, obtaining,

47) $$x\frac{dy}{dt} - y\frac{dx}{dt} = c,$$

where c is another constant of integration representing the moment of the horizontal component of the velocity about the origin and corresponding to the h of § 20.

It will be convenient to introduce polar coordinates such that (Fig. 12)

48) $$\begin{aligned} x &= l\sin\vartheta\cos\varphi,\\ y &= l\sin\vartheta\sin\varphi,\\ z &= l\cos\vartheta. \end{aligned}$$

Differentiating we have

$$\begin{aligned} dx &= l(\cos\vartheta\cos\varphi\, d\vartheta - \sin\vartheta\sin\varphi\, d\varphi),\\ dy &= l(\cos\vartheta\sin\varphi\, d\vartheta + \sin\vartheta\cos\varphi\, d\varphi),\\ dz &= -l\sin\vartheta\, d\vartheta,\\ dx^2 + dy^2 + dz^2 &= l^2(d\vartheta^2 + \sin^2\vartheta\, d\varphi^2),\\ xdy - ydx &= l^2\sin^2\vartheta\, d\varphi. \end{aligned}$$

Thus our first integrals 44) and 47) become

49) $$l^2\left\{\left(\frac{d\vartheta}{dt}\right)^2 + \sin^2\vartheta\left(\frac{d\varphi}{dt}\right)^2\right\} = 2(gl\cos\vartheta + h),$$

50) $$l^2\sin^2\vartheta\frac{d\varphi}{dt} = c.$$

Substituting the value of $\frac{d\varphi}{dt}$ derived from 50) in 49) we have

51) $$\left(\frac{d\vartheta}{dt}\right)^2 = 2\left(\frac{g}{l}\cos\vartheta + \frac{h}{l^2}\right) - \frac{c^2}{l^4\sin^2\vartheta},$$

from which

$$dt = \pm \frac{l^2 \sin \vartheta \, d\vartheta}{\sqrt{2 l^2 \sin^2 \vartheta (g l \cos \vartheta + h) - c^2}}, \tag{52}$$

or reinserting z and integrating,

$$t = \int_{z_0}^{z} \frac{l \, dz}{\pm \sqrt{2(l^2 - z^2)(gz + h) - c^2}}. \tag{53}$$

Since the integral contains the square root of a polynomial of the third degree in z, the time is given as an elliptic integral in z, or z is an *elliptic function* of the time.

Inserting the value of dt from 52) in 50) we have

$$d\varphi = \pm \frac{c \, d\vartheta}{\sin \vartheta \sqrt{2 l^2 \sin^2 \vartheta (g l \cos \vartheta + h) - c^2}}$$

or in terms of z

$$\varphi = \pm \int_{z_0}^{z} \frac{c l \, dz}{(l^2 - z^2) \sqrt{2(l^2 - z^2)(gz + h) - c^2}}. \tag{54}$$

22. Plane Pendulum. If $c = 0$, by 50) $\varphi = const.$ and we have plane motion of a pendulum. The integral 49) then reduces to

$$\left(\frac{d\vartheta}{dt}\right)^2 = 2\left(\frac{g}{l} \cos \vartheta + \frac{h}{l^2}\right). \tag{55}$$

Differentiating this gives

$$2 \frac{d^2 \vartheta}{dt^2} \frac{d\vartheta}{dt} = -\frac{2g}{l} \sin \vartheta \frac{d\vartheta}{dt}, \text{ or}$$

$$\frac{d^2 \vartheta}{dt^2} + \frac{g}{l} \sin \vartheta = 0,$$

the differential equation of plane pendular motion, which might have been directly obtained for this particular case.

If now during the motion ϑ always remains so small that its square may be neglected in comparison wit unity, we may put

$$\sin \vartheta = \vartheta,$$

so that

$$\frac{d^2 \vartheta}{dt^2} + \frac{g}{l} \vartheta = 0.$$

The integral of this is (cf. § 19, 8)

$$\vartheta = a \sin\left(\sqrt{\frac{g}{l}}\, t - \alpha\right),$$

representing a harmonic motion with period

$$T = 2\pi \sqrt{\frac{l}{g}}. \tag{56}$$

The time taken for a single swing is one half this. Here we have an example of a small vibration, which is harmonic, as stated in § 19.

In order to find the degree of approximation of this solution we must examine the exact equation. Determining the constant of integration h by putting for the velocity when $\vartheta = \vartheta_0$,

$$v_0 = l\left(\frac{d\vartheta}{dt}\right)_0,$$

we have

$$\frac{v_0^2}{l^2} = 2\left(\frac{g}{l}\cos\vartheta_0 + \frac{h}{l^2}\right).$$

Subtracting this from 55) we obtain

$$\left(\frac{d\vartheta}{dt}\right)^2 - \frac{v_0^2}{l^2} = 2\frac{g}{l}(\cos\vartheta - \cos\vartheta_0),$$

from which it follows,

$$57)\qquad t = \pm\int\limits_{\vartheta_0}^{\vartheta}\frac{l\,d\vartheta}{\sqrt{v_0^2 + 2gl(\cos\vartheta - \cos\vartheta_0)}}.$$

There are three cases according as $\frac{v_0^2 - 2gl\cos\vartheta_0}{2gl}$ is less than, equal to, or greater than 1. We shall consider the first.

Let us write in 57) $\cos\vartheta - \cos\vartheta_0 = 2\left(\sin^2\frac{\vartheta_0}{2} - \sin^2\frac{\vartheta}{2}\right)$.

$$58)\qquad t = \pm\int\limits_{\vartheta_0}^{\vartheta}\frac{l\,d\vartheta}{\sqrt{v_0^2 + 4gl\sin^2\frac{\vartheta_0}{2} - 4gl\sin^2\frac{\vartheta}{2}}}.$$

If we put

$$59)\qquad \frac{v_0^2 + 4gl\sin^2\frac{\vartheta_0}{2}}{4gl} = k^2 < 1$$

we have

$$60)\qquad t = \pm\sqrt{\frac{l}{g}}\int\limits_{\vartheta_0}^{\vartheta}\frac{d\left(\frac{\vartheta}{2}\right)}{\sqrt{k^2 - \sin^2\frac{\vartheta}{2}}}.$$

Let us now introduce a new variable ψ such that $k\sin\psi = \sin\frac{\vartheta}{2}$, then

$$k\cos\psi\,d\psi = \cos\frac{\vartheta}{2}\,d\left(\frac{\vartheta}{2}\right),$$

$$t = \pm\sqrt{\frac{l}{g}}\int\limits_{\psi_0}^{\psi}\frac{d\psi}{\sqrt{1 - k^2\sin^2\psi}}.$$

The integral $\int\limits_0^{\psi} \frac{d\psi}{\sqrt{1-k^2\sin^2\psi}}$ is an *elliptic integral* in Legendre's normal form and is denoted by $F(\psi, k)$. In this notation

61) $$t = \pm\sqrt{\frac{l}{g}}\,(F[\psi, k] - F[\psi_0, k]),$$

and the upper or lower sign is to be taken according as the particle is rising or falling at ϑ_0. For call t_m the time of reaching the lowest point. When $\vartheta = 0$, $\psi = 0$ and since $F(0, k) = 0$, we hbve

62) $$t_m = \mp\sqrt{\frac{l}{g}}\,F(\psi_0, k).$$

Since this is to be positive, we see that the lower sign is to be taken in 61) if the particle is falling at ϑ_0. Subtracting 61) from 62) we have

63) $$t_m - t = \sqrt{\frac{l}{g}}\,F(\psi, k)$$

as the time of falling from any inclination ϑ to the lowest point.

The particle swings by the lowest point and continues with negative ϑ until $\frac{d\vartheta}{dt} = 0$, that is until $k^2 - \sin^2\frac{\vartheta}{2} = 0$,

$$\sin\frac{\vartheta}{2} = -k, \quad \sin\psi = -1, \quad \psi = -\frac{\pi}{2}.$$

If the time on reaching the highest point is t_h we have by 63)

$$t_m - t_h = \sqrt{\frac{l}{g}}\,F\left(-\frac{\pi}{2}, k\right) \quad \text{or}$$

$$t_h - t_m = \sqrt{\frac{l}{g}}\,F\left(\frac{\pi}{2}, k\right).$$

The integral

$$F\left(\frac{\pi}{2}, k\right) = \int\limits_0^{\frac{\pi}{2}} \frac{d\psi}{\sqrt{1-k^2\sin^2\psi}}$$

is called the *complete* elliptic integral, and depending only on the parameter k is denoted by $K(k)$. Tables of values of F and K are given in Legendre's *Théorie des Fonctions Elliptiques*. The period of a double oscillation is $4(t_h - t_m)$,

64) $$T = 4\sqrt{\frac{l}{g}}\,K(k).$$

We may develop K in a series, for since

$$\frac{1}{\sqrt{1-k^2\sin^2\psi}} = 1 + \frac{1}{2}k^2\sin^2\psi + \frac{1\cdot 3}{2\cdot 4}k^4\sin^4\psi + \frac{1\cdot 3\cdot 5}{2\cdot 4\cdot 6}k^6\sin^6\psi + \cdots,$$

we have

$$K = \int_0^{\frac{\pi}{2}} d\psi + \frac{1}{2}k^2\int_0^{\frac{\pi}{2}}\sin^2\psi\, d\psi + \frac{1\cdot 3}{2\cdot 4}k^4\int_0^{\frac{\pi}{2}}\sin^4\psi\, d\psi + \cdots$$

Now since

$$\int_0^{\frac{\pi}{2}}\sin^{2n}\psi\, d\psi = \frac{1\,.\,3\,.\,5\ldots 2n-1}{2\,.\,4\,.\,6\,..\;2n}\,\frac{\pi}{2},$$

$$K = \frac{\pi}{2}\left\{1 + \left(\frac{1}{2}\right)^2 k^2 + \left(\frac{1\,.\,3}{2\,.\,4}\right)^2 k^4 + \left(\frac{1\,.\,3\,.\,5}{2\,.\,4\,.\,6}\right)^2 k^6 + \cdots\right\}.$$

If α be the maximum value of ϑ, for which $\psi = \frac{\pi}{2}$, $k = \sin\frac{\alpha}{2}$, and the period is given by

$$65)\quad T = 2\pi\sqrt{\frac{l}{g}}\left\{1 + \left(\frac{1}{2}\right)^2\sin^2\frac{\alpha}{2} + \left(\frac{1\,.\,3}{2\,.\,4}\right)^2\sin^4\frac{\alpha}{2} + \left(\frac{1\,.\,3\,.\,5}{2\,.\,4\,.\,6}\right)^2\sin^6\frac{\alpha}{2} + \cdots\right\}.$$

This is the formula which is used to correct our result 56) for finite oscillations. If α is 1^0 the correction is less than one part in fifty thousand, and if $\alpha = 5^0$ it is less than one in two thousand.

23. Spherical Pendulum. Let us now return to equations 53) and 54), which we will write

$$66)\qquad t = \pm\int_{z_0}^{z}\frac{l\,dz}{\sqrt{\Phi(z)}},\quad\text{and}$$

$$67)\qquad \varphi = \pm\int_{z_0}^{z}\frac{c\,l\,dz}{(l^2-z^2)\sqrt{\Phi(z)}},$$

where $\Phi(z) = 2(l^2 - z^2)(gz + h) - c^2$.

As the integrals are real $\Phi(z)$ must be positive for all values of z that occur in the motion.

Substituting successively for z, $-\infty$, $-l$, z_0, $+l$ we find

$$68)\quad \Phi(-\infty) = +\infty,\quad \Phi(-l) = -c^2,\quad \Phi(z_0) > 0,\quad \Phi(l) = -c^2.$$

Accordingly the polynomial $\Phi(z)$ has *three real* roots. If we call these α, β, γ in the order of magnitude, they lie so that

$$l > \alpha > z_0 > \beta > -l > \gamma,$$

Fig. 13 is the graph of $\Phi(z)$ as ordinate, with z as abscissa.

Since

$$\Phi(z) = -2gz^3 - 2hz^2 + 2gl^2z + 2hl^2 - c^2$$
$$= -2g(z-\alpha)(z-\beta)(z-\gamma),$$

we have, equating the coefficients of z,

$$2gl^2 = -2g(\alpha\beta + \beta\gamma + \alpha\gamma),$$

from which

$$\gamma = -\frac{l^2 + \alpha\beta}{\alpha + \beta}.$$

Since α and β both lie between $-l$ and l the numerator is positive irrespective of the sign of either, and since γ is negative, the denominator must be positive, or $\alpha + \beta > 0$. Since between β and γ $\Phi(z)$ is negative, z cannot in the motion lie in this region (for $\sqrt{\Phi(z)}$ must be real). Now since, 66)

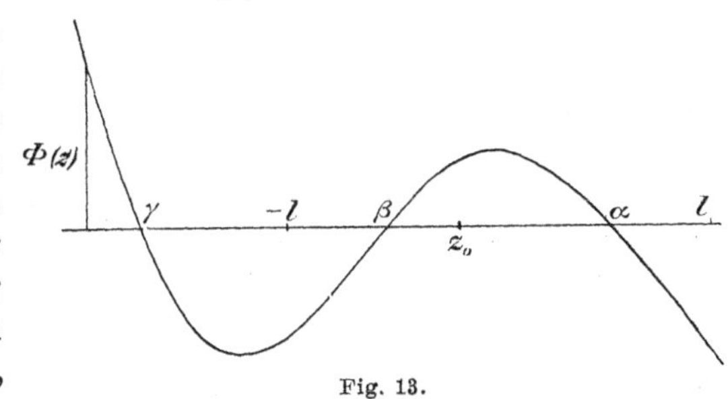

Fig. 13.

$$l^2\left(\frac{dz}{dt}\right)^2 = \Phi(z),$$

when $\Phi(z)$ vanishes z is a maximum or minimum, hence the motion takes place between two horizontal circles at depths $z=\alpha$ and $z=\beta$ below the origin. Although β may be negative, yet since $\beta+\alpha>0$ the mean position of the particle is below the center of the sphere.

Since by 50) $\frac{d\varphi}{dt} = \frac{c}{l^2 - z^2}$, φ always varies in the same sense and when z equals α or β the path has a horizontal tangent, for $\frac{dz}{dt} = 0$, while $\frac{d\varphi}{dt}$ is not equal to zero.

If z_0 is a root, that is if the particle was originally on one of the limiting circles, we must take the positive sign for the radical in the integrals if $z_0 = \beta$ (so that $\frac{dz}{dt}$ may be positive and z increase), the negative sign if $z_0 = \alpha$.

The time of passing from the highest to the lowest point is

$$T = \int_\beta^\alpha \frac{l\,dz}{\sqrt{\Phi(z)}}.$$

The meridian planes passing through the points of tangency with the parallels α and β are planes of symmetry for the path.

For if we consider two points P, P' of azimuths φ, φ' lying on the same parallel and on opposite sides of a vertex A (Fig. 14),

$$\varphi_A - \varphi = \int_z^\alpha \frac{cl\,dz}{(l^2 - z^2)\sqrt{\Phi(z)}},$$

and since the radical changes sign on passing through a vertex,

$$\varphi' - \varphi_A = -\int_\alpha^z \frac{cl\,dz}{(l^2 - z^2)\sqrt{\Phi(z)}} = \varphi_A - \varphi.$$

Therefore the points P, P' are symmetrical about A and the times of traveling the arcs PA and AP' are equal to $\int_z^\alpha \frac{l\,dz}{\sqrt{\Phi(z)}}$. In like manner it can be shown that the path is symmetrical about an upper vertex B. The path is accordingly composed of equal parts continually repeated. It of course is not generally true that the path will be reentrant after going once around the sphere.

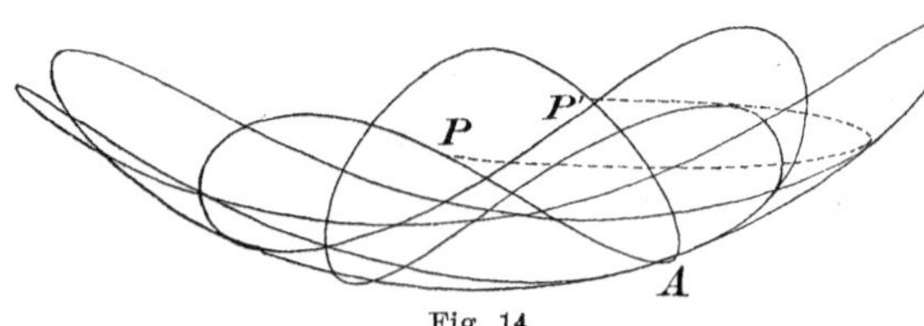

Fig. 14.

We will now consider the horizontal projection of the path.

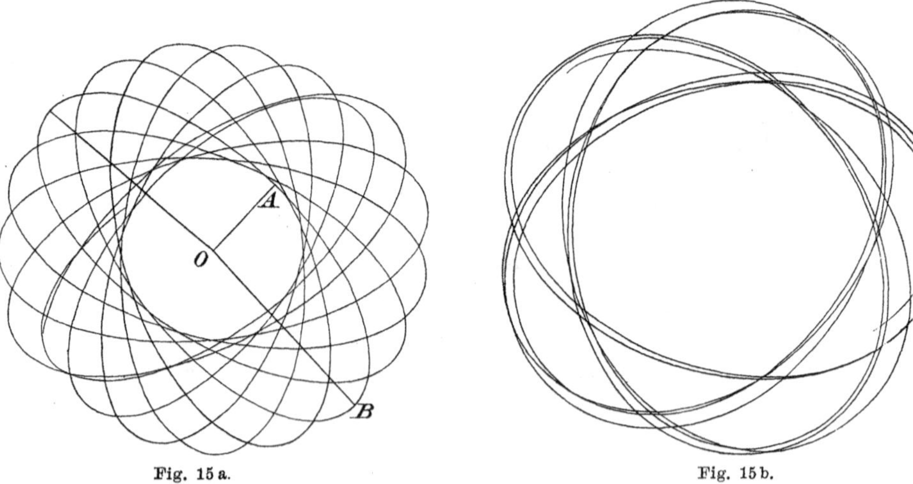

Fig. 15 a. Fig. 15 b.

1°. Suppose both limiting parallels are below the equator, the projection of the circle $z = \alpha$ is within that of $z = \beta$, and the path is similar to Fig. 15, the angle subtended at the center by two successive points of tangency being greater than a right angle, as we shall see.

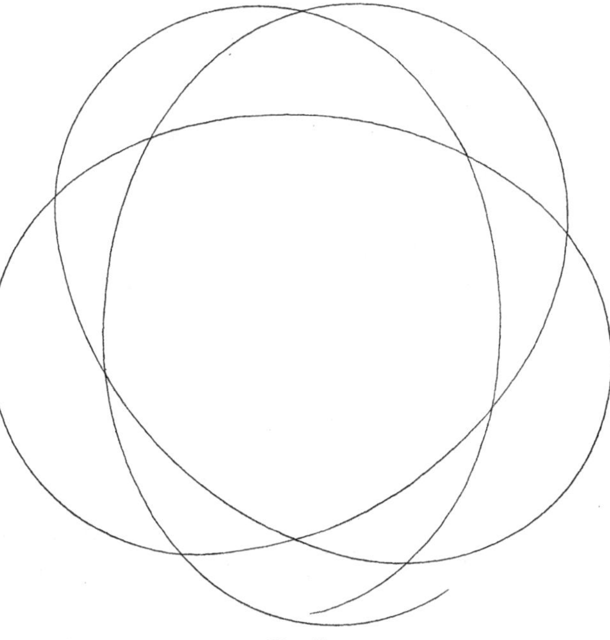
Fig. 15 c.

Figs. 14, 15a, 15b, 15c, 16a, are reproductions of photographs of actual swings of a pendulum. A brass ball was swung by a string attached to a screw-eye, and carried a small incandescent lamp. On the floor below, and at one side were placed cameras with open shutters, in a dark room. When the ball was swung, the light was turned on for a sufficient number of swings, and the path registered on the photographic plate. On the photograph Fig. 15c, the maximum and minimum radii were measured, from which could be calculated the roots α, β, and thence γ. Then from equa-

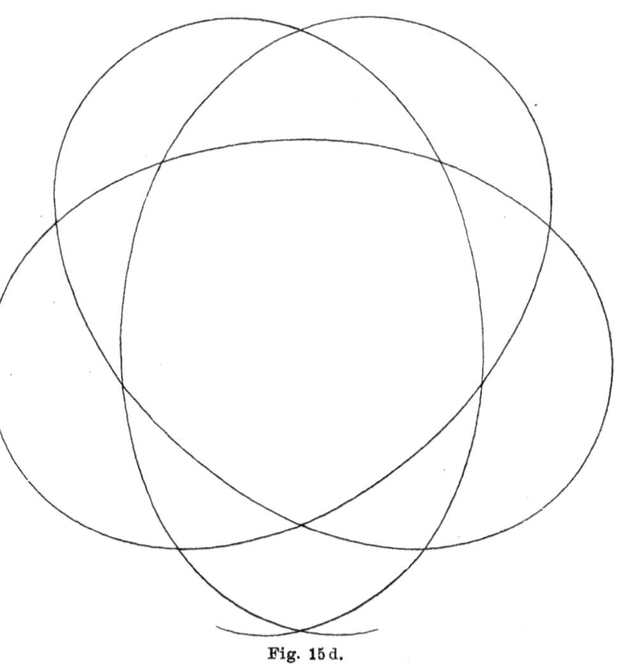
Fig. 15 d.

tion 67), by an arithmetical approximate quadrature, φ was calculated for a number of values of z, from which with the polar coordinates φ, $r = \sqrt{l^2 - z^2}$, the horizontal projection Fig. 15d was drawn. It will be observed that it almost exactly coincides with the *observed* curve Fig. 15c. From the projection and the values of z the perspective Fig. 16b was constructed, which in like manner nearly coincides with the observed Fig. 16a. The eye is *below* the shaded square in the figure. Figures 15d, 16b were constructed by Mr. Joseph G. Coffin.

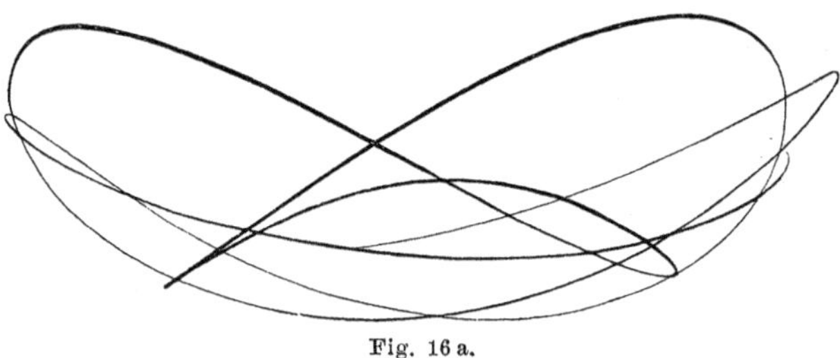

Fig. 16 a.

2°. If β is negative, the projection of the circle $z = \alpha$ is still within that of $z = \beta$, for since $\alpha + \beta > 0$, the lower circle is farther

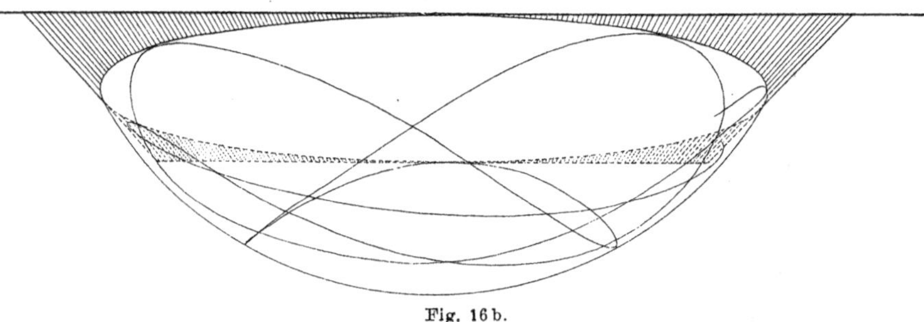

Fig. 16 b.

from the center than the upper. The projection of the path is also tangent to the equator.

The angle AOB in Fig. 15a has the value

$$69)\qquad \Psi = \int_{\beta}^{\alpha} \frac{c\,l\,dz}{(l^2 - z^2)\sqrt{\Phi(z)}}.$$

Inserting the value $\gamma = -\dfrac{l^2 + \alpha\beta}{\alpha + \beta}$ in

$$\varphi(z) = -2g(z - \alpha)(z - \beta)(z - \gamma),$$

we have

$$\Phi(z) = \frac{2g}{\alpha+\beta}(\alpha - z)(z - \beta)\{z(\alpha+\beta) + l^2 + \alpha\beta\}.$$

Now putting $z = l$,

$$\Phi(l) = -c^2 = \frac{2g}{\alpha+\beta}(\alpha - l)(l - \beta)(l + \alpha)(l + \beta),$$

writing

$$A = \sqrt{(l-\alpha)(l-\beta)},$$

$$B = \sqrt{(l+\alpha)(l+\beta)},$$

we have

$$c = AB\sqrt{\frac{2g}{\alpha+\beta}}$$

70) $$\Psi = \int_\beta^\alpha \frac{lAB\,dz}{(l^2 - z^2)\sqrt{(\alpha - z)(z-\beta)}\,\{z(\alpha+\beta)+\alpha\beta+l^2\}}.$$

In order to find limits between which this integral lies for all possible values of α and β, we notice that the coefficient of z in the last factor is positive, and that the value of the factor, varying always in the same sense as z, necessarily lies between the two values it would have when z had its extreme values, l and $-l$. But these are B^2 and A^2, so that

$$B^2 > z(\alpha+\beta) + l^2 + \alpha\beta > A^2.$$

Substituting in the radical a value that is too great or too small will make the integral have an error in the opposite sense, therefore

$$A\int_\beta^\alpha \frac{l\,dz}{(l^2 - z^2)\sqrt{(\alpha - z)(z-\beta)}} < \Psi < B\int_\beta^\alpha \frac{l\,dz}{(l^2 - z^2)\sqrt{(\alpha - z)(z-\beta)}}.$$

The polynomial under the radical being now of only the second degree, the integral can be easily calculated, as follows.

$$\int_\beta^\alpha \frac{l\,dz}{(l^2 - z^2)\sqrt{(\alpha - z)(z-\beta)}} = \frac{1}{2}\int_\beta^\alpha \frac{dz}{(z+l)\sqrt{(\alpha - z)(z-\beta)}} - \frac{1}{2}\int_\beta^\alpha \frac{dz}{(z-l)\sqrt{(\alpha - z)(z-\beta)}}$$

$$= \frac{1}{2}\left(\frac{\pi}{\sqrt{\alpha+l}\sqrt{\beta+l}} + \frac{\pi}{\sqrt{\alpha-l}\sqrt{\beta-l}}\right)$$

$$= \frac{\pi}{2}\left(\frac{1}{B} + \frac{1}{A}\right).$$

Accordingly we have

$$\frac{\pi}{2}\left(1+\frac{A}{B}\right) < \Psi < \frac{\pi}{2}\left(1+\frac{B}{A}\right),$$

therefore $\Psi > \frac{\pi}{2}$, as above stated.[1])

If in the integral 70) we substitute for the factor

$$\{z(\alpha+\beta) + l^2 + \alpha\beta\}$$

the greatest and least values that it takes during the motion, namely $z = \alpha$ and $z = \beta$, we shall get closer limits between which ψ lies. If we then make α and β approach l, Ψ will approach a right angle, so that the horizontal projection tends to be a closed curve.

This case may also be treated directly. Our equations 40) were

$$\frac{d^2x}{dt^2} = \lambda x, \quad \frac{d^2y}{dt^2} = \lambda y, \quad \frac{d^2z}{dt^2} = \lambda z + g.$$

Now we have $z = \sqrt{l^2 - (x^2+y^2)} = l\left(1 - \frac{x^2+y^2}{l^2}\right)^{\frac{1}{2}}$ and developing by the binomial theorem,

$$z = l\left\{1 - \frac{x^2+y^2}{2l^2} + \cdots\right\}.$$

If now x and y are small with respect to l and we neglect small quantities of the second order, z is constant. Then $\frac{d^2z}{dt^2} = 0$, and from the third equation above,

$$\lambda = -\frac{g}{l}.$$

Inserting this value of λ in the first two gives

$$\frac{d^2x}{dt^2} = -\frac{g}{l}x, \quad \frac{d^2y}{dt^2} = -\frac{g}{l}y,$$

the integrals of which are

$$x = a\sin\left(\sqrt{\frac{g}{l}}\,t - \alpha\right),$$

$$y = b\sin\left(\sqrt{\frac{g}{l}}\,t - \beta\right),$$

where a, b, α, β are arbitrary constants, giving elliptic harmonic motion of the same period as that of the small plane harmonic motion.

Another important case is that in which the two roots α and β are equal. We then have z and ϑ constant, and

$$z = z_0 = \alpha = \beta.$$

---

1) This treatment is taken from Appell, *Mécanique Rationelle.* The proof that $\Psi > \frac{\pi}{2}$ is due to Puiseux.

The condition for equal roots is that $\Phi(z)$ and $\Phi'(z)$ have a common root. Now

$$\Phi(z) = -2gz^3 - 2hz^2 + 2gl^2z + 2hl^2 - c^2,$$
$$\Phi'(z) = -6gz^2 - 4hz + 2gl^2.$$

If then $\Phi'(z_0) = 0$, we have

$$2h = \frac{gl^2 - 3gz_0^2}{z_0},$$

together with

$$\Phi(z_0) = 2(gz_0 + h)(l^2 - z_0^2) - c^2 = 0,$$

from which

$$c^2 = 2(gz_0 + h)(l^2 - z_0^2) = g\frac{(l^2 - z_0^2)^2}{z_0},$$
$$c = \pm\sqrt{\frac{g}{z_0}}(l^2 - z_0^2).$$

We accordingly have for the value of

$$\frac{d\varphi}{dt} = \frac{c}{l^2 - z^2} = \pm\sqrt{\frac{g}{z_0}}.$$

We thus obtain for the time of revolution

$$T = 2\pi\sqrt{\frac{z_0}{g}}.$$

The time of revolution of a conical pendulum is the same as that of a complete oscillation of a plane pendulum of length z_0 performing small vibrations.

As ϑ approaches a right angle, z_0 and therefore T approaches zero, that is the velocity increases without limit. We have in this case

$$v = l\sin\vartheta\frac{d\varphi}{dt} = \sqrt{\frac{g}{z_0}(l^2 - z_0^2)}.$$

Now the centripetal acceleration in the circular motion is (§ 10),

$$\frac{v^2}{\sqrt{l^2 - z_0^2}} = g\frac{\sqrt{l^2 - z_0^2}}{z_0} = g\tan\vartheta.$$

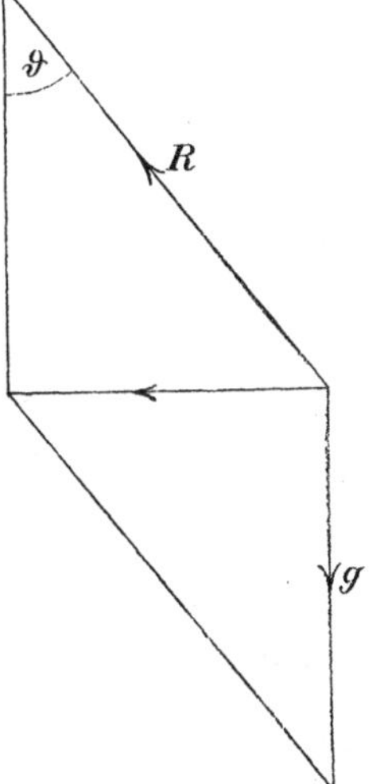

Fig. 17.

An acceleration g directed downward together with the reaction R directed toward the center of the sphere will compound into an acceleration $g\tan\vartheta$ in a horizontal direction (Fig. 17). Accordingly if the particle is projected horizontally with the velocity v, it will describe a circle.

CHAPTER III.

GENERAL PRINCIPLES. WORK AND ENERGY.

24. Work. If a point be displaced in a straight line, under the action of a force which is constant in magnitude and direction, the product of the length of the displacement and the component of the force in the direction of the displacement, that is, the *geometric product* of the force and the displacement § 4, 10), is called the *work* done by the force in producing the displacement. If the components of the force F are X, Y, Z, and those of the displacement s are s_x, s_y, s_z, the work W is

$$1)\qquad W = sF\cos(Fs) = Xs_x + Ys_y + Zs_z.$$

It is at once evident that if a force is resolved into components, the sum of the works of the components is equal to the work of the resultant, for if

$$X = X_1 + X_2,$$
$$Y = Y_1 + Y_2,$$
$$Z = Z_1 + Z_2,$$
$$W_1 = X_1 s_x + Y_1 s_y + Z_1 s_z,$$
$$W_2 = X_2 s_x + Y_2 s_y + Z_2 s_z,$$
$$W = (X_1 + X_2)s_x + (Y_1 + Y_2)s_y + (Z_1 + Z_2)s_z = W_1 + W_2.$$

Since work is defined as force$\times$distance, we have for its dimensions,

$$[\text{Work}] = [L]\left[\frac{ML}{T^2}\right] = [ML^2T^{-2}].$$

The C. G. S. unit of work is the work done when a force of one dyne produces a displacement of one centimeter in its own direction. This unit is called the *erg* = gm · cm² · sec⁻².

If the displacement be not in a straight line, and the force be not constant, the work done in an infinitesimal displacement ds is

$$2)\qquad dW = \left(X\frac{dx}{ds} + Y\frac{dy}{ds} + Z\frac{dz}{ds}\right)ds,$$

and the work done in a displacement along any path AB is the line integral

$$3)\qquad W_{AB} = \int_A^B \left(X\frac{dx}{ds} + Y\frac{dy}{ds} + Z\frac{dz}{ds}\right)ds.$$

The components of the force are supposed to be given as functions of s and the derivatives $\frac{dx}{ds}, \frac{dy}{ds}, \frac{dz}{ds}$ are known as functions of s from the equations of the path.

Understanding this, we may write

$$4)\qquad W_{AB} = \int_A^B (X\,dx + Y\,dy + Z\,dz).$$

25. Statics. Virtual Work. Suppose that we have a system of n material points. If they are entirely free to move, they require $3n$ coordinates for their specification. If however they are subjected to geometrical constraint, as explained in § 21 for a single particle, there must be certain relations satisfied by their coordinates. Let these *equations of condition* or *constraint* be

$$5)\qquad \begin{aligned} &\varphi_1(x_1, y_1, z_1, x_2, y_2, z_2, \ldots x_n, y_n, z_n) = 0,\\ &\varphi_2(x_1, y_1, z_1, x_2, y_2, z_2, \ldots x_n, y_n, z_n) = 0,\\ &\ldots\ldots\ldots\ldots\ldots\ldots\\ &\ldots\ldots\ldots\ldots\ldots\ldots\\ &\varphi_k(x_1, y_1, z_1, x_2, y_2, z_2, \ldots x_n, y_n, z_n) = 0. \end{aligned}$$

Such constraints may be caused in a great variety of ways. Particles may be caused to lie on certain fixed or moving surfaces, may be connected by inextensible strings which may pass over pulleys, or by rigid links variously jointed.

For instance, if two particles 1 and 2 are connected by a rigid rod of length l, either particle must move on a sphere of radius l of which the other is the center, and we have the equation of condition

$$\varphi \equiv (x_1 - x_2)^2 + (y_1 - y_2)^2 + (z_1 - z_2)^2 - l^2 = 0.$$

(We might have constraints defined by inequalities, e. g., if a particle were obliged to stay on or within a spherical surface of radius l the constraint would be only from without, and we should have

$$(x - a)^2 + (y - b)^2 + (z - c)^2 - l^2 \leqq 0.$$

We shall assume that the constraint is toward both sides, and is defined by an equation.)

If any particle at x_r, y_r, z_r is displaced by a small amount so that it has the coordinates

$$x_r + \delta x_r, \quad y_r + \delta y_r, \quad z_r + \delta z_r,$$

in order that the constraints may hold we must have for each φ,

$$6)\qquad \begin{aligned} &\varphi(x_r, y_r, z_r, \ldots) = 0,\\ &\varphi(x_r + \delta x_r, y_r + \delta y_r, z_r + \delta z_r, \ldots) = 0, \end{aligned}$$

and if φ be a continuous function, developing by Taylor's Theorem,

$$\varphi(x_r + \delta x_r, y_r + \delta y_r, z_r + \delta z_r, \ldots) = \varphi(x_r, y_r, z_r, \ldots) + \delta x_r \frac{\partial \varphi}{\partial x_r} + \delta y_r \frac{\partial \varphi}{\partial y_r} + \delta z_r \frac{\partial \varphi}{\partial z_r} + \cdots + \frac{1}{2}\delta x_r{}^2 \frac{\partial^2 \varphi}{\partial x_r^2} + \cdots$$

Accordingly, taking account only of the terms of the first order in the small quantities δx_r, δy_r, δz_r, and using equations 6) we have

$$7)\qquad \frac{\partial \varphi}{\partial x_r}\delta x_r + \frac{\partial \varphi}{\partial y_r}\delta y_r + \frac{\partial \varphi}{\partial z_r}\delta z_r = 0.$$

If a number of particles are displaced, we must take the sum of expressions like the above for all the particles, or

$$8)\qquad \sum_{r=1}^{r=n}\left\{\frac{\partial \varphi}{\partial x_r}\delta x_r + \frac{\partial \varphi}{\partial y_r}\delta y_r + \frac{\partial \varphi}{\partial z_r}\delta z_r\right\} = 0,$$

as the conditions which must be satisfied by all the displacements δx_r, δy_r, δz_r. There must be one such equation for each function φ. Such displacements, which are purely arbitrary, except that they satisfy the equations of condition, are called *virtual*, being possible, as opposed to the displacements that actually take place in a motion of the system. If the equations of constraint contain the time, t is supposed to be kept constant during the virtual displacement.

The number of independent coordinates possessed by a system is called the number of *degrees of freedom* of the system, which may be otherwise defined as the number of data necessary to fully specify its position. Between the $3n$ changes δx, δy, δz, occurring in an equation, there are k linear equations, hence only $3n - k$ of them may be taken arbitrarily, and this is the number of degrees of freedom of the system.

It has long been customary to make a subdivision of the subject of Dynamics entitled *Statics* which deals with only those problems in which forces produce equilibrium. A system is in equilibrium when the impressed forces upon its various particles together with the constraints balance each other in such a way that there is no tendency toward motion of any part of the system. The Principle of Virtual Work is the most general analytical statement of the conditions of equilibrium of a system. It was used in a very simple form by Galileo, but its generality and its utility for the solution of problems in statics was first recognized by Jean Bernoulli, and it was made by Lagrange the foundation of statics.[1])

If the system consists of a single free particle, in order for it to be in equilibrium the resultant of all the forces applied to it, whose components are $X = \Sigma X_r$, $Y = \Sigma Y_r$, $Z = \Sigma Z_r$, must vanish,

$$9)\qquad X = Y = Z = 0.$$

1) For the history of the principle see Lagrange, *Mécanique Analytique*, I^{e} Partie, Section I, §§ 16 and 17.

If we multiply these equations respectively by the arbitrary small quantities δx, δy, δz, and add, we get

$$10)\qquad X\delta x + Y\delta y + Z\delta z = 0,$$

which states that the work done in an infinitesimal displacement of a point from its position of equilibrium vanishes. The equation 10) is equivalent to the equations 9), for since the quantities δx, δy, δz are arbitrary, if X, Y, Z are different from zero, we may take δx, δy, δz respectively of the same sign as X, Y, Z, — each product will then be positive, and the sum will not vanish. If the sum is to vanish for *all possible choices* of δx, δy, δz, X, Y, Z must vanish.

If the particle is not free, but constrained to lie on a surface $\varphi = 0$, δx, δy, δz are not entirely arbitrary, but must satisfy

$$7)\qquad \frac{\partial \varphi}{\partial x}\delta x + \frac{\partial \varphi}{\partial y}\delta y + \frac{\partial \varphi}{\partial z}\delta z = 0.$$

Let us multiply this by a quantity λ and add it to 10), obtaining

$$11)\qquad \left(X + \lambda\frac{\partial \varphi}{\partial x}\right)\delta x + \left(Y + \lambda\frac{\partial \varphi}{\partial y}\right)\delta y + \left(Z + \lambda\frac{\partial \varphi}{\partial z}\right)\delta z = 0.$$

We may no longer conclude that the coefficients of δx, δy, δz must vanish, for δx, δy, δz are not arbitrary, being connected by the equation 7). Two of them are however arbitrary, say δy and δz, λ has not yet been fixed — suppose it determined so that

$$X + \lambda\frac{\partial \varphi}{\partial x} = 0.$$

Then we have

$$\left(Y + \lambda\frac{\partial \varphi}{\partial y}\right)\delta y + \left(Z + \lambda\frac{\partial \varphi}{\partial z}\right)\delta z = 0,$$

in which δy and δz are perfectly arbitrary, it therefore follows of necessity that the coefficients vanish.

$$Y + \lambda\frac{\partial \varphi}{\partial y} = 0,\quad Z + \lambda\frac{\partial \varphi}{\partial z} = 0.$$

By the introduction of the multiplier λ we are accordingly enabled to draw the same conclusion as if δx, δy, δz were arbitrary. If X, Y, Z refer to the resultant of the impressed forces only, not including the reaction, equations 9) do not hold, but if we suppose 10) to hold, we shall obtain the conditions for equilibrium. Eliminating λ from the above three equations we get

$$\frac{X}{\frac{\partial \varphi}{\partial x}} = \frac{Y}{\frac{\partial \varphi}{\partial y}} = \frac{Z}{\frac{\partial \varphi}{\partial z}}.$$

Now the direction cosines of the normal to the surface $\varphi = 0$ are proportional to $\frac{\partial \varphi}{\partial x}$, $\frac{\partial \varphi}{\partial y}$, $\frac{\partial \varphi}{\partial z}$, consequently, the components X, Y, Z being proportional to these direction cosines, the resultant is in the direction of the normal to the surface. But owing to the constraint the motion can be only tangential, consequently the particle cannot move, and the applied forces together with the reaction produce equilibrium.

The Principle of Virtual Work is as follows. If any system of as many bodies or particles as we please, each acted upon by any forces whatsoever, is in equilibrium, and a small arbitrary virtual displacement is given to each point of the system, the work done by all the forces will vanish (at least to the first order of small quantities). For instance a particle placed on a smooth surface under the action of gravity experiences a force mg vertically downward. If we displace it a distance ds the work done by the force will be $mg\,dz$, if the z coordinate is taken positively downward. We may write this

$$dW = mg\frac{dz}{ds}ds,$$

and if this vanishes whatever the value of ds for all directions of displacement on the surface $\frac{dz}{ds}$ must be zero, that is the tangent plane to the surface is horizontal. But the particle is in equilibrium at such a point.

Conversely, if the surface is not horizontal, dW will not vanish for all possible displacements, neither will the particle be in equilibrium. (It is to be noticed that in the neighbourhood of a point where the tangent plane is horizontal dz is proportional to ds^2, so that the work, although vanishing to the first order, does not vanish to the second, z is in this case a maximum or minimum.)

Simple illustrations of the principle of virtual work are furnished by the so-called mechanical powers. Consider in particular the pulley. The mechanical advantage or multiplying power as regards force, that is the ratio of the force sustained by the movable block to the tension on the cord, is equal to n, the number of cords coming from the movable block, for the fundamental assumption is that the tension of the cord is everywhere the same. If the end of the cord is displaced a small distance in its own direction, the block is displaced $1/n^{th}$ of that distance, consequently the work of the two equilibrating forces is equal in absolute magnitude, but one being positive and the other negative, their sum is zero.

By means of this principle Lagrange gave a simple general proof of the principle of virtual work. He supposed each force applied to a point of the system to be replaced by a pull of a block of pulleys, the number of pulleys in each block being so chosen that the proper force could be produced by the tension of a single cord passing over all the pulleys and fastened to a weight at one end, the other being fixed. If now the forces are in equilibrium, an arbitrary small displacement of all the blocks will neither raise nor lower the weight at the end of the string. Thus by the application of the property of the pulley the principle was proved.[1])

We shall not here undertake to give a more formal proof of the principle, which may be given by an analysis of the various kinds of constraint, such a proof is found in Appell, *Traité de Mécanique Rationelle*, Tom. I, Chap. 7.

If the forces X_1, Y_1, Z_1 act upon the particle 1, X_2, Y_2, Z_2 upon the particle 2, etc., the condition of equilibrium is

$$12)\quad X_1\delta x_1 + Y_1\delta y_1 + Z_1\delta z_1 + X_2\delta x_2 + Y_2\delta y_2 + Z_2\delta z_2 + \cdots Z_n\delta z_n = 0,$$

or as we may write it,

$$13)\quad \sum(X\delta x + Y\delta y + Z\delta z) = 0.$$

This is the analytical expression of the Principle of Virtual Work.

If the particles satisfy the equations of constraint 5) the displacements must satisfy the equations

$$14)\quad \begin{aligned} &\frac{\partial\varphi_1}{\partial x_1}\delta x_1 + \frac{\partial\varphi_1}{\partial y_1}\delta y_1 + \frac{\partial\varphi_1}{\partial z_1}\delta z_1 + \frac{\partial\varphi_1}{\partial x_2}\delta x_2 + \frac{\partial\varphi_1}{\partial y_2}\delta y_2 + \cdots \frac{\partial\varphi_1}{\partial z_n}\delta z_n = 0,\\ &\frac{\partial\varphi_2}{\partial x_1}\delta x_1 + \frac{\partial\varphi_2}{\partial y_1}\delta y_1 + \frac{\partial\varphi_2}{\partial z_1}\delta z_1 + \frac{\partial\varphi_2}{\partial x_2}\delta x_2 + \frac{\partial\varphi_2}{\partial y_2}\delta y_2 + \cdots \frac{\partial\varphi_2}{\partial z_n}\delta z_n = 0,\\ &\cdots\cdots\cdots\cdots\cdots\cdots\cdots\cdots\\ &\cdots\cdots\cdots\cdots\cdots\cdots\cdots\cdots\\ &\frac{\partial\varphi_k}{\partial x_1}\delta x_1 + \frac{\partial\varphi_k}{\partial y_1}\delta y_1 + \frac{\partial\varphi_k}{\partial z_1}\delta z_1 + \frac{\partial\varphi_n}{\partial x_2}\delta x_2 + \frac{\partial\varphi_k}{\partial y_2}\delta y_2 + \cdots \frac{\partial\varphi_k}{\partial z_n}\delta z_n = 0.\end{aligned}$$

Multiplying the equations 14) respectively by $\lambda_1, \lambda_2, \ldots \lambda_k$, and adding to 12) we have

$$15)\quad \begin{aligned} &\left(X_1 + \lambda_1\frac{\partial\varphi_1}{\partial x_1} + \lambda_2\frac{\partial\varphi_2}{\partial x_1} + \cdots \lambda_k\frac{\partial\varphi_k}{\partial x_1}\right)\delta x_1\\ &+\left(X_2 + \lambda_1\frac{\partial\varphi_1}{\partial x_2} + \lambda_2\frac{\partial\varphi_2}{\partial x_2} + \cdots \lambda_k\frac{\partial\varphi_k}{\partial x_2}\right)\delta x_2\\ &\cdots\cdots\cdots\cdots\cdots\cdots\\ &+\left(Y_1 + \lambda_1\frac{\partial\varphi_1}{\partial y_1} + \lambda_2\frac{\partial\varphi_2}{\partial y_1} + \cdots \lambda_k\frac{\partial\varphi_k}{\partial y_1}\right)\delta y_1\\ &\cdots\cdots\cdots\cdots\cdots\cdots\\ &\cdots\cdots\cdots\cdots\cdots\cdots\\ &+\left(Z_n + \lambda_1\frac{\partial\varphi_1}{\partial z_n} + \lambda_2\frac{\partial\varphi_2}{\partial z_n} + \cdots \lambda_k\frac{\partial\varphi_k}{\partial z_n}\right)\delta z_n = 0.\end{aligned}$$

Of the $3n$ quantities $\delta x_1, \ldots \delta z_n$, only $3n - k$ are arbitrary, we may however determine the k multipliers λ so that the coefficients

1) Ibid. § 18.

of the k other δ's vanish, then the coefficients of the $3n-k$ *arbitrary* δ's *must* vanish, so that we get the $3n$ equations

16)
$$\begin{aligned}
X_1 + \lambda_1 \frac{\partial \varphi_1}{\partial x_1} + \lambda_2 \frac{\partial \varphi_2}{\partial x_1} + \cdots \lambda_k \frac{\partial \varphi_k}{\partial x_1} &= 0,\\
X_2 + \lambda_1 \frac{\partial \varphi_1}{\partial x_2} + \lambda_2 \frac{\partial \varphi_2}{\partial x_2} + \cdots \lambda_k \frac{\partial \varphi_k}{\partial x_2} &= 0,\\
\cdots\cdots\cdots\cdots\cdots\cdots\\
\cdots\cdots\cdots\cdots\cdots\cdots\\
Z_n + \lambda_1 \frac{\partial \varphi_1}{\partial z_n} + \lambda_2 \frac{\partial \varphi_2}{\partial z_n} + \cdots \lambda_k \frac{\partial \varphi_k}{\partial z_n} &= 0.
\end{aligned}$$

Eliminating from these the k quantities λ, we have $3n-k$ equations expressing the conditions of equilibrium, being as many as the system has *degrees of freedom.*

The equations 16) were given by Lagrange[1]), to whom the principle of the use of the indeterminate multipliers λ is due.[2]) One great advantage of the principle of virtual work is that it enables us to dispense with the calculation of the reactions, for in a displacement compatible with the constraints the work of the reactions vanishes.

As an example let us find the position of equilibrium of two heavy particles of mass m_1 and m_2, connected by a rigid bar without weight, of length l, and placed inside of a smooth sphere of radius r. The equations of constraint are

$$\begin{gathered}
x_1^2 + y_1^2 + z_1^2 - r^2 = 0,\\
x_2^2 + y_2^2 + z_2^2 - r^2 = 0,\\
(x_1 - x_2)^2 + (y_1 - y_2)^2 + (z_1 - z_2)^2 - l^2 = 0.
\end{gathered}$$

The equation of virtual work is

$$m_1 g\,\delta z_1 + m_2 g\,\delta z_2 = 0,$$

where δz_1 and δz_2 satisfy the equations

$$\begin{gathered}
x_1\,\delta x_1 + y_1\,\delta y_1 + z_1\,\delta z_1 = 0,\\
x_2\,\delta x_2 + y_2\,\delta y_2 + z_2\,\delta z_2 = 0,\\
(x_1 - x_2)(\delta x_1 - \delta x_2) + (y_1 - y_2)(\delta y_1 - \delta y_2) + (z_1 - z_2)(\delta z_1 - \delta z_2) = 0.
\end{gathered}$$

These are four linear equations between the six quantities

$$\delta x_1,\quad \delta y_1,\quad \delta z_1,\quad \delta x_2,\quad \delta y_2,\quad \delta z_2.$$

We may therefore take any two of them arbitrarily. Suppose we assume $\delta y_1 = \delta y_2 = 0$. We then have four linear equations in $\delta x_1, \delta z_1, \delta x_2, \delta z_2$, and in order that they may be satisfied for values

1) Lagrange, *Mécanique Analytique,* tom. I, p. 79.

2) See Note II.

of the δ's other than 0, the determinant of the coefficients must vanish

$$\begin{vmatrix} 0\ , & m_1\ , & 0\ , & m_2 \\ x_1\ , & z_1\ , & 0\ , & 0 \\ 0\ , & 0\ , & x_2\ , & z_2 \\ x_1 - x_2, & z_1 - z_2, & x_2 - x_1, & z_2 - z_1 \end{vmatrix} = 0$$

or reducing,

$$(m_1 x_1 + m_2 x_2)(x_2 z_1 - x_1 z_2) = 0.$$

The solution that applies is given by the vanishing of the first factor, that is,

$$m_1 x_1 + m_2 x_2 = 0.$$

In like manner if we had assumed $\delta x_1 = \delta x_2 = 0$, we should have obtained

$$m_1 y_1 + m_2 y_2 = 0.$$

This equation with the preceding gives by the elimination of m_1, m_2,

$$\frac{x_1}{y_1} = \frac{x_2}{y_2}.$$

Hence the points lie in a vertical plane containing the center of the sphere. The two equations express the fact that a point dividing the line connecting the particles in the inverse ratio of their masses is vertically below the center of the sphere. The azimuth of the plane containing the particles is indeterminate on account of the symmetry about the vertical.

26. D'Alembert's Principle. The equations of motion of a particle may be written

$$\text{17)}\qquad \begin{aligned} X_r - m_r \frac{d^2 x_r}{dt^2} &= 0, \\ Y_r - m_r \frac{d^2 y_r}{dt^2} &= 0, \\ Z_r - m_r \frac{d^2 z_r}{dt^2} &= 0. \end{aligned}$$

Multiplying these equations respectively by the arbitrary quantities δx_r, δy_r, δz_r, adding, and taking the sum for all values of the suffix r, belonging to the different particles of a system,

$$\text{18)}\qquad \sum_r \left\{ \left(X_r - m_r \frac{d^2 x_r}{dt^2}\right)\delta x_r + \left(Y_r - m_r \frac{d^2 y_r}{dt^2}\right)\delta y_r + \right.$$
$$\left. + \left(Z_r - m_r \frac{d^2 z_r}{dt^2}\right)\delta z_r \right\} = 0.$$

This equation may be called the fundamental equation of dynamics, and is the analytical statement of what is known as *d'Alembert's*

Principle. Lagrange made it the basis of the entire subject of dynamics.[1]) We may interpret 18) in terms of the principle of virtual work by means of the introduction of the conception of *effective forces* due to d'Alembert.

If a system of particles is not free, when acted on by certain impressed forces it will not take on the same motion as if there were no constraint, the reactions causing it to deviate from this natural motion. Having found the actual motion, we know the system of forces that would produce it, if there were no constraints. These are termed the effective forces and if we represent them by X_r', Y_r', Z_r', they are given by the equations

$$X_r' = m_r \frac{d^2 x_r}{dt^2}, \quad Y_r' = m_r \frac{d^2 y_r}{dt^2}, \quad Z_r' = m_r \frac{d^2 z_r}{dt^2}.$$

The equation 18) accordingly states that the reversed effective forces, $-X'$, $-Y'$, $-Z'$ together with the impressed forces, X, Y, Z, will form a system in equilibrium.

We may regard the principle from another point of view. When a body is set in motion with an acceleration, it reacts on the agent which produces the motion, and this *kinetic reaction* has the properties of any force whatsoever. For instance if the accelerating agency is due to contact with a second moving body, the second body is retarded by a force, and this force is the reaction of the first. This kinetic reaction is measured by the components

$$-m\frac{d^2x}{dt^2}, \quad -m\frac{d^2y}{dt^2}, \quad -m\frac{d^2z}{dt^2},$$

and is thus in the opposite direction to the acceleration experienced by the body. The reaction is often termed the *Force of Inertia*, a very expressive term, representing in tangible form the fundamental property of inertia, possessed by all matter, this property being that matter reacts against, or in ordinary language *resists*, being put in motion. (By the use of the term resists we in no wise mean prevention of motion — the use of the term has been objected to, and Maxwell[2] has jokingly remarked that we might as well say that a cup of tea resists being sweetened, because it does not become sweet until we add sugar. The meaning here is precisely similar — we mean that matter does not move until it is moved by some agent external to itself. It is hardly likely that confusion can be caused by the use of such common phrases, which indeed seem to attribute volition to matter — we shall accordingly make no attempt to avoid them.) We may thus define *matter as that which can exert forces of*

1) Lagrange, *Mécanique Analytique,* t. 1, p. 267. The equation 18) although first explicitly given by Lagrange, will be referred as "d'Alembert's equation", as briefer than "Lagrange's equation of d'Alembert's Principle".

2) Maxwell, *Scientific Papers,* Vol. II, p. 779.

inertia. This is the only universal definition of matter now possible. (It is to be noticed that this definition includes the luminiferous ether.)

We may then state d'Alembert's Principle in these words: The impressed forces, together with the forces of inertia, form a system in equilibrium. Thus the principle is not new, but merely expresses Newton's third law of motion, embodying at the same time the other two, in the expression of the forces. The great seryice done by d'Alembert was in reducing the statement of a problem in motion to that of a statical problem.

A practical advantage frequently of great use in applications is similar to that possessed by the principle of virtual work, namely, that the reactions of the constraints do no work, and may therefore be omitted from the equation 18), for it is evident thet the reactions due to all constraints between bodies act equally in opposite directions on both, so that the work done in the motion of their common point of application vanishes.

As a simple example of the meaning of force of inertia consider two locomotives pulling in opposite directions at the ends of a train, the pulls being transmitted by spring dynamometers. If the train remains at rest, the pull recorded on both dynamometers will be the same. If now one locomotive be given more steam, so that the train begins to move, the indications of the dynamometers will be found to be unequal, the greater pull being that of the locomotive on the side toward which the train is moving, the difference being found to be exactly equal (disregarding friction) to the product of the mass of the train by the acceleration which it gains. Thus the difference of pull is balanced by the force of inertia, or kinetic reaction.

Again, consider a person standing in a street-car, when the car starts. An acceleration is impressed on his body in the direction of the motion of the car. The kinetic reaction is thus directed horizontally to the rear. The force of weight of the person being vertically downwards, the remaining force, namely, the static reaction of the floor of the car, must be such as to equilibrate these two, and is found by the triangle of vectors to be directed upwards and inclined forwards. Thus the person must lean forward in order to preserve equilibrium. Similarly when the car stops, the acceleration being directed the other way, he must lean backward. This application of d'Alembert's Principle is a matter of common knowledge, where electric railroads are common.

27. Energy. Conservative Systems. Impulse. If in the equation of d'Alembert's principle, 18), we put for δx, δy, δz the displacements which take place in the actual motion of the system in the time dt,

$$\delta x_r = \frac{dx_r}{dt}dt, \quad \delta y_r = \frac{dy_r}{dt}dt, \quad \delta z_r = \frac{dz_r}{dt}dt,$$

we obtain

$$19)\quad \sum_r \left\{ m_r \left(\frac{d^2 x_r}{dt^2}\frac{dx_r}{dt} + \frac{d^2 y_r}{dt^2}\frac{dy_r}{dt} + \frac{d^2 z_r}{dt^2}\frac{dz_r}{dt} \right) - X_r \frac{dx_r}{dt} - Y_r \frac{dy_r}{dt} - Z_r \frac{dz_r}{dt} \right\} dt = 0.$$

Now since

$$m_r \frac{d^2 x_r}{dt^2}\frac{dx_r}{dt} = \frac{1}{2}\frac{d}{dt}\left\{ m_r \left(\frac{dx_r}{dt}\right)^2 \right\},$$

the sum of the first three terms is the derivative of the sum

$$\frac{1}{2}\sum_r m_r \left\{ \left(\frac{dx_r}{dt}\right)^2 + \left(\frac{dy_r}{dt}\right)^2 + \left(\frac{dz_r}{dt}\right)^2 \right\},$$

and the equation may be written, omitting the factor dt,

$$20)\quad \frac{d}{dt}\left[\frac{1}{2}\sum_r m_r \left\{ \left(\frac{dx_r}{dt}\right)^2 + \left(\frac{dy_r}{dt}\right)^2 + \left(\frac{dz_r}{dt}\right)^2 \right\} \right] = \sum_r \left\{ X_r \frac{dx_r}{dt} + Y_r \frac{dy_r}{dt} + Z_r \frac{dz_r}{dt} \right\}.$$

The expression

$$\frac{1}{2}\sum_r m_r \left\{ \left(\frac{dx_r}{dt}\right)^2 + \left(\frac{dy_r}{dt}\right)^2 + \left(\frac{dz_r}{dt}\right)^2 \right\} = \frac{1}{2}\Sigma_r m_r v_r^2,$$

the half-sum of the products of the mass of each particle by the square of its velocity, is called the *Kinetic Energy* of the system. It is one of the most important dynamical quantities. If we denote it by T, equation 20) has on the left $\frac{dT}{dt}$. Since

$$X_r dx_r + Y_r dy_r + Z_r dz_r$$

is the work done upon the rth particle, the terms under the summation sign on the right denote the total work done by the impressed forces in unit time, or the *Activity*[1]) of the forces. The equation 20) is called the equation of activity, and states that the rate of increase of kinetic energy of the system is equal to the activity of the impressed forces.

1) The word "*actio*" is used by Newton, in a scholium on the third law, where he says, "If the activity of an agent (force) be measured by its amount and its velocity conjointly; . . . activity and counteractivity, in all combinations of machines, will be equal and opposite." The activity will sometimes be denoted by $\frac{dA}{dt}$.

Integrating equation 20) with respect to t between the limits t_0 and t_1,

$$21)\qquad \left[\frac{1}{2}\sum_r m_r\left\{\left(\frac{dx_r}{dt}\right)^2+\left(\frac{dy_r}{dt}\right)^2+\left(\frac{dz_r}{dt}\right)^2\right\}\right]_{t_0}^{t_1} = \sum_r\int_{t_0}^{t_1}\left(X_r\frac{dx_r}{dt}+Y_r\frac{dy_r}{dt}+Z_r\frac{dz_r}{dt}\right)dt.$$

The square brackets with the affixes t_0, t_1 denote that the value of the expression in brackes for $t = t_0$ is to be subtracted from the value for $t = t_1$.

The integral on the right of 21), which may be written

$$\int X_r\,dx_r + Y_r\,dy_r + Z_r\,dz_r,$$

denotes the work done by the forces of the system on the particle m_r during the motion from t_0 to t_1, and the sum of such integrals denotes the total work done by the forces acting on the system during the motion. The equation 21) thus becomes

$$22)\qquad T_{t_1} - T_{t_0} = \sum_r\int_{t_0}^{t_1}(X_r\,dx_r + Y_r\,dy_r + Z_r\,dz_r).$$

This is called the *equation of energy*, and states that the gain of kinetic energy is equal to the work done by the forces during the motion.

The equation of energy assumes an important form in the particular case that the forces acting on the yarticles depend only on the positions of the particles, and that the components may be represented by the partial derivatives of a single function of the coordinates,

$$U(x_1, y_1, z_1, x_2, y_2, z_2, \ldots z_n)$$

$$23)\qquad X_r = \frac{\partial U}{\partial x_r},\quad Y_r = \frac{\partial U}{\partial y_r},\quad Z_r = \frac{\partial U}{\partial z_r}.$$

In this case the expression

$$\Sigma_r\{X_r\,dx_r + Y_r\,dy_r + Z_r\,dz_r\} = \sum_r\left\{\frac{\partial U}{\partial x_r}dx_r + \frac{\partial U}{\partial y_r}dy_r + \frac{\partial U}{\partial z_r}dz_r\right\},$$

is the exact differential of the function U, and the integral

$$\int_{t_0}^{t_1}\Sigma_r(X_r\,dx_r + Y_r\,dy_r + Z_r\,dz_r) = U_{t_1} - U_{t_0},$$

that is the work done in the motion, does not depend upon the paths described by the various particles, but only on the initial and final configurations of the system, since

$$U_{t_1} = U(x_1, y_1, z_1, \ldots) \text{ and}$$
$$U_{t_0} = U(x_1^0, y_1^0, z_1^0, \ldots),$$

where the affix 0 denotes the value of the coordinate at the time t_0.

The equation of energy then is

24) $$T_{t_1} - T_{t_0} = U_{t_1} - U_{t_0}.$$

The function U is called the force-function, and its negative $W = -U$ is called the *Potential Energy* of the system. Inserting W in 24) we have

25) $$T_{t_1} + W_{t_1} = T_{t_0} + W_{t_0}.$$

The sum of the kinetic and potential energies of a system possessing a force-function depending only on the coordinates is the same for all instants of time. This is the Principle of *Conservation of Energy.*

Systems for which the conditions 23) are satisfied are accordingly called *conservative systems.*

The potential energy, being defined by its derivatives, contains an arbitrary constant. The functions T and W have one essential difference, namely, T contains only the velocities, $\frac{dx_r}{dt}, \frac{dy_r}{dt}, \frac{dz_r}{dt}, \ldots,$ while W does not contain the velocities, but only the coordinates. One important consequence of the equation of Conservation of Energy is that if at any time in the course of a motion, all the points of the system pass simultaneously through positions that they have occupied at a previous instant, the kinetic energy will be the same as at that instant, irrespective of the directions in which the particles may be moving, for $T + W$ is constant during the whole motion, and W depends only on the coordinates, consequently when all the coordinates resume their former values, the kinetic energy does the same.

In other words, the work done on the system has been *stored up* or *conserved,* to the amount W, and may be got out again by bringing the system back to its former configuration.

For instance, a particle thrown vertically upward, or a pendulum swinging, have the same velocity when passing a given point whether rising or falling.

As an example, consider a particle acted upon by gravity. We have

26) $$X = 0, \quad Y = 0, \quad Z = -wg,$$

so that $U = -mgz + \text{const}.$

The equation of energy is

27) $$\frac{1}{2} m (v^2 - v_0^2) = -mg(z - z_0),$$

or the velocity depends only on the vertical height fallen. Accordingly a particle, descending from a point A to another B, constrained to follow any curve, reaches B always with the same velocity, although the time occupied in the descent may be very different from one curve to another. This theorem was discovered by Galileo.

The equation 27) might have been applied to immediately give us the integral equation 44) § 21. (In that equation, the Z-axis is drawn positively *downward.*)

The principle of virtual work, § 25, may evidently be expressed by saying that for equilibrium the potential energy of the system is a maximum or minimum, and a little consideration shows that for *stable* equilibrium it is a minimum.[1])

For instance in the above example the potential energy

$$W = mgz + \text{const.},$$

z being measured positively upward. If the particle is in equilibrium on a surface concave upwards, z and with it W is a minimum, the equilibrium being stable. If the concavity is downwards, the equilibrium is unstable and W is a maximum.[2]) The question of stability of equilibrium will be discussed in § 45.

It is possible to have a force-function defined by equations 23), which contains the time as well as the coordinates. The system is not then conservative, and it is not customary to speak of its potential energy. We have now

$$\frac{dU}{dt} = \frac{\partial U}{\partial t} + \sum^r \left\{ \frac{\partial U}{\partial x_r}\frac{dx_r}{dt} + \frac{\partial U}{\partial y_r}\frac{dy_r}{dt} + \frac{\partial U}{\partial z_r}\frac{dz_r}{dt} \right\},$$

so that our equation of activity 20) is in this case

$$\text{28)} \qquad \frac{dT}{dt} = \frac{dU}{dt} - \frac{\partial U}{\partial t}.$$

In certain cases we may be able to assign the term $\frac{dU}{dt}$ to a potential energy, as $-\frac{dW}{dt}$.

In the forces depend on the velocities or on anything beside the coordinates, the system is not conservative. Such a case is that of motion with friction, where the friction, being a force that always tends to *retard* the motion, not only changes sign with the velocities but also depends upon the magnitudes of the velocities in such resisting media as the air and liquids.

The dynamical theory of heat accounts for the energy that apparently disappears in non-conservative systems.

---

1) Dirichlet, *Über die Stabilität des Gleichgewichts.* Crelle's Journal, Bd. 32, p. 85 (1846).

2) See Kirchhoff, *Mechanik,* p. 34.

We shall see later, that whereas positional forces are usually conservative, and motional forces not, there are certain conservative motional forces.

Kinetic energy being defined as $\Sigma \frac{1}{2} m v^2$ is of the dimensions $\left[\frac{ML^2}{T^2}\right]$, the same as those of work. Potential Energy is defined as work. The C. G. S. unit of energy is, therefore, the *erg*.

We have in this chapter been concerned with the line integral of the force exerted on a moving point resolved in the direction of the motion of the point of application. This has been called the work of the force, and is physically a quantity of fundamental importance. We have occasionally to consider the time-integral of a force, that is, if F be a force always in the same direction, the quantity

$$I = \int_{t_0}^{t_1} F dt,$$

which has received the name of the *impulse* of the force during the time from t_0 to t_1. The effect of a force may be measured either by the work or by the impulse, but it is to be observed that the information obtained when one or the other of these two quantities is given is of a quite different nature. Supposing the force is constant in magnitude and direction, the work done is equal to the force times the distance moved, and a knowledge of the work tells us how far the point of application will be moved by the given force, while the impulse is equal to the force times the interval of time, and tells us how long the point will move under the application of the given force. If the force is variable, considering the signification of a definite integral as a mean[1]), we may say that the work is the mean with respect to distance of the force multiplied by the length of the path, while the impulse is the mean with respect to the time multiplied by the duration of the motion. Thus the work answers the question "how far", while the impulse answers the question "how long". The work is a scalar quantity, its element being the geometric product of the force and the displacement. For the element of impulse, however, we have, using equation 7), § 3,

$$F dt = X dt \cos(Fx) + Y dt \cos(Fy) + Z dt \cos(Fz),$$

thus the element is the component in the direction of the force of the vector whose components are

$$dI_x = X dt, \quad dI_y = Y dt, \quad dI_z = Z dt.$$

1) See footnote, § 34, p. 98.

For the whole impulse we may then take as definitions

$$I_x = \int_{t_0}^{t_1} X\,dt, \quad I_y = \int_{t_0}^{t_1} X\,dt, \quad I_z = \int_{t_0}^{t_1} Z\,dt,$$

so that the impulse is a vector quantity. We thus lose the relation to the direction of the path, or of the force, in the case of a variable force, but on comparing with equations 43), § 13,

$$\frac{dM_x}{dt} = X, \quad \frac{dM_y}{dt} = Y, \quad \frac{dM_z}{dt} = Z,$$

we have by integration

$$I_x = M_x(t_1) - M_x(t_0), \quad I_y = M_y(t_1) - M_y(t_0), \quad I_z = M_z(t_1) - M_z(t_0),$$

so that the impulse of a force acting on a single particle for a certain interval of time is equal to the vector increase of momentum during that interval.

The case in which the impulse of a force is of most importance is that of what are known as impulsive forces, which arise where actions take place between bodies in such a brief interval that the bodies do not appreciably change their positions during the action, although sensible changes of momenta take place. If in the equations above, the length of the interval $t_1 - t_0$ decreases indefinitely, while the force-components X, Y, Z increase indefinitely, the integrals may still approach finite limits

$$I_x = \lim_{t_1 = t_0} \int_{t_0}^{t_1} X\,dt, \quad I_y = \lim_{t_1 = t_0} \int_{t_0}^{t_1} Y\,dt, \quad I = \lim_{t_1 = t_0} \int_{t_1}^{t_0} Z\,dt.$$

In this case we can not investigate the forces in the ordinary manner for the accelerations have been infinite, but the velocities and momenta have received finite changes in the vanishing interval. The work done is in like manner finite, though the distance moved vanish. The impulse and work of all ordinary, that is finite forces acting at the same time may thus be neglected, since the integral of a finite integrand over a vanishing range of integration vanishes.

On account of the third law, the action and reaction being equal during the operation, the impulses of the forces on the two bodies are equal and opposite, so that what one gains in momentum the other loses. It is in this manner that the impact of two billiard balls, or the action of a shot on a ballistic pendulum, is to be dealt with. Many instruments used in electrical measurements act on this principle, that the momeutum suddenly communicated to a body at rest, which afterwards proceeds to execute an observed swing, measures the time-integral of an impulsive force.[1])

In order to find the work done by a given impulse, let us make use of the equation of work and energy, 22), which says that the work done is equal to the increase of the kinetic energy. The latter may be written, bearing in mind the definition of momentum,

$$T = \frac{1}{2} m (v_x^2 + v_y^2 + v_z^2) = \frac{1}{2} (M_x v_x + M_y v_y + M_z v_z).$$

Suppose now the particle set in motion by an impulsive force, from rest. The kinetic energy acquired, and accordingly the work done, is then one-half the geometric prodnct of the impulse and the velocity generated, or in orther words, the geometric product of the impulse and the average value of the velocity at the beginning and the end of the impulsive action. This may be otherwise shown, whether the particle start from rest or not, by the following considerations.[2]) Since the interval of time and the distance moved are infinitely small, we may consider the motion as rectilinear. Suppose the initial velocity to be v_0, and the final value v_1, and let ε be a parameter which during the interval runs rapidly through all values from 0 to 1, so that at any part of the interval

$$v = v_0 + \varepsilon (v_1 - v_0).$$

But as the momentum always increases at a rate proportional to the increase of velocity, we have also

$$M = M_0 + \varepsilon (M_1 - M_0) = M_0 + \varepsilon I,$$

---

1) Suppose that a body which swings according to the law of the pendulum, or equation 8), § 19, receives, when in its position of equilibrum, an impulse I. It swings out according to the equation

$$x = a \sin nt, \quad \frac{dx}{dt} = an \cos nt$$

during a time $t = \pi/2n$ to a maximum excursion a, at which its velocity vanishes, and it turns back. If its mass is m, the momentum communicated to it while at rest was

$$I = \left(m \frac{dx}{dt}\right)_{t=0} = man,$$

so that if we know m, a, aud $n = 2\pi/\text{period}$, we can measure the impulse of the impulsive force. This is the mode of use of the ballistic galvanometer and electrometer, as well as of the ballistic pendulum formerly used in gunnery. The same formula applies (see Chapter X), to the heeling of a ship when firing a shot from a cannon.

2) Thomson and Tait, § 308.

where I is the total impulse. From the equation of motion we have

$$F = \frac{dM}{dt} = I\frac{d\varepsilon}{dt},$$

so that we obtain for the work

$$W = \int_{t_0}^{t_1} F\,ds = \int_{t_0}^{t_1} F v\,dt = I\int_0^1 [v_0 + \varepsilon(v_1 - v_0)]\,d\varepsilon = \frac{1}{2} I(v_1 + v_0).$$

Thus we find as before for the work of an impulsive force the product of the impulse by the average velocity at the beginning and end of the action. It is evident that the same is true for the infinitesimal work done by an ordinary, that is finite force, during an infinitesimal interval. This conception of the impulse will be useful to us hereafter, in connection with the following. For a system of particles, we have for the kinetic energy,

$$T = \frac{1}{2}\Sigma m v^2 = \frac{1}{2}\Sigma_r m_r (v_{xr}^2 + v_{yr}^2 + v_{zr}^2).$$

Now the kinetic energy is known when we know the velocities of every particle of the system, as well as their masses, no matter what their positions. If we consider T as a function of the velocities, we have accordingly

$$M_{xr} = m_r v_{xr} = \frac{\partial T}{\partial v_{xr}}, \quad M_{yr} = m_r v_{yr} = \frac{\partial T}{\partial v_{yr}}, \quad M_{zr} = m_r v_{zr} = \frac{\partial T}{\partial v_{zr}},$$

or the momentum components of any particle are the partial derivatives of the kinetic energy of the system, considered as a function of all the velocities of the particles, by the respective velocity-components. Thus we may write

$$T = \frac{1}{2}\sum_r \left(v_{xr}\frac{\partial T}{\partial v_{xr}} + v_{yr}\frac{\partial T}{\partial v_{yr}} + v_{zr}\frac{\partial T}{\partial v_{zr}}\right),$$

which by the theorem of Euler is true for any homogeneous quadratic function.

28. Particular Case of Force-function. The conditions necessary for the existence of a force-function being 23), we must have, since

$$\frac{\partial^2 U}{\partial x\,\partial y} = \frac{\partial^2 U}{\partial y\,\partial x}, \quad \frac{\partial^2 U}{\partial y\,\partial z} = \frac{\partial^2 U}{\partial z\,\partial y}, \quad \frac{\partial^2 U}{\partial z\,\partial x} = \frac{\partial^2 U}{\partial x\,\partial z},$$

29) $$\frac{\partial Y_r}{\partial x_r} = \frac{\partial X_r}{\partial y_r}, \quad \frac{\partial Z_r}{\partial y_r} = \frac{\partial Y_r}{\partial z_r}, \quad \frac{\partial X_r}{\partial z_r} = \frac{\partial Z_r}{\partial x_r}.$$

It will be shown below (§ 31) that these conditions are also sufficient.

In the particular case in which the only forces acting on the system are attractions or repulsions by the several particles directed along the lines joining them and depending only on their mutual distances, a force-function always exists.

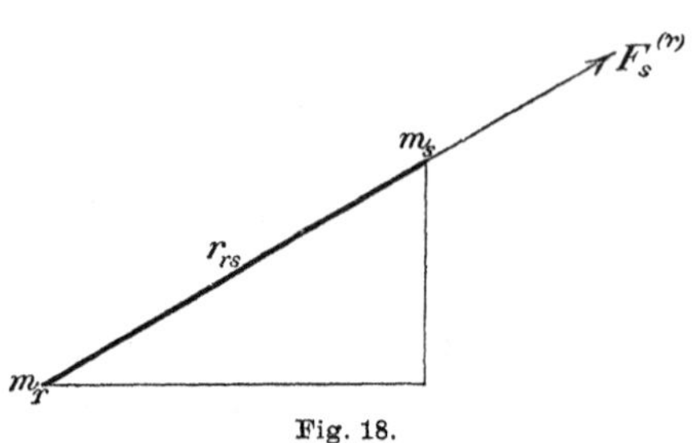

Fig. 18.

For let the force between two particles m_r and m_s at a distance apart r_{rs} be

$$F = \varphi(r_{rs}).$$

It will be convenient to consider F positive if the force is a repulsion.

Consider now the force $F_s^{(r)}$ acting on m_s and acting in the direction from m_r to m_s. Its direction cosines are those of the vector r_{rs}.

$$(30) \qquad \frac{X_s^{(r)}}{F_s^{(r)}} = \frac{x_s - x_r}{r_{rs}}, \quad \frac{Y_s^{(r)}}{F_s^{(r)}} = \frac{y_s - y_r}{r_{rs}}, \quad \frac{Z_s^{(r)}}{F_s^{(s)}} = \frac{z_s - z_r}{r_{rs}}.$$

Now since

$$r_{rs}^2 = (x_s - x_r)^2 + (y_s - y_r)^2 + (z_s - z_r)^2,$$

differentiating partially by x_s,

$$2r_{rs}\frac{\partial r_{rs}}{\partial x_s} = 2(x_s - x_r),$$

$$31) \qquad \frac{\partial r_{rs}}{\partial x_s} = \frac{x_s - x_r}{r_{rs}}, \quad \frac{\partial r_{rs}}{\partial y_{rs}} = \frac{y_s - y_r}{r_s}, \quad \frac{\partial r_{rs}}{\partial z_s} = \frac{z_s - z_r}{r_{rs}},$$

and accordingly

$$\frac{X_s^{(r)}}{F_s^{(r)}} = \frac{\partial r_{rs}}{\partial x_s}, \quad \frac{Y_s^{(r)}}{F_s^{(r)}} = \frac{\partial r_{rs}}{\partial y_s}, \quad \frac{Z_s^{(r)}}{F_s^{(r)}} = \frac{\partial r_{rs}}{\partial z_s};$$

$$X_s^{(r)} = F_s^{(r)}\frac{\partial r_{rs}}{\partial x_s} = \varphi(r_{rs})\frac{\partial r_{rs}}{\partial x_s},$$

$$Y_s^{(r)} = F_s^{(r)}\frac{\partial r_{rs}}{\partial y_s} = \varphi(r_{rs})\frac{\partial r_{rs}}{\partial y_s},$$

$$Z_s^{(r)} = F_s^{(r)}\frac{\partial r_{rs}}{\partial z_s} = \varphi(r_{rs})\frac{\partial r_{rs}}{\partial z_s}.$$

If we put U_{rs} such a function of r_{rs} that

$$\frac{dU_{xs}}{dr_{rs}} = \varphi(r_{rs}),$$

$$X_s^{(r)} = \frac{dU_{rs}}{dr_{rs}} \frac{\partial r_{rs}}{\partial x_s} = \frac{\partial U_{rs}}{\partial x_s},$$

$$Y_s^{(r)} = \frac{dU_{rs}}{dr_{rs}} \frac{\partial r_{rs}}{\partial y_s} = \frac{\partial U_{rs}}{dy_s},$$

$$Z_s^{(r)} = \frac{dU_{rs}}{dr_{rs}} \frac{\partial r_{rs}}{\partial z_s} = \frac{\partial U_{rs}}{\partial z_s}.$$

If now we find the resultant F_s of all the forces acting on m_s due to the repulsions by all the particles m_r, we shall have

32)
$$X_s = \frac{\partial U_{1s}}{\partial x_s} + \frac{\partial U_{2s}}{\partial x_s} + \cdots + \frac{\partial U_{ns}}{\partial x_s} = \frac{\partial U_s}{\partial x_s},$$
$$Y_s = \frac{\partial U_{1s}}{\partial y_s} + \frac{\partial U_{2s}}{\partial y_s} + \cdots + \frac{\partial U_{ns}}{\partial y_s} = \frac{\partial U_s}{\partial y_s},$$
$$Z_s = \frac{\partial U_{1s}}{\partial z_s} + \frac{\partial U_{2s}}{\partial z_s} + \cdots + \frac{\partial U_{ns}}{\partial z_s} = \frac{\partial U_s}{\partial z_s},$$

if we write $U_s = U_{1s} + U_{2s} \cdots + U_{ns}$. Thus U_s satisfies the conditions for a force-function as far as concerns the point m_s. In the summations s does not occur as the first index.

It is evident that the function U_{rs} serves the same purpose for m_r as for m_s. For the force $F_r^{(s)}$ exerted on m_r by m_s is equal and opposite to that exerted on m_s by m_r. But r_{rs} is the same function of $(-x_r)$ that it is of x_s, therefore

$$\frac{\partial r_{rs}}{\partial x_r} = -\frac{\partial r_{sr}}{\partial x_s},$$

and

$$X_r^{(s)} = \varphi_{rs} \frac{\partial r_{rs}}{\partial x_r} = \frac{\partial U_{rs}}{\partial x_r} = -\frac{\partial U_{rs}}{\partial x_s}.$$

We may add to U_s terms independent of x_s, y_s, z_s, without affecting the values of X_s, Y_s, Z_s. If we make U a symmetrical function of all the coordinates, containing x_s, y_s, z_s, as U_s does, then U will serve as the force-function for all the coordinates.

In particular, let the force of repulsion vary according to the Newtonian law of gravitation. Then

33)
$$\varphi(r_{rs}) = \gamma \frac{m_r m_s}{r_{rs}^2}, \quad U_{rs} = -\gamma \frac{m_r m_s}{r_{rs}},$$

34)
$$U_s = -\gamma\left\{\frac{m_1 m_s}{r_{1s}} + \frac{m_2 m_s}{r_{2s}} + \cdots + \frac{m_n m_s}{r_{ns}}\right\},$$

and the symmetrical function U will be

35)
$$\begin{aligned} U = -\frac{1}{2}\gamma\left\{\frac{m_1 m_2}{r_{12}} + \frac{m_1 m_3}{r_{13}} + \cdots + \frac{m_1 m_n}{r_{1n}}\right. \\ + \frac{m_2 m_1}{r_{21}} + \frac{m_2 m_3}{r_{23}} + \cdots + \frac{m_2 m_n}{r_{2n}} \\ + \frac{m_3 m_1}{r_{31}} + \frac{m_3 m_2}{r_{32}} + \cdots + \frac{m_3 m_n}{r_{3n}} \\ \cdots\cdots\cdots\cdots\cdots \\ \left. + \frac{m_n m_1}{r_{n1}} + \frac{m_n m_2}{r_{n2}} + \cdots + \frac{m_n m_{n-1}}{r_{n,n-1}}\right\}, \end{aligned}$$

or more briefly,

$$U = -\frac{1}{2}\gamma\sum_{r=1}^{r=n}\sum_{s=1}^{s=n}\frac{m_r m_s}{r_{rs}},$$

understanding that terms in which $r = s$ are to be omitted.

The factor $\frac{1}{2}$ is introduced because in the above summation every term appears twice. But in U each pair of particles is to appear only once.

If no constant be added to U as defined above, both it and the potential energy

36)
$$W = \frac{1}{2}\gamma\sum_{r=1}^{r=n}\sum_{s=1}^{s=n}\frac{m_r m_s}{r_{rs}},$$

will vanish when every r_{rs} is infinite, that is when no two particles are within a finite distance of each other. This furnishes a convenient zero configuration for the potential energy, and is the one generally adopted. We may accordingly define the potential energy of the system in any given configuration as the work that must be done against the mutual repulsions or attractions of the particles in order to bring them from a state of infinite dispersion to the given configuration. In the case of *attracting* forces like those of gravitation, we shall, with the notation of this section, put γ negative[1]), so that the potential energy of finite systems is negative, or in the terminology of Thomson and Tait, the *exhaustion* of potential energy, $-W$, is positive.

---

1) For the reason for the adoption of this convention see § 119.

29. Calculus of Variations. Brachistochrone. The question concerning the necessary and sufficient conditions that a line integral

$$\int_A^B (Xdx + Ydy + Zdz)$$

shall be independent of the path of integration, depending only on the terminal points A and B, though purely a question of the calculus, is of so great importance in various parts of mathematical physics that it will be considered here. For the purpose of this treatment we shall make use of the calculus of variations, which on account of the great use made of it in mechanics will now be briefly treated.

In the differential calculus, we have to consider questions of maxima and minima of functions. A function of one variable has a maximum or minimum value at a certain value of the variable if the change in the function is of the same sign for any change in the variable, provided the latter change is small enough. Since if $f(x)$ and all its derivatives are continuous at x,

$$f(x+h) = f(x) + hf'(x) + \frac{h^2}{2!}f''(x) + \cdots$$

$$f(x+h) - f(x) = hf'(x) + \frac{h^2}{2!}f''(x) + \cdots$$

If h is small enough, the expression on the right will have the sign of the first term, which will change sign with h. Accordingly the necessary condition for a maximum or minimum is

$$f'(x) = 0.$$

Suppose on the other hand that we change the *form* of the function — such a change may be made to take place gradually. For instance suppose we have a curve given by the parametric representation,

$$x = F_1(t), \quad y = F_2(t), \quad z = F_3(t),$$

where the F's are any uniform and continuous functions of an independent variable t. If we change the form of the F's we shall change the curve — suppose we change to

$$x = G_1(t), \quad y = G_2(t), \quad z = G_3(t).$$

To every value of t corresponds one point on each curve, consequently to each point on one curve corresponds a definite point on the other. Such a change from one curve to the other is called a transformation of the curve. The change may be made gradually, for example,

$$x = F_1(t) + \varepsilon\{G_1(t) - F_1(t)\},$$
$$y = F_2(t) + \varepsilon\{G_2(t) - F_2(t)\},$$
$$z = F_3(t) + \varepsilon\{G_3(t) - F_3(t)\}.$$

For every value of ε we shall have a particular curve — for $\varepsilon = 0$ we shall have the original curve, for $\varepsilon = 1$ the final curve, and for intervening values of ε other curves. A small change in ε will cause a small change in the curve, and if ε is infinitesimal we shall call the transformation an infinitesimal transformation. The changes in the values of x, y, z, or of any functions thereof, for an infinitesimal change ε, are called the *variations* of the functions, and are denoted by the sign δ.

Suppose we denote derivatives by the independent variable t,

$$\frac{dx}{dt}, \quad \frac{d^2x}{dt^2}, \dots \frac{d^kx}{dx^k} \quad \text{etc.}$$

by the letters

$$x', \quad x'', \quad x^{(k)},$$

and by φ any function of the independent variable, of the dependent variables, and of their derivatives up to the mth order

$$\varphi(t, x, y, z, x', y', z', \dots x^{(k)}, y^{(k)}, z^{(k)}, \dots x^{(m)}, y^{(m)}, z^{(m)}),$$

and consider the change in φ made by an infinitesimal transformation, where we replace x, y, z by

$$x + \varepsilon\xi(t),$$
$$y + \varepsilon\eta(t),$$
$$z + \varepsilon\zeta(t),$$

where ξ, η, ζ are *arbitrary* continuous functions of t.

Then $\frac{dx}{dt}$ or x' is replaced by $\frac{dx}{dt} + \varepsilon\frac{d\xi}{dt}$ and $\frac{d^{(k)}x}{dt^k}$ by $\frac{d^kx}{dt^k} + \varepsilon\frac{d^k\xi}{dt^k}$, i. e., by

$$x^{(k)} + \varepsilon\xi^{(k)}.$$

Hence φ becomes

$$\varphi(t, x + \varepsilon\xi, y + \varepsilon\eta, z + \varepsilon\zeta, x' + \varepsilon\xi', y' + \varepsilon\eta', \dots z^{(m)} + \varepsilon\zeta^{(m)}),$$

which developed by Taylor's theorem for any number of variables, gives on collecting terms according to powers of ε

$$\varphi(t, x, y, z, x', \dots) + \varepsilon\varphi_1 + \frac{\varepsilon^2}{2!}\varphi_2 \dots + \frac{\varepsilon^k}{k!}\varphi_k + \cdots,$$

where

$$\varphi_1 = \xi\frac{\partial\varphi}{\partial x} + \eta\frac{\partial\varphi}{\partial y} + \zeta\frac{\partial\varphi}{\partial z} + \xi'\frac{\partial\varphi}{\partial x'} + \eta'\frac{\partial\varphi}{\partial y'} + \zeta'\frac{\partial\varphi}{\partial z'} + \cdots$$

$$\varphi_2 = \xi^2\frac{\partial^2\varphi}{\partial x^2} + \eta^2\frac{\partial^2\varphi}{\partial y^2} + \zeta^2\frac{\partial^2\varphi}{\partial z^2} + 2\xi\eta\frac{\partial^2\varphi}{\partial x\,\partial y} \dots + \xi'^2\frac{\partial^2\varphi}{\partial x'^2} + \cdots$$

The terms $\varepsilon\varphi_1$, $\varepsilon^2\varphi_2$, $\varepsilon^k\varphi_k$ are called the first, second and k^{th} variations of φ and are denoted by

$$\varepsilon\varphi_1 = \delta\varphi, \quad \varepsilon^2\varphi_2 = \delta^2\varphi, \quad \varepsilon^k\varphi_k = \delta^k\varphi.$$

If for φ we put successively $x, y, z, x', y', z', \ldots$ we get

$$\begin{array}{llll}
\delta x = \varepsilon\xi, & \delta y = \varepsilon\eta, & \delta z = \varepsilon\zeta, & \delta^2 x = \delta^2 y = \delta^2 z = 0, \\
\delta x' = \varepsilon\xi', & \delta y' = \varepsilon\eta', & \delta z' = \varepsilon\zeta', & \delta^2 x' = \delta^2 y' = \delta^2 z' = 0, \\
\cdots & \cdots & \cdots & \cdots \\
\delta x^{(k)} = \varepsilon\xi^{(k)}, & \delta y^{(k)} = \varepsilon\eta^{(k)}, & \delta z^{(k)} = \varepsilon\zeta^{(k)}, & \delta^2 x^{(k)} = \delta^2 y^{(k)} = \delta^2 z^{(k)} = 0.
\end{array}$$

We thus see that the variations of x, y, z are infinitesimal *arbitrary*[1]) functions of t, the independent variable, and from the last equation

$$\delta\left(\frac{d^k x}{d t^k}\right) = \varepsilon\frac{d^k \xi}{d t^k} = \frac{d^k}{d t^k}(\varepsilon\xi) = \frac{d^k}{d t^k}\delta x,$$

that is, the operation of *differentiation* by the independent variable t and *variation* are commutative, for the variables x, y, z.

If we consider φ as a function of the variable ε, the development by Taylor's theorem for *one* variable shows that we have for all values of k

$$\varphi_k = \left(\frac{\partial^k\varphi}{\partial\varepsilon^k}\right)_{\varepsilon=0},$$

so that

$$\delta^k\varphi = \varepsilon^k\left(\frac{\partial^k\varphi}{\partial\varepsilon^k}\right)_{\varepsilon=0}.$$

Now the two variables ε and t are totally independent of each other, which may be indicated when necessary by writing the derivatives with respect to t as partial derivatives. Now since we may (subject to the usual limitations as to continuity) permute the order of differentation, we have

$$\frac{\partial^i}{\partial t^i}\frac{\partial^k\varphi}{\partial\varepsilon^k} = \frac{\partial^k}{\partial\varepsilon^k}\frac{\partial^i\varphi}{\partial t^i}.$$

Multiplying by ε^k after having put $\varepsilon = 0$ after differentiation, this becomes

$$\frac{d^i}{d t^i}\delta^k\varphi = \delta^k\frac{d^i\varphi}{d t^i},$$

so that the operations of differentiation by the independent variable, and of variation, are commutative for any function. (It is to be distinctly noted that this holds only for derivatives by the independent variable, that is the one whose variation is assumed to be zero. If

1) The functions are arbitrary because the functions G_1, G_2, G_3 are quite independent of F_1, F_2, F_3 being taken *entirely at pleasure.*

some of the variables $x, y, z, \ldots$, should depend on others, we might require the variation of some of their derivatives, for instance $\delta\left(\frac{dy}{dx}\right)$. We must then, since both functions x and y are varied, introduce the independent, or unvaried variable t, writing

$$\frac{dy}{dx} = \frac{dy}{dt} \Big/ \frac{dx}{dt} = \frac{y'}{x'},$$

and performing the operation of variation on the quotient y'/x',

$$\delta\left(\frac{dy}{dx}\right) = \delta\left(\frac{y'}{x'}\right) = \frac{\delta y'}{x'} - \frac{y' \, \delta x}{x'^2}.$$

But for derivatives by t we have

$$\delta y' = \delta\left(\frac{dy}{dt}\right) = \frac{d}{dt}\delta y, \quad \delta x' = \delta\left(\frac{dx}{dt}\right) = \frac{d}{dt}\delta x,$$

so that we may write

$$\delta\left(\frac{dy}{dx}\right) = \frac{d\delta y}{dt} \Big/ \frac{dx}{dt} - \frac{dy}{dt}\frac{d\delta x}{dt} \Big/ \left(\frac{dx}{dt}\right)^2$$

or, once more removing t from explicit appearance,

$$\delta\left(\frac{dy}{dx}\right) = \frac{d\delta y}{dx} - \frac{dy}{dx}\frac{d\delta x}{dx}.$$

(If x is the independent variable, $\delta x = 0$, so that we have the same formular as before.)

Let us now find the variation of the integral

$$I = \int_{t_0}^{t_1} \varphi(t, x, y, z, x', y', z', \ldots)\, dt.$$

Changing x to $x + \delta x$, y to $y + \delta y$, x' to $x' + \delta x'$, etc.,

$$I + \delta I + \frac{1}{2}\delta^2 I + \cdots = \int_{t_0}^{t_1} \left(\varphi + \delta\varphi + \frac{1}{2}\delta^2\varphi + \cdots\right) dt,$$

and the variations are

$$\delta I = \int_{t_0}^{t_1} \delta\varphi\, dt, \quad \delta^k I = \int_{t_0}^{t_1} \delta^k \varphi\, dt,$$

that is, the operations of variation and integration are commutative. (The limits have been supposed given, that is unvaried.) These two principles of commutativity of δ with d and $\int$ form the basis of the subject of the Calculus of Variations.

(As in the case of derivatives, it may happen that we wish to examine the integral with respect to a variable whose variation does not vanish. We must as before introduce the independent variable, writing

$$\delta\int\varphi\,dx=\delta\int\varphi x'\,dt=\int(\delta\varphi\cdot x'+\varphi\,\delta x')\,dt$$

$$=\int(\delta\varphi\cdot dx+\varphi d\delta x).$$

It may, on occasion, be more convenient to use these more general formulae, not supposing the variation of any variable to vanish.) .

If the limits are varied, we have, indicating the part of the change in I due to the change in either limit by a suffix,

$$\delta_1 I=\int_{t_0}^{t_1+\delta t_1}\varphi\,dt-\int_{t_0}^{t_1}\varphi\,dt=\int_{t_1}^{t_1+\delta t_1}\varphi\,dt=\varphi(t_1)\delta t_1,$$

$$\delta_2 I=\int_{t_0+\delta t_0}^{t_1}\varphi\,dt-\int_{t_0}^{t_1}\varphi\,dt=-\int_{t_0}^{t_0+\delta t_0}\varphi\,dt=-\varphi(t_0)\delta t_0,$$

which are to added to the part already found.

In the application of the calculus of variations, we often encounter problems involving a number of independent variables, so that we deal with partial derivatives, and multiple integrals. The principles here given will however suffice for the treatment of all the usual questions.

As a celebrated mechanical example of the use of the Calculus of Variations let us consider the question: What is that curve along which a particle must be constrained to descend under the influence of gravity in order to pass from one point to another in the least possible time?

Since $v=\frac{ds}{dt}$, we have for the time of descent $t=\int\frac{ds}{v}$, or making use of the equation of energy § 27, 27),

$$t=\int\frac{ds}{\sqrt{v_0{}^2-2g(z-z_0)}}.$$

Let us take for the independent variable corresponding to t above the vertical coordinate z. We suppose the motion to take place in a vertical plane. We have then

$$ds=\sqrt{1+x'^2}\cdot dz,\quad x'=\frac{dx}{dz}.$$

If now we make an arbitrary infinitesimal variation of the curve, if t is to be a minimum we must have the term of the first order in ε vanish,

$$\delta t=0.$$

Now

$$t = \int_{z_0}^{z_1} \sqrt{\frac{1 + x'^2}{v_0^2 - 2g(z - z_0)}}\, dz.$$

For any particular curve x is a given function of z. Giving it a variation δx we have

$$\delta t = \int_{z_0}^{z_1} \frac{x'\,\delta x'\,dz}{\sqrt{(1 + x'^2)(v_0^2 - 2g[z - z_0])}}.$$

Making use of $\delta x' = \frac{d\,\delta x}{dz}$ and integrating by parts[1]),

$$\delta t = \frac{x'\,\delta x}{\sqrt{(1 + x'^2)(v_0^2 - 2g[z - z_0])}} \Bigg/_{z_0}^{z_1} - \int_{z_0}^{z_1} \delta x \frac{d}{dz}\left(\frac{x'}{\sqrt{(1 + x'^2)(v_0^2 - 2g[z - z_0])}}\right) dz.$$

If the ends of the curve are fixed δx vanishes for both limits z_0 and z_1, hence the integrated part vanishes. Consequently for a minimum the integral must vanish.

Now since the function δx is purely arbitrary if the other factor of the integrand did not vanish for any points of the curve we might take δx of the same sign as that factor at each point. Thus the integrand would be positive everywhere and the integral would not vanish, consequently the factor multiplying δx must vanish for each point of the curve, or

$$\frac{d}{dz}\left(\frac{x'}{\sqrt{(1 + x'^2)(v_0^2 - 2g[z - z_0])}}\right) = 0.$$

This is the differential equation of the curve of quickest descent, or *brachistochrone.*

Integrating we have

$$\frac{x'}{\sqrt{(1 + x'^2)(v_0^2 - 2g[z - z_0])}} = c, \quad \text{an arbitrary constant.}$$

Squaring and solving for x'^2 we obtain

$$x'^2 = \frac{c^2(v_0^2 - 2g[z - z_0])}{1 - c^2(v_0^2 - 2g[z - z_0])}.$$

Let us put $a = \frac{v_0^2}{2g} + z_0$, $b = \frac{1}{2gc^2} - \frac{v_0^2}{2g} - z_0$ (b is arbitrary, since it involves c), then we have

$$x' = \frac{dx}{dz} = \pm\sqrt{\frac{a - z}{b + z}}.$$

1) The bar / signifies that we are to subtract the value of the expression before it at the lower limit z_0 from the value at the upper limit z_1.

If we introduce a new variable ϑ such that

$$z = \frac{a-b}{2} - \frac{a+b}{2}\cos\vartheta,$$

we have

$$dz = \frac{a+b}{2}\sin\vartheta\, d\vartheta,$$

$$a - z = \frac{a+b}{2}(1+\cos\vartheta),$$

$$b + z = \frac{a+b}{2}(1-\cos\vartheta).$$

Thus our differential equation becomes

$$\frac{dx}{dz} = \sqrt{\frac{1+\cos\vartheta}{1-\cos\vartheta}} = \frac{1+\cos\vartheta}{\sin\vartheta}.$$

Consequently

$$dx = dz\left(\frac{1+\cos\vartheta}{\sin\vartheta}\right) = \frac{a+b}{2}(1+\cos\vartheta)\,d\vartheta.$$

Integrating,

$$x + d = \frac{a+b}{2}(\vartheta + \sin\vartheta),$$

where d is an arbitrary constant.

Combining this with

$$z + b = \frac{a+b}{2}(1-\cos\vartheta),$$

we have the equations of the curve in terms of ϑ, a parameter which may be eliminated from the two equations.

If a vertical circle of radius $\frac{a+b}{2} = A$ roll under a horizontal straight line (Fig. 19), and ϑ be the angle made with the downward vertical by a radius fixed in the rolling circle, the distance moved

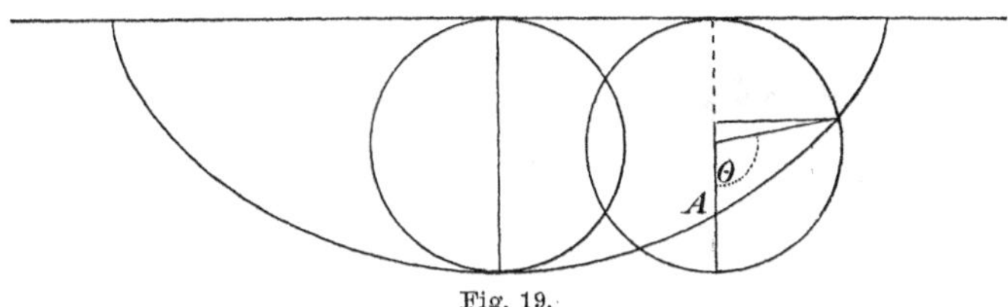

Fig. 19.

by the center of the circle from the position in which $\vartheta = 0$ is equal to the arc rolled over, $A\vartheta$. A point at the extremity of the given radius lies then at a horizontal distance $A\sin\vartheta$ farther, so that its horizontal coordinate is

$$x = A(\vartheta + \sin\vartheta).$$

Its vertical coordinate measured from its initial position $\vartheta = 0$ is

$$z = A(1-\cos\vartheta).$$

The curve described by a point on the circumference of the rolling circle is called a cycloid. Obviously from the symmetry of the case the curve is symmetrical about a vertical axis on which the point lies when $\vartheta = 0$. Thus our equations of the brachistochrone show that it is a cycloid with vertical axis.

The arbitrary constants of integration, b and d, are determined by the two points through which the curve is to pass. The discovery that the cycloid is the brachistochrone for gravity is due to Jean Bernoulli.

30. Dependence of Line Integral on Path. Stokes's Theorem. Curl. Consider now our line-integral,

$$I = \int_A^B (X\,dx + Y\,dy + Z\,dz).$$

We have first to introduce an independent variable corresponding to the t of the previous section, variation of which shall cause the point of integration to move along a given curve. Let us call this s, which to fix the ideas may (though this is unnecessary) be considered as the distance measured along the curve from the point A. Thus we write

$$I = \int_A^B \left(X\frac{dx}{ds} + Y\frac{dy}{ds} + Z\frac{dz}{ds}\right) ds.$$

The functions X, Y, Z, being given for every point x, y, z, the integral I will in general depend on the form of the curve AB. If we make an infinitesimal transformation of the curve, the integral will change, and we shall now seek an expression for the variation. We have

$$\delta I = \int \left(\delta X \cdot \frac{dx}{ds} + X\delta\frac{dx}{ds} + \delta Y \cdot \frac{dy}{ds} + Y\delta\frac{dy}{ds} + \delta Z\frac{dz}{ds} + Z\delta\frac{dz}{ds}\right) ds.$$

Now

$$\delta X = \frac{\partial X}{\partial x}\delta x + \frac{\partial X}{\partial y}\delta y + \frac{\partial X}{\partial z}\delta z,$$

and

$$\delta\frac{dx}{ds} = \frac{d(\delta x)}{ds}.$$

We may perform upon the term

$$\int_A^B X\frac{d(\delta x)}{ds}\,ds$$

an integration by parts,

$$\int_A^B X\frac{d(\delta x)}{ds}ds = X\delta x\Big/_A^B - \int_A^B \delta x\frac{dX}{ds}ds.$$

Now we have

$$\frac{dX}{ds} = \frac{\partial X}{\partial x}\frac{dx}{ds} + \frac{\partial X}{\partial y}\frac{dy}{ds} + \frac{\partial X}{\partial z}\frac{dz}{ds}.$$

Performing similar operations on the other terms we have

$$\begin{aligned}\delta I = (X\delta x + Y\delta y + Z\delta z)\Big/_A^B + \int\Big[&\left(\frac{\partial X}{\partial x}\delta x + \frac{\partial X}{\partial y}\delta y + \frac{\partial X}{\partial z}\delta z\right)\frac{dx}{ds} \\ &+ \left(\frac{\partial Y}{\partial x}\delta x + \frac{\partial Y}{\partial y}\delta y + \frac{\partial X}{\partial z}\delta z\right)\frac{dy}{ds} + \left(\frac{\partial Z}{\partial x}\delta x + \frac{\partial Z}{\partial y}\delta y + \frac{\partial Z}{\partial z}\delta z\right)\frac{dz}{ds} \\ &- \delta x\left(\frac{\partial X}{\partial x}\frac{dx}{ds} + \frac{\partial X}{\partial y}\frac{dy}{ds} + \frac{\partial X}{\partial z}\frac{dz}{ds}\right) \\ &- \delta y\left(\frac{\partial Y}{\partial x}\frac{dx}{ds} + \frac{\partial Y}{\partial y}\frac{dy}{ds} + \frac{\partial Y}{\partial z}\frac{dz}{ds}\right) \\ &- \delta z\left(\frac{\partial Z}{\partial x}\frac{dx}{ds} + \frac{\partial Z}{\partial y}\frac{dy}{ds} + \frac{\partial Z}{\partial z}\frac{dz}{ds}\right)\Big]ds.\end{aligned}$$

Now if in the variation the ends of the curve A and B are fixed, δx, δy, δz vanish for A and B, and the integrated part $X\delta x + Y\delta y + Z\delta z\Big/_A^B$ vanishes. Collecting those terms under the sign of integration that do not cancel, and removing the factor ds we have

$$\begin{aligned}\delta I = \int\Big[(\delta y\,dz - \delta z\,dy)\left(\frac{\partial Z}{\partial y} - \frac{\partial Y}{\partial z}\right) + (\delta z\,dx - \delta x\,dz)\left(\frac{\partial X}{\partial z} - \frac{\partial Z}{\partial x}\right) \\ + (\delta x\,dy - \delta y\,dx)\left(\frac{\partial Y}{\partial x} - \frac{\partial X}{\partial y}\right)\Big].\end{aligned}$$

Now the determinant

$$\delta y\,dz - \delta z\,dy$$

is the area of the parallelogram in the YZ-plane the projection of whose sides on the Y- and Z-axes are dy, dz, δy, δz. That is, if we consider the infinitesimal parallelogram whose vertices are the points s, $s + ds$, and their transformed positions, the above determinant is the area of its projection on the YZ-plane. If the area of the parallelogram is dS and n is the direction of its normal, we have

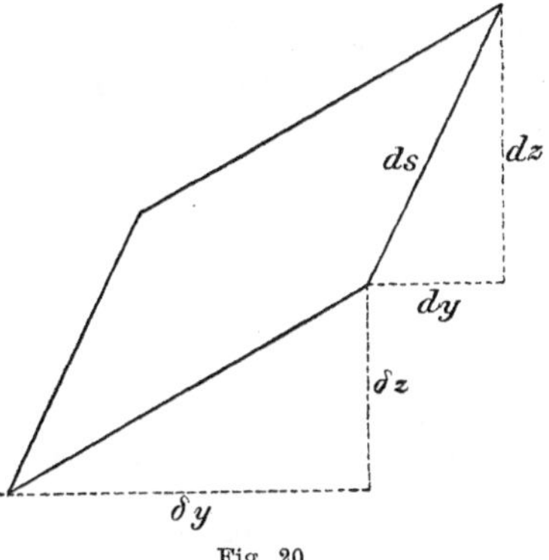

Fig. 20.

$$\delta y\, dz - \delta z\, dy = dS \cos(nx),$$
$$\delta z\, dx - \delta x\, dz = dS \cos(ny),$$
$$\delta x\, dy - \delta y\, dx = dS \cos(nz),$$

and

$$\delta I = \int \left\{ \left(\frac{\partial Z}{\partial y} - \frac{\partial Y}{\partial z}\right) \cos(nx) + \left(\frac{\partial X}{\partial z} - \frac{\partial Z}{\partial x}\right) \cos(ny) + \left(\frac{\partial Y}{\partial x} - \frac{\partial X}{\partial y}\right) \cos(nz) \right\} dS,$$

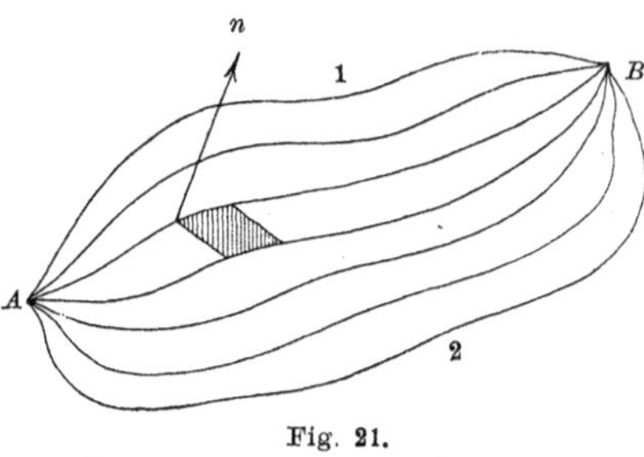

Fig. 21.

which is in the form of a *surface integral* over the strip of infinitesimal width bounded by the two curves of integration.

If we again make an infinitesimal transformation, and so continue until the path has swept over any finite portion of a surface S, and sum all the variations of I, we get for the final result that the difference in I for the two extreme paths 1 and 2 is the surface integral

$$\lim \Sigma \delta I = I_2 - I_1 = \iint \left\{ \left(\frac{\partial Z}{\partial y} - \frac{\partial Y}{\partial z}\right) \cos(nx) + \left(\frac{\partial X}{\partial z} - \frac{\partial Z}{\partial x}\right) \cos(ny) + \left(\frac{\partial Y}{\partial x} - \frac{\partial X}{\partial y}\right) \cos(nz) \right\} dS$$

taken over the portion of the surface bounded by the paths 1 and 2 from A to B. Now $-I_1$ may be considered the integral from B to A along the path 1, so that $I_2 - I_1$ is the integral around the closed path which forms the contour of the portion of surface S. We accordingly get the following, known as

STOKES'S THEOREM.[1]) The line integral, around any closed contour, of the tangential component of a vector R, whose components are X, Y, Z, is equal to the surface integral over any portion of surface bounded by the contour, of the normal component of a vector ω, whose components ξ, η, ζ are related to X, Y, Z by the relations

$$\xi = \frac{\partial Z}{\partial y} - \frac{\partial Y}{\partial z},$$
$$\eta = \frac{\partial X}{\partial z} - \frac{\partial Z}{\partial x},$$
$$\zeta = \frac{\partial Y}{\partial x} - \frac{\partial X}{\partial y}.$$

1) The proof here given is from the author's notes on the lectures of Professor von Helmholtz. A similar treatment is given by Picard, *Traité d'Analyse*, Tom. I, p. 73.

The normal must be drawn toward that side of the surface that shall make the rotation of a right-handed screw advancing along the normal agree with the direction of traversing the closed contour of integration.

37) $$\int R\cos(R, ds)\,ds = \int X\,dx + Y\,dy + Z\,dz = \int\!\!\int \omega\cos(\omega n)\,dS$$
$$= \int\!\!\int \{\xi\cos(nx) + \eta\cos(ny) + \zeta\cos(nz)\}\,dS.$$

The vector ω related to the vector point-function R by the differential equations above is called the *rotation*, *spin* (Clifford), or *curl* (Maxwell and Heaviside) of R. Such vectors are of frequent occurrence in mathematical physics. (See Part III.)

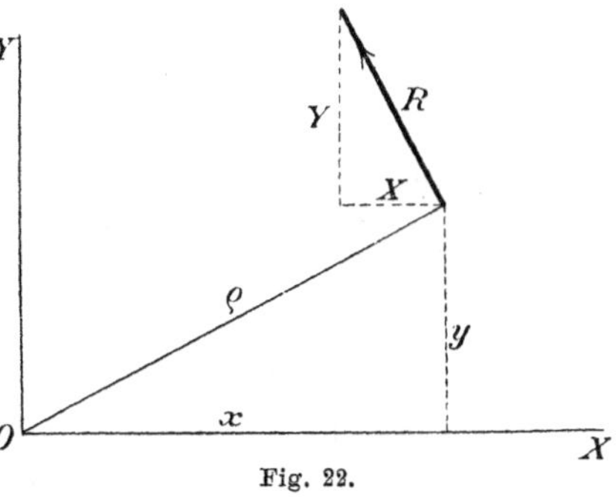

Fig. 22.

The significance of the geometrical term *curl* can be seen from the physical example in which the vector R represents the velocity of a point instantaneously occupying the position x, y, z in a rigid body turning about the Z-axis with an angular velocity ω. Then the vector $R = \omega\varrho$ is perpendicular to the radius ϱ and its components are (Fig. 22),

$$X = R\cos(Rx) = -R\sin(\varrho x) = -R\frac{y}{\varrho} = -y\omega,$$

$$Y = R\cos(Ry) = \quad R\cos(\varrho x) = \quad R\frac{x}{\varrho} = \quad x\omega,$$

where ω is constant, and

$$\frac{\partial Y}{\partial x} - \frac{\partial X}{\partial y} = 2\omega.$$

So that the z-component of the curl of the linear velocity is twice the angular velocity about the Z-axis. Further examples are presented to us in the theory of fluid motion.

31. Lamellar Vectors. In finding the variation of the integral I in the previous section, since the variations $\delta x, \delta y, \delta z$ are perfectly arbitrary functions of s, if the integral is to be independent of the path, δI must vanish, which can happen for all possible choices of $\delta x, \delta y, \delta z$, only if

38) $$\frac{\partial Z}{\partial y} - \frac{\partial Y}{\partial z} = \frac{\partial X}{\partial z} - \frac{\partial Z}{\partial x} = \frac{\partial Y}{\partial x} - \frac{\partial X}{\partial y} = 0,$$

that is if the curl of R vanishes everywhere. In case this condition is satisfied, I depends only on the positions of the limiting points A

and B, and not on the path of integration. Consequently, as stated without proof in § 28, the conditions 38) are sufficient as well as necessary.

If A is given, I is a point-function[1]) of its upper limit B, let us say φ. If B is displaced a distance s in a given direction to B', the change in the function φ is

$$\varphi_{B'} - \varphi_B = \int_B^{B'} (X\,dx + Y\,dy + Z\,dz),$$

and the limit of the ratio of the change to the distance,

38a) $$\lim_{s=0} \frac{\varphi_{B'} - \varphi_B}{s} = \frac{\partial \varphi}{\partial s} = X\frac{dx}{ds} + Y\frac{dy}{ds} + Z\frac{dz}{ds},$$

is the *derivative of φ in the direction s.*

If we take s successively in the directions of the axes of coordinates,

38b) $$\frac{\partial \varphi}{\partial x} = X, \quad \frac{\partial \varphi}{\partial y} = Y, \quad \frac{\partial \varphi}{\partial z} = Z.$$

A vector whose components are thus derived from a single scalar function φ is called the *vector differential parameter* or *gradient* of φ.

Accordingly the three equations of condition 38), equivalent to curl $R = 0$, are simply the conditions that X, Y, Z may be represented as the derivatives of a scalar point-function. In this case the expression

$$X\,dx + Y\,dy + Z\,dz = \frac{\partial \varphi}{\partial x}dx + \frac{\partial \varphi}{\partial y}dy + \frac{\partial \varphi}{\partial z}dz = d\varphi$$

is called a *perfect differential.*

From the definition of the parameter of a scalar point-function 38b), we see that the components of the vector R at any point are proportional to the direction cosines of the normal to the surface $\varphi = const.$ passing through the point in question, that is R is perpendicular to the surface. A surface for which a scalar point-function is constant is called a *level surface* of that function. Since

$$\frac{dx}{ds}, \quad \frac{dy}{ds}, \quad \frac{dz}{ds}$$

are the direction cosines of the tangent to the arc ds, we see that equation 38a) states that the derivative of φ in any direction is the projection of its vector parameter on that direction. Since a vector is the maximum value of any of its projections, we see that the direction of the normal to the level surface of φ at any point is the direction of fastest increase of φ at that point. Also if we take for $\varphi_{B'}$ and φ_B in 38a) the constant values belonging to two infinitely

---

1) A function of the coordinates of a point.

near level surfaces, we see that, the numerator being constant, the derivative in the direction of the normal, that is the value of the vector parameter R, is inversely proportional to the normal distance between the two infinitely near level surfaces of the function φ.

Such a pair of surfaces will be called a thin *level sheet* or lamina. For this reason a vector point-function that may be represented everywhere in a certain region as the vector parameter of a scalar point-function will be called a *laminar*, or *lamellar* vector (Maxwell).

The scalar function φ (or its negative) is termed the *potential* of the vector R.

32. Motion of the Center of Mass. Suppose that a system of particles is under the influence only of forces acting between the particles and depending on their mutual distances, and that the constraints, if there be any, are such as to permit of a virtual displacement which is represented by equal vectors for all the particles. Then in our equation of d'Alembert's principle (18) let us put each δx_r equal to the same quantity λ, each δy_r equal to μ, and each δz_r equal to ν.

Supposing the system to be conservative, and using equations 23) we have

$$39) \quad \sum_r \left\{ m_r \left(\lambda \frac{d^2 x_r}{dt^2} + \mu \frac{d^2 y_r}{dt^2} + \nu \frac{d^2 z_r}{dt^2} \right) \right\} = \lambda \sum_r \frac{\partial U}{\partial x_r} + \mu \sum_r \frac{\partial U}{\partial y_r} + \nu \sum_r \frac{\partial U}{\partial z_r}.$$

Now as the forces depend only on the mutual distances of the particles, and therefore only on the differences of their coordinates, if we put

$$\begin{array}{lll} \xi_1 = x_1 - x_n, & \eta_1 = y_1 - y_n, & \zeta_1 = z_1 - z_n, \\ \xi_2 = x_2 - x_n, & \eta_2 = y_2 - y_n, & \zeta_2 = z_2 - z_n, \\ \cdots & \cdots & \cdots \\ \cdots & \cdots & \cdots \\ \xi_{n-1} = x_{n-1} - x_n, & \eta_{n-1} = y_{n-1} - y_n, & \zeta_{n-1} = z_{n-1} - z_n, \end{array}$$

we shall have

$$U = U(\xi_1, \eta_1, \zeta_1, \ldots \zeta_{n-1}).$$

Accordingly

$$\frac{\partial U}{\partial x_1} = \frac{\partial U}{\partial \xi_1}\frac{\partial \xi_1}{\partial x_1} + \frac{\partial U}{\partial \xi_2}\frac{\partial \xi_2}{\partial x_1} \cdots + \frac{\partial U}{\partial \eta_1}\frac{\partial \eta_1}{\partial x_1} + \cdots = \frac{\partial U}{\partial \xi_1}$$

and likewise

$$\frac{\partial U}{\partial x_2} = \frac{\partial U}{\partial \xi_2},$$

$$\cdots\cdots\cdots$$

$$\frac{\partial U}{\partial x_{n-1}} = \frac{\partial U}{\partial \xi_{n-1}}.$$

On the other hand

$$40)\qquad \frac{\partial U}{\partial x_n} = -\frac{\partial U}{\partial \xi_1} - \frac{\partial U}{\partial \xi_2} - \cdots - \frac{\partial U}{\partial \xi_{n-1}}.$$

Consequently

$$\frac{\partial U}{\partial x_1} + \frac{\partial U}{\partial x_2} + \cdots \frac{\partial U}{\partial x_n} = \sum_r \frac{\partial U}{\partial x_r} = 0.$$

Thus 39) becomes

$$41)\qquad \lambda \sum_r m_r \frac{d^2 x_r}{dt^2} + \mu \sum m_r \frac{d^2 y_r}{dt^2} + \nu \sum m_r \frac{d^2 z_r}{dt^2} = 0.$$

But since λ, μ, ν are perfectly arbitrary this is equivalent to the three equations

$$42)\qquad \sum_r m_r \frac{d^2 x_r}{dt^2} = 0, \quad \sum m_r \frac{d^2 y_r}{dt^2} = 0, \quad \sum m_r \frac{d^2 z_r}{dt^2} = 0.$$

Since the m's are independent of the time, we may differentiate outside of the summation and write the above

$$43)\qquad \frac{d^2}{dt^2} \Sigma_r m_r x_r = 0, \quad \frac{d^2}{dt^2} \Sigma_r m_r y_r = 0, \quad \frac{d^2}{dt^2} \Sigma_r m_r z_r = 0.$$

If we define the coordinates of a point $\bar{x}$, $\bar{y}$, $\bar{z}$ by the equations

$$\bar{x}\Sigma_r m_r = \Sigma_r m_r x_r, \quad \bar{y}\Sigma_r m_r = \Sigma_r m_r y_r, \quad \bar{z}\Sigma_r m_r = \Sigma_r m_r z_r$$

and if we consider a mass m to consist of m particles of unit mass,

$$\bar{x} = \frac{\Sigma_r m_r x_r}{\Sigma_r m_r}$$

being the sum of the x-coordinates of the whole number of unit particles divided by their number is the arithmetical mean of the x-coordinates. If m is not an integer, by the method of limits we extend the motion of the mean in the usual manner. The point $\bar{x}, \bar{y}, \bar{z}$, the mean mass point thus defined is called the *center of mass* of the system. (The common term *center of gravity* is poorly adapted to express the idea here involved and had better be avoided. We shall see in the chapter on Newtonian Attractions that bodies in general do not possess centers of gravity.)

The equations 43) thus become

$$44)\qquad \frac{d^2 \bar{x}}{dt^2} = 0, \quad \frac{d^2 \bar{y}}{dt^2} = 0, \quad \frac{d^2 \bar{z}}{dt^2} = 0.$$

Therefore the center of mass of a system whose parts exert forces upon each other depending only on their mutual distances moves with constant velocity in a straight line. This is the Principle of Conservation of Motion of the Center of Mass. It evidently applies to the solar system. What the absolute velocity of the center of mass of the solar system is or what its velocity with respect to the so-called fixed stars we do not at present know with exactness.

Returning to the equations 39), whether there is a force-function or not, λ, μ, ν, being the same for each term of the summation, may be taken out from under the summation sign and being arbitrary, the equation 39) is equivalent to the three

$$45)\quad \sum_r m_r \frac{d^2 x_r}{dt^2} = \Sigma_r X_r, \quad \sum_r m_r \frac{d^2 y_r}{dt^2} = \Sigma_r Y_r, \quad \sum_r m_r \frac{d^2 z_r}{dt^2} = \Sigma_r Z_r,$$

or as before

$$46)\quad \frac{d^2\bar{x}}{dt^2}\Sigma_r m_r = \Sigma_r X_r, \quad \frac{d^2\bar{y}}{dt^2}\Sigma_r m_r = \Sigma_r Y_r, \quad \frac{d^2\bar{z}}{dt^2}\Sigma_r m_r = \Sigma_r Z_r,$$

that is: The center of mass of any system of the kind specified moves as if all the forces applied to its various parts were applied at the center of mass to a single particle whose mass is equal to the mass of the whole system.

This principle of the motion of the center of mass reduces the problem of the motion of the system to that of finding the motion of a single particle together with that of the motion of the parts of the system with respect to the center of mass.

A rigid body is a system of particles coming under the case here treated, since the only constraints are such as render all the mutual distances of individual points constant. Therefore the only new principles required in order to treat the motion of a rigid body are such as determine its motion relatively to its center of mass.

If the center of mass is to remain at rest or move uniformly, we must have

$$47)\quad \Sigma_r X_r = 0, \quad \Sigma_r Y_r = 0, \quad \Sigma_r Z_r = 0.$$

This will always be the case as shown above for mutually attracting particles, since to every action there is an equal and opposite reaction. The three equations 47) furnish three necessary conditions for the equilibrium of a rigid body.

If we introduce the relative coordinates of the particles with respect to the center of mass into the expression for kinetic energy it assumes a remarkable form. Let us put

$$x_r = \bar{x} + \xi_r, \quad y_r = \bar{y} + \eta_r, \quad z_r = \bar{z} + \zeta_r,$$

then

$$\frac{dx_r}{dt} = \frac{d\bar{x}}{dt} + \frac{d\xi_r}{dt},$$

$$\frac{dy_r}{dt} = \frac{d\bar{y}}{dt} + \frac{d\eta_r}{dt},$$

$$\frac{dz_r}{dt} = \frac{d\bar{z}}{dt} + \frac{d\zeta_r}{dt}.$$

48)
$$T = \frac{1}{2}\sum_r m_r\left\{\left(\frac{dx_r}{dt}\right)^2 + \left(\frac{dy_r}{dt}\right)^2 + \left(\frac{dz_r}{dt}\right)^2\right\}$$
$$= \frac{1}{2}\sum_r m_r\left\{\left(\frac{d\overline{x}}{dt}\right)^2 + \left(\frac{d\overline{y}}{dt}\right)^2 + \left(\frac{d\overline{z}}{dt}\right)^2\right.$$
$$+ 2\frac{d\overline{x}}{dt}\frac{d\xi_r}{dt} + 2\frac{d\overline{y}}{dt}\frac{d\eta_r}{dt} + 2\frac{d\overline{z}}{dt}\frac{d\zeta_r}{dt}$$
$$\left. + \left(\frac{d\xi_r}{dt}\right)^2 + \left(\frac{d\eta_r}{dt}\right)^2 + \left(\frac{d\zeta_r}{dt}\right)^2\right\},$$

and taking outside of the sign of summation the factors common to all the terms involving $\frac{d\overline{x}}{dt}, \frac{d\overline{y}}{dt}, \frac{d\overline{z}}{dt}$,

49)
$$T = \frac{1}{2}\left[\left\{\left(\frac{d\overline{x}}{dt}\right)^2 + \left(\frac{d\overline{y}}{dt}\right)^2 + \left(\frac{d\overline{z}}{dt}\right)^2\right\}\Sigma_r m_r\right.$$
$$+ \sum_r m_r\left\{\left(\frac{d\xi_r}{dt}\right)^2 + \left(\frac{d\eta_r}{dt}\right)^2 + \left(\frac{d\zeta_r}{dt}\right)^2\right\}$$
$$\left. + 2\frac{d\overline{x}}{dt}\sum_r m_r\frac{d\xi_r}{dt} + 2\frac{d\overline{y}}{dt}\sum_r m_r\frac{d\eta_r}{dt} + 2\frac{d\overline{z}}{dt}\sum_r m_r\frac{d\zeta_r}{dt}\right].$$

Now in the last three terms we may write

$$\sum_r m_r\frac{d\xi_r}{dt} = \frac{d}{dt}\Sigma_r m_r\xi_r = \frac{d}{dt}(\overline{\xi}\,\Sigma_r m_r),$$

if $\overline{\xi}$ is the x-coordinate of the center of mass in the ξ, η, ζ system. But since the center of mass in the origin of the relative coordinates ξ, η, ζ, this is equal to zero. Similarly for the terms in η_r and ζ_r. Thus we have remaining if we write M for the mass of the whole system,

50)
$$T = \frac{1}{2}M\left\{\left(\frac{d\overline{x}}{dt}\right)^2 + \left(\frac{d\overline{y}}{dt}\right)^2 + \left(\frac{d\overline{z}}{dt}\right)^2\right\}$$
$$+ \frac{1}{2}\sum_r m_r\left\{\left(\frac{d\xi_r}{dt}\right)^2 + \left(\frac{d\eta_r}{dt}\right)^2 + \left(\frac{d\zeta_r}{dt}\right)^2\right\}.$$

The first term is the kinetic energy of a particle whose mass is equal to the total mass of the system placed at the center of mass, while the second is the relative kinetic energy of the system with respect to the center of mass. Thus the absolute kinetic energy is always greater than its relative kinetic energy with respect to the center of mass (unless the center of mass be at rest). The center of mass is the only point for which such a decomposition of the kinetic energy is generally possible.

If the principle of the conservation of motion of the center of mass holds we have

$$\frac{d\overline{x}}{dt} = a, \quad \frac{d\overline{y}}{dt} = b, \quad \frac{d\overline{z}}{dt} = c,$$

and inserting these in the equation of energy for a conservative system, $T + W = h$,

$$51)\quad \frac{1}{2}\sum_r m_r\left\{\left(\frac{d\xi_r}{dt}\right)^2+\left(\frac{d\eta_r}{dt}\right)^2+\left(\frac{d\zeta_r}{dt}\right)^2\right\}+W=h-\frac{1}{2}M\{a^2+b^2+c^2\}=h'.$$

In this case accordingly the principle of conservation of energy holds also for the *relative* kinetic energy, the constant h being changed. Inasmuch as we know of no absolutely fixed system of axes of reference it is obvious that the kinetic energy of any system contains an indeterminate part. But in virtue of the above principle if we consider the center of mass of the solar system to be at rest all our conclusions with regard to energy will hold good. The effect in general of referring motions to systems of axes which are not at rest will be dealt with in Chapter VII.

As a simple example of the above principle let us consider the case of a rigid sphere or circular cylinder, with axis horizontal, rolling without sliding down an inclined plane under the action of gravity. If the distance that the center of the body has moved parallel to the plane be s, the first part of T is $\frac{1}{2}M\left(\frac{ds}{dt}\right)^2$. If the angle that a plane through the horizontal axis parallel to the inclined plane makes with the normal to the inclined plane be ϑ (Fig. 23), the velocity of a particle with respect to the center is $r\frac{d\vartheta}{dt}$, where r is its distance from the horizontal axis. The relative kinetic energy is thus

$$T' = \frac{1}{2}\Sigma_r m_r r_r^2\left(\frac{d\vartheta}{dt}\right)^2,$$

or since $\frac{d\vartheta}{dt}$, the angular velocity of rolling is the same for all terms of the summation,

$$52)\qquad T' = \frac{1}{2}\left(\frac{d\vartheta}{dt}\right)^2\Sigma_r m_r r_r^2.$$

The factor $\Sigma_r m_r r_r^2$ is called the *moment of inertia* of the system about the horizontal axis through the center of mass and will be denoted by K. Thus we have

$$53)\qquad T = \frac{1}{2}\left\{M\left(\frac{ds}{dt}\right)^2 + K\left(\frac{d\vartheta}{dt}\right)^2\right\}.$$

If the rolling takes place without sliding we have the geometrical condition of constraint,

$$R\frac{d\vartheta}{dt} = \frac{ds}{dt},$$

where R is the radius of the rolling body.

The loss of potential energy is Mg times the vertical distance fallen, $s \sin \alpha$, where α is the angle of inclination of the plane to the horizontal. Our equation thus becomes

$$\text{54)} \qquad \frac{1}{2}\left\{M\left(\frac{ds}{dt}\right)^2 + \frac{K}{R^2}\left(\frac{ds}{dt}\right)^2\right\} - Mgs \sin \alpha = \text{const.}$$

If $\frac{ds}{dt} = V$ when $s = 0$, determining the constant we have

$$\text{55)} \qquad \left(\frac{ds}{dt}\right)^2 - V^2 = \frac{2 Mgs \sin \alpha}{M + \frac{K}{R^2}}.$$

Thus the motion is the same (cf. § 18) as that of a particle falling freely with the acceleration diminished in the ratio $\frac{\sin \alpha}{1 + \frac{K}{MR^2}}$.

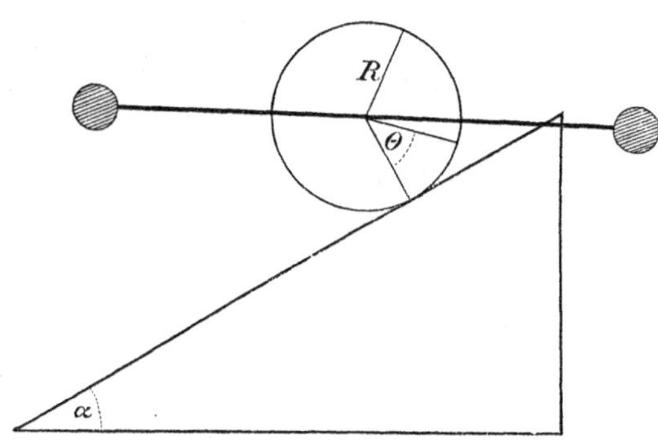

Fig. 23.

Thus by increasing K, which may be done by symmetrically attaching heavy masses to a bar fastened to the cylinder in such a way as not to interfere with the rolling of the cylinder (Fig. 23), we may make the motion as slow as we please and thus study the laws of constant acceleration.

33. Moment of Momentum. Under the supposition that the equations of constraint were compatible with the displacement of the system parallel to itself and that the force-function was thereby unchanged we obtained the principle of the conservation of motion of the center of mass. We will now suppose that the equations of constraint are compatible with a rotation of the system about the axis of X and that the force-function is thereby unaffected. This will be the case in a rigid system or in a free system left to its own internal forces (if conservative).

If we put

$$\text{56)} \qquad \begin{aligned} y_r &= r_r \cos \omega_r, \\ z_r &= r_r \sin \omega_r, \end{aligned}$$

such a displacement is obtained by changing all the ω_r's by the same amount $\delta\omega$, leaving the r's unchanged. We have then

$$\text{57)} \qquad \begin{aligned} &\delta x_r = 0, \quad \delta y_r = - r_r \sin \omega_r \delta\omega = - z_r \delta\omega, \\ &\delta z_r = r_r \cos \omega_r \delta\omega = y_r \delta\omega. \end{aligned}$$

Inserting these values in d'Alembert's equation we obtain

$$58)\quad \delta\omega \sum_r m_r \left\{ -z_r \frac{d^2 y_r}{dt^2} + y_r \frac{d^2 z_r}{dt^2} \right\} = \delta\omega\, \Sigma_r (y_r Z_r - z_r Y_r) = \delta U.$$

If U depends only on the mutual distances of the particles of the system it is unchanged in the displacement, $\delta U = 0$.

We then have

$$59)\quad \sum_r m_r \left(y_r \frac{d^2 z_r}{dt^2} - z_r \frac{d^2 y_r}{dt^2} \right) = 0.$$

As was mentioned in § 11 the quantity within the parenthesis is an exact derivative, so that

$$\sum_r m_r \frac{d}{dt} \left(y_r \frac{dz_r}{dt} - z_r \frac{dy_r}{dt} \right) = 0,$$

or differentiating outside of the sign of summation

$$\frac{d}{dt} \sum_r m_r \left(y_r \frac{dz_r}{dt} - z_r \frac{dy_r}{dt} \right) = 0.$$

Integrating we obtain

$$60)\quad \sum_r m_r \left(y_r \frac{dz_r}{dt} - z_r \frac{dy_r}{dt} \right) = H_x, \quad \text{an arbitrary constant.}$$

The expression $m\left(y\frac{dz}{dt} - z\frac{dy}{dt}\right) = ymv_z - zmv_y$ is the *moment of momentum* [42), § 13] about the X-axis of the mass m, or it is the product of twice the mass by the sectorial velocity $\frac{dS_x}{dt}$ (§ 8). The theorem consequently states that the moment of momentum of the whole system with respect to the X-axis is constant.

Under similar conditions for the other two axes we obtain

$$60)\quad \begin{aligned} \sum_r m_r \left(z_r \frac{dx_r}{dt} - x_r \frac{dz_r}{dt} \right) &= H_y, \\ \sum_r m_r \left(x_r \frac{dy_r}{dt} - y_r \frac{dx_r}{dt} \right) &= H_z. \end{aligned}$$

The vector H, whose components are H_x, H_y, H_z, is the resultant moment of momentum of the whole system, and if the above equations 60) hold it is constant both in magnitude and direction. This is the case for the solar system and we accordingly have an unvarying direction in space characteristic of the system. This direction was called by Laplace that of the Invariable Axis and the plane through the sun perpendicular to it the Invariable Plane. It may be defined as that plane for which the sum of the masses of each particle multiplied by the projection of its sectorial velocity on that plane is a maximum. Such a plane furnishes a natural plane of coordinates for the solar system.

The principle expressed by equations 60) will be referred to as the Principle of Conservation of Moment of Momentum. On account of the connection with the sectorial velocity it has received the shorter and more euphonious title of the Principle of Areas.

In case δU does not vanish, going back to equation 58), we may divide out $\delta\omega$ and instead of 60) now obtain

$$
\begin{aligned}
\frac{dH_x}{dt} &= \Sigma_r(y_r Z_r - z_r Y_r),\\
\frac{dH_y}{dt} &= \Sigma_r(z_r X_r - x_r Z_r),\\
\frac{dH_z}{dt} &= \Sigma_r(x_r Y_r - y_r X_r),
\end{aligned}
\tag{61}
$$

where H_x, H_y, H_z have the same meaning as the left-hand members of equation 60), but are not now constant. Stating in words: The time derivative of the moment of momentum of any system with respect to any point is equal to the resultant moment of all the forces of the system about the same point.

The equations 46) and 61) furnish us the six equations of motion of a rigid body. Geometrically, we may say that the radius vector of the hodograph (§ 6) of the vector *moment of momentum* of a system is parallel to the resultant moment of the forces acting on the system at each instant of time, this statement being the complement to the statement that the radius vector of the hodograph of the *velocity* of the center of mass is parallel to the resultant of the forces acting on the system.

The three principles which we have now treated, the Principle of Energy, the Principle of Motion of the Center of Mass, and the Principle of Moment of Momentum, in the cases of *conservation*, give us the first integrals of the equations of motion, and suffice for the treatment of all mechanical problems. In the next chapter we shall deal with a principle which is more general than any of these in that it enables us to *deduce* the equations of motion and thus embraces a statement of all the laws of Dynamics.

CHAPTER IV.

PRINCIPLE OF LEAST ACTION. GENERALIZED EQUATIONS OF MOTION.

34. Hamilton's Principle. We shall now consider a principle that differs from those of the last chapter in that it does not immediately furnish us with an integral of the equations of motion. On the other hand, like d'Alembert's principle it enables us to embody the laws of motion in a simple mathematical expression from which we can deduce the equations of motion, not only in the simple form hitherto used employing rectangular coordinates, but also in a form involving any coordinates whatsoever. This statement, employing the language of the calculus of variations, permits us to enunciate the principle in the convenient form that a certain integral is a minimum. The so-called Principle of Least Action was first propounded by Maupertuis[1]) on the basis of certain philosophical or religious arguments, quite other than those upon which it is now based.

We shall first treat it in the form given by Hamilton. If in d'Alembert's equation

$$\sum\left\{\left(m\frac{d^2x}{dt^2}-X\right)\delta x+\left(m\frac{d^2y}{dt^2}-Y\right)\delta y+\left(m\frac{d^2z}{dt^2}+Z\right)\delta z\right\}=0,$$

we consider δx, δy, δz arbitrary variations consistent with the equations of condition, we have

$$\begin{aligned}\frac{d^2x}{dt^2}\delta x &= \frac{d}{dt}\left(\frac{dx}{dt}\delta x\right)-\frac{dx}{dt}\frac{d\delta x}{dt}\\ &= \frac{d}{dt}\left(\frac{dx}{dt}\delta x\right)-\frac{dx}{dt}\frac{\delta dx}{dt}\\ &= \frac{d}{dt}\left(\frac{dx}{dt}\delta x\right)-\delta\frac{1}{2}\left(\frac{dx}{dt}\right)^2.\end{aligned}$$

Treating each term in this manner, taking the sum, and removing the sign of differentiation outside that of summation,

$$1)\quad \frac{d}{dt}\sum\left\{m\left(\frac{dx}{dt}\delta x+\frac{dy}{dt}\delta y+\frac{dz}{dt}\delta z\right)\right\}$$
$$=\delta\frac{1}{2}\sum\left\{m\left[\left(\frac{dx}{dt}\right)^2+\left(\frac{dy}{dt}\right)^2+\left(\frac{dz}{dt}\right)^2\right]\right\}+\Sigma(X\delta x+Y\delta y+Z\delta z).$$

---

1) Mém. de l'Acad. de Paris, 1740. Also: *Des lois de mouvement et de repos déduites d'un principe métaphysique,* Berlin, Mém. de l'Acad. 1745, p. 286.

If there is a force-function U we have

$$\Sigma(X\,\delta x + Y\,\delta y + Z\,\delta z) = \delta U,$$

consequently the right-hand member of 1) is

$$\delta T + \delta U.$$

The left-hand member being an exact derivative we may integrate with respect to t, between any two instants t_0 and t_1,

2) $$\left[\sum\left\{m\left(\frac{dx}{dt}\delta x + \frac{dy}{dt}\delta y + \frac{dz}{dt}\delta z\right)\right\}\right]_{t_0}^{t_1}$$

$$= \int_{t_0}^{t_1}\delta(T + U)\,dt = \delta\int_{t_0}^{t_1}(T + U)\,dt.$$

If the positions are given for t_0 *and* t_1, that is if the variations δx, δy, δz vanish for t_0 and t_1, then the integrated parts vanish, and

$$\delta\int_{t_0}^{t_1}(T + U)\,dt = 0,$$

or

3) $$\delta\int_{t_0}^{t_1}(T - W)\,dt = 0.$$

This is known as *Hamilton's Principle.*[1]) It may be stated by saying that if the configuration of the system is given at two instants t_0 and t_1, then the value of the time-integral of $T + U$ is *stationary* (that is less or greater) for the paths actually described in the natural motion than in any other[2]) infinitely near motion having the same terminal configurations.

Considering the signification of a definite integral as a mean[3]) we may state equation 3) in words as follows: The time mean of

1) Hamilton. *On a General Method in Dynamics. Phil. Trans.* 1834.

2) It is understood that both the natural and the varied paths are *smooth* curves, that is without sharp corners.

3) The arithmetical mean of a number of quantities is defined as their sum divided by their number. A definite integral is defined as the limit of a sum of a number of quantities as their number increases indefinitely. If we divide the interval ab into n parts of length δ_s and if we denote by f_s the value of a function $f(x)$ when x lies at some point within the interval δ_s, we define

$$\int_a^b f(x)\,dx \quad \text{as} \quad \lim_{n=\infty}\sum_1^n \delta_s f_s.$$

the difference of kinetic and potential energies is a minimum for the actual path between given configurations as compared with infinitely near paths which might be described (for instance under constraints) in the same time between the same configurations; or more freely: Nature tends to equalize the mean potential and kinetic energies during a motion.

Hamilton's principle is broader than the principle of energy, inasmuch as U may contain the time as well as the coordinates. It is true even for non-conservative systems (where a force-function U does not exist or where U contains the time), if we write instead of δU,

$$\Sigma(X\,\delta x + Y\,\delta y + Z\,\delta z).$$

We have then

$$4)\qquad \int_{t_0}^{t_1} \{\delta T + \Sigma(X\,\delta x + Y\,\delta y + Z\,\delta z)\}\,dt = 0.$$

35. Principle of Least Action. It is to be noted that in the statement of Hamilton's principle the infinitely near motion with which the actual motion is compared is perfectly arbitrary (except that it satisfies the equations of condition), so that to make the system actually move according to the supposed varied motion might require work to be done upon it by other forces. The paths described by the various particles are not necessarily geometrically different

---

It is proved in the integral calculus that the manner of subdivision into the intervals δ_s is immaterial. We may accordingly put them all equal so that $\delta_s = \frac{b-a}{n}$, then dividing by $(b-a)$ we have

$$\frac{1}{b-a}\int_a^b f(x)\,dx = \lim_{n=\infty}\frac{1}{b-a}\sum_1^n \frac{b-a}{n} f_s = \lim_{n=\infty}\frac{1}{n}\sum_1^n f_s,$$

that is the definite integral of a function in a given interval divided by the magnitude of the interval represents the limit of the arithmetical mean of all the values of the function taken at *equidistant values of that variable* throughout the interval when the number of values taken is increased indefinitely. The specification of the variable with respect to which the values are equally distributed is of the first importance. For instance suppose that we change to a new variable such that $x = \varphi(y)$, $y = \varphi^{-1}(x)$ then

$$\int_a^b f(x)\,dx = \int_{\varphi^{-1}(a)}^{\varphi^{-1}(b)} f(x)\,\varphi'(y)\,dy.$$

The integral may now be interpreted as the mean of the function $f(x)\varphi'(y)$ multiplied by the interval through which y varies, *for equally distributed values of y*. Thus we deal above with *time* means and *space* means.

from those of the actual motion, but if not they are traversed with different velocities, so that at any rate the coordinates of the various points are different functions of the time in the varied and in the actual motion.

In case the system is conservative, Hamilton's principle is equivalent to another, less general, but historically older. In this form of statement of the principle, we compare two infinitely near motions, but the second is not completely arbitrary, for instead of associating together pairs of points $x, y, z, x+\delta x, y+\delta y\ z+\delta z$, reached at the same instant in the two motions, and making $\delta x, \delta y, \delta z$ perfectly arbitrary, we assume that the variation takes place in accordance with the equation of energy,

$$5) \qquad T + W = h,$$

so that we are to put

$$\delta W = -\delta T.$$

But if the equation of energy is to hold on the varied path as well as on the unvaried, the kinetic energy of the system in any configuration is determined, and thus the system may not be in that configuration at any time we please, as that would involve arbitrary velocities, and there is a restriction on the velocities due to the determination of the kinetic energy for every configuration in the motion. We will therefore give up the assumption that pairs of points compared are reached at the same instant of time, in other words we shall no longer assume that $\delta t = 0$. No matter what the independent variable may be, as functions of which we may express t and all the coordinates, so as to compare the motion point by point, we may use the principles explained in parenthesis on p. 80, which will cause a certain modification of our result. If δt is not zero, we can no longer put in the preceding demonstration,

$$\frac{d\delta x}{dt} = \delta\frac{dx}{dt},$$

but must write, as explained on p. 80,

$$\frac{d\delta x}{dt} = \delta\frac{dx}{dt} + \frac{dx}{dt}\frac{d\delta t}{dt},$$

$$\frac{d\delta y}{dt} = \delta\frac{dy}{dt} + \frac{dy}{dt}\frac{d\delta t}{dt},$$

$$\frac{d\delta z}{dt} = \delta\frac{dz}{dt} + \frac{dz}{dt}\frac{d\delta t}{dt}.$$

We have thus to add to the right-hand member of equation 1), the term

$$\sum\left\{m\left(\frac{dx}{dt}\right)^2 + \left(\frac{dy}{dt}\right)^2 + \left(\frac{dz}{dt}\right)^2\right\}\frac{d\delta t}{dt} = 2T\frac{d\delta t}{dt},$$

so that instead of 2) we have

6) $$\left[\sum\left\{m\left(\frac{dx}{dt}\delta x + \frac{dy}{dt}\delta y + \frac{dz}{dt}\delta z\right)\right\}\right]_{t_0}^{t_1} = \int_{t_0}^{t_1}\left(\delta T + \delta U + 2T\frac{d\,\delta t}{dt}\right)dt.$$

Although the times of arriving at corresponding configurations are not the same, so that t_0 and t_1 are not the same as before, the terminal positions are still given, so that the integrated parts still vanish. Now introducing our new assumption, of variation according to the equation of energy, we obtain

$$\delta U = -\delta W = \delta T,$$

7) $$0 = \int_{t_0}^{t_1} 2(\delta T\, dt + T d\,\delta t),$$

that is,

8) $$\delta\int_{t_0}^{t_1} 2T\, dt = 0.$$

The integral $A = \int_{t_0}^{t_1} 2T\, dt$, which is twice the mean kinetic energy for equal intervals of time multiplied by the time occupied in the motion, is called the *Action.*

Accordingly the principle stated in equation 8) is known as the Principle of Least Action.

The definition of action is usually given otherwise, for since

$$T = \frac{1}{2}\sum_r m_r\left(\frac{ds_r}{dt}\right)^2 = \frac{1}{2}\sum_r m_r v_r \frac{ds_r}{dt},$$

9) $$A = \int_{t_0}^{t_1} 2T\, dt = \sum_r \int_{s_r^{(0)}}^{s_r^{(1)}} m_r v_r\, ds_r,$$

which exhibits the action as a sum of line integrals of the momentum of the particles. We may thus define the action as the sum for all the particles of the mean momentum for equal *distances* multiplied by the distance traversed by each particle.

In the enumeration of the conditions there is now a difference — the initial and final configurations of the system (positions of all the points) are given as before, but instead of prescribing the duration of the motion, $t_1 - t_0$, we prescribe the initial energy h. Thus in the variation of the paths the energy is supposed to be unchanged. In forming the integral t is supposed to be *eliminated* and all the velocities to be expressed in terms of the coordinates alone. To effect this we make use of the equation of energy,

$$T = \frac{1}{2} \sum_r m_r \left(\frac{ds_r}{dt}\right)^2 = h - W,$$

from which

10) $$dt = \sqrt{\frac{\Sigma_r m_r ds_r^2}{2(h-W)}} \quad \text{and}$$

11) $$A = \int \sqrt{2(h-W)\Sigma_r m_r ds_r^2}.$$

In order to fix the ideas we may explicitly introduce a new independent variable in the integral, supposing the equations of motion to have been integrated and all the coordinates to be expressed as functions of a single parameter q, which for example may be one of the coordinates. That is for each value that is assigned to the parameter q we suppose the position of every point in the system completely known.

Writing now $ds_r = \frac{ds_r}{dq} dq$ the integral is

12) $$A = \int_{q_0}^{q_1} \sqrt{2(h-W) \sum_r m_r \left(\frac{ds_r}{dq}\right)^2}\, dq.$$

The proper statement of the principle of least action then is that the variation of this integral vanishes, given the initial and final configurations and the total constant energy. We have now completely get rid of the variable t, and are not embarassed by the question whether its variation is zero or not.

As the simplest possible example consider the case of a single free particle acted on by no forces, then $W = 0$ and the action is

$$A = \sqrt{2hm} \int_{s_0}^{s_1} ds = mv \int_{s_0}^{s_1} ds,$$

and the action is proportional to the distance traversed.

If this is a minimum the path will be a straight line, the principle of least action accords with Newton's first law.

Suppose that the particle instead of being free is constrained to lie on a given surface. The path described must then be an arc of a shortest or *geodesic* line of the surface. The calculus of variations enables us to find the differential equations of such a line. Suppose the surface is a sphere, then if the particle is started from a point P (Fig. 24) with a given velocity v in any direction, it may be made to arrive at Q by the introduction of certain constraints, for instance, suppose it obliged to move on a plane passing through P and Q. The principle of least action says that in the natural or unconstrained motion it will go from P to Q along the shortest path, that is, an arc of a great circle. Of all possible paths there are two natural ones by which the particle travels from P to Q along a great circle, but leaving P in opposite directions. It is only for the shorter of the two paths that the action is a minimum. This is an example of a frequent occurrence in the calculus of variations, namely, that an integral possesses the minimum or maximum property only when its limits are sufficiently close together.

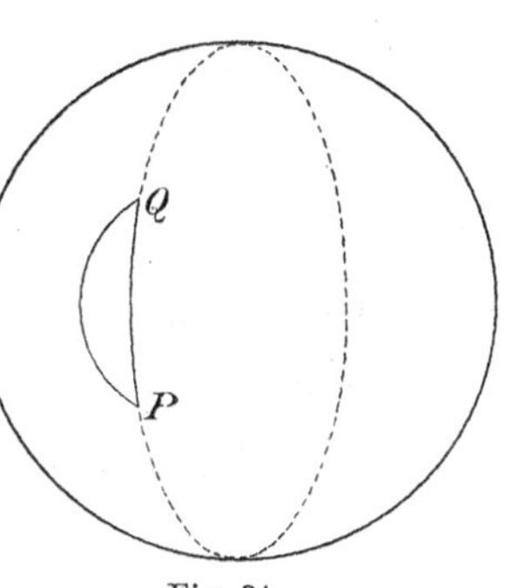

Fig. 24.

We will illustrate this by a less simple example. Consider the problem of shooting at a target, or the ideal case of a single particle acted on only by gravity, which has been treated in § 18.

Suppose the particle projected from the point x_0, z_0 with the velocity v_0, so as to reach the point x_1, z_1. If t be the time of flight, we have by § 18, 3)

$$V_x = \frac{x_1 - x_0}{t}, \quad V_z = \frac{z_1 - z_0 + gt^2/2}{t},$$

from which

$$v_0^2 = V_x^2 + V_z^2 = \frac{(x_1 - x_0)^2 + (z_1 - z_0 + gt^2/2)^2}{t^2},$$

or otherwise

13) $$\frac{1}{4} g^2 t^4 + \{g(z_1 - z_0) - v_0^2\} t^2 + (x_1 - x_0)^2 + z_1 - z_0)^2 = 0,$$

a quadratic in t^2 to determine the time of transit in terms of the given constants x_0, x_1, z_0, z_1, v_0. Introducing the following letters for the range, its horizontal and vertical projections,

$$r = \sqrt{(x_1 - x_0)^2 + (z_1 - z_0)^2}, \quad l = x_1 - x_0, \quad h = z_1 - z_0,$$

and solving the quadratic,

14) $$t^2 = \frac{2}{g^2}\left\{v_0^2 - gh \pm \sqrt{(v_0^2 - gh)^2 - g^2 r^2}\right\}.$$

If the radical is real — which will be the case if the initial velocity is great enough — since the absolute value of the term outside the radical is greater than that under it, both values of t^2 will be positive, even if the lower sign is used, therefore there will be two real possible positive values of t.

To determine α, the angle of elevation, we have

$$V_x = v_0 \cos\alpha, \quad V_z = v_0 \sin\alpha, \quad \tan\alpha = \frac{V_z}{V_x} = \frac{h + gt^2/2}{l},$$

and inserting the two values of t we get two possible elevations. Thus we find that the aim is completely determined (though not uniquely in this case) by the terminal positions and the velocity of projection.

For the action we obtain

$$\text{15)} \quad A = \int_0^t 2T\,dt = m\int_0^t \{v_0^2 + 2g(z_0 - z)\}\,dt$$

$$= m\int_0^t \{v_0^2 + g^2t^2 - 2gV_z t\}\,dt$$

$$= m\left\{v_0^2 t + \frac{1}{3}g^2t^3 - gV_z t^2\right\}.$$

Using the values of V_z and t found above we obtain two values of the action different for the two paths Thus there are two possible natural paths, differing from each other by finite distances, for only one of which is the action least. Both however have the property that between two points sufficiently near together the action is less than for any infinitely near path.

In case the radical in 14) vanishes, that is

$$\text{16)} \quad \{v_0^2 - g(z_1 - z_0)\}^2 - g^2\{(x_1 - x_0)^2 + (z_1 - z_0)^2\} = 0,$$

or

$$g^2(x_1 - x_0)^2 + 2gv_0^2(z_1 - z_0) - v_0^4 = 0,$$

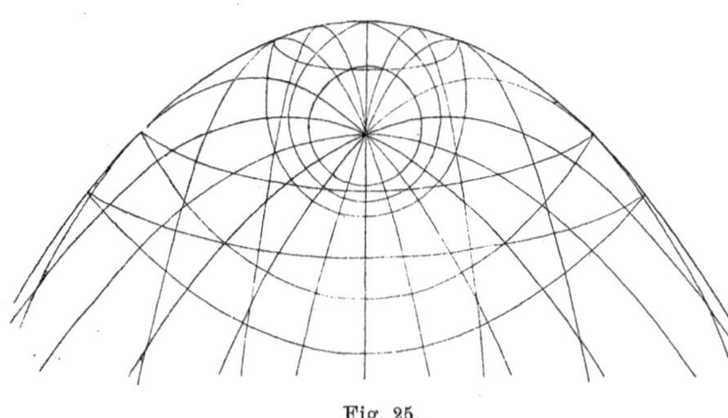

Fig. 25.

the two roots t^2 are equal and there is only one course. The terminal point x_1, z_1 then lies on a parabola whose vertex is vertically above the point of projection (Fig. 25). It is easy to see that this parabola is the envelope of all possible paths in this vertical plane starting from the same initial point x_0, z_0 with the same velocity v_0. For it is the locus of the intersection of courses whose angles of elevation α differ infinitely little. If the second point x_1, z_1 lie without this enevelope it cannot be reached under the given conditions. If upon it it can be reached by one path, and if within it by two paths. In that case the course that reaches x_1, z_1 *before* touching the envelope has the less action. A point at which two infinitely near courses from a given point with equal energy intersect is called a *kinetic focus* of the starting point, and if on any course the terminal configuration is reached before the kinetic focus on that course, the action will be a minimum. If the kinetic focus is first reached it will not.

Thus in the problem of motion on a sphere under no forces, the point diametrically opposite the initial point is a kinetic focus. Evidently a particle may reach the kinetic focus starting in any direction from the orginal point, for all great circles through a point intersect in its opposite point. The envelope of all the great circles or courses from a point in this case reduces to a point, which is the kinetic focus.

For the treatment of the difficult subject of kinetic foci, which belongs to the calculus of variations, the reader is referred to Thomson and Tait, *Principles of Natural Philosophy*, § 358, and Poincaré, *Les Méthodes Nouvelles de la Mécanique Céleste*, Tome III, p. 261, also to Kneser, *Lehrbuch der Variationsrechnung*, Bolza, *The Calculus of Variations*, and Hadamard, *Leçons sur le Calcul des Variations.*

From the principle of least action we may deduce the equations of motion. Of course the principle was itself derived from these equations, therefore, as is always the case, we obtain by mathematical transformations no new facts. It is however instructive to see how by assuming the principle of least action as a general principle we may obtain the equations from it.

Let us put in equation 12)

$$ds_r^2 = dx_r^2 + dy_r^2 + dz_r^2,$$

$$\frac{dx_r}{dq} = x'_r, \quad \frac{dy_r}{dq} = y'_r, \quad \frac{dz_r}{dq} = z'_r,$$

giving

$$17) \qquad \delta \int_{q_0}^{q_1} \sqrt{2(h-W)\{\Sigma_r m_r (x_r'^2 + y_r'^2 + z_r'^2)\}} \cdot dq = 0.$$

If we put

$$2(h - W) = M,$$

$$18) \qquad \Sigma_r m_r (x_r'^2 + y_r'^2 + z_r'^2) = N, \quad \sqrt{MN} = P,$$

since P involves all the coordinates and velocities $x_r, y_r, z_r, x'_r, y'_r, z'_r,$

$$19)\quad \delta A = \int_{q_0}^{q_1} \delta P\,dq = \int_{q_0}^{q_1} \sum_r \left\{ \frac{\partial P}{\partial x_r}\delta x_r + \frac{\partial P}{\partial y_r}\delta y_r + \frac{\partial P}{\partial z_r}\delta z_r + \frac{\partial P}{\partial x'_r}\delta x'_r + \frac{\partial P}{\partial y'_r}\delta y'_r + \frac{\partial P}{\partial z'_r}\delta z'_r \right\} dq = 0.$$

Now the term

$$\int_{q_0}^{q_1} \frac{\partial P}{\partial x'_r}\delta x'_r\,dq = \int \frac{\partial P}{\partial x'_r}\frac{d\,\delta x_r}{dq}\,dq$$

may be integrated by parts, giving

$$\frac{\partial P}{\partial x'_r}\delta x_r \Bigg|_{q_0}^{q_1} - \int_{q_0}^{q_1} \delta x_r \frac{d}{dq}\left(\frac{\partial P}{\partial x'_r}\right) dq.$$

The terms in $\delta y'_r$ and $\delta z'_r$ are to be treated in like manner. Since the variations of the coordinates vanish at the limits the integrated terms disappear, leaving

$$\int_{q_0}^{q_1} \sum_r \left[\left\{ \frac{\partial P}{\partial x_r} - \frac{d}{dq}\left(\frac{\partial P}{\partial x'_r}\right) \right\} \delta x_r + \cdots \right] dq = 0.$$

Now in virtue of 18) since N does not contain the coordinates

$$\frac{\partial P}{\partial x_r} = \frac{1}{2}\sqrt{\frac{N}{M}}\frac{\partial M}{\partial x_r} = -\sqrt{\frac{N}{M}}\frac{\partial W}{\partial x_r}.$$

Also since M does not contain x'_r,

$$\frac{\partial P}{\partial x'} = \frac{1}{2}\sqrt{\frac{M}{N}}\frac{\partial N}{\partial x'_r} = \sqrt{\frac{M}{N}}\, m_r x'_r,$$

and consequently

$$\frac{\partial P}{\partial x_r} - \frac{d}{dq}\left(\frac{\partial P}{\partial x'_r}\right) = -\sqrt{\frac{N}{M}}\frac{\partial W}{\partial x_r} - \frac{d}{dq}\left(m_r \sqrt{\frac{M}{N}}\frac{dx_r}{dq}\right).$$

The equation of energy,

$$\sum_r m_r \left\{ \left(\frac{dx_r}{dt}\right)^2 + \left(\frac{dy_r}{dt}\right)^2 + \left(\frac{dz_r}{dt}\right)^2 \right\} = 2(h - W),$$

gives

$$\sum_r m_r (x_r'^2 + y_r'^2 + z_r'^2)\left(\frac{dq}{dt}\right)^2 = 2(h - W),$$

or according to 18),

$$N\left(\frac{dq}{dt}\right)^2 = M,$$

from which we get

$$dq = \sqrt{\frac{M}{N}}\,dt.$$

Inserting this value of dq gives

$$\frac{\partial P}{\partial x_r} - \frac{d}{dq}\left(\frac{\partial P}{\partial x_r'}\right) = -\sqrt{\frac{N}{M}}\frac{\partial W}{\partial x_r} - \frac{d}{dq}\left(m_r \frac{d x_r}{dt}\right)$$
$$= -\sqrt{\frac{N}{M}}\left(\frac{\partial W}{\partial x_r} + m_r \frac{d^2 x_r}{dt^2}\right).$$

Accordingly we have

$$20) \quad \delta A = \int_{q_0}^{q_1} \delta P\, dq = -\int_{q_0}^{q_1} \sqrt{\frac{N}{M}} \sum_r \left\{ \left(\frac{\partial W}{\partial x_r} + m_r \frac{d^2 x_r}{dt^2}\right)\delta x_r \right.$$
$$\left. + \left(\frac{\partial W}{\partial y_r} + m_r \frac{d^2 y_r}{dt^2}\right)\delta y_r + \left(\frac{\partial W}{\partial z_r} + m_r \frac{d^2 z_r}{dt^2}\right)\delta z_r \right\} dq = 0.$$

In order that this may vanish for arbitrary variations, $\delta x_r, \delta y_r, \delta z_r$, the coefficient of each variation must vanish, so that we must have

$$m_r \frac{d^2 x_r}{dt^2} + \frac{\partial W}{\partial x_r} = 0, \quad \text{or} \quad m_r \frac{d^2 x_r}{dt^2} = -\frac{\partial W}{\partial x_r} = X_r,$$
$$21) \quad m_r \frac{d^2 y_r}{dt^2} + \frac{\partial W}{\partial y_r} = 0, \qquad m_r \frac{d^2 y_r}{dt^2} = -\frac{\partial W}{\partial y_r} = Y_r,$$
$$m_r \frac{d^2 z_r}{dt^2} + \frac{\partial W}{\partial z_r} = 0, \qquad m_r \frac{d^2 z_r}{dt^2} = -\frac{\partial W}{\partial z_r} = Z_r,$$

which are the ordinary equations of motion for a free system.

The variations $\delta x_r, \delta y_r, \delta z_r$ are arbitrary only if all the particles are free. If there are constraints the variations must be compatible with the equations of condition,

$$\varphi_1 = 0,$$
$$\varphi_2 = 0,$$
$$\cdots\cdots$$
$$\cdots\cdots$$
$$\varphi_k = 0,$$

that is we must have the k linear relations between the δ's, Chapter III equations 14). We may then as in § 25 multiply the equations between the δ's by undetermined factors $\lambda_1, \lambda_2, \ldots \lambda_k$ and add them to the integrand. We shall then have

$$22) \quad \int_{q_0}^{q_1} \sqrt{\frac{N}{M}} \sum_r \left\{ \left(- m_r \frac{d^2 x_r}{dt^2} - \frac{\partial W}{\partial x_r} + \lambda_1 \frac{\partial \varphi_1}{\partial x_r} + \lambda_2 \frac{\partial \varphi_2}{\partial x_r} + \cdots \right.\right.$$
$$\left.\left. + \lambda_k \frac{\partial \varphi_k}{\partial x_r}\right)\delta x_r + \cdots \right\} dq = 0.$$

We may now determine the k factors $\lambda_1, \lambda_2, \ldots \lambda_k$, so that k of the factors multiplying the variations vanish identically. Then the coefficients of the remaining $3n - k$ arbitrary variations being put equal to zero with these k give the differential equations

$$
23) \qquad \begin{aligned}
m_r \frac{d^2 x_r}{dt^2} &= X_r + \lambda_1 \frac{\partial \varphi_1}{\partial x_r} + \lambda_2 \frac{\partial \varphi_2}{\partial x_r} + \cdots + \lambda_k \frac{\partial \varphi_k}{\partial x_r}, \\
m_r \frac{d^2 y_r}{dt^2} &= Y_r + \lambda_1 \frac{\partial \varphi_1}{\partial y_r} + \lambda_2 \frac{\partial \varphi_2}{\partial y_r} + \cdots + \lambda_k \frac{\partial \varphi_k}{\partial y_r}, \\
m_r \frac{d^2 z_r}{dt^2} &= Z_r + \lambda_1 \frac{\partial \varphi_1}{\partial z_r} + \lambda_2 \frac{\partial \varphi_2}{\partial z_r} + \cdots + \lambda_k \frac{\partial \varphi_k}{\partial z_r}.
\end{aligned}
$$

From the $3n$ equations 23) we may eliminate the k multipliers $\lambda_1, \lambda_2, \ldots \lambda_k$ and obtain $3n - k$ equations of motion, which is the number of degrees of freedom of the system.

The equations 23) are known as Lagrange's differential equations in the first form. They can evidently be deduced from equations 16) of Chapter III by d'Alembert's principle, replacing X_r by

$$X_r - m_r \frac{d^2 x_r}{dt^2}, \quad \text{etc.}$$

36. Generalized Coordinates. Lagrange's Equations. In many investigations in dynamics where constraints are introduced, instead of denoting the positions of particles by rectangular coordinates (not all of which are independent) it is advantageous to specify the positions by means of certain parameters whose number is just equal to the number of degrees of freedom of the system, so that they are all independent variables. For instance if a particle is constraind to move on the surface of a sphere of radius l, we may specify its position by giving its longitude φ and colatitude ϑ, as in § 23 These are two independent variables.

The potential energy depending only on position will be expressed in terms of φ and ϑ. The kinetic energy will depend upon the expression for the length of the arc of the path in therms of φ and ϑ. Now we have, if l be the radius of the sphere,

$$ds^2 = l^2 (d\vartheta^2 + \sin^2 \vartheta \, d\varphi^2).$$

Dividing by dt^2 and writing $\vartheta' = \frac{d\vartheta}{dt}$, $\varphi' = \frac{d\varphi}{dt}$, we have

$$24) \qquad T = \frac{1}{2} m l^2 (\vartheta'^2 + \sin^2 \vartheta \, \varphi'^2).$$

The parameters ϑ and φ are coordinates of the point, since when they are known the position of the point is fully specified. Their time-derivatives ϑ', φ' being time-rates of change of coordinates may be termed velocities, and when they together with ϑ and φ are know, the velocity of the particle may be calculated. The kinetic energy in this case involves *both* the coordinates ϑ and φ *and* the velocities ϑ' and φ'. Inasmuch as the particle in any given position may have any given velocity, the variables ϑ, φ, ϑ', φ' are to be considered in this sense as independent, although *in any given actual motion* they will all be functions of a single variable t.

The form of the function T is worthy of attention. It is a homogeneous quadratic function of the velocities ϑ' and φ', the coefficients of their squares being functions of the coordinates ϑ, φ, the product term in $\vartheta'\varphi'$ being absent in this case. We may prove that if a point moves on any surface the kinetic energy is always of this form.

In the geometry of surfaces it is convenient to express the coordinates of a point in terms of two parameters q_1 and q_2. Suppose

$$x = f_1(q_1 q_2), \quad y = f_2(q_1 q_2), \quad z = f_3(q_1 q_2),$$

from these three equations we can eliminate the two parameters q_1, q_2, obtaining a single equation between x, y, z, the equation of the surface. The parameters q_1 and q_2 may be called the coordinates of a point on the surface, for when they are given its position is known. If q_1 is constant and q_2 is allowed to vary, the point x, y, z describes a certain curve on the surface. This curve changes as we change the constant value q_1. In like manner putting q_2 constant we obtain a family of curves. The two families of curves,

$$q_1 = \text{const}, \quad q_2 = \text{const},$$

may be called parametric or coordinate lines on the surface, any point being determined by the intersection of two lines, for one of which q_1 has a given value, for the other, q_2.

We may obtain the length of the infinitesimal arc of any curve in terms of q_1 and q_2. We have

$$\begin{aligned} dx &= \frac{\partial x}{\partial q_1} dq_1 + \frac{\partial x}{\partial q_2} dq_2, \\ dy &= \frac{\partial y}{\partial q_1} dq_1 + \frac{\partial y}{\partial q_2} dq_2, \\ dz &= \frac{\partial z}{\partial q_1} dq_1 + \frac{\partial z}{\partial q_2} dq_2. \end{aligned} \tag{25}$$

Squaring and adding,

$$ds^2 = dx^2 + dy^2 + dz^2 = E dq_1^2 + 2F dq_1 dq_2 + G dq_2^2, \tag{26}$$

where

$$\begin{aligned} E &= \left(\frac{\partial x}{\partial q_1}\right)^2 + \left(\frac{\partial x}{\partial q_1}\right)^2 + \left(\frac{\partial z}{\partial q_1}\right)^2, \\ F &= \frac{\partial x}{\partial q_1}\frac{\partial x}{\partial q_2} + \frac{\partial y}{\partial q_1}\frac{\partial y}{\partial q_2} + \frac{\partial z}{\partial q_1}\frac{\partial z}{\partial q_2}. \\ G &= \left(\frac{\partial x}{\partial q_2}\right)^2 + \left(\frac{\partial y}{\partial q_2}\right)^2 + \left(\frac{\partial z}{\partial q_2}\right)^2. \end{aligned} \tag{27}$$

Thus the square of the length of any infinitesimal arc is a homogeneous quadratic function of the differentials of the coordinates q_1 and q_2, the coefficients E, F, G being functions of the coordinates q_1, q_2 themselves. If the curve is one of the lines $q_1 = \text{const}$, we have, since $dq_1 = 0$,

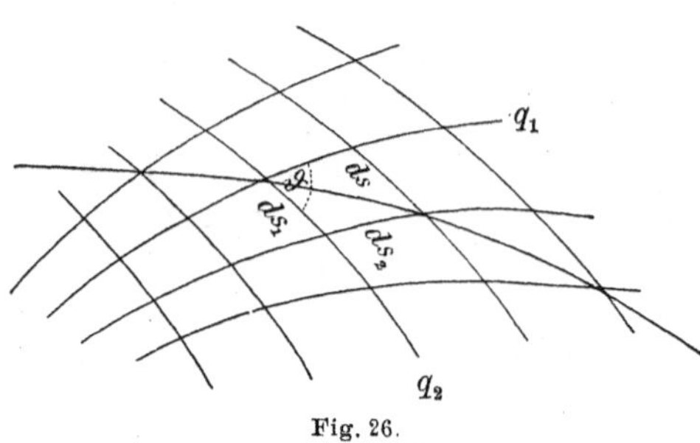

Fig. 26.

$$ds_2^2 = G dq_2^2,$$

if it is one of the curves $q_2 = \text{const}$, we have

$$ds_1^2 = E dq_1^2.$$

Considering any arc ds as the diagonal of an infinitesimal parallelogram with sides ds_1 and ds_2 including an angle ϑ (Fig. 26), we have by trigonometry,

$$ds^2 = ds_1^2 + 2 ds_1 ds_2 \cos\vartheta + ds_2^2.$$

Making use of the above values of ds_1 and ds_2 and comparing with the expression 26), we find

$$\cos\vartheta = \frac{F}{\sqrt{EG}}.$$

If the coordinate lines cut each other everywhere at right angles we shall have $\cos\vartheta = 0$, so that

$$ds^2 = E dq_1^2 + G dq_2^2. \tag{28}$$

The coordinates q_1, q_2 are then said to be *orthogonal* curvilinear coordinates. In the example above[1]) ϑ and φ are orthogonal, the lines of constant ϑ and φ being parallels and meridians intersecting at right angles and the product term in $d\vartheta\, d\varphi$ therefore disappearing.

Employing the expression 26) for the length of the arc, dividing by dt^2 and writing

---

1) We have the equations of change of coordinates,

$$x = l \sin\vartheta \cos\varphi,$$
$$y = l \sin\vartheta \sin\varphi,$$
$$z = l \cos\vartheta,$$

from which

$$\frac{\partial x}{\partial\vartheta} = l\cos\vartheta\cos\varphi, \quad \frac{\partial y}{\partial\vartheta} = l\cos\vartheta\sin\varphi, \quad \frac{\partial z}{\partial\vartheta} = -l\sin\vartheta,$$

$$\frac{\partial x}{\partial\varphi} = -l\sin\vartheta\sin\varphi, \quad \frac{\partial y}{\partial\varphi} = l\sin\vartheta\cos\varphi, \quad \frac{\partial z}{\partial\varphi} = 0,$$

$$E = l^2, \qquad F = 0, \qquad G = l^2\sin^2\vartheta.$$

$$\frac{dq_1}{dt} = q_1', \quad \frac{dq_2}{dt} = q_2',$$

we find for the kinetic energy,

$$T = \tfrac{1}{2} m (E q_1'^2 + 2 F q_1' q_2' + G q_2'^2). \tag{29}$$

This is a typical example of the employment of the *generalized coordinates* introduced by Lagrange, q_1 and q_2 being the coordinates, q_1', q_2' the velocities corresponding, and T being a homogeneous quadratic function or quadratic *form* in the velocities q_1', q_2', the coefficients of the squares and products of the velocities being functions of the *coordinates* alone. We shall show that this is a characteristic property of the kinetic energy for any system depending upon any number of variables.

In the case of a single free particle we may express the coordinates x, y, z, in terms of three parameters q_1, q_2, q_3, and we shall then have as in 25) and 26)

$$\begin{aligned} ds^2 = E_{11} dq_1^2 + E_{22} dq_2^2 + E_{33} dq_3^2 + 2 E_{12} dq_1 dq_2 \\ + 2 E_{23} dq_2 dq_3 + 2 E_{31} dq_3 dq_1, \end{aligned} \tag{30}$$

where

$$E_{rs} = \frac{\partial x}{\partial q_r}\frac{\partial x}{\partial q_s} + \frac{\partial y}{\partial q_r}\frac{\partial y}{\partial q_s} + \frac{\partial z}{\partial q_r}\frac{\partial z}{\partial q_s}. \tag{31}$$

Thus the kinetic energy has the same property as before.

Proceeding now to the general case of any number of particles, whether constrained or not, let us express all the coordinates as functions of m independent parameters, $q_1, q_2, \ldots q_m$, the generalized coordinates of the system,

$$\begin{aligned} x_r &= x_r(q_1, q_2, \ldots q_m), \\ y_r &= y_r(q_1, q_2, \ldots q_m), \\ z_r &= z_r(q_1, q_2, \ldots q_m). \end{aligned}$$

Differentiating we have

$$\begin{aligned} dx_r &= \frac{\partial x_r}{\partial q_1} dq_1 + \frac{\partial x_r}{\partial q_2} dq_2 + \cdots + \frac{\partial x_r}{\partial q_m} dq_m, \\ dy_r &= \frac{\partial y_r}{\partial q_1} dq_1 + \frac{\partial y_r}{\partial q_2} dq_2 + \cdots + \frac{\partial y_r}{\partial q_m} dq_m, \\ dz_r &= \frac{\partial z_r}{\partial q_2} dq_1 + \frac{\partial z_r}{\partial q_2} dq_2 + \cdots + \frac{\partial z_r}{\partial q_m} dq_m. \end{aligned} \tag{32}$$

The derivatives $\frac{\partial x_r}{\partial q_s}, \cdots$ are all functions of all the q's. Squaring and adding we obtain

33) $$ds_r^2 = E_{11}^{(r)} dq_1^2 + E_{21}^{(r)} dq_2^2 + \cdots + E_{mm}^{(r)} dq_m^2 + 2 E_{12}^{(r)} dq_1\, dq_2 + 2 E_{13}^{(r)} dq_1\, dq_3 + \cdots,$$

where

34) $$E_{st}^{(r)} = \frac{\partial x_r}{\partial q_s}\frac{\partial x_r}{\partial q_t} + \frac{\partial y_r}{\partial q_s}\frac{\partial y_r}{\partial q_t} + \frac{\partial z_r}{\partial q_s}\frac{\partial z_r}{\partial q_t}.$$

Thus the square of each infinitesimal arc is a quadratic form in the differentials of all the coordinates q Dividing by dt^2, denoting the time derivatives by accents as before, multiplying by $\frac{1}{2} m_r$, and taking the sum for all the particles, we obtain

35) $$T = \frac{1}{2} Q_{11} q_1'^2 + \frac{1}{2} Q_{22} q_2'^2 + \cdots + \frac{1}{2} Q_{mm} q_m^2 + Q_{12} q_1' q_2' + Q_{13} q_1' q_3' + \cdots = \frac{1}{2} \sum_{r=1}^{r=m} \sum_{s=1}^{s=m} Q_{sr}\, q_r'\, q_s',$$

where

36) $$Q_{rs} = \sum_{p=1}^{p=n} m_p E_{rs}^{(p)}.$$

(In the double sum the factor $\frac{1}{2}$ is introduced because there occur both a term in Q_{rs} and one in Q_{sr}, both being equal.) Thus the kinetic energy possesses the characteristic property mentioned above of being a quadratic form in the generalized *velocities* q', the coefficients Q_{rs} being functions of only the generalized *coordinates* q. They must satisfy the necessary conditions that for all assignable values of the q''s T shall be *positive.* Of the form of these functions no general statement can be made. They are linear functions of the masses of the particles of the system and depend upon the choice of the parameters q used to denote the configuration. We may cull them *coefficients of inertia.* It is evident from 36) that every Q_{ss} is positive, for $E_{ss}^{(p)}$ is a sum of squares. If no product terms occur we may by analogy with 28) call the coordinates orthogonal.

It is sometimes convenient to employ the language of multi-dimensional geometry. This signifies nothing more than that when we speak of a point as being in m dimensional space we mean that it requires m parameters to determine its position. Inasmuch as in motion along a curve, that is in a space of one dimension we have for the length of arc

$$ds^2 = \left(\frac{ds}{dq}\right)^2 dq^2,$$

on a surface, that is in a space of two dimensions,

$$ds^2 = Edq_1^2 + 2\,Fdq_1dq_2 + Gdq_2^2,$$

in space of three dimensions,

$$ds^2 = \sum_{r=1}^{s=3}\sum_{s=1}^{s=3} E_{rs}dq_r dq_s,$$

so by analogy, in space of m dimensions,

37)
$$ds^2 = \sum_{r=1}^{r=m}\sum_{s=1}^{s=m} E_{rs}dq_r dq_s.$$

That is to say a quadratic form in m differentials may be interpreted as the square of an arc in m dimensional space. Thus we may assimilate our system depending upon m coordinates to a single point moving in space of m dimensions, characterized by the expression for the element of arc,

$$ds^2 = \frac{1}{M}\sum_{r=1}^{r=m}\sum_{s=1}^{s=m} Q_{rs}dq_r dq_s.$$

To each possible position of this point corresponds a possible configuration of our system. No matter what be taken as the mass of the point, M, its kinetic energy, $\frac{1}{2}M\left(\frac{ds}{dt}\right)^2$, is equal to the kinetic energy of our system, the coefficients in the quadratic form for ds^2 and T being proportional.[1]) The advantage of this mode of speaking (for it is no more) may easily be seen from the many analogies that arise, connecting the dynamical theory of least action with the purely geometrical theory of geodesic lines. This method is adopted by Hertz in his *Prinzipien der Mechanik* and is worked out in a most interesting manner by Darboux in his *Théorie des Surfaces*, Tom. II. The ideas involved were first set forth by Beltrami.[2])

---

1) Since by the nature of above transformation, we have

$$ds^2 = \frac{1}{M}\sum_{r=1}^{r=n} m_r ds_r^2,$$

if as in § 32 we consider each mass m_r to be the sum of m_r unit mass-points, and

$$M = \sum_{r=1}^{r=n} m_r,$$

then ds is the *quadratic mean*, or square root of the mean square of the displacements of all the particles.

2) Beltrami, *Sulla teorica generale dei parametri differenziali* (Memorie dell' Accademia delle Scienze dell' Istituto di Bologna, Serie 2, t. VIII, p. 549; 1869.

In virtue of the homogeneity of T as a function of the q''s, we have by Euler's theorem for homogeneous functions,

$$38)\qquad \sum_{r=1}^{r=m} q'_r \frac{\partial T}{\partial q'_r} = 2T,$$

a property of which frequent use will be made.

The potential energy, if the system is conservative, on the other hand, depends only on the configuration of the system, that is on the coordinates q, the q''s not appearing. For instance in the problem of § 23,

$$W = -mgz = -mgr\cos\vartheta.$$

Whether the system is conservative or not the element of work

$$39)\qquad dA = \sum_{r=1}^{r=n}(X_r dx_r + Y_r dy_r + Z_r dz_r)$$

$$= \sum_{r=1}^{r=n}\sum_{s=1}^{s=m}\left(X_r \frac{\partial x_r}{\partial q_s} dq_s + Y_r \frac{\partial y_r}{\partial q_s} dq_s + Z_r \frac{\partial z_r}{\partial q_s} dq_s\right),$$

is a homogeneous *linear* function in the dq's which we will write

$$40)\qquad dA = P_1 dq_1 + P_2 dq_2 + \cdots + P_m dq_m.$$

By analogy with rectangular coordinates we shall call P_r the generalized force-component corresponding to the coordinate q_r and velocity q'_r.

If the system is conservative, since

$$41)\qquad dW = -dA, \quad P_r = -\frac{\partial W}{\partial q_r},$$

and in any case

$$42)\qquad P_s = \sum_{r=1}^{r=n}\left(X_r \frac{\partial x_r}{\partial q_s} + Y_r \frac{\partial y_r}{\partial q_s} + Z_r \frac{\partial z_r}{\partial q_s}\right).$$

We may now make use of Hamilton's Principle to deduce the equations of motion in terms of the generalized coordinates q.

Performing the operation of variation upon the integral occurring in Hamilton's Principle, both the q's and q''s being varied, we obtain

$$43)\qquad \int_{t_0}^{t_1}\left[\sum_{s=1}^{s=m}\left\{\frac{\partial (T-W)}{\partial q_s}\delta q_s + \frac{\partial (T-W)}{\partial q'_s}\delta q'_s\right\}\right]dt = 0,$$

and since

$$\delta q'_s = \delta \frac{dq_s}{dt} = \frac{d}{dt}\delta q_s,$$

we may integrate the second term by parts. Since the initial and final configuration of the system is supposed given, the δq's vanish at $t = t_0$ and $t = t_1$, so that the integrated part vanishes, and

$$44)\qquad \int_{t_0}^{t_1}\left[\sum_{s=1}^{s=m}\left\{\frac{\partial (T-W)}{\partial q_s} - \frac{d}{dt}\left(\frac{\partial (T-W)}{\partial q_s'}\right)\right\}\delta q_s\right] dt = 0.$$

Now if all the δq's are arbitrary, the integral vanishes only if the coefficient of every δq_s is equal to zero. Therefore we must have

$$45)\qquad \frac{\partial (T-W)}{\partial q_s} - \frac{d}{dt}\left(\frac{\partial (T-W)}{\partial q_s'}\right) = 0,$$

or if we write L for the Lagrangian function $T - W$,

$$46)\qquad \frac{d}{dt}\left(\frac{\partial L}{\partial q_s'}\right) = \frac{\partial L}{\partial q_s}.$$

Since the potential energy depends only on the coordinates, $\frac{\partial W}{\partial q_s'} = 0$, and we may write the equation 45)

$$47)\qquad \frac{d}{dt}\left(\frac{\partial T}{\partial q_s'}\right) - \frac{\partial T}{\partial q_s} = -\frac{\partial W}{\partial q_s} = P_s.$$

There are m of these equations, on for each q. These are *Lagrange's equations of motion* in generalized coordinates, generally referred to by German writers as Lagrange's equations in the second form. Their discovery constitutes one of the principal improvements in dynamical methods and we shall refer to them simply as Lagrange's equations.[1])

If the system is not conservative, by § 34, 4) we must write

$$48)\qquad \int_{t_0}^{t_1}(\delta T + \delta A)\, dt = \int_{t_0}^{t_1}(\delta T + \Sigma_s P_s \delta q_s)\, dt = 0$$

from which we easily obtain 47), except that P_s is not now derived from an energy function.

37. Lagrange's Equations by direct Transformation. Various Reactions. On account of the very great importance of Lagrange's equations, it is advantageous to consider them carefully, from as many points of view as possible. The deduction from Hamilton's principle is one of the simplest, but does not perhaps appeal as strongly to our physical sense as is desirable. Of course as Hamilton's principle is completely equivalent to d'Alembert's, and that to the equations of motion of Newton, we might have derived

1) Lagrange, *Mécanique Analytique,* Tom. I, p. 334.

the equations from either. This we will now do. It is important every time that a new quantity appears in dynamics, to have a clear conception of its physical nature. We should make free use of all analogies that our science may offer us, and here geometry aids us readily. The notion of the geometric product and the terminology of multidimensional geometry here furnish us valuable aid. The geometric product of two vectors in three dimensional space, defined by their components X, Y, Z, X', Y', Z',

$$XX' + YY' + ZZ',$$

is a scalar quantity, symmetrical with respect to both vectors, such that the geometric product of the resultants of two sets of components is the arithmetical sum of the products of all the pairs of corresponding components. If one of the vectors is an infinitesimal displacement dx, dy, dz, the geometric product is

$$Xdx + Ydy + Zdz,$$

and the multiplier of the change dx is called the *component* of the vector in the direction of the coordinate x. In like manner let us speak of a quantity defined by components $P_1, P_2, \dots P_m$ as a vector in m dimensional space. The geometric product of two such, of which the second is an infinitesimal displacement compatible with the constraints, and defined by the quantities $dq_1, dq_2, \dots dq_m$, may be, by analogy, defined as

$$P_1 dq_1 + P_2 dq_2 + \cdots + P_m dq_m.$$

If now the vector $P_1, \dots P_m$ is equivalent to the *system* of vectors X_r, Y_r, Z_r, we have equations 39), 40), 42), and the latter,

$$42)\qquad P_s = \sum_{r=1}^{r=m}\left(X_r \frac{\partial x_r}{\partial q_s} + Y_r \frac{\partial y_r}{\partial q_s} + Z_r \frac{\partial z_r}{\partial q_s}\right)$$

serves to define the component of the vector-system with reference to the coordinate q_s. Thus we have spoken of P_s as the force-component of the system for the coordinate q_s. It is to be observed that we do not insist here on the idea of direction, and that our terminology is merely a convenient mode of speaking. Nevertheless, the notion of work gives a means of realizing by the senses the meaning of our term *component*, for, if we move the system in such a way that all the q's except one q_s are unchanged the work done in a change of the coordinate dq_s will be $P_s dq_s$.[1])

Let us now find the component of our velocity-system according to our generalized coordinates. We have, according to our equation

1) For a further elucidation of the nature of the geometric product, in connection with multidimensional geometry, see Note III.

of definition 42), for the component of the velocity of the rth particle according to q_s,

$$x'_r \frac{\partial x_r}{\partial q_s} + y'_r \frac{\partial y_r}{\partial q_s} + z'_r \frac{\partial z_r}{\partial q_s}. \tag{49}$$

Now we have by 32), dividing by dt,

$$\begin{aligned} x'_r &= \frac{\partial x_r}{\partial q_1} q'_1 + \frac{\partial x_r}{\partial q_2} q'_2 + \cdots + \frac{\partial x_r}{\partial q_m} q'_m, \\ y'_r &= \frac{\partial y_r}{\partial q_1} q'_1 + \frac{\partial y_r}{\partial q_2} q'_2 + \cdots + \frac{\partial y_r}{\partial q_m} q'_m, \\ z'_r &= \frac{\partial z_r}{\partial q_1} q'_1 + \frac{\partial z_r}{\partial q_2} q'_2 + \cdots + \frac{\partial z_r}{\partial q_m} q'_m. \end{aligned} \tag{50}$$

The derivatives $\frac{\partial x_r}{\partial q_s}$ contain only the coordinates q, not the velocities q', which we see enter linearly, accordingly

$$\frac{\partial x'_r}{\partial q'_s} = \frac{\partial x_r}{\partial q_s}, \quad \frac{\partial y'_r}{\partial q'_s} = \frac{\partial y_r}{\partial q_s}, \quad \frac{\partial z'_r}{\partial q'_s} = \frac{\partial z_r}{\partial q_s}. \tag{51}$$

Making use of this relation, the expression 49) becomes

$$x'_r \frac{\partial x'_r}{\partial q'_s} + y'_r \frac{\partial y'_r}{\partial q_s} + z'_r \frac{\partial z'_r}{\partial q'_s} = \frac{1}{2} \frac{\partial}{\partial q'_s} (x'^2_r + y'^2_r + z'^2_r).$$

Thus we find that the component of the velocity of any particle according to the coordinate q_s is equal to one-half the rate of change of the square of its velocity as we change the velocity q'_s.[1]) This result is not of itself of great physical importance, but leads us to one that is. Inasmuch as the momentum is the important dynamical quantity, multiplying by the mass of the particle we find

$$m_r x'_r \frac{\partial x_r}{\partial q_s} + m_r y'_r \frac{\partial y_r}{\partial q_s} + m_r z'_r \frac{\partial z_r}{\partial q_s} = \frac{\partial}{\partial q'_s} \left[\frac{1}{2} m_r (x'^2_r + y'^2_r + z'^2_r) \right]$$

or the component of the momentum of a particle according to any coordinate is the rate of change of its kinetic energy as we change the corresponding velocity. Summing for the whole system,

$$\begin{aligned} &\sum_r \left(m_r x'_r \frac{\partial x_r}{\partial q_s} + m_r y'_r \frac{\partial y_r}{\partial q_s} + m_r z'_r \frac{\partial z_r}{\partial q_s} \right) \\ &\qquad = \frac{\partial}{\partial q'_s} \sum_r \frac{1}{2} m_r (x'^2_r + y'^2_r + z'^2_r) = \frac{\partial T}{\partial q'_s}, \end{aligned} \tag{52}$$

that is, the component of the momentum of a system according to any generalized coordinate q_s is the rate of change of kinetic energy

1) It is to be observed that this "component" is not what we have called the velocity q'_s.

with respect to the corresponding velocity. The equation 38) now says that the kinetic energy is one-half the geometric product of the velocity and momentum systems. Thus we have perfect analogy with the last two equations of § 27.

We shall hereafter denote the momentum belonging to q_s by p_s and effecting the differentiation of 35) we have

$$53)\qquad p_s = \frac{\partial T}{\partial q'_s} = Q_{1s}q'_1 + Q_{2s}q'_2 + \cdots + Q_{ms}q'_m,$$

or every generalized momentum-component is a linear function of the velocities, the coefficients being the inertia-coefficients Q_{rs}.

Let us now find the component of the effective forces according to q_s, the effective forces being defined by the system of products, for each particle, of mass by acceleration,

$$m_r\frac{dx'_r}{dt},\quad m_r\frac{dy'_r}{dt},\quad m_r\frac{dz'_r}{dt}.$$

We have

$$54)\qquad m_r\frac{dx'_r}{dt}\frac{\partial x_r}{\partial q_s} = m_r\left[\frac{d}{dt}\left(x'_r\frac{\partial x_r}{\partial q_s}\right) - x'_r\frac{d}{dt}\left(\frac{\partial x_r}{\partial q_s}\right)\right],$$

to transform which we make use not only of 51), but of a relation obtained as follows. Differentiating 50) by q_s,

$$\frac{\partial x'_r}{\partial q_s} = \frac{\partial^2 x_r}{\partial q_1\partial q_s}q'_1 + \frac{\partial^2 x_r}{\partial q_2\partial q_s}q'_2 + \cdots + \frac{\partial^2 x_r}{\partial q_m\partial q_s}q'_m = \frac{d}{dt}\left(\frac{\partial x_r}{\partial q_s}\right).$$

Using these results in 54), we obtain for the right-hand member,

$$m_r\left[\frac{d}{dt}\left(x'_r\frac{\partial x'_r}{\partial q'_s}\right) - x'_r\frac{\partial x'_r}{\partial q_s}\right] = \frac{d}{dt}\left\{\frac{\partial}{\partial q'_s}\left(\frac{1}{2}m_r x'^2_r\right)\right\} - \frac{\partial}{\partial q_s}\left(\frac{1}{2}m_r x'^2_r\right),$$

and with similar results for y and z, summing for all the particles, we have for the component of the effective forces of the system,

$$55)\qquad \sum_r\left\{m_r\frac{d^2x_r}{dt^2}\frac{\partial x_r}{\partial q_s} + m_r\frac{d^2y_r}{dt^2}\frac{\partial y_r}{\partial q_s} + m_r\frac{d^2z_r}{dt^2}\frac{\partial z_r}{\partial q_s}\right\}$$

$$= \frac{d}{dt}\left\{\frac{\partial}{\partial q'_s}\sum_r\frac{1}{2}m_r(x'^2_r + y'^2_r + z'^2_r)\right\} - \frac{\partial}{\partial q_s}\sum_r\frac{1}{2}m_r(x'^2_r + y'^2_r + z'^2_r)$$

$$= \frac{d}{dt}\left(\frac{\partial T}{\partial q'_s}\right) - \frac{\partial T}{\partial q_s}.$$

Putting the effective force equal to the applied force we have Lagrange's equation 47) by direct transformation. The equation of d'Alembert's principle thus becomes in generalized coordinates

$$56)\qquad \sum_{s=1}^{s=m}\left\{\frac{d}{dt}\left(\frac{\partial T}{\partial q'_s}\right) - \frac{\partial T}{\partial q_s} - P_s\right\}\delta q_s = 0.$$

If we had begun with d'Alembert's principle we should evidently have gone through precisely the same process that we have here followed, and assuming all the displacements δx, δy, δz to be virtual all the δq's would have been independent, so that from the transformed equation 56) would have followed the individual equations 47). This was in fact the mode of deduction followed by Lagrange.

We have a noteworthy difference between generalized and rectangular coordinates, in that the effective force-component is not generally equal to the time-derivative of the momentum $\frac{dp_s}{dt}$, but contains in addition the term $-\frac{\partial T}{\partial q_s}$. This we may accordingly call the non-momental part of the effective force. Thus in general, even though the momentum p_s is unchanging, a force P_s must be impressed in order to balance the kinetic reaction $\frac{\partial T}{\partial q_s}$. As an example, let us take the case of polar coordinates in a plane. We then have for a single particle, for the coordinates q_1, q_2 the distance r from the origin, and the angle φ subtended by the radius vector and a fixed radius. The kinetic energy is

$$T = \frac{m}{2}(r'^2 + r^2\varphi'^2),$$

from which we have the momenta,

$$p_r = \frac{\partial T}{\partial r'} = mr', \quad p_\varphi = \frac{\partial T}{\partial \varphi'} = mr^2\varphi'.$$

Thus if the momentum p_r is constant, which is the case when the radial velocity r' is constant, we still have to impress a radial component of force

$$P_r = -\frac{\partial T}{\partial r} = -mr\varphi'^2.$$

The kinetic reaction $-P_r = mr\varphi'^2$ is called the *centrifugal force*, a name to which it is as much entitled as any sort of reaction is to the term force.

By analogy we might in general call the non-momental parts of the reversed effective forces or forces of inertia the centrifugal forces of the system. These non-momental parts may be absent for some coordinates. For instance in the present example φ does not appear in the kinetic energy, but only its velocity φ'. We have then $\frac{\partial T}{\partial \varphi} = 0$, so that force need be impressed to change φ only to change the momentum p_φ. Accordingly if no such force is impressed, the momentum p_φ is conserved. Thus in the case of a central force, the momentum $p_\varphi = mr^2\varphi'$ is constant. But this is the theorem of areas, or of conservation of moment of momentum. In fact we see that the generalized component of momentum with regard to the angular coordinate φ is the moment of momentum of the particle [cf. § 8, 23), 24)]. Systems in which there are coordinates having the property that their non-momental part of the kinetic reaction vanishes have peculiar properties, and are treated in § 48.

If we perform the differentiation of p_s by the time, differentiating equation 53), and remembering that the Q's depend only on the q's, we find for the momental part of the effective force

57)
$$\begin{aligned}\frac{dp_s}{dt} &= Q_{1s}q_1'' + Q_{2s}q_2'' + \cdots + Q_{ms}q_m'' \\ &+ q_1'\left\{\frac{\partial Q_{1s}}{\partial q_1}q_1' + \frac{\partial Q_{2s}}{\partial q_2}q_2' + \cdots + \frac{\partial Q_{1s}}{\partial q_m}q_m'\right\} \\ &\quad \cdots\cdots\cdots\cdots\cdots \\ &+ q_m'\left\{\frac{\partial Q_{ms}}{\partial q_1}q_1' + \frac{\partial Q_{ms}}{\partial q_2}q_2' + \cdots + \frac{\partial Q_{ms}}{\partial q_m}q_m'\right\},\end{aligned}$$

of which the first line, which we will call $F_s^{(1)}$, is a linear function of the generalized accelerations q_r''. Here again our generalized coordinates differ from rectangular, in that there is a part of the *momental* force which is independent of the accelerations q_r'', but which is a homogeneous quadratic function of the velocities,

58)
$$F_s^{(2)} = \sum_{r=1}^{r=m}\sum_{t=t}^{t=m}\frac{\partial Q_{ts}}{\partial q_r}q_r'q_t'.$$

Consequently if at any instant of the motion we can change the signs of all the velocities, and at the same time of all the accelerations, the accelerational part of the momental force $F_s^{(1)}$ will change its sign, while the non-accelerational part $F_s^{(2)}$ will be unchanged. We may thus experimentally discriminate between the two.

Effecting the differentiation in the case of the non-momental force, we find

59)
$$-\frac{\partial T}{\partial q_s} = -\frac{1}{2}\sum_{r=1}^{r=m}\sum_{t=1}^{t=m}\frac{\partial Q_{rt}}{\partial q_s}q_r'q_t' = F_s^{(3)},$$

which is also a homogeneous quadratic function of the velocities, and thus possesses similar properties to $F_s^{(2)}$. Thus it is difficult to discriminate experimentally between these two, unless we have some experimental means of recognizing when the momentum p_s remains constant. In the simple example which we have used above, since

$$\frac{dp_r}{dt} = mr'', \qquad \frac{dp_\varphi}{dt} = m\,(r^2\varphi'' + 2r\cdot r'\varphi')$$

the non-accelerational part of the momental force belonging to r disappears, while the centrifugal or non-momental does not, while for φ

although the non-momental part $F_\varphi^{(3)} = \frac{\partial T}{\partial \varphi}$ disappears, we have the non-accelerational part $F_\varphi^{(2)} = 2mr \cdot r'\varphi'$. Experimentally this means that, if a particle move with constant radial and angular velocities, we shall have to apply to it not only a radial force $F_r^{(3)} = -mr\varphi'^2$ to balance the centrifugal force, but also a turning force $2mr \cdot r'\varphi'$. This may be done by means of a varying constraint, say by making a particle move upon a rod turning with angular velocity φ'.[1]) The particle will then react upon the rod, to which the turning moment $2mr \cdot r'\varphi'$ must be applied, for if it were not applied, owing to the conservation of angular momentum, as the particle got farther from the center its angular velocity would be less. To keep it constant the particle must be pushed around.

We have now carefully analysed the effective forces, when expressed in terms of our generalized coordinates. It is to be carefully borne in mind that all these parts come from real accelerations impressed on the particles of the system, although the accelerations of the generalized coordinates may disappear. This will depend on our choice of such coordinates. The analysis that we have made is however by no means devoid of physical significance, as we can not usually observe all the bodies with which we have to do so as to find their real motions and determine their accelerations, but are obliged to become acquainted with them in a more or less roundabout way, through the reactions that they present to various operations upon them. From this point of view it is of interest to catalogue the various reactions that we meet in dynamics. In our equation of d'Alembert's principle 56), we have called the P's which are equated to the effective forces, the impressed forces, or forces of the system. If the system is conservative, the forces of the system are derivable from a potential energy, as we have assumed in 47), while if not, part of the forces may still be derived from such a function. It will be useful to consider not the forces of the system, but the forces which must be impressed from outside in order to counterbalance all the reactions of the system. In other words, if we write $F_s^{(4)}$ for the non-conservative part not yet dealt with,

60) $$F_s = F_s^{(1)} + F_s^{(2)} + F_s^{(3)} + F_s^{(4)} + F_s^{(5)}$$
$$= \frac{dp_s}{dt} - \frac{\partial T}{\partial q_s} + F_s^{(4)} + \frac{\partial W}{\partial q_s} = F_s,$$

F_s is the force necessary to be impressed on the system from *outside* under any circumstances whatever, or $-F_s$ is the *reaction* of the system, exerted through the coordinate q_s.

1) The centrifugal force may be balanced by a spring.

If the system is left to itself, uninfluenced by other systems, then every F_s is zero, and we have equation 47) with

$$P_s = -\frac{\partial W}{\partial q_s} - F_s^{(4)}.$$

If two systems are coupled together, so that any change of the coordinate q_s is accompanied by an equal change of the corresponding coordinate of another system, then the F_s's of the two systems are equal and opposite, which is the law of action and reaction. According to what happens to the system, the effect of F_s is of different kinds. For instance, if the system it at rest, or moves very slowly, all the $F_s^{(r)}$ terms vanish except the last, and we have the static reaction

$$-F_s = -F_s^{(5)} = -\frac{\partial W}{\partial q_s} = P_s.$$

The work that is then done by the external forces,

$$\sum_s F_s\, dq_s = \sum_s \frac{\partial W}{\partial q_s} dq_s = d\,W,$$

is stored up as potential energy in the system. If there is no possibility of statical storage, and if there is no non-conservative reaction, we have only the kinetic reactions already dealt with.

As a simple example of what is meant, suppose the system to consist of a mass attached to a spring tending to draw it to the right. If the mass is at rest, it must be held by a force applied from outside, to keep the spring stretched, and the static reaction of the spring P_s is toward the right. If the mass is let go, it begins to move toward the right, and the kinetic accelerational reaction is toward the left, balancing the static reaction, or internal impressed force of the system, according to d'Alembert's principle.

If there is no inertia, so that the effective forces vanish, and no storage, the work done upon the system is not stored, but is said to be *dissipated*. The reaction $-F_s^{(4)}$ does not, in the cases that exist in nature, appear except when there is motion, that is, the reaction $-F_s^{(4)}$ is a kinetic reaction, though not due to inertia. This work dissipated,

$$\Sigma_s F_s^{(4)}\, dq_s$$

is always positive, in other words, non-conservative reactions are always such as to oppose the motion. A case of frequent occurrence is that where there are non-conservative forces proportional to the first powers of the velocities q', so that any $F_s^{(4)} = \varkappa_s q_s'$. We may then form a function F which is, like T, a homogeneous quadratic function of the velocities,

$$F = \frac{1}{2}\sum_s \varkappa_s q_s'^2, \qquad F_s^{(4)} = \frac{\partial F}{\partial q_s'},$$

and since in this case the work dissipated in unit time is

$$61) \qquad \sum_s F_s^{(4)} \frac{dq_s}{dt} = \sum_s F_s^{(4)} q_s' = \sum_s q_s' \frac{\partial F}{\partial q_s'} = 2F,$$

F represents one-half the time rate of loss, or *dissipation* of energy. F is called the Dissipation Function, or the Dissipativity.[1]) It was introduced by Lord Rayleigh, and is of use in the theory of motions of viscous media, and in the dynamical treatment of electric currents. Beside this case we have dissipative forces not capable of representation by a dissipation function.

We will now place our various reactions in a table showing their grouping in various classes and sub-classes.

- Reactions
  - Positional $F^{(5)}$
  - Motional or Kinetic
    - Inertial
      - Momental
        - Accelerational $F^{(1)}$
        - Non-accelerational $F^{(2)}$
      - Non-momental or Centrifugal $F^{(3)}$
    - Non-conservative $F^{(4)}$
      - Having Dissipation-function
      - Others

The advantage of this complete classification is as follows. Suppose that a certain system or apparatus is presented to us for dynamical examination. Its parts are concealed from our view by coverings or cases, but at certain points there protrude handles, cranks, or other *driving points,* upon which we may operate, and which will exert certain reactions. All that we can learn of the system will become known to us by a study of the reactions. Maxwell[2]) compares such a system to a set of bell-ropes hanging from holes in a roof, which are to be pulled by a number of bell ringers. If when one rope is pulled none of the others are affected, we conclude that that rope has no connection with the others. If however, when one rope is pulled, a number of others are set in motion, we conclude that there is some sort of connection between the corresponding bells. What the connection is we can find out by studying the motions. In general, if when we move one driving point, and let it go, it remains where we put it, we conclude that it is not attached to anything, but is a mere blind member. If when we push it, it

1) A case of perhaps equal importance is one in which the dissipation function contains the squares of *differences* of the velocities.

2) Maxwell, *Scientific Papers,* Vol. II, p. 783.

returns to its former position, we infer that it is connected with something of the nature of a spring, and that the system can store potential energy. If when we push it it keeps on going after we release it, we conclude that it is connected with a system possessing inertia, and capable of storing kinetic energy. If its motion dies away, we conclude that there is dissipation, and so on. By experimenting in turn, or simultaneously, on all the driving points, we may conclude how many degrees of freedom the system has, how the inertia is distributed, and how the parts of the system are connected. The means of doing this will be discussed later, and we shall find that in this manner we may learn much of a system, but that our knowledge will not always be complete. This is the nature of the process by which the physicist proceeds in the attempt to explain recondite phenomena, such as those of heat or electricity, by reducing them to the simpler phenomena of motion. The parts of the systems, be they made of molecules of matter, or of the ether, are concealed from him, but he may operate upon them in certain experimental ways, and draw definite conclusions from the results. One of the greatest triumphs of this method was Maxwell's dynamical theory of electricity.

Impulsive forces are dealt with by Lagrange's equations in the usual manner. Integrating equations 47) with respect to the time throughout a vanishing interval $t_1 - t_0$, since the velocities are finite, the non-momental forces $-\frac{\partial T}{\partial q_s}$ are by 58) finite, so that the integral of the second term vanishes, and we have

$$p_s^{(t_1)} - p_s^{(t_0)} = \lim_{t_1 = t_0} \int_{t_0}^{t_1} P_s \, dt.$$

Thus the momentum generated measures the impulse, as in the case of rectangular coordinates, § 27.

As a further example of the use of Lagrange's equations let us take the problem of the spherical pendulum, which we used to introduce the subject. We had

$$24) \qquad \begin{aligned} T &= \frac{ml^2}{2}\{\vartheta'^2 + \sin^2\vartheta \cdot \varphi'^2\}, \\ W &= -\, mgl \cos\vartheta. \end{aligned}$$

We have for the momenta p_ϑ and p_φ

$$62) \qquad \begin{aligned} p_\vartheta &= \frac{\partial T}{\partial \vartheta'} = ml^2 \vartheta', \\ p_\varphi &= \frac{\partial T}{\partial \varphi'} = ml^2 \sin^2\vartheta \cdot \varphi', \end{aligned}$$

and our differential equations are

$$63)\quad \begin{aligned} &\frac{d}{dt}(ml^2\vartheta') - \frac{\partial T}{\partial \vartheta} = -\frac{\partial W}{\partial \vartheta} = -mgl\sin\vartheta, \\ &\frac{d}{dt}(ml^2\sin^2\vartheta\cdot\varphi') - \frac{\partial T}{\partial \varphi} = -\frac{\partial W}{\partial \varphi} = 0. \end{aligned}$$

Now since m and l are constant the equation for ϑ becomes

$$64)\quad \frac{d^2\vartheta}{dt^2} - \sin\vartheta\cos\vartheta\cdot\varphi'^2 = -\frac{g}{l}\sin\vartheta,$$

in which the centrifugal force-component according to ϑ is

$$ml^2\sin\vartheta\cos\vartheta\cdot\varphi'^2.$$

The equation for φ (φ has no centrifugal part),

$$65)\quad \frac{d}{dt}(l^2\sin^2\vartheta\cdot\varphi') = 0$$

may at once be integrated, giving

$$66)\quad l^2\sin^2\vartheta\cdot\varphi' = c,$$

which is the integral equation 50), § 21.

Substituting in 64) the value of φ' derived from the integral equation 66), we obtain the differential equation for ϑ, which is the same as the derivative of equation 51), § 21. The remainder of the solution is accordingly the same as in § 21.

38. Equation of Activity. Integral of Energy. Let us multiply each of Lagrange's equations by the corresponding velocity q' and add the results for all values of r, obtaining

$$67)\quad \sum_r\left\{\frac{d}{dt}\left(\frac{\partial T}{\partial q_r'}\right) - \frac{\partial T}{\partial q_r}\right\} q_r' = \sum_r P_r q_r'.$$

The expression on the right, otherwise written

$$\sum_r P_r\frac{dq_r}{dt} = \frac{dA}{dt},$$

represents the time-rate at which the applied forces do work on the system. The equation 67) is accordingly the equation of activity, § 27, 20), in generalized coordinates.

By means of the property of T expressed in equation 38), § 36, we may transform the left-hand side of the equation, for, since T depends upon both the q's and q''s, both of which in an actual motion depend upon t, differentiating totally,

$$68)\quad \frac{dT}{dt} = \sum_r\left(\frac{\partial T}{\partial q_r}\frac{dq_r}{dt} + \frac{\partial T}{\partial q_r'}\frac{dq_r'}{dt}\right).$$

Now differentiating 38) totally

$$69)\qquad 2\frac{dT}{dt}=\sum_r\left\{\frac{\partial T}{\partial q'_r}\frac{dq'_r}{dt}+q'_r\frac{d}{dt}\left(\frac{\partial T}{\partial q'_r}\right)\right\}.$$

Subtracting equation 68) from 69) the terms $\frac{\partial T}{\partial q'_r}\frac{dq'_r}{dt}$ cancel and we have left

$$70)\qquad \frac{dT}{dt}=\sum_r\left\{q'_r\frac{d}{dt}\left(\frac{\partial T}{\partial q'_r}\right)-\frac{\partial T}{\partial q_r}\frac{dq_r}{dt}\right\}.$$

But this is exactly the left-hand member of the equation of activity 67). Thus if the system is conservative, since

$$\frac{dA}{dt}=-\frac{dW}{dt},\quad \frac{dT}{dt}=-\frac{dW}{dt},$$

so that the equation of conservation of energy is always an integral of Lagrange's equations.

39. Hamilton's Canonical Equations. Although the equations of Lagrange are by all odds those most frequently used in dynamical problems, yet in many theoretical investigations a transformation introduced by Hamilton is of importance.

The kinetic energy being a quadratic form in the velocities q' [equation 35)], the momenta p_r being the derivatives of T by the q'_r's are, as we have seen, linear forms in the q'_r's.

$$53)\qquad \begin{aligned} p_1 &= \frac{\partial T}{\partial q'_1} = Q_{11}q'_1 + Q_{12}q'_2 + \cdots + Q_{1m}q'_m,\\ &\cdots\cdots\cdots\cdots\cdots\\ &\cdots\cdots\cdots\cdots\cdots\\ p_m &= \frac{\partial T}{\partial q'_m} = Q_{m1}q'_1 + Q_{m2}q'_2 + \cdots + Q_{mm}q'_m. \end{aligned}$$

These linear equations may be solved for the q'_r's, obtaining any q'_r as a linear function of the p_r's, say,

$$71)\qquad q'_r = R_{r1}p_1 + R_{r2}p_2 + \cdots + R_{rm}p_m,$$

the R's being minors of the determinant,

$$D=\begin{vmatrix} Q_{11}, & Q_{12}, & \cdots & Q_{1m}\\ \cdot & \cdot & \cdot & \cdot\\ \cdot & \cdot & \cdot & \cdot\\ Q_{m1}, & Q_{m2}, & \cdots & Q_{mm}\end{vmatrix},$$

divided by D itself.

The R's accordingly, like the Q's, are functions of only the coordinates q. Maxwell calls them coefficients of *mobility.* The solution of the equations assumes that the determinant D does not vanish. This is always the case, being one of the conditions that T is an essentially positive function.

Let us now introduce into T the variables p in place of the variables q', so that T is expressed as a function of all the q's and p's. Since

38) $$2T = \sum_r q'_r \frac{\partial T}{\partial q'_r} = \sum_r q'_r p_r,$$

inserting the values of q'_r in terms of the p's gives

72) $$T = \frac{1}{2} \sum_{r=1}^{r=m} \sum_{s=1}^{s=m} R_{r\,s} p_r p_s,$$

that is T is now expressed as a quadratic form in the p's. We will distinguish T when expressed in terms of the p's by the suffix p, T_p. We now have by Euler's Theorem,

73) $$2T_p = \sum_r p_r \frac{\partial T_p}{\partial p_r}.$$

Since T_p is identically equal to T, comparing with equations 38), above we have by symmetry,

74) $$q'_r = \frac{\partial T_p}{\partial p_r},$$

thus the q'_r's are linear forms in the p's given by 71). The two identically equal functions, T, T_p, having the properties

75) $$\frac{\partial T}{\partial q'_r} = p_r, \quad \frac{\partial T_p}{\partial p_r} = q'_r,$$

are said to be reciprocal functions.[1])

The expressions for the forces and potential energy are left unaltered. Let us now make use of Hamilton's principle with this choice of variables. Before performing the variation it will be advantageous to introduce in the integral to be varied instead of the Lagrangian function, $L = T - W$, the Hamiltonian function, $H = T + W$, by means of the relation

76) $$L = 2T - H.$$

T and H are both to be expressed as functions of the variables q and p, both of which depend upon the time t in a manner to be found by integrating the differential equations of motion.

Hamilton's principle then takes the form

77) $$\delta \int_{t_0}^{t_1} (2T - H)\, dt = \delta \int_{t_0}^{t_1} \left(\sum_r (p_r q'_r - H) \right) dt$$

$$= \int_{t_0}^{t_1} \sum_r \left(q'_r \delta p_r + p_r \delta q'_r - \frac{\partial H}{\partial p_r} \delta p_r - \frac{\partial H}{\partial q_r} \delta q_r \right) dt = 0.$$

---

1) Webster, *Theory of Electricity and Magnetism*, § 63, 64

The term $p_r \delta q_r' = p_r \frac{d}{dt} \delta q_r$ being integrated by parts and the δq's put equal to 0 at the limits, we have

$$\int_{t_0}^{t_1} \sum_r \left\{ \left(q_r' - \frac{\partial H}{\partial p_r} \right) \delta p_r - \left(\frac{d p_r}{dt} + \frac{\partial H}{\partial q_r} \right) \delta q_r \right\} dt = 0.$$

Now since W does not depend upon the momentum p_r,

$$\frac{\partial H}{\partial p_r} = \frac{\partial T}{\partial p_r} = q_r',$$

therefore the coefficients of the δp's all vanish. If the δq's are all arbitrary, their coefficients must accordingly vanish so that we have

$$78) \qquad a) \ \frac{d p_r}{dt} = -\frac{\partial H}{\partial q_r}, \qquad b) \ \frac{d q_r}{dt} = \frac{\partial H}{\partial p_r},$$

the first equation being the equation of motion, the second defining $q_r' = \frac{d q_r}{dt}$. These equations 78) were introduced by Hamilton and on account of their peculiarly simple and symmetrical form they are often referred to as the *canonical equations of dynamics.* In practical problems they are generally not more convenient than Lagrange's equations.

We may recapitulate Hamilton's method as follows:

Form the Hamiltonian function H, representing the total energy of the system as a function of the $2m$ independent variables q and p, the coordinates and momenta. Then the time derivative of any coordinate q is equal to the partial derivative of H with respect to the corresponding momentum p, while the time derivative of any momentum is equal to minus the partial derivative of H with respect to the corresponding coordinate. A direct deduction of the equations of Hamilton without the use of Hamilton's Principle will be found in the author's *Theory of Electricity and Magnetism* § 64.

The equation of activity is most simply deduced from Hamilton's equations, for by cross multiplication of equations 78), after transposing and summing for all the coordinates we get

$$79) \qquad \sum_r \left(\frac{\partial H}{\partial q_r} \frac{d q_r}{dt} + \frac{\partial H}{\partial p_r} \frac{d p_r}{dt} \right) = 0.$$

But this is equal to the total derivative of H by t,

$$\frac{dH}{dt} = 0,$$

which being integrated gives

$$H = h,$$

a constant. But since $H = T + W$, this is the equation of energy.

If the system is not conservative, there may be still some forces which are derivable from a potential energy function. In that case the Hamiltonian function is to be formed with that energy, but we must add to the right of equation 78a) the non-conservative force $-F_r^{(4)}$. Thus our equations become

80) $$\frac{dp_r}{dt} = -\frac{\partial H}{\partial q_r} - F_r^{(4)}, \quad \frac{dq_r}{dt} = \frac{\partial H}{\partial p_r}.$$ [1]

The equation of activity then becomes

81) $$\sum_r \left(\frac{\partial H}{\partial q_r}\frac{dq_r}{dt} + \frac{\partial H}{\partial p_r}\frac{dp_r}{dt}\right) = -\sum_r F_r^{(4)}\frac{dq_r}{dt}$$

or

$$\frac{dH}{dt} = -2F,$$

if there is a dissipation function.

39a. Varying Constraint. It may happen that the equations of constraint contain the time explicitly, that is

82) $$\begin{aligned} &\varphi_1(t, x_1, y_1, z_1, \ldots x_n, y_n, z_n) = 0, \\ &\varphi_2(t, x_1, y_1, z_1, \ldots x_n, y_n, z_n) = 0, \\ &\cdots\cdots\cdots\cdots \\ &\cdots\cdots\cdots\cdots \\ &\varphi_k(t, x_1, y_1, z_1, \ldots x_n, y_n, z_n) = 0. \end{aligned}$$

Such a case is that of a particle constrained to move on a surface which is itself in motion, say a sphere whose center moves with a prescribed motion. The constraint is then said to be variable, and the work done by the constraint no longer vanishes, for the surface has generally a normal component in its motion, which causes the reaction to do work. The variability of the constraint has an important effect on the equations of motion. We can then no longer determine the position of the system by means of a set of independent parameters, but must give not only their values, but also the time. We may put

83) $$\begin{aligned} x_r &= x_r(t, q_1, q_2, \ldots q_m), \\ y_r &= y_r(t, q_1, q_2, \ldots q_m), \\ z_r &= z_r(t, q_1, q_2, \ldots q_m), \end{aligned}$$

from which, by the elimination of the q's, we may obtain equations 82).

---

1) cf. § 37, 60).

Differentiating now totally, we have

$$
\begin{aligned}
dx_r &= \frac{\partial x_r}{\partial t}dt + \frac{\partial x_r}{\partial q_1}dq_1 + \frac{\partial x_r}{\partial q_2}dq_2 + \cdots + \frac{\partial x_r}{\partial q_m}dq_m,\\
84)\qquad dy_r &= \frac{\partial y_r}{\partial t}dt + \frac{\partial y_r}{\partial q_1}dq_1 + \frac{\partial y_r}{\partial q_2}dq_2 + \cdots + \frac{\partial y_r}{\partial q_m}dq_m,\\
dz_r &= \frac{\partial z_r}{\partial t}dt + \frac{\partial z_r}{\partial q_1}dq_1 + \frac{\partial z_r}{\partial q_2}dq_2 + \cdots + \frac{\partial z_r}{\partial q_m}dq_m,
\end{aligned}
$$

or on dividing through by dt,

$$
\begin{aligned}
x_r' &= \frac{\partial x_r}{\partial t} + \sum_{s=1}^{s=m}\frac{\partial x_r}{\partial q_s}q_s',\\
85)\qquad y_r' &= \frac{\partial y_r}{\partial t} + \sum_{s=1}^{s=m}\frac{\partial y_r}{\partial q_s}q_s',\\
z_r' &= \frac{\partial z_r}{\partial t} + \sum_{s=1}^{s=m}\frac{\partial z_r}{\partial q_s}q_s'.
\end{aligned}
$$

We have now in each x', y', z', beside the linear function of q_1', q_2', $\ldots$ q_m', a term independent of the q''s, but which may be expressed in terms of the coordinates q and t On squaring there are accordingly not only quadratic terms in the q''s, but also terms of the first and zero orders, On forming the kinetic energy

$$
\begin{aligned}
86)\qquad T = \frac{1}{2}\sum_{r=1}^{r=n} m_r \Bigg\{ & \sum_{s=1}^{s=m}\sum_{t=1}^{t=m}\left(\frac{\partial x_r}{\partial q_s}\frac{\partial x_r}{\partial q_t} + \frac{\partial y_r}{\partial q_s}\frac{\partial y_r}{\partial q_t} + \frac{\partial z_r}{\partial q_s}\frac{\partial z_r}{\partial q_t}\right)q_s' q_t'\\
& + 2\frac{\partial x_r}{\partial t}\sum_{s=1}^{s=m}\frac{\partial x_r}{\partial q_s}q_s' + 2\frac{\partial y_r}{\partial t}\sum_{s=1}^{s=m}\frac{\partial y_r}{\partial q_s}q_s' + 2\frac{\partial z_n}{\partial t}\sum_{s=1}^{s=m}\frac{\partial z_r}{\partial q_t}q_s'\\
& + \left(\frac{\partial x_r}{\partial t}\right)^2 + \left(\frac{\partial y_r}{\partial t}\right)^2 + \left(\frac{\partial z_r}{\partial t}\right)^2 \Bigg\},
\end{aligned}
$$

we accordingly find that instead of being, as before, a homogeneous function of the q''s, it contains not only quadratic terms, but also terms linear in and others independent of the q''s. The effect of these linear terms in the kinetic energy, whatever be their origin, will be discussed in § 50.

40. Hamilton's Principle the most general dynamical principle. We have seen in this chapter how by means of Hamilton's Principle we may deduce the general equations of motion, and from these the principle of Conservation of Energy. As

Hamilton's Principle holds whether the system is conservative or not, it is more general than the principle of Conservation of Energy, which it includes. The principle of energy is not sufficient to deduce the equations of motion. If we know the Lagrangian function we can at once form the equations of motion by Hamilton's Principle, and without forming them we may find the energy. For we have

$$L = T - W,$$

$$E = T + W.$$

Accordingly

$$87)\qquad E = 2T - L = \sum_s q_s' \frac{\partial T}{\partial q_s'} - L = \sum_s q_s' \frac{\partial L}{\partial q_s'} - L,$$

so that the energy is given in terms of L and its partial derivatives. If on the other hand the energy E is given a function of the coordinates and velocities, the Lagrangian function must be found by integrating the partial differential equation 87), the integration involving an arbitrary function. In fact if F be a homogeneous linear function of the velocities, the equation 87) will be satisfied not only by L but also by $L + F$. For, F being homogeneous of degree one,

$$F = \sum_s q_s' \frac{\partial F}{\partial q_s'}.$$

Consequently a knowledge of the energy is not sufficient to find the motion, while a knowledge of the Lagrangian function is. The attempt has been made by certain writers to found the whole of physics upon the principle of energy. The fact that the principle of energy is but one integral of the differential equations, and is not sufficient to deduce them, should be sufficient to show the futility of this attempt. It is the infinite order of variability of the motion involved in the variations occurring in Hamilton's Principle that makes it embrace what the Principle of Energy does not

41. Principle of Varying Action. We shall now deal with a principle, likewise due to Hamilton, somewhat broader than that which we have hitherto called Hamilton's Principle or Principle of Least Action, and furnishing a means of integrating the equations of motion. In the principle of least action a certain integral, belonging to a motion naturally described by a system under the action of certain forces according to the differential equations of motion, has been compared with the value of the same integral for a slightly different motion between the same terminal configurations, but not a natural motion and therefore violating the equations of motion. Under these circumstances the principle states that the integral is less for the natural motion than for the other. The new principle, on the other hand, compares the integrals *always taken for a natural motion* satisfying the differential equations, but the terminal configurations are varied from one motion to another. The principle is therefore known as the Principle of Varying Action.

In the process of § 34 equation 2) we cannot now put the integrated part equal to zero, but instead of 3) we shall have

$$(88) \quad \left[\sum_r \left\{ m_r \left(\frac{dx_r}{dt} \delta x_r + \frac{dy_r}{dt} \delta y_r + \frac{dz_r}{dt} \delta z_r \right) \right\} \right]_{t_0}^{t_1} = \delta \int_{t_0}^{t_1} (T - W)\, dt.$$

The integrated part, which is the sum of the geometric products of the momenta and the variations of the corresponding positions at the end of the motion minus the corresponding sum at the beginning, may now be transformed into generalized coordinates. The integral

$$S = \int_{t_0}^{t_1} (T - W)\, dt,$$

where T and W are expressed as functions of the time, appropriate to any given motion (whether natural or not) depends upon the terminial configurations, and is called by Hamilton the Principal Function. The terminal configurations being given we had $\delta S = 0$. Let us now find an expression for δS in generalized coordinates corresponding to the expression above in rectangular coordinates.

Proceeding as in § 36 equation 43) we obtain

$$89) \quad \delta S = \int_{t_0}^{t_1} \left[\sum_r \left\{ \frac{\partial (T-W)}{\partial q_r} \delta q_r + \frac{\partial (T-W)}{\partial q'_r} \delta q'_r \right\} \right] dt$$

$$= \sum_r \frac{\partial (T-W)}{\partial q'_r} \delta q_r \Bigg|_{t_0}^{t_1} + \int_{t_0}^{t_1} \left[\sum_r \left\{ \frac{\partial (T-W)}{\partial q_r} - \frac{d}{dt}\left(\frac{\partial (T-W)}{\partial q'_r} \right) \right\} \delta q_r \right] dt.$$

Since the various motions are all natural ones satisfying the differential equations of motions, the factor of every δq in the integrand vanishes, so that the integral vanishes of itself, and δS is accordingly expressed as a linear function of the variations of the initial and terminal coordinates. Since W is independent of the q''s and $\frac{\partial T}{\partial q'_r} = p_r$, making use of the affixes 0 and 1 for the limits t_0 and t_1, we may write

$$90) \qquad \delta S = \Sigma_r p_r^1 \delta q_r^1 - \Sigma_r p_r^0 \delta q_r^0,$$

an equation which could have been obtained from the considerations regarding geometric products at the beginning of § 37. This expression for the variation of S is of great importance, for by means of it we can obtain a method of integrating the equations of motion, and obtaining the coordinates q and momenta p at any time t_1. As we are now to consider the upper limit t_1 as variable it will be convenient to drop the subscript 1.

Suppose we have integrated the differential equations of motion completely so as to obtain every coordinate as a function of the time t, involving $2m$ arbitrary constants, $c_1, c_2, \ldots c_{2m}$, the number necessarily introduced in integrating the m Lagrangian equations of the second order or the $2m$ Hamiltonian equations of the first order. Let the integrals be

91)
$$\begin{aligned} q_1 &= f_1(t, c_1, c_2, \ldots c_{2m}), \\ q_2 &= f_2(t, c_1, c_2, \ldots c_{2m}), \\ &\cdots\cdots\cdots \\ &\cdots\cdots\cdots \\ q_m &= f_m(t, c_1, c_2, \ldots c_{2m}). \end{aligned}$$

Differentiating these by t we obtain

92)
$$q_1' = \frac{df_1}{dt}, \quad q_2' = \frac{df_2}{dt}, \ldots q_m' = \frac{df_m}{dt},$$

from which by equation 53) we may find the p's as functions of t,

93)
$$p_r = \varphi_r(t, c_1, c_2, \ldots c_{2m}).$$

These equations with 91) constitute $2m$ integral equations of the system.

Inserting the particular value t_0 in our integral equations we have

91′)
$$q_r^0 = f_r(t_0, c_1, c_2, \ldots c_{2m}),$$

93′)
$$p_r^0 = \varphi_r(t_0, c_1, c_2, \ldots c_{2m}).$$

We accordingly have the $4m+1$ variables,

$$t, \; q_1, \ldots q_m, \; p_1, \ldots p_m, \; q_1^0, \ldots q_m^0, \; p_1^0, \ldots p_m^0,$$

connected by $2m$ integral equations. We may thus choose any $2m+1$ of them as variables in terms of which to express the remaining $2m$.

For instance in the problem of shooting at a target § 35 we saw that the motion was completely determined by the coordinates of the initial and final positions and the initial velocity. The latter determined the time of transit t, so that it together with the initial coordinates, $q_1^0, \ldots q_m^0$, and the final coordinates, $q_1, \ldots q_m$, may be taken as independent variables in terms of which everything may be expressed.

Thus the integral

$$S = \int_{t_0}^{t} (T - W)\, dt \tag{94}$$

is supposed to be expressed in terms of these $2m + 1$ variables. Now if the initial and final coordinates are varied without varying the time of transit $t - t_0$ (t the upper limit of the integral) we have

$$\delta S = \sum_r \frac{\partial S}{\partial q_r} \delta q_r + \sum_r \frac{\partial S}{\partial q_r^0} \delta q_r^0. \tag{95}$$

We have however proved that under these conditions we have

$$\delta S = \Sigma_r p_r \delta q_r - \Sigma_r p_r^0 \delta q_r^0. \tag{90}$$

Since these expressions must be equal for arbitrary variations of the q's and q^0's we must have

$$\frac{\partial S}{\partial q_1} = p_1, \qquad \frac{\partial S}{\partial q_2} = p_2, \ldots \frac{\partial S}{\partial q_m} = p_m, \tag{96}$$

$$\frac{\partial S}{\partial q_1^0} = -p_1^0, \quad \frac{\partial S}{\partial q_2^0} = -p_2^0, \ldots \frac{\partial S}{\partial q_m^0} = -p_m^0. \tag{97}$$

We may now, if we please, regard the initial coordinates $q_1^0, \ldots q_m^0$, and the initial momenta, $p_1^0, \ldots p_m^0$, as $2m$ arbitrary constants replacing the $c_1, c_2, \ldots c_{2m}$ of equations 91) and 93). Then the equations 97) will be the general integrals of the equations of motion, for if the form of the function S is known in terms of $t, q_1, \ldots q_m, q_1^0, \ldots q_m^0$, the equations 97) are m equations involving $q_1, \ldots q_m$ without their derivatives, which may be solved to obtain the q's as functions of t and $2m$ arbitrary constants $q_1^0, \ldots q_m^0, p_1^0, \ldots p_m^0$, as in equations 91).

It has appeared as if in order to find S it were necessary to integrate the equations of motion, so to obtain $T - W$ as a function of the time, which being integrated would give S. If this were so the statement just made would be of little interest. But this is not necessary, for Hamilton showed that the function S, which he called the Principal Function, satisfies a certain partial differential equation, a solution of which being obtained, the whole problem is solved.

The function S is a function of the variables q, the constants q^0 and the time t, which thus occurs explicitly and implicitly. Differentiating by t we have therefore

$$\frac{dS}{dt} = \frac{\partial S}{\partial t} + \sum_r \frac{\partial S}{\partial q_r} \frac{dq_r}{dt}. \tag{98}$$

Differentiating 94) by t, the upper limit, gives however

$$\frac{dS}{dt} = T - W.$$

Equating the two values,

$$T - W = \frac{\partial S}{\partial t} + \Sigma_r p_r q_r',$$

$$= \frac{\partial S}{\partial t} + 2T,$$

by 38).

Transposing and writing $T + W = H$,

$$\frac{\partial S}{\partial t} + H = 0.$$

The function H, the sum of the energies, depends upon the coordinates q_r and the momenta, $p_r = \frac{\partial S}{\partial q_r}$. If the force-function depends upon the time H will also contain t explicitly. Thus we have the partial differential equation

$$99)\qquad \frac{\partial S}{\partial t} + H\left(t, q_1, \dots q_m, \frac{\partial S}{\partial q_1}, \dots \frac{\partial S}{\partial q_m}\right) = 0.$$

The equation is of the first order since only first derivatives of S appear, and, from the way in which T contains the momenta [equation 72)], is of the second degree in the derivatives $\frac{\partial S}{\partial q_r}$. Since S appears only through its derivatives an arbitrary constant may be added to it.

Thus we have the theorem due to Hamilton: If $q_1, \dots q_m$ expressed as integrals of the differential equations in terms of t and $2m$ arbitrary constants $q_1^0, \dots q_m^0$, $p_1^0, \dots p_m^0$, are introduced into the integral 94), and the result is expressed in terms of t, $q_1, \dots q_m$, $q_1^0, \dots q_m^0$, then S is a solution of the partial differential equation 99).

The converse of the proposition was proved by Jacobi, namely, that if we take any solution of the equation 99) containing m arbitrary constants, $q_1^0, \dots q_m^0$ (other than the one which may always be added), the equations 97) obtained by putting the derivatives of S by the m arbitrary constants equal to other arbitrary constants, $p_1^0, \dots p_m^0$ will be integrals of the differential equations of motion For the proof of this the reader is referred to Jacobi, *Vorlesungen über Dynamik*, XX.

Before giving examples of the utility of this method we shall show that the arbitrary constants by which we differentiate need not be the q^0's, but may be *any* m constants appearing in the integral equations.

Suppose that in equations 91) we vary m of the arbitrary constants $c_1, \dots c_m$. We then have

$$\delta q_r = \frac{\partial f_r}{\partial c_1}\delta c_1 + \frac{\partial f_r}{\partial c_2}\delta c_2 + \cdots + \frac{\partial f_r}{\partial c_m}\delta c_m,$$

and putting $t = 0$,

$$\delta q_r^0 = \left(\frac{\partial f_r}{\partial c_1}\right)_0 \delta c_1 + \left(\frac{\partial f_r}{\partial c_2}\right)_0 \delta c_2 + \cdots + \left(\frac{\partial f_r}{\partial c_m}\right)_0 \delta c_m.$$

Then equation 90) becomes

100) $$\delta S = \Sigma_r p_r \delta q_r - \Sigma_r C_r \delta c_r,$$

where

$$C_r = \sum_s p_s^0 \frac{\partial f_s}{\partial c_r}.$$

Then comparing with

$$\delta S = \sum_r \frac{\partial S}{\partial q_r}\delta q_r + \sum_r \frac{\partial S}{\partial c_r}\delta c_r,$$

we have corresponding to equations 96) and 97)

101) $$\frac{\partial S}{\partial q_r} = p_r,$$

102) $$\frac{\partial S}{\partial c_r} = - C_r.$$

Thus we may differentiate S with respect to the m arbitrary constants, no matter how they may appear in the solution of 99), putting the result equal to other arbitrary constants.

Hamilton's equation 99) assumes a somewhat simpler form when the force-function and consequently H are independent of the time, that is when the system is conservative. We may then advantageously replace the principal function S by another function called by Hamilton the Characteristic Function, which represents the action A, § 35. Making use of the equation of energy, $T + W = h$, to eliminate W, we have

$$S = \int_{t_0}^{t} (T - W)\, dt = \int_{t_0}^{t} 2T dt - h(t - t_0) = A - h(t - t_0)$$

or

103) $$A = \int_{t_0}^{t} 2T\, dt = S + h(t - t_0).$$

If now the function A does not contain the time explicitly we have differentiating partially

104) $$\frac{\partial S}{\partial t} = - h, \quad \frac{\partial S}{\partial q_r} = \frac{\partial A}{\partial q_r},$$

and our partial differential equation 99) becomes merely

105) $$H\left(q_1, \ldots q_m, \frac{\partial A}{\partial q_1}, \ldots \frac{\partial A}{\partial q_m}\right) = h.$$

The arbitrary constant of energy, h, takes the place of one of the constants c.

The variation of the action on changing the terminal configurations will cause a change in the energy necessary, h, if the time of transit, $t - t_0$, is unchanged. Accordingly

$$106)\quad \delta A = \delta S + (t - t_0)\,\delta h = \sum_{r=1}^{r=m} p_r\,\delta q_r - \sum_{r=1}^{r=m-1} C_r\,\delta c_r + (t - t_0)\,\delta h,$$

by equation 100). We have therefore

$$107)\quad p_r = \frac{\partial A}{\partial q_r},$$

$$108)\quad - C_r = \frac{\partial A}{\partial c_r},$$

$$109)\quad t - t_0 = \frac{\partial A}{\partial h}.$$

As examples of the use of the method of Hamilton we will now solve a few problems that have been already treated.

First, let us take the case of any number of free particles. We have

$$T = \frac{1}{2}\sum_r m_r\,(x_r'^2 + y_r'^2 + z_r'^2).$$

Putting

$$110)\quad \begin{aligned} p_{xr} &= \frac{\partial T}{\partial x_r'} = m_r x_r', \\ p_{yr} &= \frac{\partial T}{\partial y_r'} = m_r y_r', \\ p_{zr} &= \frac{\partial T}{\partial z_r'} = m_r z_r', \end{aligned}$$

this becomes

$$T = \frac{1}{2}\sum_r \frac{1}{m_r}(p_{xr}^2 + p_{yr}^2 + p_{zr}^2).$$

By 107)

$$111)\quad p_{xr} = \frac{\partial A}{\partial x_r},\quad p_{yr} = \frac{\partial A}{\partial y_r},\quad p_{zr} = \frac{\partial A}{\partial z_r},$$

and equation 105) then is

$$112)\quad H = \frac{1}{2}\sum_r \frac{1}{m_r}\left\{\left(\frac{\partial A}{\partial x_r}\right)^2 + \left(\frac{\partial A}{\partial y_r}\right)^2 + \left(\frac{\partial A}{\partial z_r}\right)^2\right\} + W = h.$$

In the case of a single particle comparing equations 110) and 111) we have

$$113)\quad m x' = \frac{\partial A}{\partial x},\quad m y' = \frac{\partial A}{\partial y},\quad m z' = \frac{\partial A}{\partial z}.$$

In other words if the action A is expressed in terms of the coordinates x, y, z, the momentum of a particle describing any path under the action of the given forces with the constant energy h is the vector differential parameter of the action A, and therefore, by the properties of lamellar vectors (§ 31), the velocity of a particle moving in this manner is normal to all the surfaces of constant action, and is inversely proportional to the distance between two infinitely near surfaces of constant action. Otherwise expressed, if from all points of any surface particles be projected normally with the same energy h, their paths will always be normal to a set of surfaces, and the action from one surface to another will be the same for all the particles. This theorem is due to Thomson and Tait.[1])

Suppose first there are no forces acting, then equation 112) becomes

$$114)\qquad \left(\frac{\partial A}{\partial x}\right)^2 + \left(\frac{\partial A}{\partial y}\right)^2 + \left(\frac{\partial A}{\partial z}\right)^2 = 2mh,$$

which is satisfied by the linear function

$$115)\qquad A = ax + by + cz,$$

if

$$116)\qquad a^2 + b^2 + c^2 = 2mh.$$

In virtue of this last equation only three of the constants a, b, c, h are arbitrary. Suppose we take a, b, h, then we have

$$117)\qquad A = ax + by + \sqrt{2mh - (a^2 + b^2)} \cdot z.$$

Then equations 107) or 113) are

$$118)\qquad mx' = a, \quad my' = b, \quad mz' = \sqrt{2mh - (a^2 + b^2)},$$

which are first integrals of the equations of motion, showing that the motion of the point is uniform. Equations 108) and 109) are

$$119)\qquad \begin{aligned} \frac{\partial A}{\partial a} &= x - \frac{az}{\sqrt{2mh - (a^2 + b^2)}} = a_1, \\ \frac{\partial A}{\partial b} &= y - \frac{bz}{\sqrt{2mh - (a^2 + b^2)}} = b_1, \\ \frac{\partial A}{\partial h} &= \frac{mz}{\sqrt{2mh - (a^2 + b^2)}} = t - t_0. \end{aligned}$$

The first two of these equations are the equations of the path, showing it to be a straight line, while the last gives the time. By means of it we may find z as a function of the time, and from the first two x and y. Thus 119) are the integral equations of the motion.

Corresponding to this solution, a and b being constants, the surfaces of constant action are parallel planes. The path of any

1) *Natural Philosophy,* § 332.

particle projected normally to one of these planes with the energy (kinetic) h is a straight line normal to these planes, and the velocity is constant.

In order to find solutions suited to surfaces of equal action having other forms, we should require to find other particular solutions of equation 114), which would take us too far into the subject of partial differential equations. Whatever the nature of the surfaces, since the velocity in all the motions considered is constant, the action is proportional to the distance traversed and consequently if we measure off on the normals to a surface of constant action equal distances, the locus of the points thus obtained will be another surface of equal action, or all the surfaces of equal action are so-called *parallel* surfaces.

Next, suppose we have a single particle of mass unity under the action of gravity. Then

$$W = gz,$$

and our equation is

$$H = \frac{1}{2}\left\{\left(\frac{\partial A}{\partial x}\right)^2 + \left(\frac{\partial A}{\partial y}\right)^2 + \left(\frac{\partial A}{\partial z}\right)^2\right\} + gz = h. \tag{120}$$

We may find a solution

$$A = ax + by + \varphi(z),$$

where

$$a^2 + b^2 + [\varphi'^2(z)]^2 + 2(gz - h) = 0,$$

or

$$\varphi(z) = \int \sqrt{2(h - gz) - (a^2 + b^2)}\, dz.$$

Making use of this value we have

$$A = ax + by + \int \sqrt{2(h - gz) - (a^2 + b^2)}\, dz. \tag{121}$$

Equations 107) become

$$\begin{aligned} x' &= \frac{\partial A}{\partial x} = a, \quad y' = \frac{\partial A}{\partial y} = b, \\ z' &= \frac{\partial A}{\partial z} = \sqrt{2(h - gz) - (a^2 + b^2)}, \end{aligned} \tag{122}$$

giving the velocities in terms of the position of the point. These are first integrals of the equations of motion. Equations 108) become

$$\begin{aligned} \frac{\partial A}{\partial a} &= x + \int \frac{-a\,dz}{\sqrt{2(h-gz)-(a^2+b^2)}} = x + \frac{a}{g}\sqrt{2(h - gz) - (a^2 + b^2)} = a_1, \\ \frac{\partial A}{\partial b} &= y + \int \frac{-b\,dz}{\sqrt{2(h-gz)-(a^2+b^2)}} = y + \frac{b}{g}\sqrt{2(h - gz) - (a^2 + b^2)} = b_1. \end{aligned} \tag{123}$$

These are the equations of the path completely integrated, showing that it is a parabola in a vertical plane. The equation 109) is

$$124)\quad \frac{\partial A}{\partial h} = \int \frac{dz}{\sqrt{2(h-gz)-(a^2+b^2)}} = -\frac{1}{g}\sqrt{2(h-gz)-(a^2+b^2)} = t - t_0,$$

giving the time.

From the last equation we may obtain z in terms of t, and, from the two preceding, x and y. Thus the problem is completely solved, the constants, a, b, and h, being determined by the terminal conditions.

Suppose we put $b = 0$, then 121) and 123) give

$$125)\quad A = ax - \frac{1}{3g}\{2(h-gz) - a^2\}^{\frac{3}{2}},$$

$$126)\quad x + \frac{a}{g}\{2(h-gz) - a^2\}^{\frac{1}{2}} = a_1,$$

or

$$(x - a_1)^2 = \frac{a^2}{g^2}\{2(h-gz) - a^2\}.$$

If we consider motions for which a is a constant, but a_1 has different values for the different motions, all the parabolas are obtained from a single one by displacing it horizontally. The curves of constant action,

$$127)\quad \{3g(ax - A)\}^2 = \{2(h-gz) - a^2\}^3,$$

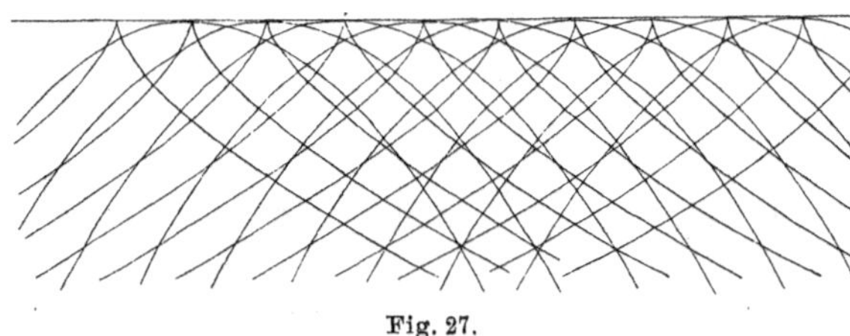
Fig. 27.

are semicubical parabolas, similarly obtained by displacing a single one horizontally, and cut the parabolas at right angles (Fig. 27).

The same solution of the differential equation may be adapted to the treatment of other problems. If we put $a_1 = 0$, x and A will vanish simultaneously, or one of the curves of equal action will be the vertical line $x = 0$. Thus we have a solution of the following problem. Particles are projected horizontally in a vertical plane from points on the same vertical line with such velocities that the total energy is equal to h, the same for all. The different parabolic paths contain the parameter a which changes from one to another. The action along any path contains the same parameter. Eliminating a between equations 125) and 126) we obtain the action as a function of the coordinates x and z, whichever path is described,

$$128) \quad A = \sqrt{\frac{2g}{3}}\left\{\left(\frac{h}{g} - z + x\right)^{\frac{3}{2}} - \left(\frac{h}{g} - z - x\right)^{\frac{3}{2}}\right\}.$$

Putting A equal to a constant we obtain the curves of equal action which cut the parabolic paths orthogonally. This problem is treated in Tait's *Dynamics*, § 219.

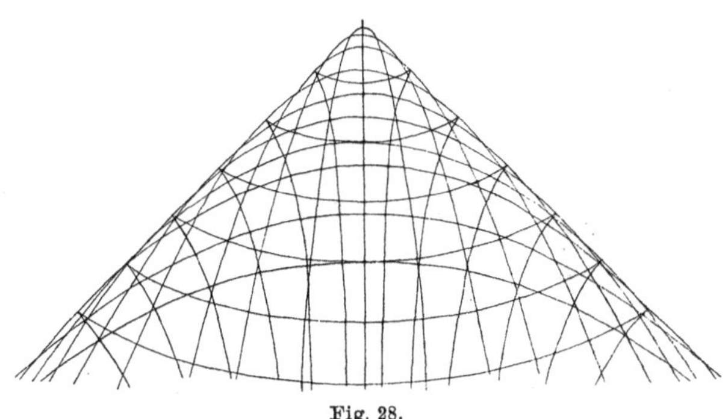

Fig. 28.

This problem is geometrically equivalent to that of streams of water issuing from holes in a vertical side of a tank, for it will be proved in Chapter XI that the velocity of water so issuing varies with the height in precisely the manner above prescribed, the parabolic paths corresponding to the jets of water. It is easy to show that all these parabolas touch a common line making and angle of forty-five degrees with the vertical, and that the curves of equal action have cusps on this line. (Fig. 28.)

As a further example, let us add an arbitrary constant to the value of A (which may always be done) writing,

$$129) \quad A = ax - \frac{1}{3g}\left[\{2(h - gz) - a^2\}^{\frac{3}{2}} - \{2(h - gz_0) - a^2\}^{\frac{3}{2}}\right],$$

$$130) \quad x + \frac{a}{g}\left[\{2(h - gz) - a^2\}^{\frac{1}{2}} - \{2(h - gz_0) - a^2\}^{\frac{1}{2}}\right] = 0.$$

If now $x = 0$, $z = z_0$, A vanishes, thus one of the curves of equal action shrinks to a point. The problem is then that of particles projected in a vertical plane from the same point $(0, z_0)$ with the same velocity. The equations of the various paths and the action corresponding contain the parameter a. Eliminating a we obtain the action in terms of the position,

$$131) \quad A = \frac{1}{3g}\left[\left\{2h - g(z + z_0) + g\sqrt{x^2 + (z - z_0)^2}\right\}^{\frac{3}{2}} - \left\{2h - g(z + z_0) - g\sqrt{x^2 + (z - z_0)^2}\right\}^{\frac{3}{2}}\right].$$

The various paths here treated have a parabolic envelope as described in § 35. The curves of equal action here again have cusps on the envelope, Fig. 25. The hydrodynamical illustration of this problem is a lawn sprinkler or fountain from a ball pierced with holes.

Let us now treat the motion of a planet about the sun, using the coordinates r, ϑ, φ, defined as in § 21. Since r is not constant, we have to use the element of arc

$$132)\qquad ds^2 = dr^2 + r^2 d\vartheta^2 + r^2 \sin^2\vartheta\, d\varphi^2.$$

Form which

$$133)\qquad T = \frac{m}{2}\{r'^2 + r^2\vartheta'^2 + r^2\sin^2\vartheta\cdot\varphi'^2\},$$

also

$$W = -\frac{\gamma M m}{r},$$

by § 28 (γ being taken positive). Let us for simplicity take the mass of the planet as unity and write $\gamma M = k^2$, we have then

$$134)\qquad \begin{aligned} p_r &= \frac{\partial T}{\partial r'} = r' = \frac{\partial A}{\partial r},\\ p_\vartheta &= \frac{\partial T}{\partial \vartheta'} = r^2\vartheta' = \frac{\partial A}{\partial \vartheta},\\ p_\varphi &= \frac{\partial T}{\partial \varphi'} = r^2\sin^2\vartheta\cdot\varphi' = \frac{\partial A}{\partial \varphi}. \end{aligned}$$

Accordingly our differential equation becomes

$$135)\qquad H = \frac{1}{2}\left[\left(\frac{\partial A}{\partial r}\right)^2 + \frac{1}{r^2}\left(\frac{\partial A}{\partial \vartheta}\right)^2 + \frac{1}{r^2\sin^2\vartheta}\left(\frac{\partial A}{\partial \varphi}\right)^2\right] - \frac{k^2}{r} = h.$$

Let us undertake to find a solution in the form

$$136)\qquad A = R(r) + F(\vartheta, \varphi),$$

where the functions R and F contain only the variables indicated. Then

$$\frac{\partial A}{\partial r} = \frac{dR}{dr},\quad \frac{\partial A}{\partial \vartheta} = \frac{\partial F}{\partial \vartheta},\quad \frac{\partial A}{\partial \varphi} = \frac{\partial F}{\partial \varphi}.$$

Substituting in equation 135) we have

$$137)\qquad \frac{1}{2}\left[\left(\frac{dR}{dr}\right)^2 + \frac{1}{r^2}\left(\frac{\partial F}{\partial \vartheta}\right)^2 + \frac{1}{r^2\sin^2\vartheta}\left(\frac{\partial F}{\partial \varphi}\right)^2\right] - \frac{k^2}{r} = h.$$

Multiplying by r^2 and transposing

$$138)\qquad \frac{1}{2}r^2\left(\frac{dR}{dr}\right)^2 - k^2 r - hr^2 = -\frac{1}{2}\left[\left(\frac{\partial F}{\partial \vartheta}\right)^2 + \frac{1}{\sin^2\vartheta}\left(\frac{\partial F}{\partial \varphi}\right)^2\right].$$

On one side of this equation we have functions of r alone, on the other functions of ϑ and φ alone. Since r, ϑ and φ are independent variables this cannot hold identically unless each side reduces to a constant. The partial differential equation thus falls apart into the two

$$139)\qquad \frac{1}{2}r^2\left(\frac{dR}{dr}\right)^2 - k^2 r - hr^2 = -\beta,$$

$$140)\qquad \left(\frac{\partial F^2}{\partial \vartheta}\right) + \frac{1}{\sin^2\vartheta}\left(\frac{\partial F}{\partial \varphi}\right)^2 = 2\beta.$$

The first of these is an ordinary differential equation for R giving

$$141)\qquad R=\int\sqrt{\frac{2k^2}{r}-\frac{2\beta}{r^2}+2h}\cdot dr.$$

The equation 140) is to be treated in a similar manner writing

$$142)\qquad F=\Theta(\vartheta)+\Phi(\varphi).$$

Proceeding as before, multiplying by $\sin^2\vartheta$ and transposing,

$$143)\qquad \sin^2\vartheta\left(\frac{d\Theta}{d\vartheta}\right)^2-2\beta\sin^2\vartheta=-\left(\frac{d\Phi}{d\varphi}\right)^2.$$

From which, as before, we must have

$$144)\qquad \left(\frac{d\Phi}{d\varphi}\right)^2=\gamma,$$

$$145)\qquad \left(\frac{d\Theta}{d\vartheta}\right)^2-2\beta+\frac{\gamma}{\sin^2\vartheta}=0,$$

whose integrals are $\Phi=\sqrt{\gamma}\cdot\varphi$,

$$146)\qquad \Theta=\int\sqrt{2\beta-\frac{\gamma}{\sin^2\vartheta}}\cdot d\vartheta.$$

Substituting the values of R, Θ, Φ,

$$147)\qquad A=\int\sqrt{\frac{2k^2}{r}-\frac{2\beta}{r^2}+2h}\cdot dr+\int\sqrt{2\beta-\frac{\gamma}{\sin^2\vartheta}}\cdot d\vartheta+\sqrt{\gamma}\cdot\varphi.$$

This solution contains the three arbitrary constants, β, γ, h. Differentiating by them we obtain the integrals

$$148)\qquad \begin{aligned}\frac{\partial A}{\partial\beta}&=-\int\frac{dr}{r^2\sqrt{\frac{2k^2}{r}-\frac{2\beta}{r^2}+2h}}+\int\frac{d\vartheta}{\sqrt{2\beta-\frac{2\gamma}{\sin^2\vartheta}}}=\beta',\\ \frac{\partial A}{\partial\gamma}&=-\int\frac{d\vartheta}{2\sqrt{2\beta\sin^4\vartheta-\gamma\sin^2\vartheta}}+\frac{\varphi}{2\sqrt{\gamma}}=\gamma',\\ \frac{\partial A}{\partial h}&=\int\frac{dr}{\sqrt{\frac{2k^2}{r}-\frac{2\beta}{r^2}+2h}}=t-t_0.\end{aligned}$$

If we put $\gamma=0$, necessitating according to the second equation $\varphi=0$, the first equation becomes

$$149)\qquad \frac{\vartheta}{\sqrt{2\beta}}+\int\frac{d\left(\frac{1}{r}\right)}{\sqrt{2h+\frac{2k^2}{r}-\frac{2\beta}{r^2}}}=\beta',$$

the equation of the path, which, on performing the integration indicated, takes the form obtained in § 20 equation 23).

CHAPTER V.

OSCILLATIONS AND CYCLIC MOTIONS.

42. Tautochrone for Gravity. A curve along which a particle will descend under the action of gravity to a fixed point from a variable point in the same time is called a *tautochrone curve.* If the particle is dropped from rest we have the equation of energy

1) $$v^2 = \left(\frac{ds}{dt}\right)^2 = 2g(z_0 - z),$$

and the time of falling to the level $z = 0$ is

2) $$t = \frac{1}{\sqrt{2g}}\int_0^{z_0}\frac{ds}{\sqrt{z_0 - z}}.$$

Let the length of the arc s measured from the fixed point be $\varphi(z)$, then

3) $$t = \frac{1}{\sqrt{2g}}\int_0^{z_0}\frac{\varphi'(z)\,dz}{\sqrt{z_0 - z}}.$$

If the curve is to be a tautochrone this must be independent of z_0 or

$$\frac{dt}{dz_0} = 0.$$

Let us change the variable by putting $z = z_0 u$, then

$$t = \frac{1}{\sqrt{2g}}\int_0^1\frac{\sqrt{z_0}\,\varphi'(z_0 u)\,du}{\sqrt{1-u}},$$

$$\frac{dt}{dz_0} = \frac{1}{\sqrt{2g}}\int_0^1\frac{du}{2\sqrt{z_0(1-u)}}\{\varphi'(z_0 u) + 2z_0 u\,\varphi''(z_0 u)\},$$

or changing the variable back to z,

$$\frac{dt}{dz_0} = \frac{1}{\sqrt{2g}}\int_0^{z_0}\frac{dz}{2z_0\sqrt{z_0 - z}}\{\varphi'(z) + 2z\varphi''(z)\}.$$

If this is to vanish for all values of the limit z_0 the integrand must vanish, or

4) $$\varphi'(z) + 2z\varphi''(z) = 0,$$

which is the differential equation of the curve. Writing this

$$\frac{\varphi''(z)}{\varphi'(z)} = -\frac{1}{2z},$$

we may integrate, obtaining

$$\log \varphi'(z) = -\frac{1}{2} \log z + const.$$

Taking the antilogarithm,

5) $$\varphi'(z) = \frac{c}{\sqrt{z}} = \frac{ds}{dz},$$

since $s = \varphi(z)$. Integrating again,

6) $$s = 2c\sqrt{z} + d,$$

where c and d are arbitrary constants. This is the equation of the curve. In order to recognize its nature let us square equation 5) writing

7) $$\left(\frac{ds}{dz}\right)^2 = \left(\frac{dx}{dz}\right)^2 + 1 = \frac{2a}{z},$$

where $a = \frac{c^2}{2}$ is an arbitrary constant. Solving for $\frac{dz}{dx}$,

8) $$\frac{dz}{dx} = \sqrt{\frac{z}{2a - z}}.$$

If we put $z = a(1 - \cos\vartheta)$ this becomes

9) $$\frac{dz}{dx} = a \sin\vartheta \frac{d\vartheta}{dx} = \sqrt{\frac{1 - \cos\vartheta}{1 + \cos\vartheta}} = \frac{\sin\vartheta}{1 + \cos\vartheta},$$

from which

$$a(1 + \cos\vartheta)\, d\vartheta = dx.$$

Integrating,

$$x = a(\vartheta + \sin\vartheta) + const.,$$

or, if x and z vanish together,

10) $$x = a(\vartheta + \sin\vartheta),$$
$$z = a(1 - \cos\vartheta).$$

These are the equations of a cycloid, § 29 p. 83, accordingly the cycloid is not only a brachistochrone for gravity, but also a tautochrone. For a particular cycloid the time of descent is by 2) and 7)

11) $$T = \sqrt{\frac{a}{g}} \int_0^{z_0} \frac{dz}{\sqrt{z_0 z - z^2}},$$

or putting $z = z_0 u$,

12) $$T = \sqrt{\frac{a}{g}} \int_0^1 \frac{du}{\sqrt{u - u^2}}.$$

Putting $u = \sin^2\vartheta$, we easily obtain

13) $$T = \pi \sqrt{\frac{a}{g}}.$$

If the particle be allowed to ascend after passing the lowest point of the cycloid it will rise to the same height from which it fell and the motion being repeated, the time of a complete to- and fro-oscillation is

$$4T = 2\pi\sqrt{\frac{4a}{g}}.$$

Thus the time of an oscillation of whatever amplitude on a cycloid is the same as that of the infinitesimal oscillation of a circular pendulum of length $4a$ or twice the diameter of the rolling circle which generates the cycloid. Since the time of oscillation is independent of its amplitude we are led to the question of whether the motion is harmonic.

We may more generally inquire whether an isochronous or tautochronous vibration is necessarily harmonic, that is: Is the elongation of a particle, performing a vibration whose period is independent of the amplitude, necessarily represented by a sine or cosine function of the time?

Let the distance along the path from the point to which the motion is tautochronous be s. Then if the system is conservative the force will be a function of s. Suppose

$$14)\qquad \frac{d^2 s}{dt^2} = f(s) = \frac{dF(s)}{ds}.$$

Multiplying by $\frac{ds}{dt}$ and integrating we obtain the equation of energy

$$15)\qquad \frac{1}{2}\left(\frac{ds}{dt}\right)^2 = \int_{s_0}^{s} f(s)\,ds = F(s) - F(s_0),$$

where s_0 is the initial value from which the particle started from rest. The time of the motion from $s = s_0$ to $s = 0$ will be

$$16)\qquad t = \int_{s_0}^{0} \frac{ds}{\sqrt{2[F(s) - F(s_0)]}},$$

or putting $s = s_0 u$,

$$t = \frac{1}{\sqrt{2}}\int_{1}^{0} \frac{s_0\,du}{\sqrt{F(s_0 u) - F(s_0)}}.$$

Differentiating by s_0,

$$17)\qquad \frac{dt}{ds_0} = \frac{1}{\sqrt{2}}\int_{1}^{0} \frac{du}{F(s_0 u) - F(s_0)}\left\{\sqrt{F(s_0 u) - F(s_0)} - \frac{s_0[uF'(s_0 u) - F'(s_0)]}{2\sqrt{F(s_0 u) - F(s_0)}}\right\}$$

$$= \frac{1}{\sqrt{2}}\int_{s_0}^{0} \frac{ds}{s_0[F(s) - F(s_0)]}\left\{\sqrt{F(s) - F(s_0)} - \frac{sF'(s) - s_0 F'(s_0)}{2\sqrt{F(s) - F(s_0)}}\right\}.$$

Since this is to vanish for all values of the limit s_0 we must have

$$\sqrt{F(s) - F(s_0)} - \frac{sF'(s) - s_0 F'(s_0)}{2\sqrt{F(s) - F(s_0)}} = 0,$$

or

18) $$2[F(s) - F(s_0)] - sF'(s) + s_0 F'(s_0) = 0,$$

which is a differential equation for $F(s)$.

Let us put $y = F(s)$,

$$2F(s_0) - s_0 F'(s_0) = c,$$

then 18) becomes

19) $$s\frac{dy}{ds} - 2y + c = 0.$$

a linear equation of the first order. An integrating factor is $\frac{1}{s^3}$, multiplying by which the equation becomes

$$d\left(\frac{y}{s^2}\right) + c\frac{ds}{s^3} = 0.$$

Integrating,

$$\frac{y}{s^2} - \frac{c}{2s^2} = b,$$

20) $$y = bs^2 + \frac{c}{2} = F(s).$$

From this we obtain

$$f(s) = \frac{dF}{ds} = 2bs,$$

so that the equation of tautochronous motion 14) must be

21) $$\frac{d^2s}{dt^2} = 2bs,$$

accordingly the motion must be harmonic, and evidently b must be negative.

We have seen that the cycloid is a tautochronous curve and that a tautochronous vibration must be harmonic.

By equation 6) the length of the arc of the cycloid measured from $z = 0$ is

22) $$s = 2\sqrt{2az},$$

from which

$$z = \frac{s^2}{8a},$$

or inserting in equation 1),

23) $$\frac{ds}{dt} = \sqrt{2g\left(z_0 - \frac{s^2}{8a}\right)}.$$

Differentiating,

24) $$\frac{d^2s}{dt^2} = -\frac{gs}{4a\sqrt{2g\left(z_0 - \frac{s^2}{8a}\right)}}\frac{ds}{dt} = -\frac{gs}{4a}.$$

Thus the differential equation of the cycloidal motion is

$$\frac{d^2 s}{d t^2} + \frac{g}{4a} s = 0, \tag{25}$$

showing that the motion is harmonic with the period $2\pi\sqrt{\frac{4a}{g}}$ as found above. The cycloid is isochronous for all arcs, the circle only for infinitely small arcs. The circle having the same curvature as the cycloid at its vertex is less steep than the cycloid (Fig. 29) and therefore the time of descent on the circle is greater for larger arcs, as shown in § 22. The evolute or envelope of the normals of a cycloid is an equal cycloid, hence the cycloidal pendulum may be realized. If two material half-cycloids be constructed tangent at O (Fig. 30), where the string is attached, and the string be allowed to wind itself against them, if its length is that of the half-cycloid, its end will describe a cycloid. This pendulum was constructed by Huygens.[1])

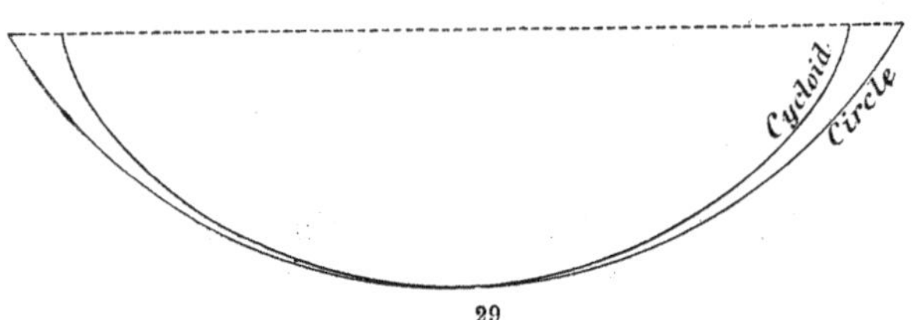

29

The length being $4a$ agrees with the above. On account of the motion on the cycloid being harmonic Thomson and Tait call harmonic motions *cycloidal.*

43. Damped Oscillations. Let us now consider a particle under the influence of a force proportional to its displacement from a certain point and directed toward the position of equilibrium, the motion being resisted by a non-conservative force proportional to the first power of the velocity. Calling the accelerations produced by the *positional* (conservative) force $-h^2 s$, and the *motional* (non-conservative) force

$$-\varkappa \frac{ds}{dt},$$

the equation of motion is

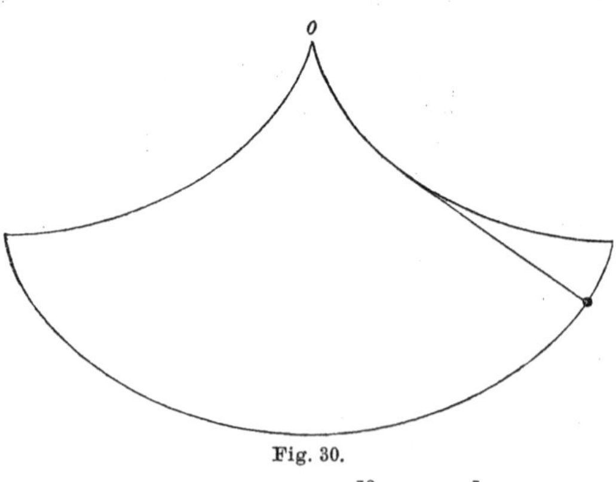

Fig. 30.

$$\frac{d^2 s}{dt^2} + \varkappa \frac{ds}{dt} + h^2 s = 0. \tag{26}$$

1) Huygens, *Horologium oscillatorium,* Paris 1673.

This equation is linear with constant coefficients, and is a type of those that appear in the theory of oscillations. The fundamental property of such equations is that any solution multiplied by a constant is a solution, and that the sum of two solutions is a solution. In order to find a particular solution we put

$$s = e^{\lambda t}.$$

Differentiating we have

$$\frac{ds}{dt} = \lambda e^{\lambda t}, \quad \frac{d^2 s}{dt^2} = \lambda^2 e^{\lambda t}.$$

Substituting in the differential equation we may divide out the factor $e^{\lambda t}$, obtaining

27) $$\lambda^2 + \varkappa\lambda + h^2 = 0,$$

a quadratic to determine the constant λ.

Calling its roots λ_1, λ_2 we have

28) $$\lambda_1 = -\frac{\varkappa}{2} + \frac{1}{2}\sqrt{\varkappa^2 - 4h^2}, \quad \lambda_2 = -\frac{\varkappa}{2} - \frac{1}{2}\sqrt{\varkappa^2 - 4h^2}.$$

The general solution is obtained by multiplying the particular solutions $e^{\lambda_1 t}$ and $e^{\lambda_2 t}$ by arbitrary constants and adding. Thus we obtain

29) $$s = Ae^{\frac{1}{2}\left(-\varkappa + \sqrt{\varkappa^2 - 4h^2}\right)t} + Be^{\frac{1}{2}\left(-\varkappa - \sqrt{\varkappa^2 - 4h^2}\right)t}$$

We have to consider two cases,

$$\text{I.} \quad \varkappa^2 > 4h^2,$$

$$\text{II.} \quad \varkappa^2 < 4h^2.$$

In case I the radical is real, and since its absolute value is less than $\varkappa$ both λ_1 and λ_2 are negative and s eventually decreases as the time goes on, vanishing when $t = \infty$. We have

$$\frac{ds}{dt} = \lambda_1 Ae^{\lambda_1 t} + \lambda_2 Be^{\lambda_2 t}.$$

This vanishes when

$$\frac{-\lambda_2 B}{\lambda_1 A} = e^{(\lambda_1 - \lambda_2)t}$$

or

30) $$t = \frac{1}{\lambda_1 - \lambda_2} \log\left(\frac{-\lambda_2 B}{\lambda_1 A}\right).$$

Consequently if B and A are of opposite signs s will increase to a maximum and then continually die away. If they are of the same sign the motion dies away from the start. Both cases are shown in Fig. 31, where t is the abscissa and s the coordinate.

In case II the radical is imaginary and both λ_1 and λ_2 are complex. Then writing

$$\lambda_1 = \mu + i\nu, \quad \lambda_2 = \mu - i\nu,$$

and making use of the fundamental formula of imaginaries,

31) $$e^{i\nu t} = \cos \nu t + i \sin \nu t,$$

and the principle that both the real part and the coefficient of i in the imaginary part of a solution are particular solutions, we obtain the two particular solutions

$$e^{\mu t} \cos \nu t \quad \text{and} \quad e^{\mu t} \sin \nu t.$$

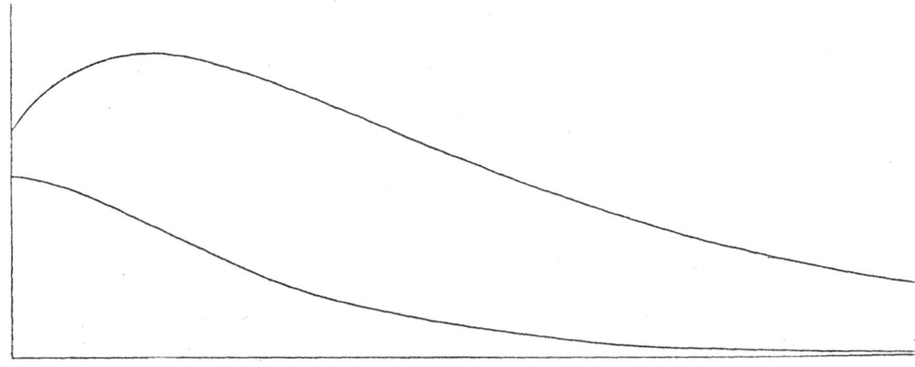

Fig. 31.

We thus obtain the general solution

32) $$s = e^{\mu t}(A \cos \nu t + B \sin \nu t)$$

$$= e^{-\frac{\varkappa}{2}t}\left(A \cos \frac{1}{2}\sqrt{4h^2 - \varkappa^2} \cdot t + B \sin \frac{1}{2}\sqrt{4h^2 - \varkappa^2} \cdot t\right)$$

(A and B being *new* arbitrary constants),

or as in § 19 equation 10),

33) $$s = a e^{-\frac{\varkappa}{2}t} \cos\left(\frac{1}{2}\sqrt{4h^2 - \varkappa^2} \cdot t - \alpha\right).$$

The trigonometric factor represents a simple harmonic oscillation, which on account of the continually decreasing exponential factor dies away as the time increases (Fig. 32). Such a motion is called a *damped* oscillation, and $\varkappa$ is a measure of the amount of damping.

The extreme elongation occurs when

34) $$\frac{ds}{dt} = a e^{-\frac{\varkappa}{2}t}\left\{-\frac{1}{2}\sqrt{4h^2 - \varkappa^2} \cdot \sin\left(\frac{1}{2}\sqrt{4h^2 - \varkappa^2} \cdot t - \alpha\right) - \frac{\varkappa}{2} \cos\left(\frac{1}{2}\sqrt{4h^2 - \varkappa^2} \cdot t - \alpha\right)\right\} = 0,$$

that is when

35) $$\tan\left(\frac{1}{2}\sqrt{4h^2 - \varkappa^2} \cdot t - \alpha\right) = -\frac{\varkappa}{\sqrt{4h^2 - \varkappa^2}}.$$

The smaller the damping $\varkappa$, the more nearly does the time of the maximum coincide with that of the maximum of the cosine factor in 33). In any case successive maxima follow each other at intervals equal to the period of the oscillation,

36) $$T = \frac{2\pi}{\nu} = \frac{4\pi}{\sqrt{4h^2 - \varkappa^2}}.$$

At two successive maxima on the same side, s_1 and s_2, the cosine term will have the same value, therefore the ratio of the elongations will be that of the exponential factors, or

$$\frac{s_1}{s_2} = e^{\frac{\varkappa}{2}T}.$$

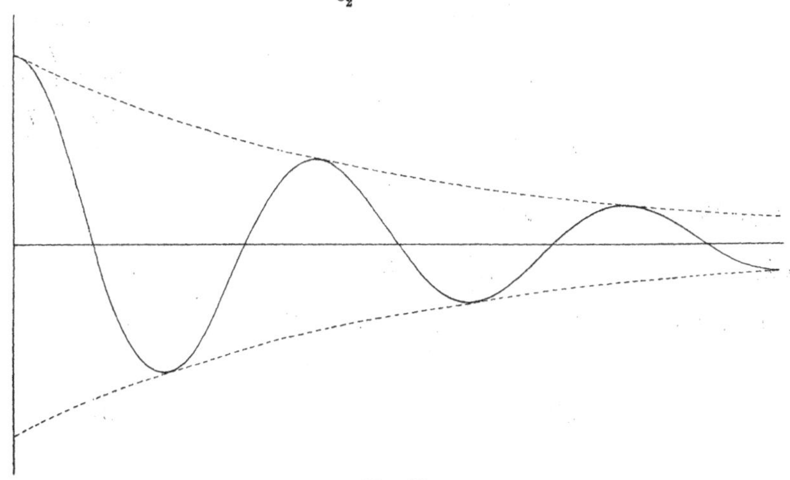

Fig. 32.

The logarithm of the ratio,

37) $$\delta = \log\frac{s_1}{s_2} = \frac{1}{2}\varkappa T = \frac{2\pi}{\sqrt{\frac{4h^2}{\varkappa^2} - 1}},$$

is accordingly constant, and by means of observations on the logarithmic decrement we may determine the damping. We see that the decrement depends on and increases with the ratio of the square of the coefficient of damping $\varkappa$ to the coefficient of "*stiffness*" h^2.

If there were no damping, $\varkappa = 0$, we should have for the period,

$$T_0 = \frac{2\pi}{h}.$$

Introducing these values of T_0 and δ, we may write

38) $$T = T_0\sqrt{1 + \frac{\delta^2}{4\pi^2}} = T_0\left(1 + \frac{\delta^2}{8\pi^2} + \cdots\right),$$

so that if the damping is small, as is usually the case, it affects the period only by small quantities of the second order.

As has been shown in § 38 we have here an instance of the use of a dissipation function

39) $$F = \frac{1}{2} \varkappa \left(\frac{ds}{dt}\right)^2,$$

and the energy is dissipated at a rate proportional to the exponential $e^{-\varkappa t}$.

44. Forced Vibrations. Resonance. The motion considered in the last section being that of a system left to itself is called a free oscillation or vibration. We shall now consider a problem of a different sort from any yet treated and involving a force depending upon the time, and thus introducing or withdrawing energy from the system. Let us suppose a particle to be subject to the same conditions as above, but in addition to be acted upon by an extraneous force varying according to a harmonic function of the time,

40) $$F = E \cos pt,$$

so that the differential equation of motion is

41) $$\frac{d^2 s}{dt^2} + \varkappa \frac{ds}{dt} + h^2 s = E \cos pt.$$

We may find a particular solution by putting

$$s = a \cos (pt - \alpha),$$

42) $$\frac{ds}{dt} = - ap \sin (pt - \alpha), \quad \frac{d^2 s}{dt^2} = - ap^2 \cos (pt - \alpha).$$

Substituting in the differential equation, we have

43) $$a(h^2 - p^2) \cos (pt - \alpha) - a\varkappa p \sin (pt - \alpha) = E \cos pt$$
$$= E\{\cos \alpha \cos (pt - \alpha) - \sin \alpha \sin (pt - \alpha)\}.$$

This can be identically true for all values of t only if the coefficients of the sine and cosine of the variable angle $(pt - \alpha)$ are respectively equal on both sides of the equation, accordingly we must have

44) $$a\varkappa p = E \sin \alpha,$$
$$a(h^2 - p^2) = E \cos \alpha,$$

from which eliminating first E and then α,

45) $$\tan \alpha = \frac{\varkappa p}{h^2 - p^2}.$$
$$E = a \sqrt{(h^2 - p^2)^2 + \varkappa^2 p^2},$$

from which we obtain the amplitude

46) $$a = \frac{E}{\sqrt{(h^2 - p^2)^2 + \varkappa^2 p^2}}.$$

Thus our solution is

$$47)\qquad s = \frac{E}{\sqrt{(h^2-p^2)^2+\varkappa^2p^2}}\cos(pt-\alpha).$$

The motion represented by this solution is called the *forced* vibration, for the system is forced to assume the same period as that of the extraneous force F, namely $\frac{2\pi}{p}$, of frequency $\frac{p}{2\pi}$, while the frequency of the *free* or natural vibration would be $\frac{1}{2\pi}\sqrt{h^2-\frac{\varkappa^2}{4}}$, or without damping $\frac{h}{2\pi}$.

The displacement is not in phase with the force, lagging behind it by less than a quarter-period if $\tan\alpha$ is positive, that is, if h is greater than p, in other words if the natural frequency is greater than the forced. If on the contrary the natural frequency is less than the forced, $\tan\alpha$ is negative, and since $\sin\alpha$ is positive, the displacement is between a quarter and a half-period behind the force. If the frequencies of the forced and free vibrations coincide, $\tan\alpha$ becomes infinite, the lag is a quarter period, so that the displacement is a maximum when the force is zero and vice versa. Then 47) becomes

$$48)\qquad s = \frac{E}{p\varkappa}\sin ht,$$

and if the damping $\varkappa$ is small, the amplitude is very large. This is the case in the phenomenon of *resonance*, of great importance in various parts of physics, including acoustics, electricity, and dispersion in optics. The equation shows how a very small force may produce a very large vibration if the period coincides nearly enough with the natural one, and explains the danger to bridges from the accumulated effect of the measured step of soldiers, the heavy rolling of ships caused by waves of proper period, and kindred phenomena

Although in the phenomenon of resonance the excursion and consequently the kinetic energy becomes very large, it is of course not to be supposed that this energy comes from nothing as has been frequently contended by inventive charlatans proposing to obtain vast stores of energy from sound vibrations.[1])

If we form the equation of activity, by multiplying 41) by $\frac{ds}{dt}$,

$$49)\qquad \frac{d(T+W)}{dt}+\varkappa\left(\frac{ds}{dt}\right)^2 = \frac{1}{2}\frac{d}{dt}\left\{\left(\frac{ds}{dt}\right)^2+h^2s^2\right\}+\varkappa\left(\frac{ds}{dt}\right)^2 = F\frac{ds}{dt}$$

$$= \frac{-E^2p}{\sqrt{(h^2-p^2)^2+\varkappa^2p^2}}\sin(pt-\alpha)\cos pt$$

$$= \frac{E^2p}{\sqrt{(h^2-p^2)^2+\varkappa^2p^2}}(-\cos\alpha\sin pt\cos pt+\sin\alpha\cos^2 pt),$$

1) Of these the United States has produced more than its share. The ignorance of the above mentioned principle enabled John Keely to abstract in the neighborhood of a million dollars from intelligent (!) American shareholders.

we see that energy is being alternately introduced into and withdrawn from the system by the extraneous force. On the average however, as we find by integrating the trigonometric terms with respect to the time,

$$\int_0^T \sin pt \cos pt\, dt = 0,$$

$$\int_0^T \cos^2 pt\, dt = \frac{\pi}{p}, \quad T = \frac{2\pi}{p},$$

the time average of the activity depends upon the last term containing $\sin\alpha$, and this is always positive, consequently the extraneous force is on the whole continually doing work on the system, which is being dissipated at the rate $\varkappa\left(\frac{ds}{dt}\right)^2$. This work is a maximum when $\alpha = \frac{\pi}{2}$, when the system is in complete resonance. Thus the mechanical effects producible by resonance are shown to be commensurate with the causes acting, and the impossibility of the common story of the fiddler fiddling down a bridge is demonstrated.

The exactness of "tuning", or approach to exact coincidence of period necessary for resonance is shown in Fig. 33, which is the graph of the curve

$$y = \frac{1}{\sqrt{(1-x^2)^2 + \alpha^2 x^2}},$$

where $y = a\Big/\frac{E}{h^2}$ is the ratio of the actual amplitude of equation 46) to the steady statical displacement $\frac{E}{h^2}$ produced by a constant force E (that is when $p = 0$), $x = \frac{p}{h}$ is the ratio of the frequencies of forced and free vibration, and $\alpha^2 = \frac{\varkappa^2}{h^2}$.[1]) Thus curves are drawn for values of the parameter α^2 equal to ·01, ·05, ·10, ·15, ·20. Thus the magnitude of the resonance for any particular case can be seen by a glance at the figure. The resonance is sharper the smaller α. The maximum amplitude is not for *perfect* tuning, but for $x = \sqrt{1 - \frac{\alpha^2}{2}}$. The value of the maximum is nearly equal to $\frac{1}{\alpha}$.

If there is no friction, for $p = h$ the vibration becomes infinite, which means simply that in this case friction *must* be taken into account. If there is no friction we have by 44),

$$\sin\alpha = 0, \quad \cos\alpha = 1$$

50)
$$s = \frac{E}{h^2 - p^2}\cos pt,$$

1) This parameter α is not the *angle* α above.

and the displacement is in the same or opposite phase with the force, according as h is greater than or less than p. In the latter case the excursion is a maximum in one direction when the force is exerting a maximum pull in the opposite direction. This need not appear paradoxical, for consider the limiting case of a system with very little stiffness in proportion to its inertia, that is h very small and the natural period very great. Then the excursion is always opposite in phase to the force on account of the inertia of the system. In the opposite case of a system with very little inertia in proportion to the stiffness, h is very large, and the excursion is in phase with the force. In this case (that of complete agreement) we have what is called the *equilibrium theory* of oscillation, the displacement being the same as if the problem were one in statics $\left(s = \frac{F}{h^2}\right)$, except that the force and displacement are varying together. Such a theory was given by Newton for the tides, which consist of a forced vibration of the water covering the earth under the periodic force due to the moon's attraction. The more accurate theory taking account of inertia was given by Lagrange. The relation of the dynamical to the equilibrium theory is shown in Fig. 33.

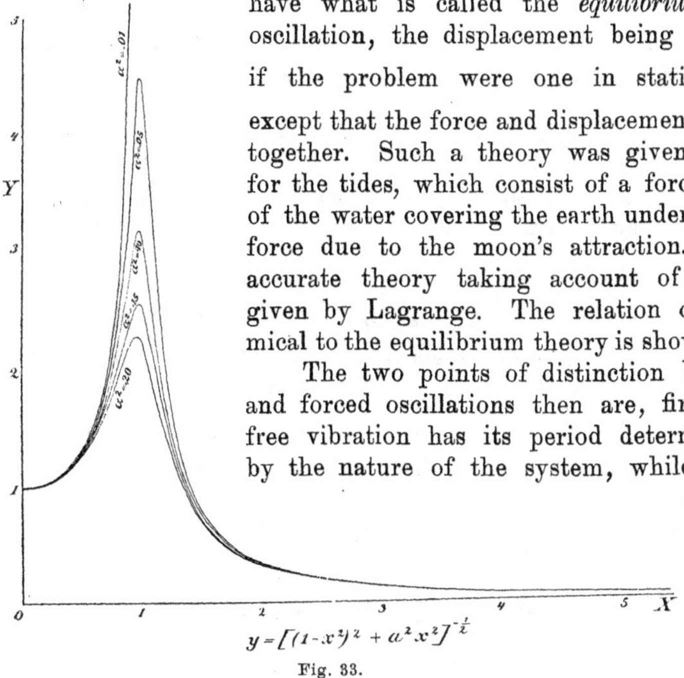

Fig. 33.

The two points of distinction between free and forced oscillations then are, first, that the free vibration has its period determined solely by the nature of the system, while the forced vibration takes the period of the force, and secondly, that if there is damping, the free vibration dies away, while the forced vibration persists unchanged.

The theory of the forced vibration which we have given does not take account of the gradual production of the motion from a state of rest, but refers only to the motion after the steady state has been reached. We may now complete the treatment and take account of the motion at the start. Our previous solution is merely a particular solution. According to the theory of linear differential equations in order to obtain the general solution we must add to the particular solution just obtained the solution of the equation 41) when the second member is equal to zero, or in physical terms the forced and free vibrations exist superimposed. Accordingly we have

$$51) \quad s = \frac{E}{\sqrt{(h^2 - p^2)^2 + \varkappa^2 p^2}} \cos(pt - \alpha) + Ae^{-\frac{1}{2}\varkappa t} \cos\left(\sqrt{h^2 - \frac{\varkappa^2}{4}} \cdot t - \beta\right).$$

If the system starts from rest we must determine A and β so that when $t = 0$, s and $\frac{ds}{dt}$ are equal to zero. These conditions will be very nearly satisfied, if p and h are nearly equal and $\varkappa$ small, by

$$52) \qquad s = a\left\{\cos(pt - \alpha) - e^{-\frac{1}{2}\varkappa t} \cos\left(\sqrt{h^2 - \frac{\varkappa^2}{4}} \cdot t - \alpha\right)\right\}.$$

The simultaneous existence of two harmonic vibrations of nearly equal frequencies gives rise to the phenomenon known as *beats.* Suppose

$$53) \qquad s = a\cos(pt - \alpha) + b\cos\{(p + \varDelta p)t - \beta\},$$

where $\varDelta p$ is a small quantity, equal to 2π times the difference of frequencies. We may write the last term

$$\begin{aligned} &b\cos\{(pt - \alpha) + \varDelta p \cdot t + \alpha - \beta\} \\ &= b\{\cos(pt - \alpha)\cos(\varDelta p \cdot t + \alpha - \beta) \\ &\quad - \sin(pt - \alpha)\sin(\varDelta p \cdot t + \alpha - \beta)\}, \end{aligned}$$

so that

$$54) \qquad \begin{aligned} s = \{a + b\cos(\varDelta p \cdot t + \alpha - \beta)\}\cos(pt - \alpha) \\ - b\sin(\varDelta p \cdot t + \alpha - \beta)\sin(pt - \alpha), \end{aligned}$$

or if we write

$$\begin{aligned} a + b\cos(\varDelta p \cdot t + \alpha - \beta) &= A \\ - b\sin(\varDelta p \cdot t + \alpha - \beta) &= B, \end{aligned}$$

$$s = A\cos(pt - \alpha) + B\sin(pt - \alpha),$$

$$55) \qquad s = D\cos(pt - \alpha - \varepsilon),$$

where

$$\varepsilon = \tan^{-1}\frac{B}{A},$$

and

$$D = \sqrt{A^2 + B^2} = \{a^2 + b^2 + 2ab\cos(\varDelta p \cdot t + \alpha - \beta)\}^{\frac{1}{2}}.$$

Accordingly the compound vibration may be considered as a harmonic motion of variable amplitude and phase, the amplitude varying from $a + b$ to $a - b$, with the period $\frac{2\pi}{\varDelta p}$ and frequency $\frac{\varDelta p}{2\pi}$ equal to the difference of the frequencies of the two constituents. The phenomenon of beats or interferences is represented graphically in Fig. 34.

In the case of free and forced vibrations coexisting [equation 52)], we have at the beginning beats which gradually die away owing to the factor $e^{-\frac{1}{2}\varkappa t}$ in the free vibration, leaving only the forced vibration.

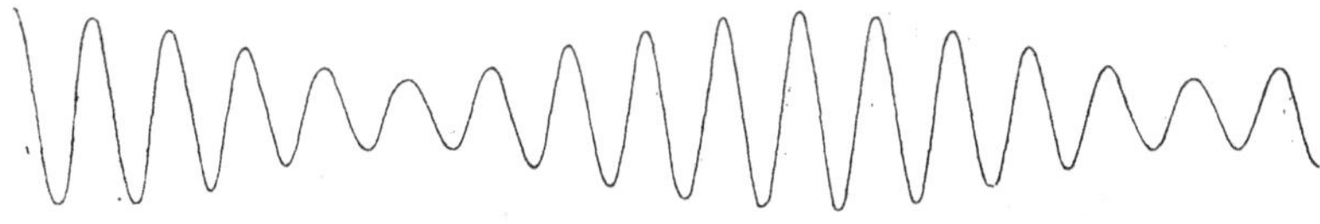

Fig. 34.

This is shown in an interesting manner by a tuning fork electrically excited by another fork not quite in unison with it, the phenomenon of a single driven fork apparently producing beats with itself being very striking (Fig. 35). It will be noticed that the first maximum is greater than the steady amplitude.

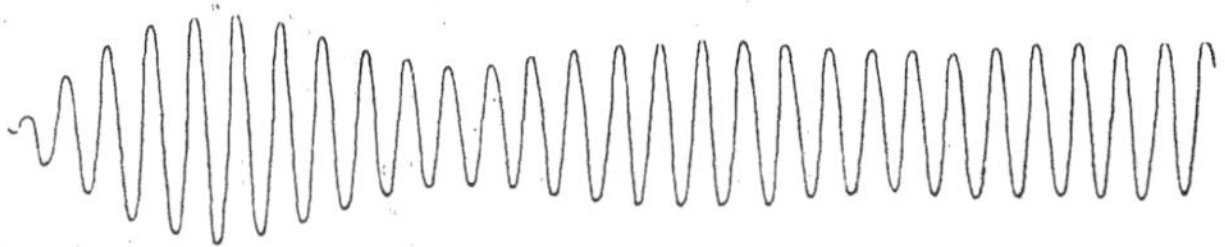

Fig. 35.

The greater part of this section and the preceding is taken from Rayleigh's *Theory of Sound.*

45. General Theory of small Oscillations. Having now set forth the general characteristics of vibrations excuted by systems possessing one degree of freedom, we will now treat the problem of the small vibrations of any system about a configuration of equilibrium after the manner of Lagrange, wo first investigated it.

Suppose a system is defined by n parameters $q_1, q_2, \ldots q_n$. Its potential energy will depend only on the coordinates q, and developing by Taylor's Theorem,

$$56)\qquad W = W_0 + \sum_{r=1}^{r=n} q_r \left(\frac{\partial W}{\partial q_r}\right)_0 + \frac{1}{2}\sum_{r=1}^{r=n}\sum_{s=1}^{s=n} q_r q_s \left(\frac{\partial^2 W}{\partial q_r \partial q_s}\right)_0 + \cdots,$$

where the suffix zero denotes the value when all the q's are zero. Suppose that this is a configuration of equilibrium, then W is a minimum or maximum and every $\left(\frac{\partial W}{\partial q_r}\right)_0$ equals zero. Thus $W - W_0$ begins with a quadratic function of the q's. If the motion is small enough we may neglect the terms of higher orders of small quantities. Accordingly, neglecting the constant W_0 (for the potential energy always contains an arbitrary constant which does not affect the motion), we shall put W a homogeneous quadratic function of the q's with *constant* coefficients,

$$57)\qquad W = \frac{1}{2}\sum_{r=1}^{r=n}\sum_{s=1}^{s=n} c_{rs} q_r q_s.$$

If the equilibrium is stable the potential energy must be a minimum so that the constants c_{rs} will be such that the quadratic function W is positive for all possible values of the variables q.

The kinetic energy will be a quadratic function of the time derivatives, $q_1', q_2', \dots q_n'$,

$$T = \frac{1}{2} \sum_{r=1}^{r=n} \sum_{s=1}^{s=n} a_{rs} q_r' q_s', \tag{58}$$

where the a's are functions of the coordinates q alone. We may develop the functions a_{rs} in series, thus one term of the sum becomes

$$a_{rs} q_r' q_s' = q_r' q_s' \left\{ a_{rs}^0 + \sum_{t=1}^{t=n} q_t \left(\frac{\partial a_{rs}}{\partial q_t} \right)_0 + \cdots \right\}, \tag{59}$$

and since the velocities q' are small at the same time as the coordinates q, we may neglect all the terms within the braces except that of lowest order a_{rs}^0, therefore we may consider the a's as constants. If we have besides the conservative forces of restitution, arising from the potential energy W, non-conservative resistances which are linear functions of the velocities, we may make use of a dissipation function F, § 39, such that the dissipative force corresponding to the coordinate q_r will be $-\frac{\partial F}{\partial q_r'}$. We thus have the three homogeneous quadratic functions with constant coefficients,

$$\begin{aligned} T &= \frac{1}{2} \sum_{r=1}^{r=n} \sum_{s=1}^{s=n} a_{rs} q_r' q_s', \\ F &= \frac{1}{2} \sum_{r=1}^{r=n} \sum_{s=1}^{s=n} \varkappa_{rs} q_r' q_s', \\ W &= \frac{1}{2} \sum_{r=1}^{r=n} \sum_{s=1}^{s=n} c_{rs} q_r q_s. \end{aligned} \tag{60}$$

Each of these has the property of being positive for all possible values of the variables of which it is a function. The a's may be called coefficients of inertia, the c's, coefficients of stiffness, and the $\varkappa$'s, coefficients of viscosity or resistance. We may now form Lagrange's equations for any coordinate q_r.

$$\frac{dp_r}{dt} - \frac{\partial T}{\partial q_r} = P_r = -\frac{\partial W}{\partial q_r} - \frac{\partial F}{\partial q_r'}, \tag{61}$$

where

$$\frac{\partial T}{\partial q_r} = 0,$$

$$p_r = \frac{\partial T}{\partial q_r'} = a_{r1}q_1' + a_{r2}q_2' + \cdots + a_{rn}q_n',$$

62)
$$\frac{\partial W}{\partial q_r} = c_{r1}q_1 \;\; + c_{r2}q_2 \;\; + \cdots + c_{rn}q_n,$$

$$\frac{\partial F}{\partial q_r'} = \varkappa_{r1}q_1' + \varkappa_{r2}q_2' + \cdots + \varkappa_{rn}q_n'.$$

Inserting these values in our differential equation we have, performing the differentiation by t,

63)
$$\begin{aligned} & a_{r1}\frac{d^2q_1}{dt^2} + a_{r2}\frac{d^2q_2}{dt^2} + \cdots + a_{rn}\frac{d^2q_n}{dt^2} \\ + & \varkappa_{r1}\frac{dq_1}{dt} + \varkappa_{r2}\frac{dq_2}{dt} + \cdots + \varkappa_{rn}\frac{dq_n}{dt} \\ + & c_{r1}q_1 \;\;\; + c_{r2}q_2 \;\;\; + \cdots + c_{rn}q_n = 0, \end{aligned}$$

a linear differential equation of the second order with constant coefficients. We have one such equation for each coordinate q_r. As in the case of one variable we may find a particular solution by assuming for each coordinate

$$q_r = A_r e^{\lambda t},$$

the λ being the same constant for all the q's. Inserting these values in 63) we obtain, after dividing by the common factor $e^{\lambda t}$,

64)
$$\begin{aligned} & A_1(a_{11}\lambda^2 + \varkappa_{11}\lambda + c_{11}) + \cdots + A_n(a_{1n}\lambda^2 + \varkappa_{1n}\lambda + c_{1n}) = 0, \\ & A_1(a_{21}\lambda^2 + \varkappa_{21}\lambda + c_{21}) + \cdots + A_n(a_{2n}\lambda^2 + \varkappa_{2n}\lambda + c_{2n}) = 0, \\ & \cdots\cdots\cdots\cdots\cdots\cdots\cdots\cdots\cdots \\ & \cdots\cdots\cdots\cdots\cdots\cdots\cdots\cdots\cdots \\ & A_1(a_{n1}\lambda^2 + \varkappa_{n1}\lambda + c_{n1}) + \cdots + A_n(a_{nn}\lambda^2 + \varkappa_{nn}\lambda + c_{2n}) = 0. \end{aligned}$$

These are linear equations in the A's and suffice to determine their ratios when λ is known. In order that they may be satisfied by other than zero values of the A's, the determinant of the coefficients must vanish, namely,

65)
$$\begin{vmatrix} a_{11}\lambda^2 + \varkappa_{11}\lambda + c_{11}, & \cdots & a_{1n}\lambda^2 + \varkappa_{1n}\lambda + c_{1n} \\ \cdot\;\cdot\;\cdot & \cdot\;\cdot\;\cdot & \cdot\;\cdot\;\cdot \\ \cdot\;\cdot\;\cdot & \cdot\;\cdot\;\cdot & \cdot\;\cdot\;\cdot \\ a_{n1}\lambda^2 + \varkappa_{n1}\lambda + c_{n1}, & \cdots & a_{nn}\lambda \;\; + \varkappa_{nn}\lambda + c_{nn} \end{vmatrix} = 0.$$

This is Lagrange's determinantal equation for λ. It is of degree $2n$ and does not contain odd powers of λ if $F = 0$. We shall denote its roots by $\lambda_1, \lambda_2, \ldots \lambda_{2n}$, which we shall suppose are all different. We shall first prove that none of the roots are real and positive.

If we multiply the rth equation 64), which may be written

$$64)\qquad \lambda^2\sum_{s=1}^{s=n} a_{rs}A_s + \lambda\sum_{s=1}^{s=n}\varkappa_{rs}A_s + \sum_{s=1}^{s=n}c_{rs}A_s = 0$$

by A_r and take the sum for all the r's, we obtain

$$66)\quad \lambda^2\sum_{r=1}^{r=n}\sum_{s=1}^{s=n} a_{rs}A_rA_s + \lambda\sum_{r=1}^{r=n}\sum_{s=1}^{s=n}\varkappa_{rs}A_rA_s + \sum_{r=1}^{r=n}\sum_{s=1}^{s=n}c_{rs}A_rA_s = 0.$$

The double sum by which λ^2 is multiplied is the value of the function $2T$ when for every q_r' is substituted A_r. We shall denote this by $2T(A)$. Similarly the coefficient of λ is $2F(A)$, and the constant term or that independent of λ is $2W(A)$. But by the fundamental property of the three functions each must be positive for every set of values of its variables. This equation 66) thus written,

$$67)\qquad \lambda^2 T(A) + \lambda F(A) + W(A) = 0$$

shows at once that λ cannot be real and positive, for that would involve the sum of three positive terms being zero.

Secondly if $F = 0$, that is if there is no dissipation,

$$\lambda^2 = -\frac{W(A)}{T(A)}$$

which is negative and λ is a pure imaginary. In this case $e^{\lambda t}$ and $e^{-\lambda t}$ are replaced as above by trigonometric functions representing an undamped harmonic oscillation of the same period for all the parameters q.

Thirdly, if F is large enough λ can be real and negative. In this case each parameter q gradually dies away to zero, the rate of dying away being the same for all. This corresponds to case I of the preceding section.

Fourthly, in the limiting case of a system devoid either of inertia or of stiffness, so that T or W is zero, F not zero, instead of a pair of roots we have a single one which is real and negative, so that the motion dies away.

Fifthly, in other cases, that is when neither T, F, nor W vanish and F is not too large, λ is complex. This is the most frequent case in practice.

We shall prove that the real part of λ is negative. When the value of any root λ is determined, the equations 64) determine the quantities A_r except for a common factor. If complex values enter, since any equation which involves i will also hold good if i be changed to $-i$, changing any root λ to its conjugate λ' causes every A to change to its conjugate A'. Let us accordingly write

$$68)\qquad \begin{aligned} \lambda &= \mu + i\nu, & \lambda' &= \mu - i\nu, \\ A_r &= \alpha_r + i\beta_r, & A'_r &= \alpha_r - i\beta_r. \end{aligned}$$

Let us now apply the method that gave us equation 66), except that we multiply the equations 64) containing λ by the A''s corresponding to λ', obtaining

$$69)\qquad \lambda^2\sum_{r=1}^{r=n}\sum_{s=1}^{s=n} a_{rs} A_r A'_s + \lambda\sum_{r=1}^{r=n}\sum_{s=1}^{s=n} \varkappa_{rs} A_r A'_s + \sum_{r=1}^{r=n}\sum_{s=1}^{s=n} c_{rs} A_r A'_s = 0.$$

In this equation any coefficient a_{rs} appears in the two terms for which $r = p$, $s = t$ and also $r = t$, $s = p$, so that the sum is

$$a_{pt}\{A_p A'_t + A_t A'_p\},$$

or substituting the values of the A's,

$$a_{pt}\{(\alpha_p + i\beta_p)(\alpha_t - i\beta_t) + (\alpha_t + i\beta_t)(\alpha_p - i\beta_p)\}$$
$$= 2a_{pt}(\alpha_p\alpha_t + \beta_p\beta_t).$$

Accordingly using a notation similar to that before employed equation 69) is

$$70)\qquad \lambda^2[T(\alpha) + T(\beta)] + \lambda[F(\alpha) + F(\beta)] + W(\alpha) + W(\beta) = 0.$$

Now performing the same process on equations 64) with the root λ' and multiplying by the A's we obtain

$$71)\qquad \lambda'^2[T(\alpha) + T(\beta)] + \lambda'[F(\alpha) + F(\beta)] + W(\alpha) + W(\beta) = 0.$$

Then λ and λ' are roots of the same quadratic. We therefore have their sum

$$72)\qquad \lambda + \lambda' = 2\mu = -\frac{F(\alpha) + F(\beta)}{T(\alpha) + T(\beta)},$$

so that μ is negative. The solution therefore represents a damped vibration, as in case II § 43, the period and damping being the same for all the q's. For another treatment of Lagrange's determinant see Note V.

Having obtained all the roots λ, by substitution of any one λ_r in the equations 64) we obtain the ratios $A_1 : A_2 : \cdots : A_n$. For each value of λ_r we obtain a different set of ratios. We will distinguish the values belonging to λ_r by an upper affix r, so that A^r_s means the coefficient of $e^{\lambda_r t}$ in the coordinate q_s.

The theory of linear differential equations shows us that for the general solution we must take the sum of the particular solutions $A^r_s e^{\lambda_r t}$ for all the roots λ_r, so that we obtain

$$
73)\qquad
\begin{aligned}
q_1 &= A_1^1 e^{\lambda_1 t} + A_1^2 e^{\lambda_2 t} + \cdots + A_1^{2n} e^{\lambda_{2n} t},\\
q_2 &= A_2^1 e^{\lambda_1 t} + A_2^2 e^{\lambda_2 t} + \cdots + A_2^{2n} e^{\lambda_{2n} t},\\
&\cdots\cdots\cdots\cdots\cdots\cdots\\
&\cdots\cdots\cdots\cdots\cdots\cdots\\
q_n &= A_n^1 e^{\lambda_1 t} + A_n^2 e^{\lambda_2 t} + \cdots + A_n^{2n} e^{\lambda_{2n} t}.
\end{aligned}
$$

It is to be noticed that the ratios of the A's in any column have been determined by the linear equations 64), so that there is a factor which is still arbitrary for each column, that is to say, $2n$ in all. We may now replace the exponentials by trigonometric terms. The appearance of the terms with conjugate imaginaries

$$A_r e^{\lambda t} + A_r' e^{\lambda' t} = 2 e^{\mu t} (\alpha_r \cos \nu t - \beta_r \sin \nu t)$$

leads to the disappearance of imaginaries from the result. Changing the notation we will accordingly write

$$
74)\qquad
\begin{aligned}
q_1 &= B_1^1 e^{\mu_1 t} \cos(\nu_1 t - \varepsilon_1) + B_1^2 e^{\mu_2 t} \cos(\nu_2 t - \varepsilon_2) + \cdots\\
&\qquad + B_1^n e^{\mu_n t} \cos(\nu_n t - \varepsilon_n),\\
q_2 &= B_2^1 e^{\mu_1 t} \cos(\nu_1 t - \varepsilon_1) + B_2^2 e^{\mu_2 t} \cos(\nu_2 t - \varepsilon_2) + \cdots\\
&\qquad + B_2^n e^{\mu_n t} \cos(\nu_n t - \varepsilon_n),\\
&\cdots\cdots\cdots\cdots\cdots\cdots\\
&\cdots\cdots\cdots\cdots\cdots\cdots\\
q_n &= B_n^1 e^{\mu_1 t} \cos(\nu_1 t - \varepsilon_1) + B_n^2 e^{\mu_2 t} \cos(\nu_2 t - \varepsilon_2) + \cdots\\
&\qquad + B_n^n e^{\mu_n t} \cos(\nu_n t - \varepsilon_n).
\end{aligned}
$$

If these be substituted in the differential equations it will be found that the B's satisfy the same linear equations as the A's.[1]) Each column then contains an arbitrary constant as before in the B's and a second arbitrary constant in the ε belonging to the column. We may therefore state the general result: — The motion of any system, possessing n degrees of freedom, slightly displaced from a position of stable equilibrium may be described as follows: Each coordinate performs the resultant of n damped harmonic oscillations of different periods. The phase and damping factor of any simple oscillation of a particular period are the same for all the coordinates. The absolute value of the amplitude for any particular coordinate is arbitrary, but the ratios of the amplitudes for a particular period for the different coordinates are determined solely by the nature of the system, that is, by its inertia, stiffness and resistance coefficients. The $2n$ arbitrary constants determining the n amplitudes and phases are found from the values of the n coordinates q and velocities q' for a particular instant of time.

1) In 64) put $\lambda = \mu + i\nu$ and take the real part.

We further notice that, since the different periods depending upon ν are derived from the roots of an algebraic equation, they are not in general commensurable, so that the motion is not as a whole generally periodic. For instance in the case of Lissajous's curves described in § 19, unless the two periods are commensurable the curve will never close. In the case, however, of the spherical pendulum performing small oscillations the periods of the two coordinates were equal, so that the path became a closed curve, an ellipse.

There is one set of coordinates of peculiar importance. For simplicity let us suppose there is no dissipation, $F = 0$. Let us make a linear transformation with constant coefficients, putting

75)
$$\begin{aligned} q_1 &= \gamma_{11}\,\varphi_1 + \gamma_{12}\,\varphi_2 + \cdots + \gamma_{1n}\varphi_n, \\ q_2 &= \gamma_{21}\,\varphi_1 + \gamma_{22}\,\varphi_2 + \cdots + \gamma_{2n}\varphi_n, \\ &\cdots\cdots\cdots\cdots\cdots \\ &\cdots\cdots\cdots\cdots\cdots \\ q_n &= \gamma_{n1}\,\varphi_1 + \gamma_{n2}\varphi_2 + \cdots + \gamma_{nn}\varphi_n. \end{aligned}$$

Differentiating by t the q''s are obtained from the φ''s by the same substitution. It is shown by algebra that we may determine the coefficients γ in such a way that the two quadratic functions T and W are simultaneously transformed to sums of squares, the product terms being absent. Supposing this done we have

76)
$$\begin{aligned} T &= \frac{1}{2}\left(a_1\varphi_1'^2 + a_2\varphi_2'^2 + \cdots + a_n\varphi_n'^2\right), \\ W &= \frac{1}{2}\left(c_1\varphi_1^2 + c_2\varphi_2^2 + \cdots + c_n\varphi_n^2\right). \end{aligned}$$

Then we have

$$p_r = a_r\varphi_r'$$

and our differential equations 63) are

77)
$$a_r\frac{d^2\varphi_r}{dt^2} + c_r\varphi_r = 0,$$

and the integrals,

78)
$$\varphi_r = A_r\cos\left(\sqrt{\frac{c_r}{a_r}}\cdot t - \varepsilon_r\right).$$

In other words, each coordinate φ appears in its own differential equation entirely separate from the rest, and performs a harmonic vibration independent of the others with its own period. The φ's are called *normal* or *principal coordinates.* The q's being linear functions of the φ's describe compound harmonic oscillations. A vibration in which all the normal coordinates but one vanish is called a *normal vibration.* The effect of this on the q's is, if every φ equals zero except φ_s, to make

79)
$$\begin{aligned} q_1 &= \gamma_{1s}\varphi_s, \\ q_2 &= \gamma_{2s}\varphi_s, \\ &\;\cdots\cdots \\ &\;\cdots\cdots \\ q_n &= \gamma_{ns}\varphi_s, \end{aligned}$$

that is in a normal vibration all the coordinates of whatever sort are in ratios constant throughout the motion, or the solutions 74) are reduced to a single column. The motion is then completely periodic, all the coordinates passing through the equilibrium values simultaneously. We may thus describe the general motion as the resultant of n normal oscillations. The normal coordinates have the property that the energy of any vibration is the sum of the energies of the separate normal vibrations, for substituting 78) in 76) we have

80) $$T + W = \frac{1}{2}(c_1 A_1^2 + c_2 A_2^2 + \cdots + c_n A_n^2).$$

46. Vibration of a String of Beads. Continuous String. As an example of the preceding theory let us consider the problem, solved by Lagrange[1]), of the motion of a string on which are fastened a number of beads of equal mass equidistant from each other and from the ends of the string, the mass of the string being neglected in comparison. Let the number of beads be n, the mass of each, m, the distances apart, a, and the length of the string, $l = (n+1)a$. Suppose for simplicity the motion of each bead takes place in a straight line at right angles to the stretched string, all the displacements y_r being in the same plane. Then the kinetic energy is

81) $$T = \frac{m}{2}(y_1'^2 + y_2'^2 + \cdots + y_n'^2).$$

The coefficients of inertia are the same for all, equal to the mass of any bead. The displacements being small quantities, the length of the string connecting any two beads is equal to a plus small quantities of the second order which will be neglected. The tension of the string will thus be considered constant and equal to S. Neglecting the weight of the beads the only forces acting on a bead are the components of the tension of the two adjacent portions of the string in the direction of the

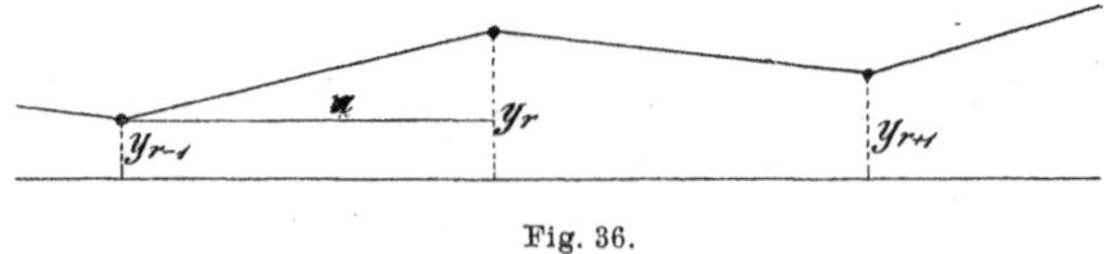

Fig. 36.

displacement y, Fig. 36. To find the component we have to multiply the tension by the cosine of the angle made by the displacement

1) Lagrange, *Mécanique Analytique,* Tom. I, p. 390.

with the segment of the string, which is the difference of two consecutive displacements divided by a. Accordingly the force on the rth bead is

$$82)\qquad \frac{S}{a}\{(y_{r-1}-y_r)+(y_{r+1}-y_r)\}=-\frac{\partial W}{\partial y_r},$$

from which we obtain the potential energy,

$$83)\qquad W=\frac{S}{2a}\{y_1^2+(y_2-y_1)^2+(y_3-y_2)^2+\cdots+y_n^2\}.$$

It will be seen that the y's are not normal coordinates since product terms appear in W.

Forming the differential equations of motion we obtain

$$84)\qquad \begin{aligned} &m\frac{d^2y_1}{dt^2}+\frac{S}{a}(y_1-0+y_1-y_2)=0,\\ &m\frac{d^2y_2}{dt^2}+\frac{S}{a}(y_2-y_1+y_2-y_3)=0,\\ &\cdot\quad\cdot\quad\cdot\quad\cdot\quad\cdot\quad\cdot\quad\cdot\quad\cdot\quad\cdot\quad\cdot\quad\cdot\quad\cdot\\ &m\frac{d^2y_n}{dt^2}+\frac{S}{a}(y_n-y_{n-1}+y_n-0)=0.\end{aligned}$$

Now putting $y_r=A_re^{\lambda t}$ and collecting according to the A's we obtain

$$86)\qquad \begin{aligned} &\left(m\lambda^2+\frac{2S}{a}\right)A_1-\frac{S}{a}A_2=0,\\ &-\frac{S}{a}A_1+\left(m\lambda^2+\frac{2S}{a}\right)A_2-\frac{S}{a}A_3=0,\\ &\cdot\quad\cdot\quad\cdot\quad\cdot\quad\cdot\quad\cdot\quad\cdot\quad\cdot\quad\cdot\quad\cdot\quad\cdot\quad\cdot\quad\cdot\,,\end{aligned}$$

or, dividing through by $\frac{S}{a}$ and putting

$$2+\frac{ma\lambda^2}{S}=C,$$

$$87)\qquad \begin{aligned} &CA_1-A_2+0+\cdots\cdots\cdots\cdots=0,\\ &-A_1+CA_2-A_3+0-\cdots\cdots\cdots=0,\\ &0-A_2+CA_3-A_4+0+\cdots\cdots=0,\\ &\cdot\quad\cdot\quad\cdot\quad\cdot\quad\cdot\quad\cdot\quad\cdot\quad\cdot\quad\cdot\quad\cdot\quad\cdot\quad\cdot\end{aligned}$$

The determinantal equation for λ is

$$88)\qquad D_n=\begin{vmatrix} C, & -1, & 0, & 0, & 0, & \dots\\ -1, & C, & -1, & 0, & 0, & \dots\\ 0, & -1, & C, & -1, & 0, & \dots\\ 0, & 0, & -1, & C, & -1, & \dots\\ \cdot & \cdot & \cdot & \cdot & \cdot & \cdot \end{vmatrix}=0,\quad n \text{ rows}.$$

Expanding the determinant in terms of its first minors we have

89) $$D_n = CD_{n-1} - D_{n-2}.$$

This equation between three consecutive determinants of the same form suggests a trigonometric relation, namely, making use of the relation

$$\sin(a+b) + \sin(a-b) = 2 \sin a \cos b,$$

with $b = \vartheta$, $a = n\vartheta$, we have

$$\sin(n+1)\vartheta + \sin(n-1)\vartheta = 2 \sin n\vartheta \cos \vartheta.$$

Comparing this with the formula 89),

$$D_n + D_{n-2} = CD_{n-1},$$

we see that they are identical if we put

$$C = 2\cos\vartheta, \qquad D_n = c \sin(n+1)\vartheta,$$

where c is independent of n. To find it put $n = 1$,

$$D_1 = C = 2\cos\vartheta = \frac{\sin 2\vartheta}{\sin\vartheta}.$$

Accordingly

90) $$D_n = \frac{\sin(n+1)\vartheta}{\sin\vartheta}.$$

If this is to vanish we must have

$$(n+1)\vartheta = k\pi,$$

where k is any integer (not a multiple of $n+1$, to prevent $\sin\vartheta$ in the denominator from vanishing). Introducing the values of ϑ thus found we obtain

91) $$C = 2 + \frac{ma\lambda^2}{S} = 2\cos\vartheta = 2\cos\frac{k}{n+1}\pi,$$

from which

$$-\lambda^2 = \nu^2 = \frac{2S}{ma}\left(1 - \cos\frac{k}{n+1}\cdot\pi\right)$$

92) $$\nu_k = 2\sqrt{\frac{S}{ma}}\cdot\sin\frac{k}{2(n+1)}\cdot\pi.$$

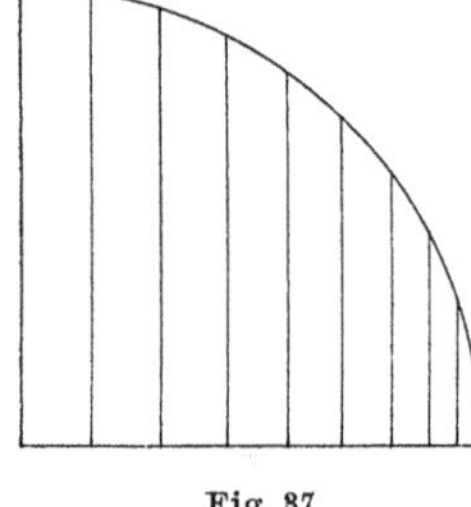

Fig. 37.

Letting $k = 1, 2, 3, \ldots n$, we obtain n different frequencies proportional to the ordinates of points dividing a quadrant into $(n+1)$ equal parts, Fig. 37. Giving k other values not multiples of $(n+1)$, we shall merely repeat these frequencies. There are accordingly n different frequencies for the vibrations.

We may arrive at the same result by noticing that the linear equations for the A's,

87) $$-A_{r-1} + CA_r - A_{r+1} = 0,$$

are satisfied by

$$A_s = P \sin s\vartheta,$$

where P is a constant, making use of the same trigonometric formula as before. Accordingly let us substitute in the differential equation

84) $$-y_{r-1} + \frac{ma}{S}\frac{d^2 y_r}{dt^2} + 2y_r - y_{r+1} = 0$$

the solution

93) $$y_r = P \sin r\vartheta \cos(\nu t - \varepsilon).$$

Every term will contain the same cosine, so that dividing out we have

$$-\sin(r-1)\vartheta + 2\left(1 - \frac{ma\nu^2}{2S}\right)\cdot \sin r\vartheta - \sin(r+1)\vartheta = 0,$$

which is an identity if

$$1 - \frac{ma\nu^2}{2S} = \cos\vartheta,$$

giving

$$\nu^2 = \frac{2S}{ma}(1 - \cos\vartheta),$$

as before, 92). The complete solution is then

94) $$y_r = \sum_{s=1}^{s=n} P_s \sin\frac{rs\pi}{n+1}\cos(\nu_s t - \alpha_s),$$

with the $2n$ arbitrary constants P_s, α_s to be determined by the initial displacements and velocities.

Consider the case first in order of simplicity, n equals 2. Then

95) $$\nu_1 = 2\sqrt{\frac{S}{ma}}\sin\frac{\pi}{6} = \sqrt{\frac{S}{ma}},$$
$$\nu_2 = 2\sqrt{\frac{S}{ma}}\sin\frac{\pi}{3} = \sqrt{\frac{3S}{ma}}.$$

Thus the frequency of the higher pitched vibration is in the ratio of $\sqrt{3}:1 = 1.732$ to that of the lower, — somewhat more than the musical interval of a sixth. In this particular case it is easy to find the normal coordinates. Writing

96) $$\varphi_1 = \frac{1}{2}(y_1 + y_2), \quad y_1 = \varphi_1 + \varphi_2, \quad y_1' = \varphi_1' + \varphi_2',$$
$$\varphi_2 = \frac{1}{2}(y_1 - y_2), \quad y_2 = \varphi_1 - \varphi_2, \quad y_2' = \varphi_1' - \varphi_2',$$

we obtain

97) $$T = \frac{m}{2}(y_1'^2 + y_2'^2) = m(\varphi_1'^2 + \varphi_2'^2),$$
$$W = \frac{S}{2a}[y_1^2 + (y_1 - y_2)^2 + y_2^2] = \frac{S}{a}(\varphi_1^2 + 3\varphi_2^2),$$

so that the φ's are normal coordinates. Forming the differential equations,

$$98)\qquad \begin{aligned} m\frac{d^2\varphi_1}{dt^2}+\frac{S}{a}\varphi_1=0,\\ m\frac{d^2\varphi_2}{dt^2}+\frac{3S}{a}\varphi_2=0,\end{aligned}$$

the integrals of which are

$$99)\qquad \begin{aligned} \varphi_1=A_1\cos\left(\sqrt{\frac{S}{ma}}\cdot t-\alpha_1\right),\\ \varphi_2=A_2\cos\left(\sqrt{\frac{3S}{ma}}\cdot t-\alpha_2\right),\end{aligned}$$

agreeing with the above result.

The two normal vibrations are found, the first by putting $\varphi_2=0$, — in this case $y_1=y_2$ and the two beads swing together, the second normal vibration by $\varphi_1=0$, $y_2=-y_1$, and the two beads swing in opposite directions with a frequency $\sqrt{3}$ times as great as before. The middle point of the string is now at rest, or forms a *node*.

The general case above treated is very interesting when we pass to the limit as the number of beads is increased, giving us the case of a continuous string, of the greatest importance in the theory of musical instruments.

Let us introduce in equation 94) the distance of the bead from one end of the string,

$$x=r\cdot a=\frac{rl}{n+1}.$$

Accordingly 94) becomes

$$100)\qquad y(x)=\sum_{s=1}^{s=n}P_s\sin\frac{s\pi x}{l}\cos(\nu_s t-\alpha_s).$$

A glance at Fig. 37 shows us that, as we increase n, the ratios at least of the smaller frequencies approach those of the integers, 1, 2, 3, ... By passage to the limit we may demonstrate that this is exactly true for all the frequencies.

If ϱ be the line density of matter of the continuous string, that is, the mass per unit length, we have

$$\varrho l=\lim_{n=\infty}(n+1)m.$$

Accordingly since

$$l=(n+1)\,a,$$

we have in the limit

$$am=\frac{\varrho l^2}{(n+1)^2}.$$

Introducing this into the value of ν_s, 92),

$$101)\qquad \nu_s=2\sqrt{\frac{S}{\varrho}}\lim_{n=\infty}\frac{n+1}{l}\sin\frac{s\pi}{2(n+1)}.$$

As n increases without limit y preserves its form, while ν_s approaches the limit

$$102) \qquad \nu_s = \frac{s\pi}{l}\sqrt{\frac{S}{\varrho}}. \text{ [1]}$$

We have therefore for the continuous string,

$$103) \qquad y = \sum_{s=1}^{s=\infty} P_s \sin\frac{s\pi x}{l}\cos\left(\frac{s\pi}{l}\sqrt{\frac{S}{\varrho}}\cdot t - \alpha_s\right).$$

The frequencies of the different terms of the series are in the ratios of the integers. Such partial vibrations are called *harmonics* or *overtones* of the lowest or fundamental, for which $s = 1$. Since, if we consider a single term of the series, the excursions of all the particles are in the same ratios throughout the motion, we see that the harmonics are normal vibrations. On account of the factor depending upon x the sth harmonic has nodes for

$$x = \frac{l}{s}, \frac{2l}{s}, \ldots \frac{(s-1)l}{s},$$

or at any instant the string has the form of a sine curve and is divided by nodes into s segments vibrating oppositely, generally known as ventral segments.

In order to show how rapidly the string of beads approximates to the motion of a continuous string, the following table from Rayleigh's *Theory of Sound* is inserted. It is to be noticed that it does not give exactly the ratios of the frequencies on account of the variable factor s under the sine in ν_s, but it approximately does so, and for the fundamental, $s = 1$, it gives exactly the ratio of frequency for n beads to that of the continuous string.

n	1	2	3	4	9	19	39
$\frac{2(n+1)}{\pi}\sin\frac{\pi}{2(n+1)}$	·9003	·9549	·9745	·9836	·9959	·9990	·9997

By means of an extension of the above method, Pupin has treated the problem of the vibrations of a *heavy* string loaded with beads,

1) Writing the factor of $\frac{2}{l}\sqrt{\frac{S}{\varrho}}$ in the form

$$\frac{s\pi}{2}\left(\sin\frac{s\pi}{2(n+1)} \Big/ \frac{s\pi}{2(n+1)}\right),$$

since

$$\lim_{x=0}\frac{\sin x}{x} = 1,$$

we obtain the result.

both for free and forced vibrations, and by an electrical application has solved a very important telephonic problem.[1])

On account of the importance and typical nature of the problem of the continuous string, we shall also solve it by means of Hamilton's Principle. Replacing the length of a segment a by the differential dx, writing ϱdx for the mass m, and for y', $\frac{\partial y}{\partial t}$ (partial derivative because y depends upon both t and x), and for the sum, the definite integral, we have the kinetic energy

$$104)\qquad T = \frac{1}{2}\int_0^l \varrho \left(\frac{\partial y}{\partial t}\right)^2 dx.$$

Similarly in the potential energy the limit of the term

$$\left(\frac{y_r - y_{r-1}}{a}\right)^2 a \quad \text{is} \quad \left(\frac{\partial y}{\partial x}\right)^2 dx,$$

so that the potential energy becomes

$$105)\qquad W = \frac{1}{2}\int_0^l S\left(\frac{\partial y}{\partial x}\right)^2 dx.$$

As the number of degrees of freedom is now infinite we are not able to use Lagrange's equations, but we can use Hamilton's Principle, which includes them.

$$106)\qquad \begin{aligned} \delta\int_{t_0}^{t_1}(T - W)\,dt &= \frac{1}{2}\int_{t_0}^{t_1} dt \cdot \delta\int_0^l \left\{\varrho\left(\frac{\partial y}{\partial t}\right)^2 - S\left(\frac{\partial y}{\partial x}\right)^2\right\} dx \\ &= \int_0^{t_1} dt \cdot \int_0^l \left\{\varrho\,\frac{\partial y}{\partial t}\frac{\partial \delta y}{\partial t} - S\,\frac{\partial y}{\partial x}\frac{\partial \delta y}{\partial x}\right\} dx = 0. \end{aligned}$$

Integrating the first term partially with respect to t and the second with respect to x,

$$107)\qquad \begin{aligned} &\int_0^l \left[\varrho\,\frac{\partial y}{\partial t}\,\delta y\right]_{t_0}^{t_1} dx - \int_{t_0}^{t_1}\left[S\,\frac{\partial y}{\partial t}\,\delta y\right]_0^l dt \\ &- \int_{t_0}^{t_1} dt \int_0^l dx \left\{\varrho\,\frac{\partial^2 y}{\partial t^2} - S\,\frac{\partial^2 y}{\partial x^2}\right\}\delta y = 0. \end{aligned}$$

The variation δy is as usual to be put equal to zero at the time limits, and, as the ends of the string are fixed, δy equals zero at

1) Pupin, *Wave Propagation over non-uniform Electrical Conductors.* Trans. American Mathematical Society, I, p. 259, 1900.

the limits for x also, consequently we must have the factor of the arbitrary δy vanish, that is,

108)
$$\varrho \frac{\partial^2 y}{\partial t^2} - S \frac{\partial^2 y}{\partial x^2} = 0.$$

Putting $\frac{S}{\varrho} = a^2$ we have the partial differential equation for the motion of the continuous string,

109)
$$\frac{\partial^2 y}{\partial t} = a^2 \frac{\partial^2 y}{\partial x^2},$$

which may also be obtained from the ordinary differential equations 84) by passage to the limit in an obvious manner.

The passage from n ordinary differential equations to a single partial differential equation when n is infinite is worth noting as a type of a phenomenon of frequent occurrence. At the same time the notion of normal vibrations gives rise to that of normal *functions*. To find a normal vibration let us find a particular solution of 109),

110)
$$y = X(x) \cdot \varphi(t),$$

where the two functions contain only the variables indicated. This satisfies the definition of a normal vibration, since the ratios of displacements of the different points are the same throughout the motion. Inserting in the differential equation we obtain

111)
$$X \frac{d^2 \varphi}{dt^2} = a^2 \varphi \cdot \frac{d^2 X}{dx^2}.$$

Dividing by $X\varphi$ we have

$$\frac{1}{\varphi} \frac{d^2 \varphi}{dt^2} = \frac{a^2}{X} \frac{d^2 X}{dt^2}.$$

Since one side depends only on x and the other only on t, which are independent variables, this can hold only if either member is constant, say $-\nu^2 a^2$, where ν is arbitrary. Thus we have the two equations

112)
$$\frac{d^2 \varphi}{dt^2} + \nu^2 a^2 \varphi = 0,$$
$$\frac{d^2 X}{dx^2} + \nu^2 X = 0.$$

The first of these shows, like 77), that φ is a normal coordinate. Its integral is

113)
$$\varphi = C \cos(\nu a t - \alpha),$$

the integral of the second is

114)
$$X = A \cos \nu x + B \sin \nu x.$$

The normal vibration is accordingly represented by

115)
$$y = (A \cos \nu x + B \sin \nu x) \cos(\nu a t - \alpha),$$

the arbitrary constant C being merged in A and B.

Since for all values of t, $y = 0$ for $x = 0$, we must have $A = 0$, and since $y = 0$ for $x = l$, we must also have $B \sin \nu l = 0$, that is

$$\nu l = s\pi, \tag{116}$$

where s is any integer, accordingly we obtain for the sth normal vibration,

$$\nu_s = \frac{s\pi}{l}, \tag{117}$$

and the vibration is given by

$$y = B_s \sin \frac{s\pi x}{l} \cos \left(\frac{s\pi a t}{l} - \alpha_s\right). \tag{118}$$

The general solution is therefore represented as an infinite series of normal vibrations,

$$y = \sum_{s=1}^{s=\infty} B_s \sin \frac{s\pi x}{l} \cos \left(\frac{s\pi a t}{l} - \alpha_s\right), \tag{103}$$

the arbitrary constants, B_s, α_s, being determined by the initial displacements and velocities. In order to determine them let us make use of the other fundamental property of normal coordinates, namely, that the energy functions do not contain product terms. Let us write

$$y = \sum_{s=1}^{s=\infty} X_s \varphi_s, \tag{119}$$

then

$$T = \frac{1}{2}\int_0^l \varrho \left(\frac{\partial y}{\partial t}\right)^2 dx = \frac{\varrho}{2}\int_0^l \left[\sum_{s=1}^{s=\infty} \varphi_s' X_s\right]^2 dx \tag{120}$$

$$= \frac{\varrho}{2} \varphi_1'^2 \int_0^l X_1^2\, dx + \frac{\varrho}{2} \varphi_2'^2 \int_0^l X_2^2\, dx + \cdots$$

$$+ \varrho \varphi_1' \varphi_2' \int_0^l X_1 X_2\, dx + \cdots.$$

Inasmuch as product terms in the φ''s are not to appear we must have

$$\int_0^l X_r X_s\, dx = 0 \qquad (r \neq s). \tag{121}$$

Putting X_r equal to $\sin \frac{r\pi x}{l}$ this result is at once verified by integration. The property of normal functions expressed by equation 121) is of fundamental importance in the theory of developments in infinite series. We may now make use of it to obtain the constants in the series above. Putting $t = 0$ and writing $B_s \cos \alpha_s = A_s$, we have

122) $$y(0) = \sum_{s=1}^{s=\infty} A_s \sin \frac{s\pi}{l} x = f(x).$$

The problem is, f being an arbitrary function of x to find the coefficients in the development in the trigonometric series. To find the coefficients A_r multiply the equation 122) by the rth normal function and integrate from 0 to l, giving

123) $$\int_0^l f(x) \sin \frac{r\pi x}{l} dx = \sum_{s=1}^{s=\infty} A_s \int_0^l \sin \frac{r\pi x}{l} \sin \frac{r\pi x}{l} dx,$$

and by the property just found the integral on the right vanishes in every term except that in which $r = s$. But

$$\int_0^l \sin^2 \frac{r\pi x}{l} dx = \frac{l}{2}.$$

Therefore we have the value of the coefficient

124) $$A_r = \frac{2}{l} \int_0^l f(x) \sin \frac{r\pi x}{l} dx.$$

We are thus led to a particular case of the remarkable trigonometric series associated with the name of Fourier. Such series were first considered by Daniel Bernoulli in connection with this very problem of a vibrating string. This determination of the coefficients was given by Euler in 1777. The importance of the series in analysis was first brought out by Fourier who insisted that such a series was capable of representing an arbitrary function, as had been maintained by Bernoulli, but doubted by Euler and Lagrange.

47. Forced Vibrations of General System. Let us now briefly consider the question of forced vibrations of the general system of § 45.

Suppose that there is impressed upon each coordinate a harmonically varying force,

$$F_r = E_r \cos pt,$$

the period and phase being the same for all, the amplitude E_r being taken at pleasure. The equations are most easily dealt with if, instead of proceeding as we did in treating equations 41) and 42) we make use of the principle that, in an equation involving complex quantities, the real and the imaginary parts must be equated separately. Let us therefore put instead of the above value of F_r the value $E_r e^{ipt}$, whose real part agrees with the above, and having found a particular solution of the differential equation, let us retain its real part only. Thus we have instead of equations 63) n equations of which the rth is

125)
$$a_{r1}\frac{d^2 q_1}{dt^2} + a_{r2}\frac{d^2 q_2}{dt^2} + \cdots + a_{rn}\frac{d^2 q_n}{dt^2}$$
$$+ \varkappa_{r1}\frac{dq_1}{dt} + \varkappa_{r2}\frac{dq_2}{dt} + \cdots + \varkappa_{rn}\frac{dq_n}{dt}$$
$$+ c_{r1}q_1 \quad + c_{r2}q_2 \quad + \cdots + c_{rn}q_n = E_r e^{ipt}.$$

Guided by the result of § 44, assuming

$$q_r = A_r e^{ipt},$$

these become

126)
$$(-a_{11}p^2 + \varkappa_{11}\,ip + c_{11})A_1 + \cdots + (-a_{1n}p^2 + \varkappa_{1n}\,ip + c_{1n})A_n = E_1,$$
$$\cdots\cdots\cdots\cdots\cdots\cdots$$
$$\cdots\cdots\cdots\cdots\cdots\cdots$$
$$(-a_{n1}p^2 + \varkappa_{n1}\,ip + c_{n1})A_1 + \cdots + (-a_{nn}p^2 + \varkappa_{nn}\,ip + c_{nn})A_n = E_n.$$

If we call the determinant of equation 65) $D(\lambda)$ and the minor of the element of the rth column and s^{th} row $D_{rs}(\lambda)$, we have as the solution of 126)

126)
$$A_r = \frac{\sum_{s=1}^{s=n} D_{rs}(ip)\,E_s}{D(ip)}.$$

Since $D(\lambda) = 0$ is the determinantal equation 65) for the free vibration, whose roots are $\lambda_1, \lambda_2, \ldots \lambda_{2n}$, we have

128)
$$D(\lambda) = C(\lambda - \lambda_1)(\lambda - \lambda_2)\cdots(\lambda - \lambda_{2n}) = C\prod_{s=1}^{s=n}(\lambda - \lambda_s)$$

where C is the proper constant.

Accordingly the denominator $D(ip)$ is

129)
$$D(ip) = C\prod_{s=1}^{s=n}(ip - \lambda_s) = C\prod_{s=1}^{s=n}\{-\mu_s + i(p - \nu_s)\}.$$

The minors $D_{rs}(ip)$ are polynomials in ip and the numerators are therefore complex quantities, which however reduce to real ones if the $\varkappa$'s are zero. We may write

130)
$$\sum_{s=1}^{s=n} D_{rs}(ip)\,E_s = B_r e^{i\vartheta_r},$$

where B_r and ϑ_r are real and ϑ_r vanishes with the $\varkappa$'s, and is small if they are small. We thus have

131) $$A_r = \frac{B_r e^{i\vartheta_r}}{C\prod_{s=1}^{s=n}(-\mu_s + i(p-\nu_s))} = \frac{B_r e^{i\left(\vartheta_r - \sum_{s=1}^{s=n}\alpha_s\right)}}{C\prod_{s=1}^{s=n} P_s},$$

where

$$P_s = \sqrt{\mu_s^2 + (p-\nu_s)^2}, \qquad \tan\alpha_s = -\frac{p-\nu_s}{\mu_s}.$$

Retaining now only the real parts, we have for our solution,

132) $$q_r = \frac{B_r \cos\left(pt + \vartheta_r - \sum_{s=1}^{s=n}\alpha_s\right)}{C\prod_{s=1}^{s=n}\sqrt{\mu_s^2 + (p-\nu_s)^2}}.$$

Thus if the damping coefficients $\varkappa$ are small, all the oscillations are in nearly the same phase. If the frequency of the impressed force coincides with that of any one of the free oscillations, $p - \nu_s = 0$, and one factor of the denominator reduces to μ_s, so that if the damping of that oscillation is small, the amplitude is very large, or infinite if there is no damping. This is the case of resonance. (Resonance may also be defined in a slightly different manner as occurring when ip is one of the roots of the equation $D(\lambda) = 0$ in which all the $\varkappa$'s have been put equal to zero. This corresponds with our example in § 44. In practical cases the difference is very small.)

48. Cyclic Motions. Ignoration of Coordinates. In certain large classes of motions some of the coordinates do not appear in the expression for the kinetic energy, although their velocities may. For instance in the case of rectangular coordinates,

$$T = \frac{1}{2} m\,(x'^2 + y'^2 + z'^2),$$

the coordinates themselves x, y, z do not appear. In spherical coordinates, § 41, 133),

$$T = \frac{1}{2} m\,(r'^2 + r^2\vartheta'^2 + r^2\sin^2\vartheta\cdot\varphi'^2),$$

φ does not appear while both r and ϑ do. Further examples are furnished in the case of systems in which throughout the motion the place of one particle is immediately taken by another equal particle moving with the same velocity, as for instance in the case of the system of balls in a ball-bearing (bicycle) or better in the case of a continuous chain passing over pulleys, or through a tube of any form, or by the particles of water circulating through a tube.

In order that this condition may be permanent it is evidently necessary that the path traversed by the successive particles shall be reentrant, or that they shall circulate. Under the conditions supposed it is evident that the absolute position of any particle does not affect the kinetic energy, for throughout the motion at any point on the path of the particles there is always a particle moving with the same definite velocity. On account of the character of these examples the term *cyclic coordinates* has been applied by Helmholtz to coordinates which do not appear in the kinetic energy. We shall when necessary distinguish cyclic coordinates by a bar, thus

$$133)\qquad \frac{\partial T}{\partial \overline{q}} = 0$$

is the condition that $\overline{q}$ is cyclic. This of course involves that every

$$134)\qquad \frac{\partial Q_{rs}}{\partial \overline{q}} = 0,$$

that is the coefficients of inertia do not depend upon the cyclic coordinates. Thus a cyclic coordinate is characterized by the fact that the corresponding reaction is wholly *momental.* Examples of cyclic coordinates are found in x, y, z, φ, above, and φ in the case of plane polar coordinates.

Inserting equation 133) in Lagrange's equations we have

$$135)\qquad \frac{d}{dt}\left(\frac{\partial T}{\partial \overline{q}_r'}\right) = \frac{d\overline{p}_r}{dt} = \overline{P}_r,$$

or the fundamental property of a cyclic coordinate is that the force corresponding goes entirely to increasing the corresponding cyclic momentum. If the cyclic force $\overline{P}_r$ vanished, we have

$$136)\qquad \frac{d}{dt}\left(\frac{\partial T}{\partial \overline{q}_r'}\right) = 0,$$

and integrating,

$$137)\qquad \frac{\partial T}{\partial \overline{q}_r'} = \overline{p}_r = c_r.$$

In this case we may with advantage employ a transformation introduced by Routh[1]) and afterwards by Helmholtz[2]), which is analogous to that invented by Hamilton and described in § 39. By means of equations 53) and 71) § 39, we have expressed the velocities as linear functions of the momenta with coefficients R_{rs}, which were functions of the coordinates, and have thus introduced the momenta into the kinetic energy in place of the velocities. We have thus been led to use instead of the Lagrangian function $L = T - W$,

1) Routh, *Stability of Motion,* 1877.

2) Helmholtz, *Studien zur Statik monocyclischer Systeme,* 1884. Ges. Abh. III, p. 119.

whose variation appears in Hamilton's Principle, the Hamiltonian function $H = T + W$. The transformation of Routh and Helmholtz, instead of eliminating *all* the m velocities q', eliminates a certain number, which we will choose so as to replace those having the suffixes $1, 2, \ldots r$, by the corresponding momenta, but to retain the velocities with suffixes $r + 1, \ldots m$, in the equations. This transformation, while it may be made in the general case, is of particular advantage where the eliminated velocities are cyclic and the corresponding momenta constant, as in the case just described.

The equations 53) § 39 for the elimination become by transposition

$$138)\quad \begin{aligned} &Q_{11}q_1' + Q_{12}q_2' + \cdots + Q_{1r}q_r' = p_1 - (Q_{1,r+1}q_{r+1}' + \cdots + Q_{1m}q_m'), \\ &\cdots\cdots\cdots\cdots\cdots\cdots\cdots\cdots\cdots \\ &\cdots\cdots\cdots\cdots\cdots\cdots\cdots\cdots\cdots \\ &Q_{r1}q_1' + Q_{r2}q_2' + \cdots + Q_{rr}q_r' = p_r - (Q_{r,r+1}q_{r+1}' + \cdots + Q_{rm}q_m'). \end{aligned}$$

It will be convenient to write the right hand members above,

$$p_1, - S_1, \ldots p_r - S_r.$$

Let the solutions of equations 138) be

$$139)\quad \begin{aligned} &q_1' = R_{11}(p_1 - S_1) + R_{12}(p_2 - S_2) + \ldots + R_{1r}(p_r - S_r), \\ &\cdots\cdots\cdots\cdots\cdots\cdots\cdots\cdots\cdots \\ &\cdots\cdots\cdots\cdots\cdots\cdots\cdots\cdots\cdots \\ &q_r' = R_{r1}(p_1 - S_1) + R_{r2}(p_2 - S_2) + \ldots + R_{rr}(p_r - S_r), \end{aligned}$$

where the R's are the quotients of the corresponding minors of the determinant

$$\begin{vmatrix} Q_{11}, & Q_{12}, & \ldots & Q_{1r} \\ \cdot & \cdot & \cdot & \cdot \\ \cdot & \cdot & \cdot & \cdot \\ Q_{r1}, & Q_{r2}, & \ldots & Q_{rr} \end{vmatrix}$$

by the determinant itself, and, like the Q's, are functions of the coordinates only.

Introducing the values 139) for the q''s into the kinetic energy, the latter becomes a function of the velocities $q_{r+1}', \ldots q_m'$ and of the momenta $p_1, \ldots p_r$. It is a homogeneous quadratic function of *all* these variables, but not of the p's or q''s considered separately on account of product terms, such as $p_s q_t'$ which are linear in terms of either the p's or q''s. The function T thus transformed has lost its utility for Langrage's equations, but may be replaced by a new function, as follows.

Let us call the function T expressed in terms of the new variables T'. We have thus identically

$$140)\quad \begin{aligned} &T(q_1, q_2, \ldots q_m, q_1', q_2', \ldots q_m') \\ &\equiv T'(q_1, q_2, \ldots q_m, p_1, p_2, \ldots p_r, q_{r+1}', \ldots q_m'). \end{aligned}$$

It is to be noticed that since the coordinates q appear in the coefficients R of equations 139) they are introduced into T' in a way in which they do not appear in T, so that we do not have

$$\frac{\partial T}{\partial q_s} = \frac{\partial T'}{\partial q_s},$$

but since q enters in T' both explicitly and implicitly through equations 139), we have for $s = r + 1, r + 2, \ldots m$,

141) $$\frac{\partial T'}{\partial q_s} = \frac{\partial T}{\partial q_s} + \sum_{t=1}^{t=r} \frac{\partial T}{\partial q'_t} \frac{\partial q'_t}{\partial q_s} \qquad (s = r + 1, \ldots m),$$

and since the same may be said of the velocities,

142) $$\frac{\partial T'}{\partial q'_s} = \frac{\partial T}{\partial q'_s} + \sum_{t=1}^{t=r} \frac{\partial T}{\partial q'_t} \frac{\partial q'_t}{\partial q'_s} \qquad (s = r + 1, m).$$

Now if the eliminated velocities with suffixes $1, 2, \ldots r$ are cyclic and the corresponding forces vanish, we have

$$\frac{\partial T}{\partial \bar{q}'_s} = \bar{p}_s = c_s.$$

Accordingly equations 141/142) become

143) $$\frac{\partial T'}{\partial q_s} = \frac{\partial T}{\partial q_s} + \sum_{t=1}^{t=r} c_t \frac{\partial \bar{q}'_t}{\partial q_s},$$
$$\frac{\partial T'}{\partial q'_s} = \frac{\partial T}{\partial q'_s} + \sum_{t=1}^{t=r} c_t \frac{\partial \bar{q}'_t}{\partial q'_s},$$

or transposing and differentiating outside of the sign of summation.

143) $$\frac{\partial T}{\partial q_s} = \frac{\partial}{\partial q_s}\left(T' - \sum_{t=1}^{t=r} c_t \bar{q}'_t\right),$$
$$\frac{\partial T}{\partial q'_s} = \frac{\partial}{\partial q'_s}\left(T' - \sum_{t=1}^{t=r} c_t \bar{q}'_t\right).$$

We may therefore use with the coordinates whose velocities remain in the equations Lagrange's equations, except that instead of the kinetic energy T we use the function

$$T' - \sum_{t=1}^{t=r} c_t q'_t,$$

and instead of the Lagrangian function $L = T - W$, we use the function

$$\Phi = T' - \sum_{t=1}^{t=r} c_t q'_t - W.$$

Equations 45) or 46) § 36 accordingly become

$$\frac{d}{dt}\left(\frac{\partial \Phi}{\partial q_s'}\right) - \frac{\partial \Phi}{\partial q_s} = 0, \tag{145}$$

(or equal to P_s if this represents an extraneous force not included in the potential energy, or any dissipative force). The function Φ is called by Routh the *modified Lagrangian function*, and its negative by Helmholtz the *kinetic potential*. It is to be understood that Φ is to be expressed in terms of the velocities, $q'_{r+1}, \ldots q'_m$ by means of equations 139) in which $p_1, \ldots p_r$ have been replaced by $c_1, \ldots c_r$. The important thing to notice about Φ is that it contains linear terms in the velocities, as well as a homogeneous quadratic function of the c's whose coefficients depend only on the coordinates $q_{r+1}, \ldots q_m$, like the Q's from which they are derived. The terms of the latter sort in $-\frac{\partial \Phi}{\partial q_s}$ cause precisely the same effect as if they were added to the potential energy. The effect of cyclic motions in a system is accordingly partly represented by an apparent change of potential energy, so that a system devoid of potential energy would seem to possess it, if we were in ignorance of the existence of the cyclic motions in it. The effect of the linear terms in Φ is quite different and will be discussed in § 50.

A system is said to contain concealed masses, when the coordinates which become known to us by observation do not suffice to define the positions of all the masses of the system. The motions of such bodies are called *concaled* motions. It is often possible to solve the problem of the motions of the visible bodies of a system, even when there are concealed motions going on. For it may be possible to form the kinetic potential of the system for the visible motions, not containing the concealed coordinates, and in this case we may use Lagrange's equations, as in the case just treated, for all visible coordinates, while the coordinates of the concealed masses may be *ignored*. Such problems are incomplete, inasmuch as they tell us nothing of the concealed motions, but very often we are concerned only with the visible motions. Such concealed motions enable us to explain the forces acting between visible systems by means of concealed motions of systems connected with them.

The process of eliminating the cyclic coordinates of the concealed motions as above described is termed by Thomson and Tait *ignoration of coordinates*.[1])

Examples of the process may be obtained in any desired number from the theory of the motion of rigid bodies rotating freely about

1) Thomson and Tait, *Natural Philosophy*, Part I, § 319, example *G*.

axes pivoted in bearings fastened to bodies themselves in motion. Such motions will be treated in § 94.

A very simple case of the above process is encountered in treating the motion of a particle m sliding on a horizontal rod, revolving about a vertical axis, at a distance r from the axis. Let the angle made by the rod with a fixed horizontal line be φ, then the velocity perpendicular to the rod is $r\varphi'$. The velocity along the rod being r', the kinetic energy of the body m is

$$146)\qquad T = \frac{1}{2} m (r^2 \varphi'^2 + r'^2).$$

Since φ does not appear in T, φ is a cyclic coordinate. If there is no force tending to change the angle φ we have

$$147)\qquad p_\varphi = \frac{\partial T}{\partial \varphi'} = m r^2 \varphi' = c,$$

from which we obtain

$$148)\qquad \varphi' = \frac{c}{m r^2}$$

to eliminate φ'. Thus we get

$$149)\qquad T' = \frac{1}{2}\left\{\frac{c^2}{m r^2} + m r'^2\right\}.$$

Supposing that there is no potential energy we have

$$150)\qquad \Phi = T' - c\varphi' = \frac{1}{2}\left\{m r'^2 - \frac{c^2}{m r^2}\right\},$$

illustrating the general property of Φ mentioned, $\frac{1}{2} m r'^2$ being the quadratic function of the remaining velocity r' and $-\frac{1}{2}\frac{c^2}{m r^2}$ being the quadratic function of the constant c, which contains as a coefficient a function of the coordinate r. We may now, ignoring the coordinate φ, use the differential equation for r,

$$151)\qquad \frac{d}{dt}\left(\frac{\partial \Phi}{\partial r'}\right) = \frac{\partial \Phi}{\partial r}, \quad \text{or}$$

$$m \frac{d^2 r}{d t^2} = \frac{\partial \Phi}{\partial r} = \frac{c^2}{m r^3}.$$

We accordingly see that the system acts as if, there being no rotation, it possessed an amount of potential energy $-\Phi$, producing the force $\frac{c^2}{m r^3}$ directed from the center. This example accordingly illustrates the effect of ignored cyclic motions in producing an apparent potential energy, but it does not illustrate the effect of linear terms in φ, for they disappear in this example, which is chosen on account of its very simplicity. The example hardly seems to illustrate the case of concealed motions, for the fact of there being a rotation φ' could with difficulty be concealed. Nevertheless this is exactly what happens to bodies at rest relatively to the surface of the earth. To the unsophisticated mind they seem at rest, the constant rotation of the earth being concealed. The rotation however produces an effect in altering the weight of the body by the vertical component of the force just found, which is known as the centrifugal force. (In the case of the rotation of the earth it is to be observed that it is φ' that is constant rather than p_φ, nevertheless, if in the case above r is constant, φ' will be constant at the same time with p_φ.) The effect of the motion of the earth will be treated in detail in § 104.

49. Example. Three Degrees of Freedom. General Case. As a further example of the process of ignoration of coordinates we will work out the case of a system with three degrees of freedom, one of whose coordinates is cyclic, this being the simplest example which typifies the behavior of the general system. We have then,

152) $$T = \frac{1}{2} Q_{11} q_1'^2 + \frac{1}{2} Q_{22} q_2'^2 + \frac{1}{2} Q_{33} q_3'^2 + Q_{12} q_1' q_2' + Q_{13} q_1' q_3' + Q_{23} q_2' q_3'.$$

If q_3 is the cyclic coordinate, all the Q's are independent of q_3, and if the corresponding force P_3 vanishes, we have the constant momentum,

153) $$p_3 = Q_{13} q_1' + Q_{23} q_2' + Q_{33} q_3' = c_3.$$

From this we determine the cyclic velocity,

154) $$q_3' = \frac{c_3 - Q_{13} q_1' - Q_{23} q_2'}{Q_{23}},$$

inserting which in the kinetic energy gives, on combining terms,

155) $$T' = \frac{1}{2}\left(\frac{Q_{11} Q_{33} - Q_{13}^2}{Q_{33}}\right) q_1'^2 + \frac{1}{2}\left(\frac{Q_{22} Q_{33} - Q_{23}^2}{Q_{33}}\right) q_2'^2 + \frac{Q_{12} Q_{33} - Q_{13} Q_{23}}{Q_{33}} q_1' q_2' + \frac{1}{2} \frac{c_3^2}{Q_{33}}.$$

It is noticeable that the linear terms in q_1', q_2' have cancelled each other. It will be proved below that this always happens. But when we form the kinetic potential, which is to be used instead, they reappear. We have

156) $$\begin{aligned} \Phi &= T' - c_3 q_3' \\ &= \frac{1}{2}\left(\frac{Q_{11} Q_{33} - Q_{13}^2}{Q_{33}}\right) q_1'^2 + \frac{1}{2}\left(\frac{Q_{22} Q_{33} - Q_{23}^2}{Q_{33}}\right) q_2'^2 + \frac{Q_{12} Q_{33} - Q_{13} Q_{23}}{Q_{33}} q_1' q_2' \\ &\quad + \frac{c_3 Q_{13}}{Q_{33}} q_1' + \frac{c_3 Q_{23}}{Q_{33}} q_2' - \frac{1}{2} \frac{c_3^2}{Q_{33}}. \end{aligned}$$

Thus the effect of the cyclic motion, which may itself be concealed from us, is made evident to our observation by the presence of the fourth and fifth terms, which are linear in q_1', q_2'. The apparent coefficients of inertia, that is the coefficients of $q_1'^2, q_2'^2, q_1' q_2'$, are changed from their real values (unless $Q_{13} = Q_{23} = 0$), while there appears the term $-\frac{c_3^{\,2}}{2\,Q_{33}}$ independent of the velocities, depending on the coordinates q_1, q_2. This is, since it gives rise to a conservative positional redaction, undistinguishable in its effect from potential energy. In reality, the reaction to which it gives rise is motional, instead of positional, as it appears to be. If we could explain all potential energy in this manner, namely as due to concealed cyclic motions, we should have solved the chief mystery of dynamics. In his remarkable work on dynamics, Hertz treats all energy from this kinetic point of view. In order to have a successful model for this representation of potential energy, which needs in order to be perfect no linear terms, we must have $Q_{13} = Q_{23} = 0$.

We can now see why the simple example of § 48 showed no linear terms, since by putting all the Q's with one suffix 2 equal to zero we pass to the case of a system with two degrees of freedom. If at the same time the coordinates are orthogonal, $Q_{13} = 0$, so that the single linear term disappears. This was the case above.

Let us now pass to the general case. We have for the momenta the equations 53) § 37 and, for the first r, 137) which are written out,

$$157)\quad \begin{array}{l}
p_1 = Q_{11}\, q_1' + Q_{12}\, q_2' \cdots + Q_{1r}\, q_r' + Q_{1\,r+1}\, q_{r+1}' \cdots + Q_{1m}\, q_m = c_1 \\
p_2 = Q_{21}\, q_1' + Q_{22}\, q_2' \cdots + Q_{2r}\, q_r' + Q_{2\,r+1}\, q_{r+1}' \cdots + Q_{2m}\, q_m' = c_2 \\
\cdots\cdots\cdots\cdots\cdots\cdots \\
\cdots\cdots\cdots\cdots\cdots\cdots \\
p_r = Q_{r1}\, q_1' + Q_{r2}\, q_2' \cdots + Q_{rr}\, q_r' + Q_{r\,r+1}\, q_{r+1}' \cdots + Q_{rm}\, q_m' = c_r \\
\hline
p_{r+1} = Q_{r+1\,1}\, q_1' + Q_{r+1\,2}\, q_2' \cdots + Q_{r+1\,r}\, q_r' + Q_{r+1\,r+1}\, q_{r+1}' \cdots + Q_{r+1\,m}\, q_m' \\
\cdots\cdots\cdots\cdots\cdots\cdots \\
\cdots\cdots\cdots\cdots\cdots\cdots \\
p_m = Q_{m1}\, q_1' + Q_{m2}\, q_2' \cdots + Q_{mr}\, q_r' + Q_{m\,r+1}\, q_{r+1}' \cdots + Q_{mm}\, q_m'
\end{array}$$

Let us now form the kinetic energy from the definition, § 36, 38),

$$158)\qquad 2\,T = \sum_{s=1}^{s=m} p_s\, q_s'.$$

Multiplying the above equations, the sth line by q_s', and adding, we obtain from the first r lines on the right,

$$\sum_{s=1}^{s=r} c_s\, q_s'.$$

The terms coming from the last $m - r$ lines, and the first r columns, as marked off by the dotted lines, are found to be, on collecting according to *columns*,

$$\sum_{s=1}^{s=r} q_s' S_s,$$

on referring to the definition of the definitions of the $S'_s s$, 138),

159) $$S_s = Q_{s,r+1} q'_{r+1} + Q_{s,r+2} q'_{r+2} + \cdots + Q_{s,m} q_m.$$

Finally the terms from the lower right hand square, of $m - r$ rows and columns gives us a quadratic function of the last $m - r$ velocities, namely that part of $2T$ which originally depended on these velocities and no others. This part we will call $2T_a$. We have therefore

160) $$2T = 2T_a + \sum_{s=1}^{s=r} q_s' (S_s + c_s).$$

Now if we form the quadratic functions, with the coefficients R from the determinant of equations 139),

161) $$S = \frac{1}{2}\sum_{s=1}^{s=r}\sum_{t=1}^{t=r} R_{st} S_s S_t,$$
$$C = \frac{1}{2}\sum_{s=1}^{s=r}\sum_{t=1}^{t=r} R_{st} c_s c_t,$$

we may write equations 139) as

162) $$q_s' = \frac{\partial C}{\partial c_s} - \frac{\partial S}{\partial S_s} \qquad (s = 1, 2, \ldots r),$$

so that we may write

163) $$2T = 2T_a + \sum_{s=1}^{s=r} (c_s + S_s)\left(\frac{\partial C}{\partial c_s} - \frac{\partial S}{\partial S_s}\right).$$

But since C, S are homogeneous functions of c_s, S_s respectively,

$$\sum_{s=1}^{s=r} c_s \frac{\partial C}{\partial c_s} = 2C, \quad \sum_{s=1}^{s=r} S_s \frac{\partial S}{\partial S_s} = 2S,$$

so that the above becomes,

164) $$2T = 2T_a + 2C - 2S + \sum_{s=1}^{s=r} S_s \frac{\partial C}{\partial c_s} - \sum_{s=1}^{s=r} c_s \frac{\partial S}{\partial S_s}.$$

But we also have

$$\sum_{s=1}^{s=r} S_s \frac{\partial C}{\partial c_s} = \sum_{s=1}^{s=r} S_s \sum_{t=1}^{t=r} R_{st} c_t = \sum_{t=1}^{t=r} c_t \sum_{s=1}^{s=r} R_{st} S_s = \sum_{t=1}^{t=r} c_t \frac{\partial S}{\partial S_s},$$

so that the sums in 164) destroy each other, and there remains

165) $$T' = T_a - S + C.$$

But S is a homogeneous quadratic function of the S_s's, which are themselves homogeneous linear functions of the q_s''s, so that S, like T_a, is a homogeneous quadratic function of the non-eliminated velocities. Thus we have proved that the linear terms disappear from the kinetic energy. At the same time we have obtained the general value of the part independent of the velocities. Forming the function Φ for the kinetic potential,

$$166)\quad \Phi = T' - \sum_{s=1}^{s=r} c_s q_s' = T_a - S + C - \sum_{s=1}^{s=r} c_s \left(\frac{\partial C}{\partial c_s} - \frac{\partial S}{\partial S_s}\right)$$

$$= T_a - S - C + \sum_{s=1}^{s=r} c_s \frac{\partial S}{\partial S_s},$$

so that the part C which imitates the potiential energy is a homogeneous quadratic function of the momenta c_s of the concealed cyclic motions. The terms under the sign of summation are linear in the remaining velocities.

50. Effect of Linear Terms in Kinetic Potential. Gyroscopic Forces. We will now examine the effect of terms linear in the velocities in the kinetic potential, arising from any cause whatever. We have seen that such terms arise from variable constraints, and from ignored cyclic motions. We shall find a third case when we treat of relative motion, § 103.

Suppose now that the kinetic potential contains the linear part

$$167)\quad \Phi_1 = L_1 q_1' + L_2 q_2' + \cdots L_m q_m',$$

where the coefficients L are functions of the coordinates, and may also involve the time explicitly. Let the part of the force P_s that must be applied on account of the part Φ_1 be denoted by $P_s^{(1)}$, so that

$$168)\quad \frac{d}{dt}\left(\frac{\partial \Phi_1}{\partial q_s'}\right) - \frac{\partial \Phi_1}{\partial q_s} = P_s^{(1)}.$$

Now

$$\frac{\partial \Phi_1}{\partial q_s'} = L_s$$

and differentiating,

$$169)\quad \frac{d}{dt}\left(\frac{\partial \Phi_1}{\partial q_s'}\right) = \frac{dL_s}{dt} = \frac{\partial L_s}{\partial t} + \frac{\partial L_s}{\partial q_1} q_1' + \frac{\partial L_s}{\partial q_2} q_2' + \cdots + \frac{\partial L_s}{\partial q_s} q_m'.$$

We have also

$$\frac{\partial \Phi_1}{\partial q_s} = \frac{\partial L_1}{\partial q_s} q_1' + \frac{\partial L_2}{\partial q_s} q_2' + \cdots + \frac{\partial L_m'}{\partial q_s} q_m.$$

Using these values in 168), we obtain for the force,

$$170)\quad P_s^{(1)} = \left(\frac{\partial L_s}{\partial q_1} - \frac{\partial L_1}{\partial q_s}\right) q_1' + \cdots + \left(\frac{\partial L_s}{\partial q_m} - \frac{\partial L_m}{\partial q_s}\right) q_m' + \frac{\partial L_s}{\partial t}$$

$$= G_{s1}\, q_1' + G_{s2}\, q_2' + \cdots + G_{sm}\, q_m' + \frac{\partial L_s}{\partial t},$$

where the G's are functions of the coordinates deined b—

$$171)\qquad G_{st} = \frac{\partial L_s}{\partial q_t} - \frac{\partial L_t}{\partial q_s}.$$

For the force applied to change a coordinate q_t we have a similar form, with coefficients such that

$$172)\qquad G_{ts} = \frac{\partial L_t}{\partial q_s} - \frac{\partial L_s}{\partial q_t} = - G_{st}.$$

We have then the result that the terms linear in the velocities in the kinetic potential give rise to reactions linear in the velocities, with the property that the coefficient of q_t' in the reaction P_s is equal and of *opposite* sign to the coefficient of q_s' in the reaction P_t. Such reactions are called *gyroscopic forces* by Thomson and Tait[1]), since we have examples of them where gyrostats, or symmetrical bodies spinning about axes attached to parts of systems, act as concealed cyclic motions.

If we find the activity of the gyroscopic forces,

$$173)\qquad \frac{dA_1}{dt} = \sum_{s=1}^{s=m} P_s^{(1)} q_s' = \sum_{s=1}^{s=m} \sum_{t=1}^{t=m} G_{st}\, q_s'\, q_t',$$

we find that the part $P_s^{(1)} q_s'$ we have the term $G_{st}\, q_s'\, q_t'$ while in the part $P_t^{(1)} q_t'$ we have the term $G_{ts}\, q_s'\, q_t'$, and since $G_{ts} = - G_{st}$, these two terms destroy each other. Accordingly the gyrostatic forces disappear from the equation of activity. These forces are consequently conservative motional forces. They are however perfectly distinguishable by their effects from the conservative motional forces arising from the term C which imitates potential energy, and they in no wise imitate potential energy, as we shall see by an example. A system containing gyrostatic members behaves in such a peculiar manner that their presence is easily inferred. The theory of gyrostats will be treated in Chapter VII. In the mean time the following simple example will illustrate the theory, and at the same time serve to prepare for the general theory of the gyrostat, of which it constitutes a special case.

---

1) Thomson and Tait, *Nat. Phil.* § 345VI.

Let four equal masses, $\frac{m}{4}$, be fastened to the ends of two mutually perpendicular arms of length $2l$ and of negligible mass (Fig. 38), which are fastened rigidly where they cross, at their middle points, to an axis perpendicular to them both, about which they turn. Let the point of crossing of the three arms be fixed while the system can spin about the axis OP, which can move in any manner.

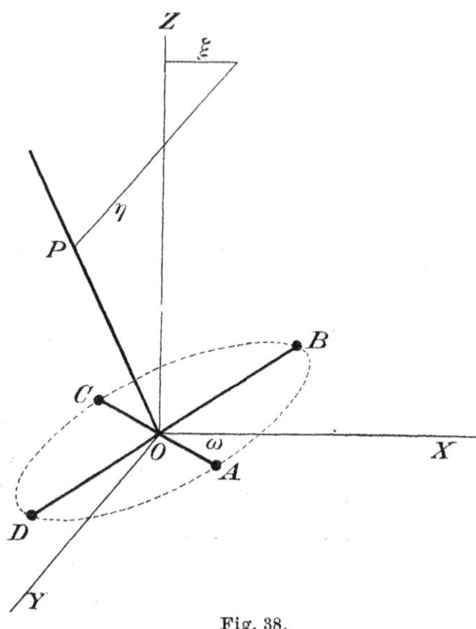

Fig. 38.

We will suppose that during the motion the axis OP makes with the Z-axis a small angle whose square can be neglected in comparison with unity. Let the position of the axis be determined by the coordinates ξ, η, of the point in which it intersects a plane perpendicular to the Z-axis at unit distance from the origin. The squares and products of ξ, η, are consequently to be neglected. Let us further specify the position of the system by the angle φ that the projection of the arm OA on the XY-plane makes with the X-axis. Thus the three coordinates ξ, η, φ determine the position of the whole system.

If the coordinates of the point A are x, y, z, since it lies in a plane whose normal passes through the point ξ, η, 1, we have

$$174) \qquad z + \xi x + \eta y = 0.$$

But since OA always makes a small angle with the XY-plane, the projection of OA on this plane differs from it in length only by a quantity of the second order, which we neglect. We therefore have

$$x = l\cos\varphi, \quad dx = -y\,d\varphi,$$
$$y = l\sin\varphi, \quad dy = \quad x\,d\varphi.$$

Differentiating 174),

$$\begin{aligned} -dz &= \xi dx + \eta dy + x d\xi + y d\eta \\ &= x d\xi + y d\eta + (\eta x + \xi y)\, d\varphi, \end{aligned}$$

so that we have

$$\begin{aligned} dx^2 + dy^2 + dz^2 = {} & (l^2 + \eta^2 x^2 + \xi^2 y^2 - 2\xi\eta x y)\, d\varphi^2 \\ & + x^2 d\xi^2 + y^2 d\eta^2 + 2xy\, d\xi\, d\eta \\ & + 2(\eta x^2 - \xi x y)\, d\xi\, d\varphi \\ & + 2(\eta x y - \xi y^2)\, d\eta\, d\varphi, \end{aligned}$$

and the part contributed by the particle A to the kinetic energy is

175) $$\frac{1}{2}\frac{m}{4}\{(l^2+\eta^2x^2+\xi^2y^2-2\xi\eta xy)\varphi'^2+x^2\xi'^2+y^2\eta'^2+2xy\xi'\eta' + 2(\eta x^2-\xi xy)\xi'\varphi'+2(\eta xy-\xi y^2)\eta'\varphi'\}.$$

The opposite particle C, for which x^2, y^2, xy have the same values, contributes the same amount. The other pair of particles, for which the values of x^2, y^2 are respectively those of y^2, x^2, for the first pair, and the values of xy the negatives of the values for the first pair, consequently contributes an amount of energy which, added to that already found, makes the terms in xy disappear, and replaces each term in x^2, y^2, by the same term with l^2 written in the place of x^2 or y^2. Neglecting then ξ^2, η^2, we have finally

176) $$T=\frac{ml^2}{2}\left\{\varphi'^2+\frac{1}{2}(\xi'^2+\eta'^2)+\eta\xi'\varphi'-\xi\eta'\varphi'\right\}.$$

We accordingly see that φ is a cyclic coordinate for the system, so that if the system is spinning without any force tending to change φ, we are dealing with a case of the example in § 49. We have, proceeding as there,

177) $$\frac{\partial T}{\partial\varphi'}=ml^2\left\{\varphi'+\frac{1}{2}(\eta\xi'-\xi\eta')\right\}=c,$$

and eliminating φ',

178) $$T'=\frac{ml^2}{2}\left\{\frac{1}{2}(\xi'^2+\eta'^2)-\frac{1}{4}(\eta\xi'-\xi\eta')^2\right\}+\frac{c^2}{2ml^2},$$

from which we form

$$\Phi=\frac{ml^2}{2}\left\{\frac{1}{2}(\xi'^2+\eta'^2)-\frac{1}{4}(\eta\xi'-\xi\eta')^2\right\}+\frac{c}{2}(\eta\xi'-\xi\eta')-\frac{c^2}{2ml^2}.$$

In order to form the differential equations for the motion of ξ, η, we have by differentiation

179) $$\frac{\partial\Phi}{\partial\xi'}=\frac{ml^2}{2}\left\{\xi'-\frac{1}{2}\eta(\eta\xi'-\xi\eta')\right\}+\frac{c\eta}{2},$$
$$\frac{\partial\Phi}{\partial\xi}=\frac{ml^2}{2}\left\{\frac{\eta'}{2}(\eta\xi'-\xi\eta')\right\}-\frac{c\eta'}{2},$$

and neglecting the squares and products of the small quantities ξ, η and ξ', η', which are small at the same time,

180) $$\frac{\partial\Phi}{\partial\xi'}=\frac{ml^2}{2}\xi'+\frac{c\eta}{2},$$
$$\frac{\partial\Phi}{\partial\xi}=-\frac{c\eta'}{2}.$$

Proceeding in the same manner for η, we have with the same degree of approximation

181)
$$\frac{\partial \Phi}{\partial \eta'} = \frac{m l^2}{2} \eta' - \frac{c \xi}{2},$$
$$\frac{\partial \Phi}{\partial \eta} = \frac{c}{2} \xi'.$$

If W is the potential energy (there being no apparent potential energy due to the cyclic motion, since the part C is here constant), the equations of motion are accordingly,

182)
$$\frac{m l^2}{2} \xi'' + c \eta' + \frac{\partial W}{\partial \xi} = 0,$$
$$\frac{m l^2}{2} \eta'' - c \xi' + \frac{\partial W}{\partial \eta} = 0.$$

Thus the gyroscopic terms in c have the property proved in 172).

If there is no potential energy, the gyroscopic forces cause the motion to be of such a nature that

$$\xi'' \xi' + \eta'' \eta' \mp 0, \quad \sqrt{\xi''^2 + \eta''^2} = \frac{2c}{m l^2} \sqrt{\xi'^2 + \eta'^2},$$

that is the acceleration is perpendicular to the velocity, and proportional to it. Under these circumstances the motion is uniform circular motion. In fact the equations are satisfied by

183)
$$W = 0, \quad \begin{array}{l} \xi = A \cos p t, \\ \eta = A \sin p t, \end{array} \quad p = \frac{2c}{m l^2}.$$

Thus the circle, whatever its size, is described in the same time $\frac{\pi m l^2}{c}$, which is inversely proportional to the momentum of the cyclic motion. We may describe the effect of the gyroscopic forces in general for a system with two degrees of freedom by saying that they tend to cause a point to veer out from its path always toward the same side. This effect is characteristic, and cannot be imitated by any arrangement of potential energy whatever. By the aid of this principle all the motions of tops and gyrostats may be explained.

51. Cyclic Systems. A system in which the kinetic energy is represented with sufficient approximation by a homogeneous quadratic function of its cyclic velocities is ealled a Cyclic System. Of course the rigid expression of the kinetic energy contains the velocities of every coordinate of the system, cyclic or not, for no mass can be moved without adaing a certain amount of kinetic energy. Still if certain of the coordinates change so slowly that their velocities may be neglected in comparison with the velocities of the cyclic coordinates, the approximate condition will be fulfilled. These coordinates define the position of the cyclic systems, and may be called the *positional coordinates* or *parameters* of the system. In the example of § 48 if we suppose the radial motion to be so slow that we may neglect r'^2 in comparison with $r^2\varphi'^2$ we have

184) $$T = \frac{1}{2} m r^2 \varphi'^2,$$

and the system is cyclic, r being the positional, φ the cyclic coordinate. In the case of a liquid circulating through an endless rubber tube, the positional coordinates would specify the shape and position of the tube. The positional coordinates will be distinguished from the cyclic coordinates by not being marked with a bar. The analytical conditions for a cyclic system will accordingly be, for all coordinates, either

185) $$\frac{\partial T}{\partial \bar{q}_s} = 0 \quad \text{or} \quad \frac{\partial T}{\partial q_s'} = p_s = 0,$$

or if we use the Hamiltonian equations 78) § 39 with the value of T obtained by replacing the velocities by the momenta, which we shall denote by T_p, since the non-cyclic momenta vanish

186) $$\frac{\partial T_p}{\partial p_s} = 0, \quad \text{and} \quad \frac{\partial T_p}{\partial \bar{q}_s} = 0,$$

for the cyclic coordinates, as before. We accordingly have for the external impressed forces tending to increase the positional coordinates, by § 37, 60) § 39, 80) respectively, the first term vanishing,

187) $$F_s = P_s = -\frac{\partial (T - W)}{\partial q_s} = \frac{\partial (T_p + W)}{\partial q_s},$$ [1]

and for the cyclic coordinates

$$\bar{P}_s = \frac{d}{dt}\left(\frac{\partial T}{\partial \bar{q}_s'}\right) = \frac{d\bar{p}_s}{dt}.$$

A motion in which there are no forces tending to change the cyclic coordinates is called an *adiabatic* motion, since in it no energy enters or leaves the system through the *cyclic* coordinates. (It may do so through the positional coordinates.) Accordingly in such a motion the cyclic *momenta* remain constant. The case worked out above was such a motion.

In adiabatic motions the cyclic *velocities* do not generally remain constant. In the above example, for instance, the cyclic velocity φ' was given by

$$\varphi' = \frac{c}{mr^2}.$$

A motion in which the cyclic *velocities* remain constant is called *isocyclic*.

---

1) That $-\frac{\partial T}{\partial q_s} = \frac{\partial T_p}{\partial q_s}$ may be seen by putting $r = m$ in 144), when the parenthesis becomes $T' - 2T = -T_p$.

The motion of a particle relatively at rest upon the surface of the earth is isocyclic, taking account of the earth's rotation.

In such a motion the cyclic momenta do not generally remain constant, but forces have to be applied.

In the example of the bead on the revolving rod if r varied forces would have to be applied to the rod to keep the rotation φ' constant.

If the motion is isocyclic, the only *variables* appearing in T are the q's, the positional coordinates. The positional forces, 187), are then derivable from a force-function $W - T$[1]), so that even if the system possessed no potential energy, it would appear to possess an amount of potential energy $-T$. If the motion on the other hand is adiabatic, the energy in the form T_p again contains as variables only the coordinates q_s, and the positional forces are now derivable from the force-function $T_p + W$, so that in this case a system without potential energy would appear to contain the amount of potential energy $+T_p$. In this manner we are enabled to explain potential energy as kinetic energy of concealed cyclic motions, thus adding materially to our conceptions of the nature of force For it is to be noted that kinetic energy is an entity depending only on the property of inertia, which is possessed by all bodies, while potential energy is a term employed only to disguise our ignorance of the nature of force. Accordingly when we are able to proceed to an explanation of a static force by means of kinetic phenomena, we have made a distinct advance in our knowledge of the subject. A striking example is furnished by the kinetic theory of gases, by means of which we are enabled to pass from the bare statement that all gases press against their confining vessels to the statement that this pressure is due to the impact of the molecules of the gas against the walls of the vessel.

52. Properties of Cyclic Systems. Reciprocal Relations. Since by the properties of the kinetic energy we have three different kinds of quanties represented by partial derivatives of one or the other of two functions,

$$189)\quad P_s = -\frac{\partial T}{\partial q_s},\quad \bar{p}_s = \frac{\partial T}{\partial \bar{q}_s'},\quad P_s = \frac{\partial T_p}{\partial q_s},\quad \bar{q}_s' = \frac{\partial T_p}{\partial \bar{p}_s},$$

applying the principle that a derivative by two variables is independent of the order of the differentiations we obtain six reciprocal theorems. We shall throughout suppose that there is no potential energy.

1) The reason for the appearance of W with the *positive* sign is that as explained in § 37, P_s denotes the *external* impressed forces, which in the case of equilibrium, are equal and *opposite* to the internal forces given by W.

Ia. In an adiabatic motion if an increase in one positional coordinate q_r causes an increase in the impressed force P_s belonging to another positional coordinate q_s at a certain rate, then an increase in the positional coordinate q_s causes an increase in the impressed force P_r at the same rate. For

$$190)\qquad \frac{\partial P_s}{\partial q_r} = \frac{\partial^2 T_p}{\partial q_r \partial q_s} = \frac{\partial P_r}{\partial q_s}.$$

Ib. In an isocyclic motion we have the same property as above. For

$$191)\qquad \frac{\partial P_s}{\partial q_r} = -\frac{\partial^2 T}{\partial q_r \partial q_s} = \frac{\partial P_r}{\partial q_s}.$$

IIa. If in any motion an increase of any cyclic momentum p_r, the positional coordinates being unchanged, causes an increase in a cyclic velocity q_s' at a certain rate, then an increase in the momentum $\bar{p}_s$, the positional coordinates being unchanged, causes an increase in the velocity $\bar{q}_r'$ at the same rate. For

$$192)\qquad \frac{\partial \bar{q}_s'}{\partial p_r} = \frac{\partial^2 T_p}{\partial \bar{p}_r \partial \bar{p}_s} = \frac{\partial \bar{q}_r}{\partial \bar{p}_s}.$$

IIb. If in any motion an increase in any cyclic velocity $\bar{q}_r'$, the positional coordinates being unchanged, causes an increase in a cyclic momentum $\bar{p}_s$, then an increase in the velocity $\bar{q}_s'$ causes an increase in the momentum $\bar{p}_r$ at the same rate. For

$$193)\qquad \frac{\partial \bar{p}_s}{\partial \bar{q}_r'} = \frac{\partial^2 T}{\partial \bar{q}_r' \partial \bar{q}_s'} = \frac{\partial \bar{p}_r}{\partial \bar{q}_s'}.$$

IIIa. If an increase in one of the cyclic momenta $\bar{p}_r$, the positional coordinates being unchanged, causes an increase in the impressed force P_s necessary to be applied to one of the positional coordinates q_s (in order to prevent its changing), then an adiabatic increase of the positional coordinate q_s will cause the cyclic velocity $\bar{q}_r'$ to increase at the same rate. For

$$194)\qquad \frac{\partial P_s}{\partial \bar{p}_r} = \frac{\partial^2 T_p}{\partial \bar{p}_r \partial q_s} = \frac{\partial \bar{q}_r'}{\partial q_s}.$$

IIIb. If an increase in one of the cyclic velocities $\bar{q}_r'$, the positional coordinates being unchanged, causes an increase in the impressed force P_s necessary to be applied to one of the positional coordinates q_s (in order to prevent its changing), then an isocyclic increase of the positional coordinate q_s will cause the cyclic momentum $\bar{p}_r$ to *decrease* at the same rate. For

$$195)\qquad \frac{\partial P_s}{\partial \bar{q}_r'} = -\frac{\partial^2 T}{\partial \bar{p}_r' \partial q_s} = -\frac{\partial \bar{p}_r}{\partial q_s}.$$

53. Work done by the Cyclic and Positional Forces.

I. In an isocyclic motion, the work done *by* the cyclic forces is double the work done by the system *against* the positional forces. In such motions the energy of the system accordingly increases by one-half the work done by the cyclic forces, the other half being given out against the positional forces. For if we use the energy in the form

$$T = \frac{1}{2}\sum_s \bar{q}_s' \bar{p}_s,$$

we have in any change

196) $$\delta T = \frac{1}{2}\sum_s (\bar{q}_s' \,\delta \bar{p}_s + \bar{p}_s \,\delta \bar{q}_s'),$$

and in an isocyclic change, every $\delta \bar{q}_s'$ vanishing,

197) $$\delta T = \frac{1}{2}\sum_s \bar{q}_s' \,\delta \bar{p}_s.$$

But since

198) $$\frac{d\bar{p}_s}{dt} = \bar{P}_s,\ \delta \bar{p}_s = \bar{P}_s \delta t, \quad \text{and since} \quad \bar{q}_s' = \frac{d\bar{q}_s}{dt},\ \bar{q}_s' \,\delta t = \delta \bar{q}_s,$$

and the above expression for the gain of energy becomes

199) $$\delta T = \frac{1}{2}\sum_s \bar{q}_s' \bar{P}_s \delta t = \frac{1}{2}\sum_s \bar{P}_s \delta \bar{q}_s.$$

But the work done by the cyclic forces is

200) $$\delta \bar{A} = \sum_s \bar{P}_s \delta \bar{q}_s = 2\,\delta T.$$

Therefore the last part of the theorem is proved. Again, in any motion,

201) $$\delta T = \sum_s \frac{\partial T}{\partial \bar{q}_s'} \,\delta \bar{q}_s' + \sum_s \frac{\partial T}{\partial q_s} \,\delta q_s,$$

and in an isocyclic motion,

202) $$\delta T = \sum_s \frac{\partial T}{\partial q_s} \,\delta q_s.$$

But since the work of the positional forces is

203) $$\delta A = \sum_s P_s \,\delta q_s = -\sum \frac{\partial T}{\partial q_s} \,\delta q_s = -\,\delta T,$$

the first part of the proposition is also proved.

II. In an adiabatic motion, the cyclic velocities will in general be changed.

Then they change in such a way that the positional forces caused by the change of cyclic velocities oppose the motion, that is, do a positive amount of work. For since for any positional force

$$P = -\frac{\partial T}{\partial q_s},$$

the change due to the motion is

204) $$\delta P_s = -\frac{\partial \delta T}{\partial q_s} = -\sum_r \frac{\partial^2 T}{\partial q_s \partial q_r} \delta q_r - \sum_r \frac{\partial^2 T}{\partial q_s \partial \bar{q}_r'} \delta \bar{q}_r'.$$

Of this the part due to the change in the cyclic velocities is

205) $$\delta_{\bar{q}'} P_s = -\sum_r \frac{\partial^2 T}{\partial q_s \partial \bar{q}_r'} \delta \bar{q}_r' = -\sum_r \frac{\partial \bar{p}_r}{\partial q_s} \delta \bar{q}_r',$$

and the work done by these forces is

206) $$\delta_{\bar{q}'} A = \sum_s \delta_{\bar{q}'} P_s \delta q_s = -\sum\sum_r \frac{\partial \bar{p}_r}{\partial q_s} \delta q_s \delta \bar{q}_r'.$$

Now we have for any motion

207) $$\delta \bar{p}_r = \sum_s \frac{\partial \bar{p}_r}{\partial q_s} \delta q_s + \sum_s \frac{\partial \bar{p}_r}{\partial \bar{q}_s} \delta \bar{q}_s',$$

and in an adiabatic motion this is zero, so that

208) $$\sum_s \frac{\partial \bar{p}_r}{\partial q_s} \delta q_s = -\sum_s \frac{\partial \bar{p}_r}{\partial \bar{q}_s} \delta \bar{q}_s'.$$

Substituting this in the double sum 206), we get

209) $$\delta_{\bar{q}'} A = \sum_s \sum_r \frac{\partial \bar{p}_r}{\partial \bar{q}_s'} \delta q_s' \delta \bar{q}_r' = \sum_s \sum_r Q_{rs} \delta \bar{q}_s' \delta q_r'.$$

But this expression represents [§ 36, 35)] twice the energy of a possible motion in which the velocities would be $\delta \bar{q}_s'$, and must therefore be positive for all values of $\delta \bar{q}_s'$, $\delta \bar{q}_r'$.

Accordingly $\delta_{\bar{q}'} A > 0$.

The interpretation of this theorem for electrodynamics is known as Lenz's Law[1]), namely, an electrical current being represented by a cyclic velocity, and the shape and relative position of the circuits by positional coordinates, if in any system of conductors carrying currents, the relative positions of the conductors are changed, the induced currents due to the motion of the conductors are so directed as by their magnetic action to oppose the motion.

54. Examples of Cyclic Systems. Let us consider the example of equation 184) as illustrating the previous theorems.

We have for the momenta

$$p_r = \frac{\partial T}{\partial r'} = 0, \quad \bar{p}_\varphi = \frac{\partial T}{\partial \varphi'} = m r^2 \varphi',$$

1) These Theorems are all given by Hertz, *Prinzipien der Mechanik*, §§ 568—583.

and introducing these instead of the velocities

$$T_p = \frac{1}{2mr^2}\bar{p}_\varphi^2. \tag{210}$$

We have for the positional force

$$P_r = -\frac{\partial T}{\partial r} = -mr\varphi'^2 = \frac{\partial T_p}{\partial r} = -\frac{1}{mr^3}\bar{p}_\varphi^2. \tag{211}$$

This being negative denotes that a force P_r toward the axis must be impressed on the mass m order to maintain the cyclic state. This may be accomplished by means of a geometrical constraint, or by means of a spring. The force or reaction $-P_r$ which the mass m exerts in the direction from the axis in virtue of the rotation is the so-called centrifugal force. We see that if the motion is isocyclic, the positional force increases with r, while if it is adiabatic, as in the case worked out above, it decreases when r increases. The verification of the theorems of § 52 is obvious. The cyclic force

$$\overline{P}_\varphi = \frac{d\bar{p}_\varphi}{dt} = m\frac{d}{dt}(r^2\varphi)$$

vanishes when the rotation is uniform, and the radius constant. If, the motion being isocyclic, that is, one of uniform angular velocity, the body moves farther from the axis, $\overline{P}_\varphi$, the cyclic force is positive, that is, unless a positive force $\overline{P}_\varphi$ is applied, the angular velocity will diminish. In moving out from r_1 to r_2 work will be done against the positional force P_r of amount

$$-A = -\int_{r_1}^{r_2} P_r dr = m\varphi'^2 \int_{r_1}^{r_2} r\,dr = \frac{m\varphi'^2}{2}(r_2^2 - r_1^2), \tag{212}$$

while the energy *increases* by the same amount.

Thus the first theorem of § 53 is verified. If the motion is adiabatic,

$$\bar{p}_\varphi = mr^2\varphi' = c.$$

If the body moves from the axis, φ' will accordingly decrease, so that

$$r^2\delta\varphi' + 2r\varphi'\delta r = 0. \tag{213}$$

The change in P_r due to a displacement δr is, by 211)

$$\delta P_r = -m(\varphi'^2\delta r + 2r\varphi'\delta\varphi') \tag{214}$$

of which the part containing $\delta\varphi'$

$$\delta_{\varphi'} P_r = -2mr\varphi'\delta\varphi' \tag{215}$$

does the work

216) $$\delta_{\varphi'} A = \delta_{\varphi'} P_r \delta r = -2 m r \varphi' \delta \varphi' \delta r,$$
or by 213),
217) $$\delta_{\varphi'} A = m r^2 \delta \varphi'^2,$$
which is positive, illustrating Theorem II, § 53.

A further example is found in the motion of the following system. Two particles of equal masses m are fastened to a rod of length $2a$ pivoted at its central point upon an axis fastened to the horizontal rod of the previous example at a distance b from the axis of rotation in such a way that the two masses can move in the vertical plane containing the axis of rotation. The inclination of the pivoted rod to the vertical being ϑ, the distances of the particles from the axis of rotation are respectively

$$r_1 = b + a \sin\vartheta, \quad r_2 = b - a \sin\vartheta.$$

The system is fully specified by the coordinates ϑ and φ, the latter having the same meaning as before.

It is evident that the kinetic energy is given by

218) $$T = \frac{m}{2}\{r_1^2 \varphi'^2 + r_2^2 \varphi'^2 + 2 a^2 \vartheta'^2\} = m\{(b^2 + a^2 \sin^2\vartheta)\varphi'^2 + a^2\vartheta'^2\}$$

so that φ is again the cyclic coordinate.[1])

To find the change of ϑ we have

$$\frac{\partial T}{\partial \vartheta'} = 2 m a^2 \vartheta', \quad \frac{\partial T}{\partial \vartheta} = 2 m a^2 \varphi'^2 \sin\vartheta \cos\vartheta,$$

giving us the differential equation,

219) $$2 m a^2 \frac{d^2\vartheta}{dt^2} - 2 m a^2 \varphi'^2 \sin\vartheta \cos\vartheta = 0,$$
$$\frac{d^2\vartheta}{dt^2} = \varphi'^2 \sin\vartheta \cos\vartheta.$$

If the motion is isocyclic φ is constant, and since the angular acceleration $\frac{d^2\vartheta}{dt^2}$ vanishes when ϑ equals zero or $\frac{\pi}{2}$, we see that the rod carrying the particles will remain at rest relatively to the horizontal rod in either a vertical or horizontal position. It is easy to see that the vertical position is one of unstable equilibrium, for, writing the equation 219)

220) $$\frac{d^2(2\vartheta)}{dt^2} = \varphi'^2 \sin 2\vartheta,$$

we see that if ϑ be slightly different from zero, ϑ will tend to become still greater in absolute value. Writing however $\vartheta = \frac{\pi}{2} - \vartheta'$ the equation becomes

221) $$\frac{d^2(2\vartheta)}{dt^2} = -\varphi'^2 \sin 2\vartheta'.$$

1) The *system* is cyclic if we neglect $a^2\vartheta'^2$.

If ϑ' is slightly different from zero, it will accordingly tend to approach the value zero, so that the horizontal position is stable.

A body moving according to the differential equation 221) is called by Thomson and Tait[1]) a *quadrantal pendulum*, since ϑ changes "according to the same law with reference to a quadrant on each side of its positions of equilibrium as the common pendulum with reference to a half-circle on each side", or in other words, in the ordinary pendulum the acceleration is proportional to the sine of the angle of deviation from equilibrium, and in the quadrantal to the sine of twice the angle. The *small* oscillation performed by the bar will be harmonic with the frequency $\frac{\varphi'}{2\pi}\cdot$ Here we have an excellent example of an apparent potential energy which is really kinetic.

1) Thomson and Tait, *Nat. Phil.* § 322.

PART II

DYNAMICS OF RIGID BODIES

CHAPTER VI.

SYSTEMS OF VECTORS. DISTRIBUTION OF MASS. INSTANTANEOUS MOTION.

55. Translations and Rotations. A rigid body or system of material particles is one in which the distance of each point of the system from every other is invariable. Its position is known when the positions of any three of its points are known, for every point is determined by its distances from three given points. These three points have each three coordinates, but, since there are three conditions between them, defining their mutual distances, there are only six *independent* coordinates. Thus, a rigid body has *six* coordinates.

A rigid body may evidently be displaced in such a manner that the displacement of every point is represented by equal vectors, that is equal in length and parallel. Such a displacement is called a *translation*, and, being represented by a *free* vector, has three coordinates.

A rigid body may also evidently be displaced so that two given points in it, A and B, remain fixed. Since any point P must move on a sphere of radius BP about B, and also on a sphere of radius AP about A, the locus of its positions is the intersection of the two spheres, that is a circle whose plane is perpendicular to the line AB, and whose radius CP is the perpendicular distance from P to the line AB. If this is zero, the point does not move, therefore all points on the line AB remain fixed. The displacement is called a *rotation* and the line AB, the axis of rotation. The rotation is specified if we know the situation of the line AB and the magnitude of the angle PCP', or the angle of rotation.

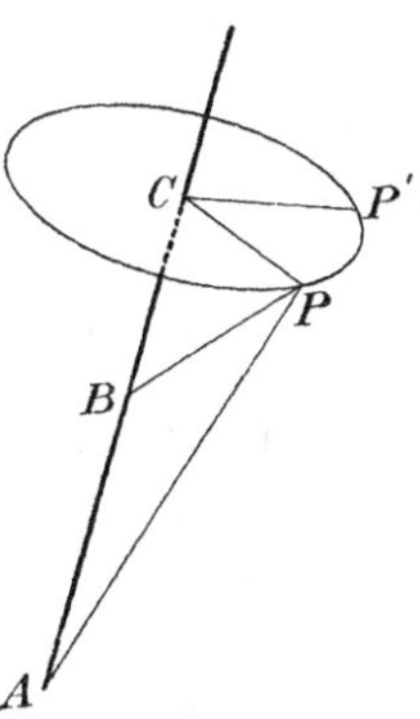

Fig. 39.

A line may be specified by giving the two pairs of coordinates of the points in which it intersects two of the coordinate planes. A line has thus four coordinates, and a rotation, five — the four of the axis together with the magnitude of the angle.

Any displacement of a rigid body may be brought about in an indefinite number of ways. Let three points ABC (Fig. 40) be displaced to $A'B'C'$. We may first give the body a translation defined by the vector AA'. This will bring B to B_1 and C to C_1. Then through A' pass an axis perpendicular to the plane $B_1A'B'$, and rotate the body about this axis through the angle $B_1A'B'$. This brings B_1 to B' and C_1 to a new position C_2. Finally rotate the body about $A'B'$ until C_2 arrives at C'. We have thus brought about the given displacement by means of a succession of translation and rotations. Evidently the order of these may be varied. Accordingly,

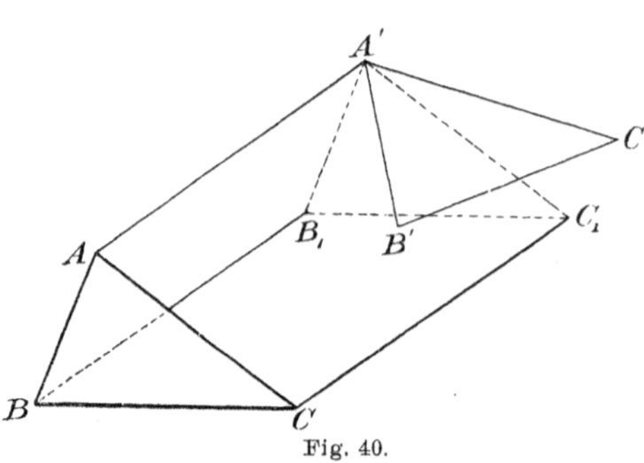

Fig. 40.

Any displacement of a rigid body may be reduced to a succession of translations and rotations.

We have seen that a translation may be represented by a free vector, a rotation, by a vector that must give the axis and the angle. If we agree to draw the vector *in* the axis, and make its length numerically equal to the angle of rotation, it will completely specify the rotation, if we adopt a convention about the direction of rotation. This shall be that, if the rotation is in the direction of the hands of a watch, the vector shall point from face to back of the watch. Vector and rotation correspond then to the translation and rotation in the motion of a cork-screw, or any right-handed screw. As the vector may be placed anywhere along the axis, but not out of it, it has five coordinates, and may be characterized as a *sliding vector*.

Translations are compounded by the law of addition of vectors. The resultant of two rotations about the *same axis* is evidently the algebraic sum of the individual rotations. The resultant of translation and rotation is evidently independent of the order in which they take place.

The resultant of a rotation and a translation perpendicular to its axis is equivalent to a rotation about a parallel axis, for it is evident that all points move in planes perpendicular to the axis, and that the motions of all such planes are alike, or the motion is *uniplanar*.

Now the motions of any two points in a plane determine the motion of the plane parallel to itself.

From O (Fig. 41) lay off the translation vector OO' of length τ and find a point C on the perpendicular bisecting OO' which makes the angle OCO' equal to ω, the angle of rotation, and in the right sense. Then if OC be rotated about O through the angle ω to C' and then C' be moved by the translation it will return to C. Therefore the point C remains fixed, and is the center of rotation, and thus the rotation ω about C is equivalent to the equal rotation about O together with the translation,

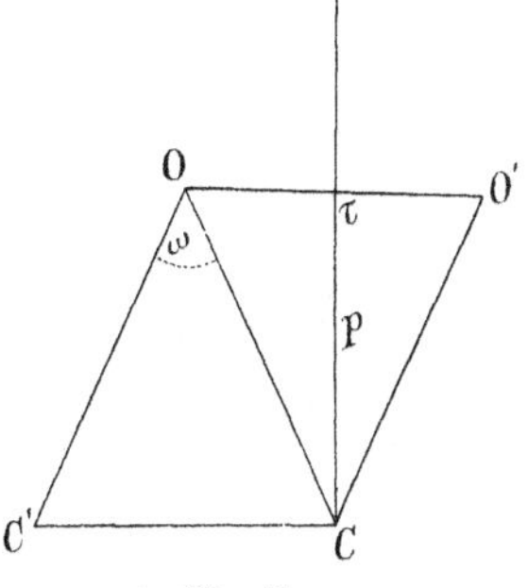

Fig. 41.

1) $$\tau = 2\,\overline{OC}\sin\frac{\omega}{2},$$

and if p is the perpendicular from C to OO',

2) $$p = OC\cos\frac{\omega}{2} = \frac{\tau}{2}\cot\frac{\omega}{2}.$$

56. Rotations about two Parallel Axes. As before the motion is uniplanar and is specified by two points. Let A and B (Fig. 42) be the intersections of the axes with the plane of the paper perpendicular to them. Turn about A through the angle ω_1, bringing B to B'. Then turn about B' through the angle ω_2, bringing A to A'. Bisect ω_1 by AC. B could be brought to B' by rotation about any point of AC, since all such points are equidistant from BB'. Bisect ω_2 by $B'D$. A could be brought to A' by rotation about any point in $B'D$. Therefore the motion of A and B could be produced by a rotation about O, the intersection of AC and $B'D$. Triangle AOA' is isosceles.

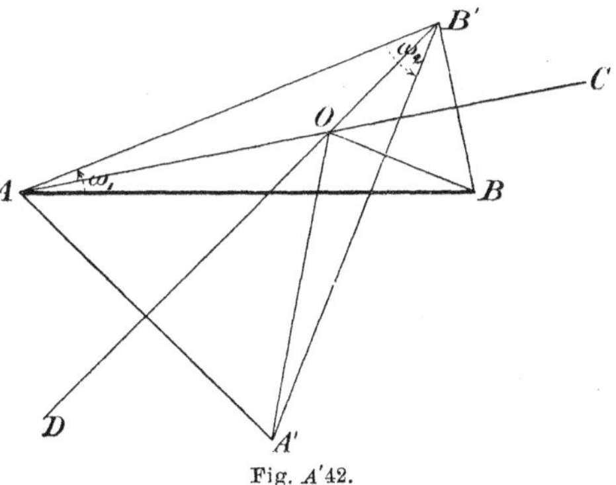

Fig. A'42.

$$\text{Angle } AOD = \text{angle } OAB' + \text{angle } AB'O = \frac{\omega_1}{2} + \frac{\omega_2}{2},$$

$$\text{Angle } AOA' = 2 \cdot \text{angle } AOD = \omega_1 + \omega_2,$$

that is, two rotations about parallel axes compound into a rotation equal to their algebraic sum about a parallel axis. To find the position of this axis we have

3) $$\frac{OB'}{\sin\frac{\omega_1}{2}} = \frac{OA}{\sin\frac{\omega_2}{2}} = \frac{AB}{\sin\frac{\omega_1+\omega_2}{2}}.$$

If the order of rotation is changed we obtain a *different* result.

If the rotations ω_1 and ω_2 are of opposite signs and of equal magnitudes, the intersecton of the two bisectors is at infinity and the axis of rotation is thus at infinity. A motion about an infinitely distant axis is a translation. The direct proof is as follows.

Let A be the center of rotation ω, bringing B to B'. Then rotate about B' through an equal angle in the opposite direction, bringing A to A'. Triangles ABB' and $AA'B'$ have AB' common, and $AB = A'B'$ and the included angles equal, therefore AA' and BB' are equal and parallel and two points — consequently all points — have moved parallel to each other the same distance. The motion is therefore a translation of magnitude,

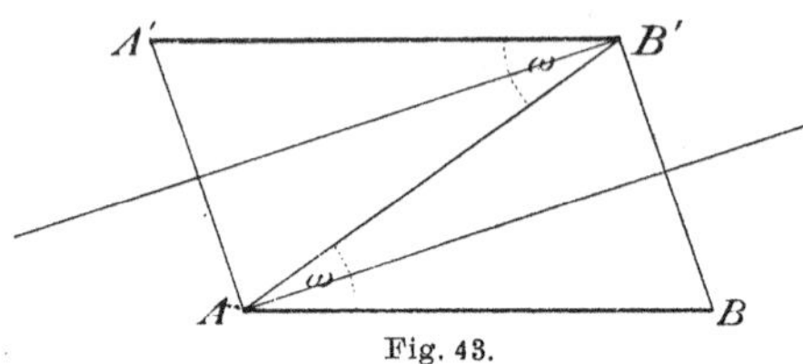

Fig. 43.

$$\tau = 2AB \sin\frac{\omega}{2}. \tag{4}$$

Accordingly every translation may be decomposed into rotations, and we may reduce all displacements to rotations.

57. Rotations about Intersecting Axes. Infinitesimal Rotations. Let OA and OB be two intersecting axes about which we revolve the body through the angles ω_1 and ω_2 respectively. Describe a sphere with the center O. Let the rotation ω_1 about A bring B to B', and ω_2 about B' bring A to A'. Pass planes through the vertices bisecting the angles ω_1 and ω_2, then, as in § 56, the displacement just given is equivalent to a rotation about the line of intersection CO of these planes. The order of the rotations affects the result.

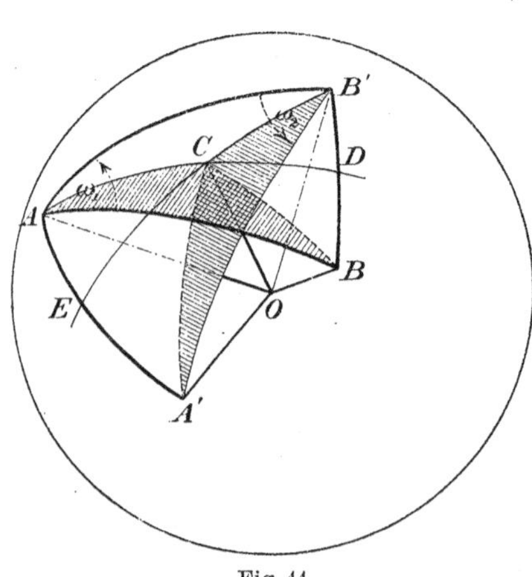

Fig. 44.

Since AC bisects the angle BAB' and the spherical triangle BAB' is isosceles,

$$\text{angle } ABC = \text{angle } AB'C = \frac{\omega_2}{2}.$$

Thus the resultant rotation, $\omega = \text{angle } ACA' = \text{angle } BCB'$

$$\text{Angle } ACE = \text{angle } B'CD = \text{angle } DCB = \frac{\omega}{2}.$$

In the spherical triangle ABC we have

$$5) \qquad \frac{\sin\frac{\omega_1}{2}}{\sin COB} = \frac{\sin\frac{\omega_2}{2}}{\sin COA} = \frac{\sin\frac{\omega}{2}}{\sin AOB}.$$

The preceding results are much simplified if the rotations are infinitely small.

We shall first prove that two equal infinitely small rotations in the same sense about axes infinitely near each other may be regarded as equal. Suppose the axes first parallel, and perpendicular to the paper which they cut in A and B. Let a point P be rotated about A through the angle $d\omega$ to P', and through the same angle about B to P''. The arcs PP' and PP'' differ by the amount $dr\,d\omega$, if dr is the difference between AP and BP. They are inclined to each other at an infinitely small angle BPA, and as the sides PP' and PP'' are infinitely small, and differ by an infinitely small quantity of the second order, $P'\,P''$ is of the second order. If the axes are inclined to each other at an infinitesimal angle, there is a third component perpendicular to $P'\,P''$, which is likewise of the second order. Therefore the theorem is proved.

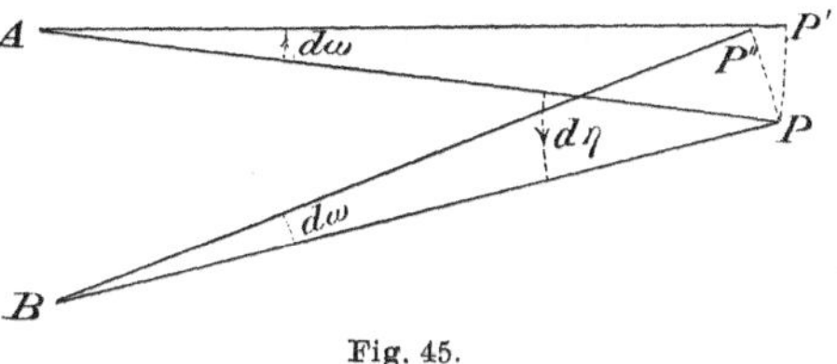

Fig. 45.

The theorem of rotations about intersecting axes may then be stated. Two infinitesimal rotations about intersecting axes are equivalent to a rotation about an axis in their plane, the order of rotations being immaterial. To find the position of the axis of the resultant rotation, we have, by 5)

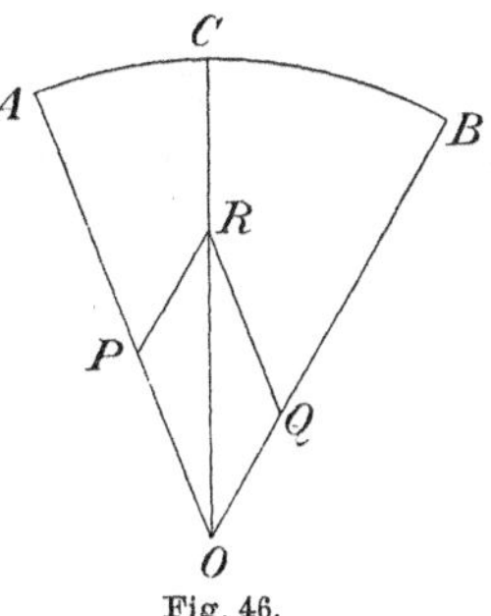

Fig. 46.

$$6) \qquad \frac{\frac{d\omega_1}{2}}{\sin COB} = \frac{\frac{d\omega_2}{2}}{\sin COA} = \frac{\frac{d\omega}{2}}{\sin AOB}.$$

If we lay off on the axes OA and OB (Fig. 46) lengths OP and OQ proportional to the rotations $d\omega_1$ and $d\omega_2$, the above equations show that OC is in the direction of the diagonal of a parallelogram constructed on OP and OQ as sides and the resultant rotation $d\omega$ is proportional to the diagonal OR.

Therefore the resultant of two infinitesimal rotations whose axes intersect is found by the parallelogram construction, or by the law of addition of vectors. This process may be extended to any number of infinitesimal rotations whose axes intersect.

The theorem regarding rotations about parallel axes becomes: Infinitesimal rotations about two parallel axes compound into a rotation about a parallel axis lying in their plane. We have for its position by 3),

$$7)\qquad \frac{OB}{\omega_1} = \frac{OA}{\omega_2} = \frac{AB}{\omega_1 + \omega_2},$$

showing that the point of application of the resultant is at the center of mass of masses proportional to the component rotations placed at their points of application.

If vectors representing ω_1 and ω_2 are laid off anywhere on their axes, the position of the axis O may be found by the following construction. At A a point on the axis of rotation ω_1 lay off $AR = \omega_2$ and at B at a point on the axis of rotation ω_2. in the *opposite direction* $BS = \omega_1$. Join R and S, and where this straight line RS cuts AB, draw OT parallel to AR, BS equal in length to $\omega_1 + \omega_2$. For

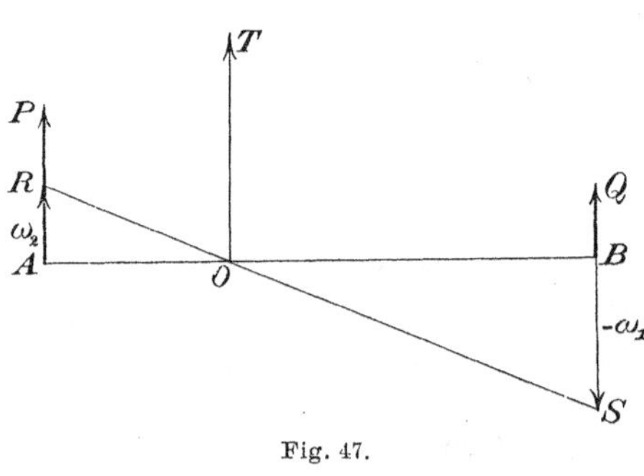

Fig. 47.

$$\frac{OA}{OB} = \frac{AR}{BS} = \frac{\omega_2}{\omega_1},$$

as required by 7).

The construction (Fig. 47) shows that if ω_1 and ω_2 have the same sign, the resultant $\omega_1 + \omega_2$ has its axis O between A and B.

If ω_1 and ω_2 are of opposite signs the same construction may be used (Fig. 48), but O is on AB produced and on the side of the greater rotation. If $\omega_1 = -\omega_2$ evidently O is at infinity and $\omega = 0$. The resultant is then a translation perpendicular to the plane of the two axes, and its magnitude τ is by 4) equal to $\omega\omega_1$ times the perpendicular distance between the axes.

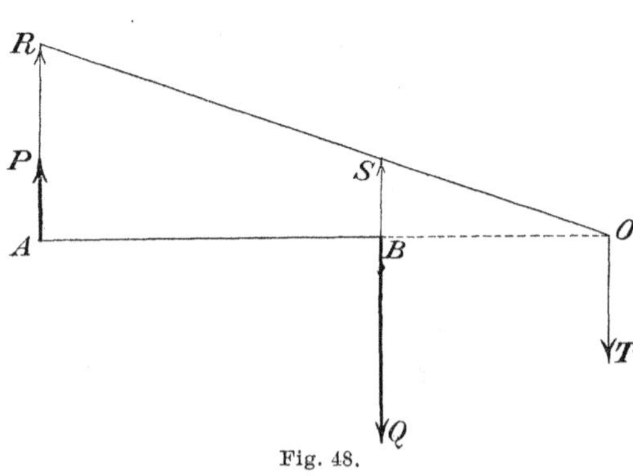

Fig. 48.

58. Vector-couples. A pair of equal, parallel, oppositely directed, sliding vectors will be called a *vector-couple*. A rotation vector-couple is thus equivalent to a translation perpendicular to its plane, equal to the product of the length of either vector by the perpendicular distance between their lines, or the *arm* of the couple. This product is called the *moment* of the couple.

Two couples whose planes are parallel give rise to parallel translations, and if their moments are equal, to equal translations. Therefore a rotation-couple may be displaced without altering its effect, if its plane is kept parallel to itself and its moment is unchanged.

A vector-couple may then be represented by a single vector perpendicular to its plane, whose length is equal to the moment of the couple. Its direction will be governed by the same convention as before, namely, the vector moment is to be drawn in such a direction that rotation in the direction of the couple and translation in that of the moment correspond to the motion of a right-handed screw.

Moments will be represented by heavy vectors. The moment of a vector-couple is a *free* vector, hence the composition of couples is simpler than that of the slide-vectors themselves.

We may now state the theorem of the general infinitely small displacement of a body as follows: *The infinitely small displacement of a body may be reduced to a translation and a rotation, or in other words to a rotation and a rotation-couple.* The choice of components may be made in an infinite number of ways.

59. Statics of a Rigid Body. Two equal, parallel, oppositely directed forces applied to a rigid body in the same line are in equilibrium. For otherwise they can produce only distortion or motion. Distortion is excluded according to the definition of a rigid body. They satisfy the conditions of equilibrium, § 32, for if applied at the center of mass they are in equilibrium, and their moments about *any* point are equal and opposite. Accordingly a force applied to a rigid body may be applied at *any* point in its line of direction without change of effect. Thus forces applied to a rigid body are not free, but are sliding vectors (five coordinates). (This is not a property of forces, but of rigid bodies.) Forces, whose lines of direction intersect, may be applied at the point of intersection and compounded by the rule of vector addition.

59a. Parallel Forces. Force-couples. Let AP and BQ (Fig. 49) represent two parallel forces applied to a rigid body at A and B. Introduce at A and B two equal and opposite forces AR and BS of any magnitude in the line AB. These being in equilibrium do not affect the system. Find the resultant of AP and AR by the parallelogram, giving AC, also of BQ and BS giving BD. All these forces are coplanar, therefore the lines AC and BD will meet at E, if produced. Slide AC and BD to E, and then resolve into components parallel to the original ones. We get EH and EJ equal and opposite (being equal to AR and BS), and EK equal to AP and EL to BQ applied at E. Therefore the resultant of two parallel forces is a parallel force equal to their algebraic sum, and applied on a line EO, whose position is to be found as follows.

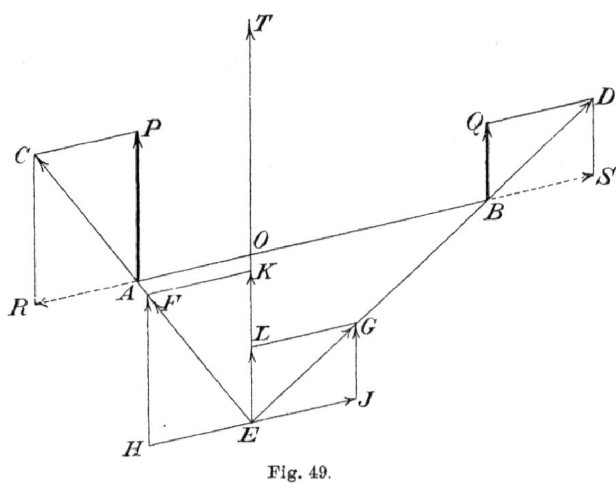

Fig. 49.

From the similar triangles,

$$\frac{AO}{FK} = \frac{OE}{KE} = \frac{OE}{AP},$$

$$\frac{BO}{GL} = \frac{OE}{LE} = \frac{OE}{BQ}.$$

By division, since $FK = GL$,

$$\frac{AO}{BO} = \frac{BQ}{AP}.$$

Thus the position of the resultant of parallel forces is to be found by the same construction as the resultant of two rotations about parallel axes, Fig. 47.

If the two forces are oppositely directed (Fig. 50), O is on AB produced, and if the forces are equal O lies at infinity. Accordingly there is no force that can replace two equal, parallel and oppositely directed forces not along the same line, or *force-couple*. The distance between the lines of direction is the *arm*, and the product of either force by the arm is the *moment of the couple.*

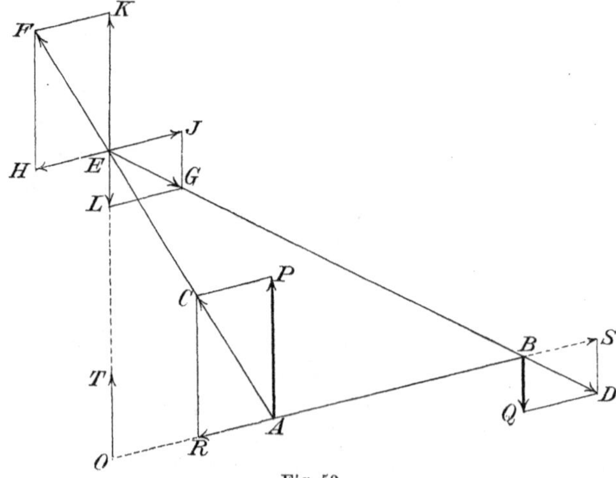

Fig. 50.

We shall prove the following theorems.

Theorem I. A couple may be transported parallel to itself either in its own or a parallel plane without changing its effect.

Consider the forces P_1 and P_2 both equal to P, applied perpendicularly at the ends of AB (Fig. 51). At the ends of an equal and parallel line $A'B'$ apply four equal and opposite forces P_3, P_4, P_5, P_6, each equal to P, which are in equilibrium. The resultant of the equal parallel forces P_1, P_6 is a force $2P$ applied half-way between A and B'. The resultant of P_2 and P_5 is a force $2P$ in the opposite direction applied half-way between A' and B. Since $ABB'A'$ is a parallelogram these two points of application coincide and the two resultants neutralize each other. We have left the couple P_3P_4 equivalent in effect to the original couple.

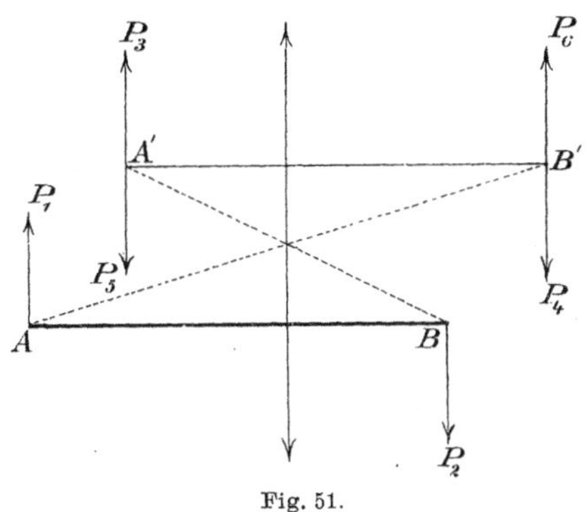

Fig. 51.

Theorem II. A couple may be turned in its plane about its center of symmetry without changing its effect.

Let $A'B'$ be a line of the same length and with the same center O as AB, the arm of the couple, and in the plane of the couple (Fig. 52). Apply at A' and B' four equal and opposite forces in equilibrium, each equal to P, and perpendicular to $A'B'$ and in the plane of the couple. Consider P_1 and P_5 applied at C, their point of intersection, and by symmetry their resultant will be along OC. Similarly the resultant of P_2 and P_6 is an equal force along OD in the opposite direction. These two resultants neutralize each other, leaving the couple P_3P_4 which has the same effect as the original couple.

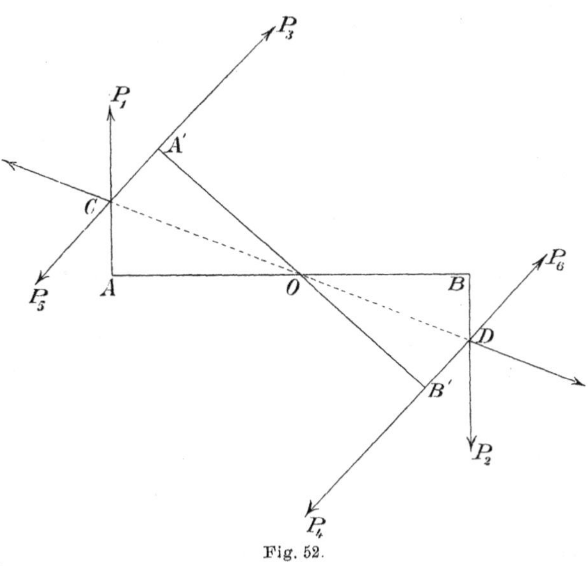

Fig. 52.

Theorem III. A couple may be replaced by another in the same plane having equal moment.

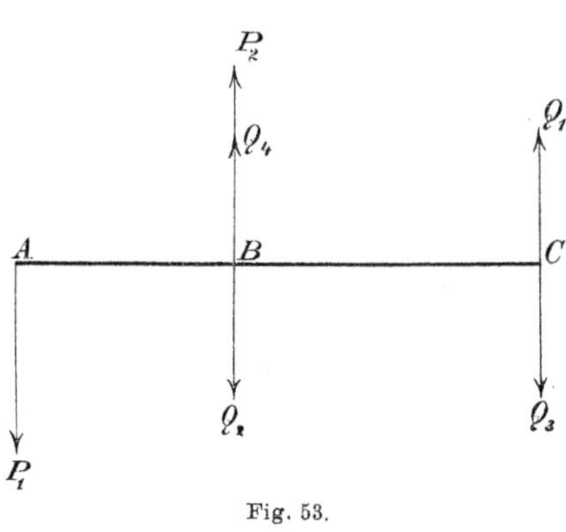

Fig. 53.

Let the couple be $P_1 P_2$ and the arm be AB (Fig. 53). At C on AB produced and at B apply four equal and opposite forces Q of such magnitude that

$$\frac{Q}{P} = \frac{AB}{BC}.$$

The resultant of the parallel forces, P_1, Q_3, is equal to P_3 plus Q_3 applied at B on account of the above equation. This is counterbalanced by the forces P_2 and Q_4 a pplied at B, leaving the couple $Q_1 Q_2$ of moment

$$Q \cdot BC = P \cdot AB,$$

equivalent to the original couple.

A force-couple is determined therefore by its plane and moment, and may be represented by a *free* vector perpendicular to its plane and of length equal to the moment.

Theorem IV. Composition of Couples. Suppose the two couples are in different planes. By turning each in its own plane bring all the four forces into directions perpendicular to the intersection of their planes, and then by varying one of the couples cause them to have the same arm AB. The forces $Q_1 P_1$ applied at A compound by the parallelogram into R_1. P_2 and Q_2 applied at B compound into R_2 equal and opposite to R_1. The arm of all these couples is the same, therefore their moments are proportional to P, Q and R. The vectors representing the moments are perpendicular to AB and to P, Q and R respectively, thus they form the sides and diagonal of a parallelogram similar to that of P, Q, R. Therefore couples are compounded by compounding their moments by the law of addition of vectors.

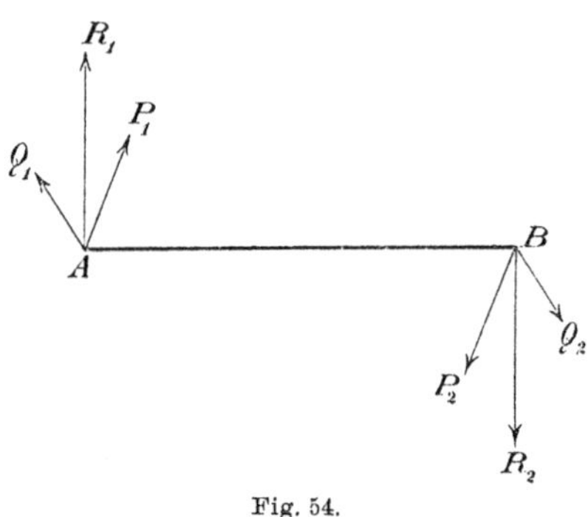

Fig. 54.

60. Reduction of Groups of Forces. Dualism. Suppose we have any number of forces applied to various points of a rigid body. Let one such be P applied at A. At any point O apply two equal and opposite forces equal and parallel to P. One of these P_2 forms a couple with P. The other is equal and parallel to P. The moment of the couple is perpendicular to this force.

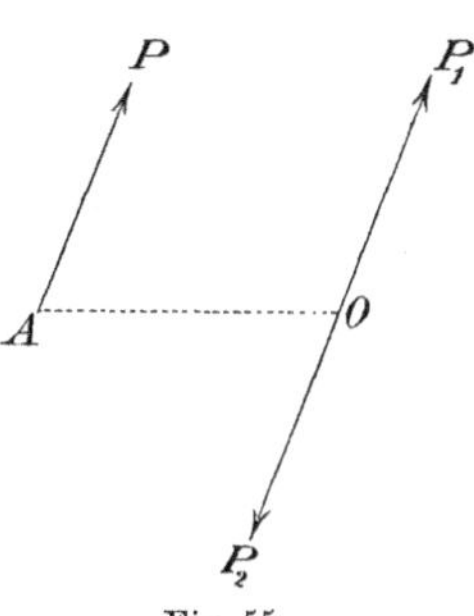

Fig. 55.

In this manner the points of application of all the forces may be brought to O, where they can then be compounded into a single resultant R. For each force thus transferred there remains a couple, and all the couples may be compounded into a single one. Therefore all the forces applied to a rigid body may be replaced by a single force and a single couple.

We may now state the following dualism existing between infinitesimal rotations and forces:

Infinitesimal rotations are sliding vectors.	Forces applied to a rigid body are sliding vectors.

When their axes intersect they are compounded by the vector law.

Parallel infinitesimal rotations	Parallel forces

have a resultant parallel and equal to their algebraic sum, placed at the center of mass of their points of application.

Two equal and opposite parallel rotations form a rotation-couple represented by its moment, a free vector.	Two equal and opposite parallel forces form a couple, represented by its moment, a free vector.
Every displacement of a rigid body may be reduced to a rotation and a rotation-couple.	Every combination of forces applied to a rigid body may be reduced to a force and force-couple.

The theory of couples is due to Poinsot.

61. Variation of the Elements of the Reduction. Central Axis. Null-System. We have seen that any system of slide-vectors may be reduced to the resultant of a single vector and a single moment applied at any point whatever. We have now to examine the variation of the pair of elements, vector R and moment S, as we vary the point of application O. R is invariable. As we move O along the line of R there is no change since R may be applied at any point of its axis, and S may be moved parallel to itself. If we make the resolution at any other point, O', the couple to be compounded with S at O', is perpendicular to R and OO', so that if S has any component parallel to R it cannot be neutralized by the new couple. Accordingly in order that the couple may vanish for *any* point O', the couple S must be perpendicular to R at all other points. As a change of O introduces only a component of S perpendicular to R, the component *parallel* to R is unchanged. Therefore the projection of S on R is the same for all points O,

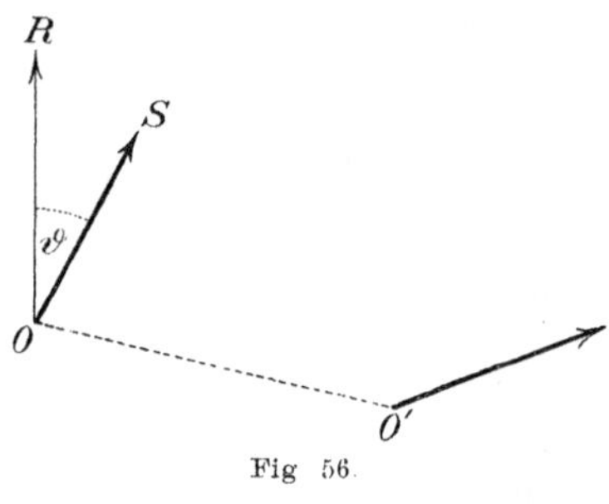

Fig. 56.

$$S \cos \vartheta = S_0. \tag{8}$$

Although in general R and S have different directions, we may find points O' for which they have the same direction. Let S and R include the angle ϑ at O. Resolve S into $S_0 = S \cos \vartheta$ parallel to R, and $S_1 = S \sin \vartheta$ perpendicular to R. If we take O' on a line perpendicular to SR at a distance d such that $d \cdot R = S \sin \vartheta$ in the positive direction of translation corresponding to a rotation from R to S, the component S_1 will be neutralized, and we shall have at O', R and $S' = S_0$ in the same direction. This property holds for all points on the line of R through O'. This line is called *Poinsot's central axis.*

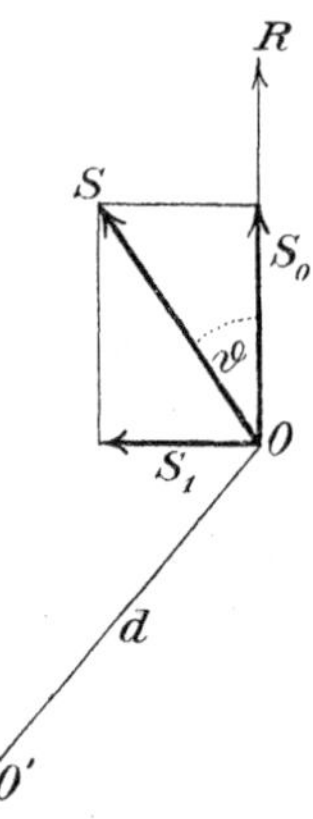

Fig. 57.

In order to consider the resolution at any point O we may refer it to the central axis. Drop a perpendicular from O (Fig. 58) on the central axis, and take this perpendicular for the axis of X, the central axis for the axis of Z.

Then as above

$$S \cos \vartheta = S_0, \tag{8}$$

$$S \sin \vartheta = d \cdot R, \tag{9}$$

and if xyz are the coordinates of the end of S, we have

$$z = S_0, \quad y = Rx, \quad \tan \vartheta = x \frac{R}{S_0}, \tag{10}$$

and for any point on the line of S,

$$\frac{z}{y} = \frac{S_0}{Rx}, \quad \text{or} \quad zx = \frac{S_0}{R} y, \tag{11}$$

that is the line of S lies on a hyperbolic paraboloid.

It is evident that if we slide the whole of Fig. 58 along or turn it around the central axis nothing is changed, consequently if we suppose the vector S laid off at every point of space O, and consider the assemblage of couples thus formed, the assemblage remains unchanged if we rotate it about or slide it along the central axis.

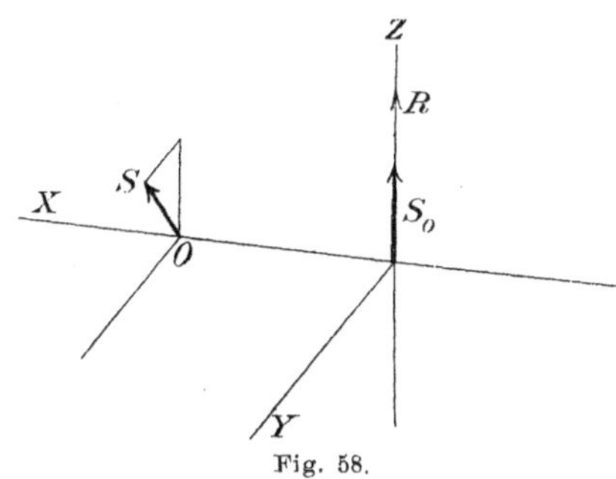

Fig. 58.

Every S is tangent to a certain helix, or locus of a point which moves on a circular cylinder in a path making a constant angle with its generators (Fig. 59). This angle is less as the diameter of the cylinders is less, so that

$$10) \qquad \tan\vartheta = x\frac{R}{S_0}.$$

All these helices have however one constant in common, namely the distance traversed parallel to the central axis for each turn. If $d\tau$ be the translation for a rotation $d\omega$, we have

$$\frac{x\,d\omega}{d\tau} = \tan\vartheta = x\frac{R}{S_0},$$

$$\frac{d\omega}{d\tau} = \frac{R}{S_0}, \quad \tau = \frac{S_0}{R}\omega.$$

Then

$$12) \qquad p = 2\pi\frac{S_0}{R}$$

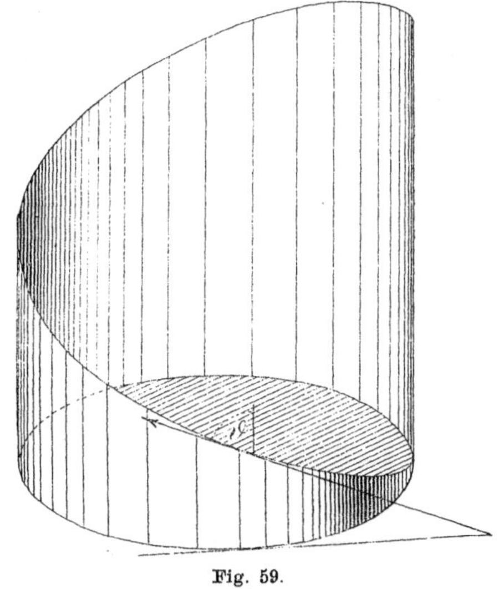
Fig. 59.

is the traverse for each turn, and is called the *pitch* of the helix. Every helix lies on a ruled screw-surface, made by the revolution of a line perpendicular to the central axis, which slides along it a distance proportional to the angle of rotation, the pitch of the screw being $p = 2\pi\frac{S_0}{R}$. The lines of the assemblage of moments have every direction in space — there are a triple infinity of lines of the system (one for each point in space), but only a double infinity of directions — therefore every plane cutting all these lines has for its points (a double infinity), every possible direction for S. For *one* point only is this perpendicular to the plane. This point is called the *focus* of the plane. Let the plane cut the central axis in A. Through A draw a plane perpendicular to the central axis, intersecting the given plane in AO. As we go along the line AO, S turns about it, and for one point has the direction of the normal to the given plane.

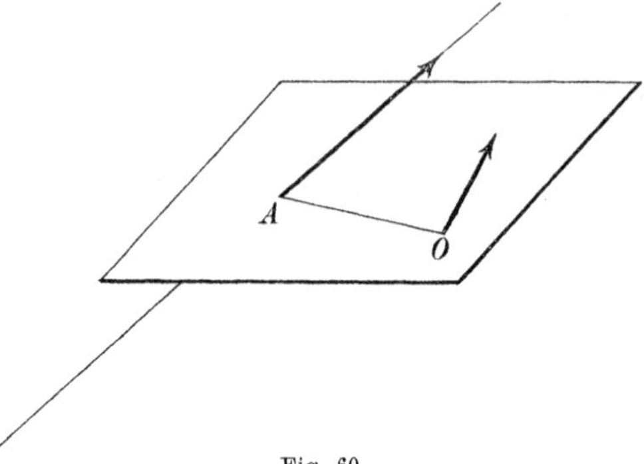

Fig. 60.

Accordingly to every point in space there corresponds one plane, and to every plane one point The correspondence was discovered by Chasles, and the system of points and planes was called a *Null-System* by Möbius.

62. Vector-cross. Besides the reduction to the screw-type we may reduce the system of vectors to two vectors not lying in the same plane, without a couple. This reduction may be made in an infinite number of ways, and the line of one of the vectors may

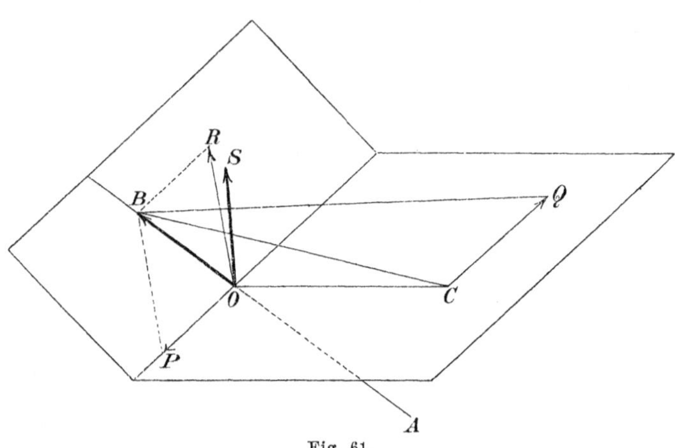

Fig. 61.

be given. Let AB (Fig. 61) be the given line. At any point O on AB let R be the resultant vector, S the resultant couple. (R and S will not in general lie in a plane with AB.) At O pass a plane perpendicular to S, intersecting the plane of R and AB in OP. Resolve S into the pair of vectors OP and CQ so taken that the resultant of R and OP shall lie in AB. The length of OP is thus determined, and the distance between its line and that of CQ is determined by S. Thus the line AB determines the line CQ.

The lengths OB and CQ are determined as soon as the line AB is given. Two such non-parallel and non-coplanar vectors OB, CQ will be termed a *vector-cross*. The crossing will degenerate to intersection only when $S = 0$ and to parallelism when $R = 0$.

As any line may be taken for AB, and as there are a quadruple infinity of lines in space, there are a quadruple infinity of vector-crosses. They all posses a property in common, namely, that the tetrahedron formed by joining the four ends of a vector-cross has a constant volume. Let OB, CQ (Fig. 61) be the vector-cross, and let us reverse the preceding resolution. The volume of a tetrahedron is equal to one-third the product of its altitude by the area of its base. The area of the base OCQ in one-half the moment of CQ about O, or $\frac{1}{2} S$, while the altitude is the projection of OB on the perpendicular to OCQ, that is, on S. But since BR is parallel to the plane OCQ, OR has the same projection on S as OB, namely $R \cos \vartheta$, consequently

$$V = \frac{1}{3} R \cos \vartheta \cdot \frac{1}{2} S = \frac{1}{6} RS \cos \vartheta.$$

But by 8),

$$S \cos \vartheta = S_0,$$

therefore

$$V = \frac{1}{6} RS_0. \tag{13}$$

This theorem is due to Chasles.

Corresponding lines of vector-crosses possess a remarkable relation to the null-system. Let AB and CQ (Fig. 62) be the two lines of the vector-cross. Through CQ pass any plane cutting AB in O. The moment of CQ is perpendicular to the plane OCQ, and the other vector has no moment about O, since it passes through it. Accordingly O is the focus of the plane OCQ. Thus, if a plane turns about a line, its focus traverses another line, and these two *conjugate* lines are lines of a vector-cross.

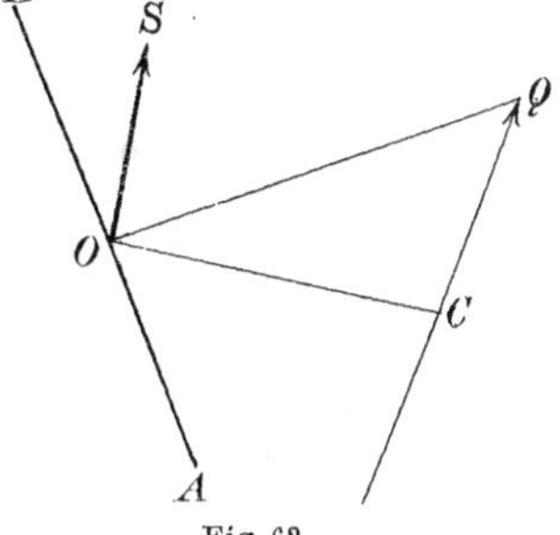

Fig. 62.

We have here shown the intermediate nature of a line between a point and a plane, in the dual role as generated by the motion of a point and by the rotation of a plane. In the first relation the line is spoken of as a *ray*, in the second as an *axis*.

If two conjugate lines are at right angles, pass a plane through one, AB, perpendicular to the other, CD (Fig. 63). By the preceding theorem, the point of intersection of the plane with CD is the focus of the plane. Resolving at any point P in AB, the moment of OD, being perpendicular to OD and OP, lies in the plane OAB.

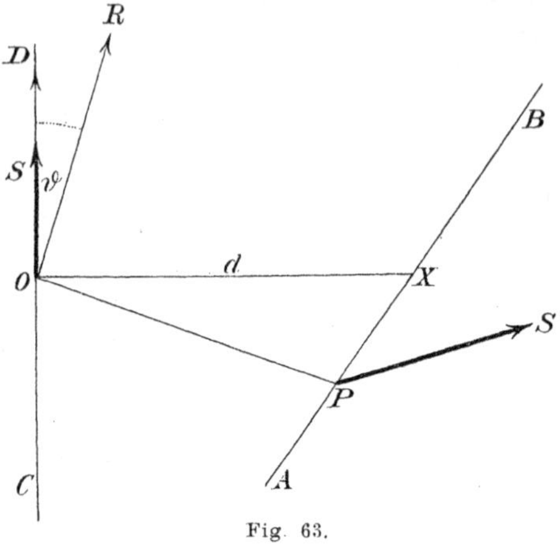

Fig. 63.

That line in a plane which has the property that for all its points the resultant moment lies in the plane is called the *characteristic* of the plane, or of its focus. Its distance $OX = d$ from the focus is such that[1])

14) $dR \sin \vartheta = S.$

The line OX, of length

$$d = \frac{S}{R \sin \vartheta},$$

is perpendicular to the plane of R and S, and drawn toward the side corresponding to the motion of a right-handed screw when rotated in the direction from R to S. If we should go from O in the direction OX a distance $d' = \frac{S \sin \vartheta}{R}$ we should reach the central axis, and

15) $$dd' = \frac{S^2}{R^2}.$$

63. Complex of Double-lines. If a plane 1 pass through the pole of a plane 2, then the plane 2 passes through the pole of the plane 1. Let P (Fig. 64) be the pole of the plane 1, and let PO be any line in 1 through P. The moment of R about O is perpendicular to PO, and so is S, hence so is their resultant. Thus the moment at O is perpendicular to OP, and the polar plane of O contains the line OP, that is, if O, the pole of 2 lies in 1, then P, the pole of 1 lies in 2.

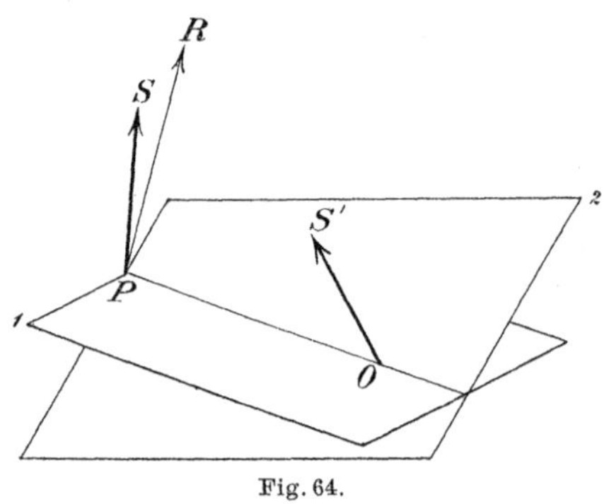

Fig. 64.

In this case the two poles lie in the line of intersection of the planes, and we see that if a plane turns about a line through its pole, its pole traverses that line. Such a double line is conjugate

1) For the component in AB, $R \sin \vartheta$, has the moment S about O.

to itself. The necessary and sufficient condition that a line is self-conjugate is that the pole (focus) of a plane through the line falls in the line. For then as the plane rotates about the double line as axis, the focus describes the line as a ray. Hence the double lines lying in a particular plane all pass through the pole of that plane, and conversely, all the double lines passing through a point lie in the polar plane of the point. Such a system of lines is called by Plücker a line *complex* of the first degree. There are in all a double infinity of lines passing through any point in space, but of these only a single infinity belong to the complex. Therefore lines belonging to the complex have one less degrees of freedom than lines in general, or a complex contains a triple infinity of lines. A complex may be represented analytically by a single relation between the four parameters determining a line. If we mark off on a line any length R, and give its projections on a set of rectangular axes X, Y, Z, and the projections L, M, N of its moment about an origin O, the line is completely determined. For its direction is given and giving the moment $S=\sqrt{L^2+M^2+N^2}$ gives the plane through O containing R, and the distance from the line, if the length of R is given, but this is given by $R=\sqrt{X^2+Y^2+Z^2}$.

As the determination of the line is independent of the length of R, the ratios of the six quantities determine the line. But these five ratios are not independent, for since by § 5, 12),

16)
$$\begin{aligned} L &= yZ - zY, \\ M &= zX - xZ, \\ N &= zY - yX, \end{aligned}$$

we have the identical relation,

17)
$$LX + MY + NZ = 0,$$

expressing the fundamental property that the moment of a vector is perpendicular to it. The coordinates $LMNXYZ$ are known as Plücker's line-coordinates.

Thus there remain four independent quantities to determine a line. A relation between these denotes a complex, and in particular a *linear* relation,

18)
$$aX + bY + cZ + dL + eM + fN = 0,$$

denotes a complex of the *first* degree.

Since the double lines of the null-system are the loci of points which are the poles of planes containing the double-lines, at every point of a double-line the resultant moment is perpendicular to it, or double lines are lines of no moment. In the kinematical application, points on a double-line experience no translation along it.

If a double-line cuts one of a pair of conjugate lines, it cuts the other. Let PQ be a double-line cutting the line AB. Then the pole of the plane BPQ lies in the line conjugate to AB. But since PQ is a double-line, the pole of BPQ lies on PQ. Hence PQ cuts the conjugate to AB. Conversely, every line cutting two conjugates is a double-line.

The complex of double-lines is symmetrical with respect to the central axis. Let AB (Fig. 65) be a line of the complex, and let OX be the common perpendicular to it and the central axis. Now AB is perpendicular to the moment S at X, but S is perpendicular to OX, and the distance OX is $d = \frac{S_0}{R} \tan \vartheta$. If φ is the angle that the line AB makes with the central axis we have

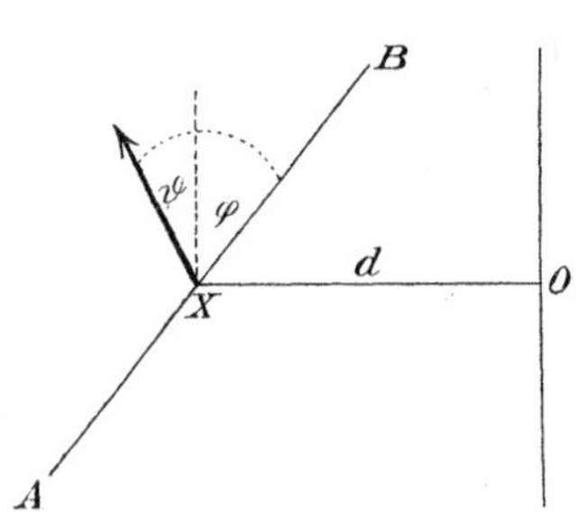

Fig. 65.

$$19)\qquad \tan \varphi = \operatorname{ctn} \vartheta = \frac{S_0}{Rd}.$$

This equation shows that the double-lines constituting the complex are tangent to an infinite number of helices, which become less steep as d decreases, so that the double-lines cutting the central axis are perpendicular to it, and those at infinity are parallel to it. For the pitch p of any helix tangent to lines of the complex we have

$$20)\qquad \frac{p}{2\pi d} = \operatorname{ctn} \varphi = \frac{Rd}{S_0}.$$

Thus the pitch is not constant, but varies as d^2.

This construction shows the triple infinity of complex-lines. In a plane perpendicular to the central axis every point x is on one complex line AB. There is a double infinity of such points. But there is a single infinity of such planes, and therefore in all a triple infinity of complex lines. It is evident that the complex is unchanged if we rotate it about, or slide it along the axis.

64. Composition of Screws. Suppose we have two systems of vectors, each reduced to the type of a screw. (The combination of forces of this type, namely a force, and a couple tending to cause rotation about its line of direction, is called a *wrench*. R is called the intensity of the wrench, or the amplitude of the rotation.) The resultant of both systems may also be reduced to a screw, and we may find its position.

Let us first suppose that the axes of the component screws intersect at rigth angles, and let us take them for axes of X and Y. Let their pitches be

$$p_x = 2\pi \frac{S_x}{R_x}, \quad p_y = 2\pi \frac{S_y}{R_y}.$$

Let R (Fig. 66), the resultant of R_x and R_y, make an angle α with the X-axis, and let S, the resultant of S_x and S_y, make an angle β with the same axis. Then the central axis is parallel to R, and cuts the Z-axis at a distance from the origin

$$OZ = \frac{S}{R} \sin(\beta - \alpha).$$

The resultant moment along this line is

$$S_0 = S \cos(\beta - \alpha).$$

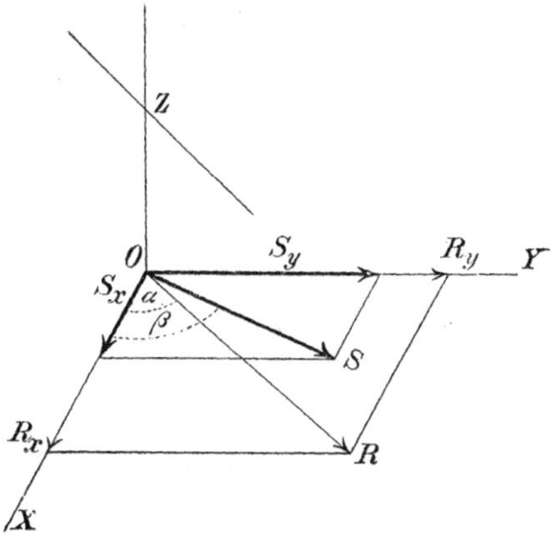

Fig. 66.

We have now

$$S \cos\beta = S_x = \frac{p_x}{2\pi} R_x = \frac{p_x}{2\pi} R \cos\alpha,$$

$$S \sin\beta = S_y = \frac{p_y}{2\pi} R_y = \frac{p_y}{2\pi} R \sin\alpha,$$

21) $$OZ = \frac{S}{R} \sin(\beta - \alpha) = \frac{p_y - p_x}{2\pi} \sin\alpha \cos\alpha,$$

$$S_0 = S \cos(\beta - \alpha) = \frac{R}{2\pi} (p_x \cos^2\alpha + p_y \sin^2\alpha)$$

For the pitch of the resultant screw we obtain

$$\frac{p}{2\pi} = \frac{S_0}{R} = \frac{1}{2\pi} (p_x \cos^2\alpha + p_y \sin^2\alpha),$$

22) $$p = p_x \cos^2\alpha + p_y \sin^2\alpha.$$

The equations of the central axis are

$$y = x \tan\alpha,$$

23) $$z = \frac{p_y - p_x}{2\pi} \sin\alpha \cos\alpha.$$

Inserting the values

$$\cos\alpha = \frac{x}{\sqrt{x^2 + y^2}}, \quad \sin\alpha = \frac{y}{\sqrt{x^2 + y^2}},$$

we have

$$z = \frac{(p_x - p_y)xy}{2\pi(x^2 + y^2)},$$

or

24) $$z(x^2 + y^2) - \frac{p_y - p_x}{2\pi} xy = 0,$$

as the equation of the ruled surface of the third order in which the resultant screw must lie, whatever the values of R_x, R_y. This surface is called the *Cylindroid.*

Since

23) $$z = \frac{p_y - p_x}{4\pi} \sin 2\alpha$$

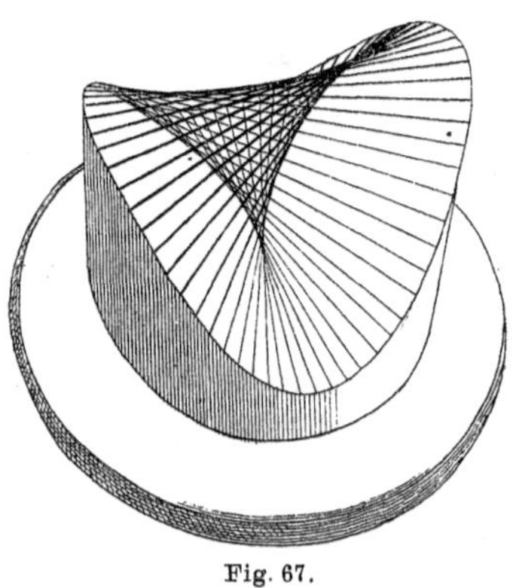

Fig. 67.

the surface is the locus of a line which, always intersecting a fixed line at right angles, revolves about it, and makes a harmonic oscillation along it, making *two* complete oscillations for each rotation. In this manner the model shown in Fig. 67 was constructed.

For every screw lying on the cylindroid there is a definite pitch, given by the equation 22). If we lay off the square roots of the reciprocals of the pitches on lines making angles α with the X-axis in the plane of XY, and call the coordinates of their ends xy, we have

$$x = \frac{1}{\sqrt{p}} \cos \alpha, \quad y = \frac{1}{\sqrt{p}} \sin \alpha,$$

and our equation is

$$p = p_x p x^2 + p_y p y^2,$$

25) $$p_x x^2 + p_y y^2 = 1,$$

representing a conic section, such that the pitch belonging to the direction of any radius vector is inversely proportional to the square of the length of the radius vector. This is called the pitch-conic. If p_x and p_y are of the same sign, the pitch-conic is an ellipse, if of opposite signs it is an hyperbola. In the latter case, there are two lines of zero pitch, given by the asymptotes. In other words, if one screw is right-handed, while the other is left-handed, there are two screws on the cylindroid representing merely rotation.

Any two screws determine a cylindroid. Let their pitches be $p_1 p_2$, let them make an angle γ and let the length of their common perpendicular be h. Then if they lie on a cylindroid we must have, by 23), 22),

$$z_2 - z_1 = h, \quad \alpha_2 - \alpha_1 = \gamma,$$

26)
$$z_1 = \frac{p_y - p_x}{4\pi} \sin 2\alpha_1, \quad z_2 = \frac{p_y - p_x}{4\pi} \sin 2\alpha_2,$$
$$p_1 = p_x \cos^2 \alpha_1 + p_y \sin^2 \alpha_2,$$
$$p_2 = p_x \cos^2 \alpha_2 + p_y \sin^2 \alpha_2,$$

as six equations to determine p_x, p_y, z_1, z_2, α_1, α_2. We have by elimination

$$z_2 - z_1 = \frac{p_y - p_x}{4\pi}(\sin 2\alpha_2 - \sin 2\alpha_1)$$
$$= \frac{p_y - p_x}{2\pi} \cos(\alpha_2 + \alpha_1) \sin(\alpha_2 - \alpha_1),$$

or using the first two equations,

27)
$$h = \frac{p_y - p_x}{2\pi} \cos(\alpha_2 + \alpha_1) \sin \gamma,$$

$$z_1 + z_2 = \frac{p_y - p_x}{4\pi}(\sin 2\alpha_2 + \sin 2\alpha_1)$$
$$= \frac{p_y - p_x}{2\pi} \sin(\alpha_2 + \alpha_1) \cos(\alpha_2 - \alpha_1)$$

28)
$$= \frac{p_y - p_x}{2\pi} \sin(\alpha_2 + \alpha_1) \cos \gamma,$$

$$p_2 - p_1 = p_x(\cos^2 \alpha_2 - \cos^2 \alpha_1) + p_y(\sin^2 \alpha_2 - \sin^2 \alpha_1)$$
$$= (p_x - p_y)(\cos^2 \alpha_2 - \cos^2 \alpha_1)$$
$$= (p_x - p_y) \sin(\alpha_2 + \alpha_1) \sin(\alpha_2 - \alpha_1),$$

29)
$$p_2 - p_1 = (p_x - p_y) \sin(\alpha_2 + \alpha_1) \sin \gamma,$$

30)
$$p_2 + p_1 = p_x(\cos^2 \alpha_2 + \cos^2 \alpha_1) + p_y(\sin^2 \alpha_2 + \sin^2 \alpha_1)$$
$$= p_x + p_y + (p_x - p_y)(\cos^2 \alpha_2 - \sin^2 \alpha_1)$$
$$= p_x + p_y + (p_x - p_y) \cos(\alpha_1 + \alpha_2) \cos \gamma.$$

From 27) and 29) we obtain

$$4\pi^2 h^2 + (p_2 - p_1)^2 = (p_y - p_x)^2 \sin^2 \gamma,$$

31)
$$p_y - p_x = \frac{\sqrt{4\pi^2 h^2 + (p_2 - p_1)^2}}{\sin \gamma}.$$

From 27) and 30),

$$p_y + p_x = p_2 + p_1 - 2\pi h \operatorname{ctn} \gamma.$$

From 29) and 31),

$$\sin(\alpha_1 + \alpha_2) = \frac{p_2 - p_1}{\sqrt{4\pi^2 h^2 + (p_1 - p_2)^2}},$$
$$\cos(\alpha_1 + \alpha_2) = \frac{2\pi h}{\sqrt{4\pi^2 h^2 + (p_1 - p_2)^2}},$$

32)
$$\tan(\alpha_1 + \alpha_2) = \frac{p_2 - p_1}{2\pi h}.$$

From 32) with $\alpha_2 - \alpha_1 = \gamma$,

$$33)\quad \begin{aligned} \alpha_2 &= \frac{1}{2}\left(\tan^{-1}\frac{p_2 - p_1}{2\pi h} + \gamma\right), \\ \alpha_1 &= \frac{1}{2}\left(\tan^{-1}\frac{p_2 - p_1}{2\pi h} - \gamma\right). \end{aligned}$$

Since the cylindroid is thus determined by 31) and 33), a twist about p_1 can be resolved into a twist about p_x and one about p_y. A twist about p_2 can be likewise resolved. The two components about p_x add together, so do those about p_y, and since the resultant of any twists about p_x and p_y lies on the cylindroid, the resultant of p_1 and p_2 does. Its direction can be found, since the amplitudes R of the two twists about $p_1 p_2$ compound by the parallelogram law, hence the angle made by the resultant with the axes is known. The pitch is then found from the pitch-conic.

65. Work of Wrench in Producing a Twist. Let us find an expression for the work done during a twist of amplitude R_k about a screw of pitch p_k by a wrench of intensity R_f about another screw of pitch p_f. We already know the work done by a force in a translation, namely, it is equal to the product of the magnitudes by the cosine of the included angle. If the force is R_f and the translation (rotation-couple) is S_k, we have

$$W = R_f S_k \cos(R_f S_k).$$

Notice that the vector of one system is multiplied by the vector-couple in the other.

We can find the work done by the force-couple in a rotation about its axis. Apply the couple so that one of its members P passes through the axis of rotation. In a rotation this member does no work, for its point of application is at rest, while that of the other member Q moves in a rotation a distance $d\omega$, where d is the arm of the couple. Accordingly the work is $W = Pd\omega$ which is equal to the product of the twist by the moment of the couple. Here again we multiply the vector of one system by the vector-couple of the other.

If the axis of rotation is perpendicular to the axis of the couple, the motion is perpendicular to the force, and no work is done. Hence we must take the resolved part of the couple on the vector, as before.

We can now find the work of a wrench during a twist. The work of the force in the displacement S_k is $R_f S_k \cos\alpha$, α being the angle between the two screws. The work of the couple $S_f = \frac{p_f}{2\pi} R_f$ in the rotation R_k is

$$S_f R_k \cos\alpha = \frac{p_f}{2\pi} R_f R_k \cos\alpha.$$

But when R_f is changed to the origin of R_k it gives rise to a moment perpendicular to R_f equal to $R_f d$, d being the perpendicular distance between the screws. This moment therefore makes with R_k the angle $\alpha + \frac{\pi}{2}$, and the work done by it in the rotation R_k is

$$d R_f R_k \cos\left(\alpha + \frac{\pi}{2}\right) = -d R_f R_k \sin\alpha.$$

Thus the whole work is

$$34) \qquad W = R_f R_k \left(\frac{p_f + p_k}{2\pi} \cos\alpha - d\sin\alpha\right).$$

It is symmetrical with respect to both screws, hence the wrench and twist might have been interchanged.

The geometrical quantity in parentheses is called the virtual coefficient of the two screws, and if it vanishes no work is done, that is, a body free to twist only about a particular screw is in equilibrium under a wrench about another screw if the virtual coefficient of the two screws is zero. The two screws are then said to be *reciprocal.*

66. Analytical Representation. Line Coordinates. In Plücker's line coordinates referred to any origin, since each component of vector does work on the corresponding component of couple in the other system,

$$35) \quad W = X_f L_k + Y_f M_k + Z_f N_k + L_f X_k + M_f Y_k + N_f Z_k.$$

If a screw is reciprocal to two screws on a cylindroid, it is evidently reciprocal to all the screws on it.

For two screws to be reciprocal, the condition is,

$$36) \quad X_1 L_2 + Y_1 M_2 + Z_1 N_2 + L_1 X_2 + M_1 Y_2 + N_1 Z_2 = 0.$$

If the coordinates of one of the screws be constant, while those of the other be variable, this is the equation 18) of a complex of the first degree, so that all the screws reciprocal to a given screw form such a complex.

Since between the six coordinates $X_1\, Y_1\, Z_1\, L_1\, M_1\, N_1$ there is always the identical relation

$$X_1 L_1 + Y_1 M_1 + Z_1 N_1 = 0,$$

we may always make them satisfy five equations like the above, that is, we may always find a screw reciprocal to *five* arbitrarily given screws.

Suppose the coordinates of the system of vectors for an origin O are $XYZLMN$, being the projections of R and S at O. Let

$XYZL'M'N'$ denote the same for a point O' whose coordinates are xyz. Then

$$37)\qquad \begin{aligned} L &= L' + yZ - zY, \\ M &= M' + zX - xZ, \\ N &= N' + xY - yX. \end{aligned}$$

In order that the point O' may lie on the central axis, the direction of resultant and couple must coincide, or

$$\frac{L'}{X} = \frac{M'}{Y} = \frac{N'}{Z},$$

hence the equations of the central axis in Cartesian coordinates are

$$38)\qquad \frac{L - yZ + zY}{X} = \frac{M - zX + xZ}{Y} = \frac{N - xY + yX}{N}.$$

The equation of the focal or polar plane to a point $x'\,y'\,z'$ is, since it is perpendicular to $L'\,M'\,N'$,

$$39)\qquad (x - x')\,L' + (y - y')\,M' + (z - z')\,N' = 0$$

and inserting the values of $L'\,M'\,N'$,

$$(x - x')\,(L - y'\,Z + z'Y) + (y - y')\,(M - z'X + x'Z) \\ + (z - z')\,(N - x'Y + y'X) = 0,$$

or, more symmetrically arranged,

$$40)\qquad L(x - x') + M(y - y') + N(z - z') + X\,(zy' - yz') \\ + Y(xz' - zx') + Z\,(yx' - xy') = 0.$$

This equation is symmetrical with respect to xyz, $x'y'z'$, hence if $x'y'z'$ is fixed, xyz is on its polar plane, or if xyz is considered fixed, $x'y'z'$ is on its polar plane, showing the reciprocal relation of pole and polar.

If the vector system is to reduce to a single vector, the resultant and couple ot any point must be perpendicular, or

$$41)\qquad LX + MY + NZ = 0.$$

We must have in general, at any point, $S\cos\vartheta = S_0$ that is,

$$42)\qquad \frac{LX + MY + NZ}{R = \sqrt{X^2 + Y^2 + Z^2}} = S_0,$$

and the pitch p is given by

$$43)\qquad \frac{p}{2\pi} = \frac{S_0}{R} = \frac{LX + MY + NZ}{X^2 + Y^2 + Z^2}.$$

The volume of the tetrahedron on a vector-cross is

$$44)\qquad \frac{1}{6}RS_0 = \frac{1}{6}(LX + MY + NZ),$$

and this, like the last expression, is independent of the choice of origin or axes, that is, is an invariant

Suppose that the two members of a vector-cross have Plücker's coordinates

$$X_1 Y_1 Z_1 L_1 M_1 N_1 \quad \text{and} \quad X_2 Y_2 Z_2 L_2 M_2 N_2$$

with the identical relations,

$$L_1 X_1 + M_1 Y_1 + N_1 Z_1 = 0, \quad L_2 X_2 + M_2 Y_2 + N_2 Z_2 = 0.$$

Their resultant has components

$$\begin{aligned} X &= X_1 + X_2, & L &= L_1 + L_2, \\ X &= Y_1 + Y_2, & M &= M_1 + M_2, \\ Z &= Z_1 + Z_2, & N &= N_1 + N_2, \end{aligned}$$

and the volume of the tetrahedron is one sixth of

$$LX + MY + NZ =$$
$$(L_1 + L_2)(X_1 + X_2) + (M_1 + M_2)(Y_1 + Y_2) + (N_1 + N_2)(Z_1 + Z_2),$$

which in virtue of the two identities is

45) $$L_1 X_2 + M_1 Y_2 + N_1 Z_2 + L_2 X_1 + M_2 Y_1 + N_2 Z_1.$$

If any two lines are given by their Plücker's coordinates, the condition that they shall intersect is that the above expression shall vanish.

We may now find the equation of the complex of double-lines. We have seen that every line meeting two conjugate lines is a double-line. Let the coordinates of the two conjugate lines be $X_1 \ldots N_1$, $X_2 \ldots N_2$, satisfying the conditions

46) $$\begin{aligned} X_1 + X_2 &= X_0, & L_1 + L_2 &= L_0, \\ Y_1 + Y_2 &= Y_0, & M_1 + M_2 &= M_0, \\ Z_1 + Z_2 &= Z_0, & N_1 + N_2 &= N_0, \end{aligned}$$

where $X_0 Y_0 Z_0 L_0 M_0 N_0$ define the vector-system. Let the coordinates of a double-line be $XYZLMN$. The condition that it meets the line $X_1 Y_1 Z_1 L_1 M_1 N_1$ is

$$L_1 X + M_1 Y + N_1 Z + X_1 L + Y_1 M + Z_1 N = 0,$$

and that it meets $X_2 Y_2 Z_2 L_2 M_2 N_2$,

$$L_2 X + M_2 Y + N_2 Z + X_2 L + Y_2 M + Z_2 N = 0.$$

Adding these equations, and using the conditions 46) we obtain,

47) $$L_0 X + M_0 Y + N_0 Z + X_0 L + Y_0 M + Z_0 N = 0,$$

as the equation of the complex, that is, *any* linear relation in Plücker's coordinates represents a linear complex, as stated in § 63.

It is to be noticed that the equation 47) does not signify that the line $XYZLMN$ cuts the line $X_0 Y_0 Z_0 L_0 M_0 N_0$ unless the latter *are* the coordinates of a *line* (not of a general system of vectors), that is fulfill the relation

$$L_0 X_0 + M_0 Y_0 + N_0 Z_0 = 0.$$

If they do, then every line of the complex cuts the line $X_0 Y_0 Z_0 L_0 M_0 N_0$, and the equation may be considered the equation in Plücker's coordinates of the line $X_0 Y_0 Z_0 L_0 M_0 N_0$ (see Clebsch, Geometrie, Vol. II, p. 51). For further information on this subject, the reader may consult, Ball, *Theory of Screws.*

67. Momentum Screw. Dynamics. The previous sections have shown how to combine systems of vectors having different points of application, provided they are unchanged if slid along their lines of direction. As one particular system to which the operation is applicable we have had the various rotation-velocities of a rigid body, as another, sets of forces applied to a rigid body. That these vectors are susceptible of such treatment may be considered as due to properties of a rigid body, rather than of the vectors themselves. We have however previously dealt with two other sorts of vectors which may be dealt with in similar fashion, on account of their physical nature, and independently of the nature of the bodies in which their points of application lie. By means of these properties we are able to connect the kinematical aspect of a rigid body, as expressed by its instantaneous screw motion, with its dynamical aspect, as expressed by an applied wrench about another screw.

If for each point of the system we consider the momentum, whose six coordinates (one being redundant), in the sense of § 66 are,

$$m v_x, \quad m v_y, \quad m v_z, \quad m(y v_z - z v_y), \quad m(z v_x - x v_z), \quad m(x v_y - y v_x),$$

and form the general resultant, we obtain a system whose coordinates are

$$48) \qquad \begin{aligned} M_x &= \Sigma m v_x, & H_x &= \Sigma m (y v_z - z v_y), \\ M_y &= \Sigma m v_y, & H_y &= \Sigma m (z v_x - x v_z), \\ M_z &= \Sigma m v_z, & H_z &= \Sigma m (x v_y - y v_x), \end{aligned}$$

which represent the momentum of the system, the three projections M_x, M_y, M_z, being more particularly characterized as the linear momentum, the others H_x, H_y, H_z, as the *angular momentum* or moment of momentum with respect to the origin.

We have now by the general principles of dynamics, as shown in § 32, 45), § 33, 61), the fact that the time-derivatives of these six components of momentum are equal to the corresponding components of the resultant wrench,

$$X = \Sigma X, \quad Y = \Sigma Y, \quad Z = \Sigma Z,$$
$$L = \Sigma(yZ - zY), \quad M = \Sigma(zX - xZ), \quad N = \Sigma(xX - yX),$$

applied to the system. That is,

$$49) \qquad \begin{aligned} \frac{dM_x}{dt} &= X, & \frac{dM_y}{dt} &= Y, & \frac{dM_z}{dt} &= Z, \\ \frac{dH_x}{dt} &= L, & \frac{dH_y}{dt} &= M, & \frac{dH_z}{dt} &= N. \end{aligned}$$

Integrating these equations with respect to the time,

$$50) \qquad \begin{aligned} M_x &= \int_0^t X\,dt, & M_y &= \int_0^t Y\,dt, & M_z &= \int_0^t Z\,dt, \\ H_x &= \int_0^t L\,dt, & H_y &= \int_0^t M\,dt, & H_z &= \int_0^t N\,dt, \end{aligned}$$

we may, in the sense of § 27, call the momentum the *impulsive wrench* of the system. Physically, then, the momentum that a system possesses at any instant is equal to the impulsive wrench necessary to suddenly communicate to it when at rest the velocity-system that it actually possesses. As a prelude to the dynamics of a rigid body we must accordingly study the properties of the momentum or impulsive wrench of a body possessing a given instantaneous twist-velocity.

All the systems of vectors in question may be reduced to the screw type, and their respective screws are in general all different. Thus we may speak of the instantaneous velocity-screw and instantaneous axis, the momentum screw, and the force-screw. As the body moves, all these screws change both their pitch and position in the body, describing ruled surfaces both in the body and in space. The integration of the differential equations of motion 49) will enable us to find these surfaces. The kinematical description of the motion will be complete if we know the two ruled surfaces described in space and in the body by the instantaneous axis, together with such data as will give their mutual relations at each instant of time.

68. Momentum of Rigid Body. The properties of the momentum of a rigid body are conveniently investigated by the consideration of the velocity-system as an instantaneous screw-motion. Let V be the velocity of translation, and ω of rotation. Then every particle of mass m has one component of momentum parallel to the axis of the instantaneous twist (which we will take for Z-axis), equal to $mv_z = mV$ and the resultant for all is

$$51) \qquad M_z = \Sigma mV = V\Sigma m = MV,$$

where M is the total mass of the body. By the construction of §§ 57, 59 the resultant of parallel vectors P and Q is applied at the center of mass of masses proportional to P and Q placed at their points of application. Consequently the various elements being proportional to the masses m, this component of the momentum is applied at the center of mass of the body.

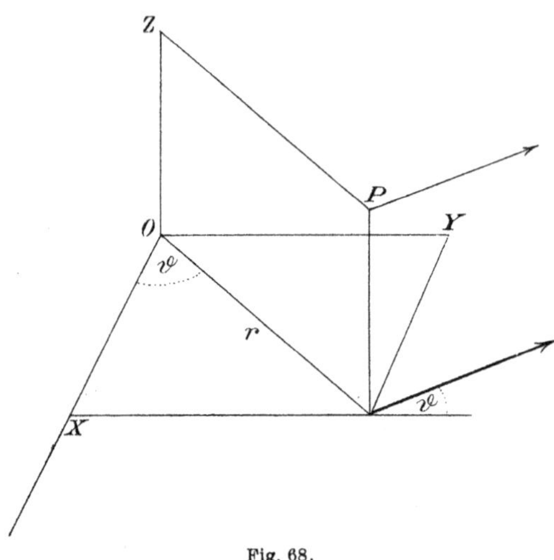

Fig. 68.

There remains the component of momentum perpendicular to the instantaneous axes. Let OZ (Fig. 68) be the instantaneous axis, and let r be the perpendicular distance from it of any point P, and let the angle made by r with the X-axis be ϑ. Now P is moving parallel to the XY-plane with the velocity $v = r\omega$ perpendicular to r, so that the projections of this velocity are

$$v_x = -v \sin\vartheta = -\omega r \sin\vartheta = -\omega y,$$
$$v_y = v \cos\vartheta = \omega r \cos\vartheta = \omega x.$$

Thence we obtain the components of momentum

$$\text{52)} \quad \begin{aligned} M_x &= -\Sigma m\omega y = -\omega\Sigma m y = -M\omega\bar{y}, \\ M_y &= \Sigma m\omega x = \omega\Sigma m x = M\omega\bar{x}, \end{aligned}$$

where $\bar{x}$, $\bar{y}$ are the coordinates of the center of mass. The resultant momentum is accordingly equal and parallel to the momentum that the body would have if concentrated at the center of mass, but its point of application is different, for the components M_x, M_y are not applied at the center of mass, inasmuch as their elements are proportional, not to m but to my and mx. The magnitude of the resultant momentum being given by M_x, M_y, M_z, we may find its axis by obtaining its three remaining coordinates, representing the angular momentum. We have

$$\text{53)} \quad \begin{aligned} H_x &= \Sigma(ymv_z - zmv_y) = V\Sigma my - \omega\Sigma mzx = MV\bar{y} - \omega\Sigma mzx, \\ H_y &= \Sigma(zmv_x - xmv_z) = -\omega\Sigma myz - V\Sigma mx = -MV\bar{x} - \omega\Sigma myz, \\ H_z &= \Sigma(xmv_y - ymv_x) = \omega\Sigma m(x^2 + y^2) = \omega\Sigma mr^2. \end{aligned}$$

Of these the terms in V are the moments of the vector MV in the direction of the Z-axis applied at the center of mass, while the terms in ω are applied elsewhere. The equations of the central axis of momentum are, by § 66, 38), $x'y'z'$ being the running coordinates,

$$54)\qquad \frac{H_x - y' M_z + z' M_y}{M_x} = \frac{H_y - z' M_x + x' M_z}{M_y} = \frac{H_z - x' M_y + y' M_x}{M_z},$$

or inserting the values,

$$55)\qquad \frac{MV\bar{y} - \omega\Sigma mxz - y' MV + z' M\omega\bar{x}}{-M\omega\bar{y}}$$
$$= \frac{-MV\bar{x} - \omega\Sigma myz + z' M\omega\bar{y} + x' MV}{M\omega\bar{x}}$$
$$= \frac{\omega\Sigma m(x^2 + y^2) - x' M\omega\bar{x} - y' M\omega\bar{y}}{MV}.$$

This does not pass through the center of mass unless, putting $x' = \bar{x}$, $y' = \bar{y}$, $z' = \bar{z}$,

$$56)\qquad \frac{-\omega\Sigma mxz + M\omega\bar{z}\bar{x}}{-M\omega\bar{y}} = \frac{-\omega\Sigma myz + M\omega\bar{y}\bar{z}}{M\omega\bar{x}}$$
$$= \frac{\omega\Sigma m(x^2 + y^2) - M\omega(\bar{x}^2 + \bar{y}^2)}{MV}.$$

We see that the resultant momentum involves the various sums

$$\Sigma mx, \quad \Sigma my, \quad \Sigma mxz, \quad \Sigma myz, \quad \Sigma mr^2,$$

the axis of Z being the instantaneous axis. These sums are constants for the rigid body, depending on the distribution of mass in it. The first two represent the mass of the body multiplied by the coordinates of the center of mass. The last represents the sum of the mass of each particle multiplied by the square of its distance from the Z-axis and is what has been called the *moment of inertia* of the body with respect to that axis. We are thus led to consider the sums

$$A = \Sigma m(y^2 + z^2), \quad B = \Sigma m(z^2 + x^2), \quad C = \Sigma m(x^2 + y^2),$$
$$D = \Sigma myz, \qquad E = \Sigma mzx, \qquad F = \Sigma mxy.$$

Of these the last three, D, E, F, are termed the *products of inertia* with respect to the respective pairs of axes.

In the case of a continuous distribution of mass, we must divide the body up into infinitesimal elements of volume $d\tau$, and if the density is ϱ, the element of mass is $dm = \varrho\, d\tau$ and the six sums become the definite integrals

$$A = \iiint \varrho(y^2 + z^2)\,d\tau, \quad B = \iiint \varrho(z^2 + x^2)\,d\tau, \quad C = \iiint \varrho(x^2 + y^2)\,d\tau,$$
$$D = \iiint \varrho yz\,d\tau, \qquad E = \iiint \varrho zx\,d\tau, \qquad F = \iiint \varrho xy\,d\tau.$$

The determination of these quantities is then, like that of centers of mass, a subject belonging to the integral calculus.

The six constants A, B, C, D, E, F together with the mass M and coordinates $\bar{x}, \bar{y}, \bar{z}$, of the center of mass, completely characterize the body for dynamical purposes, since when we know their values and the instantaneous twist, the momentum or impulsive wrench is completely given. The body may therefore be replaced by any other having the same mass, center of mass, and moments and products of inertia, and the new body will, when acted upon by the same forces, describe the same motion.

69. Centrifugal Forces. As the body moves, its different parts exercise forces of inertia upon each other, so that there is a resultant tending to change the instantaneous screw in the body. Let us suppose the translation to vanish, and examine the kinetic reactions developed by the rotation, or the centrifugal forces. The instantaneous axis being again taken as the axis of Z, a particle P experiences the centripetal acceleration $\frac{v^2}{r} = r\omega^2$ towards the axis, and the centrifugal force is $R_c = mr\omega^2$ (see p. 119) directed along the radius r from the axis OZ, and having the projections

57)
$$\begin{aligned} X_c &= R_c \frac{x}{r} = mx\omega^2, \\ Y_c &= R_c \frac{y}{r} = my\omega^2, \\ Z_c &= 0. \end{aligned}$$

For the moment of the centrifugal force we have

58)
$$\begin{aligned} L_c &= yZ_c - zY_c = -myz\omega^2, \\ M_c &= zX_c - xZ_c = mxz\omega^2, \\ N_c &= xY_c - yX_c = 0, \end{aligned}$$

so that the coordinates of the *resultant* centrifugal force and couple are

59)
$$\begin{aligned} X_c &= \omega^2 \Sigma mx = \omega^2 M\bar{x}, \\ Y_c &= \omega^2 \Sigma my = \omega^2 M\bar{y}, \\ Z_c &= 0, \\ L_c &= -\omega^2 \Sigma myz = -D\omega^2, \\ M_c &= \omega^2 \Sigma mzx = E\omega^2, \\ N_c &= 0. \end{aligned}$$

Thus the centrifugal force is equal and parallel to that of a mass placed at the center of mass, and moving as the latter point does. It vanishes when the center of mass lies in the axis. The system of centrifugal forces is however, as in the case of the momentum, not to be replaced by a single force placed *at* the center of mass, for the couple is not equal to what its value would be in that case, unless $\frac{D}{E} = \frac{\bar{y}}{\bar{x}}$. If the center of mass lies on the axis, although the centrifugal force R_c vanishes, the centrifugal couple S_c does not, unless $D = E = 0$.

The centrifugal forces then in general tend to change the instantaneous twist, unless the axis of the latter passes through the center of mass, and for it $D = E = 0$. Such axes are called *principal axes* of inertia of the body at the center of mass, and are characterized by the property that if the body be moving with an instantaneous twist about such an axis, it will remain twisting about it, unless acted on by external forces. In order to examine the effect of the distribution of mass of the body, we are led to interrupt the consideration of dynamics in order to consider the purely geometrical relations among moments and products of inertia.

70. Moments of Inertia. Parallel Axes. Consider the moments of inertia of a body about two parallel axes. Let the perpendicular distances from a point P on the two axes be p_1 and p_2 and let the distance apart of the axes be d. Let A and B (Fig. 69) be the intersections of the axes with the plane of p_1 and p_2. If we take AB for the X-axis, A for origin, we have

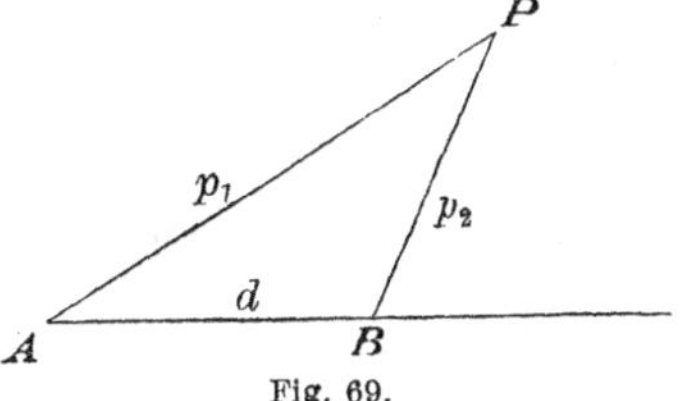

Fig. 69.

$$p_2^2 = p_1^2 + d^2 - 2p_1 d \cos(p_1 x),$$

60) $$\Sigma m p_2^2 = \Sigma m p_1^2 + Md^2 - 2d\Sigma m p_1 \cos(p_1 x) = \Sigma m p_1^2 + Md^2 - 2d\,\Sigma m x.$$

The last term is equal to $-2dM\bar{x}$ and vanishes if the axis 1 passes through the center of mass. Consequently the moment of inertia about any axis is equal to the moment of inertia about a parallel axis through the center of mass plus the moment of inertia of a particle of mass equal to that of the body placed at the center of mass, about the original axis. Consequently of all moments of inertia about parallel axes, that about an axis through the center of mass is the least. In virtue of this theorem the study of moments of inertia is reduced to the study of moments of inertia about axes in different directions passing through the same point.

71. Moments of Inertia at a Point. Ellipsoid of Inertia. Consider now moments of inertia about different axes all passing through the same point O. Let α, β, γ be the direction cosines of any axis. Let p be the perpendicular distance of a point P from the axis, r its distance from O, and q the distance from O of the foot of the perpendicular. Now since q is the projection of r on the axis,

$$61)\qquad q = \alpha x + \beta y + \gamma z,$$

and we have

$$62)\qquad \begin{aligned} p^2 = r^2 - q^2 &= x^2 + y^2 + z^2 - (\alpha x + \beta y + \gamma z)^2 \\ &= x^2(1-\alpha^2) + y^2(1-\beta^2) + z^2(1-\gamma^2) \\ &\quad - 2(\beta\gamma yz + \gamma\alpha zx + \alpha\beta xy). \end{aligned}$$

Now since we have

$$\alpha^2 + \beta^2 + \gamma^2 = 1,$$

$$1-\alpha^2 = \beta^2 + \gamma^2, \quad 1-\beta^2 = \gamma^2 + \alpha^2, \quad 1-\gamma^2 = \alpha^2 + \beta^2,$$

and replacing in 62),

$$63)\qquad \begin{aligned} p^2 &= \alpha^2(y^2+z^2) + \beta^2(z^2+x^2) + \gamma^2(x^2+y^2) \\ &\quad - 2(\beta\gamma yz + \gamma\alpha zx + \alpha\beta xy), \\ \Sigma m p^2 &= \alpha^2 \Sigma m(y^2+z^2) + \beta^2 \Sigma m(z^2+x^2) + \gamma^2 \Sigma m(x^2+y^2) \\ &\quad - 2\beta\gamma \Sigma myz - 2\gamma\alpha \Sigma mzx - 2\alpha\beta \Sigma mxy. \end{aligned}$$

Thus the moment of inertia K about any axis whose direction cosines are α, β, γ, is given by

$$64)\qquad K = A\alpha^2 + B\beta^2 + C\gamma^2 - 2D\beta\gamma - 2E\gamma\alpha - 2F\alpha\beta \equiv F(\alpha, \beta, \gamma),$$

as a homogeneous quadratic function of the direction cosines of the axis.

The sum of products of the mass of each particle multiplied by the square of its distance from a given plane is called the moment of inertia of the system with respect to the plane. Although it has no physical significance it will be convenient to consider it. For a plane normal to the preceding axis we have

$$65)\qquad \begin{aligned} Q = \Sigma m q^2 &= \alpha^2 \Sigma m x^2 + \beta^2 \Sigma m y^2 + \gamma^2 \Sigma m z^2 \\ &\quad + 2\beta\gamma \Sigma myz + 2\gamma\alpha \Sigma mzx + 2\alpha\beta \Sigma mxy, \end{aligned}$$

and if we put

$$A' = \Sigma m x^2, \quad B' = \Sigma m y^2, \quad C' = \Sigma m z^2,$$

we have

$$66)\qquad Q = A'\alpha^2 + B'\beta^2 + C'\gamma^2 + 2D\beta\gamma + 2E\gamma\alpha + 2F\alpha\beta \equiv F'(\alpha, \beta, \gamma).$$

The six quantities, A, B, C, A', B', C', being sums of squares, are all positive. We have evidently

$$A = B' + C', \qquad B = C' + A', \qquad C = A' + B',$$

$$67)\qquad B + C = A + 2A', \quad C + A = B + 2B', \quad A + B = C + 2C',$$

so that the sum of any two of the moments A, B, C is greater than the third.

If we lay off on the axis a length ϱ and call the coordinates of the point P so determined ξ, η, ζ we have

$$\xi = \varrho\alpha, \quad \eta = \varrho\beta, \quad \zeta = \varrho\gamma,$$

$$68) \qquad F'(\xi, \eta, \zeta) = \varrho^2 F'(\alpha, \beta, \gamma) = \varrho^2 Q.$$

If we now make the length of OP vary in such a manner that $\varrho^2 Q = 1$, we obtain for the coordinates of P the equation

$$69) \quad F'(\xi, \eta, \zeta) = A'\xi^2 + B'\eta^2 + C'\zeta^2 + 2D\eta\zeta + 2E\zeta\xi + 2F\xi\eta = 1,$$

or P lies on a *central quadric surface.* Since $\varrho = \frac{1}{\sqrt{Q}}$ is always real, this is an ellipsoid. It possesses the property that the moment Q with respect to any plane through its center is inversely proportional to the square of a radius vector perpendicular to it. It will be termed the *fundamental ellipsoid* of inertia at the point O. It was discovered by Binet.

In a similar manner the moments of inertia about the various axes are inversely proportional to the square of the radii vectores in their direction of another ellipsoid

$$70) \quad F(\xi, \eta, \zeta) = A\xi^2 + B\eta^2 + C\zeta^2 - 2D\eta\zeta - 2E\zeta\xi - 2F\xi\eta = 1.$$

This is known as Poinsot's ellipsoid of inertia at the point O.

Since a central quadric always has three principal axes perpendicular to each other (see Note IV), we find that there are at any point in a body three mutually perpendicular directions, namely those of the axes of the two ellipsoids of inertia, characterized by the property that for them the products of inertia D, E, F, are equal to zero. These are termed the principal axes of inertia of the body at the point in question. They have, as shown in § 69, the property that if the body be rotating about one of them the centrifugal couple vanishes, so that if the center of mass lies on the axis the body remains rotating about the same axis, unless acted on by external forces.

The moments A, B, C about these axes are called principal moments of inertia.

It is important to notice that as we pass along a line which is a principal axis at one of its points, the directions of the axes of the ellipsoids at successive points are not the same, so that in general a line is a principal axis of inertia at only one of its points. We are thus led to study the relative directions of the principal axes at different points of the body.

72. Ellipsoid of Gyration. The moment of inertia about any axis may be considered equal to that of a particle whose mass is that of the body placed at a distance k from the axis, such that $K = Mk^2$. k is called the *radius of gyration* for this axis. The radii of gyration about the principal axes of inertia at any point are called the principal radii of gyration for that point. If we call their lengths a, b, c we have

$$A = Ma^2, \quad B = Mb^2, \quad C = Mc^2,$$

and 70) becomes

71) $$k^2 = \alpha^2 a^2 + \beta^2 b^2 + \gamma^2 c^2.$$

Another ellipsoid besides Poinsot's, which referred to its axes is

72) $$F(x, y, z) = Ax^2 + By^2 + Cz^2 = 1$$

is sometimes convenient. If at any point x, y, z on Poinsot's ellipsoid we draw the tangent plane, and from the center let fall a perpendicular upon it, its length p will be the projection of the radius vector r on a line parallel to the normal,

73) $$p = x \cos(nx) + y \cos(ny) + z \cos(nz).$$

But since

$$\cos(nx) = \frac{\partial F}{\partial x} \Big/ \left\{ \left(\frac{\partial F}{\partial x}\right)^2 + \left(\frac{\partial F}{\partial y}\right)^2 + \left(\frac{\partial F}{\partial z}\right)^2 \right\}^{\frac{1}{2}} = \frac{Ax}{\sqrt{A^2x^2 + B^2y^2 + C^2z^2}},$$

74) $$\cos(ny) = \frac{\partial F}{\partial y} \Big/ \left\{ \left(\frac{\partial F}{\partial x}\right)^2 + \left(\frac{\partial F}{\partial y}\right)^2 + \left(\frac{\partial F}{\partial z}\right)^2 \right\}^{\frac{1}{2}} = \frac{By}{\sqrt{A^2x^2 + B^2y^2 + C^2z^2}},$$

$$\cos(nz) = \frac{\partial F}{\partial z} \Big/ \left\{ \left(\frac{\partial F}{\partial x}\right)^2 + \left(\frac{\partial F}{\partial y}\right)^2 + \left(\frac{\partial F}{\partial z}\right)^2 \right\}^{\frac{1}{2}} = \frac{Cz}{\sqrt{A^2x^2 + B^2y^2 + C^2z^2}},$$

this gives for the ellipsoid

$$p = \frac{x\frac{\partial F}{\partial x} + y\frac{\partial F}{\partial y} + z\frac{\partial F}{\partial z}}{\sqrt{\left(\frac{\partial F}{\partial x}\right)^2 + \left(\frac{\partial F}{\partial y}\right)^2 + \left(\frac{\partial F}{\partial z}\right)^2}} = \frac{1}{\sqrt{A^2x^2 + B^2y^2 + C^2z^2}}.$$

Thus the direction cosines of p are, by 74),

75) $$\begin{aligned} \alpha' &= \cos(nx) = Apx = Apr\alpha, \\ \beta' &= \cos(ny) = Bpy = Bpr\beta, \\ \gamma' &= \cos(nz) = Cpz = Cpr\gamma. \end{aligned}$$

If on the perpendicular we mark off a point P', at a distance $OP' = r' = \frac{R^2}{p}$, and call its coordinates x', y', z', we have

$$x' = r'\,\alpha' = Apr'\,x = R^2 Ax,$$
76) $$y' = r'\,\beta' = Bpr'\,y = R^2 By,$$
$$z' = r'\,\gamma' = Cpr'\,z = R^2 Cz,$$

from which we obtain

77) $$x = \frac{x'}{AR^2}, \quad y = \frac{y'}{BR^2}, \quad z = \frac{z'}{CR^2},$$

and by 72)

78) $$\frac{x'^2}{A} + \frac{y'^2}{B} + \frac{z'^2}{C} = R^4.$$

Accordingly the locus of P' is an ellipsoid, whose axes are inversely proportional to those of the original ellipsoid. It is called the *inverse ellipsoid.* If we take $R^4 = \frac{1}{M}$ we have

79) $$\frac{x'^2}{a^2} + \frac{y'^2}{b^2} + \frac{z'^2}{c^2} = 1$$

and the semi-axes of the inverse ellipsoid are equal to the principal radii of gyration a, b, c.

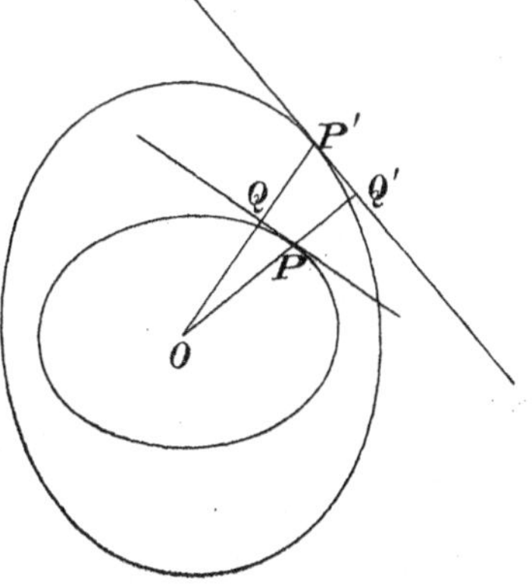

Fig. 70.

Since the two ellipsoids have the directions of their principal axes coincident (namely the directions in which p and r coincide), the relations are evidently reciprocal, and OP is perpendicular to the tangent plane at P'. Let the length of the perpendicular in this direction be p'. Then since the triangles OPQ, $OP'Q'$ (Fig. 70) are similar,

80) $$\frac{r}{p} = \frac{r'}{p'}, \quad p'r = r'p = R^2.$$

Since the moment of inertia about OP is

81) $$K = Mk^2 = \frac{1}{r^2} = \frac{p'^2}{R^4} = Mp'^2,$$

we have

$$k = p',$$

and the property of the inverse ellipsoid is that the radius of gyration about any line is equal to the part intercepted by a plane perpendicular to it tangent to the inverse ellipsoid. The inverse ellipsoid is accordingly called the *ellipsoid of gyration.*

It is evident that the direct ellipsoid more nearly resembles the given body in shape than the inverse ellipsoid, for if the body is spread out much about any particular axis the inertia and radius of gyration about that axis are large, so that the inverse ellipsoid has a large dimension in the direction of that axis, while the direct ellipsoid, like the body, has a small one.

73. Ellipsoidal Coordinates. The equation of a central quadric referred to its axes may be written,

$$\frac{x^2}{a_1} + \frac{y^2}{a_2} + \frac{z^2}{a_3} = 1, \tag{82}$$

where a_1, a_2, a_3, may be positive or negative. If they are all negative, the surface is imaginary, for the equation is not satisfied by any real values of x, y, z.

1⁰. Suppose one is negative, say

$$a_3 = -c^2,$$

while

$$a_1 = a^2, \quad a_2 = b^2.$$

Let

$$a > b > c.$$

The equation now is

$$\frac{x^2}{a^2} + \frac{y^2}{b^2} - \frac{z^2}{c^2} = 1.$$

The surface is cut by the XY-plane in the ellipse

$$\frac{x^2}{a^2} + \frac{y^2}{b^2} = 1,$$

whose semi-axes are a and b, and whose foci are at distances from the center

$$\sqrt{a^2 - b^2} = \sqrt{a_1 - a_2}$$

on the X-axis.

The section by the ZX-plane is the hyperbola

$$\frac{x^2}{a^2} - \frac{z^2}{c^2} = 1,$$

with semi-axes a, c, and foci at distances $\sqrt{a^2 + c^2} = \sqrt{a_1 - a_3}$ on the X-axis. The section by the YZ-plane is the hyperbola,

$$\frac{y^2}{b^2} - \frac{z^2}{c^2} = 1,$$

with semi-axes b, c, and foci at distances $\sqrt{b^2 + c^2} = \sqrt{a_2 - a_3}$ on the Y-axis. The surface is an hyperboloid of one sheet.

2⁰. Let two of the constants a_1, a_2, a_3, be negative, say

$$a_2 = -b^2, \quad a_3 = -c^2.$$

The equation is

$$\frac{x^2}{a^2} - \frac{y^2}{b^2} - \frac{c^2}{z^2} = 1.$$

The sections by the coordinates planes and their focal distances are

$$XY \quad \frac{x^2}{a^2} - \frac{y^2}{b^2} = 1 \quad \text{Hyperbola,} \quad \sqrt{a^2 + b^2} = \sqrt{a_1 - a_2} \text{ on } X\text{-axis,}$$

$$ZX \quad \frac{x^2}{a^2} - \frac{z^2}{c^2} = 1 \quad \text{Hyperbola,} \quad \sqrt{a^2 + c^2} = \sqrt{a_1 - a_3} \text{ on } X\text{-axis,}$$

$$YZ \quad \frac{y^2}{b^2} + \frac{z^2}{c^2} = -1 \quad \text{Imaginary Ellipse,} \quad \sqrt{-(b^2 - c^2)} = \sqrt{a_2 - a_3}.$$

The surface is an hyperboloid of two sheets.

3°. If a_1, a_2, a_3 are all positive, the sections are all ellipses, and the surface is an ellipsoid. In all three cases, the squares of the focal distances are the differences of the constants a_1, a_2, a_3. Consequently if we add to the three the same number, we get a surface whose principal sections have the same foci as before, or a surface *confocal* with the original. Accordingly

$$\text{83)} \qquad \frac{x^2}{a^2+\varrho} + \frac{y^2}{b^2+\varrho} + \frac{z^2}{c^2+\varrho} = 1$$

represents a quadric confocal with the ellipsoid

$$\frac{x^2}{a^2} + \frac{y^2}{b^2} + \frac{z^2}{c^2} = 1,$$

for any real value of ϱ.

If $a > b > c$ and $\varrho > -c^2$, the surface is an ellipsoid. If $-c^2 > \varrho > -b^2$, the surface is an hyperboloid of one sheet, and if $-b^2 > \varrho > -a^2$, an hyperboloid of two sheets. If $\varrho < -a^2$, the surface is imaginary.

Suppose we attempt to pass through a given point x, y, z, a quadric confocal with the ellipsoid

$$\frac{x^2}{a^2} + \frac{y^2}{b^2} + \frac{z^2}{c^2} = 1 \qquad (a > b > c).$$

Its equation is 83), where the parameter ϱ is to be determined. Clearing of fractions, the equation is

$$\text{84)} \quad f(\varrho) = (a^2 + \varrho)(b^2 + \varrho)(c^2 + \varrho) - x^2(b^2 + \varrho)(c^2 + \varrho) - y^2(c^2 + \varrho)(a^2 + \varrho) - z^2(a^2 + \varrho)(b^2 + \varrho) = 0$$

a cubic in ϱ. But this is easily shown to have three real roots. Putting successively ϱ equal to $\infty, -c^2, -b^2, -a^2$ and observing signs of $f(\varrho)$,

$$\begin{array}{lll} \varrho = \infty, & f(\varrho) = \infty & + \\ \varrho = -c^2, & f(\varrho) = -z^2(a^2 - c^2)(b^2 - c^2) & - \\ \varrho = -b^2, & f(\varrho) = -y^2(c^2 - b^2)(a^2 - b^2) & + \\ \varrho = -a^2, & f(\varrho) = -x^2(b^2 - a^2)(c^2 - a^2) & - \end{array}$$

Call the roots in order of magnitude λ, μ, ν. The changes of sign above show that λ lies in the interval $\lambda > -c^2$ necessary in order that the surface shall be an ellipsoid, μ in the interval $-c^2 > \mu > -b^2$ that it may be an hyperboloid of one sheet, and ν in the interval $-b^2 > \nu > -a^2$ that it may be an hyperboloid of two sheets. There pass therefore through every point in space one surface of each of the three kinds. If we call

$$85)\qquad F(\lambda, x, y, z) = \frac{x^2}{a^2+\lambda} + \frac{y^2}{b^2+\lambda} + \frac{z^2}{c^2+\lambda} - 1,$$

the equation $F = 0$ defines λ as a function of x, y, z. The normal to the surface $\lambda(xyz) = const.$ has direction cosines proportional to

$$\frac{\partial \lambda}{\partial x}, \quad \frac{\partial \lambda}{\partial y}, \quad \frac{\partial \lambda}{\partial z}.$$

Now since identically $F = 0$, differentiating totally,

$$dF = \frac{\partial F}{\partial x} dx + \frac{\partial F}{\partial y} dy + \frac{\partial F}{\partial z} dz + \frac{\partial F}{\partial \lambda} d\lambda = 0,$$

and we have

$$\frac{\partial \lambda}{\partial x} = \left(\frac{d\lambda}{dx}\right)_{\substack{dy=0\\dz=0}} = -\frac{\partial F}{\partial x} \Big/ \frac{\partial F}{\partial \lambda},$$

for the required partial derivative of λ with respect to x, when y and z are constant.

Therefore

$$\frac{\partial \lambda}{\partial x} = \frac{2x}{a^2+\lambda} \Big/ \left\{\frac{x^2}{(a^2+\lambda)^2} + \frac{y^2}{(b^2+\lambda)^2} + \frac{z^2}{(c^2+\lambda)^2}\right\} = -\frac{2x}{(a^2+\lambda)F'(\lambda)}.$$

Similarly

$$86)\qquad \begin{aligned} \frac{\partial \lambda}{\partial y} &= -\frac{2y}{(b^2+\lambda)F'(\lambda)}, \\ \frac{\partial \lambda}{\partial z} &= -\frac{2z}{(c^2+\lambda)F'(\lambda)}. \end{aligned}$$

The sum of the squares of the derivatives being called h_λ^2, we have

$$87)\qquad \begin{aligned} h_\lambda^2 &= \left(\frac{\partial \lambda}{\partial x}\right)^2 + \left(\frac{\partial \lambda}{\partial y}\right)^2 + \left(\frac{\partial \lambda}{\partial z}\right)^2 = \frac{4}{\{F'(\lambda)\}^2}\left\{\frac{x^2}{(a^2+\lambda)^2} + \frac{y^2}{(b^2+\lambda)^2} + \frac{z^2}{(c^2+\lambda)^2}\right\} \\ &= -\frac{4}{F'(\lambda)}, \\ h_\lambda &= \pm \frac{2}{\sqrt{-F'(\lambda)}}. \end{aligned}$$

Now the direction cosines of the normal to the surface $\lambda = const.$ are

$$88)\qquad \begin{aligned} \cos(n_\lambda x) &= \frac{1}{h_\lambda}\frac{\partial \lambda}{\partial x} = \pm \frac{\sqrt{-F'(\lambda)}}{2} \cdot \frac{2x}{(a^2+\lambda)F'(\lambda)} = \pm \frac{x}{(a^2+\lambda)\sqrt{-F'(\lambda)}}, \\ \cos(n_\lambda y) &= \pm \frac{y}{(b^2+\lambda)\sqrt{-F'(\lambda)}}, \\ \cos(n_\lambda z) &= \pm \frac{z}{(c^2+\lambda)\sqrt{-F'(\lambda)}}. \end{aligned}$$

Similarly for the normals to the surface $\mu = const.$,

$$89)\quad \begin{aligned} \cos(n_\mu x) &= \pm \frac{x}{(a^2+\mu)\sqrt{-F'(\mu)}}, \\ \cos(n_\mu y) &= \pm \frac{y}{(b^2+\mu)\sqrt{-F'(\mu)}}, \\ \cos(n_\mu z) &= \pm \frac{z}{(c^2+\mu)\sqrt{-F'(\mu)}}. \end{aligned}$$

The angle between the normals to λ and μ is given by

$$90)\quad \cos(n_\lambda n_\mu) = \left\{ \frac{x^2}{(a^2+\lambda)(a^2+\mu)} + \frac{y^2}{(b^2+\lambda)(b^2+\mu)} + \frac{z^2}{(c^2+\lambda)(c^2+\mu)} \right\} \frac{1}{\sqrt{F'(\lambda)F'(\mu)}}.$$

Now by subtracting from the equation

$$\frac{x^2}{a^2+\lambda} + \frac{y^2}{b^2+\lambda} + \frac{z^2}{c^2+\lambda} = 1,$$

the equation

$$\frac{x_2}{a^2+\mu} + \frac{y^2}{b^2+\mu} + \frac{z^2}{c^2+\mu} = 1,$$

we obtain

$$91)\quad x^2\left\{\frac{1}{a^2+\lambda} - \frac{1}{a^2+\mu}\right\} + y^2\left\{\frac{1}{b^2+\lambda} - \frac{1}{b^2+\mu}\right\} + z^2\left\{\frac{1}{c^2+\lambda} - \frac{1}{c^2+\mu}\right\} = 0$$

or

$$92)\quad (\lambda - \mu)\left\{\frac{x^2}{(a^2+\lambda)(a^2+\mu)} + \frac{y^2}{(b^2+\lambda)(b^2+\mu)} + \frac{z^2}{(c^2+\lambda)(c^2+\mu)}\right\} = 0.$$

Accordingly, unless $\lambda = \mu$, $\cos(n_\lambda n_\mu) = 0$, and the two normals are at right angles. Similarly for the other pairs of surfaces. Accordingly the three surfaces of the confocal system passing through any point cut each other at right angles.

If we give the values of λ, μ, ν we determine completely the ellipsoid and two hyperboloids, and hence the point of intersection x, y, z. To be sure there are the seven symmetrical points in the other quadrants which have the same values of λ, μ, ν, but if we specify which quadrant is to be considered this will cause no ambiguity. Thus the point is specified by the three quantities λ, μ, ν, which are called the *ellipsoidal* or *elliptic coordinates* of the point.

74. Axes of Inertia at Various Points. Let $K = Mk^2$ be the moment of inertia about an axis whose direction cosines are α, β, γ, at a point O whose coordinates with respect to the principal axes at the center of mass G are xyz. Let p be the distance of the axis at O from a parallel axis through G, and q the distance of the foot of the perpendicular from G. Then by the two theorems of § 70 and § 71,

$$93)\quad \begin{aligned} K &= A\alpha^2 + B\beta^2 + C\gamma^2 + Mp^2, \\ k^2 &= a^2\alpha^2 + b^2\beta^2 + c^2\gamma^2 + p^2. \end{aligned}$$

Now

$$94)\quad \begin{aligned} p^2 &= r^2 - q^2 = r^2 - (\alpha x + \beta y + \gamma z)^2, \\ k^2 &= a^2\alpha^2 + b^2\beta^2 + c^2\gamma^2 + r^2 - q^2. \end{aligned}$$

In order to find the principal axes at O we must make this a maximum or minimum with respect to α, β, γ subject to the condition,

$$\alpha^2 + \beta^2 + \gamma^2 = 1.$$

Multiplying this by a constant σ, subtracting from 94), and differentiating

$$95)\quad \begin{aligned} \frac{1}{2}\frac{\partial}{\partial \alpha}\{k^2 - \sigma(\alpha^2 + \beta^2 + \gamma^2)\} &= a^2\alpha - qx - \sigma\alpha = 0, \\ \frac{1}{2}\frac{\partial}{\partial \beta}\{k^2 - \sigma(\alpha^2 + \beta^2 + \gamma^2)\} &= b^2\beta - qy - \sigma\beta = 0, \\ \frac{1}{2}\frac{\partial}{\partial \beta}\{k^2 - \sigma(\alpha^2 + \beta^2 + \gamma^2)\} &= c^2\gamma - qz - \sigma\gamma = 0. \end{aligned}$$

Multiplying these equations respectively by α, β, γ and adding,

$$96)\quad \begin{aligned} a^2\alpha^2 + b^2\beta^2 + c^2\gamma^2 - q(\alpha x + \beta y + \gamma z) - \sigma &= 0, \\ k^2 - r^2 - \sigma &= 0. \end{aligned}$$

Thus σ is determined as

$$97)\quad \sigma = k^2 - r^2.$$

Inserting this value in 95) we have

$$98)\quad \begin{aligned} (a^2 + r^2 - k^2)\,\alpha &= qx, \\ (b^2 + r^2 - k^2)\,\beta &= qy, \\ (c^2 + r^2 - k^2)\,\gamma &= qz. \end{aligned}$$

Multiplying these equations respectively by

$$\frac{x}{a^2 + r^2 - k^2}, \quad \frac{y}{b^2 + r^2 - k^2}, \quad \frac{z}{c^2 + r^2 - k^2},$$

and adding, we get, since q divides out,

$$99)\quad \frac{x^2}{a^2 + r^2 - k^2} + \frac{y^2}{b^2 + r^2 - k^2} + \frac{z^2}{c^2 + r^2 - k^2} = 1.$$

If we now put $r^2 - k^2 = \varrho$, this is the same cubic as 83) to determine ϱ, and gives three real roots for k^2,

$$\begin{aligned} k_1^2 &= r^2 - \lambda, \\ k_2^2 &= r^2 - \mu, \\ k_3^2 &= r^2 - \nu. \end{aligned}$$

The direction cosines are then given, according to 95) and 98), by

$$\frac{\alpha_1}{x}(a^2+r^2-k_1^2)=\frac{\beta_1}{y}(b^2+r^2-k_1^2)=\frac{\gamma_1}{z}(c^2+r^2-k_1^2),$$

100) $$\frac{\alpha_2}{x}(a^2+r^2-k_2^2)=\frac{\beta_2}{y}(b^2+r^2-k_2^2)=\frac{\gamma_2}{z}(c^2+r^2-k_2^2),$$

$$\frac{\alpha_3}{x}(a^2+r^2-k_3^2)=\frac{\beta_3}{y}(b^2+r^2-k_3^2)=\frac{\gamma_3}{z}(c^2+r^2-k_3^2),$$

that is

$$\alpha_1:\beta_1:\gamma_1=\frac{x}{a^2+\lambda}:\frac{y}{b^2+\lambda}:\frac{z}{c^2+\lambda},\quad \text{etc.}$$

Hence the principal axes of inertia at any point O are normal to the three surfaces through O confocal with the ellipsoid of gyration at the center of mass. This theorem is due to Binet.

Since $\lambda>\mu>\nu$, the least moment of inertia is about the normal to the ellipsoid, the greatest about the two-sheeted hyperboloid, and the mean about the normal to the one-sheeted hyperboloid.

We have

$$k_1^2+k_2^2+k_3^2=3r^2-(\lambda+\mu+\nu).$$

But the sum of the three roots is the negative of the coefficient of ϱ^2 in the cubic 83),

$$\lambda+\mu+\nu=x^2+y^2+z^2-(a^2+b^2+c^2),$$

101) $$k_1^2+k_2^2+k_3^2=2r^2+a^2+b^2+c^2.$$

Thus the sum of the principal moments of inertia is the same for all points lying at equal distances from the center of mass.

It is now easy to see that any given line is a principal axis for only one of its points, unless it passes through the center of mass, when it is such for all of its points. It is also evident that not every line in space can be a principal axis.

If the central ellipsoid of gyration is a sphere, all the ellipsoids of the confocal system are spheres, and all the hyperboloids cones. Every ellipsoid of inertia is a prolate ellipsoid of revolution, with its axis passing through the center of mass.

If the central ellipsoid has two equal axes, the ellipsoids of inertia for points on the axis of revolution are also of revolution. If the distance of a point on this line from the center of mass is d, and the moment of inertia about it is Mk_1^2

$$k_1^2=a^2,$$

$$k_2^2=b^2+d^2,$$

$$k_3^2=b^2+d^2.$$

If $b < a$ there are two points for which the ellipsoids of inertia are spheres, namely where $d = \pm\sqrt{a^2 - b^2}$. This is the only case, except the above, where there are spheres.

If we look for ellipsoids of revolution in the general case when a, b, c are unequal, we must distinguish between prolate and oblate ellipsoids of gyration.

1°. *Prolate.* The two equal radii of gyration are the two smaller k_1 and k_2. For these to be equal, we must have $\lambda = \mu$. But as λ and μ are separated by $-c^2$, if they are equal they must be equal to $-c^2$. In this case the axis of the ellipsoid and one-sheeted hyperboloid are both zero, and the ellipsoid becomes the elliptical *disk* with axes $\sqrt{a^2 - c^2}$, $\sqrt{b^2 - c^2}$, forming part of the XY-plane, and the hyperboloid all the rest of the XY-plane. Points lying on both surfaces lie on the ellipse whose axes are $\sqrt{a^2 - c^2}$, $\sqrt{b^2 - c^2}$, which passes through the four foci of the system lying on the X- and Y-axes, and is accordingly called the focal ellipse of the confocal system. (We saw by 92] that if $\lambda = \mu$ the two surfaces were not necessarily orthogonal.) All points lying on this ellipse have prolate ellipsoids of gyration, the axes of rotation lying in the plane of the ellipse.

2°. *Oblate ellipsoids of gyration.* In this case we have

$$k_2 = k_3, \quad \mu = \nu = -b^2.$$

The Y-axes of the two hyperboloids now vanish. That of one sheet becomes the part of the XZ-plane lying within the hyperbola

$$\frac{x^2}{a^2 - b^2} - \frac{z^2}{b^2 - c^2} = 1,$$

and that of two sheets the remaining parts. The points common to both are those lying on the hyperbola, whose axes are $\sqrt{a^2 - b^2}$, $\sqrt{b^2 - c^2}$ and which passes through the remaining two foci of the system, and is called the focal hyperbola. The axes of revolution of the ellipsoids of gyration lie in the plane of the hyperbola.

75. Calculation of Moments of Inertia. In the case of a continuous solid, the sums all become definite integrals, as stated in § 68. All the preceding theorems of course are unaltered. If the body is homogeneous all the integrals are proportional to the density. Since the mass is likewise, the radii of gyration are independent of the density. We will therefore put $\varrho = 1$.

Rectangular Parallelepiped, of dimensions $2a$, $2b$, $2c$.

$$A' = \int_{-a}^{a}\int_{-b}^{b}\int_{-c}^{c} x^2\,dx\,dy\,dz = \frac{8}{3}a^3bc,$$

$$B' = \int_{-a}^{a}\int_{-b}^{b}\int_{-c}^{c} y^2\,dx\,dy\,dz = \frac{8}{3}b^3ca, \qquad (M = 8abc)$$

$$C' = \int_{-a}^{a}\int_{-b}^{b}\int_{-c}^{c} z^2\,dx\,dy\,dz = \frac{8}{3}c^3ab,$$

$$A = B' + C' = \frac{8}{3}abc(b^2 + c^2),$$

102) $$B = C' + A' = \frac{8}{3}abc(c^2 + a^2),$$

$$C = A' + B' = \frac{8}{3}abc(a^2 + b^2),$$

Thus the radii of gyration a_0, b_0, c_0 are

103) $$a_0 = \sqrt{\frac{b^2 + c^2}{3}}, \quad b_0 = \sqrt{\frac{c^2 + a^2}{3}}, \quad c_0 = \sqrt{\frac{a^2 + b^2}{3}}.$$

Sphere, with radius R.

$$A' = \iiint x^2\,dx\,dy\,dz, \quad B' = \iiint y^2\,dx\,dy\,dz, \quad C' = \iiint z^2\,dx\,dy\,dz,$$

the limits of integration being given by the inequality $x^2 + y^2 + z^2 < R^2$.

$$A' + B' + C' = \iiint (x^2 + y^2 + z^2)\,dx\,dy\,dz.$$

Changing to polar coordinates,

$$A' + B' + C' = \int_0^R 4\pi r^4\,dr = \frac{4}{5}\pi R^5,$$

$$A' = B' = C' = \frac{4}{15}\pi R^5,$$

$$A = B' + C' = \frac{8}{15}\pi R^5, \quad M = \frac{4}{3}\pi R^3,$$

104) $$A = B = C = \frac{2}{5}MR^2, \quad k = R\sqrt{\frac{2}{5}}.$$

Ellipsoid, with semi-axes a, b, c.

$$A' = \iiint x^2\,dx\,dy\,dz,$$

the limits of integration being given by the inequality

$$\frac{x^2}{a^2} + \frac{y^2}{b^2} + \frac{z^2}{c^2} < 1.$$

The integration is most easily performed by a change of variable. If we put

$$\frac{x}{a} = x', \quad \frac{y}{b} = y', \quad \frac{z}{c} = z',$$

$$A' = \iiint a^2 x'^2 a\,dx'\,b\,dy'\,c\,dz' = a^3 bc \iiint x'^2\,dx'\,dy'\,dz',$$

the integral being taken for values corresponding to points within the sphere

$$x'^2 + y'^2 + z'^2 < 1.$$

Now the moment of inertia of the sphere with respect to a diametral plane is $\frac{4}{15}\pi$, hence

$$A' = \frac{4}{15}\pi a^3 bc, \quad B' = \frac{4}{15}\pi b^3 ca, \quad C' = \frac{4}{15}\pi c^3 ab.$$

$$A = B' + C' = \frac{4}{15}\pi abc(b^2 + c^2),$$

105) $$B = C' + A' = \frac{4}{15}\pi abc(c^2 + a^2), \qquad \left(M = \frac{4}{3}\pi abc\right)$$

$$C = A' + B' = \frac{4}{15}\pi abc(a^2 + b^2),$$

106) $$a_0 = \sqrt{\frac{b^2 + c^2}{5}}, \quad b_0 = \sqrt{\frac{b^2 + a^2}{5}}, \quad c_0 = \sqrt{\frac{a^2 + b^2}{5}}.$$

Thin Circular Disk normal to Z-axis

$$A' = \iint x^2\,dx\,dy, \quad B' = \iint y^2\,dx\,dy, \quad C' = 0,$$

$$A' + B' = \iint (x^2 + y^2)\,dx\,dy = \int_0^R 2\pi r^3\,dr = \frac{1}{2}\pi R^4,$$

$$M = \pi R^2, \quad A' = B' = \frac{1}{4}MR^2, \quad C' = 0,$$

107) $$A = B = \frac{1}{4}MR^2, \quad C = \frac{1}{2}MR^2.$$

The moment about the normal to the disk is double that about a diameter.

Circular Cylinder of radius R, length $2l$.

The moment about the axis of rotation, is, as for the disk,

$$C = \frac{1}{2}MR^2, \quad A' = B' = \frac{1}{4}MR^2,$$

$$C' = \int_{-l}^{l} \pi R^2 z^2\,dz = \frac{2}{3}\pi R^2 l^3 = \frac{1}{3}Ml^2,$$

108) $$A = B = M\left(\frac{R^2}{4} + \frac{l^2}{3}\right), \quad C = \frac{1}{2}MR^2.$$

We have $A = B = C$ if

$$\frac{R^2}{4} + \frac{l^2}{3} = \frac{R^2}{2},$$

$$l = R\frac{\sqrt{3}}{2}.$$

Then the cylinder is dynamically equivalent to a sphere, as is also the case for a cube.

These examples furnish the means of treating the cases that usually appear in practice.

76. Analytical Treatment of Kinematics of a Rigid System. Moving Axes. In §§ 55—57 we have treated the general motion of a rigid system, from the purely geometrical point of view, without analysis. We shall now give the analytical treatment of the same subject. Let us refer the position of a point in the system to two different sets of coordinates. Let x', y', z' be its coordinates with respect to a set of axes fixed in space, and let x, y, z be its coordinates with respect to a set of axes moving in any manner. The position of the moving axes is defined by the position of their origin, whose coordinates referred to the fixed axes are ξ, η, ζ, and by the nine direction cosines of one set of axes with respect to the other. Let these be given by the following table

	X	Y	Z
X'	α_1	β_1	γ_1
Y'	α_2	β_2	γ_2
Z'	α_3	β_3	γ_3

The equations for the transformation of coordinates are then,

$$109)\qquad \begin{aligned} x' &= \xi + \alpha_1 x + \beta_1 y + \gamma_1 z, \\ y' &= \eta + \alpha_2 x + \beta_2 y + \gamma_2 z, \\ z' &= \zeta + \alpha_3 x + \beta_3 y + \gamma_3 z. \end{aligned}$$

Since α_1, α_2, α_3 are the direction cosines of the X-axis with respect to X', Y', Z', we have

$$110)\qquad \alpha_1^2 + \alpha_2^2 + \alpha_3^2 = 1,$$

and similarly

$$110)\qquad \begin{aligned} \beta_1^2 + \beta_2^2 + \beta_3^2 &= 1, \\ \gamma_1^2 + \gamma_2^2 + \gamma_3^2 &= 1. \end{aligned}$$

Since the axes Y, Z are perpendicular, their direction cosines satisfy the conditions,

$$\beta_1\gamma_1 + \beta_2\gamma_2 + \beta_3\gamma_3 = 0, \tag{111}$$

and similarly,

$$\begin{aligned} \gamma_1\alpha_1 + \gamma_2\alpha_2 + \gamma_3\alpha_3 &= 0, \\ \alpha_1\beta_1 + \alpha_2\beta_2 + \alpha_3\beta_3 &= 0. \end{aligned} \tag{111}$$

Thus the nine cosines are not independent, but, satisfying six conditions, may be expressed in terms of three parameters. These, with the three ξ, η, ζ, show the *six* degrees of freedom possessed by a rigid system.

By interchanging the rôles of the axes, and considering the direction cosines of X', Y', Z' with respect to X, Y, Z we find the equivalent conditions

$$\begin{aligned} \alpha_1^2 + \beta_1^2 + \gamma_1^2 &= 1, \\ \alpha_2^2 + \beta_2^2 + \gamma_2^2 &= 1, \\ \alpha_3^2 + \beta_3^2 + \gamma_3^2 &= 1, \end{aligned} \tag{112}$$

$$\begin{aligned} \alpha_1\alpha_2 + \beta_1\beta_2 + \gamma_1\gamma_2 &= 0, \\ \alpha_2\alpha_3 + \beta_2\beta_3 + \gamma_2\gamma_3 &= 0, \\ \alpha_3\alpha_1 + \beta_3\beta_1 + \gamma_3\gamma_1 &= 0. \end{aligned} \tag{113}$$

If we now differentiate the first of equations 109), supposing x, y, z to be constant, we obtain

$$\begin{aligned} v'_x &= \frac{dx'}{dt} = x\frac{d\alpha_1}{dt} + y\frac{d\beta_1}{dt} + z\frac{d\gamma_1}{dt} + \frac{d\xi}{dt}, \\ v'_y &= \frac{dy'}{dt} = x\frac{d\alpha_2}{dt} + y\frac{d\beta_2}{dt} + z\frac{d\gamma_2}{dt} + \frac{d\eta}{dt}, \\ v'_z &= \frac{dz'}{dt} = x\frac{d\alpha_3}{dt} + y\frac{d\beta_3}{dt} + z\frac{d\gamma_3}{dt} + \frac{d\zeta}{dt}, \end{aligned} \tag{114}$$

for the components in the directions of the fixed axes of the velocity of a point fixed to the moving axes.

Let us now resolve the velocity in the direction which is at a given instant that of one of the moving axes. To resolve in the direction of the X-axis we have

$$\begin{aligned} v_x = \alpha_1 v'_x + \alpha_2 v'_y + \alpha_3 v'_z &= \alpha_1\frac{d\xi}{dt} + \alpha_2\frac{d\eta}{dt} + \alpha_3\frac{d\zeta}{dt} \\ &+ x\left(\alpha_1\frac{d\alpha_1}{dt} + \alpha_2\frac{d\alpha_2}{dt} + \alpha_3\frac{d\alpha_3}{dt}\right) + y\left(\alpha_1\frac{d\beta_1}{dt} + \alpha_2\frac{d\beta_2}{dt} + \alpha_3\frac{d\beta_3}{dt}\right) \\ &+ z\left(\alpha_1\frac{d\gamma_1}{dt} + \alpha_2\frac{d\gamma_2}{dt} + \alpha_3\frac{d\gamma_3}{dt}\right). \end{aligned} \tag{115}$$

The coefficient of x in this expression is the derivative of the left-hand member of the first of equations 110), and is accordingly equal to zero. If we now denote the coefficients of y and z by single letters, and compare them with the results of differentiating equations 111), writing

$$\begin{aligned} p &= \gamma_1 \frac{d\beta_1}{dt} + \gamma_2 \frac{d\beta_2}{dt} + \gamma_3 \frac{d\beta_3}{dt} = -\left(\beta_1 \frac{d\gamma_1}{dt} + \beta_2 \frac{d\gamma_2}{dt} + \beta_3 \frac{d\gamma_3}{dt}\right), \\ q &= \alpha_1 \frac{d\gamma_1}{dt} + \alpha_2 \frac{d\gamma_2}{dt} + \alpha_3 \frac{d\gamma_3}{dt} = -\left(\gamma_1 \frac{d\alpha_1}{dt} + \gamma_2 \frac{d\alpha_2}{dt} + \gamma_3 \frac{d\alpha_3}{dt}\right), \\ r &= \beta_1 \frac{d\alpha_1}{dt} + \beta_2 \frac{d\alpha_2}{dt} + \beta_3 \frac{d\alpha_3}{dt} = -\left(\alpha_1 \frac{d\beta_1}{dt} + \alpha_2 \frac{d\beta_2}{dt} + \alpha_3 \frac{d\beta_3}{dt}\right), \end{aligned} \tag{116}$$

we obtain

$$\begin{aligned} v_x &= \alpha_1 \frac{d\xi}{dt} + \alpha_2 \frac{d\eta}{dt} + \alpha_3 \frac{d\zeta}{dt} + qz - ry, \\ v_y &= \beta_1 \frac{d\xi}{dt} + \beta_2 \frac{d\eta}{dt} + \beta_3 \frac{d\zeta}{dt} + rx - pz, \\ v_z &= \gamma_1 \frac{d\xi}{dt} + \gamma_2 \frac{d\eta}{dt} + \gamma_3 \frac{d\zeta}{dt} + py - qx. \end{aligned} \tag{117}$$

These equations express the fact that the velocity of a point attached to the moving axes is the resultant of two vectors, one of which, V, is the same for all points of the system, being independent of x, y, z, and having the components in the direction of x', y', z' equal to $\frac{d\xi}{dt}, \frac{d\eta}{dt}, \frac{d\zeta}{dt}$, and in the direction of x, y, z, equal to

$$\begin{aligned} V_x &= \alpha_1 \frac{d\xi}{dt} + \alpha_2 \frac{d\eta}{dt} + \alpha_3 \frac{d\zeta}{dt}, \\ V_y &= \beta_1 \frac{d\xi}{dt} + \beta_2 \frac{d\eta}{dt} + \beta_3 \frac{d\zeta}{dt}, \\ V_z &= \gamma_1 \frac{d\xi}{dt} + \gamma_2 \frac{d\eta}{dt} + \gamma_3 \frac{d\zeta}{dt}. \end{aligned} \tag{118}$$

This part of the motion is accordingly a *translation.*

The other part of the velocity, whose components in the direction of the instantaneous positions of the X, Y, Z-axes are given by

$$\begin{aligned} v_x &= qz - ry, \\ v_y &= rx - pz, \\ v_z &= py - qx, \end{aligned} \tag{119}$$

being the vector product of a vector ω whose components are p, q, r, and of the position vector ϱ of the point, is perpendicular to both these vectors and is in magnitude equal to $\omega \varrho \sin(\omega \varrho)$. It accordingly represents a motion due to a rotation of the body with angular velocity ω about an axis in the direction of the vector ω. Thus we have an analytical demonstration of the vector nature of angular velocity. If we take as a position of the fixed axes one which coincides with that of the moving axes at some particular instant of time, the direction cosines vanish with the exception of $\alpha_1, \beta_2, \gamma_3$, which are equal to unity. We then have

$$120)\quad p = \frac{d\beta_3}{dt} = -\frac{d\gamma_2}{dt}, \quad q = \frac{d\gamma_1}{dt} = -\frac{d\alpha_3}{dt}, \quad r = \frac{d\alpha_2}{dt} = -\frac{d\beta_1}{dt}.$$

But since $\beta_3 = \cos(yz')$, $\gamma_2 = \cos(zy')$, we have on differentiating

$$\frac{d\beta_3}{dt} = -\sin(yz')\frac{d(yz')}{dt}, \quad \frac{d\gamma_2}{dt} = -\sin(zy')\frac{d(zy')}{dt},$$

and since

$$\sin(yz') = \sin(zy') = 1, \quad \frac{d\beta_3}{dt} = -\frac{d(yz')}{dt},$$

$$p = -\frac{d(yz')}{dt} = \frac{d(zy')}{dt}.$$

Thus it is clearly seen that p, q, r are angular velocities, being the rates of increase of the angles zy', xz', yx', or in other words, the angular velocities with which the moving axes X, Y, Z are turning about *each other*.

It is to be noticed that p, q, r, though angular velocities, are *not* time-derivatives of any functions of the coordinates, which might be taken for three generalized Lagrangian coordinates q.

They are merely linear functions of the derivatives of the nine cosines, which latter may *themselves* be expressed in terms of three q's.

If we seek to find those points of the body whose actual velocity is a minimum, we must differentiate the quantity,

$$121)\quad v^2 = (V_x + qz - ry)^2 + (V_y + rx - pz)^2 + (V_z + py - qx)^2$$

with respect to x, y, z, and equate the derivatives to zero. We thus obtain

$$122)\quad \begin{aligned} r(V_y + rx - pz) - q(V_z + py - qx) &= 0, \\ p(V_z + py - qx) - r(V_x + qz - ry) &= 0, \\ q(V_x + qz - ry) - p(V_y + rx - pz) &= 0, \end{aligned}$$

which are equivalent to the two independent equations,

$$123)\quad \frac{V_x + qz - ry}{p} = \frac{V_y + rx - pz}{q} = \frac{V_z + py - qx}{r}.$$

These are the equations of a line in the body, namely of the central axis, as found in § 66, 38).

Calling the value of the common ratio λ, clearing of fractions, multiplying by p, q, r, and adding, we obtain the value of λ,

$$\lambda(p^2 + q^2 + r^2) = pV_x + qV_y + rV_z,$$

$$124)\quad \lambda = \frac{pV_x + qV_y + rV_z}{p^2 + q^2 + r^2}.$$

Making use of this value of λ with equations 121), 123), we obtain for v for points on the central-axis

$$v^2 = \lambda^2 (p^2 + q^2 + r^2) = \frac{(pV_x + qV_y + rV_z)^2}{p^2 + q^2 + r^2},$$

125) $$v = \frac{pV_x + qV_y + rV_z}{\sqrt{p^2 + q^2 + r^2}},$$

agreeing with 42).

If the velocity of points on the central axis is to be zero, we must have

126) $$pV_x + qV_y + rV_z = 0,$$

when the motion reduces to a rotation, as in 41).

77. Relative Motion. In forming equations 114) and the following, we have supposed the point in question fixed in the body, so that x, y, z were constants. If this is not the case we have to add to the right hand members of 114) the quantities

127) $$\begin{aligned} &\alpha_1 \frac{dx}{dt} + \beta_1 \frac{dy}{dt} + \gamma_1 \frac{dz}{dt}, \\ &\alpha_2 \frac{dx}{dt} + \beta_2 \frac{dy}{dt} + \gamma_2 \frac{dz}{dt}, \\ &\alpha_3 \frac{dx}{dt} + \beta_3 \frac{dy}{dt} + \gamma_3 \frac{dz}{dt}, \end{aligned}$$

which, on being multiplied by the proper cosines, will appear in equations 117) as $\frac{dx}{dt}$, $\frac{dy}{dt}$, $\frac{dz}{dt}$, so that we have for the components of the actual velocity in the direction of the axes X, Y, Z at the instant in question, if the *origin* of the latter is fixed,

128) $$\begin{aligned} v_x &= \frac{dx}{dt} + qz - ry, \\ v_y &= \frac{dy}{dt} + rx - pz, \\ v_z &= \frac{dz}{dt} + py - qx. \end{aligned}$$

These equations are of very great importance, for by means of them we may express the velocity components in directions coinciding with the instantaneous direction of the moving axes of the end of *any* vector x, y, z. If for x, y, z we put the components of the velocity v, we obtain the acceleration-components (§ 103), if the components of angular momentum H we have a dynamical result treated in § 84.

If we apply these equations to a point fixed in space, for which v_x, v_y, v_z vanish, we obtain

$$129)\quad \frac{dx}{dt} = ry - qz, \quad \frac{dy}{dt} = pz - rx, \quad \frac{dz}{dt} = qx - py.$$

Taking a point on the X'-axis at unit distance from the origin, we have $x = \alpha_1$, $y = \beta_1$, $z = \gamma_1$,

$$130)\quad \frac{d\alpha_1}{dt} = \beta_1 r - \gamma_1 q, \quad \frac{d\beta_1}{dt} = \gamma_1 p - \alpha_1 r, \quad \frac{d\gamma_1}{dt} = \alpha_1 q - \beta_1 p,$$

and in like manner taking points on the Y' and Z'-axes,

$$130)\quad \begin{aligned} \frac{d\alpha_2}{dt} &= \beta_2 r - \gamma_2 q, & \frac{d\beta_2}{dt} &= \gamma_2 p - \alpha_2 r, & \frac{d\gamma_2}{dt} &= \alpha_2 q - \beta_2 p, \\ \frac{d\alpha_3}{dt} &= \beta_3 r - \gamma_3 q, & \frac{d\beta_3}{dt} &= \gamma_3 p - \alpha_3 r, & \frac{d\gamma_3}{dt} &= \alpha_3 q - \beta_3 p. \end{aligned}$$

Using these values of the derivatives of the cosines, we find that they identically satisfy equations 116).

78. Angular Acceleration. If we call p', q', r' the components of the angular velocity on the fixed axes X', Y', Z', we have

$$131)\quad \begin{aligned} p' &= \alpha_1 p + \beta_1 q + \gamma_1 r, \\ q' &= \alpha_2 p + \beta_2 q + \gamma_2 r, \\ r' &= \alpha_3 p + \beta_3 q + \gamma_3 r. \end{aligned}$$

The time derivatives of these quantities will be called the angular accelerations about the axes X', Y', Z'. Differentiating the first,

$$\frac{dp'}{dt} = \alpha_1 \frac{dp}{dt} + \beta_1 \frac{dq}{dt} + \gamma_1 \frac{dr}{dt} + p \frac{d\alpha_1}{dt} + q \frac{d\beta_1}{dt} + r \frac{d\gamma_1}{dt},$$

and substituting the values of $\frac{d\alpha_1}{dt}, \frac{d\beta_1}{dt}, \frac{d\gamma_1}{dt}$ from 130), we have

$$132)\quad \begin{aligned} \frac{dp'}{dt} &= \alpha_1 \frac{dp}{dt} + \beta_1 \frac{dq}{dt} + \gamma_1 \frac{dr}{dt}, \\ \frac{dq'}{dt} &= \alpha_2 \frac{dp}{dt} + \beta_2 \frac{dq}{dt} + \gamma_2 \frac{dr}{dt}, \\ \frac{dr'}{dt} &= \alpha_3 \frac{dp}{dt} + \beta_3 \frac{dq}{dt} + \gamma_3 \frac{dr}{dt}. \end{aligned}$$

Thus the angular acceleration is obtained by resolving a vector whose components about the axes X, Y, Z are $\frac{dp}{dt}, \frac{dq}{dt}, \frac{dr}{dt}$. In other words, the time derivatives of the components p, q, r, of the angular velocity in the directions of the moving axes at any instant are equal to the angular accelerations of the motion about axes fixed in space coinciding in direction with the position of the moving axes at the given instant. This theorem, which is by no means self-evident, is of great importance, as is the similar property of the angular velocity, of which we have already made use.

79. Kinetic Energy and Momentum due to Rotation. From equations 119) we find for the part of the kinetic energy of the rigid body due to the rotation, supposing $\frac{d\xi}{dt} = \frac{d\eta}{dt} = \frac{d\zeta}{dt} = 0$,

$$
\begin{aligned}
133) \quad T &= \frac{1}{2} \Sigma m (v_x^2 + v_y^2 + v_z^2) \\
&= \frac{1}{2} p^2 \Sigma m (y^2 + z^2) + \frac{1}{2} q^2 \Sigma m (z^2 + x^2) + \frac{1}{2} r^2 \Sigma m (x^2 + y^2) \\
&- qr \Sigma myz - rp \Sigma mzx - pq \Sigma mxy \\
&= \frac{1}{2} Ap^2 + \frac{1}{2} Bq^2 + \frac{1}{2} Cr^2 - Dqr - Erp - Fpq,
\end{aligned}
$$

and for the angular momentum, introducing 119) in 48),

$$
\begin{aligned}
& H_x = \Sigma m [y(py - qx) - z(rx - pz)] = \quad Ap - Fq - Er, \\
134) \quad & H_y = \Sigma m [z(qz - ry) - x(py - qx)] = - Fp + Bq - Dr, \\
& H_z = \Sigma m [x(rx - pz) - y(qz - ry)] = - Ep - Dq + Cr,
\end{aligned}
$$

the last column being what we obtained in § 68, 53).

It is evident that

$$
135) \qquad H_x = \frac{\partial T}{\partial p}, \quad H_y = \frac{\partial T}{\partial q}, \quad H_z = \frac{\partial T}{\partial r},
$$

so that in this respect p, q, r, H_x, H_y, H_z have the relation of Lagrangian generalized velocities and momenta.

Since we have

$$
136) \quad T = \frac{1}{2} \left(p \frac{\partial T}{\partial p} + q \frac{\partial T}{\partial q} + r \frac{\partial T}{\partial r} \right) = \frac{1}{2} (pH_x + qH_y + rH_z),
$$

the kinetic energy is one-half the geometric product of the angular velocity and angular momentum.

CHAPTER VII.

DYNAMICS OF ROTATING BODIES.

80. Dynamics of Body moving about a Fixed Axis. The simplest case of motion of a rigid body next to that of translation is a movement of rotation with one degree of freedom, namely a motion about a fixed axis. The centrifugal force exerted by the body on the axis is $Md\omega^2$ where d is the distance from the axis of the center of mass of the body, and since this is in the direction of d, which is continually changing, if a body is to run rapidly in bearings the center of mass should be in the axis, otherwise the bearings are subjected to periodically varying forces. At the same time, even if this condition is fulfilled, there will be a centrifugal couple, also tending to tear the body from its bearings, unless the axis is a principal axis of inertia. It is worth noticing that the first condition may be obtained in practice by statical means, by making the axis horizontal, and attaching weights until the body is in equilibrium in any position, but that the second condition is only obtained by experiments on the body in motion. For this reason the former condition is generally fulfilled in such pieces of machinery as the armatures of dynamos, while the latter is not especially provided for.

Let us consider the motion of a heavy body about a horizontal axis. The resultant of all the parallel force acting on its various particles is by § 59a equal to a single force equal to the weight of the body Mg applied at the center of mass. The position of the body is determined by a single coordinate which we will take as the angle ϑ made with the vertical by the perpendicular from the center of mass on the axis. If the length of the perpendicular is h the work done in turning the body from the position of equilibrium is

$$1)\qquad W = Mgh(1 - \cos\vartheta).$$

The kinetic energy is

$$2)\qquad T = \frac{1}{2} K\omega^2 = \frac{1}{2} K\left(\frac{d\vartheta}{dt}\right)^2.$$

The equation of energy accordingly is

$$3)\qquad \frac{1}{2} K\left(\frac{d\vartheta}{dt}\right)^2 + Mgh(1 - \cos\vartheta) = \text{const.}$$

But this is the equation of motion of a simple pendulum of length

4) $$l = \frac{K}{Mh} = \frac{k^2}{h}.$$

The body, which is often called a compound pendulum, accordingly moves like a simple pendulum of length l. This is called the equivalent simple pendulum. It is to be noticed that in virtue of the constraint of rigidity, points at distances from the axis less than l move more slowly than they would if moving alone in the same paths, while those at greater distances move faster, and those at distance l move just as they would if free to move in the same circular paths.

Let k_g be the radius of gyration of the body about a parallel axis through the center of gravity. Then by § 70,

4) $$k^2 = k_g^2 + h^2, \quad l = \frac{k_g^2}{h} + h,$$

so that l is always greater than h. If we take a parallel axis O' at a distance $h' = l - h$ beyond the center of mass G, so that it, G, and the original axis are in the same plane, we have

5) $$hh' = k_g^2.$$

If now the axis O' be made the axis of suspension, the equivalent simple pendulum has a length

6) $$l' = \frac{k'^2}{h'} = \frac{k_g^2 + h'^2}{h'} = h + h' = l.$$

The axis O' is called the axis of oscillation, and we have the theorem that the axes of suspension and oscillation are interchangeable and separated by the distance equal to the length of the equivalent simple pendulum. This is the principle of Kater's reversible pendulum, used to determine the acceleration of gravity. The pendulum is furnished with two knife edges, so that it may be swung with either end down. Movable masses attached to the pendulum are so adjusted that the time of vibration is the same in both positions, and then the distance between the knife-edges gives the length l from which $g = \frac{4\pi^2 l}{T^2}$. The present example also includes the metronome and the beam of the ordinary balance. The masses of the pans may be regarded as concentrated at the knife-edges.

If the fixed axis is not horizontal, the modification in the result is very simple. Suppose the axis makes an angle α with the vertical. Let us take two sets of fixed axes, Z' vertical, Z the axis of rotation, Y' horizontal in the plane of Z and Z', Y in the same plane, and X and X' coincident. Then we have for the transformation of coordinates

$$z' = -y \sin \alpha + z \cos \alpha,$$

and determining the position of the system by the angle ϑ made by the perpendicular from the center of mass on the axis of rotation with the Y-axis,

$$y = r \cos \vartheta,$$

$$z' = -r \sin \alpha \cos \vartheta + z \cos \alpha.$$

The potential energy is as before

$$W = Mg\bar{z}' = -Mgh \sin \alpha \cos \vartheta + const.$$

thus the equation of energy is

$$7) \qquad \frac{1}{2} K \left(\frac{d\vartheta}{dt}\right)^2 - Mgh \sin \alpha \cos \vartheta = const.$$

Thus the equation is the same as before, except that the length of the equivalent simple pendulum is increased in the ratio of $1 : \sin \alpha$. This example includes the case of a swinging gate and of the important physical instrument, the horizontal pendulum of Zöllner. The mode of action of the latter depends on the fact that the moment of the force required to produce a given deflection ϑ,

$$\Theta = -\frac{\partial W}{\partial \vartheta} = Mgh \sin \alpha \sin \vartheta,$$

may be made as small as we please by decreasing α, which is observed in practice by making the time of vibration long.

81. Motion of a Rigid Body about a Fixed Point. Kinematics. We shall now consider one of the most important and interesting cases of the motion of a rigid body, namely that of a body one of whose points is fixed, and which thus possesses three degrees of freedom. This case was dealt with very fully by Poinsot, in his celebrated memoir "*Théorie nouvelle de la rotation des corps*", in the Journal de Liouville, tom. XVI, 1851. On account of the instructive nature of his processes, which are entirely geometrical, we shall present his method first. The treatment of the properties of the moment of inertia, which is contained in the same paper, has already been given in §§ 70—72.

If one point of the body remains fixed, the instantaneous axis must at all times pass through that point. The motion is completely described if we know at all times the position of the instantaneous axis in the body and in space, and the angular velocity about it.

Let O, Fig. 71, be the fixed point, and let OI_1 be the instantaneous axis at a given instant. During the time Δt suppose a line OI_2 moves to the position OI_2', and during the next interval Δt let the body turn about this line as instantaneous axis. During this interval let another line OI_3 move to OI_3' which then becomes the instantaneous axis, and so on. We have thus obtained two pyramids, one $OI_1\,OI_2\,OI_3\ldots$ fixed in space, the other $OI_1\,OI_2'\,OI_3'\ldots$ fixed in the body, and we may evidently describe the motion by saying that one pyramid *rolls* upon the other. As we pass to the limit, making Δt infinitely small, the pyramids evidently become cones, and the generator of tangency is the instantaneous axis at any instant.

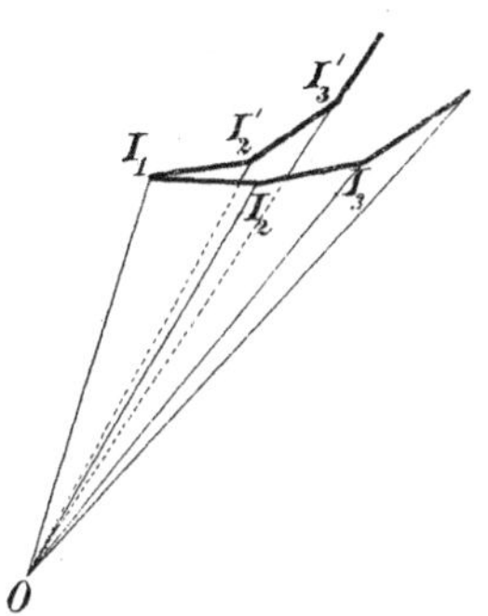

Fig. 71.

The rolling cone may be external or internal to the fixed one. In the former case, Fig 72a, the instantaneous axis moves around the fixed cone in the same direction in which the body rotates, and the motion is said to be progressive, in the second case, Fig. 72c it goes in the opposite direction, and the motion is said to be regressive or retrograde. It is to be noticed that it makes no difference whether the rolling cone is convex (Fig. 72a) or concave (Fig. 72b) toward the fixed cone. (In the figures, in which merely for convenience the cones are shown circular, C_1 denotes the fixed, C_2 the rolling cone.)

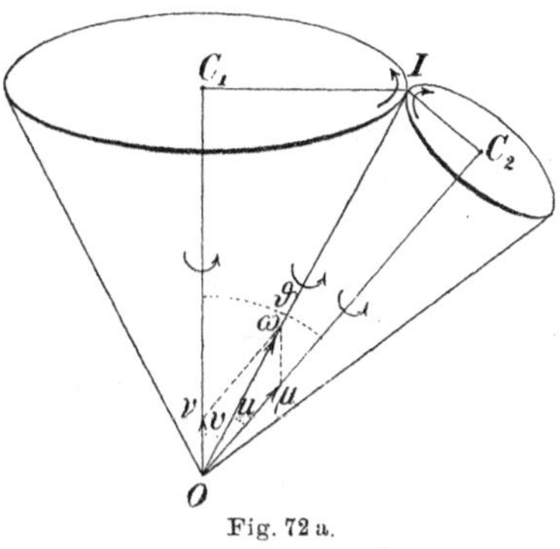

Fig. 72 a.

If one of the cones closes up to a line, upon which the other rolls, it always remains in contact with the same generator, that is, the instantaneous axis does not move. Accordingly if either cone degenerates to a line, the other does also, and the instantaneous axis remains fixed in space and in the body. This case has been already treated.

If we lay a plane perpendicular to the instantaneous axis at a distance R from O, Fig. 73, and if the radii of curvature of its intersections with the fixed and rolling cones be ϱ_1 and ϱ_2 (taken with the same sign if they lie on the same side of the common tangent), and the angles made by consecutive tangents at the ends of corresponding arcs ds_1 and ds_2 are $d\tau_1$, $d\tau_2$, we have $ds_1 = \varrho_1\,d\tau_1$, $ds_2 = \varrho_2\,d\tau_2$.

The angle turned by the body in rolling the arc ds_2 on its equal ds_1 is $d\tau_2 - d\tau_1$, and the angular velocity

$$8)\qquad \omega = \frac{1}{dt}(d\tau_2 - d\tau_1) = \frac{ds}{dt}\left(\frac{1}{\varrho_2} - \frac{1}{\varrho_1}\right).$$

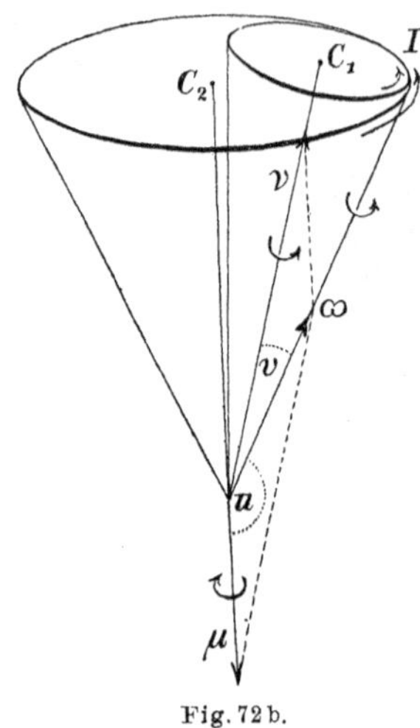

Fig. 72 b.

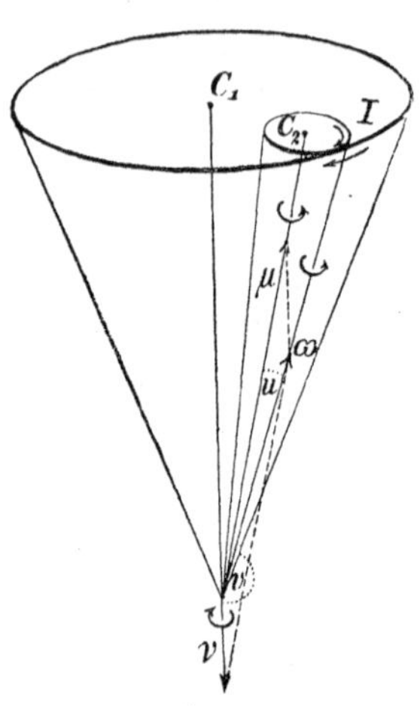

Fig. 72 c.

Now if w denote the angular velocity with which the instantaneous axis is turning about an axis through O perpendicular to the common tangent plane to the cones, we have

$$Rw = \frac{ds}{dt},$$

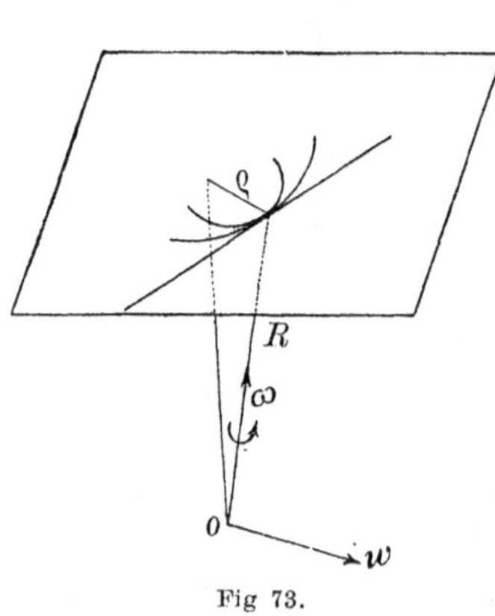

Fig 73.

which inserted in 8) gives

$$9)\qquad \frac{\omega}{w} = R\left(\frac{1}{\varrho_1} - \frac{1}{\varrho_2}\right).$$

If the cones have external contact, ϱ_2 is negative, and if we consider the absolute values, we must take the sign plus.

$$8')\qquad \omega = \frac{ds}{dt}\left(\frac{1}{|\varrho_2|} + \frac{1}{|\varrho_1|}\right),$$

$$9')\qquad \frac{\omega}{w} = R\left(\frac{1}{|\varrho_1|} + \frac{1}{|\varrho_2|}\right).$$

Consequently if we give the values at every instant of three of the quantities, ω, the angular velocity about the instantaneous axis, w, the angular velocity of change of the instantaneous axis, and ϱ_1

and ϱ_2, the radii of curvature of the sections of the fixed and rolling cones, the fourth, and consequently the whole motion, are determined. This corresponds to the fact that the body has three degrees of freedom.

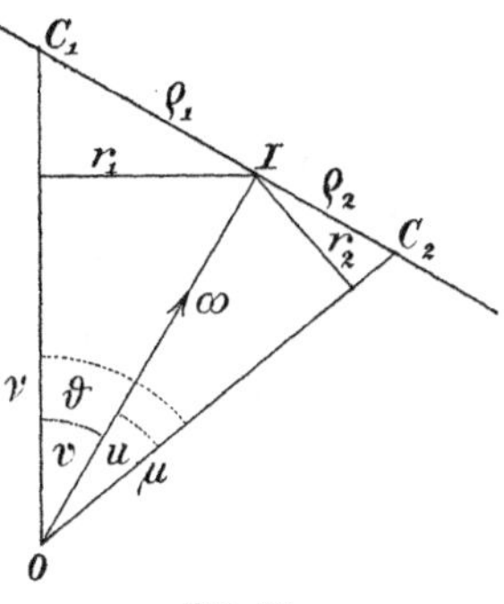

Fig. 74.

If OI, Fig. 72, is the instantaneous axis, OC_1 and OC_2 the lines of centers of curvature, the point I may be considered to be travelling around the cones with the velocity $\frac{ds}{dt}$, and about the lines OC_1 and OC_2 as axes with the angular velocities ν, μ, of which ω is the resultant. This was proved by Poinsot as follows. Let Fig. 74 represent a section of Fig. 72a in the plane of the axes OC_1, OI, OC_2. If r_1, r_2 are the perpendiculars from I on OC_1, OC_2, and v and u are the angles C_1OI, C_2OI, we have, considering absolute values only,

$$10)\quad \begin{aligned} &\frac{ds_1}{dt} = \nu r_1 = \nu \varrho_1 \cos v = \frac{ds_2}{dt} = \mu r_2 = \mu \varrho_2 \cos u, \\ &\frac{1}{\varrho_1}\frac{ds_1}{dt} = \nu \cos v, \quad \frac{1}{\varrho_2}\frac{ds_2}{dt} = \mu \cos u. \end{aligned}$$

Inserting in 8′)

$$11)\qquad \omega = \mu \cos u + \nu \cos v,$$

also since

$$\frac{r_1}{\sin v} = \frac{r_2}{\sin u}, \quad \text{and} \quad \nu r_1 = \mu r_2,$$

we have

$$12)\qquad \mu \sin u = \nu \sin v.$$

That is, since the three axes are in the same plane, and ω is the sum of the components of μ and ν in its direction, while their components in the perpendicular direction are equal and opposite, ω is the resultant of μ and ν. Figs. 72a, 72b, 72c show the three cases, where the cones have external contact, where the fixed cone is internal, and the rolling cone is internal, respectively. The parallelogram construction is shown on the figures, and the direction of rotation is shown by the arrows representing the vector rotations. It will be noticed that in each case the arrow on the figure showing the direction in which I is travelling around the *rolling* cone is *opposite* to the direction of rotation μ about C_2.

The rotation about OC_1 is known as *precession.* If both fixed and rolling cones are cones of revolution, and μ, ν, ω constant, the precession is called *regular*. If we call ϑ the angle C_1OC_2 between the axes of the cones, we have

$$13)\qquad \frac{\mu}{\sin v} = \frac{\nu}{\sin u} = \frac{\omega}{\sin\vartheta},$$

$$14)\qquad \omega^2 = \mu^2 + \nu^2 + 2\mu\nu\cos\vartheta.$$

An important case of a regular precession is furnished us in the motion of the earth, which, disregarding nutation (§ 93), describes a cone with $\vartheta = 23^0\,27'\,32''$ in the time 25,868 years, the motion being retrograde, Fig. 72c. We thus have

$$\sin u = \frac{\sin 23^0\,27'\,32''}{25{,}868 \times 365.256} = \sin 0'',0087,$$

so that the pole of the earth describes a circular cone whose half angle is $0'',0087$, an angle too small to be perceived by astronomical means, the radius of the circle cut by this cone on the surface of the earth being only 27 centimeters.

82. Dynamics. Motion under no Forces. We have already found, §§ 68, 69, following Poinsot, the expressions for the momentum and the centrifugal forces for the general motion of a rigid body. If the fixed point be the center of mass, both the linear momentum and the centrifugal resultant vanish, and we have to deal only with the angular momentum and the centrifugal couple. At the same time the resultant of the effect of gravity passes through the fixed point, and is neutralized by the reaction of the support. Let us then consider the motion of a body turning about its center of mass, or more generally, the motion of a body under the action of no forces. Such a motion will be called a Poinsot-motion.

Let OZ be the instantaneous axis. Then we have from § 68, 53)

$$15)\qquad \begin{aligned} H_x &= -E\omega, \\ H_y &= -D\omega, \\ H_z &= \quad C\omega. \end{aligned}$$

Let us call the resultant of H_x and H_y, H_2, Fig. 75. We have for the centrifugal couple S_c, from § 69, 59),

$$16)\qquad \begin{aligned} L_c &= -D\omega^2, \\ M_c &= \quad E\omega^2, \\ N_c &= \quad 0. \end{aligned}$$

Since N_c is zero, the axis of the centrifugal couple is perpendicular to the instantaneous axis. But since

$$17) \qquad H_x L_c + H_y M_c + H_z N_c = 0,$$

it is also perpendicular to the angular momentum. Consequently the axis of the centrifugal couple is perpendicular to the plane containing the instantaneous axis and the axis of angular momentum, and is drawn in such a direction that if S_c be turned through a right angle in the direction of the body's motion it will coincide in direction with H_2. Also since the components of S_c are equal in absolute value to the components of H_2 multiplied by ω, we have

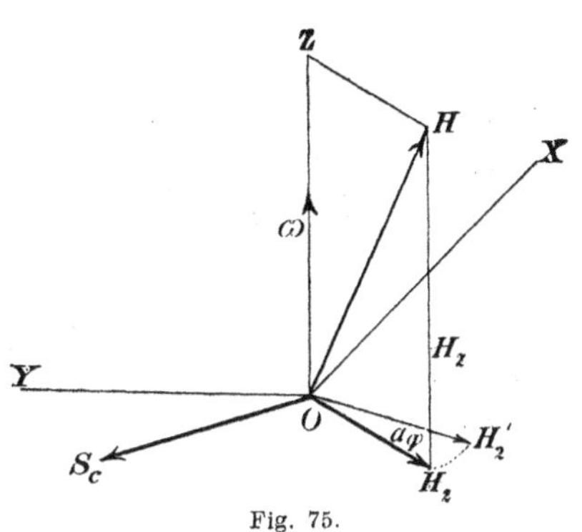

Fig. 75.

$$18) \qquad S_c = H_2 \omega = \omega H \sin (H \omega).$$

Thus the centrifugal couple is equal to the vector product of the angular momentum by the angular velocity. In case the instantaneous axis is a principal axis at O, the direction of H coincides with that of ω and the centrifugal couple vanishes. The body will then remain permanently turning about the same axis. This property of a heavy body turning about its center of gravity about a principal axis of maintaining the direction of that axis fixed in space was utilized by Foucault in his gyroscope, the axis of which points in a fixed direction while the earth turns, and thus the motion of the earth is made observable. The same principle is utilized practically in the Obry steering gear contained in the Whitehead automobile torpedo, in which a rapidly rotating gyroscope is made to give the direction to the torpedo, and by acting on the steering gear to make it return to its course if it accidentally leaves it.

Suppose on the other hand that the instantaneous axis is not a principal axis. The centrifugal couple then tends to generate an angular momentum whose axis is in its own direction, and this new momentum compounds with that which the body already possesses. Let us consider two successive positions of the body. Suppose that in the time dt the body turns about the instantaneous axis through an angle $d\varphi = \omega\, dt$. At the end of that time the vector H_2 (Fig. 75), would have turned through the angle $d\varphi$ into the position H_2', the length of the infinitesimal vector $H_2 H_2'$ being $H_2\, d\varphi = H_2 \omega\, dt$. But during this time the centrifugal couple S_c has given rise to the angular momentum

$$S_c\, dt = H_2 \omega\, dt.$$

This vector, being parallel to S_c and thus perpendicular to H_2' gives, when compounded with H_2' a resultant exactly equal to H_2. The component of H parallel to ω being unchanged by the motion, we find, geometrically, that the angular momentum remains constant throughout the motion, as we have found by a general theorem in § 33.

As we now wish to follow the motion of the body from one instant to another, it will be convenient to free ourselves from the choice of axes which made the instantaneous axis the Z-axis. Let us take for axes the principal axes of the body at O. Let the components of the angular velocity ω on the axes be p, q, r. Then the angular momentum, being the resultant of the three angular momenta due to the three angular velocities p, q, r, is by § 68, 53) or § 79, 134),

$$19)\qquad H_x = Ap, \quad H_y = Bq, \quad H_z = Cr.$$

If we draw any radius vector to the ellipsoid of inertia at the fixed center of mass

$$F \equiv Ax^2 + By^2 + Cz^2 = 1,$$

the perpendicular δ on the tangent plane at the point x, y, z has direction cosines proportional to Ax, By, Cz.

If we draw the radius vector ϱ in the direction of the instantaneous axis, so that

$$20)\qquad \frac{p}{x} = \frac{q}{y} = \frac{r}{z} = \frac{\omega}{\varrho} = n,$$

equations 19) give

$$21)\qquad H_x = nAx, \quad H_y = nBy, \quad H_z = nCz,$$

or the angular momentum vector bears to the angular velocity vector the relation, as to direction, of the perpendicular on the tangent plane to the radius vector. Otherwise, if the angular momentum is given, the instantaneous axis is the diameter conjugate to the diametral plane of the ellipsoid perpendicular to the angular momentum.

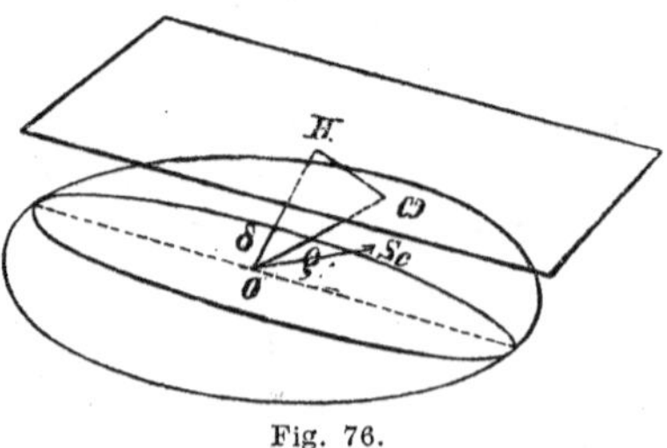

Fig. 76.

The centrifugal couple being perpendicular to the plane of δ and ϱ, lies *in* the diametral plane conjugate to ϱ. It produces in the time dt an angular momentum $S_c\,dt$ whose axis is in the same direction. To find the axis of the angular velocity corresponding thereto we must find the diameter conjugate to the plane perpendicular to S_c, that is the plane $\varrho\delta$. But the diameter conjugate to a plane is conjugate to all diameters in it, hence the required diameter is conjugate to ϱ and lies in the plane conjugate to ϱ, that is, parallel to the tangent plane at x, y, z. Consequently, if we compound with the velocity ω about ϱ the velocity corresponding to $S_c\,dt$ parallel to the tangent plane, the resultant has the same component perpendicular to the tangent plane as ω. In other words the component $\omega\cos(\omega, H)$ is constant throughout the motion.

Now we have found that H is constant in magnitude and direction, hence, multiplying by the constant $\omega\cos(\omega H)$,

22) $$H\omega\cos(\omega H) = \text{const.}$$

But $H\cos(H\omega)$ is that component of the angular momentum which is parallel to the instantaneous axis, and is accordingly equal, by § 68, 53) to the product of the angular velocity by the moment of inertia about the instantaneous axis.

23) $$H\cos(H\omega) = K\omega.$$

Accordingly 22) becomes

24) $$K\omega^2 = \text{const.}$$

But this is equal to twice the kinetic energy. Accordingly we obtain geometrically the integral of energy. Thus for a rigid body this principle follows from that of the conservation of angular momentum.

In the ellipsoid of inertia we have, § 71,

$$K = \frac{1}{\varrho^2}.$$

Accordingly

25) $$K\omega^2 = \frac{\omega^2}{\varrho^2} = n^2,$$

and the equation of energy shows that n is constant during the motion, or during the whole motion the angular velocity is proportional to the radius vector to the ellipsoid of inertia in the direction of the instantaneous axis. But since $\omega\cos(H\omega)$ is constant, $\varrho\cos(\varrho\delta) = \delta$ must be constant, and therefore the tangent plane is at a constant distance from the center during the motion. But since the direction of the line δ is constant in space, and its length is also constant, the tangent plane must be a fixed plane in space. As the point where it is touched by the ellipsoid of inertia is on the instantaneous axis the ellipsoid must be turning about this radius vector, and hence *rolling* without sliding on the fixed tangent plane. The motion of the body is thus completely described, and we see that the problem of a Poinsot-motion is equivalent to the geometrical one of the rolling of an ellipsoid whose center is fixed on a fixed tangent plane, together with the kinematical statement that the angular velocity of rolling is proportional to the radius vector to the point of tangency. Before taking up the discussion of this result, as given by Poinsot, we will consider the analytical method of establishing the result.

84. Euler's Dynamical Equations. If H_x', H_y', H_z' represent the angular momentum about the fixed X', Y', Z'-axes, L', M', N', the moment of the applied couple, the equations of § 67, 49) are

$$26)\qquad \frac{dH_x'}{dt} = L', \quad \frac{dH_y'}{dt} = M', \quad \frac{dH_z'}{dt} = N',$$

where (cf. § 76)

$$27)\qquad \begin{aligned} H_x' &= \alpha_1 H_x + \beta_1 H_y + \gamma_1 H_z, \\ H_y' &= \alpha_2 H_x + \beta_2 H_y + \gamma_2 H_z, \\ H_z' &= \alpha_3 H_x + \beta_3 H_y + \gamma_3 H_z. \end{aligned}$$

Differentiating we have, after making use of § 77, 130),

$$28)\qquad \begin{aligned} \frac{dH_x'}{dt} &= \alpha_1 \frac{dH_x}{dt} + \beta_1 \frac{dH_y}{dt} + \gamma_1 \frac{dH_z}{dt} \\ &\quad + H_x(\beta_1 r - \gamma_1 q) + H_y(\gamma_1 p - \alpha_1 r) + H_z(\alpha_1 q - \beta_1 p). \end{aligned}$$

If we now choose for fixed axes the instantaneous positions of the moving axes, we have $\alpha_1 = \beta_2 = \gamma_3 = 1$, all other cosines zero, and the equations 28) become simply

$$29)\qquad \begin{aligned} \frac{dH_x}{dt} + qH_z - rH_y &= L, \\ \frac{dH_y}{dt} + rH_x - pH_z &= M, \\ \frac{dH_z}{dt} + pH_y - qH_x &= N. \end{aligned}$$

We may obtain the same results by the use of the equations § 77, 128). Let us take for the point x, y, z the end of the vector H. Its coordinates with respect to the moving axes being H_x, H_y, H_z, substituting them in equations § 77, 128) we obtain for their velocities resolved along the X, Y, Z-axes the expression on the left of 29).

We must now put for H_x, H_y, H_z the expressions § 79, 134). If now the moving axes are taken at random, the moments and products of inertia of the body with respect to them will vary with the time, so that their time-derivatives enter into the dynamical equations, which are thus too complicated to be of any use. It is therefore immediately suggested that we choose for the moving axes a set of axes fixed in the body, and moving with it. The quantities A, B, C, D, E, F are then constants. If in addition we take as axes the principal axes at the origin of the moving axes, D, E, F vanish, and then since $H_x = Ap$, $H_y = Bq$, $H_z = Cr$, the equations become simply

$$
\begin{aligned}
&A\frac{dp}{dt} + (C-B)qr = L,\\
30)\qquad &B\frac{dq}{dt} + (A-C)rp = M,\\
&C\frac{dr}{dt} + (B-A)pq = N.
\end{aligned}
$$

These are Euler's dynamical equations for the rotation of a rigid body.

In case the moments of the applied forces about the origin vanish, they become

$$
\begin{aligned}
&A\frac{dp}{dt} = (B-C)\,qr,\\
31)\qquad &B\frac{dq}{dt} = (C-A)rp,\\
&C\frac{dr}{dt} = (A-B)pq,
\end{aligned}
$$

and we see that the quantities on the right, being the vector product of the angular velocity by the angular momentum, represent the centrifugal couple, which alone acts to produce the angular acceleration, whose components appear on the left. We thus obtain the result obtained geometrically by Poinsot, the quantities on the left denoting the velocity of the end of H *in the body*.

The equations 29) may be simplified in another manner, if the ellipsoid of inertia is of revolution. If for one of the moving axes we take the axis of revolution, and for the others, any axes perpendicular to it, whether fixed in the body or not, the axes will be principal axes, and the moments of inertia constant, since the moment of inertia about all axes perpendicular to the axis of rotation in the same. Examples of this will be given in §§ 96, 106.

85. Poinsot's Discussion of the Motion. We may now integrate the equations 31) by making use of the fact that the centrifugal couple is perpendicular to the angular velocity and the angular momentum. Multiplying equations 31) respectively by p, q, r and adding

$$32)\qquad Ap\frac{dp}{dt} + Bq\frac{dq}{dt} + Cr\frac{dr}{dt} = 0,$$

which is at once integrated as

$$33)\qquad \frac{1}{2}Ap^2 + \frac{1}{2}Bq^2 + \frac{1}{2}Cr^2 = const.$$

This is the equation of energy.

Multiplying now by Ap, Bq, Cr, and adding,

34) $$A^2 p \frac{dp}{dt} + B^2 q \frac{dq}{dt} + C^2 r \frac{dr}{dt} = 0,$$

which is integrated as

35) $$A^2 p^2 + B^2 q^2 + C^2 r^2 = const.$$

But this, since the left-hand member is equal to H^2, is the equation of conservation of angular momentum. The equation alone does not show the fixity of the *direction* of H in space.

The point P in which the instantaneous axis intersects the ellipsoid of inertia at the fixed point O is called the *pole* of the instantaneous axis. Its coordinates are

$$x = \frac{\varrho p}{\omega}, \quad y = \frac{\varrho q}{\omega}, \quad z = \frac{\varrho r}{\omega}.$$

Now the length of the perpendicular δ is, since it is the projection of ϱ on the direction of the normal,

36) $$\begin{aligned}\delta &= x \cos(nx) + y \cos(ny) + z \cos(nz) \\ &= \frac{Ax^2 + By^2 + Cz^2}{\sqrt{A^2x^2 + B^2y^2 + C^2z^2}} = \frac{\varrho(Ap^2 + Bq^2 + Cr^2)}{\omega\sqrt{A^2p^2 + B^2q^2 + C^2r^2}} \\ &= \frac{\varrho}{\omega}\frac{2T}{H} = \frac{\sqrt{2T}}{H},\end{aligned}$$

since

$$2T = K\omega^2 = \frac{\omega^2}{\varrho^2}.$$

Accordingly since T and H are constant, δ is constant, and the tangent plane being perpendicular to the invariable line H is fixed in space. Poinsot called the locus of the pole of the instantaneous axis on the ellipsoid, the *polhode* (πόλος axis, ὁδός path), and its locus on the tangent plane the *herpolhode*.

The ellipsoid of inertia being

37) $$Ax^2 + By^2 + Cz^2 = 1,$$

the distance of the tangent plane at x, y, z from the center is

38) $$\delta = \frac{1}{\sqrt{A^2x^2 + B^2y^2 + C_1^2 z^2}}.$$

Since this is to be constant, this equation and that of the ellipsoid define the polhode curve. Combining the equation

39) $$A^2x^2 + B^2y^2 + C^2z^2 = \frac{1}{\delta^2}$$

with that of the ellipsoid, divided by δ^2, we obtain by subtraction

40) $$A\left(\frac{1}{\delta^2} - A\right)x^2 + B\left(\frac{1}{\delta^2} - B\right)y^2 + C\left(\frac{1}{\delta^2} - C\right)z^2 = 0.$$

This is the equation of the *cone* passing through the polhode, with its vertex at the fixed point, that is the rolling or polhode cone.

We find then that the rolling cone for a body moving under no forces is of the second order. If it is to be real, we must have

$$A \geqq \frac{1}{\delta^2} \geqq C, \tag{41}$$

that is the perpendicular must have a length intermediate between the greatest and least axes of the ellipsoid. If $\frac{1}{\delta^2} = A$ the cone is

$$B(A-B)y^2 + C(A-C)z^2 = 0 \tag{42}$$

representing a pair of imaginary planes, intersecting in the real line $y = z = 0$, the X-axis. Thus in this case the rolling cone reduces to a line, fixed both in the body and in space. If $\frac{1}{\delta^2} = C$, we have a similar result. If $\frac{1}{\delta^2} = B$, we have

$$A(A-B)x^2 - C(B-C)z^2 = 0, \tag{43}$$

representing two real planes intersecting in the Y-axis, and making an angle with the XY-plane whose tangent is

$$\frac{z}{x} = \pm\sqrt{\frac{A(A-B)}{C(B-C)}}. \tag{44}$$

These are the planes which separate the polhodes surrounding the end of the major axis from those about the minor axis. The polhodes are twisted curves of the fourth order, whose appearance is shown in perspective in Fig. 77. The separating polhodes are drawn black.

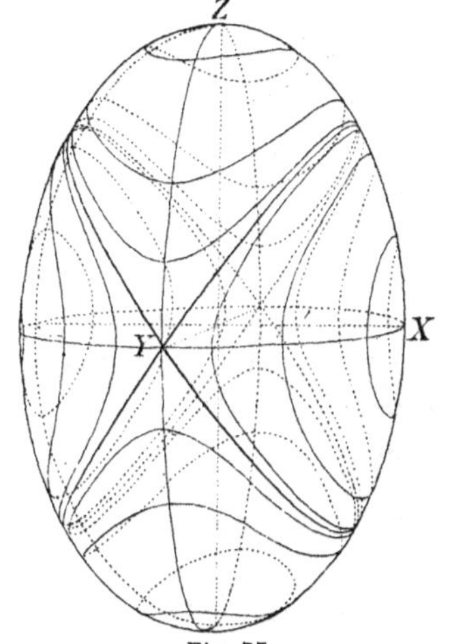

Fig. 77.

Since the polhode is a closed curve, the radius vector of a point on it oscillates between a maximum and a minimum value. If σ is the distance of a point on the herpolhode from the foot of the perpendicular δ, since $\sigma^2 = \varrho^2 - \delta^2$, σ oscillates between two constant values, and the herpolhode is tangent to two circles. Since the polhode is described periodically the various arcs of the herpolhode corresponding to repetitions of the polhode are all alike. The herpolhode is not in general a reentrant curve. The name herpolhode was given by Poinsot from the verb ἕρπειν, to creep (like a snake) from the supposedly undulating nature of the curve, it has however been proved to have no points of inflexion, and is like Fig. 78, which has been calculated for $A = 8$, $B = 5$, $C = 3$, $\frac{1}{\delta^2} = 4 \cdot 9$.

86. Stability of Axes. We have seen that the body if rotating about either of the principal axes of inertia will remain rotating about it. If the instantaneous axis be the axis of either greatest or least inertia, and be displaced a little, as the polhodes encircle the ends of these axes the instantaneous axis will travel around on a small polhode, and the herpolhode will be small, neither ever leaving the original axis by a large amount. These axes are accordingly said to be axes of stable motion. If on the other hand the mean axis be the instantaneous axis, and there is a slight displacement, the axis immediately begins to go farther and farther from the original position, and nearly reaches a point diametrically opposite before returning to the original position. The mean axis is thus said to be an axis of instability. It is however to be noticed that if either $A - B$ or $B - C$ is small with respect to the other, the separating polhode closes up about either the axis of greatest or least inertia respectively, and thus a small displacement may lead to a considerable departure from the original pole, the rotation is thus less stable. The rotation about either axis is most stable when the wedge of the separating polhode enclosing it is most open.

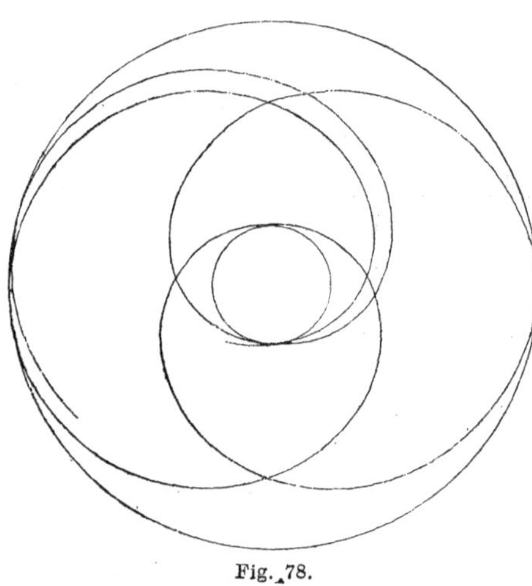

Fig. 78.

87. Projections of the Polhode. From the equations of the polhode 37), 39), we may obtain its projections on the coordinate planes by eliminating either of the coordinates. Eliminating x,

$$45) \qquad \delta^2\{B(A-B)y^2 + C(A-C)z^2\} = A\delta^2 - 1,$$

an ellipse of semi-axes,

$$\frac{1}{\delta}\sqrt{\frac{A\delta^2-1}{B(A-B)}}, \quad \frac{1}{\delta}\sqrt{\frac{A\delta^2-1}{C(A-C)}},$$

the ratio of which is

$$\sqrt{\frac{C(A-C)}{B(A-B)}},$$

a constant, so that all the projections are similar. The motion about the axis A is most stable when the small polhode is a circle, that is when the above ratio is unity, or $B = C$.

Eliminating z we obtain

46) $$\delta^2\{A(A-C)x^2 + B(B-C)y^2\} = 1 - C\delta^2,$$

an ellipse the ratio of whose axes is

$$\sqrt{\frac{B(B-C)}{A(A-C)}},$$

and for maximum stability this is unity, or $A=B$. These projections are shown in Fig. 79.

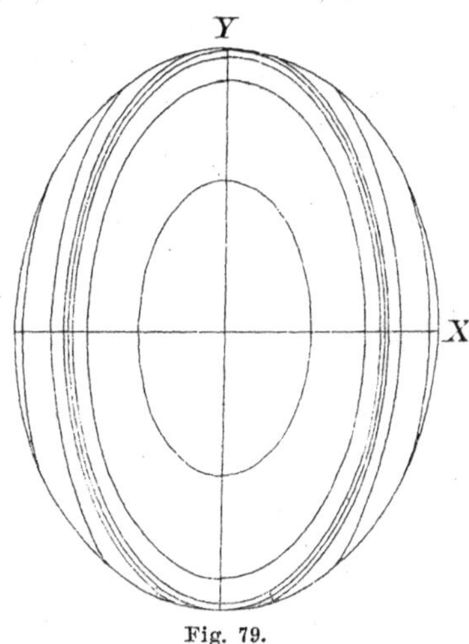

Fig. 79.

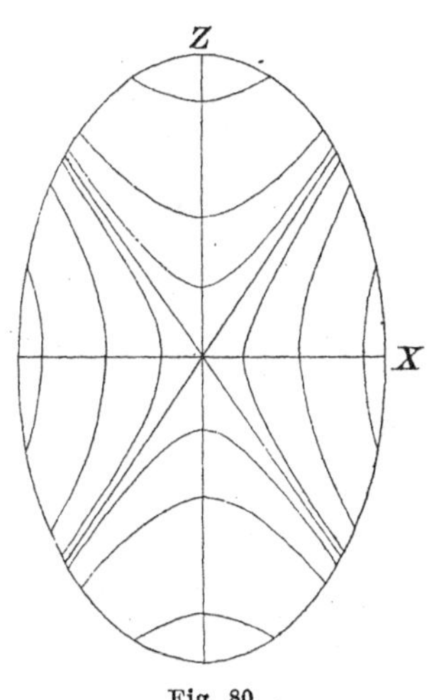

Fig. 80.

Eliminating y, we have

47) $$\delta^2\{A(A-B)x^2 - C(B-C)z^2\} = 1 - B\delta^2,$$

an hyperbola the ratio of whose axes is

$$\sqrt{\frac{C(B-C)}{A(A-B)}}.$$

All the hyperbolas have the separating polhode projections as asymptotes (Fig. 80).

88. Invariable Line. The invariable line describes a cone in the body. Its equation may be simply found from consideration of the reciprocal ellipsoid

48) $$\frac{x^2}{A} + \frac{y^2}{B} + \frac{z^2}{C} = 1,$$

whose radius in the direction of δ is $\frac{1}{\delta}$ and therefore constant. The cone of the invariable axis is accordingly the cone passing through the intersection of the ellipsoid 48) with the sphere

49) $$x^2 + y^2 + z^2 = \frac{1}{\delta^2},$$

that is

50) $$\left(\delta^2 - \frac{1}{A}\right) x^2 + \left(\delta^2 - \frac{1}{B}\right) y^2 + \left(\delta^2 - \frac{1}{C}\right) z^2 = 0.$$

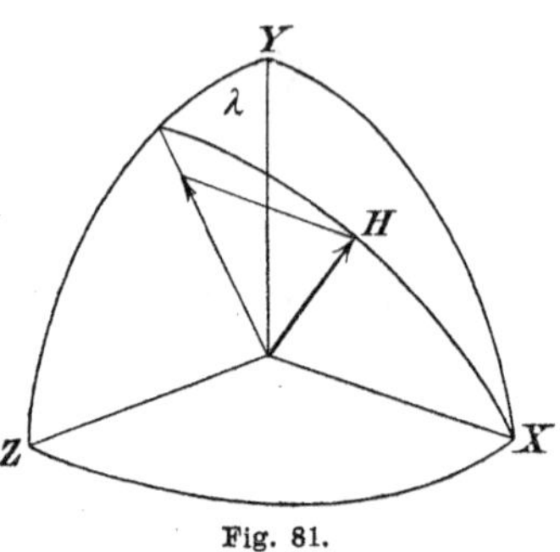

Fig. 81.

The axis of this, like that of the polhode cone, is the axis of greatest or least inertia.

Let us find how fast the invariable line revolves around one of the principal axes. Since the invariable axis is fixed in space, its relative motion is equal and opposite to the actual motion of the part of the body in which it lies. If we call λ the diedral angle between the plane of the invariable axis and the axis of X and the XY-plane, we may find $\frac{d\lambda}{dt}$. Projecting H upon the YZ-plane (Fig. 81), the projection makes with the Y-axis the angle λ, given by

51) $$\tan\lambda = \frac{H_z}{H_y} = \frac{Cr}{Bq},$$

from which

$$\sec^2\lambda = \frac{B^2 q^2 + C^2 r^2}{B^2 q^2}.$$

Differentiating,

$$\sec^2\lambda \frac{d\lambda}{dt} = \frac{C}{B}\left(q\frac{dr}{dt} - r\frac{dq}{dt}\right)\frac{1}{q^2},$$

52) $$\frac{d\lambda}{dt} = \frac{BC}{B^2 q^2 + C^2 r^2}\left(q\frac{dr}{dt} - r\frac{dq}{dt}\right).$$

Inserting from Euler's equations 31),

$$\frac{dq}{dt} = \frac{C-A}{B} rp, \quad \frac{dr}{dt} = \frac{A-B}{C} pq,$$

53) $$\frac{d\lambda}{dt} = \frac{p\{B(A-B)q^2 + C(A-C)r^2\}}{B^2 q^2 + C^2 r^2}$$

$$= p\left\{\frac{A(Bq^2 + Cr^2)}{B^2 q^2 + C^2 r^2} - 1\right\}$$

$$= p\left\{\frac{A(2T - Ap^2)}{H^2 - H_x^2} - 1\right\} = p\left\{\frac{2AT - H^2}{H^2 - H_x^2}\right\}$$

$$= p\left\{\frac{\frac{2AT}{H^2} - 1}{1 - \left(\frac{H_x}{H}\right)^2}\right\} = p\,\frac{\delta^2 A - 1}{\sin^2(Hx)}.$$

Similarly for the rotation around the Y and Z-axes,

$$54)\qquad \begin{aligned} \frac{d\mu}{dt} &= q\frac{\delta^2 B - 1}{\sin^2(Hy)},\\ \frac{d\nu}{dt} &= r\frac{\delta^2 C - 1}{\sin^2(Hz)}. \end{aligned}$$

Looking at the signs of the numerators, we see that the invariable axis rotates around the axis of greatest moment of inertia in the direction of rotation, about the least axis in the direction opposite to that of rotation, and about the mean axis according to the value of δ.

If we mark off on the invariable axis a line of unit length, its end describes a sphero-conic, the intersection of the invariable cone

$$50)\qquad \left(\delta^2 - \frac{1}{A}\right)x^2 + \left(\delta^2 - \frac{1}{B}\right)y^2 + \left(\delta^2 - \frac{1}{C}\right)z^2 = 0$$

with the sphere

$$x^2 + y^2 + z^2 = 1,$$

whose projections on planes perpendicular to the X and Z-axes are ellipses, and perpendicular to the Y-axis an hyperbola. The radius vector of the X-projection is $r_x = \sin(Hx)$ and since it turns with the angular velocity $\frac{d\lambda}{dt}$ it describes area at the rate

$$55)\qquad 2\frac{dS_x}{dt} = r_x^2\frac{d\lambda}{dt} = p(\delta^2 A - 1).$$

The time of one revolution of the body turning with the velocity p would be, if p were constant, $t = \frac{2\pi}{p}$.

The equation of the ellipse is obtained by eliminating x from the equations of the sphero-conic as

$$56)\qquad \left(\frac{1}{A} - \frac{1}{B}\right)y^2 + \left(\frac{1}{A} - \frac{1}{C}\right)z^2 = \frac{1}{A} - \delta^2,$$

whose axes are

$$\sqrt{\frac{\frac{1}{A} - \delta^2}{\frac{1}{A} - \frac{1}{B}}} = \sqrt{\frac{\delta^2 AB - B}{A - B}},$$

$$\sqrt{\frac{\frac{1}{A} - \delta^2}{\frac{1}{A} - \frac{1}{C}}} = \sqrt{\frac{\delta^2 AC - C}{A - C}},$$

and whose area is

$$57)\qquad \pi(A\delta^2 - 1)\sqrt{\frac{BC}{(A-B)(A-C)}}.$$

Now the area described in one revolution about the instantaneous axis would be, if p were constant [see 55)],

$$S_x = \frac{p}{2}(\delta^2 A - 1)t = \pi(\delta^2 A - 1), \tag{58}$$

and the number of turns the body makes for one revolution of the invariable axis about the X-axis is the area 57) divided by this, or

$$\sqrt{\frac{BC}{(A-B)(A-C)}}. \tag{59}$$

This may be made as large as we please by making A approach B or C. If $B = C$ or the ellipsoid of inertia is of revolution, about the X-axis, p is constant, and the invariable cone is circular, and described with uniform velocity, the number of revolutions of the body for one circuit of the invariable axis being $\frac{B}{(A-B)}$. The motion is direct or inverse, according as the X-axis is that of greatest or least inertia.

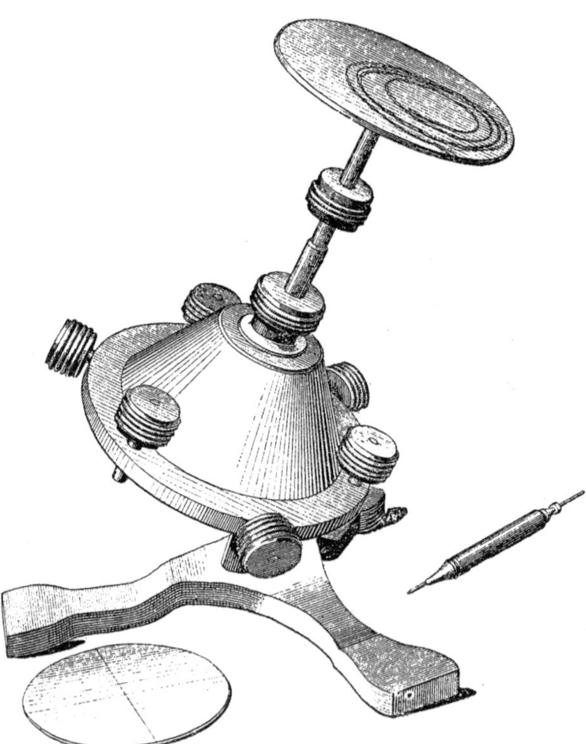

Fig. 82.

These properties may all be illustrated experimentally by means of Maxwell's Dynamical Top[1]), constructed by Maxwell for the purpose of studying the motion of the earth about its center of mass. An example of this top constructed in the workshop of the Department of Physics of Clark University is shown in Fig. 82. The six weights projecting from the bell allow the moments of inertia to be changed in a great variety of ways, while at the same time the center of

1) Maxwell, Papers, Vol. I, p. 248.

mass is constantly kept at the point of support, a sharp steel point turning in a sapphire cup. Maxwell's ingenious device for the observation of the motion of the invariable axis, is the disk, divided into four colored segments, attached to the axis of figure. The colors chosen, red, blue, yellow and green, combine into a neutral gray when the top is revolving rapidly about the axis of figure. If however the top revolves about a line passing through a point in the red sector, there will be in the center a circle of red, the diameter of which is greater as the axis is farther from the center of the disk and the boundaries of the red sector. Thus the center of the gray disk changes from one color to another as the pole moves about in the body, and by following the changes of color we can study the motion. By noticing the order of the succession of colors we can determine whether the axis of figure coincides most nearly with the axis of greatest or least inertia, and by changing the adjustments we may make it a principal axis, which is known by the disappearance of wabbling, or we may make it deviate by any desired amount from a principal axis. If the deviation is great, and the top spun about the axis of figure, and then left to itself, the top will wabble to a startling amount, but eventually the pole will reach its first position and the wabbling will cease, to be repeated periodically. The recovery of the top from its apparantly lawless gyrations is very striking. If the adjustment is such as to make the axis of figure lie near the mean axis of inertia, the top will not recover, but must be stopped in its motion before striking its support.

The path of the invariable axis has been made visible by Mr. G. F. C. Searle, of the Cavendish Laboratory, Cambridge, by attaching to the axis of figure a card, upon which ink was projected from an electrified jet. Acting upon this suggestion, the author attached to the top a disk of smoked paper, upon which a steel stylus, playing easily in a vertical support (shown in Fig. 82 lying on the table) could write with very slight friction. One easily finds by looking at the disk in its gyrations a point which remains fixed, and by applying the stylus to this point, holding it on a proper support, the path of the invariable axis is drawn, and found to be an ellipse or hyperbola. If the stylus is not held exactly on the invariable axis, small loops are formed, which enable us to count the number of turns of the top in going around the polhode, and thus to verify the theory. The results of several spins are shown in Fig. 83, reproduced from actual traces.

The loops are turned out if the principal axis at the center of the ellipse is that if greatest inertia, and in if it is the least, for the reason that in the former case the invariable axis and the herpolhode cone lie within the polhode (Fig. 83a), while in the latter they lie without (Fig. 83b) so that if we consider the relative motion, in the former case a point fixed to the herpolhode describes a sort of

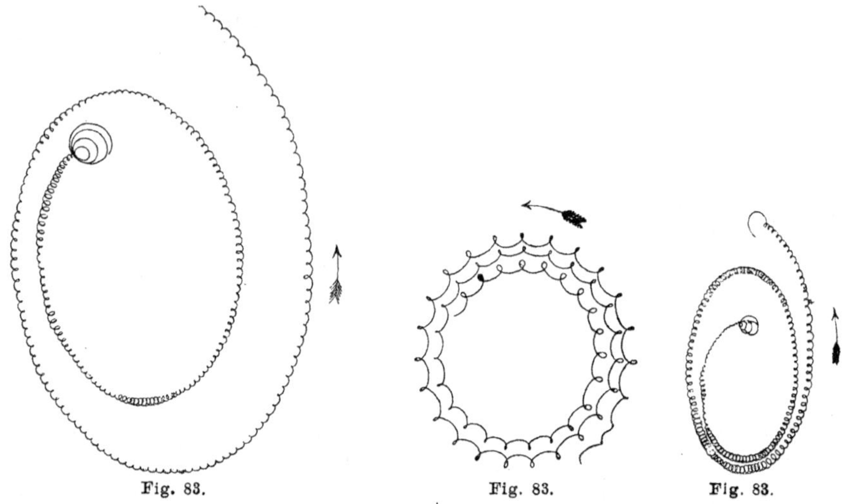

Fig. 83. Fig. 83. Fig. 83.

hypocycloid (loops out) on the card attached to the polhode, in the latter a sort of epicycloid (loops in).

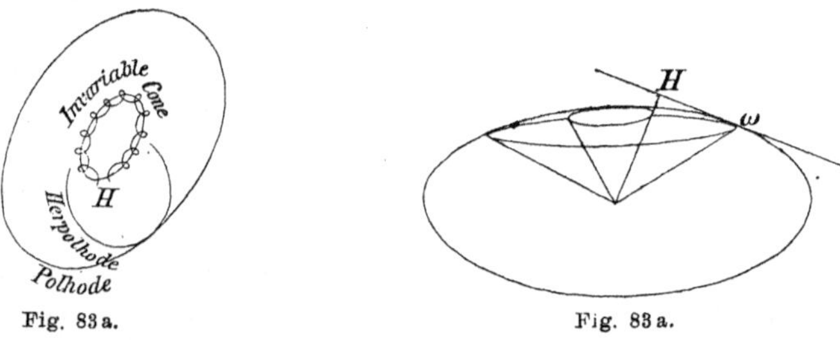

Fig. 83a. Fig. 83a.

The recent astronomical discovery of the motion of the earth's pole is probably due to a sort of *variable* Poinsot-motion, the moments of inertia of the earth being gradually varied.

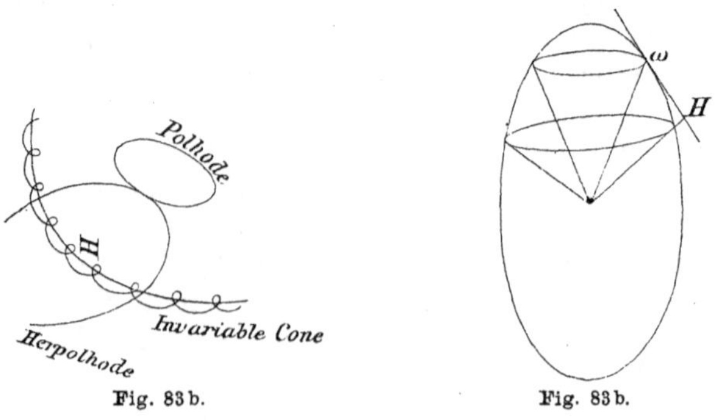

Fig. 83b. Fig. 83b.

89. Symmetrical Top. Constrained Motion. While we have in the preceding section considered the very interesting and instructive question of the motion of the most general rigid body under the action of no forces, by far the most frequent case under the practical conditions of experiment is that in which the body is dynamically symmetrical about an axis, that is, the ellipsoid of inertia is of revolution. Such a body we shall call a symmetrical top. This will include not only all ordinary tops and gyroscopes, as well as flywheels, rolling hoops, billiard balls, but even the earth and planets. Suppose such a body to be spinning under the action of no forces, about its axis of symmetry. We have seen that it will remain so spinning, and the angular momentum will have the direction of the axis of symmetry. If now the axis of symmetry OF (Fig. 84) is to move to some other position, OF', which is then to coincide with the new instantaneous axis, the angular momentum HH' must be communicated to the body, that is a couple whose axis is parallel to HH' must act on the body. This may be made evident experimentally by placing a loop of string over the axis F of a symmetrical top balanced on its center of mass (Fig. 85) and pulling on the string. The axis of the top, instead of following the direction of the pull P moves off a right angles thereto, although the string can only impart a force in its own direction. The pull of the string, together with the reaction of the point of support, constitute a couple, whose moment is perpendicular to the plane of the string and of the point of support, and it is in this direction that the end of the axis, or *apex* of the top, moves, as is required by the theory.

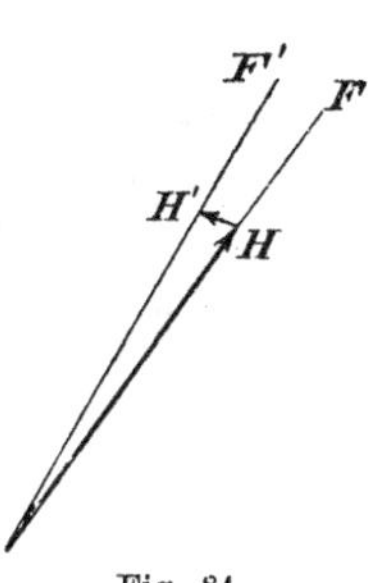

Fig. 84.

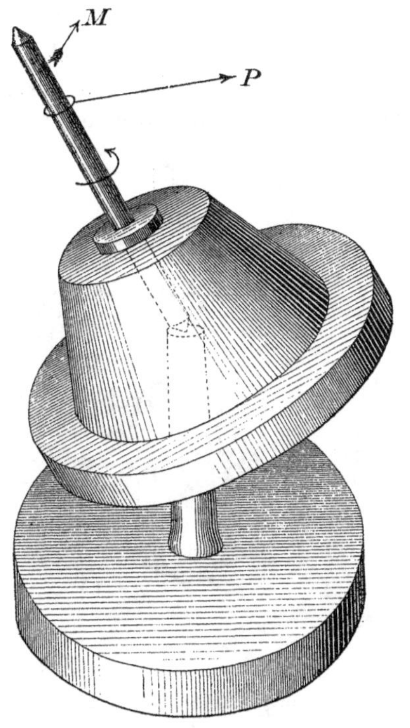

Fig. 85.

This simple experiment and the theory which it illustrates will make clear most of the apparantly paradoxical phenomena of rotation. We may describe it by saying that the kinetic reaction of a symmetrical rotating top is not in the direction of the motion of the apex, but nearly at right angles thereto. (Exactly at right angles to the motion of OH.)

An ingenious application of this principle is found in the Howell automobile torpedo, invented by Admiral Howell of the United States navy. In this the energy necessary for driving the torpedo is stored up in a heavy steel flywheel, weighing one hundred and thirty-five pounds, and turning with a speed of ten thousand turns per minute. The axis of the flywheel lies horizontally perpendicular to the axis of the torpedo (Fig 86), thus steadying the torpedo in its course. If now any force acts tending to deflect the torpedo horizontally from its course, by means of a moment about a vertical axis, the end of the axis of the disk moves vertically, causing the torpedo to roll instead of yielding to the deflecting force. The rolling is utilized, by means of a vertically hanging pendulum, to bring rudders into action, and to cause the torpedo to roll back to its original position, while maintaining its course.

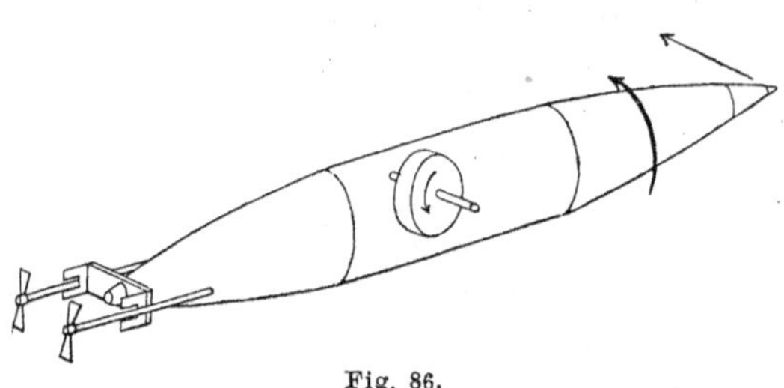

Fig. 86.

A striking example of the principle enunciated above is found in an ingenious top (Fig. 87), spinning on its center of mass, with its axis rolling on various curves constructed of metal wire. No matter what the shape of the wire, the axis of the top clings to it as if held by magnetism, no matter how sharply the curve may bend. The passing around sharp corners at a high speed, in apparant defiance of centrifugal force, is extremely remarkable. The explanation of the action is immediate, on the lines just laid down. The instantaneous axis passes though the point of support O (Fig. 88) and the point of contact of the axis of the top with the wire. The wire, in fact,

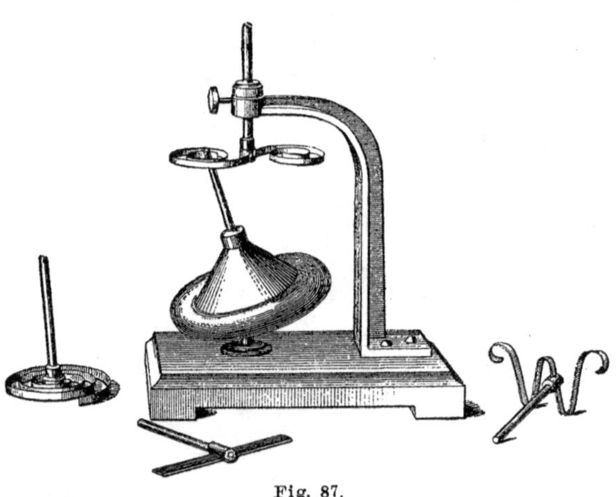

Fig. 87.

constitutes the directrix of the herpolhode cone. Since the ellipsoid of inertia is of rotation, the axis of figure OF, the instantaneous axis OI, and the axis of angular momentum OH, lie in the same plane, which is perpendicular to the tangent plane to the herpolhode cone. During the rolling, all these axes move parallel to this tangent plane, so that the vector HH', representing the change of angular momentum, is parallel to the tangent plane, and in the direction of advance of the axis of figure. The couple causing the motion accordingly due to the reaction between the wire herpolhode and the top, is always parallel to the tangent plane, and never vanishes, but always tends to press the top against the wire. Or in general, in constrained motion, the motion causes the polholde cone to press against the herpolholde cone. This seems to have been first explicitly stated by Klein and Sommerfeld, *Theorie des Kreisels*, p. 173.

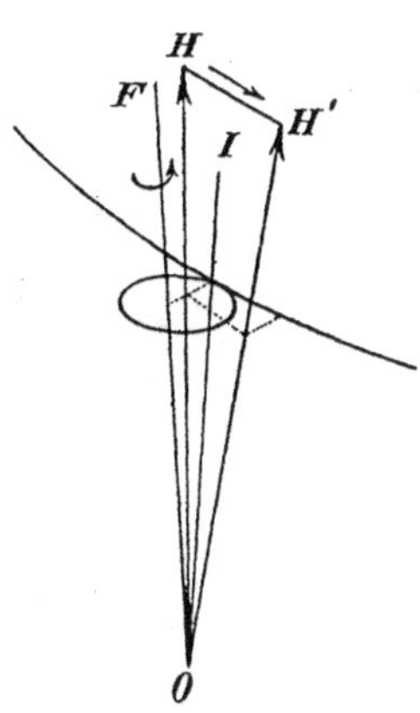

Fig. 88.

An application of the above principle on a large scale, and the only one known to the author[1]), is found in the Griffin grinding mill. A massive steel disk or roller A (Fig. 89) hangs from a vertical shaft by a universal or Hooke's joint C, in the middle of a steel ring B forming the side of a pan. If now the shaft be set rotating, the roller spins quietly about a fixed axis, with no tendency to move sidewise. If on the contrary it be brought into contact with the ring, it immediately rolls around with great velocity, pressing with great force against the steel ring or herpolhode, and grinding any material placed in the pan with great efficiency. It is interesting to note that a somewhat similar mill, in which the axis, instead of passing through a fixed point, hangs vertically from a revolving arm, and therefore is devoid of the action just

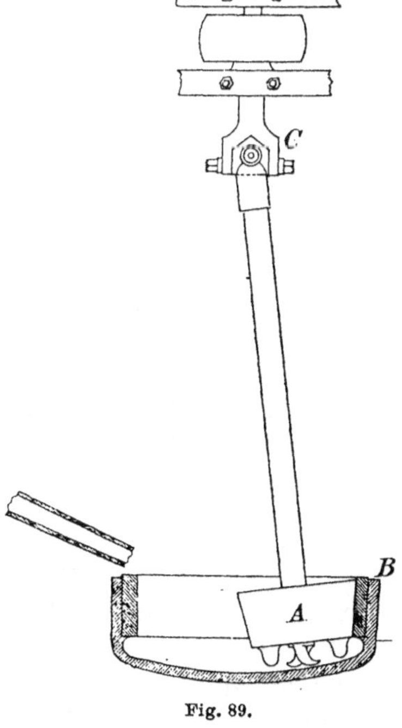

Fig. 89.

1) Since this was written the Brennan monorail car has been invented.

described, although both mills possess in common the centrifugal force due to the circular motion of the center of mass of the roller, is much less efficient. The first mill is an excellent example of the centrifugal force and centrifugal couple, while the second lacks the centrifugal couple, the instantaneous axis and the axis of angular momentum being parallel.

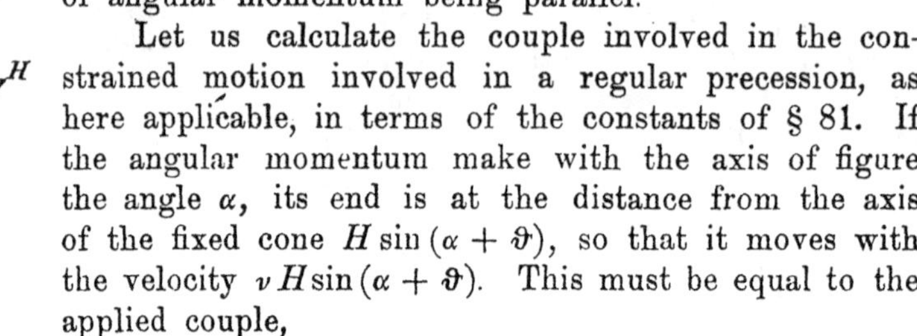

Fig. 90.

Let us calculate the couple involved in the constrained motion involved in a regular precession, as here applicable, in terms of the constants of § 81. If the angular momentum make with the axis of figure the angle α, its end is at the distance from the axis of the fixed cone $H \sin(\alpha + \vartheta)$, so that it moves with the velocity $\nu H \sin(\alpha + \vartheta)$. This must be equal to the applied couple,

60) $$K = \nu H \sin(\alpha + \vartheta).$$

Now resolving H parallel and perpendicular to the axis of figure, we have

61) $$H \cos\alpha = C\omega \cos u, \quad H \sin\alpha = A\omega \sin u,$$

so that

62) $$K = \nu\omega(A \sin u \cos\vartheta + C \cos u \sin\vartheta).$$

But we have, § 81,

13) $$\frac{\omega}{\sin\vartheta} = \frac{\nu}{\sin u},$$

14) $$\omega^2 = \mu^2 + \nu^2 - 2\mu\nu\cos\vartheta,$$

from which

$$\omega \sin u = \nu \sin\vartheta, \quad \omega \cos u = \mu - \nu\cos\vartheta,$$

so that finally

63) $$\begin{aligned} K &= \nu\{A\nu \sin\vartheta \cos\vartheta + C \sin\vartheta(\mu - \nu\cos\vartheta)\} \\ &= \nu \sin\vartheta\{C\mu + (A - C)\nu\cos\vartheta\}. \end{aligned}$$

It is to be noticed that the body will perform a regular precession under no constraint or other applied couple, if putting $K = 0$,

64) $$\cos\vartheta = \frac{\mu}{\nu}\,\frac{C}{C - A}.$$

90. Heavy Symmetrical Top. We will now take up one of the most interesting problems of the motion of a rigid body, namely the motion of a body dynamically symmetrical about an axis, on which its center of mass lies, and spinning about some other point of that axis. This is the problem of the common top or gyroscope. In order to determine the position of the top it will be convenient to introduce three coordinate parameters, namely the three angles of Euler. Let these be ϑ the angle between the Z-axis, which we take as the axis of symmetry, and the fixed vertical Z'-axis (Fig. 91). We may call the XY-plane the equator of the top. Let ON (Fig. 91) be the line of nodes, or the line in which the equator intersects the fixed $X'Y'$-plane. Let ψ be the longitude of the line of nodes, or the angle $X'ON$ measured positively from X' to Y'. Let φ be the angle from the line of nodes to the X-axis, the positive direction of increase being from X to Y. By means of the three angles ϑ, ψ, φ, we may express the nine direction cosines, and the position of the body is completely determined. The meaning of the angles is easily seen on the gyroscope in gimbals (Fig. 92). It will not be necessary for us to express the cosines, as we need only the values of p, q, r in terms of Euler's angles and their velocities,

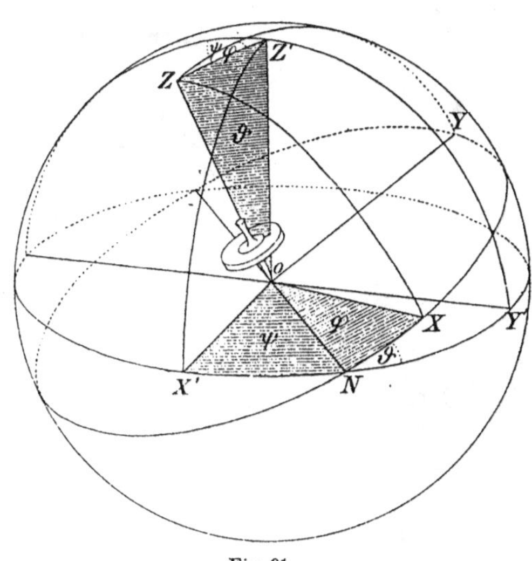

Fig. 91.

$$\frac{d\vartheta}{dt}, \quad \frac{d\psi}{dt}, \quad \frac{d\varphi}{dt}.$$

As these are the angular velocities about ON, OZ' and OZ, respectively, we need only the cosines of the angles made by these lines with the X, Y, Z-axes, which are evidently as given in the following table.

	X	Y	Z
Z'	$\sin\vartheta\sin\varphi$	$\sin\vartheta\cos\varphi$	$\cos\vartheta$
N	$\cos\varphi$	$-\sin\varphi$	0
Z	0	0	1

Resolving now in the direction of the three axes, we obtain

65)
$$p = \frac{d\psi}{dt}\sin\vartheta\sin\varphi + \frac{d\vartheta}{dt}\cos\varphi,$$
$$q = \frac{d\psi}{dt}\sin\vartheta\cos\varphi - \frac{d\vartheta}{dt}\sin\varphi,$$
$$r = \frac{d\psi}{dt}\cos\vartheta + \frac{d\varphi}{dt}.$$

These are Euler's kinematical equations. They illustrate the statement made in § 76, about p, q, r as not being time derivatives, for it is easily seen that $p\,dt$, $q\,dt$, $r\,dt$ do not satisfy the conditions of being exact differentials.

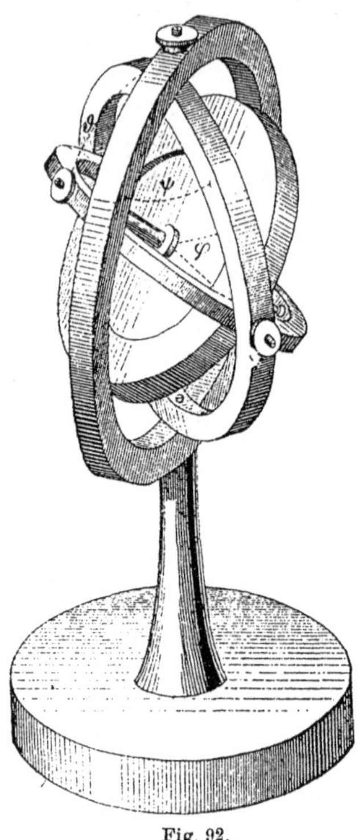
Fig. 92.

The resultant of the weight of all the parts of the body is $Mg = W$ applied at the center of mass. If this is at a distance l from the fixed point the moment of the applied force is $Wl \sin\vartheta$ about the axis ON.

$$66)\qquad \begin{aligned} L &= \quad Wl \sin\vartheta \cos\varphi, \\ M &= -Wl \sin\vartheta \sin\varphi, \\ N &= \quad 0, \end{aligned}$$

so that Euler's dynamical equations are

$$67)\qquad \begin{aligned} A\frac{dp}{dt} &= (B - C)qr + Wl \sin\vartheta \cos\varphi, \\ B\frac{dq}{dt} &= (C - A)rp - Wl \sin\vartheta \sin\varphi, \\ C\frac{dr}{dt} &= (A - B)pq. \end{aligned}$$

Multiplying respectively by p, q, r and adding, we obtain

$$68)\qquad \begin{aligned} Ap\frac{dp}{dt} + Bq\frac{dq}{dt} + Cr\frac{dr}{dt} &= Wl(p \sin\vartheta \cos\varphi - q \sin\vartheta \sin\varphi) \\ &= Wl \sin\vartheta \frac{d\vartheta}{dt}. \end{aligned}$$

Integrating we get the equation of energy,

$$69)\qquad Ap^2 + Bq^2 + Cr^2 = 2(h - Wl\cos\vartheta).$$

Since the moment of the applied forces has no vertical component, the vertical component of the angular momentum is constant, or the end of the vector H describes a plane curve in a horizontal plane. Resolving $H_x = Ap$, $H_y = Bq$, $H_z = Cr$ on the vertical OZ', we obtain

$$70)\qquad H_z' = Ap \sin\vartheta \sin\varphi + Bq \sin\vartheta \cos\varphi + Cr \cos\vartheta = const.$$

If the top is symmetrical about the Z-axis, we have $A = B$. Then the third equation 67) is

$$C\frac{dr}{dt} = 0,$$

$$71)\qquad Cr = const. = H_z.$$

The integral of energy 69) becomes

$$72)\qquad p^2+q^2=\frac{2h-Cr^2-2\,Wl\cos\vartheta}{A}=\alpha-a\cos\vartheta,$$

if we introduce the constants

$$73)\qquad \alpha=\frac{2h}{A}-\frac{H_z^2}{AC},\quad a=\frac{2\,Wl}{A}.$$

The integral of vertical angular momentum 70) becomes

$$74)\qquad \sin\vartheta\,(p\sin\varphi+q\cos\varphi)=\frac{H_z'-H_z\cos\vartheta}{A}=\beta-b\cos\vartheta,$$

putting

$$75)\qquad \beta=\frac{H_z'}{A},\quad b=\frac{H_z}{A}.$$

Inserting the values of p, q from 65),

$$76)\qquad \begin{aligned} \left(\frac{d\vartheta}{dt}\right)^2+\sin^2\vartheta\left(\frac{d\psi}{dt}\right)^2&=\alpha-a\cos\vartheta,\\ \sin^2\vartheta\frac{d\psi}{dt}&=\beta-b\cos\vartheta,\\ \frac{d\psi}{dt}\cos\vartheta+\frac{d\varphi}{dt}&=r. \end{aligned}$$

Eliminating $\frac{d\psi}{dt}$ between the first two gives

$$77)\qquad (\beta-b\cos\vartheta)^2+\sin^2\vartheta\left(\frac{d\vartheta}{dt}\right)^2=\sin^2\vartheta\,(\alpha-a\cos\vartheta),$$

which if we put for $\cos\vartheta$ the single letter z becomes simply

$$78)\qquad \left(\frac{dz}{dt}\right)^2=(1-z^2)(\alpha-az)-(\beta-bz)^2\equiv f(z).$$

From the second equation 76),

$$79)\qquad \frac{d\psi}{dt}=\frac{\beta-bz}{1-z^2},$$

and from the third,

$$80)\qquad \frac{d\varphi}{dt}=r-\frac{z(\beta-bz)}{1-z^2}.$$

The letter z represents the height above the origin of a point on the axis of symmetry, at unit distance from the fixed point. This point will be spoken of as the *apex* of the top. Equation 78) determines the rise and fall of the apex, equation 79) its horizontal motion.

91. Top Equations deduced by Lagrange's Method. Before proceeding with the discussion, let us find the equations by means of Lagrange's method. We have the kinetic energy

$$81)\qquad \begin{aligned} T&=\frac{1}{2}\{A(p^2+q^2)+Cr^2\}\\ &=\frac{1}{2}\{A(\vartheta'^2+\sin^2\vartheta\cdot\psi'^2)+C(\varphi'+\cos\vartheta\cdot\psi')^2\} \end{aligned}$$

so that it appears that φ and ψ are cyclic coordinates. The forces tending to change ϑ, ψ, φ are

$$
82)\quad \begin{aligned}
P_\vartheta &= \frac{d}{dt}\frac{\partial T}{\partial \vartheta'} - \frac{\partial T}{\partial \vartheta} = \frac{d}{dt} A\vartheta' \\
&\quad - \{A\psi'^2 \sin\vartheta \cos\vartheta - C(\varphi' + \cos\vartheta \cdot \psi')\psi' \sin\vartheta\}, \\
P_\psi &= \frac{d}{dt}\frac{\partial T}{\partial \psi'} = \frac{d}{dt}\{A\sin^2\vartheta \cdot \psi' + C(\varphi' + \cos\vartheta \cdot \psi')\cos\vartheta\}, \\
P_\varphi &= \frac{d}{dt}\frac{\partial T}{\partial \varphi'} = \frac{d}{dt}\{C(\varphi' + \cos\vartheta \cdot \psi'\}.
\end{aligned}
$$

If there is no force tending to change the spinning of the top, $P_\varphi = 0$,

$$\frac{d}{dt}\{C(\varphi' + \cos\vartheta \cdot \psi')\} = 0,$$

$$83)\quad C(\varphi' + \cos\vartheta \cdot \psi') = const. = H_z,$$

which is the integral 71). Eliminating φ' by means of this equation, and forming the kinetic potential,

$$84)\quad \varphi' = \frac{H_z}{C} - \cos\vartheta \cdot \psi',$$

$$85)\quad \Phi = T - H_z\varphi' = \frac{1}{2}A(\vartheta'^2 + \sin^2\vartheta \cdot \psi'^2) + H_z \cos\vartheta \cdot \psi' - \frac{1}{2}\frac{H_z^2}{C},$$

the second term containing ψ' in the first power. Such terms in the kinetic potential give rise to what have been called by Thomson and Tait *gyroscopic forces*, whose theory has been treated in § 50. Using this form to determine the forces, we have

$$
86)\quad \begin{aligned}
P_\vartheta &= \frac{d}{dt}\left(\frac{\partial \Phi}{\partial \vartheta'}\right) - \frac{\partial \Phi}{\partial \vartheta} = \frac{d}{dt}A\vartheta' - A\psi'^2 \sin\vartheta\cos\vartheta + H_z \sin\vartheta \cdot \psi', \\
P_\psi &= \frac{d}{dt}\left(\frac{\partial \Phi}{\partial \psi'}\right) = \frac{d}{dt}(A\sin^2\vartheta \cdot \psi' + H_z\cos\vartheta).
\end{aligned}
$$

The influence of the cyclic motion may be very simply shown if the spinning body is mounted as a balanced gyroscope in gimbals, as in Fig. 92. Suppose the vertical ring be held fixed. Then $\psi = const.$, $\psi' = 0$,

$$
87)\quad \begin{aligned}
P_\vartheta &= A\frac{d^2\vartheta}{dt^2}, \\
P_\psi &= -H_z \sin\vartheta \frac{d\vartheta}{dt}.
\end{aligned}
$$

Spinning the inner ring about the horizontal axis requires the same force P_ϑ whether the cyclic motion exists or not, whereas a force is developed tending to make the vertical ring revolve about its axis, which must be balanced by the force of the constraint, P_ψ, proportional to $\frac{d\vartheta}{dt}$. On the other hand let us hold the inner gimbal ring horizontal. Then $\vartheta = \frac{\pi}{2}$, $\vartheta' = 0$.

88)
$$P_\vartheta = H_z \frac{d\psi}{dt},$$
$$P_\psi = A\frac{d^2\psi}{dt^2},$$

and the same force is required for the rotation about the vertical as if there were no spinning, whereas a force is developed tending to turn about the horizontal axis, which must be balanced by the constraint, P_ϑ, proportional to $\frac{d\psi}{dt}$. Thus the effect of the concealed motion would be made evident, even if the disposition of the concealed rotating parts were unknown. The effect of the gyroscopic term may be described by saying that if the apex of the top be moved in any direction, the spinning tends to move it at right angles to that direction, as shown in § 50.

In our present problem, we have $P_\psi = 0$,

$$A\sin^2\vartheta \cdot \psi' + C(\varphi' + \cos\vartheta \cdot \psi')\cos\vartheta = const. = H_z',$$

or by 83),

89)
$$A\sin^2\vartheta \cdot \psi' + H_z\cos\vartheta = H_z',$$

which is the integral of 70).

The differential equation for ϑ is

90)
$$A\frac{d^2\vartheta}{dt^2} - A\psi'^2\sin\vartheta\cos\vartheta + H_z\sin\vartheta \cdot \psi' = Wl\sin\vartheta,$$

which, on replacing ψ' by the value from 89), and using the constants of 73), 75) becomes

91)
$$\frac{d^2\vartheta}{dt^2} - \frac{(\beta - b\cos\vartheta)^2\cos\vartheta}{\sin^3\vartheta} + \frac{b(\beta - b\cos\vartheta)}{\sin\vartheta} = \frac{a}{2}\sin\vartheta.$$

If we now multiply this by $2\sin^2\vartheta \cdot \frac{d\vartheta}{dt}$, it becomes an exact derivative, and integrates into 77). Thus our three integrals are immediate integrals of Lagrange's equations.

92. Nature of the Motion. Equation 78) which states that the time-derivative of z, the cosine of the inclination of the axis to the vertical, is a polynomial of the third degree in z, shows that z is an elliptic function of the time. As we do not here presuppose a knowledge of the elliptic functions, we will discuss the motion without explicitly finding the solution in terms of elliptic functions.

We see at once that the solution depends on the four arbitrary constants α, a, of the dimensions $[T^{-2}]$, which enter equation 78) linearly, and β, b, of dimensions $[T^{-1}]$, which enter homogeneously in the second degree, so that if we divide b, β by the same number and a, α by its square, while we multiply t by the same number, the two equations 78) 79) are unchanged, that is to any value of z

corresponds the same value of ψ, or the *path* of the point of the top is the same, but described at a different rate. Thus the shape of the path depends on the three ratios of the constants, or there is a triple infinity of paths. As for the meanings of the constants, a depends simply on the nature of the top, irrespective of the motion, and by comparison with § 80 is seen to be inversely proportional to the square of the time in which the top would describe small oscillations as a pendulum, if supported with its apex downwards, without spinning. If we change the top, we may obtain the same path by suitably changing α, b, β as just described. These three constants depend on the circumstances of the motion, b being proportional to the angular momentum about the axis of figure, or to the velocity of spinning, β to the angular momentum about the vertical, and α depending on both the velocity of spinning and the energy constant. Expressed in terms of the initial position and velocities they are

$$92)\qquad \begin{aligned} a &= \frac{2\,Wl}{A}, \quad b = \frac{C}{A}r, \\ \beta &= \sin^2\vartheta_0\left(\frac{d\psi}{dt}\right)_0 + b\cos\vartheta_0, \\ \alpha &= \left(\frac{d\vartheta}{dt}\right)_0^2 + \sin^2\vartheta_0\left(\frac{d\psi}{dt}\right)_0^2 + a\cos\vartheta_0. \end{aligned}$$

With the convention that we have adopted, a is positive. As it is evident that any path may be described in either direction, we shall obtain all the paths if we spin the top always in the same direction. We shall thus suppose b to be positive, while β may be positive or negative, according to the sign and magnitude of $\left(\frac{d\psi}{dt}\right)_0$ and $\cos\vartheta_0$.

Since $\frac{dz}{dt}$ is real, $f(z)$ 78) must be positive throughout the motion, except when it vanishes. Since $f(1) = -(\beta - b)^2$ and $(-1) = -(\beta + b)^2$ are both negative, $f(\infty) = \infty$ and $f(-\infty) = -\infty$, the course of the function $f(z)$ is as shown in Fig. 93. Thus it is evident that $f(z)$ has three real roots, two z_1, z_2, lying between 1 and -1, while the third, z_3 lies outside of that interval on the positive side. Thus the motion is confined to that part of the z-axis between z_1, z_2, and the apex of the top rises and falls between the two values of ϑ whose cosines are z_1 and z_2. The triple infinity of paths may be characterized by giving the three roots all possible real values, instead of giving the constants b, β, α. In practice it will be convenient to give the two roots indicating the highest and lowest points reached by the apex, and the value

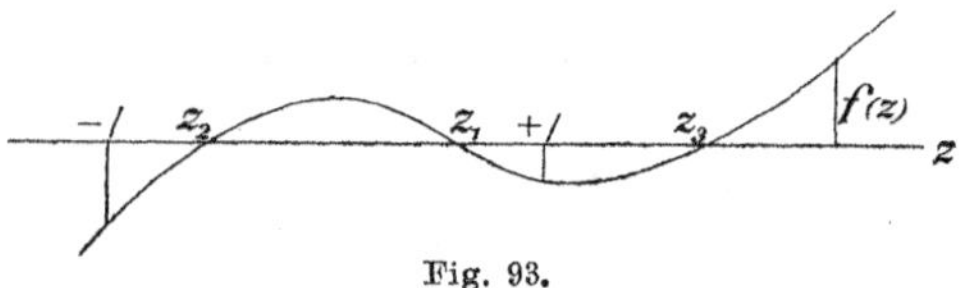

Fig. 93.

of $\frac{d\psi}{dt}$ the horizontal angular velocity at one of them, which three data completely characterize the motion.

Since z is an elliptic function of the time, the rise and fall is periodic, and after a certain time, z will have attained the same value, and so will $\frac{d\psi}{dt}$ and $\frac{d\varphi}{dt}$, accordingly during successive periods the angles ψ and φ will increase by the same amounts. The horizontal projection of the axis of the top advances at the rate $\frac{d\psi}{dt}$. This vanishes and changes sign if $z = \frac{\beta}{b}$ and then we have

$$93) \qquad \left(\frac{dz}{dt}\right)^2 = \left(\alpha - a\frac{\beta}{b}\right)\left(1 - \frac{\beta^2}{b^2}\right).$$

The second factor is positive if $\frac{\beta^2}{b^2} < 1$, which will be the case if

$$94) \qquad \left\{\frac{1}{b}\sin^2\vartheta_0\left(\frac{d\psi}{dt}\right)_0 + \cos\vartheta_0\right\}^2 < 1,$$

that is if the top is spun fast enough or $\left(\frac{d\psi}{dt}\right)_0$ is small enough. We must then have

$$\alpha - a\frac{\beta}{b} > 0,$$

that is,

$$95) \qquad \left(\frac{d\vartheta}{dt}\right)_0^2 + \sin^2\vartheta_0\left(\frac{d\psi}{dt}\right)_0\left\{\left(\frac{d\psi}{dt}\right)_0 - \frac{a}{b}\right\} > 0.$$

This will certainly be the case if $\left(\frac{d\psi}{dt}\right)_0$ is negative at the highest point of the path, and if it be positive and greater than $\frac{a}{b}$. If the top be spinning so slowly that $\frac{\beta}{b}$ is greater than unity, z cannot attain this value, and $\frac{d\psi}{dt}$ will never vanish. It is evident that when $\frac{d\psi}{dt} = 0$ the projection of the path on the horizontal plane has a radial tangent (unless $z = 0$).

The axis of the top describes a cone limited by the two circular cones of angles ϑ_1, ϑ_2, where $z_1 = \cos\vartheta_1$, $z_2 = \cos\vartheta_2$. It is in general tangent to the two cones, as may be shown as follows. The projection of the apex on the horizontal plane $X'Y'$ has the polar coordinates (Fig. 94),

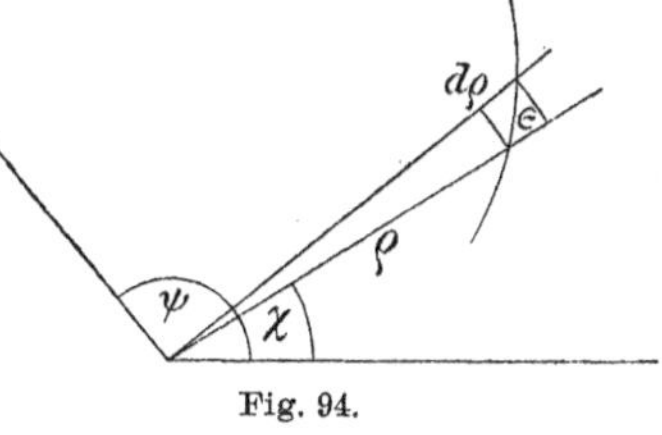

Fig. 94.

$$\rho = \sin\vartheta = \sqrt{1 - z^2}, \quad \chi = \psi - \frac{\pi}{2}.$$

If ε denote the angle made by the tangent with the radius vector,

$$\text{96)}\qquad \tan\varepsilon = \varrho\frac{d\chi}{d\varrho} = \varrho\frac{d\psi}{d\varrho} = \frac{d\psi}{dt} \Big/ \frac{1}{\varrho}\frac{d\varrho}{dt}$$

$$= \frac{\beta - bz}{1-z^2} \Big/ \frac{-z}{1-z^2}\frac{dz}{dt} = \frac{bz-\beta}{z\sqrt{f(z)}}.$$

Accordingly for z_1, z_2, roots of $f(z)$, $\tan\varepsilon = \infty$, and the axis cone is tangent to the limiting cones, unless at the same time the numerator vanishes. This can be the case for only one of the limiting values. In case the numerator vanishes, say for $z = z_1$, we have $\beta = bz_1$,

$$\text{97)}\qquad \tan\varepsilon = \frac{b\sqrt{z-z_1}}{z\sqrt{a(z-z_2)(z-z_3)}}$$

which *vanishes* for $z = z_1$, so that the cone has cuspidal edges. If the top is merely set spinning, and let go, so that $\frac{d\vartheta}{dt} = \frac{d\psi}{dt} = 0$ it evidently begins to fall vertically, so that the cusps are on the upper limiting circle, while the path is tangent to the lower. The reason

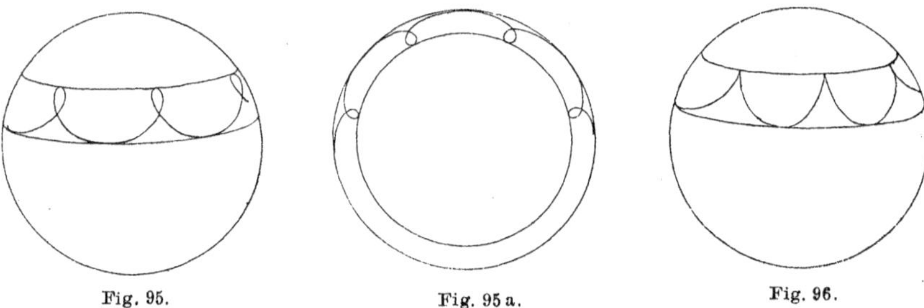

Fig. 95. Fig. 95 a. Fig. 96.

that the top starts to fall vertically is, of course, that the gyroscopic action does not begin until the velocity of falling $\frac{d\vartheta}{dt}$ begins, as shown in 87). It is to be noticed that when z is positive, as in the ordinary top, the horizontal projection has the cusps turned inward, while

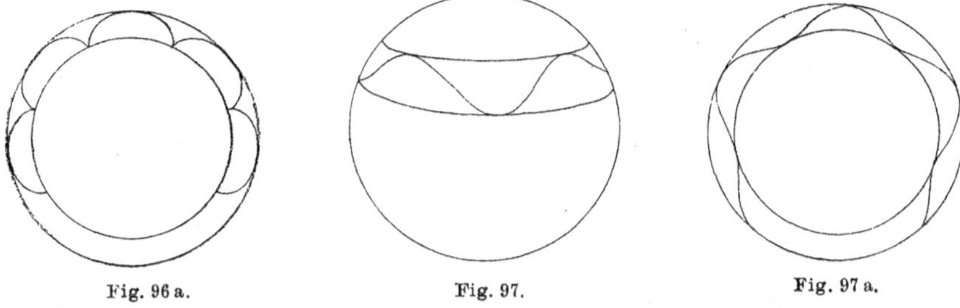

Fig. 96 a. Fig. 97. Fig. 97 a.

when z is negative, as in the gyroscopic pendulum, the cusps are turned outward. The three types of motion are shown for the first case in Figs. 95, 96, 97, with the horizontal projections marked a, while the cusped type for the second case is shown in Fig. 98.

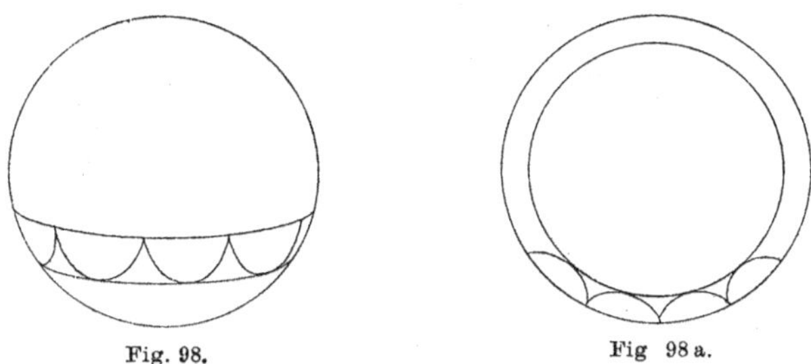

Fig. 98. Fig 98 a.

93. Precession and Nutation. It will be convenient to express the motion in therms of one root z_1 and the constants b, β. In order to eliminate α we have

$$98)\qquad \frac{f(z)}{1-z^2} = \alpha - az - \frac{(\beta - bz)^2}{1-z^2}.$$

Now since z_1 is a root, $f(z_1) = 0$ and

$$99)\qquad 0 = \alpha - az_1 - \frac{(\beta - bz_1)^2}{1-z_1^{\,2}}.$$

Subtracting from 98),

$$100)\qquad \frac{f(z)}{1-z^2} = a(z_1 - z) + \frac{(\beta - bz_1)^2(1-z^2) - (\beta - bz)^2(1-z_1^{\,2})}{(1-z_1^{\,2})(1-z^2)}$$

$$= a(z_1 - z) + \frac{2b\beta(1+z_1 z)(z-z_1) - (b^2+\beta^2)(z^2-z_1^{\,2})}{(1-z_1^{\,2})(1-z^2)}.$$

We thus find that $z - z_1$ is a factor of the expression on the right, so that, multiplying by $1-z^2$, we have $f(z)$ exhibited in the form

$$101)\qquad f(z) = (z - z_1) f_1(z),$$

where $f_1(z)$ is the polynomial of order two,

$$102)\qquad f_1(z) = -a(1-z^2) + \frac{2b\beta(1+z_1 z) - (b^2+\beta^2)(z+z_1)}{1-z_1^{\,2}},$$

so that the other two roots are found by solving the quadratic $f_1(z) = 0$. As the roots z_1, z_2 approach each other, the rise and fall decreases, and vanishes when $f(z)$ has two equal roots. The condition for this is that $f(z)$ and $f'(z) = f_1(z) + (z-z_1)f'_1(z)$ have a common root z_1, that is that

$$f_1(z_1) = 0,$$

from which

$$103)\qquad a(1-z_1^{\,2})^2 = 2b\beta(1+z_1^{\,2}) - 2z_1(b^2+\beta^2) \equiv 2(\beta - bz_1)(b - \beta z_1).$$

If z_1 and one of the constants b, β are given, this is a quadratic to determine the other. We find

104) $$\frac{d\psi}{dt} = \frac{\beta - bz_1}{1 - z_1^2},$$

which is constant, so that the motion is a regular precession, without rise and fall. There are thus, for a given velocity of spinning, and a given angle of inclination with the vertical, *two* values of the velocity of precession. We may also find these by considering the equation 90), putting ϑ constant in which gives, if $\sin\vartheta$ is not zero,

105) $$-\psi'^2 \cos\vartheta + b \cdot \psi' = \frac{a}{2}$$

a quadratic for ψ' with the roots

106) $$\psi' = \frac{b}{2\cos\vartheta}\left\{1 \pm \left(1 - \frac{2a\cos\vartheta}{b^2}\right)^{\frac{1}{2}}\right\}.$$

These values are real if $b^2 > 2a\cos\vartheta$. If the top be spun so fast that $\frac{2a\cos\vartheta}{b^2}$ is a small quantity whose square may be neglected, we find for one value of ψ'

107) $$\psi_1' = \frac{b}{\cos\vartheta}\left(1 - \frac{a}{2b^2}\cos\vartheta\right)$$

which is a large quantity of the order of b, while the other root is

108) $$\psi_2' = \frac{a}{2b}$$

which is a small quantity of the order of $\frac{1}{b}$. Of these it is the slow precsssion which is usually observed.

It is to be observed that if we put $\psi' = \nu$, $\varphi' = \mu$, the first of equations 82) gives for P_ϑ the same result as obtained for K in 63). When we make a vanish, so that the body is under the action of no impressed moment, the root ψ_2' becomes zero, so that the axis of figure stands still, while the root ψ_1' becomes $\frac{b}{\cos\vartheta}$, that is, the body performs a Poinsot-motion around the vertical as the invariable axis. Thus the effect to the impressed forces may be looked upon as a small perturbation of the Poinsot-motion.

We will now consider the motion when the condision 103) is not fulfilled. From equation 78), we have t given by the elliptic integral,

109) $$t = \int \frac{dz}{\sqrt{a(z - z_1)(z - z_2)(z - z_3)}}.$$

We may easily find two limits within which this value lies, by substituting for the factor $\sqrt{z - z_3}$ in the integrand its greatest and least values, as we did in the case of the spherical pendulum.

Since throughout the motion

$$z_1 > z > z_2, \quad z_3 - z_1 < z_3 - z < z_3 - z_2$$

we have the inequalities,

110) $$\frac{1}{\sqrt{a(z_3 - z_1)}}\int\frac{dz}{\sqrt{(z_1 - z)(z - z_2)}} > t > \frac{1}{\sqrt{a(z_3 - z_2)}}\int\frac{dz}{\sqrt{(z_1 - z)(z - z_2)}}.$$

By means of a linear substitution we may simplify the integral. Let us put

111) $$\begin{aligned} z_1 - z &= c(1 - x), \quad c = \frac{z_1 - z_2}{2}, \\ z - z_2 &= c(1 + x), \quad x = \frac{z}{c} - \frac{z_1 + z_2}{2c}, \end{aligned}$$

when the integral becomes

112) $$\int\frac{dz}{\sqrt{(z_1 - z)(z - z_2)}} = \int\frac{dx}{\sqrt{1 - x^2}} = \cos^{-1} x + const.,$$

so that we have for t,

113) $$\frac{1}{\sqrt{a(z_3 - z_1)}}\cos^{-1} x > t + const. > \frac{1}{\sqrt{a(z_3 - z_2)}}\cos^{-1} x.$$

If now the difference $z_1 - z_2 = x$ is sufficiently small in comparison with $z_3 - z_1$ and $z_3 - z_2$, we may obtain an approximate result by putting under the radical the mean of the quantities which are too great and too small respectively, so that if $z_1 + z_2 = 2z_0$ we have the approximate result

114) $$t + const. = \frac{1}{\sqrt{a(z_3 - z_0)}}\cos^{-1} x,$$

from which we obtain

$$x = \frac{z}{c} - \frac{z_0}{c} = \cos\sqrt{a(z_3 - z_0)}\cdot t,$$

115) $$z = z_0 + c\cdot\cos\sqrt{a(z_3 - z_0)}\cdot t = z_0 + \xi.$$

The arbitrary constant has been taken so that $t = 0$ when the top is at its highest, and $z = z_0 + c = z_1$.

We thus see that when the roots z_1, z_2 are nearly enough equal, the apex of the top rises and falls with a harmonic oscillation ξ of the small amplitude $c = \frac{(z_1 - z_2)}{2}$. In order to determine when the approximation is justified, we have to consider what will cause the third root z_3 to be large. Since z_2 and z_3 are the roots of the quadratic function $f_1(z)$ 102), their sum is the negative of the coefficient of z divided by that of z^2, that is

116) $$z_2 + z_3 = \frac{b^2 + \beta^2 - 2b\beta z_1}{a(1 - z_1^2)} = \frac{b^2}{a} + \frac{(\beta - bz_1)^2}{a(1 - z_1^2)}.$$

Thus we see that by making b large enough we may make z_3 as large as we please, when z_1 and z_2 are given, so that the approximation is better the faster the top spins.

Let us now consider the horizontal motion, or precession. We have

$$117)\qquad \frac{d\psi}{dt} = \frac{\beta - bz}{1 - z^2} = \frac{\beta - b(z_0 + \zeta)}{(1 - z_0^2)\left(1 - \frac{2 z_0 \zeta + \zeta^2}{1 - z_0^2}\right)}.$$

We have already supposed ζ to be a small quantity, so that if we neglect the square of $\frac{z_0 \zeta}{(1 - z_0^2)}$, we have, after developing the second factor of the denominator,

$$118)\qquad \frac{d\psi}{dt} = \frac{\beta - b z_0}{1 - z_0^2} - \frac{b\zeta}{1 - z_0^2} + \frac{2 z_0 (\beta - b z_0)}{(1 - z_0^2)^2} \cdot \zeta.$$

Now inserting the value of ζ from 115) and integrating,

$$119)\qquad \psi = \frac{\beta - b z_0}{1 - z_0^2} t - \frac{c}{\sqrt{a(z_3 - z_0)}} \left\{ \frac{b}{1 - z_0^2} - \frac{2 z_0 (\beta - b z_0)}{(1 - z_0^2)^2} \right\} \sin \sqrt{a(z_3 - z_0)} \cdot t.$$

Thus we see that ψ varies with a harmonic oscillation about the value that it would have in the regular precession at the mean height z_0, of the same period as the vertical oscillation. If we project the motion of the apex on the tangent plane to the sphere on which it moves, calling ξ the horizontal coordinate, and η the distance moved from the horizontal mean axis, we have, Fig. 99,

$$120)\qquad \xi = \psi \sin \vartheta = \sqrt{1 - z_0^2} \cdot \psi, \quad \eta = \frac{\zeta}{\sqrt{1 - z_0^2}}.$$

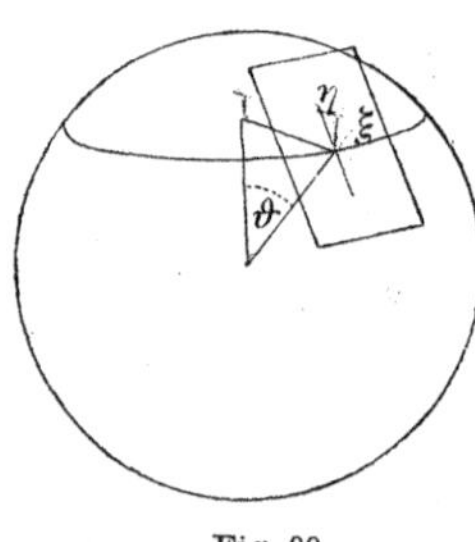

Fig. 99.

Thus we see that the second terms of 115) and 119) represent an elliptic harmonic motion of the apex of the top. This is termed *nutation*. We thus have a complete description of the motion of a top when differing by a small amount from a regular precession, as a regular precession combined with a nutation in an ellipse about the point which advances with the regular precession.

We shall now make an additional supposition with regard to the constants of the motion. We have seen from 108) that in the case of regular precession with rapid spinning, the precession was slow. Let us then suppose that

$$121)\qquad \frac{\beta - b z_0}{1 - z_0^2} = \mu$$

is a small quantity of the same order as c, so that their squares and product may be neglected. Since z_0 is the cosine of the angle between the vectors whose magnitudes are b and β, this supposition is equivalent to saying that the angular momentum makes a small angle with the axis of figure, as we see from Fig. 100, in which the distance $DE = \beta - bz_0$. Making this supposition, the last term in 116) is negligible, also that in 119). Thus we obtain from 116),

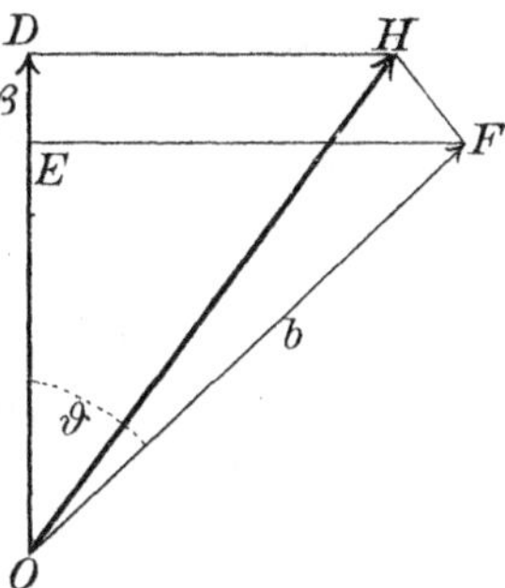

Fig. 100.

$$z_3 - z_0 = \frac{b^2}{a} - z_2 - z_0,$$

and since $\frac{b^2}{a}$ is supposed to be large we may neglect $-(z_2 + z_0)$, so that we have finally,

$$122) \qquad a(z_3 - z_0) = b^2,$$

$$123) \qquad \zeta = c \cdot \cos bt, \quad \psi = \mu t - \frac{c}{1 - z_0^2} \sin bt,$$

$$124) \qquad \xi = \frac{\beta - bz_0}{\sqrt{1 - z_0^2}} t - \frac{c}{\sqrt{1 - z_0^2}} \sin bt, \quad \eta = \frac{c}{\sqrt{1 - z_0^2}} \cos bt.$$

It is evident from Fig. 100 that $\beta - bz_2$ is positive, accordingly [cf. 117)] the apex is always moving so that $\frac{d\psi}{dt}$ is positive at the bottom of its path, and thus the average motion is in that sense. The motion at the top may be in either direction, according to the magnitude of c. We see that the motion of nutation is opposite to the motion of the clock-hands. Thus the motion of the apex, as given by 124), is that of a point at a distance $\frac{c}{\sqrt{1 - z_0^2}}$ from the center of a circle which *rolls* on a line above it with its center advancing at a velocity $\frac{(\beta - bz_0)}{\sqrt{1 - z_0^2}}$. The radius of the rolling circle is $\frac{(\beta - bz_0)}{b\sqrt{1 - z_0^2}}$.

Such a locus is called a cycloid. In the ordinary cycloid, the tracing point is on the circumference of the rolling circle, or $\beta - bz_0 = bc$. If the tracing point is an internal one, the cycloid is called *prolate*. It has no loops, nor vertical tangents, and $\frac{d\psi}{dt}$ is never zero, but it has points of inflexion. If the point is external the cycloid is called *curtate*, and has loops, but no inflexions. It is evident that this curve will be described when the apex is given a push to the left at the top of its motion, while if it be given a push to the right it describes the prolate cycloid, and if it be simply let go, it describes the ordinary cycloid with cusps. (The prolate and curtate cycloids are also called *trochoids*.) Since the height of a cycloid is to the length of its base as $1 : \pi$, the base being the distance traversed in one revolution, we see that when the top is spun rapidly, so that the precession is slow, the rise and fall is very rapid $\left(\text{for } b = r\frac{C}{A}\right)$, and very small. For this reason it is seldom noticed, and this accounts for the popular opinion, expressed in many text books, that the motion of a top is such that its axis describes a circular cone with a constant angular velocity, or a regular precession. Thus the reason of the vertical force of gravity producing a horizontal motion remains a paradox. We have seen that such a motion is the very particular exception, and not the rule, being only exhibited when the necessary horizontal velocity is imparted at the outset, so that the action of gravity is always balanced by the centrifugal couple generated by the precession. If the necessary velocity is not imparted, the top immediately begins to fall in

Fig. 101 a. Fig. 101 b.

obedience to gravity. The motion which we have just described is called by Klein and Sommerfeld a *pseudo-regular* precession, and may be called a small oscillation about a regular precession. In Fig. 101 are shown curves of the actual path obtained by photographing a small incandescent lamp attached to the axis of a gyroscope, with ϑ nearly a right angle.

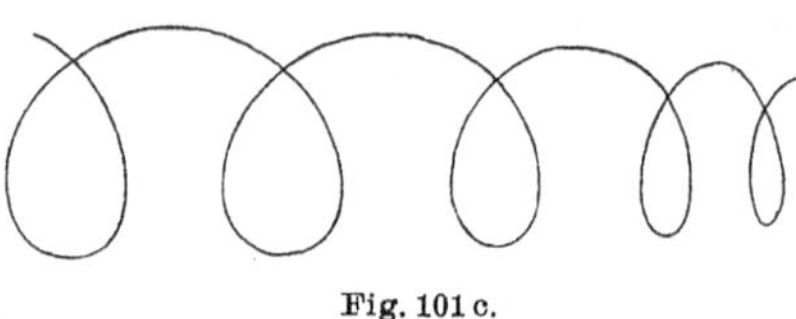

Fig. 101 c.

94. Small Oscillations about the Vertical. In the discussion which has just been given, it has been supposed that $1 - z^2$ was not a small quantity. If however in the course of the motion the axis of the top becomes nearly vertical this will no longer be true, so that for this case a special investigation is necessary. Let us suppose that ϑ and ϑ' are so small that in the kinetic potential all their powers above the second may be neglected. Let us use for coordinates the rectangular coordinates of the projection of the apex on the horizontal plane,

$$x = r\cos\psi, \quad y = r\sin\psi, \quad r = \sin\vartheta.$$

Using then the expression of 85) for the kinetic potential, with potential energy $W\cos\vartheta$,

125) $$\Phi = \frac{1}{2} A (\vartheta'^2 + \sin^2 \vartheta \cdot \psi'^2) + H_z \cos \vartheta \cdot \psi' - Wl \cos \vartheta,$$

we will convert it into terms of x, y, x', y', neglecting all terms of order higher than the second.

In the first term, since, to the order of approximation,

$$r' = \cos \vartheta \cdot \vartheta' = \vartheta',$$

we have $r'^2 + r^2 \psi'^2$, the square of the velocity in polar coordinates which is in rectangular coordinates $x'^2 + y'^2$. Also we have

$$\tan \psi = \frac{y}{x}, \quad \sec^2 \psi \cdot \psi' = \frac{xy' - yx'}{x^2},$$

126) $$\psi' = \frac{xy' - yx'}{x^2 + y^2},$$

and since

127) $$\cos \vartheta = \{1 - (x^2 + y^2)\}^{\frac{1}{2}} = 1 - \frac{x^2 + y^2}{2},$$

we have finally

128) $$\Phi = \frac{1}{2} A (x'^2 + y'^2) + H_z \left(1 - \frac{x^2 + y^2}{2}\right) \frac{xy' - yx'}{x^2 + y^2} - Wl \left(1 - \frac{x^2 + y^2}{2}\right).$$

We have then in the term in H_z an example of the gyroscopic terms of § 50, in which $x = q_1$, $y = q_2$,

129) $$L_1 = H_z \left(\frac{-y}{x^2 + y^2} + \frac{y}{2}\right), \qquad L_2 = H_z \left(\frac{x}{x^2 + y^2} - \frac{x}{2}\right).$$

Forming the equations of motion, since

$$\frac{\partial L_1}{\partial y} - \frac{\partial L_2}{\partial x} = H_z,$$

we have finally

130) $$Ax'' + H_z y' + Wlx = 0, \qquad Ay'' - H_z x' + Wly = 0,$$

or in terms of our constants,

131) $$x'' + by' + \frac{a}{2} x = 0, \qquad y'' - bx' + \frac{a}{2} y = 0.$$

These equations are a particular case of a problem that is interesting enough to be considered in full. If b were zero, they would be the equations for the small vibrations of a system of two degrees of freedom, the stiffness and inertia coefficients of which are the same for both freedoms. Let us consider the general system, for which

$$132)\qquad T = \frac{1}{2}(Ax'^2 + By'^2), \qquad W = \frac{1}{2}(Cx^2 + Dy^2),$$

into which a gyrostat, or rapidly rotating symmetrical solid, is introduced, the direction of whose axis is determined as in the present case by the coordinates x and y. (It is to be noticed that x and y are *principal* coordinates.) The equations for the small oscillations of the system are then

$$133)\qquad \begin{aligned} Ax'' + H_z y' + Cx &= 0, \\ By'' - H_z x' + Dy &= 0. \end{aligned}$$

These may be treated by the general method of § 45 for small oscillations. In order to simplify the notation, it will be convenient to put

$$134)\qquad x = \frac{\xi}{\sqrt{A}}, \quad y = \frac{\eta}{\sqrt{B}}, \quad \frac{H_z}{\sqrt{AB}} = b, \quad \frac{C}{A} = c, \quad \frac{D}{B} = d,$$

when our equations become

$$135)\qquad \begin{aligned} \xi'' + b\eta' + c\xi &= 0, \\ \eta'' - b\xi' + d\eta &= 0. \end{aligned}$$

Having solved these, we may pass to the case of our vertical top by putting $c = d$.

In accordance with the method of § 45, let us put

$$\xi = A_1 e^{\lambda t}, \quad \eta = A_2 e^{\lambda t},$$

from which we obtain

$$136)\qquad \begin{aligned} A_1(\lambda^2 + c) + A_2 b\lambda &= 0, \\ -A_1 b\lambda + A_2(\lambda^2 + d) &= 0. \end{aligned}$$

The determinantal equation is

$$137)\qquad \lambda^4 + (c + d + b^2)\lambda^2 + cd = 0,$$

whose roots are

$$138)\qquad \lambda^2 = \frac{1}{2}\left\{-(c + d + b^2) \pm \sqrt{(c + d + b^2)^2 - 4cd}\right\}.$$

If the solution is to represent oscillations, all the values of λ must be pure imaginary, thus both values of λ^2 must be real and negative.

If we call them $-\mu^2$, $-\nu^2$, we have for their sum and product the coefficients in 137),

$$139)\qquad \mu^2 + \nu^2 = c + d + b^2, \quad \mu^2\nu^2 = cd.$$

In order that μ, ν shall be real it is accordingly necessary that c, d shall be of the same sign, that is our system must be either stable for both freedoms, or unstable for both. Extracting the square root of the second equation 139), doubling, and adding to and subtracting from the first,

$$140)\qquad \begin{aligned} (\mu+\nu)^2 &= c + d + b^2 \pm 2\sqrt{cd} = b^2 + (\sqrt{c} \pm \sqrt{d})^2, \\ (\mu-\nu)^2 &= c + d + b^2 \mp 2\sqrt{cd} = b^2 + (\sqrt{c} \mp \sqrt{d})^2. \end{aligned}$$

Extracting the roots, adding and subtracting,

$$\mu + \nu = \pm \left\{b^2 + (\sqrt{c} \pm \sqrt{d})^2\right\}^{\frac{1}{2}},$$

$$\mu - \nu = \pm \left\{b^2 + (\sqrt{c} \mp \sqrt{d})^2\right\}^{\frac{1}{2}},$$

$$141)\qquad \begin{aligned} \mu &= \pm\frac{1}{2}\left[\left\{b^2 + (\sqrt{c} + \sqrt{d})^2\right\}^{\frac{1}{2}} + \left\{b^2 + (\sqrt{c} - \sqrt{d})^2\right\}^{\frac{1}{2}}\right], \\ \nu &= \pm\frac{1}{2}\left[\left\{b^2 + (\sqrt{c} + \sqrt{d})^2\right\}^{\frac{1}{2}} - \left\{b^2 + (\sqrt{c} - \sqrt{d})^2\right\}^{\frac{1}{2}}\right]. \end{aligned}$$

The inner double sign is evidently unnecessary. Since $\mu\nu = \pm\sqrt{cd}$ we have also,

$$142)\qquad \begin{aligned} \frac{1}{\mu} &= \pm\frac{\left\{b^2 + (\sqrt{c} + \sqrt{d})^2\right\}^{\frac{1}{2}} - \left\{b^2 + (\sqrt{c} - \sqrt{d})^2\right\}^{\frac{1}{2}}}{2\sqrt{cd}}, \\ \frac{1}{\nu} &= \pm\frac{\left\{b^2 + (\sqrt{c} + \sqrt{d})^2\right\}^{\frac{1}{2}} + \left\{b^2 + (\sqrt{c} - \sqrt{d})^2\right\}^{\frac{1}{2}}}{2\sqrt{cd}}. \end{aligned}$$

From the values of μ and ν it is evident that both are real if c and d are positive. If they are negative it is necessary in order to have real values that

$$b > \sqrt{-c} + \sqrt{-d}.$$

Thus we find that even if the system is unstable, sufficiently rapid spinning of the gyrostat makes it stable. This is the case of the top with its center of mass over the point of support. In order to complete the discussion we have to determine the coefficients A_1, A_2 for the various roots. If we call the roots

$$\lambda_1 = i\mu, \quad \lambda_2 = -i\mu, \quad \lambda_3 = i\nu, \quad \lambda_4 = -i\nu,$$

we have for the general solution,

143)
$$\xi = A_1^{(1)} e^{i\mu t} + A_1^{(2)} e^{-i\mu t} + A_1^{(3)} e^{i\nu t} + A_1^{(4)} e^{-i\nu t},$$
$$\eta = A_2^{(1)} e^{i\mu t} + A_2^{(2)} e^{-i\mu t} + A_2^{(3)} e^{i\nu t} + A_2^{(4)} e^{-i\nu t},$$

where we have by the first of equations 136),

144)
$$\frac{A_2^{(1)}}{A_1^{(1)}} = i\frac{c-\mu^2}{b\mu}, \quad \frac{A_2^{(2)}}{A_1^{(2)}} = -i\frac{c-\mu^2}{b\mu}, \quad \frac{A_2^{(3)}}{A_1^{(3)}} = i\frac{c-\nu^2}{b\nu}, \quad \frac{A_2^{(4)}}{A_1^{(4)}} = -i\frac{c-\nu^2}{b\nu}.$$

Introducing the values of the A_2's in terms of the A_1's, and writing

145) $A_1^{(1)} + A_1^{(2)} = \alpha,\ A_1^{(1)} - A_1^{(2)} = -i\beta,\ A_1^{(3)} + A_1^{(4)} = \alpha',\ A_1^{(3)} - A_1^{(4)} = -i\beta'$,

we have, replacing exponentials by trigonometric terms,

146)
$$\xi = \alpha \cos \mu t + \beta \sin \mu t + \alpha' \cos \nu t + \beta' \sin \nu t,$$
$$\eta = \frac{c-\mu^2}{b\mu}(\beta \cos \mu t - \alpha \sin \mu t) + \frac{c-\nu^2}{b\nu}(\beta' \cos \nu t - \alpha' \sin \nu t),$$

with the four arbitrary constants α, β, α', β', or putting

147) $\alpha = A_1 \cos \varepsilon_1,\quad \beta = A_1 \sin \varepsilon_1,\quad \alpha' = A_2 \cos \varepsilon_2,\quad \beta' = A_2 \sin \varepsilon_2,$

148)
$$\xi = A_1 \cos(\mu t - \varepsilon_1) + A_2 \cos(\nu t - \varepsilon_2),$$
$$\eta = \frac{\mu^2 - c}{b\mu} A_1 \sin(\mu t - \varepsilon_1) + \frac{\nu^2 - c}{b\nu} A_2 \sin(\nu t - \varepsilon_2).$$

Accordingly the motion may be described as the resultant of two elliptic harmonic motions of frequencies $\frac{\mu}{2\pi}$, $\frac{\nu}{2\pi}$, the directions of the axes of the ellipses being coincident, and given by the directions in which the system can make a principal oscillation when the gyrostat is not spinning. The absolute sizes of the two ellipses are arbitrary, but the ratios of the axes, and the phases, are determined by the nature of the system and the rapidity of the spinning.

Calculating the coefficients in 148) from the values of μ, $\frac{1}{\mu}$, ν, $\frac{1}{\nu}$,

149)
$$\frac{\mu^2 - c}{b\mu} = \pm \frac{1}{2b}\left[\left(1 - \sqrt{\frac{c}{d}}\right)\left\{b^2 + (\sqrt{c} + \sqrt{d})^2\right\}^{\frac{1}{2}} + \left(1 + \sqrt{\frac{c}{d}}\right)\left\{b^2 + (\sqrt{c} - \sqrt{d})^2\right\}^{\frac{1}{2}}\right],$$
$$\frac{\nu^2 - c}{b\nu} = \pm \frac{1}{2b}\left[\left(1 - \sqrt{\frac{c}{d}}\right)\left\{b^2 + (\sqrt{c} + \sqrt{d})^2\right\}^{\frac{1}{2}} - \left(1 + \sqrt{\frac{c}{d}}\right)\left\{b^2 + (\sqrt{c} - \sqrt{d})^2\right\}^{\frac{1}{2}}\right].$$

If now $c = d$, as in the case of the top, both these expressions become equal to plus or minus unity, so that both ellipses become circles. The motion of the top making small oscillations about the vertical is accordingly to be described by saying that its apex describes epicycloids (epitrochoids) or hypocycloids (hypotrochoids) upon the horizontal plane. It is to be noticed that according as we take the signs in 149) the relative sense of the rotations in the two circular motions will be alike or different. By considering which way the top tends to fall we may decide whether the cusps are turned inwards or outwards, and it will be found that if the center of mass is above the point of support the cusps or loops are turned inwards, and the curves are epicycloids, while if it is below the cusps or loops are turned outwards and the curves are hypocycloids.

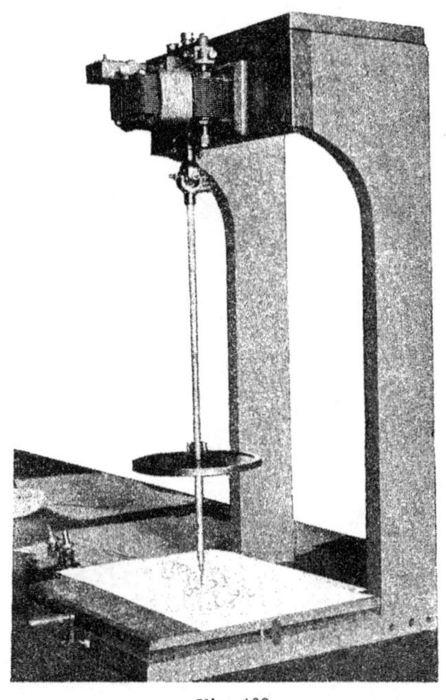
Fig. 102.

An instrument to show these properties of the motion has been constructed by the author, and is shown in Fig. 102. A heavy

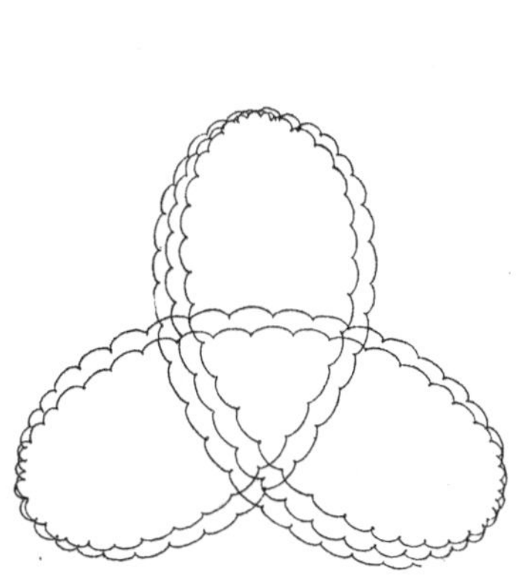
Fig. 103 a.

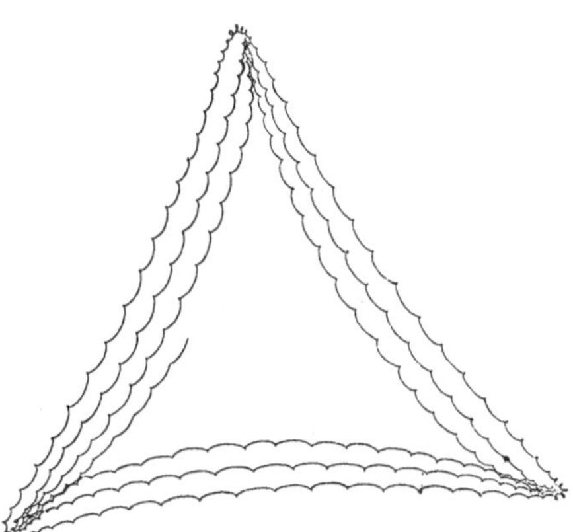
Fig. 103 b.

symmetrical disc hangs by a universal joint (Hooke's or Cardan's suspension) from a shaft which is rotated by an electric motor.

A pointed steel wire slides easily in the end of the axis of the pendulum, and draws a curve upon a plate of smoked glass which is

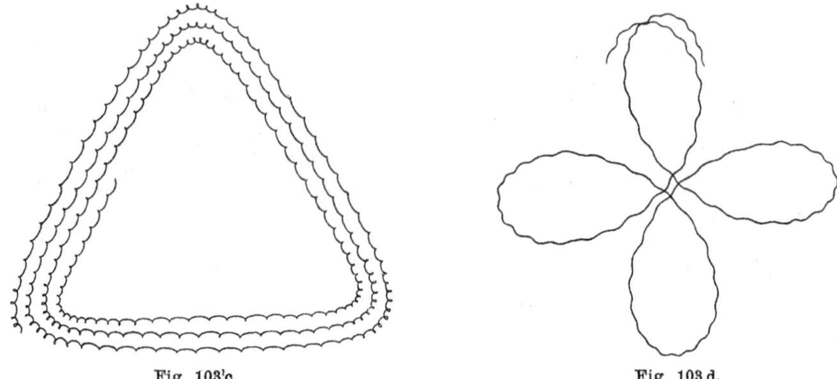

Fig. 103 c. Fig. 103 d.

brought against it by a lifting table. By means of a lantern and a

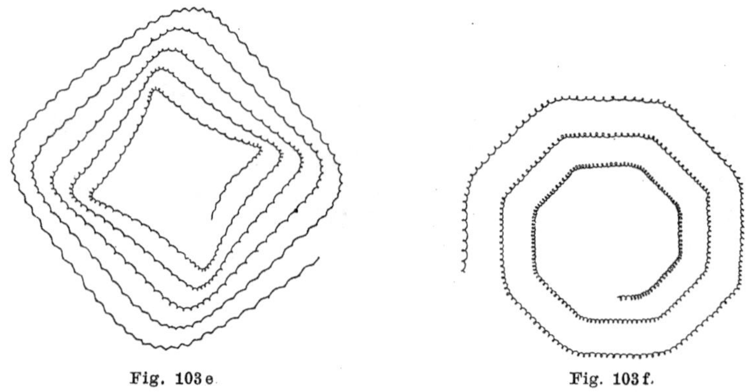

Fig. 103 e Fig. 103 f.

right angled prism the curves are projected upon a wall in the act

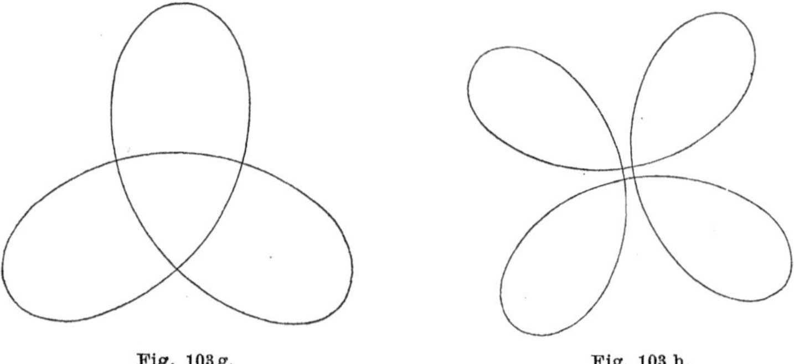

Fig. 103 g. Fig. 103 h.

of being traced. Examples of the curves obtained are shown in Fig. 103. (Figs. g, h, i are hypocycloids drawn geometrically, for comparison.) In order to compare theory with experiment, let us calculate how many revolutions in one circle go to one of the other. Let us call this ratio m. We have then from 141)

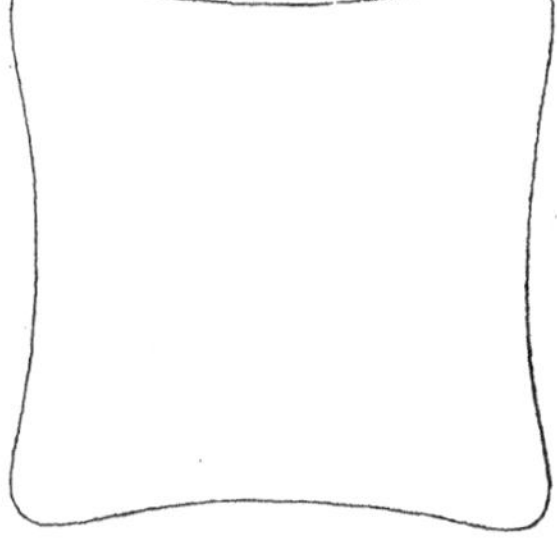

Fig. 103 i.

$$150)\qquad m = \frac{\mu}{\nu} = \pm \frac{\sqrt{b^2 + 4c} + b}{\sqrt{b^2 + 4c} - b}.$$

It is noticeable that this ratio depends only on the constants of the system and the velocity of spinning, but not on the circumstances of projection. This is shown in the figures. In each group m is made an integer, by properly adjusting the height of the disc, and the rate of spinning, which is maintained constant by stroboscopic observation. If the apex is merely drawn aside, and let go, the curves have cusps. If pushed to one side, the curves have loops, and if to the other, there are no loops, but the curve is a sort of curvilinear polygon, and if the spinning is rapid enough, there are inflexions. The ratio m is the same for the three types of curve. The slight perturbations noticeable in the figures arise from the slight looseness in the tracing point, and permit of counting the number of revolutions of the top about its axis (thus determining r),

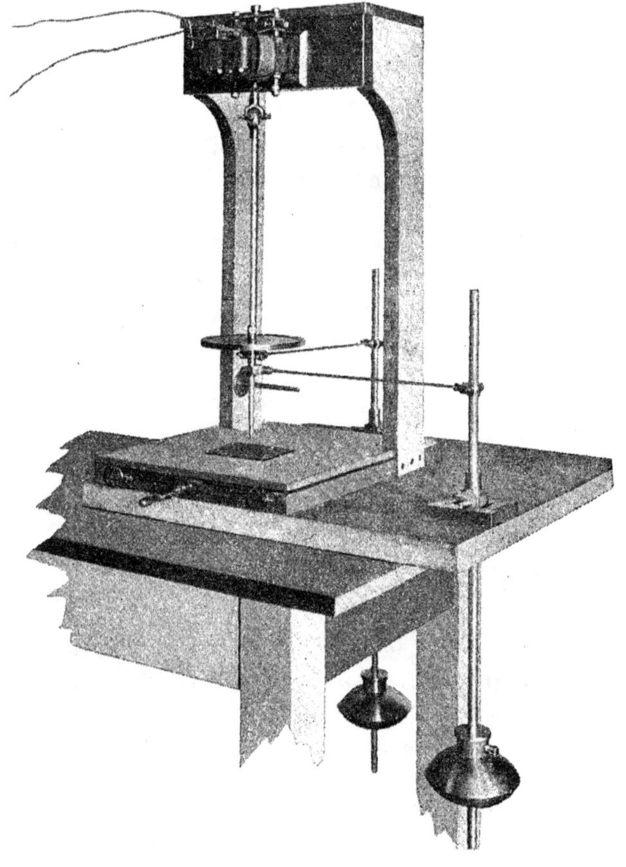

Fig. 104.

which is found to be the same for the same value of m, as may be verified on the figures.

In order to illustrate the more general case above treated, the spinning top is included in a system of two pendulums (Fig. 104), whose frequencies may be made to have any ratio to each other, so that when the top is not spinning the point describes a Lissajous's curve. The influence of the spinning on the curves is shown in Fig. 105.

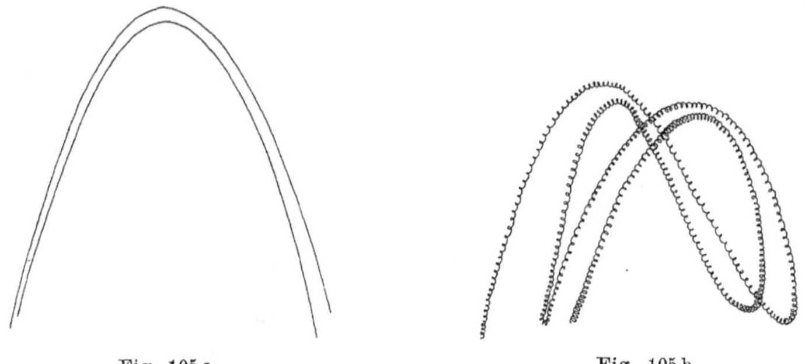

Fig. 105 a. Fig. 105 b.

An interesting application of the heavy symmetrical top is the gyroscopic horizon invented by Admiral Fleuriais of the French navy. A small top is spun upon a pivot in vacuo, in a box which is attached to a sextant. The top executes a slow movement of precession

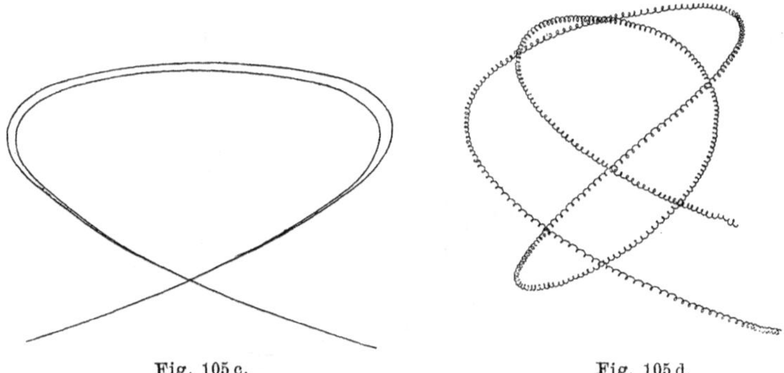

Fig. 105 c. Fig. 105 d.

about the vertical, and by means of lines ruled on two lenses which it carries, the vertical is observed, so that observations may be made when the horizon is obscured by fog.[1])

Other recent applications of the gyroscope are the Schlick anti-rolling control for ships, and the Brennan mono-rail car. For the theory of the former see Föppl, Dynamik, Bd. VI. An excellent treatment is also given in Crabtree, "The Gyroscope".

1) Schwerer. *L'horizon gyroscopique dans le vide de M. le Contre-Amiral Fleuriais.* Annales Hydrographiques. 1896.

95. Top Equations deduced by Jacobi's Method. We will conclude the treatment of the top by deducing the equations of motion by the method of Jacobi, § 41. Since we have for the kinetic energy,

$$77)\qquad T = \frac{1}{2}\{A(\vartheta'^2 + \sin^2\vartheta \cdot \psi'^2) + C(\varphi' + \cos\vartheta \cdot \psi')^2\}$$

and for the momenta

$$p_\vartheta = A\vartheta', \quad p_\psi = A\sin^2\vartheta \cdot \psi' + C\cos\vartheta\,(\varphi' + \cos\vartheta \cdot \psi'),$$
$$p_\varphi = C(\varphi' + \cos\vartheta \cdot \psi'),$$

we obtain at once

$$151)\qquad \vartheta' = \frac{p_\vartheta}{A}, \quad \psi' = \frac{p_\psi - p_\varphi \cos\vartheta}{A\sin^2\vartheta},$$
$$\varphi' = \frac{p_\varphi}{C} - \frac{\cos\vartheta\,(p_\psi - p_\varphi\cos\vartheta)}{A\sin^2\vartheta}.$$

Forming the sum of products of corresponding velocities and momenta, we obtain the energy, and also the Hamiltonian function,

$$152)\qquad H = T + W = \frac{1}{2}\left\{\frac{p_\vartheta^2}{A} + \frac{p_\varphi^2}{C} + \frac{(p_\psi - p_\varphi\cos\vartheta)^2}{A\sin^2\vartheta}\right\} + Wl\cos\vartheta.$$

From this we form the Hamiltonian equation § 41, 99),

$$153)\qquad \frac{\partial S}{\partial t} + \frac{1}{2}\left\{\frac{1}{A}\left(\frac{\partial S}{\partial \vartheta}\right)^2 + \frac{1}{C}\left(\frac{\partial S}{\partial \varphi}\right)^2 + \frac{1}{A\sin^2\vartheta}\left(\frac{\partial S}{\partial \psi} - \frac{\partial S}{\partial \varphi}\cos\vartheta\right)^2\right\}$$
$$+ Wl\cos\vartheta = 0.$$

We find, as in the problems of § 41, that this is satisfied by a linear function of t, φ, ψ, plus a function Θ of ϑ, which we will determine. We shall obtain the result in the notation of § 90 if we put

$$154)\qquad S = -ht + A(b\varphi + \beta\psi + \Theta).$$

Iuserting in the differential equation 153), we obtain

$$155)\qquad -h + \frac{1}{2}\left\{A\Theta'^2 + \frac{A^2 b^2}{C} + \frac{A}{\sin^2\vartheta}(\beta - b\cos\vartheta)^2\right\} + Wl\cos\vartheta = 0,$$

from which

$$156)\qquad \Theta'^2 = \frac{2h}{A} - \frac{Ab^2}{C} - \frac{2Wl}{A}\cos\vartheta - \frac{(\beta - b\cos\vartheta)^2}{\sin^2\vartheta} = F(\vartheta).$$

Accordingly we have the solution,

$$157)\qquad S = -ht + A\left(b\varphi + \beta\psi + \int\sqrt{F(\vartheta)}\,d\vartheta\right).$$

The integrals are obtained by differentiating by the arbitrary constants, h, b, β,

$$\frac{\partial S}{\partial h} = -t + \int \frac{d\vartheta}{\sqrt{F(\vartheta)}} = c_1,$$

$$158)\quad \frac{\partial S}{\partial \beta} = A\left(\psi - \int \frac{(\beta - b\cos\vartheta)\,d\vartheta}{\sin^2\vartheta\sqrt{F(\vartheta)}}\right) = c_2,$$

$$\frac{\partial S}{\partial b} = A\left(\varphi + \int \left\{-\frac{A b}{C} + \frac{\cos\vartheta(\beta - b\cos\vartheta)}{\sin^2\vartheta}\right\}\frac{d\vartheta}{\sqrt{F(\vartheta)}} = c_3.$$

Bearing in mind that $F(\vartheta) = \frac{f(z)}{\sin^2\vartheta}$, and that $\frac{A b}{C} = r$, we see that the first equation is the integral of equation 78), the second of 79), and the third of 80).

96. Rotation of the Earth. Precession and Nutation. Since the earth is not an exact sphere, it is not centrobaric, that is the direction of the resultant of the attraction of its various parts on a distant point does not pass through its center of mass. Or, in other words, the attraction of a distant mass-point, not passing through the center of mass of the earth, possesses a moment about it, which tends to tilt the earth's axis. The sun and moon are so nearly spherical that they may be considered as concentrated at their respective centers of mass. One of them, placed at M (Fig. 106), attracting the nearer portions of the earth more strongly than the more distant ones, tends to tip the earth's axis more nearly vertical in the figure, and it is seen that this is the same in whichever side of the earth the body lies. Thus the sun always tends to make the earth's axis more nearly perpendicular to the ecliptic, except when the sun lies on the earth's equator, that is at the equinoxes. The deflecting moment thus always tends to cause a motion of precession in the same direction, the tendency being greatest at the solstices, and disappearing at the equinoxes. The moon, which moves nearly in the plane of the ecliptic, produces a similar effect.

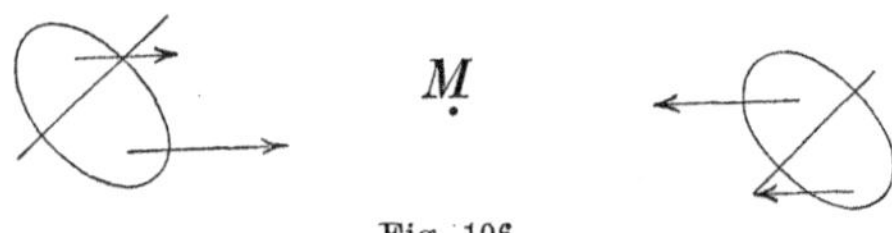

Fig. 106.

It will be shown, in § 148, that the potential of a body at a distant point, x, y, z is given very approximately by

$$159)\quad V = \frac{M}{r} + \frac{1}{2}\,\frac{(B+C-2A)x^2 + (C+A-2B)y^2 + (A+B-2C)z^2}{r^5},$$

where $r^2 = x^2 + y^2 + z^2$, and A, B, C, are the principal moments of inertia of the body. If the distant point is the center of the sun, whose mass is m, the force exerted by the earth on the sun is

$$160)\qquad X = \gamma m\frac{\partial V}{\partial x},\quad Y = \gamma m\frac{\partial V}{\partial y},\quad Z = \gamma m\frac{\partial V}{\partial z}.$$

But this is equal and opposite to the action of the sun upon the earth, the moment of which about the earth's center of mass is accordingly

161)
$$L = -(yZ - zY),\quad M = -(zX - xZ),\quad N = -(xY - yX).$$

Differentiating the expression 159), since x appears both explicitly, and implicitly in r, and $\frac{\partial r}{\partial x} = \frac{x}{r}$,

162)
$$\frac{\partial V}{\partial x} = \frac{\partial V}{\partial r}\frac{x}{r} + \frac{(B + C - 2A)x}{r^5},$$
$$\frac{\partial V}{\partial y} = \frac{\partial V}{\partial r}\frac{y}{r} + \frac{(C + A - 2B)y}{r^5},$$
$$\frac{dV}{\partial z} = \frac{\partial V}{\partial r}\frac{z}{r} + \frac{(A + B - 2C)z}{r^5},$$

and inserting in 161),

163)
$$L = \frac{3\gamma m(C - B)yz}{r^5},$$
$$M = \frac{3\gamma m(A - C)zx}{r^5},$$
$$N = \frac{3\gamma m(B - A)xy}{r^5}.$$

We may now insert these in Euler's equations, so that, if x, y, z, the coordinates of the sun, are given as functions of the time, the earth's motion may be found. Considering the earth to be symmetrical about its axis of figure, we put $A = B$, so that $N = 0$, and the third equation gives $r = const.$, as in the case of the top. It is however more convenient for our purpose to use, instead of Euler's equations a set of equations proposed by Puiseux, Résal, and Slesser, in which we take for axes, as suggested in § 84, the axis of symmetry, and two axes perpendicular to it, that is, lying in the equator, and moving in the earth. We have, since we are dealing with principal axes

164)
$$H_x = Ap, \quad H_y = Aq, \quad H_z = Cr,$$

which are to be inserted in equations 29), § 84, where we are to put the velocities with which the moving axes turn about themselves, which we will call p_0, q_0, r_0, so that our equations are

165)
$$\frac{dH_x}{dt} + q_0 H_z - r_0 H_y = L,$$
$$\frac{dH_y}{dt} + r_0 H_x - p_0 H_z = M,$$
$$\frac{dH_z}{dt} + p_0 H_y - q_0 H_x = N.$$

If we choose as X-axis the line of nodes, or intersection of the equator with the ecliptic, or plane of the sun's orbit about the earth, we have, in 65), $\varphi = 0$, so that Euler's geometric equations become simply,

166) $$p_0 = \frac{d\vartheta}{dt}, \quad q_0 = \frac{d\psi}{dt}\sin\vartheta, \quad r_0 = \frac{d\psi}{dt}\cos\vartheta.$$

We have also $p = p_0$, $q = q_0$, while r is not equal to r_0. Inserting in the third equation 165), we have $C\frac{dr}{dt} = 0$, $r = const. = \Omega$, where Ω is the angular velocity of the earth's daily rotation.

We shall content ourselves with an approximate solution of the equations, which may be obtained by neglecting the squares and products of small quantities. Observations show that $\frac{d\psi}{dt}$ and $\frac{d\vartheta}{dt}$ are small, $\left(\frac{d\psi}{dt} = 50'',37 \text{ per year}\right)$, so that we may neglect $r_0 q_0$, $r_0 p_0$. Thus our equations 165) become

167) $$A\frac{dp}{dt} + C\Omega q = L, \qquad A\frac{dq}{dt} - C\Omega p = M.$$

If the sun, or other disturbing body, did not move with respect to the axes of X, Y, Z, then L, M would be constant, and the equations would be satisfied by constant values of p, q,

168) $$p = -\frac{M}{C\Omega}, \quad q = \frac{L}{C\Omega}.$$

In order to ascertain whether these approximations are sufficient when L and M vary, let us differentiate equations 167), substituting in either the value of the first derivative of p or q from the other, obtaining

169) $$A\frac{d^2p}{dt^2} + \frac{C\Omega}{A}(C\Omega p + M) = \frac{dL}{dt}, \qquad A\frac{d^2q}{dt^2} - \frac{C\Omega}{A}(-C\Omega q + L) = \frac{dM}{dt}.$$

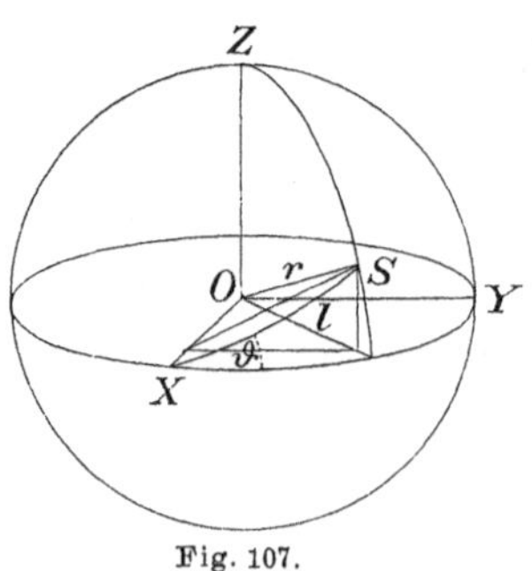

Fig. 107.

We have now to find the values of L, M in terms of the motion of the sun.

If l be the longitude of the sun, that is the angle its radius vector OS makes with the X-axis, we have, passing a plane through the sun perpendicular to the X-axis (Fig. 107),

$$x = r\cos l, \quad y = r\sin l\cos\vartheta,$$
$$z = r\sin l\sin\vartheta,$$

so that, inserting in 163),

170)
$$L = \frac{3\gamma m(C-A)}{r^3}\sin^2 l \sin\vartheta\cos\vartheta,$$
$$M = \frac{3\gamma m(A-C)}{r^3}\sin l\cos l\sin\vartheta.$$

If we suppose the sun's path relative to the earth to be a circle, described with angular velocity n, we have

$$l = l_0 + nt,$$

so t hat

171)
$$\frac{dL}{dt} = \frac{3\gamma m(C-A)}{r^3}\left\{n\cdot\sin 2l\sin\vartheta\cos\vartheta + \sin^2 l\cos 2\vartheta\frac{d\vartheta}{dt}\right\},$$
$$\frac{dM}{dt} = \frac{3\gamma m(A-C)}{r^3}\left\{n\cdot\cos 2l\sin\vartheta + \sin l\cos l\cos\vartheta\frac{d\vartheta}{dt}\right\}.$$

Now if $A = C$, there would be no motion of the earth's axis, so that $C - A$ is a small quantity of the order of $\frac{d\psi}{dt}$. The angular velocity n, though much larger, is still $365\frac{1}{4}$ times smaller than Ω, so that if we neglect its product and that of $\frac{d\vartheta}{dt}$ with $C - A$, we may neglect the right hand sides of 169). Thus the approximation 168) is justified, for differentiating, it will make $\frac{d^2p}{dt^2}$, $\frac{d^2q}{dt^2}$ negligible, so that equations 167) are satisfied. Inserting the values of p, q, L, M, in 168), we have

172)
$$\frac{d\psi}{dt} = \frac{3\gamma m}{2\Omega r^3}\frac{(C-A)}{C}\cos\vartheta(1-\cos 2l),$$
$$\frac{d\vartheta}{dt} = \frac{3\gamma m}{2\Omega r^3}\frac{(C-A)}{C}\sin\vartheta\sin 2l.$$

These are the equations for the precession and nutation. In order to integrate them approximately, we may neglect the small difference, on the right, between ϑ and its mean value, so that inserting the value of $2l = 2nt + 2l_0$, considering ϑ constant, and integrating,

173)
$$\psi = \frac{3\gamma m}{2\Omega r^3}\frac{C-A}{C}\cos\vartheta\left(t - \frac{\sin 2l}{2n}\right) + const.,$$
$$\vartheta = \frac{3\gamma m}{2\Omega r^3}\frac{C-A}{C}\sin\vartheta\frac{\cos 2l}{2n} + const.$$

We thus find the motion to be a regular precession, of amount

174)
$$\psi' = \frac{3\gamma m}{2\Omega r^3}\frac{C-A}{C}\cos\vartheta,$$

together with a nutation in an ellipse (compare § 93), whose period is one-half that of the revolution of the disturbing body.

By means of observations of the value of the precession, we may thus obtain the ratio of $\frac{(C-A)}{C}$. We see that the forces causing precession are proportional to $\frac{m}{r^3}$. On account of the nearness of the moon, therefore, and in spite of its small mass, the precession produced by the moon is somewhat greater than that due to the sun. Since the moon's orbit departs but little from the plane of the ecliptic the precession due to the moon may be calculated approximately by the above formulae, and compounded with that due to the sun.

97. Top on smooth Table. Having treated in detail the motion of a body with one point fixed, and three degrees of freedom, it remains to consider the motion of bodies which, like the ordinary top, spin upon a table or other surface. We must now consider the reaction between the body and the surface, and we have to distinguish between the ideal case of perfectly smooth, or frictionless bodies, where the reaction is normal, and bodies between which there is friction, so that the reaction is not normal. We will consider the first case. Let us examine the motion of a symmetrical top, spinning on a sharp point resting on a smooth horizontal plane. The top has five degrees of freedom, its position being defined as before by the three angles ϑ, ψ, φ, and in addition, by the coordinates x, y, of the center of mass, the z-coordinate being given by

$$z = l\cos\vartheta.$$

Since the only force which we have not already considered is the reaction, which has no horizontal component, the horizontal component of the acceleration of the center of mass vanishes, so that its motion is in a straight line with constant velocity. It therefore remains only to determine the motion of rotation. This being independent of the horizontal motion just found, we may consider the latter to vanish, so that the center of mass will be supposed to move in a vertical line. The motion thus becomes one of three freedoms, and we shall treat it by Lagrange's method as before. By the principle of § 32, 50), the kinetic energy is equal to that which the body would have if concentrated at its center of mass,

$$\frac{1}{2}Mz'^2 = \frac{1}{2}Ml^2\sin^2\vartheta\cdot\vartheta'^2,$$

plus that which it would have if it performed its motion of rotation about the center of mass supposed at rest. If then A and C denote the moments of inertia about the *center of mass* (in § 90 they were the moments about the fixed point), we have

$$175)\quad T = \frac{1}{2}[Ml^2\sin^2\vartheta\cdot\vartheta'^2 + A(\sin^2\vartheta\cdot\psi'^2 + \vartheta'^2) + C(\varphi' + \psi'\cos\vartheta)^2].$$

The potential energy is as before $Wl\cos\vartheta$. Consequently the only difference in the problem from that treated in § 90 is in the extra term in ϑ', $Ml^2\sin^2\vartheta\cdot\vartheta'^2$ in the kinetic energy. Carrying out the various steps of §§ 90, 91, we find instead of the first equation 76) the equation

$$176)\qquad \vartheta'^2\left(1+\frac{Ml^2}{A}\sin^2\vartheta\right)+\sin^2\vartheta\cdot\psi'^2=\alpha-a\cos\vartheta$$

and putting $z=\cos\vartheta$,

$$177)\qquad \left(\frac{dz}{dt}\right)^2=\frac{f(z)}{1+\frac{Ml^2}{A}(1-z^2)}=\frac{a}{-\frac{Ml^2}{A}}\,\frac{(z-z_1)(z-z_2)(z-z_3)}{(z-z_4)(z-z_5)},$$

where we denote the roots of the denominator by z_4, z_5. It is to be noticed that they lie outside the interval 1, -1, for evidently the coefficient of ϑ'^2 in 176) cannot vanish for real values of ϑ.

The square of $\frac{dz}{dt}$ being now the quotient of two polynomials in z, z is a *hyperelliptic* function of t. We may however, without a knowledge of these functions, treat the problem just as we did the former one, and we shall find that the top in general rises and falls between two of the roots of the numerator, and that the motion resembles the motion already discussed. The path of the peg has loops, cusps, or inflexions, according to the initial conditions, as before, while the regular precession and the small oscillations may be investigated as before. Whereas accordingly the functional relations involved are considerably different, physically this motion, which is that of the common top, closely resembles that already studied.

98. Effect of Friction. Rising of Top. We have now to take account of the effect of friction. Here we have in addition to the normal component of the reaction a tangential component called the force of friction, and the ordinary law assumed is that the tangential component is equal to the normal component multiplied by a constant depending on the nature of the two surfaces in contact, called the coefficient of friction. If the friction is less than a certain amount, the two surfaces will slide one upon the other, and the direction of the friction will be such as to oppose the sliding, being in the direction of the relative motion of the points instantaneously in contact. The bodies are then said to be "imperfectly rough". If the friction is greater than a definite amount, it will prevent the sliding, and there is then no relative motion of the points of contact, so that there is a constraint due to the friction, which is expressed by an equation stating that the velocities of the points of the two surfaces in contact are equal. If one of the surfaces is at rest, as is usually the case, the instantaneous axis then always passes through the point of contact. If it is in the tangent plane, the motion is said to be pure rolling, and the bodies act as if "perfectly rough". If the instantaneous angular velocity has a normal component, this is known as *pivoting*, and is also resisted by a frictional moment. The pivoting friction is however usually neglected where the surfaces are supposed to touch at a single point. The conception of perfect roughness, involving the absolute prevention of slipping under all circumstances is as far from the truth as that of perfect smoothness, nevertheless slipping may often cease in actual motions, so that motions of perfect rolling, whether or not accompanied by pivoting, are important in practice. For instance, a bicycle wheel under normal circumstances rolls and pivots, if it slips the consequences may be serious.

In the following sections we shall consider the methods of treating various cases of friction. We may however, without calculation, consider the effect of imperfect friction on the motion of the top spinning on the table. Let P (Fig. 108) represent the peg, no longer considered as a sharp point. Let OH represent the angular momentum at the center of mass O. The friction is in the direction F, opposite to the motion of the point of contact of the peg with the table. The moment of this force with respect to the center of mass is perpendicular to the plane OF, or K. Thus the end of OH moves in the direction of K, that is rises. Thus the effect of friction is to make the top rise toward a vertical position. When it has reached that, it "sleeps" and the friction has become merely pivoting friction, tending to stop the motion. We have before seen that under conservative forces, the top would never become vertical except instantaneously by oscillation.

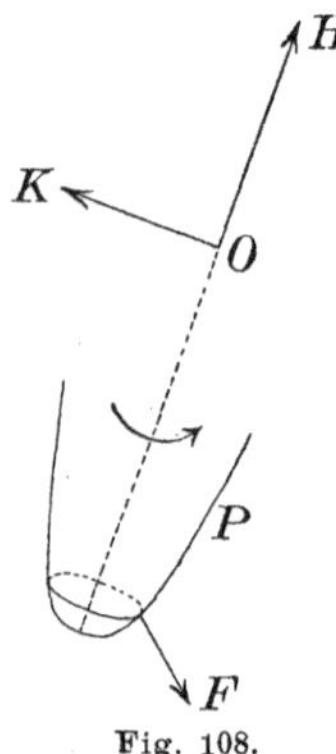

Fig. 108.

The effect of friction on the Maxwell top may be most easily seen from the fact that the friction tends to *stop* the spinning, accordingly it causes a moment which is represented by a vector opposite in direction to ω, Figs. 83a, b. Compounding this vector with H we see that the moment of momentum vector H tends to move away from the axis of the two cones in Fig. 83b while it tends towards it in Fig 83a, thus the trace of the invariable axis (as it would be but for friction), instead of being an ellipse, is a spiral winding outwards in the former case, and inwards in the latter, as is shown by the arrows in Fig. 83.

99. Motion of a Billiard Ball. We will now treat the problem of the motion of a sphere on a horizontal plane, taking account of friction. The friction of sliding is supposed to be a force of magnitude $F = \mu R$ where R is the reaction between the ball and the table, and μ the coefficient of friction. F has the direction opposite to that of the motion of the point of contact of the ball and table.

If the axes of X, Y are taken horizontal, Z vertical, we have for the motion of the center of mass the equations

178)
$$\begin{aligned} M\frac{d^2x}{dt^2} &= X = F\cos(Fx) = \mu R\cos(Fx), \\ M\frac{d^2y}{dt^2} &= Y = F\cos(Fy) = \mu R\cos(Fy), \\ M\frac{d^2z}{dt^2} &= Z = R - Mg, \end{aligned}$$

and since z is constant, $R = Mg$.

Euler's dynamical equations are, since $A = B = C$,

179)
$$\begin{aligned} A\frac{dp}{dt} &= aY, \\ A\frac{dq}{dt} &= -aX, \\ A\frac{dr}{dt} &= 0, \end{aligned}$$

a being the radius of the sphere. To determine the direction of F we have

180)
$$\frac{X}{Y} = \frac{v_x}{v_y},$$

where v_x, v_y are the velocities of the lowest point of the sphere,

181)
$$v_x = \frac{dx}{dt} - aq, \quad v_y = \frac{dy}{dt} + ap.$$

Differentiating these equations, and making use of 179),

182)
$$\begin{aligned} \frac{dv_x}{dt} &= \frac{d^2x}{dt^2} - a\frac{dq}{dt} = \frac{X}{M} + \frac{a^2}{A}X, \\ \frac{dv_y}{dt} &= \frac{d^2y}{dt^2} + a\frac{dp}{dt} = \frac{Y}{M} + \frac{a^2}{A}Y. \end{aligned}$$

Dividing one of these by the other, and using 180),

183)
$$\frac{dv_x}{dv_y} = \frac{X}{Y} = \frac{v_x}{v_y},$$

from which

$$\frac{dv_x}{v_x} = \frac{dv_y}{v_y}.$$

Integrating we have

184) $$\frac{v_x}{v_y} = const = \frac{X}{Y}.$$

Thus we find that F makes a constant angle with the axes of coordinates, and since it has the constant magnitude μMg the center of the sphere experiences a constant acceleration, and describes a parabola.

If the center of the sphere starts to move with the velocities V_x, V_y and with a "twist", whose components are p_0, q_0, r_0, we have, integrating 179), since X, Y are constant,

185) $$p = p_0 + \frac{aY}{A} t, \quad q = q_0 - \frac{aX}{A} t, \quad r = r_0.$$

Integrating the equations for the center of mass

186) $$\frac{dx}{dt} = V_x + \frac{X}{M} t, \quad \frac{dy}{dt} = V_y + \frac{Y}{M} t.$$

Inserting in 181) we find for v_x, v_y

187) $$v_x = V_x + \frac{X}{M} t - a q_0 + \frac{a^2}{A} X t,$$
$$v_y = V_y + \frac{Y}{M} t + a p_0 + \frac{a^2}{A} Y t,$$
$$\frac{v_x}{v_y} = \frac{X}{Y} = \frac{V_x - a q_0 + \frac{X}{M}\left(1 + \frac{a^2}{A}\right) t}{V_y + a p_0 + \frac{Y}{M}\left(1 + \frac{a^2}{A}\right) t}.$$

Accordingly,

188) $$\frac{X}{Y} = \frac{V_x - a q_0}{V_y + a p_0}, \quad \sqrt{X^2 + Y^2} = \mu M g,$$

189) $$X = -\mu M g \frac{V_x - a q_0}{\sqrt{(V_x - a q_0)^2 + (V_y + a p_0)^2}},$$
$$Y = -\mu M g \frac{V_y + q p_0}{\sqrt{(V_x - a q_0)^2 + (V_y + a p_0)^2}}.$$

Since v_x, v_y are linearly decreasing functions of the time, whose ratio is constant, they vanish at the same time

186) $$t = \frac{\sqrt{(V_x - a q_0)^2 + (V_y + a p_0)^2}}{\mu g \left(1 + \frac{M a^2}{A}\right)}.$$

The sliding then ceases. Obviously it cannot change sign, so that the above solution ceases to hold. The ball now rolls without sliding, and we have always, at subsequent instants, the equations of constraint

$$v_x = v_y = 0,$$

$$\frac{dx}{dt} = aq, \quad \frac{dy}{dt} = -ap,$$

190) $$\frac{X}{M} = \frac{d^2x}{dt^2} = a\frac{dq}{dt} = -\frac{a^2}{A}X,$$

$$\frac{Y}{M} = \frac{d^2y}{dt^2} = -a\frac{dp}{dt} = -\frac{a^2}{A}Y.$$

From this we obtain

$$X\left(\frac{1}{M} + \frac{a^2}{A}\right) = 0, \quad Y\left(\frac{1}{M} + \frac{a^2}{A}\right) = 0,$$

so that $X = Y = 0$, and the ball moves uniformly in a straight line.

In reality there is always a certain friction of *pivoting*, causing a moment about the normal, but this would only affect the rotation component r, which would not affect the motion of the center of the ball.

100. Pure Rolling. The preceding problem has illustrated both sliding friction and pure rolling. The treatment of the latter is interesting on account of a peculiarity in the nature of rolling constraint which makes the ordinary treatment of Lagrange's equations require modification. We shall accordingly first present the application of Euler's equations to this subject, but before doing so, we will treat by means of results already obtained one of the most important practical problems, which illustrates the steering of the bicycle, namely the rolling of a hoop or of a coin upon a rough horizontal plane.

As the hoop rolls, if its plane is not vertical, it tends to fall, and thus to change the direction of its axis of symmetry. The falling motion developes a gyroscopic action, which causes the hoop to pivot about the point of contact, so that the path described on the table is not straight but curved. The pivoting motion, like the precession of the top, tends to prevent the falling, and to this is added the effect of the centrifugal force due to the curvilinear motion of the center of mass. Thus the hoop automatically *steers* itself so as to prevent falling, and a bicycle left to itself does the same thing.

Let the position of the hoop be defined by the coordinates of its center of mass, and by the angles ϑ, ψ, φ of § 90, ϑ being the inclination to the vertical of the axis of symmetry, or normal to the plane of the hoop at its center. We will examine the conditions for a regular precession, in which ϑ, φ', ψ' being constant, the center of mass and the point of contact of the hoop with the table evidently describe circles. In this case we have for the moment about the center of mass of the forces tending to increase ϑ, by 82),

$$191)\quad P_\vartheta = -A\psi'^2 \sin\vartheta \cos\vartheta + C(\varphi' + \psi' \cos\vartheta)\psi' \sin\vartheta.$$

The forces which act to change ϑ are, the weight of the hoop, which has zero moment about the center of mass, and the reaction of the table. Let R, Fig. 109, (an edge view of the hoop) represent the vertical reaction, F the horizontal component due to friction, which is normal to the path of the point of contact, the tangential component disappearing on account of the assumed constancy of the velocity of rolling, as in the case of the rolling sphere. We accordingly have, taking moments,

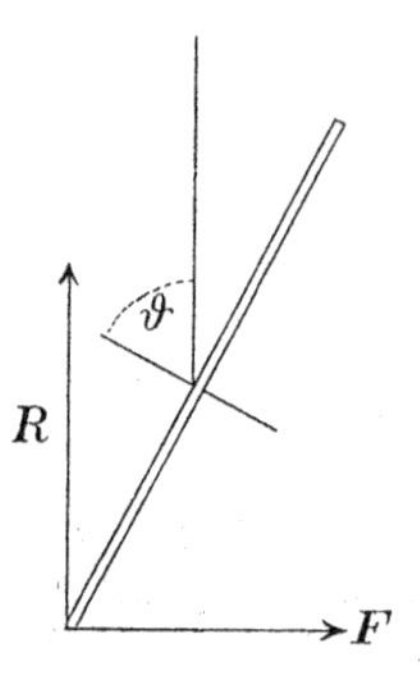

Fig. 109.

$$192)\quad P_\vartheta = Fa \sin\vartheta - Ra \cos\vartheta,$$

a being the radius of the hoop. But considering the motion of the center of mass, which is uniform circular motion, and supposing all the forces there applied, since there is no vertical motion, the resultant vertical component vanishes, or $R = Mg$ and the horizontal component balances the centrifugal force, so that

$$193)\quad F = Mb\psi'^2,$$

where b is the radius of the circle described by the center of mass.

Beside the dynamical equation we have the equation of constraint describing the rolling. Since there is no slipping, the rate at which the center of mass advances in its path is

$$194)\quad ar = a(\varphi' + \psi' \cos\vartheta).$$

But this is also, from the circular motion, equal to $-b\psi'$. From the equation of constraint,

$$195)\quad a(\varphi' + \psi' \cos\vartheta) = -b\psi',$$

we may express φ' in terms of ψ'. Doing this, aud inserting in 191), 192), we obtain,

$$-A\psi' \sin\vartheta \cos\vartheta - \frac{b}{a} C\psi'^2 \sin\vartheta = a(Mb\psi'^2 \sin\vartheta - Mg \cos\vartheta)$$

or

$$196)\quad \psi'^2\{Aa \sin\vartheta \cos\vartheta + b(C + Ma^2)\sin\vartheta\} = Mga^2 \cos\vartheta$$

as the equation for the steady motion, connecting the inclination of the hoop, the radius of the path, and the velocity of its description. In order to make the hoop roll in this manner, the proper velocity of pivoting ψ', as well as that of rolling φ', must be initially imparted to it. It is to be observed, either by considering this example, or from the results of § 90, that in order to prevent the hoop from falling, it must be steered, or given a pivoting movement, towards the side to which it tends to fall, and this is the practical manner of steering a bicycle. It is to be remarked that in steering the bicycle by the rider, the centrifugal force plays a greater part than the gyroscopic action of the wheel. For a treatment of this subject, the reader is referred to Appell, *Traité de Mécanique Rationelle*, and *Les Mouvements de Roulement en Dynamique*, and to papers by Bourlet, Carvallo, and Boussinesq.

We will now consider the general treatment of the motion of a body bounded by a surface of revolution, and dynamically symmetrical about the axis of revolution, rolling without sliding on a rough horizontal plane. We shall follow the method and notation of Appell. Let us take as axes, with origin at the center of mass of the body, as in § 90, a set of moving axes turning about themselves with angular velocities p_0, q_0, r_0, of which the Z-axis is the axis of revolution, the Y-axis the horizontal axis in the equator of the body (the line of nodes of § 90) and the X-axis directed toward the ground in the vertical plane containing the Z-axis (Fig. 110). We have accordingly to put in Euler's geometrical equations, 65), $\varphi = -\frac{\pi}{2}$, so that

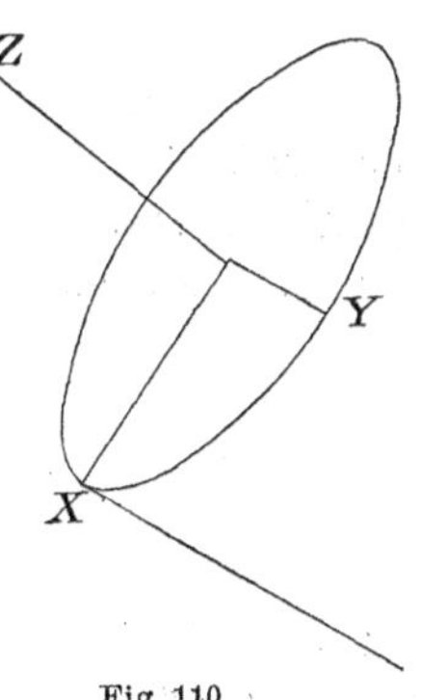

Fig. 110.

197) $$p_0 = -\psi' \sin \vartheta, \quad q_0 = \vartheta', \quad r_0 = \psi' \cos \vartheta.$$

These are connected with the rotation of the body by the relation

$$p_0 = p, \quad q_0 = q, \quad r_0 = r - \varphi'$$

(φ not being constant for the *body*).

For the motion of the center of mass, we have the components of the weight of the body

198) $$X = Mg \sin \vartheta, \quad Y = 0, \quad Z = -Mg \cos \vartheta,$$

together with the unknown components of the reaction, which we will call R_x, R_y, R_z. The resultant is to be equated to the product of the mass by the acceleration of the center of mass, using the method of § 77 for moving axes. If v_x, v_y, v_z are the components of the velocity of the center of mass along the instantaneous positions of the moving axes, we have accordingly, substituting v_x, v_y, v_z for x, y, z, in 128) § 77,

$$199)\qquad \begin{aligned} M\Big(\frac{dv_x}{dt} + q_0 v_z - r_0 v_y\Big) &= R_x + Mg\sin\vartheta,\\ M\Big(\frac{dv_y}{dt} + r_0 v_x - p_0 v_z\Big) &= R_y,\\ M\Big(\frac{dv_z}{dt} + p_0 v_y - q_0 v_x\Big) &= R_z - Mg\cos\vartheta. \end{aligned}$$

For the rotation we have, as in § 96, 165), since $p = p_0$, $q = q_0$,

$$200)\qquad \begin{aligned} A\frac{dq}{dt} + (Cr - Ar_0)q &= -zR_y,\\ A\frac{dp}{dt} - (Cr - Ar_0)p &= zR_x - xR_z,\\ C\frac{dr}{dt} &= xR_y. \end{aligned}$$

We have finally, as the conditions for rolling and pivoting, the equations stating that the velocity of the point of contact with the plane (whose coordinates are x, y, z) is at rest.

$$201)\qquad \begin{aligned} v_x + qz &= 0,\\ v_y + rx - pz &= 0,\\ v_z - qx &= 0. \end{aligned}$$

The coordinates x, z of the point of contact are obtained as known functions of ϑ from the equation of the meridian of the body. We have accordingly the eighteen equations 197—201) between the eighteen quantities

$$v_x,\ v_y,\ v_z,\ p,\ q,\ r,\ p_0,\ q_0,\ r_0,\ X,\ Y,\ Z,\ \vartheta,\ \psi,\ \varphi,\ R_x,\ R_y,\ R_z,$$

or just enough to determine them. The differential equations are all of the first order.

The reactions may be at once eliminated from the equations 199), 200). By differentiating 201) we may eliminate the derivatives of v_x, v_y v_z from 199). In doing this, however, we introduce the derivatives of x, z, which are functions of ϑ, so that in general the equations become complicated. We shall therefore confine ourselves to the case of a body rolling on a sharp edge, like a circular cylinder with a plane bottom, or a hoop or disk. We then have x, z constants,

$$x = a, \quad z = c,$$

where a is the radius of the circular edge, c the distance of the center of mass from its plane, which is zero in the case of the hoop.

The equations of the motion of the center of mass thus become

$$
\begin{aligned}
&M\left(-c\frac{dq}{dt}+aq^2+r_0(ar-cp)\right)=R_x+Mg\sin\vartheta,\\
202)\quad &M\left(\ \ c\frac{dp}{dt}-a\frac{dr}{dt}-cr_0q-apq\right)=R_y,\\
&M\left(\ \ a\frac{dq}{dt}+p(cp-ar)+cq^2\right)=R_z-Mg\cos\vartheta.
\end{aligned}
$$

We may verify that these equations are satisfied by the steady motion in which $q=0$, p and r are constants. For the case of the hoop, in which $c=0$, they thus become

$$
203)\quad
\begin{aligned}
Mar_0r&=R_x+Mg\sin\vartheta,\\
0&=R_y,\\
-Mapr&=R_z-Mg\cos\vartheta,
\end{aligned}
$$

while the second of 200) becomes

$$204)\qquad (Cr-Ar_0)p=aR_z.$$

Eliminating R_z we have the equation for the steady motion

$$205)\qquad (Cr-Ar_0)p+aM(apr-g\cos\vartheta)=0$$

which, on inserting the values of p, r, and r_0 from 197) is the same as equation 196). From the first equation 203) we may calculate R_x from which the tendency to slip, $R\cos\vartheta$ is found. If this is greater than $Mg\mu$ the hoop will slip. The slipping of the bicycle may be similarly dealt with.

In order to treat the general motion, let us eliminate R_y from the first and third of equations 200) and then from the third of 200) and the second of 202), obtaining

$$
206)\quad
\begin{aligned}
&Aa\frac{dp}{dt}+Cc\frac{dr}{dt}+(Cr-Ar_0)q=0,\\
&Ma\left\{c\frac{dp}{dt}-a\frac{dr}{dt}-(cr_0+ap)q\right\}-C\frac{dr}{dt}=0.
\end{aligned}
$$

If in these equations we substitute from 197),

$$q=\frac{d\vartheta}{dt},\quad r_0=-p\operatorname{ctn}\vartheta,$$

we find that each term contains dt in the denominator, so that we. may change the variable from t to ϑ by multiplying by $\frac{dt}{d\vartheta}$, giving

$$
207)\quad
\begin{aligned}
&Aa\frac{dp}{d\vartheta}+Cc\frac{dr}{dt}+Car+Aap\operatorname{ctn}\vartheta=0,\\
&Mac\frac{dp}{d\vartheta}-(C+Ma^2)\frac{dr}{d\vartheta}-Ma^2p+Macp\operatorname{ctn}\vartheta=0,
\end{aligned}
$$

as two equations to determine p and r as functions of ϑ. When they are thus determined, the equation of energy

208) $M(v_x^2+v_y^2+v_z^2)+A(p^2+q^2)+Cr^2=2\{h-Mg(a\sin\vartheta-c\cos\vartheta)\}$,

or by 201),

209) $$(A+Mc^2)p^2+(A+Ma^2+Mc^2)q^2+(C+Ma^2)r^2-2Macpr = 2\{h-Mg(a\sin\vartheta-c\cos\vartheta)\},$$

suffices to determine q as a function of ϑ. Thus we see that whenever ϑ returns to a former value, the circumstances of the rolling are repeated, so that the motion is periodic.

Eliminating $\frac{dp}{d\vartheta}$ from 207), we obtain

210) $$MAa^2p=-\{AC+M(Aa^2+Cc^2)\}\frac{dr}{d\vartheta}-MCacr,$$

differentiating which, we may eliminate p, obtaining for r,

211) $$\frac{d^2r}{d\vartheta^2}+\operatorname{ctn}\vartheta\frac{dr}{d\vartheta}+\frac{MCa}{AC+M(Aa^2+Cc^2)}(c\operatorname{ctn}\vartheta-a)r=0$$

a linear differential equation for r, with variable coefficients. In the case of the disk, where $c=0$, by introducing the new variable

$$x=\cos^2\vartheta,$$

we reduce the equation to the form

212) $$x(1-x)\frac{d^2r}{dx^2}+\left(\frac{1}{2}-\frac{3}{2}x\right)\frac{dr}{dx}-\frac{MCa^2r}{4A(C+Ma^2)}=0$$

which is the differential equation of Gauss,

213) $$x(1-x)\frac{d^2r}{dx^2}+[\gamma-(\alpha+\beta+1)x]\frac{dr}{dx}-\alpha\beta r=0,$$

if we put

$$\gamma=\frac{1}{2},\quad \alpha+\beta=\frac{1}{2},\quad \alpha\beta=\frac{MCa^2}{4A(C+Ma^2)}.$$

This differential equation is satisfied by the hypergeometric series

214) $$F(\alpha,\beta,\gamma,x)\equiv 1+\frac{\alpha\beta}{1\cdot\gamma}x+\frac{\alpha(\alpha+1)\beta(\beta+1)}{1\cdot2\cdot\gamma(\gamma+1)}x^2 +\frac{\alpha(\alpha+1)(\alpha+2)\beta(\beta+1)(\beta+2)}{1\cdot2\cdot3\cdot\gamma(\gamma+1)(\gamma+2)}x^3+\cdots$$

and by the theory of linear differential equations we find that the general integral is

215) $$r=c_1F(\alpha,\beta,\gamma,x)+c_2x^{1-\gamma}F(\alpha-\gamma+1,\beta-\gamma+1,2-\gamma,x),$$
$$r=c_1F\left(\alpha,\beta,\frac{1}{2},\cos^2\vartheta\right)+c_2\cos\vartheta\,F\left(\alpha+\frac{1}{2},\beta+\frac{1}{2},\frac{3}{2},\cos^2\vartheta\right),$$

where c_1 and c_2 are arbitrary constants. From this we obtain

$$p = -\frac{C + Ma^2}{Ma^2} \frac{dr}{d\vartheta},$$

and from equation 209) we obtain $q = \frac{d\vartheta}{dt}$, or $\frac{dt}{d\vartheta}$ as a function of ϑ, so that the time is given in terms of ϑ by a quadrature. The explicit completion of the solution is too complicated to be of use in investigating the motion. The equations 207) have been investigated by Carvallo by a development in series, from which the properties of the motion are investigated.

101. Lagrange's Equations applied to Rolling. Non-integrable Constraints. In the attempt to apply the method of Lagrange to the problem of rolling we are met with a peculiar difficulty, which has been the subject of researches by Vierkant and Hadamard.[1] We shall follow the treatment of the former of the rolling of a disk. Let us characterize the position of the disk by the angles ϑ, ψ, φ, as before, and in addition by the coordinates x, y with respect to a set of fixed axes in the horizontal plane, of the point of contact of the disk and plane. These five coordinates completely characterize the position of the disk, but are not all independent, on account of the constraint of rolling. If we measure ψ as the angle made by the vertical plane through the normal to the disk with the X-axis, as indicated in Fig. 111, we see that changes of ψ and ϑ do not affect the coordinates of the point of contact, but that a change $\delta\varphi$ of φ causes a shifting whose components δx, δy are given by the equations of rolling,

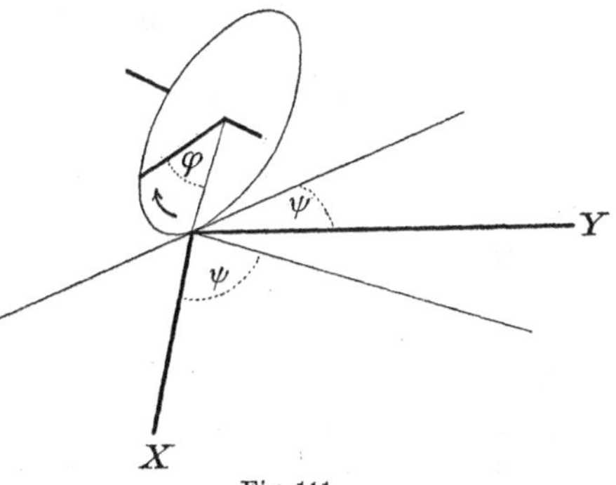

Fig. 111.

$$\text{216)} \qquad \begin{aligned} \delta x + a\,\delta\varphi \sin\psi &= 0, \\ \delta y - a\,\delta\varphi \cos\psi &= 0, \end{aligned}$$

which constitute the equations of constraint. These equations differ from any that we have heretofore met, in that they are not *integrable*, that is, they are not, like the equations 8) Chapter III, derived from equations obtained by putting certain functions of our five variables

1) Vierkant. *Über gleitende und rollende Bewegung.* Monatshefte für Math. u. Physik, 1892, p. 31.

equal to constants. We can not then, as was assumed in § 36 in deducing Lagrange's equations, use the equations of constraint to express the position of the system in terms of a less number of coordinates equal to the number of degrees of freedom, three in this case.

Moreover, we can not even use the equations connecting the velocities,

$$217)\qquad \begin{aligned} x' &= -a\varphi' \sin\psi, \\ y' &= \quad a\varphi' \cos\psi, \end{aligned}$$

to express the kinetic energy in terms of ϑ', φ', ψ' alone, as was explicitly pointed out by Vierkant. On the other hand, we must keep all the coordinates and their volocities in the expression of the kinetic energy, as if there were no constraints, and form the equation of d'Alembert's Principle as in § 37, 56), and afterwards introduce the fact that the changes of the coordinates are not all independent, by means of undetermined multipliers, as in § 25.

If ξ, η, ζ are the coordinates of the center of mass of the disk, we have for the kinetic energy,

$$218)\qquad T = \frac{1}{2} M(\xi'^2 + \eta'^2 + \zeta'^2) + \frac{1}{2} A(\vartheta'^2 + \sin^2\vartheta \cdot \psi'^2) + \frac{1}{2} C(\varphi' + \psi' \cos\vartheta)^2,$$

where

$$219)\qquad \begin{aligned} \xi &= x + a\cos\vartheta\cos\psi, \\ \eta &= y + a\cos\vartheta\sin\psi, \\ \zeta &= \quad a\sin\vartheta, \end{aligned}$$

and differentiating,

$$220)\qquad \begin{aligned} \xi' &= x' - a(\sin\vartheta\cos\psi \cdot \vartheta' + \cos\vartheta\sin\psi \cdot \psi'), \\ \eta' &= y' - a(\sin\vartheta\sin\psi \cdot \vartheta' - \cos\vartheta\cos\psi \cdot \psi'), \\ \zeta' &= \quad a\cos\vartheta \cdot \vartheta'. \end{aligned}$$

Squaring and adding, we obtain for the kinetic energy,

$$221)\qquad \begin{aligned} T = \frac{1}{2} M\{x'^2 + y'^2 + a^2(\vartheta'^2 + \cos^2\vartheta \cdot \psi'^2) \\ + 2a[-\sin\vartheta \cdot \vartheta'(x'\cos\psi + y'\sin\psi) \\ + \cos\vartheta \cdot \psi'(-x'\sin\psi + y'\cos\psi)]\} \\ + \frac{1}{2} A(\vartheta'^2 + \sin^2\vartheta \cdot \psi'^2) + \frac{1}{2} C(\varphi' + \psi'\cos\vartheta)^2. \end{aligned}$$

Forming now the equation of d'Alembert, adding equations 216) multiplied by λ and μ respectively, and equating to zero the coefficients of δx, δy, $\delta\vartheta$, $\delta\psi$, $\delta\varphi$, we obtain the equations of motion

$$222\text{a})\quad M\frac{d}{dt}\{x' + a(-\sin\vartheta\cos\psi\cdot\vartheta' - \cos\vartheta\sin\psi\cdot\psi')\} + \lambda = 0,$$

$$\text{b})\quad M\frac{d}{dt}\{y' + a(-\sin\vartheta\sin\psi\cdot\vartheta' + \cos\vartheta\cos\psi\cdot\psi')\} + \mu = 0,$$

$$\begin{aligned}\text{c})\quad &\frac{d}{dt}\{(Ma^2 + A)\vartheta' - Ma\sin\vartheta(x'\cos\psi + y'\sin\psi)\}\\ &+ (Ma^2 - A)\psi'^2\sin\vartheta\cos\vartheta\\ &+ C\sin\vartheta\cdot\psi'(\varphi' + \psi'\cos\vartheta) + Ma\cos\vartheta\cdot\vartheta'(x'\cos\psi + y'\sin\psi)\\ &+ Ma\sin\vartheta\cdot\psi'(-x'\sin\psi + y'\cos\psi) = -Mga\cos\vartheta,\end{aligned}$$

$$\begin{aligned}\text{d})\quad &\frac{d}{dt}\{(Ma^2\cos^2\vartheta + A\sin^2\vartheta)\psi' + C\cos\vartheta(\varphi' + \psi'\cos\vartheta)\\ &+ Ma\cos\vartheta(-x'\sin\psi + y'\cos\psi)\}\\ &+ Ma\sin\vartheta\cdot\vartheta'(-x'\sin\psi + y'\cos\psi)\\ &- Ma\cos\vartheta\cdot\psi'(-x'\cos\psi - y'\sin\psi) = 0,\end{aligned}$$

$$\text{e})\quad C\frac{d}{dt}(\varphi' + \psi'\cos\vartheta) + \lambda a\sin\psi - \mu a\cos\psi = 0.$$

We must observe that if we had taken account of the equations 217) in the expression 221) for the kinetic energy, before differentiating, we should have obtained quite different equations. Having performed the differentiations, however, and introduced in the equations all the reactions belonging to the different coordinates, we may now take account of the equations of constraint, thus introducing, in effect, the statement of the equilibration of some of the reactions, and causing some of the terms to drop out.

Now introducing the values of x', y', from 217) in 222), and eliminating λ, μ from 222a b, e), we obtain

$$\begin{aligned}223)\quad Ma^2\Big\{&\sin\psi\frac{d}{dt}(\sin\psi\cdot\varphi' + \sin\vartheta\cos\psi\cdot\vartheta' + \cos\vartheta\sin\psi\cdot\psi')\\ &+ \cos\psi\frac{d}{dt}(\cos\psi\cdot\varphi' - \sin\vartheta\sin\psi\cdot\vartheta' + \cos\vartheta\cos\psi\cdot\psi')\Big\}\\ &+ C\frac{d}{dt}(\varphi' + \cos\vartheta\cdot\psi') = 0,\end{aligned}$$

which, on performing the differentiations, and cancelling some of the terms, becomes

$$224)\quad (Ma^2 + C)\frac{d}{dt}(\varphi' + \cos\vartheta\cdot\psi') - Ma^2\sin\vartheta\cdot\vartheta'\psi' = 0.$$

Making corresponding simplifications in 222d) it becomes

$$\begin{aligned}225)\quad &\frac{d}{dt}\Big\{(Ma^2 + C)\cos\vartheta(\varphi' + \cos\vartheta\cdot\psi') + A\sin^2\vartheta\cdot\psi'\Big\}\\ &+ Ma^2\sin\vartheta\cdot\vartheta'\varphi' = 0.\end{aligned}$$

Since the last terms of both these equations contains ϑ', it is suggested that we change the independent variable from t to ϑ, which is done by dividing through by ϑ', giving

$$\text{226a)}\quad (Ma^2+C)\frac{d}{d\vartheta}(\varphi'+\cos\vartheta\cdot\psi') - Ma^2\sin\vartheta\cdot\psi' = 0,$$

$$\text{b)}\quad \frac{d}{d\vartheta}\{(Ma^2+C)\cos\vartheta(\varphi'+\cos\vartheta\cdot\psi') + A\sin^2\vartheta\cdot\psi'\} + Ma^2\sin\vartheta\cdot\varphi' = 0,$$

as two simultaneous equations to determine ψ' and φ' as functions of ϑ. Now observing that $\varphi'+\cos\vartheta\cdot\psi' = r$, let us multiply the first equation by $\cos\vartheta$ and subtract it from the second, obtaining, on performing the differentations and simplifying

$$\text{227)}\quad -C\sin\vartheta\cdot r + A\frac{d}{d\vartheta}(\sin^2\vartheta\cdot\psi') = 0.$$

Introducing the value of ψ' from 226a), performing the differentiations, we have finally,

$$\text{228)}\quad \frac{d^2 r}{d\vartheta^2} + \operatorname{ctn}\vartheta\frac{dr}{d\vartheta} - \frac{MCa^2}{A(Ma^2+C)}r = 0,$$

which is the same as the equation 210) obtained by Appell and Korteweg, by a totally different process.

Having obtained r, we obtain ψ' from 227), we may then obtain ϑ' from the equation of energy, and obtain the time as before.

102. Moving Axes. It is often convenient to refer the motion of a body to a set of axes which are themselves moving in space. Let us first suppose that they move parallel to themselves and that the moving origin has the coordinates ξ, η, ζ with respect to a system of parallel axes fixed in space. Let the coordinates of a point with respect to the fixed axes be x', y', z' and to the moving axes x, y, z, then

$$x' = \xi + x, \qquad y' = \eta + y, \qquad z' = \zeta + z,$$

$$\text{229)}\quad \begin{aligned} \frac{dx'}{dt} &= \frac{d\xi}{dt} + \frac{dx}{dt}, & \frac{d^2x'}{dt^2} &= \frac{d^2\xi}{dt^2} + \frac{d^2x}{dt^2},\\ \frac{dy'}{dt} &= \frac{d\eta}{dt} + \frac{dy}{dt}, & \frac{d^2y'}{dt^2} &= \frac{d^2\eta}{dt^2} + \frac{d^2y}{dt^2},\\ \frac{dz'}{dt} &= \frac{d\zeta}{dt} + \frac{dz}{dt}, & \frac{d^2z'}{dt^2} &= \frac{d^2\zeta}{dt^2} + \frac{d^2z}{dt^2}, \end{aligned}$$

showing that the velocity and acceleration of a point with respect to the fixed axes are the resultants of the velocity and acceleration of the point with respect to the moving axes, and of those of a point rigidly connected to the moving axes. We may accordingly consider the moving axes at rest, provided that, in addition to the forces impressed upon the system, we impress forces capable of producing accelerations *equal and opposite* to the actual accelerations of the origin of the moving axes.

As an example, let us consider the problem of two bodies (§ 16), which is important in the practical case of the sun and a planet, neglecting the action of the other planets. We have seen in § 32 that the center of mass of the two bodies remains at rest, while the sun moves about it, in practice however we are interested in the motion of the planets with respect to the sun. We will therefore, in order to consider the sun as at rest, apply to the whole system an acceleration equal and opposite to that possessed by the sun. Let us call the mass of the sun M, its coordinates with respect to fixed axes ξ, η, ζ, the mass of the planet m, its coordinates with respect to the fixed axes x', y', z', with respect to parallel axes through the sun x, y, z. We then have by the equations of § 16,

$$230)\qquad \begin{aligned} M\frac{d^2\xi}{dt^2} &= \gamma Mm\frac{x}{r^3} = -m\frac{d^2x'}{dt^2},\\ M\frac{d^2\eta}{dt^2} &= \gamma Mm\frac{y}{r^3} = -m\frac{d^2y'}{dt^2},\\ M\frac{d^2\zeta}{dt^2} &= \gamma Mm\frac{z}{r^3} = -m\frac{d^2z'}{dt^2}, \end{aligned}$$

while by combining these with 229) we obtain for the relative motion,

$$231)\qquad \begin{aligned} m\frac{d^2x}{dt^2} &= m\left(\frac{d^2x'}{dt^2} - \frac{d^2\xi}{dt^2}\right) = -\gamma m(M+m)\frac{x}{r^3},\\ m\frac{d^2y}{dt^2} &= m\left(\frac{d^2y'}{dt^2} - \frac{d^2\eta}{dt^2}\right) = -\gamma m(M+m)\frac{y}{r^3},\\ m\frac{d^2z}{dt^2} &= m\left(\frac{d^2z'}{dt^2} - \frac{d^2\zeta}{dt^2}\right) = -\gamma m(M+m)\frac{z}{r^3}. \end{aligned}$$

We accordingly find that the differential equations for the relative motion are the same as those for the absolute motion, except for the factor $M + m$ on the right instead of M. Thus if the sun be considered to be at rest, the first two of Kepler's laws are still valid, while the third needs the slight correction that the ratio of the cubes of the semi-axes to the squares of the times of revolution are not absolutely constant for all the planets, but proportional to the sums of masses of sun and planet. As even in the case of Jupiter, the largest, m is less than one-thousandth of M, the correction is slight.

103. Rotating Axes. Theorem of Coriolis. Suppose now that the origin is fixed; but that the moving axes revolve with an angular velocity whose projections upon their own instantaneous directions are p, q, r. Then we have found in § 77, 128) for the actual velocities of a point projected on the instantaneous directions of the axes x, y, z,

232)
$$v_x = zq - yr + \frac{dx}{dt},$$
$$v_y = xr - zp + \frac{dy}{dt},$$
$$v_z = yp - xq + \frac{dz}{dt}.$$

The first two terms, respresenting the vector-product of the angular velocity of the moving axes by the position-vector of the point, represent the components of the velocity of a point fixed to the moving axes, the last terms represent the velocity relative to the moving axes.

We might now, in order to find the components of the actual acceleration along the instantaneous positions of the moving axes, make use of equations 128), § 77, to obtain the velocity of the end of the *velocity*-vector, that is put for x, y, z the quantities v_x, v_y, v_z when on the left we should obtain a_x, a_y, a_z, as has been suggested for H in § 84 (after 29) but we shall rather choose for the sake of variety, to proceed by means of Lagrange's method to find the forces tending to increase the *relative* coordinates x, y, z. Suppose a particle of mass m to have coordinates x, y z in the moving system. Its kinetic energy is then

$$T = \frac{1}{2} m (v_x^2 + v_y^2 + v_z^2),$$

that is

233)
$$T = \frac{1}{2} m \Big[\left(\frac{dx}{dt}\right)^2 + \left(\frac{dy}{dt}\right)^2 + \left(\frac{dz}{dt}\right)^2 + 2\left\{\frac{dx}{dt}(zq - yr) + \frac{dy}{dt}(xr - zp) + \frac{dz}{dt}(yp - xq)\right\} + (zq - yr)^2 + (xr - zp)^2 + (yp - xq)^2 \Big].$$

Then the force tending to increase the coordinate x is by Lagrange's equations,

234)
$$X = \frac{d}{dt}\left(\frac{\partial T}{\partial x'}\right) - \frac{\partial T}{\partial x} = m\Big[\frac{d}{dt}\left\{\frac{dx}{dt} + (zq - yr)\right\} - r\frac{dy}{dt} + q\frac{dz}{dt} - r(xr - zp) + q(yp - xq)\Big].$$

Accordingly, the acceleration due to X is

235)
$$a_x = \frac{X}{m} = \frac{d^2x}{dt^2} + 2q\frac{dz}{dt} - 2r\frac{dy}{dt} + z\frac{dq}{dt} - y\frac{dr}{dt} - x(q^2 + r^2) + rpz + pqy.$$

This is the expression for the component of the actual acceleration of the point resolved along the instantaneous direction of the axis of X. We see that besides the relative acceleration $\frac{d^2x}{dt^2}$ it contains terms involving the relative velocities $\frac{dx}{dt}, \frac{dy}{dt}, \frac{dz}{dt}$, the angular velocities of the moving axes p, q, r and their derivatives, $\frac{dp}{dt}, \frac{dq}{dt}, \frac{dr}{dt}$.

A point *fixed* to the moving system at x, y, z would have the accelerations

$$
\begin{aligned}
a_{x0} &= z\frac{dq}{dt} - y\frac{dr}{dt} - x(q^2 + r^2) + p(qy + rz),\\
a_{y0} &= x\frac{dr}{dt} - z\frac{dp}{dt} - y(r^2 + p^2) + q(rz + px), \qquad (236)\\
a_{z0} &= y\frac{dp}{dt} - x\frac{dq}{dt} - z(p^2 + q^2) + r(px + qy).
\end{aligned}
$$

These may be called the components of acceleration of transportation (*entraînement*) or the acceleration of the *moving space.* They represent the centripetal acceleration of the transported point. (If p, q, r are constant, we have in the last two terms the ordinary expressions for centripetal acceleration, whose resultant is v^2 divided by the distance from the axis of rotation.) Beside these and the relative accelerations there are terms

$$
\begin{aligned}
J_x &= 2\left\{q\frac{dz}{dt} - r\frac{dy}{dt}\right\},\\
J_y &= 2\left\{r\frac{dx}{dt} - p\frac{dz}{dt}\right\}, \qquad (237)\\
J_z &= 2\left\{p\frac{dy}{dt} - q\frac{dx}{dt}\right\}.
\end{aligned}
$$

These are termed the components of the *compound centripetal acceleration.* We accordingly have for the total acceleration

$$
\begin{aligned}
a_x &= \frac{d^2x}{dt^2} + a_{x0} + J_x,\\
a_y &= \frac{d^2y}{dt^2} + a_{y0} + J_y, \qquad (238)\\
a_z &= \frac{d^2z}{dt^2} + a_{z0} + J_z,
\end{aligned}
$$

that is the actual acceleration of the point is the resultant of the *relative acceleration,* the acceleration of transportation, and of the compound centripetal acceleration. Accordingly we may consider the axes at rest if we add to the actual forces applied forces capable of producing an acceleration equal and opposite to the acceleration of transportation and the compound centripetal acceleration. This is known as Coriolis's theorem.

The resultant J, often known as the acceleration of Coriolis, is evidently perpendicular to the relative velocity whose components are $\frac{dx}{dt}, \frac{dy}{dt}, \frac{dz}{dt}$ and to the axis of p, q, r and is equal to twice the vector-product of the angular velocity of the axes and the relative velocity of the particle. It is interesting to notice that the acceleration of Coriolis arises from the presence of linear terms in the velocities, $\frac{dx}{dt}, \frac{dy}{dt}, \frac{dz}{dt}$ in the kinetic energy, the effect of which in introducing gyroscopic terms was explained in § 50 Thus a particle may be arranged to represent by its motions relatively to a uniformly revolving body, such as the earth, the motions of a system containing a gyrostat. This remark is due to Thomson and Tait.

104. Motion relatively to the Earth. Let us suppose the axes chosen are taken fixed in the earth, the origin at the center, the z-axis the axis of rotation. Let the earth rotate with the constant angular velocity Ω, which expressed in seconds is

$$\Omega = \frac{2\pi}{86{,}164.1 \text{ sec.}} = .0000729211 \text{ sec.}^{-1}$$

and is very small. Then $p = q = 0$, $r = \Omega$. The centripetal acceleration of transportation is then

$$\begin{aligned} \alpha_{x0} &= -x\Omega^2, \\ \alpha_{y0} &= -y\Omega^2, \\ \alpha_{z0} &= 0. \end{aligned}$$

Accordingly for a point *at rest* on the earth we may consider the earth at rest, provided we add to other applied forces a centrifugal force whose components are $m\Omega^2 x$, $m\Omega^2 y$. This centrifugal force is

$$239) \qquad m\Omega^2\sqrt{x^2+y^2} = m\Omega^2 R\cos\varphi,$$

where R is the radius of the earth and φ is the latitude. This is a subtractive part of g, the acceleration of gravity, which is consequently greatest at the poles, least at the equator. The *vertical* part of the centrifugal force is $m\Omega^2 R\cos^2\varphi$. This acceleration is common to all bodies at rest on the earth, and hence is included with gravity in our ordinary experiments. It need not then be further noticed. There is however to be considered the apparent compound centrifugal force, $-mJ_x$, $-mJ_y$, $-mJ_z$, which acts on bodies *in motion* relatively to the earth.

$$240) \qquad \begin{aligned} -mJ_x &= 2m\Omega\frac{dy}{dt}, \\ -mJ_y &= -2m\Omega\frac{dx}{dt}, \\ -mJ_z &= 0. \end{aligned}$$

The equations of motion of a body acted on by forces X, Y, Z are then

$$m\frac{d^2x}{dt^2} = X + 2m\Omega\frac{dy}{dt},$$

241) $$m\frac{d^2y}{dt^2} = Y - 2m\Omega\frac{dx}{dt},$$

$$m\frac{d^2z}{dt^2} = Z,$$

the terms in Ω having the usual property of gyroscopic forces. For a falling body we have, if the plane XZ is the vertical plane at the place of observation,

$$X = -mg\cos\varphi, \quad Z = -mg\sin\varphi, \quad Y = 0.$$

Then the equations are

$$\frac{d^2x}{dt^2} = -g\cos\varphi + 2\Omega\frac{dy}{dt},$$

242) $$\frac{d^2y}{dt^2} = \qquad\qquad - 2\Omega\frac{dx}{dt},$$

$$\frac{d^2z}{dt^2} = -g\sin\varphi.$$

Let us now introduce a set of axes ξ, η, ζ, with ζ vertical, ξ from north to south, η from west to east (Fig. 112). The direction cosines of the new axes are given by the table

	X	Y	Z
ξ	$\sin\varphi$	0	$-\cos\varphi$
η	0	1	0
ζ	$\cos\varphi$	0	$\sin\varphi$

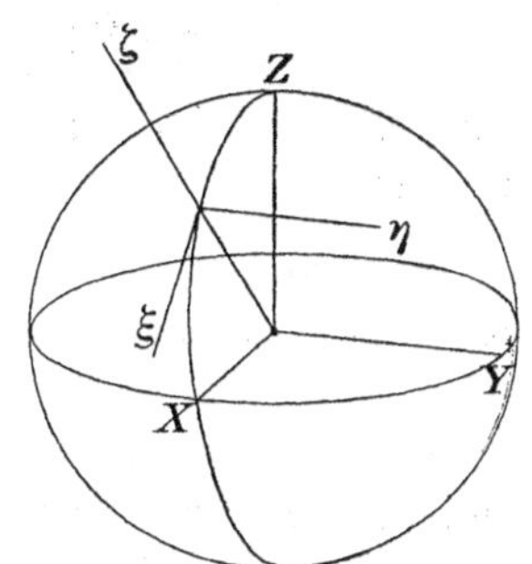

Fig. 112.

from which we have for the equations for the transformation of coordinates

243) $$\begin{aligned} x &= \xi\sin\varphi + \zeta\cos\varphi, & \xi &= x\sin\varphi - z\cos\varphi,\\ y &= \eta, & \eta &= y,\\ z &= -\xi\cos\varphi + \zeta\sin\varphi, & \zeta &= x\cos\varphi + z\sin\varphi. \end{aligned}$$

Inserting in the differential equations 242),

$$\frac{d^2\xi}{dt^2} = \quad 2\Omega\sin\varphi\frac{d\eta}{dt},$$

244) $$\frac{d^2\eta}{dt^2} = -2\Omega\left\{\frac{d\xi}{dt}\sin\varphi + \frac{d\zeta}{dt}\cos\varphi\right\},$$

$$\frac{d^2\zeta}{dt^2} = -g + 2\Omega\cos\varphi\frac{d\eta}{dt}.$$

These are linear equations with constant coefficients, and may readily be integrated. As a simple example let us consider the effect of the earth's rotation on a railway train moving with constant velocity. A railway train running with a velocity v is urged to the right with a force

$$m\frac{d^2\eta}{dt^2} = 2m\Omega v \sin\varphi.$$

The acceleration experienced by a train running 50 miles per hour in latitude 45° would be

$$\sin 45^0 \frac{4\pi \times 50 \times 160{,}933}{86{,}164.1 \times 3600}\frac{\text{cm}}{\text{sec}^2} = 0.2305\frac{\text{cm}}{\text{sec}^2} \quad \text{or a force of} \quad \frac{0.2305}{980}$$

of its weight.

Secondly consider a body falling freely. We shall assume that the body is dropped from a point in the ζ-axis with no initial velocity. Then integrating the first of equations 244) we obtain

$$\frac{d\xi}{dt} = 2\Omega \sin\varphi \cdot \eta.$$

Integrating the third

$$\frac{d\zeta}{dt} = -gt + 2\Omega \cos\varphi \cdot \eta.$$

Substituting this in the second

$$\frac{d^2\eta}{dt^2} = -2\Omega\{2\Omega \sin^2\varphi \cdot \eta - gt\cos\varphi + 2\Omega\cos^2\varphi \cdot \eta\}.$$

Integrating this, making the assumption that Ω^2 may be neglected, we have

$$\frac{d\eta}{dt} = \Omega\cos\varphi \cdot gt^2.$$

Integrating again,

$$\eta = \frac{1}{3}\Omega\cos\varphi \cdot gt^3,$$

and inserting this value in $\frac{d\zeta}{dt}$,

$$\frac{d\zeta}{dt} = -gt + \frac{2}{3}\Omega^2\cos^2\varphi \cdot gt^3.$$

Consequently we have finally

$$\zeta - \zeta_0 = -\frac{1}{2}gt^2,$$

$$\frac{d\xi}{dt} = \frac{2}{3}\Omega^2 \sin\varphi\cos\varphi \cdot gt^3,$$

$$\xi = 0,$$

to this order of approximation. We have

$$\eta = \frac{\Omega\cos\varphi}{3}\sqrt{\frac{8(\zeta_0-\zeta)^3}{g}}.$$

The particle falls to the east by an amount proportional to the square root of the cube of the height of fall and to the cosine of the latitude. This has been experimentally verified.

105. Motion of a Spherical Pendulum. We have for the pendulum the equation of constraint

$$\varphi = \frac{1}{2}(\xi^2 + \eta^2 + \zeta^2 = l^2) = 0,$$

so that to the previous equations of motion are added terms

$$\lambda \frac{\partial \varphi}{\partial \xi}, \quad \lambda \frac{\partial \varphi}{\partial \eta}, \quad \lambda \frac{\partial \varphi}{\partial \zeta},$$

giving

245)
$$\begin{aligned} \frac{d^2\xi}{dt^2} &= \quad 2\Omega \sin \varphi \frac{d\eta}{dt} + \lambda \xi, \\ \frac{d^2\eta}{dt^2} &= -2\Omega\left(\sin \varphi \frac{d\xi}{dt} + \cos \varphi \frac{d\zeta}{dt}\right) + \lambda \eta, \\ \frac{d^2\zeta}{dt^2} &= -g + 2\Omega \cos \varphi \frac{d\eta}{dt} + \lambda \zeta. \end{aligned}$$

Multiplying by $\frac{d\xi}{dt}, \frac{d\eta}{dt}, \frac{d\zeta}{dt}$ respectively and adding, then integrating, we get the equation of energy,

246)
$$\left(\frac{d\xi}{dt}\right)^2 + \left(\frac{d\eta}{dt}\right)^2 + \left(\frac{d\zeta}{dt}\right)^2 = a - 2g\zeta,$$

the gyroscopic terms disappearing, as usual. For a second integral we get as in § 23,

247)
$$\xi \frac{d^2\eta}{dt^2} - \eta \frac{d^2\xi}{dt^2} = -2\Omega \sin \varphi \left\{\xi \frac{d\xi}{dt} + \eta \frac{d\eta}{dt}\right\} - 2\Omega \cos \varphi \cdot \xi \frac{d\zeta}{dt}.$$

If we assume that the oscillations are infinitely small,

$$\zeta - l = -\frac{\xi^2 + \eta^2}{\zeta + l}$$

is infinitely small, and the last term above is of the third order and may be neglected. Integrating we have

248)
$$\xi \frac{d\eta}{dt} - \eta \frac{d\xi}{dt} = -\Omega \sin \varphi \{\xi^2 + \eta^2\} + b.$$

The equation of energy 246) becomes

249)
$$\left(\frac{d\xi}{dt}\right)^2 + \left(\frac{d\eta}{dt}\right)^2 = a + \frac{g}{l}(\xi^2 + \eta^2).$$

Inserting polar coordinates,

$$\xi = r \cos \omega,$$
$$\eta = r \sin \omega,$$

the integrals become

$$250)\qquad r^2 \frac{d\omega}{dt} = b - \Omega \sin\varphi \cdot r^2,$$
$$\left(\frac{dr}{dt}\right)^2 + r^2 \left(\frac{d\omega}{dt}\right)^2 = a + \frac{g}{l} r^2.$$

If we put

$$\omega + \Omega \sin\varphi \cdot t = \psi,$$

the first becomes

$$251)\qquad r^2 \frac{d\psi}{dt} = b,$$

and the second,

$$252)\qquad \left(\frac{dr}{dt}\right)^2 + r^2 \left(\frac{d\psi}{dt}\right)^2 - 2\Omega r^2 \sin\varphi \frac{d\psi}{dt} + r^2 \Omega^2 \sin^2\varphi = a + \frac{g}{l} r^2.$$

Introducing the above value of $r^2 \frac{d\psi}{dt}$, putting $a + 2b\Omega \sin\varphi = c$, and neglecting Ω^2,

$$253)\qquad \left(\frac{dr}{dt}\right)^2 + r^2 \left(\frac{d\psi}{dt}\right)^2 = c + \frac{g}{l} r^2.$$

This and 251) are the equations of the spherical pendulum, but we have

$$\omega = \psi - \Omega \sin\varphi \cdot t,$$

hence the axes of the ellipse described by the bob revolve around the vertical with the angular velocity $\Omega \sin\varphi$ in the direction east-south-west-north. This was verified by Foucault in his celebrated experiment made in the Panthéon at Paris in 1852.

106. Foucault's Gyroscope. Let us now consider the celebrated experiments by which, by means of a gyroscope, Foucault demonstrated the rotation of the earth. Let us consider a symmetrical gyroscope, suspended by its center of mass. If it is free to move, and is started spinning about its axis of symmetry, it will evidently by the principle of conservation of angular momentum, keep the axis of angular momentum, which is here the axis of symmetry, pointing in the same direction in space, so that this axis, while pointing always at the same star, describes a circular cone with reference to the earth. Instead of treating the general motion, which would lead to too great complications, we shall treat two important cases, in which the axis of symmetry is constrained to move either in a vertical or horizontal plane. This we shall do by making use of equations 29), § 84, following the method of Hayward, who gave those equations, in a paper in the Transactions of the Cambridge Philosophical Society, Vol. 10, read in 1856.

Suppose first the top constrained to move in a vertical plane, and take for axis of Z, the axis of figure, which makes an angle ϑ with the earth's axis, for axes of X and Y axes fixed in the meridian and at right angles to it, like the axes ξ, η, ζ of Fig. 112. We then have for the motion of the axes

$$254)\qquad p = -\Omega \sin\vartheta, \quad q = \frac{d\vartheta}{dt}, \quad r = \Omega\cos\vartheta,$$

while if ω be the velocity of rotation of the top about the Z-axis, we have for the moment of momentum, the axes being principal axes, though not fixed in the top,

$$255)\qquad H_x = -A\Omega\sin\vartheta, \quad H_y = A\frac{d\vartheta}{dt}, \quad H_z = C\omega.$$

Inserting now in the equations 29), § 84, the constraint producing a couple L,

$$256)\qquad \begin{aligned} -A\Omega\cos\vartheta\frac{d\vartheta}{dt} + C\omega\frac{d\vartheta}{dt} - A\Omega\cos\vartheta\frac{d\vartheta}{dt} &= L, \\ A\frac{d^2\vartheta}{dt^2} - A\Omega^2\sin\vartheta\cos\vartheta + C\Omega\omega\sin\vartheta &= 0, \\ C\frac{d\omega}{dt} - A\Omega\sin\vartheta\frac{d\vartheta}{dt} + A\Omega\sin\vartheta\frac{d\vartheta}{dt} &= 0. \end{aligned}$$

From the last of these equations, ω is constant, while from the second, neglecting Ω^2, we have

$$257)\qquad \frac{d^2\vartheta}{dt^2} + \frac{C}{A}\Omega\omega\sin\vartheta = 0.$$

The first equation 256) determines the constraint L. Equation 257) is the equation for the motion of a plane pendulum, § 22, so that the gyroscope will perform oscillations about a line parallel to the earth's axis, or will be in equilibrium when $\vartheta = 0$, thus affording a means of determining the latitude. The time of a small oscillation will be, $2\pi\sqrt{\frac{A}{C\Omega\omega}}$ which, on account of the smallness of Ω, will be very great unless ω be made very great. The experiment was performed with success by Foucault.

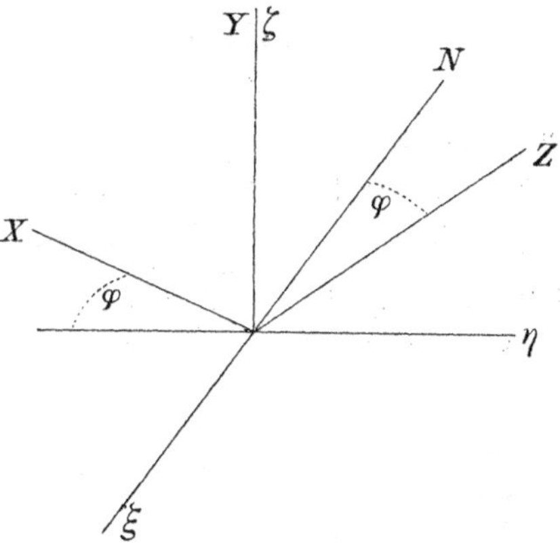

Fig. 113.

In the second case let us suppose the gyroscope constrained to move in a horizontal plane. Let us take for Y-axis the vertical, corresponding to the ζ-axis of Fig. 112, for the Z-axis the axis of figure of the top, making the variable angle φ with the north, towards the east, and for the X-axis a perpendicular to these (Fig. 113). The rotation of the earth gives the components $-\Omega\sin\vartheta$, $\Omega\cos\vartheta$ in the direction of the ξ, ζ axes respectively (ϑ being the co-latitude and not variable), which give by the use of the table of direction cosines,

	X	Y	Z
ξ	$-\sin\varphi$	0	$-\cos\varphi$
η	$-\cos\varphi$	0	$\sin\varphi$
ζ	0	1	0

the values of the rotations of the axes

$$258)\qquad \begin{aligned} p &= \Omega \sin\vartheta \sin\varphi, \\ q &= \Omega \cos\vartheta - \frac{d\varphi}{dt}, \\ r &= \Omega \sin\vartheta \cos\varphi, \end{aligned}$$

and for the angular momenta, ω being again the velocity of spinning,

$$259)\qquad \begin{aligned} H_x &= A\,\Omega \sin\vartheta \sin\varphi, \\ H_y &= A\left(\Omega \cos\varphi - \frac{d\varphi}{dt}\right), \\ H_z &= C\omega. \end{aligned}$$

Inserting in equations 29, § 84, the constraint producing the couple L,

$$260)\qquad \begin{aligned} A\,\Omega \sin\vartheta \cos\varphi \frac{d\varphi}{dt} - C\omega\left(\Omega\cos\vartheta - \frac{d\varphi}{dt}\right) \\ - A\left(\Omega\cos\vartheta - \frac{d\varphi}{dt}\right)\Omega \sin\vartheta\cos\varphi = L, \\ -A\frac{d^2\varphi}{dt^2} + A\Omega^2\sin^2\vartheta\sin\varphi\cos\varphi - C\Omega\omega\sin\vartheta\sin\varphi = 0, \\ C\frac{d\omega}{dt} + A\left(\Omega\cos\vartheta - \frac{d\varphi}{dt}\right)\Omega\sin\vartheta\sin\varphi \\ - A\Omega\left(\Omega\cos\vartheta - \frac{d\varphi}{dt}\right)\sin\vartheta\sin\varphi = 0. \end{aligned}$$

The last equation again shows that ω is constant, while from the second, neglecting Ω^2 we have

$$261)\qquad \frac{d^2\varphi}{dt^2} + \frac{C}{A}\Omega\omega\sin\vartheta\cdot\sin\varphi = 0.$$

The first equation determines the constraint L. The gyroscope again performs oscillations about the meridian, with the period

$$2\pi\sqrt{\frac{A}{C\Omega\omega\sin\vartheta}},$$

which is greater the greater the latitude, being infinite at the poles. The gyroscope in this mounting therefore constitutes a dynamical compass. This theory has been reduced to practice in a very ingenious manner in the gyroscopic compass of Anschütz.[1])

It is to be noticed in both cases that the equilibrium is stable for $\vartheta = 0$ or $\varphi = 0$ if ω is positive, and for $\vartheta = \pi$, $\varphi = \pi$ if ω is negative, in other words the gyroscope tends to set its axis as nearly as possible parallel with the earth's axis, so that its direction of rotation shall correspond with that of the earth. This was clearly stated by Foucault, although he employed no mathematics.

1) Nature, July 1911.

PART III

THEORY OF THE POTENTIAL, DYNAMICS OF DEFORMABLE BODIES

CHAPTER VIII.

NEWTONIAN POTENTIAL FUNCTION.

107. Point-Function. If for every position of a point in a region of space τ a quantity has one or more definite values assigned, it is said to be a function of the point, or point-function. This term was introduced by Lamé. If at every point it has a single value, it is a uniform function. Functions of the two or three rectangular coordinates of the point are point-functions. A point-function is continuous at a point A if we can find corresponding to any positive ε, however small, a value δ such that when B is *any* point inside a sphere of radius $< \delta$,

$$| f(B) - f(A) | < \varepsilon.$$

We may have vector as well as scalar point-functions, the length and direction of the vector being given for every point. A vector point-function is continuous if its components along the coordinate axes are continuous point-functions.

108. Level Surface of Scalar Point-Function. If V is a uniform function of the point M, continuous and without maximum or minimum in a portion of space τ, through any point M in the region τ we may construct a surface having the property that for every point on it V has the same value.

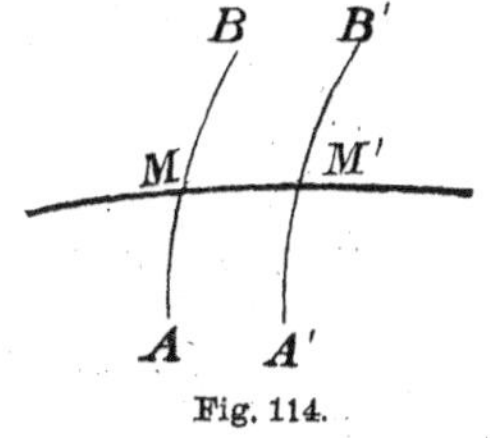

Fig. 114.

For let the value of V at M be c. Then since c is neither a maximum nor minimum, we can find in the neighbourhood of M two points A and B, such that at A, V is less, and at B, greater than c, and that in moving along a line AB through M, V continually increases. If the line AMB is displaced to the position $A'M'B'$, so that

$$| V(A) - V(A') | < c - V(A)$$

and

$$| V(B) - V(B') | < V(B) - c,$$

then $V(A') < c < V(B')$, therefore there is a point M' on the line $A'B'$ for which $V = c$.

As AB moves continuously M describes a line, and this line in its motion describes a surface, for every point of which $V = c$. Such a surface is called a *level surface* of the function V. A level surface divides space into two parts, for one of which V is greater, and for the other less, than in the surface.

As examples of point-functions we may take (1) the length of a line drawn from the point M parallel to a given line until it cuts a given plane. Its level surfaces are planes parallel to the given plane. (2) The distance of M from a fixed point O. The level surfaces are spheres with centers at O. (3) The angle that the radius vector OM makes with a fixed line OX. The level surfaces are right circular cones with OX as axis. (4) The dihedral angle made by the plane MOX with a fixed plane through OX. The level surfaces are planes through OX.

109. Coordinates. If a point is restricted to lie on a given surface S, the intersections of that surface with the level surfaces of a function V are the level lines of the function on the surface S; e. g. in examples (3) and (4) above, if S is a sphere with O as center, the level lines are parallels and meridians respectively.

A function $f(V_1, V_2, \ldots)$ of several point-functions is itself a point-function. If it is a function of one V only, its level surfaces are the same as those of V, for when V is constant, $f(V)$ is also constant.

Let q_1, q_2, q_3 be three uniform point-functions. Each has a level surface passing through the point M. If these three level surfaces do not coincide or intersect in a common curve, they determine the point M, and we may regard the point-functions q_1, q_2, q_3 as the *coordinates* of the point M. The level surfaces of q_1, q_2, q_3 are the coordinate surfaces, and the intersections of pairs $(q_1 q_2)$, $(q_2 q_3)$, $(q_3 q_1)$, are the coordinate lines. The tangents to the coordinate lines at M are called the coordinate axes at M. If at every point M the coordinate axes are mutually perpendicular, the system is said to be an orthogonal system.

110. Differential Parameter. The consideration of point-functions leads to the introduction of a particular sort of derivative. If V is a uniform point-function, continuous at a point M, and possessing there the value V, and at a point M' the value V', in virtue of continuity, when the distance MM' is infinitesimal, $V' - V = \Delta V$ is also. The ratio

$$\frac{V' - V}{MM'} = \frac{\Delta V}{\Delta s}$$

is finite, and as $MM' = \Delta s$ approaches 0, the direction of MM being given, the limit

$$\lim_{\Delta s = 0} \frac{\Delta V}{\Delta s} = \frac{\partial V}{\partial s}$$

is defined as the *derivative of V in the direction s*. We may lay off on a line through M in the direction of s a length $MQ = \frac{\partial V}{\partial s}$ and as we give s successively all possible directions, we may find the surface that is the locus of Q.

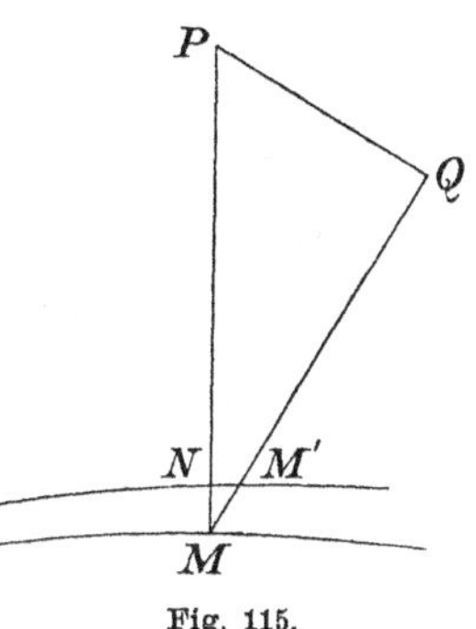

Fig. 115.

Let MN (Fig. 115) be the direction of the normal to the level surface at M, and let MP, drawn toward the side on which V is greater, represent in magnitude the derivative in that direction. Let M' and N be the intersections of the same neighboring level surface, for which $V = V'$, with MQ and MP. Then

$$\frac{\Delta V}{MM'} = \frac{\Delta V}{MN} \frac{MN}{MM'}.$$

As MM' approaches zero, we have

$$\lim_{s=0} \frac{\Delta V}{MM'} = \frac{\partial V}{\partial s}, \quad \lim \frac{\Delta V}{MN} = \frac{\partial V}{\partial n}, \quad \lim \frac{MN}{MM'} = \cos PMQ.$$

Hence

$$\frac{\partial V}{\partial s} = \frac{\partial V}{\partial n} \cos PMQ,$$

that is, the derivative in any direction at any point is equal to the projection on that direction of the derivative in the direction of the normal to the level surface at that point. Accordingly all points Q lie on a sphere whose diameter is MP.

The derivative in the direction of the normal to the level surface was called by Lamé[1]) the *first differential parameter* of the function V, and since it has not only magnitude but direction, we shall call it the *vector* differential parameter, or where no ambiguity will result, simply the parameter, denoted by $\overline{P}$ or $\overline{P}_V$. The above theorem may then be stated by saying that the derivative in any direction is the projection of the vector parameter on that direction. The theorem shows that the parameter gives the direction of the *fastest* increase of the function V.

If V is a function of a point-function q, $V = f(q)$, its level surfaces are those of q, and

$$\dot{P} = \frac{\partial V}{\partial n} = \frac{dV}{dq} \frac{\partial q}{\partial n} = f'(q) \frac{\partial q}{\partial n},$$

and if

$$\pm \frac{\partial q}{\partial n} = h, \quad P = \pm f'(q) \cdot h,$$

1) G. Lamé. *Lecons sur les coordonnées curvilignes et leurs diverses applications.* Paris, 1859, p. 6.

where the sign + is to be taken if V and q increase in the same, — if in opposite directions.

Suppose now that $V = f(q_1, q_2, q_3, \ldots)$

$$\frac{\partial V}{\partial s} = \frac{\partial V}{\partial q_1}\frac{\partial q_1}{\partial s} + \frac{\partial V}{\partial q_2}\frac{\partial q_2}{\partial s} + \frac{\partial V}{\partial q_3}\frac{\partial q_3}{\partial s} + \cdots$$

and if $h_1, h_2, \ldots$ denote the parameters of $q_1, q_2, \ldots$ the above theorem gives

$$P\cos(Ps) = \frac{\partial V}{\partial q_1} h_1 \cos(h_1 s) + \frac{\partial V}{\partial q_2} h_2 \cos(h_2 s) + \cdots$$

Now $\pm \frac{\partial V}{\partial q_i} h_i$ is the parameter of V, considered as a function of q_i, and we may call it the partial parameter P_i, and since P_i and h_i have the same sign if $\frac{\partial V}{\partial q_i} > 0$, opposite signs if $\frac{\partial V}{\partial q_i} < 0$, we have in either case

$$\frac{\partial V}{\partial q_i} h_i \cos(h_i s) = P_i \cos(P_i s),$$

and

$$P\cos(Ps) = P_1\cos(P_1 s) + P_2 \cos(P_2 s) + \cdots$$

This formula holds for *any* direction s and therefore shows that the parameter P is the geometrical sum, or resultant, of the partial parameters,

$$\overline{P} = \overline{P}_1 + \overline{P}_2 + \cdots$$

Thus we have the rule for finding the parameter of any function of several point-functions. If we know the parameters $h_1, h_2, \ldots$ of the functions $q_1, q_2, \ldots$ and the partial derivatives $\frac{\partial V}{\partial q_1}, \frac{\partial V}{\partial q_2}, \cdots$ we lay off the partial parameters

$$P_i = \pm h_i \frac{\partial V}{\partial q_i} \cdots,$$

in the directions $h_1, h_2, \ldots$ or their opposites, according as $\frac{\partial V}{\partial q_i} > 0$, or the opposite, and find the resultant of $P_1, P_2, \ldots$

If the functions $q_1, q_2, \ldots$ are three in number, and form an *orthogonal* system, the equation

$$\overline{P} = \overline{P}_1 + \overline{P}_2 + \overline{P}_3,$$

gives for the absolute, or numerical value of the parameter

$$P^2 = P_1^2 + P_2^2 + P_3^2.$$

Examples. (1) in § 108. Let the distance of M in the given direction from the plane be u. $\Delta V = \Delta u = \frac{\Delta n}{\cos\alpha}$, where α is the angle between the given direction and the given plane.

$$\frac{\Delta u}{\Delta n} = \frac{1}{\cos\alpha}, \quad P = \frac{1}{\cos\alpha}.$$

If the given direction is perpendicular to the given plane $P = 1$. Accordingly for $q_1 = x$, $q_2 = y$, $q_3 = z$, the rectangular coordinates of a point, we have $P_x = P_y = P_z = 1$, and for any function $f(x, y, z)$

$$P_1 = \pm \frac{\partial f}{\partial x}, \quad P_2 = \pm \frac{\partial f}{\partial y}, \quad P_3 = \pm \frac{\partial f}{\partial z},$$

$$P = \left[\left(\frac{\partial f}{\partial x}\right)^2 + \left(\frac{\partial f}{\partial y}\right)^2 + \left(\frac{\partial f}{\partial z}\right)^2\right]^{\frac{1}{2}}.$$

The projections of P on the coordinate axes are the partial parameters

$$P_1 = P\cos(Px) = \frac{\partial f}{\partial x}, \; P_2 = P\cos(Py) = \frac{\partial f}{\partial y}, \; P_3 = P\cos(Pz) = \frac{\partial f}{\partial z}.$$

This agrees with the definition already given in § 31.

Consequently, if $\cos(sx)$, $\cos(sy)$, $\cos(sz)$ are the direction cosines of a direction s, the derivative in that direction

$$\begin{aligned}\frac{\partial V}{\partial s} &= P_1 \cos(sx) + P_2 \cos(sy) + P_3 \cos(sx)\\ &= \frac{\partial V}{\partial x}\cos(sx) + \frac{\partial V}{\partial y}\cos(sy) + \frac{\partial V}{\partial z}\cos(sz),\end{aligned}$$

which is the same as equation 38a) of § 31.

We have in this section defined the differential parameter in a geometrical manner, not depending on the choice of axes of coordinates. If however we take as the definition of the arithmetical value of the parameter the equation

$$P = \sqrt{\left(\frac{\partial f}{\partial x}\right)^2 + \left(\frac{\partial f}{\partial y}\right)^2 + \left(\frac{\partial f}{\partial z}\right)^2}$$

and then transform to other coordinates x', y', z', by equations 109), § 76, we easily find by calculation that

$$\sqrt{\left(\frac{\partial f}{\partial x'}\right)^2 + \left(\frac{\partial f}{\partial y'}\right)^2 + \left(\frac{\partial f}{\partial z'}\right)^2}$$

is equal to P, that is, the parameter is a differential *invariant*, as is at once evident from its geometrical nature.

If $f(x, y, z)$ is a homogeneous function of degree n, by Euler's Theorem,

$$nf = x\frac{\partial f}{\partial x} + y\frac{\partial f}{\partial y} + z\frac{\partial f}{\partial z},$$

or

$$nf = P\{x\cos(Px) + y\cos(Py) + z\cos(Pz)\}.$$

Now the $\pm$ parenthesis is the distance of the tangent plane to the level surface at x, y, z from the origin. Calling this δ,

$$nf = \pm P\delta, \quad P = \pm\frac{nf}{\delta},$$

or the parameter of a homogeneous function is inversely proportional to the perpendicular from the origin to the tangent plane to the level surface. For example, if $n = 1$,

$$V = ax + by + cz,$$

$$P\cos(Px) = a, \quad P\cos(Py) = b, \quad P\cos(Pz) = c, \quad P = \sqrt{a^2 + b^2 + c^2}.$$

The level surfaces are parallel planes, and the parameter is constant,

$$P = \pm \frac{V}{\delta}, \quad V = \pm\,\delta\sqrt{a^2 + b^2 + c^2}.$$

V is proportional to the distance of the level surface from the origin.

If $n = 2$,

$$V = \frac{x^2}{a_1} + \frac{y^2}{a_2} + \frac{z^2}{a_3},$$

$$P\cos(Px) = \frac{2x}{a_1}, \quad P\cos(Py) = \frac{2y}{a_2}, \quad P\cos(Pz) = \frac{2z}{a_3},$$

$$P = 2\sqrt{\frac{x^2}{a_1{}^2} + \frac{y^2}{a_2{}^2} + \frac{z^2}{a_3{}^2}},$$

$$\delta = \pm \frac{2V}{P} = \pm \frac{V}{\sqrt{\frac{x^2}{a_1{}^2} + \frac{y^2}{a_2{}^2} + \frac{z^2}{a_3{}^2}}}.$$

For the surface, $V = 1$,

$$\delta = \pm \frac{1}{\sqrt{\frac{x^2}{a_1{}^2} + \frac{y^2}{a_2{}^2} + \frac{z^2}{a_3{}^2}}},$$

a familiar result of analytic geometry.

111. Polar Coordinates. If we call the point functions of Examples 2, 3, and 4, of § 108, r, ϑ, φ, we obtain the system of spherical, or polar coordinates ϑ and φ may be called the co-latitude and longitude. The level surfaces of r being spheres, the normal coincides with r. Accordingly

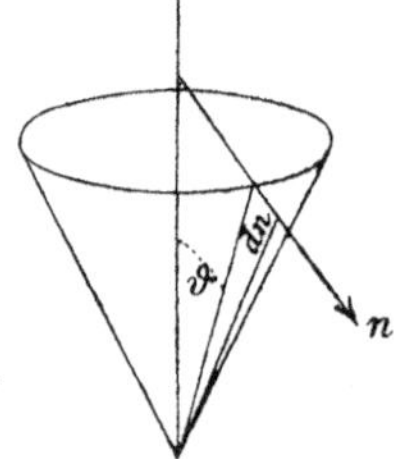

Fig. 116.

$$\frac{\partial r}{\partial n} = \frac{\partial r}{\partial r} = 1, \quad h_r = 1.$$

The level surface of ϑ is a circular cone of angular opening ϑ, (Fig. 116), and

$$dn = r\,d\vartheta, \quad \frac{\partial \vartheta}{\partial n} = \frac{d\vartheta}{r\,d\vartheta} = \frac{1}{r}, \quad h_\vartheta = \frac{1}{r}.$$

The level surfaces of φ are meridian planes through the axis of the above cones, (Fig. 117), and

$$dn = r \sin \vartheta \, d\varphi, \quad \frac{\partial \varphi}{\partial n} = \frac{d\varphi}{r \sin \vartheta \, d\varphi} = \frac{1}{r \sin \vartheta},$$

$$h_\varphi = \frac{1}{r \sin \vartheta}.$$

For any function $f(r, \vartheta, \varphi)$, the partial parameters are

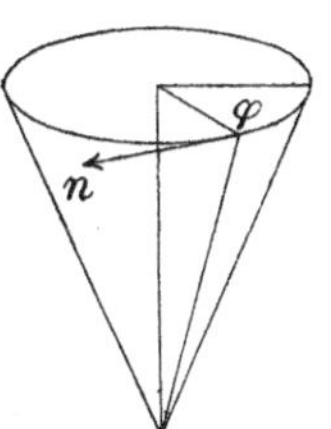

Fig. 117.

$$P_r = \pm \frac{\partial f}{\partial r} h_r = \pm \frac{\partial f}{\partial r},$$

$$P_\vartheta = \pm \frac{\partial f}{\partial \vartheta} h_\vartheta = \frac{1}{r} \frac{\partial f}{\partial \vartheta},$$

$$P_\varphi = \pm \frac{\partial f}{\partial \varphi} h_\varphi = \frac{1}{r \sin \vartheta} \frac{\partial f}{\partial \varphi}.$$

The total parameter, the resultant of these, is given by

$$P^2 = \left(\frac{\partial f}{\partial r}\right)^2 + \frac{1}{r^2}\left(\frac{\partial f}{\partial \vartheta}\right)^2 + \frac{1}{r^2 \sin^2 \vartheta}\left(\frac{\partial f}{\partial \varphi}\right)^2.$$

112. Cylindrical, or Semi-polar Coordinates. If we take the rectangular coordinate z, the perpendicular distance from the Z-axis, ϱ, and ω the longitude, or angle made by plane including the point M and the Z-axis, we have the system of semi-polar, cylindrical, or columnar coordinates, for which we have immediately,

$$h_z = 1, \quad h_\varrho = 1, \quad h_\omega = \frac{1}{\varrho}.$$

The parameter of a function $f(z, \varrho, \omega)$ is the resultant of the partial parameters

$$P_z = \pm \frac{\partial f}{\partial z}, \quad P_\varrho = \pm \frac{\partial f}{\partial \varrho}, \quad P_\omega = \pm \frac{1}{\varrho} \frac{\partial f}{\partial \omega},$$

$$P^2 = \left(\frac{\partial f}{\partial z}\right)^2 + \left(\frac{\partial f}{\partial \varrho}\right)^2 + \frac{1}{\varrho^2}\left(\frac{\partial f}{\partial \omega}\right)^2.$$

113. Ellipsoidal Coordinates. Let us now find the value of the parameter in terms of the ellipsoidal coordinates described in § 73, which are defined for a point x, y, z as the three roots of the equation

$$1) \qquad \frac{x^2}{a^2+\lambda} + \frac{y^2}{b^2+\lambda} + \frac{z^2}{c^2+\lambda} - 1 = 0.$$

The three coordinate-surfaces at any point have been proved to be mutually perpendicular at each point x, y, z. Since the equation 1) is an identity, we have, differentiating totally, that is changing x, y, z, λ,

$$2) \quad 2\left\{\frac{x\,dx}{a^2+\lambda} + \frac{y\,dy}{b^2+\lambda} + \frac{z\,dz}{c^2+\lambda}\right\} - d\lambda \left\{\frac{x^2}{(a^2+\lambda)^2} + \frac{y^2}{(b^2+\lambda)^2} + \frac{z^2}{(c^2+\lambda)^2}\right\} = 0.$$

Now if δ_λ is the perpendicular distance of the tangent plane from the origin, we have by the last formula of § 110,

$$\delta_\lambda = 1 \Big/ \sqrt{\frac{x^2}{(a^2+\lambda)^2} + \frac{y^2}{(b^2+\lambda)^2} + \frac{z^2}{(c^2+\lambda)^2}},$$

so that we may write for the direction cosines of the normal,

$$\begin{aligned} \cos(n_\lambda x) &= \frac{x\,\delta_\lambda}{a^2+\lambda}, \\ \cos(n_\lambda y) &= \frac{y\,\delta_\lambda}{b^2+\lambda}, \\ \cos(n_\lambda z) &= \frac{z\,\delta_\lambda}{c^2+\lambda}. \end{aligned} \tag{3}$$

Now as we move along the normal, we have

$$\begin{aligned} dx &= dn \cos(n_\lambda x) = \frac{x\,\delta_\lambda}{a^2+\lambda}\,dn, \\ dy &= dn \cos(n_\lambda y) = \frac{y\,\delta_\lambda}{b^2+\lambda}\,dn, \\ dz &= dn \cos(n_\lambda z) = \frac{z\,\delta_\lambda}{c^2+\lambda}\,dn. \end{aligned}$$

Inserting these values in 2),

$$\begin{aligned} 4)\quad 2\delta_\lambda \left\{\frac{x^2}{(a^2+\lambda)^2} + \frac{y^2}{(b^2+\lambda)^2} + \frac{z^2}{(c^2+\lambda)^2}\right\} dn & \\ - \left\{\frac{x^2}{(a^2+\lambda)^2} + \frac{y^2}{(b^2+\lambda)^2} + \frac{z^2}{(c^2+\lambda)^2}\right\} d\lambda &= 0, \end{aligned}$$

so that

$$h_\lambda = \frac{d\lambda}{dn} = 2\delta_\lambda = \frac{2}{\sqrt{\frac{x^2}{(a^2+\lambda)^2} + \frac{y^2}{(b^2+\lambda)^2} + \frac{z^2}{(c^2+\lambda)^2}}}. \tag{5}$$

In order to express this result in terms of the elliptic coordinates alone we may express x, y, z, in terms of λ, μ, ν. Observe that the function

$$F(\varrho) \equiv \frac{x^2}{\varrho+a^2} + \frac{y^2}{\varrho+b^2} + \frac{z^2}{\varrho+c^2} - 1$$

has as roots λ, μ, ν, and being reduced to the common denominator

$$(\varrho + a^2)(\varrho + b^2)(\varrho + c^2)$$

has a numerator of the third degree in ϱ. As this vanishes for

$$\varrho = \lambda, \quad \varrho = \mu, \quad \varrho = \nu,$$

it can only be

$$-(\varrho - \lambda)(\varrho - \mu)(\varrho - \nu).$$

Accordingly we have the identity

$$6)\qquad F(\varrho) \equiv \frac{x^2}{\varrho+a^2} + \frac{y^2}{\varrho+b^2} + \frac{z^2}{\varrho+c^2} - \equiv \frac{-(\varrho-\lambda)(\varrho-\mu)(\varrho-\nu)}{(\varrho+a^2)(\varrho+b^2)(\varrho+c^2)}.$$

Multiplying this by $\varrho + a^2$ and then putting $\varrho = -a^2$ we get

$$x^2 = \frac{(a^2+\lambda)(a^2+\mu)(a^2+\nu)}{(a^2-b^2)(a^2-c^2)},$$

and in like manner

$$7)\qquad \begin{aligned} y^2 &= \frac{(b^2+\mu)(b^2+\mu)(b^2+\nu)}{(b^2-c^2)(b^2-a^2)}, \\ z^2 &= \frac{(c^2+\lambda)(c^2+\mu)(c^2+\nu)}{(c^2-a^2)(c^2-b^2)}. \end{aligned}$$

If λ, μ, ν are contained in the intervals specified in § 73, these will all be positive, so that the point will be real.

If we insert these values in δ_λ, we shall have h_λ expressed in terms of λ, μ, ν.

This is more easily accomplished as follows.

Differentiating the above identity 6) according to ϱ,

$$8)\qquad -\left\{\frac{x^2}{(a^2+\varrho)^2} + \frac{y^2}{(b^2+\varrho)^2} + \frac{z^2}{(c^2+\varrho)^2}\right\}$$
$$\equiv \frac{-(\varrho-\lambda)(\varrho-\mu)(\varrho-\nu)}{(a^2+\varrho)(b^2+\varrho)(c^2+\varrho)}\left\{\frac{1}{\varrho-\lambda} - \frac{1}{\varrho+a^2} + \frac{1}{\varrho-\mu} - \frac{1}{\varrho+b^2} + \frac{1}{\varrho-\nu} - \frac{1}{\varrho+c^2}\right\}.$$

If we put $\varrho = \lambda$, all the terms on the right except the first, being multiplied by $\varrho - \lambda$, vanish, and we have

$$9)\qquad \frac{x^2}{(a^2+\lambda)^2} + \frac{y^2}{(b^2+\lambda)^2} + \frac{z^2}{(c^2+\lambda)^2} = \frac{(\lambda-\mu)(\lambda-\nu)}{(a^2+\lambda)(b^2+\lambda)(c^2+\lambda)}.$$

The expression on the left is $\frac{1}{\delta_\lambda^2}$. Therefore

$$10)\qquad h_\lambda = 2\delta_\lambda = 2\sqrt{\frac{(a^2+\lambda)(b^2+\lambda)(c^2+\lambda)}{(\lambda-\mu)(\lambda-\nu)}}.$$

In a similar manner we find

$$h_\mu = 2\sqrt{\frac{(a^2+\mu)(b^2+\mu)(c^2+\mu)}{(\mu-\nu)(\mu-\lambda)}},$$

$$h_\nu = 2\sqrt{\frac{(a^2+\nu)(b^3+\nu)(c^2+\nu)}{(\nu-\lambda)(\nu-\mu)}},$$

and the parameter of any function $V(\lambda, \mu, \nu)$ is

$$11)\qquad P^2 = \left(\frac{\partial V}{\partial \lambda}\right)^2 h_\lambda^2 + \left(\frac{\partial V}{\partial \mu}\right)^2 h_\mu^2 + \left(\frac{\partial V}{\partial \nu}\right)^2 h_\nu^2.$$

114. Infinitesimal Arc; Area and Volume. If we have any three point-functions q_1, q_2, q_3 forming an orthogonal system of coordinates, since their parameters are

$$h_1 = \frac{\partial q_1}{\partial n_1}, \quad h_2 = \frac{\partial q_2}{\partial n_2}, \quad h_3 = \frac{\partial q_3}{\partial n_3},$$

the normal distance between two consecutive level surfaces q_1 and $q_1 + dq_1$ is $dn_1 = \frac{dq_1}{h_1}$, consequently if we take six surfaces

$$q_1, q_1+dq_1, q_2, q_2+dq_2, q_3, q_3+dq_3,$$

the edges of the infinitesimal curvilinear rectangular parallelopiped whose edges are the intersections of the surfaces are

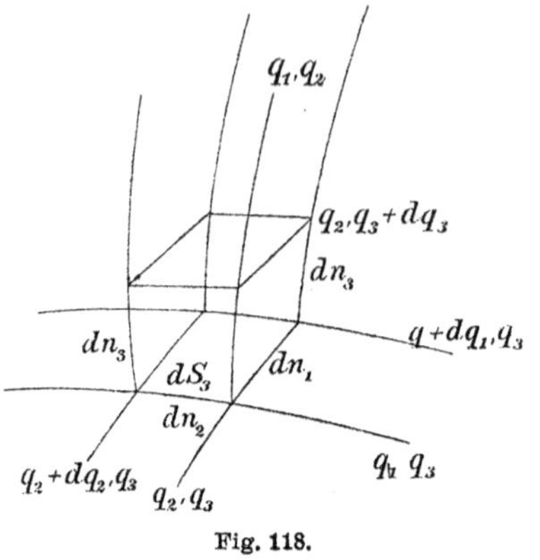

Fig. 118.

$$dn_1 = \frac{dq_1}{h_1}, \quad dn_2 = \frac{dq_2}{h_2}, \quad dn_3 = \frac{dq_3}{h_3},$$

and since the edges are mutually perpendicular, the diagonal, or element of arc is

$$ds^2 = \frac{dq_1^2}{h_1^2} + \frac{dq_2^2}{h_2^2} + \frac{dq_3^2}{h_3^2},$$

the elements of area of the surfaces q_1, q_2, q_3 are respectively

$$dS_1 = \frac{dq_2\, dq_3}{h_2\, h_3}, \quad dS_2 = \frac{dq_3\, dq_1}{h_3\; h_1}, \quad dS_3 = \frac{dq_1\; dq_2}{h_1\;\, h_2},$$

and the element of volume is

$$d\tau = \frac{dq_1\, dq_2\, dq_3}{h_1\, h_2\, h_3}.$$

Examples. Rectangular coordinates x, y, z.

$$h_x = h_y = h_z = 1,$$

$$dS_x = dy\, dz, \quad dS_y = dz\, dx, \quad dS_z = dx\, dy, \quad d\tau = dx\, dy\, dz.$$

Polar coordinates r, ϑ, φ,

$$h_r = 1, \quad h_\vartheta = \frac{1}{r}, \quad h_\varphi = \frac{1}{r \sin \vartheta},$$

12)
$$\begin{aligned} dS_r &= r^2 \sin \vartheta\, d\vartheta\, d\varphi, && \text{element of area of sphere,} \\ dS_\vartheta &= r\, \sin \vartheta\, dr\; d\varphi, && \text{element of area of cone,} \\ dS_\varphi &= r\, dr\, d\vartheta \qquad , && \text{element of area of plane,} \end{aligned}$$

$$d\tau = r^2 \sin \vartheta\, dr\, d\vartheta\, d\varphi.$$

Cylindrical coordinates, z, ϱ, ω,

$$h_z = h_\varrho = 1, \quad h_\omega = \frac{1}{\varrho},$$

13)
$$\begin{aligned} dS_z &= \varrho\, d\varrho\, d\omega, && \text{element of area of plane,} \\ dS_\varrho &= \varrho\, d\omega\, dz, && \text{element of area of cylinder,} \\ dS_\omega &= d\varrho\, dz\ , && \text{element of area of meridian plane,} \\ d\tau &= \varrho\, d\varrho\, d\omega\, dz. \end{aligned}$$

Elliptic coordinates, λ, μ, ν.

14)
$$\begin{aligned} dS_\lambda &= \frac{d\mu\, d\nu \sqrt{(\mu-\nu)(\mu-\lambda)(\nu-\lambda)(\nu-\mu)}}{4\sqrt{(a^2+\mu)(b^2+\mu)(c^2+\mu)(a^2+\nu)(b^2+\nu)(c^2+\nu)}}, && \text{ellipsoid,} \\ dS_\mu &= \frac{d\nu\, d\lambda \sqrt{(\nu-\lambda)(\nu-\mu)(\lambda-\mu)(\lambda-\nu)}}{4\sqrt{(a^2+\nu)(b^2+\nu)(c^2+\nu)(a^2+\lambda)(b^2+\lambda)(c^2+\lambda)}}, && \text{hyperboloid,} \\ dS_\nu &= \frac{d\lambda\, d\mu \sqrt{(\lambda-\mu)(\lambda-\nu)(\mu-\nu)(\mu-\lambda)}}{4\sqrt{(a^2+\lambda)(b^2+\lambda)(c^2+\lambda)(a^2+\mu)(b^2+\mu)(c^2+\mu)}}, && \text{hyperboloid,} \\ d\tau &= \frac{d\lambda\, d\mu\, d\nu\, (\lambda-\mu)(\mu-\nu)(\nu-\lambda)}{8\sqrt{(a^2+\lambda)(b^2+\lambda)(c^2+\lambda)(a^2+\mu)(b^2+\mu)(c^2+\mu)(a^2+\nu)(b^2+\nu)(c^2+\nu)}}. \end{aligned}$$

115. Connectivity of Space. Green's Theorem. We supposed in § 30 that it was possible to change the path 1 from A to B into the path 2 by continuous deformation, without passing out of the space considered. A portion of space in which any path between two points may be thus changed into any other between the same two points is said to be singly-connected. For instance, in the case of a two-dimensional space, any area bounded by a single closed contour will have this property. If, however, we consider an area bounded externally by a closed contour C, and internally by one or more closed contours I, Fig. 119, such as the surface of a lake containing islands, it will be possible to go from any point A to any other point B by two routes which cannot be continuously changed into each other without passing out of the space considered, that is traversing the shaded part.

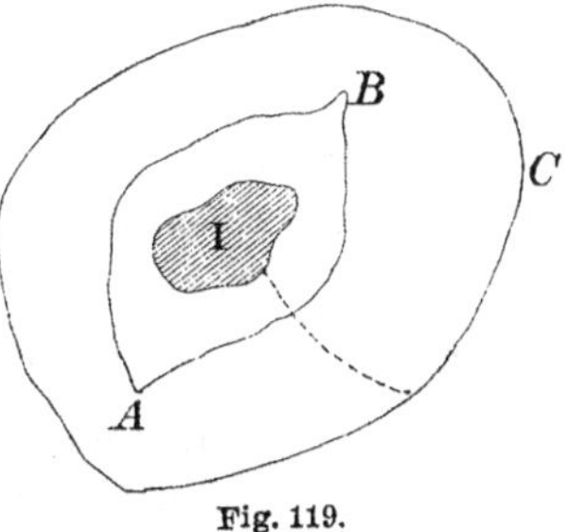

Fig. 119.

The space in Fig. 119 between the contour C and the island I is said to be doubly-connected. We may make it singly-connected by drawing a barrier connecting the island with the contour C, represented by the dotted line. If no path is allowed which crosses the barrier the space is singly-connected.

A three-dimensional space bounded externally by a single closed surface is not made doubly-connected by containing an inner closed boundary. For instance, the space lying between two concentric spheres allows all paths between two given points to be deformed into each other, avoiding the inner sphere. But the space bounded by an endless tubular surface, Fig. 120, is doubly-connected, because we may go from A to B in either direction of the tube, and the two paths cannot be deformed into each other. We may make the space singly-connected by the insertion of a barrier in the shape of a diaphragm, closing the tube so that one of the paths is inadmissible. The connectivity of a portion of space is defined as one more than the least number of barriers or diaphragms necessary to make it singly connected. Thus the space in a closed vase with three hollow handles, Fig. 121, is quadruply-connected. *We shall always suppose the spaces with which we deal in this book to be singly-connected, or to be made so by the insertion of diaphragms, unless the contrary is expressly stated.*

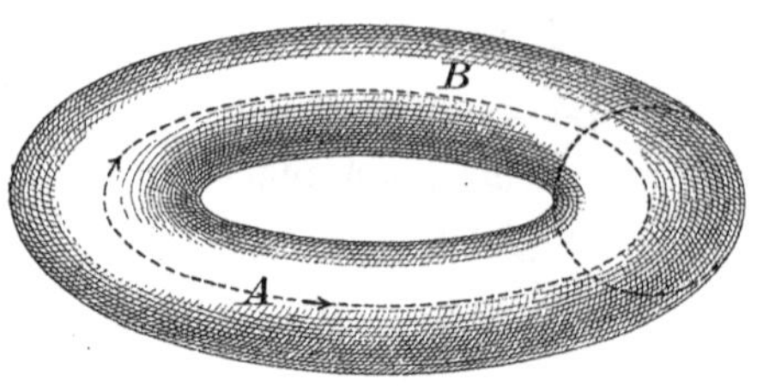

Fig. 120.

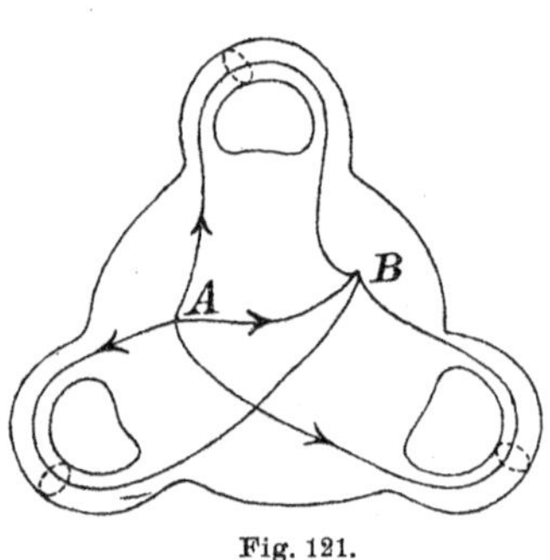

Fig. 121.

Suppose that W is a point-function which, together with its derivative in any direction, is uniform and continuous in a certain portion of space τ bounded by a closed surface S. Then its derivative $\frac{\partial W}{\partial x}$ is finite in the whole region, and if we multiply it by the element of volume $d\tau$ and integrate throughout the volume τ, the integral is finite, being less than the maximum value attained by $\frac{\partial W}{\partial x}$ in the space τ multiplied by the volume τ. We have at once

15) $$J = \iiint \frac{\partial W}{\partial x} dx\,dy\,dz = \iint dy\,dz \left[\int \frac{\partial W}{\partial x} dx \right].$$

If, keeping y and z constant, we perform the integration with respect to x, the volume is divided into elementary prisms whose sides are parallel to the X-axis, and whose bases are rectangles with sides dy, dz.

The portion of the integral due to one such prism is

$$dy\,dz \int \frac{\partial W}{\partial x} dx.$$

Now the integral is to be taken between the values of x where the edge of the elementary prism cuts into the surface S and where it cuts out from the surface. If it cuts in more than once, it will, since the surface is closed, cut out the same number of times. Let the values of x, at the successive points of cutting, be

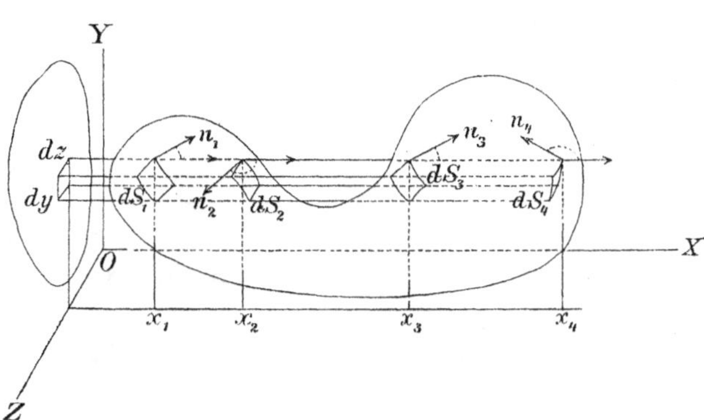

Fig. 122.

$x_1,\ x_2,\ \cdots\ x_{2n},$

then

$$\int \frac{\partial W}{\partial x}\,dx = W_2 - W_1 + W_4 + W_3 \cdots + W_{2n} - W_{2n-1},$$

W_k being the value of W for x_k, and

$$16)\quad \iiint \frac{\partial W}{\partial x}\,dx\,dy\,dz$$
$$= \iint [W_2 - W_1 + W_4 - W_3 \cdots + W_{2n} - W_{2n-1}]\,dy\,dz.$$

Now let $dS_1,\ dS_2, \cdots dS_{2n}$ denote the areas of the elements of the surface S cut out by the prism in question at $x_1,\ x_2, \cdots x_{2n}$, these all have the same projection on the YZ-plane, namely $dy\,dz$. If all these elements are considered positive, and if n be the normal always drawn *inward* from the surface S toward the space τ, at each point of cutting into the surface S, n makes an acute angle with the positive direction of the axis of X, and the projection of dS is

$$dy\,dz = dS \cos(nx),$$

but where the edge cuts out n makes an obtuse angle, with negative cosine, and therefore

$$dy\,dz = -\,dS\cos(nx).$$

We may accordingly write

$$\begin{aligned} dy\,dz\,W_1 &= W_1 \cos(n_1 x)\,dS_1,\\ -\,dy\,dz\,W_2 &= W_2 \cos(n_2 x)\,dS_2,\\ dy\,dz\,W_3 &= W_3 \cos(n_3 x)\,dS_3,\\ &\cdots\cdots\cdots\cdots\\ dy\,dz\,W_{2n} &= W_{2n} \cos(n_{2n} x)\,dS_{2n}, \end{aligned}$$

and in integrating with respect to y and z we cover the whole of the projection of the surface S on the YZ-plane. At the same time we cover the whole of the surface S, so that the volume integral is transformed into a surface integral,

$$17)\quad J=\iiint\frac{\partial W}{\partial x}\,d\tau=\iint dy\,dz\,[W_2-W_1+\cdots-W_{2n-1}]$$
$$=-\iint W\cos(nx)\,dS,$$

taken all over the surface S.

In like manner we may transform the two similar integrals

$$\iiint\frac{\partial W}{\partial y}\,d\tau=-\iint W\cos(ny)\,dS,$$
$$\iiint\frac{\partial W}{\partial z}\,d\tau=-\iint W\cos(nz)\,dS.$$

Applying this lemma to the function

$$W=U\frac{\partial V}{\partial x},$$

where both U, V and their derivatives in any direction are uniform and continuous point-functions in the space τ, we have

$$18)\qquad \iiint\frac{\partial}{\partial x}\left(U\frac{\partial V}{\partial x}\right)d\tau=-\iint U\frac{\partial V}{\partial x}\cos(nx)\,dS.$$

Similarly putting for W,

$$W=U\frac{\partial V}{\partial y},$$

and integrating with respect to y,

$$\iiint\frac{\partial}{\partial y}\left(U\frac{\partial V}{\partial y}\right)d\tau=-\iint U\frac{\partial V}{\partial y}\cos(ny)\,dS,$$

and for

$$W=U\frac{\partial V}{\partial z},$$

$$\iiint\frac{\partial}{\partial z}\left(U\frac{\partial V}{\partial z}\right)d\tau=-\iint U\frac{\partial V}{\partial z}\cos(ny)\,dS.$$

Adding these three equations, and performing the differentiations,

$$19)\quad \iiint\left[U\left(\frac{\partial^2 V}{\partial x^2}+\frac{\partial^2 V}{\partial y^2}+\frac{\partial^2 V}{\partial z^2}\right)+\frac{\partial U}{\partial x}\frac{\partial V}{\partial x}+\frac{\partial U}{\partial y}\frac{\partial V}{\partial y}+\frac{\partial U}{\partial z}\frac{\partial V}{\partial z}\right]d\tau$$
$$=-\iint U\left(\frac{\partial V}{\partial x}\cos(nx)+\frac{\partial V}{\partial y}\cos(ny)+\frac{\partial V}{\partial z}\cos(nz)\right)dS,$$

or, transposing, and denoting the symmetrical integral by J,

$$20)\quad J=\iiint\left[\frac{\partial U}{\partial x}\frac{\partial V}{\partial x}+\frac{\partial U}{\partial y}\frac{\partial V}{\partial y}+\frac{\partial U}{\partial z}\frac{\partial V}{\partial z}\right]d\tau$$
$$=-\iint U\left\{\frac{\partial V}{\partial x}\cos(nx)+\frac{\partial V}{\partial y}\cos(ny)+\frac{\partial V}{\partial z}\cos(nz)\right\}dS$$
$$-\iiint U\left\{\frac{\partial^2 V}{\partial x^2}+\frac{\partial^2 V}{\partial y^2}+\frac{\partial^2 V}{\partial z^2}\right\}d\tau.$$

This result is known as *Green's Theorem.*[1])

By the definition of differentiation in any direction (p. 333) the parenthesis in the surface integral on the right is

$$\frac{\partial V}{\partial n}=P_V\cos(P_V n),$$

if P_V is the parameter of V. In like manner the symmetrical function of U and V on the left, the integrand in J, is the geometric product of the vector parameters of U and V. This symmetric function, which we will denote by $\varDelta(U, V)$,

$$\varDelta(U, V)\equiv\frac{\partial U}{\partial x}\frac{\partial V}{\partial x}+\frac{\partial U}{\partial y}\frac{\partial V}{\partial y}+\frac{\partial U}{\partial z}\frac{\partial V}{\partial z}=P_U P_V\cos(P_U P_V)$$

is often called the *mixed* differential parameter of U and V. From its geometrical properties, or by direct calculation, it is also a differential invariant for a transformation of coordinates.

Since the integral on the left is symmetrical in U and V, we may interchange them on the right, so that

$$21)\quad J=-\iint V\frac{\partial U}{\partial n}dS-\iiint V\left\{\frac{\partial^2 U}{\partial x^2}+\frac{\partial^2 U}{\partial y^2}+\frac{\partial^2 U}{\partial z^2}\right\}d\tau.$$

Writing this equal to the former value, and transposing, we obtain

$$22)\quad \iint\left[U\frac{\partial V}{\partial n}-V\frac{\partial U}{\partial n}\right]dS=\iiint\left\{V\left(\frac{\partial^2 U}{\partial x^2}+\frac{\partial^2 U}{\partial y^2}+\frac{\partial^2 U}{\partial z^2}\right)\right.$$
$$\left.-U\left(\frac{\partial^2 V}{\partial x^2}+\frac{\partial^2 V}{\partial y^2}+\frac{\partial^2 V}{\partial z^2}\right)\right\}d\tau,$$

which will be refered to as Green's theorem in its second form.

We shall, unless the contrary is stated, always mean by n the *internal* normal to a closed surface, but if necessary we shall distinguish the normals drawn internally and externally as n_i and n_e. If we do not care to distinguish the inside from the outside we shall denote the normals toward opposite sides by n_1 and n_2.

1) *An Essay on the Application of Mathematical Analysis to the theories of Electricity and Magnetism.* Nottingham, 1828, Geo. Green, Reprint of papers, p. 25.

116. Second Differential Parameter. If for the function U we take a constant, say 1,

$$\frac{\partial U}{\partial x}=\frac{\partial U}{\partial y}=\frac{\partial U}{\partial z}=0, \quad P_U=0,$$

and we have simply

$$23)\qquad -\iint P_V \cos(P_V n)\, dS = -\iint \frac{\partial V}{\partial n}\, dS$$
$$=\iiint\left(\frac{\partial^2 V}{\partial x^2}+\frac{\partial^2 V}{\partial y^2}+\frac{\partial^2 V}{\partial z^2}\right) d\tau.$$

The function

$$\frac{\partial^2 V}{\partial x^2}+\frac{\partial^2 V}{\partial y^2}+\frac{\partial^2 V}{\partial z^2},$$

which, following the usage of the majority of writers, we shall denote by ΔV, was termed by Lamé[1]) the second differential parameter of V. As is it a scalar quantity it will be sufficiently distinguished from the first parameter if we call it the scalar parameter. We have accordingly the theorem giving the relation between the two: —

The volume integral of the scalar differential parameter of a uniform continuous point-function throughout any volume is equal to the surface integral of the vector parameter resolved along the *outward* normal to the surface S bounding the volume.

We may obtain a geometrical notion of the significance of ΔV in a number of ways. In the neighborhood of a point O, let us develop V by Taylor's theorem, calling the coordinates of neighboring points with respect to O, x, y, z, then

$$24)\qquad V=V_0+x\left(\frac{\partial V}{\partial x}\right)_0+y\left(\frac{\partial V}{\partial y}\right)_0+z\left(\frac{\partial V}{\partial z}\right)_0$$
$$+\frac{1}{2}\left\{x^2\left(\frac{\partial^2 V}{\partial x^2}\right)_0+y^2\left(\frac{\partial^2 V}{\partial y^2}\right)_0+z^2\left(\frac{\partial^2 V}{\partial z^2}\right)_0+2yz\left(\frac{\partial^2 V}{\partial y\,\partial z}\right)_0+\cdots\right\},$$

where the suffix 0 denotes the value at O.

Integrating the value of $V-V_0$ throughout the volume of a small sphere with center at O, we have

$$25)\qquad \iiint(V-V_0)\,d\tau=\left(\frac{\partial V}{\partial x}\right)_0\iiint x\,d\tau+\left(\frac{\partial V}{\partial y}\right)_0\iiint y\,d\tau$$
$$+\left(\frac{\partial V}{\partial z}\right)_0\iiint z\,d\tau+\frac{1}{2}\left(\frac{\partial^2 V}{\partial x^2}\right)_0\iiint x^2 d\tau+\frac{1}{2}\left(\frac{\partial^2 V}{\partial y^2}\right)_0\iiint y^2 d\tau$$
$$+\frac{1}{2}\left(\frac{\partial^2 V}{\partial z^2}\right)_0\iiint z^2 d\tau+\left(\frac{\partial^2 V}{\partial y\,\partial z}\right)_0\iiint yz\,d\tau+\cdots$$

---

1) G. Lamé. *Leçons sur les Coordonnées curvilignes et leurs diverses Applications.* Paris, 1859, p. 6.

The integrals in x, y, z, yz, zx, xy, being proportional to the coordinates of the center of gravity and products of inertia of a homogeneous sphere, all disappear from symmetry, while those in x^2, y^2, z^2 are all equal and represent the moment of inertia of the sphere with respect to the diametral plane, which by § 75 is $\frac{4}{15}\pi R^5$. Accordingly if $\overline{V}$ denotes the mean value of V in the sphere of radius R, the above integral equation becomes

$$26)\quad \frac{4}{3}\pi R^3(\overline{V}-V_0)=\frac{2}{15}\pi R^5\left\{\left(\frac{\partial^2 V}{\partial x^2}\right)_0+\left(\frac{\partial^2 V}{\partial y^2}\right)_0+\left(\frac{\partial^2 V}{\partial z^2}\right)_0+\cdots \text{ terms of higher order}\right\}.$$

Dividing by R^5 and taking the limit for $R=0$,

$$27)\quad \lim_{R=0}\frac{(\overline{V}-V_0)}{R^2}=\frac{1}{10}(\Delta V)_0,$$

that is, the excess of the mean value of V throughout the volume of a small sphere over the value at the center is proportional to the value of ΔV at the center and is of the second order of small quantities. This interpretation is due to Stokes.

From this point of view Maxwell calls $-\Delta V$ the *concentration* of V, since it is proportional to the excess of the value of V at a point over the values at neighboring points. It is evident from this interpretation of ΔV that if the concentration of a function vanishes throughout a certain region, then about any point in the region the values at neighboring points are partly greater and partly less than at the point itself, so that the function cannot have at any point in the region either a maximum or minimum with respect to surrounding points. A function that in a certain region is uniform, continuous, and has no concentration is said to be *harmonic* in that region. The study of such functions constitutes one of the most important parts, not only of the theory of functions, but also of mathematical physics.

Another interpretation of ΔV may be obtained as follows: we have by the rule for the derivation of any function in any direction r, with direction cosines, $\cos(rx)=\alpha$, $\cos(ry)=\beta$, $\cos(rz)=\gamma$,

$$\frac{\partial}{\partial r}=\alpha\frac{\partial}{\partial x}+\beta\frac{\partial}{\partial y}+\gamma\frac{\partial}{\partial z}.$$

Applying this to the function $\frac{\partial V}{\partial r}$ we obtain

$$\begin{aligned}28)\quad \frac{\partial^2 V}{\partial r^2}&=\frac{\partial}{\partial r}\left(\frac{\partial V}{\partial r}\right)=\left(\alpha\frac{\partial}{\partial x}+\beta\frac{\partial}{\partial y}+\gamma\frac{\partial}{\partial z}\right)\left(\alpha\frac{\partial V}{\partial x}+\beta\frac{\partial V}{\partial y}+\gamma\frac{\partial V}{\partial z}\right)\\&=\alpha^2\frac{\partial^2 V}{\partial x^2}+\beta^2\frac{\partial^2 V}{\partial y^2}+\gamma^2\frac{\partial^2 V}{\partial z^2}\\&\quad+2\beta\gamma\frac{\partial^2 V}{\partial y\partial z}+2\alpha\gamma\frac{\partial^2 V}{\partial x\partial z}+2\alpha\beta\frac{\partial^2 V}{\partial x\partial y}.\end{aligned}$$

If we consider at a point x, y, z all directions r and take the mean of $\frac{\partial^2 V}{\partial r^2}$ for these directions, inasmuch as the mean of any product of cosines is equal to zero, because for every cosine that appears the negative also appears, the terms with products disappear. Also from symmetry, denoting the mean value by the bar,

$$\overline{\alpha^2} = \overline{\beta^2} = \overline{\gamma^2}.$$

Since always $\alpha^2 + \beta^2 + \gamma^2 = 1$, we have

$$\overline{\alpha^2} + \overline{\beta^2} + \overline{\gamma^2} = 1,$$

and

$$\overline{\alpha^2} = \overline{\beta^2} = \overline{\gamma^2} = \frac{1}{3}.$$

Accordingly

$$29)\qquad \overline{\frac{\partial^2 V}{\partial r^2}} = \frac{1}{3}\left(\frac{\partial^2 V}{\partial x^2} + \frac{\partial^2 V}{\partial y^2} + \frac{\partial^2 V}{\partial z^2}\right)$$

therefore ΔV is equal to three times the mean of the second derivative of V in a definite direction for all possible direction leading from the point in question. This interpretation is due to Boussinesq.[1])

By means of this result we may obtain a third interpretation connecting the value of ΔV at a point with the mean excess of values on the surface of a small sphere, with center at the point, over the value at the center.

If V_0 denote the value at the center, the value at a distance R in any direction is given by Taylor's theorem,

$$V = V_0 + R\left(\frac{\partial V}{\partial r}\right)_0 + \frac{1}{2}R^2\left(\frac{\partial^2 V}{\partial r^2}\right)_0 + \cdots$$

Integrating over the surface of a sphere of radius R, the derivatives of V varying with the direction, since $dS = R^2 d\omega$, dividing by the constant R^2,

$$30)\qquad \iint (V - V_0)\,d\omega = R\iint\left(\frac{\partial V}{\partial r}\right)_0 d\omega + \frac{1}{2}R^2\iint\left(\frac{\partial^2 V}{\partial r^2}\right)_0 d\omega + \cdots$$

Now since

$$\left(\frac{\partial V}{\partial r}\right)_0 = \cos(rx)\left(\frac{\partial V}{\partial x}\right)_0 + \cos(ry)\left(\frac{\partial V}{\partial y}\right)_0 + \cos(rz)\left(\frac{\partial V}{\partial z}\right)_0,$$

the terms in the first integral depend upon the directions simply through the direction cosines of r, which on account of symmetry cause the integral to vanish. If $\overline{V}$ is the mean of V on the surface the equation then becomes

2) Boussinesq, *Application des Potentiels à l'étude de l'équilibre et du mouvement des solides élastiques*, p. 45.

31) $$4\pi(V - V_0) = 2\pi R^2 \left(\overline{\frac{\partial^2 V}{\partial r^2}}\right)_0,$$

where the bar over the derivative denotes the mean for all directions at 0, but this mean has just been proved toe be equal to $\frac{1}{3}\Delta V$ at 0. Consequently dividing by R^2 and taking the limit the terms of higher orders in R disappear and we have

32) $$\lim_{R=0} \frac{\overline{V} - V_0}{R^2} = \frac{1}{6}(\Delta V)_0.$$

The difference in the numerical coefficient in the two equations 27) and 32) is accounted for by the fact that in 32) we have a mean over a surface whereas in 27) we had a mean throughout a volume.

Any of the interpretations of the second differential parameter shows that it is also a differential invariant. Thus Green's theorem involves three different sorts of differential invariants.

117. Divergence. Solenoidal Vectors. If the components of the vector parameter are

33) $$\begin{aligned} P\cos(Px) &= X = \frac{\partial V}{\partial x}, \\ P\cos(Py) &= Y = \frac{\partial V}{\partial y}, \\ P\cos(Pz) &= Z = \frac{\partial V}{\partial z}, \end{aligned}$$

we have

34) $$\Delta V = \frac{\partial X}{\partial x} + \frac{\partial Y}{\partial y} + \frac{\partial Z}{\partial z},$$

and the theorem 23) becomes

35) $$-\iint P\cos(Pn)\,dS = -\iint [X\cos(nx) + Y\cos(ny) + Z\cos(nz)]\,dS = \iiint \left(\frac{\partial X}{\partial x} + \frac{\partial Y}{\partial y} + \frac{\partial Z}{\partial z}\right) d\tau.$$

If P is everywhere outward from the surface S, the integral is positive, and

$$\text{mean}\left(\frac{\partial X}{\partial x} + \frac{\partial Y}{\partial y} + \frac{\partial Z}{\partial z}\right) > 0.$$

Accordingly $\frac{\partial X}{\partial x} + \frac{\partial Y}{\partial y} + \frac{\partial Z}{\partial z}$ is called the *divergence* of the vector point-function whose components are X, Y, Z, and will be denoted by div. R.

The theorem as given in equation 35) may be stated as follows, and will be referred to as the DIVERGENCE THEOREM: *The mean value of the normal component of any vector point-function* outward *from any closed surface S within which the function is uniform and continuous, multiplied by the area of the surface, is equal to the mean value of the divergence of the vector in the space within S multiplied by its volume.* The theorem is here proved for a vector which is the parameter of a scalar point-function V, but it is evident that it may be proved directly whether this is the case or not by putting in equation 17) for W and x successively X, Y, Z and x, y, z respectively.

Let us consider the geometrical nature of a vector point-function R whose divergence vanishes in a certain region. In the neighborhood of any point, the vector will at some points be directed toward the point and at others away. We may then draw curves of such a nature that at every point of any curve the tangent is in the direction of the vector point-function R at that point. Such curves will be called *lines* of the vector function. Their differential equations are

36) $$\frac{dx}{X} = \frac{dy}{Y} = \frac{dz}{Z}.$$

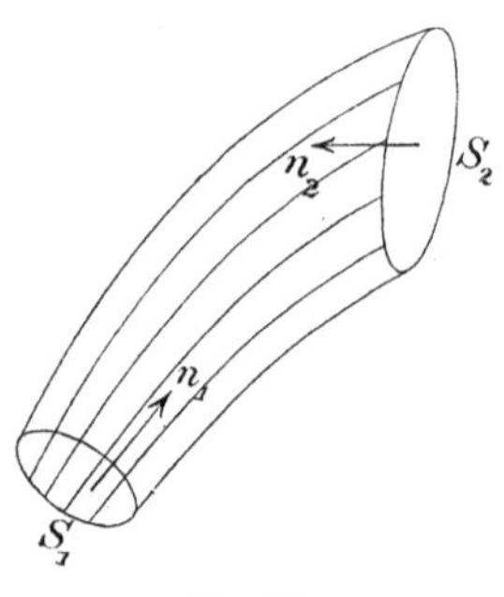

Fig. 123.

Suppose that such lines be drawn through all points of a closed curve, they will generate a tubular surface, which will be called a *tube* of the vector function. Let us now construct any two surfaces S_1 and S_2 cutting across the vector tube and apply the divergence theorem to the portion of space inclosed by the tube and the two surfaces or caps S_1 and S_2. Since at every point on the surface of the tube, R is tangent to the tube, the normal component vanishes. The only parts contributing anything to the surface integral are accordingly the caps, and since the divergence everywhere vanishes in τ, we have

37) $$\iint_{S_1} R\cos(Rn_1)\,dS_1 + \iint_{S_2} R\cos(Rn_2)\,dS_2 = 0.$$

If we draw the normal to S_2 in the other direction, so that as we move the cap along the tube the direction of the normal is continuous, the above formula becomes

38) $$\iint_{S_1} R\cos(Rn_1)\,dS_1 = \iint_{S_2} R\cos(Rn_2)\,dS_2,$$

or the surface integral of the normal component of R over *any* cap cutting the same vector tube is constant.

Such a vector will be termed *solenoidal*, or tubular, and the condition $\frac{\partial X}{\partial x} + \frac{\partial Y}{\partial y} + \frac{\partial Z}{\partial z} = 0$ will be termed the solenoidal condition (Maxwell). We may abbreviate it, div. $R = 0$. If a vector point-function R is lamellar as well as solenoidal, the scalar function V of which it is the vector parameter is harmonic, for

$$\frac{\partial X}{\partial x} + \frac{\partial Y}{\partial y} + \frac{\partial Z}{\partial z} = \text{div. } R = \Delta V = 0.$$

A solenoidal vector may be represented by its tubes, its direction being given by the tangent to an infinitesimal tube, and its magnitude being inversely proportional to its cross-section. As an example of a solenoidal vector we may take the velocity of particles of a moving liquid. If the velocity is R, with components X, Y, Z, the amount of liquid flowing through an element of surface dS in unit time is that contained in a prism of slant height R, and base dS, whose volume is

$$R \cos (Rn)\, dS.$$

The total flux, or quantity flowing in unit time through a surface S, is the surface integral

$$\iint R \cos (Rn)\, dS = \iint [X \cos (nx) + Y \cos (ny) + Z \cos (nz)]\, dS.$$

Such a surface integral may accordingly be called the *flux* of the vector R through S.

A tube of the vector R is a tube through whose sides no fluid flows, such as a material rigid tube through which a liquid flows, and the divergence theorem shows that as much liquid flows in through one cross-section as out through another, if the solenoidal condition holds. If the liquid is incompressible, this must of course be true.

As a second example of a solenoidal vector we have an y vector which is the curl of another vector, for

$$\frac{\partial}{\partial x}\left\{\frac{\partial Z}{\partial y} - \frac{\partial Y}{\partial z}\right\} + \frac{\partial}{\partial y}\left\{\frac{\partial X}{\partial z} - \frac{\partial Z}{\partial x}\right\} + \frac{\partial}{\partial z}\left\{\frac{\partial Y}{\partial x} - \frac{\partial X}{\partial y}\right\} = 0$$

identically.

The equation

$$39)\qquad \frac{\partial^2 V}{\partial x^2} + \frac{\partial^2 V}{\partial y^2} + \frac{\partial^2 V}{\partial z^2} = \Delta V = 0$$

is called Laplace's equation, and the operator

$$\Delta = \frac{\partial^2}{\partial x^2} + \frac{\partial^2}{\partial y^2} + \frac{\partial^2}{\partial z^2},$$

Laplace's operator.

The parameter ΔV is often called the *Laplacian* of V.

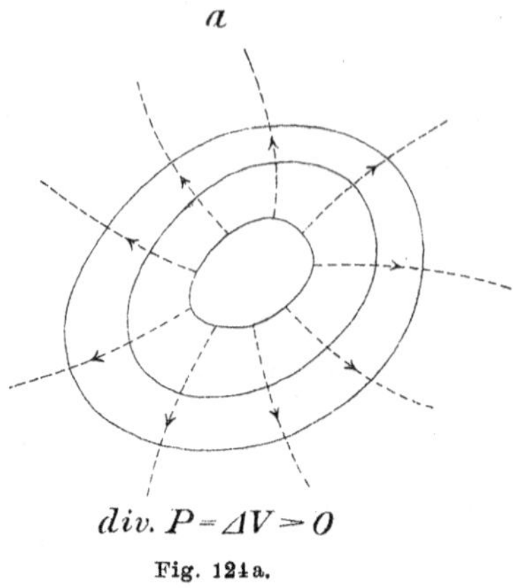

Fig. 124 a.

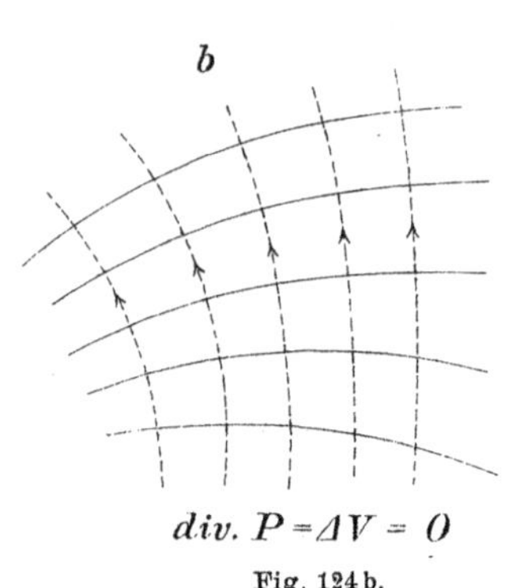

Fig. 124 b.

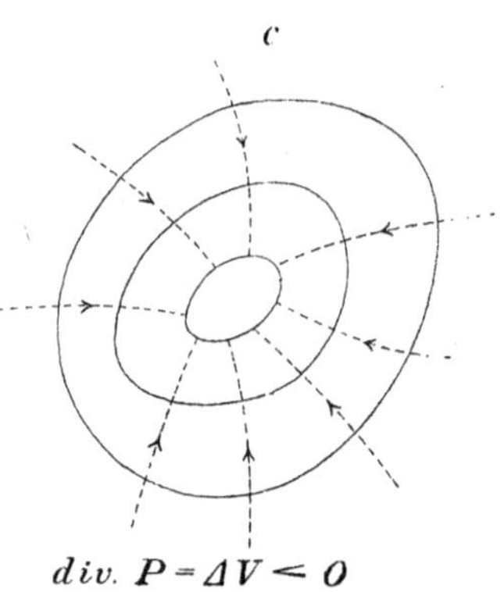

Fig. 124 c.

In Fig. 124 a, b, c, are graphically represented regions of divergent, solenoidal, and convergent vectors, with the level surfaces of the functions V of which they are the vector parameters. The arrows on the vector lines show the direction of increase of V, and it is evident that V has positive concentration (and a maximum value) where P is convergent, negative concentration (and a minimum value) where P is divergent, and no concentration (nor maximum) where P is solenoidal.

118. Reciprocal Distance. Gauss's Theorem. Consider the scalar point-function, $V = \frac{1}{r}$, where r is the distance from a fixed point or *pole* O. Then the level surfaces are spheres, and the parameter is

$$R = \pm \frac{d}{dr}\left(\frac{1}{r}\right) h_r,$$

and since $h_r = 1$,

$$|R| = \frac{1}{r^2},$$

drawn toward O (§ 110).

Consider the surface integral of the normal component of R directed into the volume bounded by a closed surface S not containing O, or as we have called it, the flux of R into S,

$$40)\qquad \iint R \cos(Rn)\, dS = -\iint \frac{1}{r^2} \cos(rn)\, dS.$$

The latter geometrical integral was reduced by Gauss. If to each point in the boundary of an element dS we draw a radius and thus get an infinitesimal cone with vertex O, and call the part of the surface of a sphere of radius r cut by this cone $d\Sigma$, $d\Sigma$ is the projection of dS on the sphere, Fig. 125, and as the normal to the sphere is in the direction of r, we have

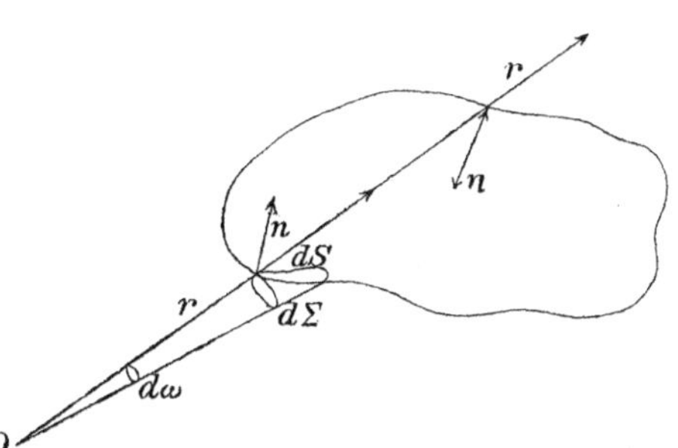

Fig. 125.

$$d\Sigma = \pm dS \cos(rn),$$

the upper sign, for r cutting in, the lower for r cutting out. If now we draw about O a sphere of radius 1, whose area is 4π, and call the portion of its area cut by the above-mentional cone $d\omega$, we have from the similarity of the right sections of the cone

$$\frac{d\Sigma}{d\omega} = \frac{r^2}{1},$$

$$d\Sigma = r^2 d\omega.$$

The ratio $d\omega$ is called the *solid angle* subtended by the infinitesimal cone.

Accordingly

41) $$\frac{dS \cos(nr)}{r^2} = \pm \frac{d\Sigma}{r^2} = \pm d\omega.$$

Now for every element $d\omega$, where r cuts into S, there is another equal one, $-d\omega$, where r cuts out, and the two annul each other. Hence for O outside S,

42) $$\iint \frac{\cos(rn)}{r^2} dS = 0.$$

If on the contrary, O lies inside S, the integral $\iint d\omega$ is to be taken over the whole of the unit sphere with the same sign, and consequently gives the area 4π. Hence for O within S,

43) $$\iint \frac{\cos(rn)}{r^2} dS = -4\pi.$$

These two results are known as Gauss's theorem, and the integral will be called Gauss's integral.[1])

1) Gauss, *Theoria Attractionis Corporum Sphaeroidicorum Ellipticorum homogeneorum Methodo nova tractata.* Werke, Bd. V, p. 9.

These results could have been obtained as direct results of the divergence theorem. For the tubes of the vector function R are cones with vertex O. If O is outside S, R is continuous in every point within S, and since the area of any two spheres cut out by a cone are proportional to the squares of the radii of the spheres, we have the normal flux of

$$R = \frac{1}{r^2}$$

equal for all spherical caps. Consequently R is solenoidal, and the flux through *any* closed surface is zero. If O is within S, R is solenoidal in the space between S and any sphere with centre O lying entirely within S, and the flux through S is the same as the flux through the sphere, which is evidently -4π.

The fact that R is solenoidal and V harmonic may be directly shown by differentiation. If the coordinates of O are a, b, c,

$$44)\qquad r^2 = (x-a)^2 + (y-b)^2 + (z-c)^2,$$
$$\frac{\partial (r^2)}{\partial x} = 2r\frac{\partial r}{\partial x} = 2(x-a);$$

$$45)\qquad \frac{\partial r}{\partial x} = \frac{x-a}{r}, \quad \frac{\partial r}{\partial y} = \frac{y-b}{r}, \quad \frac{\partial r}{\partial z} = \frac{z-c}{r};$$

$$46)\qquad \frac{\partial}{\partial x}\left(\frac{1}{r}\right) = -\frac{1}{r^2}\frac{\partial r}{\partial x} = -\frac{x-a}{r^3};$$

$$47)\qquad \frac{\partial^2}{\partial x^2}\left(\frac{1}{r}\right) = -\frac{1}{r^3} + \frac{3(x-a)}{r^4}\frac{\partial r}{\partial x} = \frac{3(x-a)^2 - r^2}{r^5},$$
$$\frac{\partial^3}{\partial y^2}\left(\frac{1}{r}\right) = \frac{3(y-b)^2 - r^2}{r^5}, \quad \frac{\partial^2}{\partial z^2}\left(\frac{1}{r}\right) = \frac{3(z-c)^2 - r^2}{r^5};$$

$$48)\qquad \Delta\left(\frac{1}{r}\right) = \frac{\partial^2\left(\frac{1}{r}\right)}{\partial y^2} + \frac{\partial^2\left(\frac{1}{r}\right)}{\partial y^2} + \frac{\partial^2\left(\frac{1}{r}\right)}{\partial z^2}$$
$$= \frac{3\left\{(x-a)^2 + (y-b)^2 + (z-c)^2\right\} - 3r^2}{r^5} = 0,$$

and $\frac{1}{r}$ is harmonic, except where $r = 0$.

119. Definition and fundamental Properties of Potential. We have seen in § 28, 34) that if we have any number of material particles m repelling or attracting according to the Newtonian Law of the inverse square of the distance, the function

$$U_s = -\gamma\left\{\frac{m_1 m_s}{r_1} + \frac{m_2 m_s}{r_2} + \cdots + \frac{m_n m_s}{r_n}\right\}.$$

where $r_1, r_2, \ldots r_n$ are the distances from the repelling points, is the force-function for all the forces acting upon the particle m_s. If we put the mass m_s equal to unity, at a point P whose coordinates are x, y, z the function

$$49)\qquad V = \frac{m_1}{r_1} + \frac{m_2}{r_2} + \cdots + \frac{m_n}{r_n} = \sum \frac{m}{r}$$

is called the *potential function* at the point P of the field of force due to the actions of the particles $m_1, m_2, \ldots m_n$, and γ times its *negative* vector parameter,

$$50)\qquad X = -\gamma \frac{\partial V}{\partial x}, \quad Y = -\gamma \frac{\partial V}{\partial y}, \quad Z = -\gamma \frac{\partial V}{\partial z},$$

is the *strength of the field*, that is, the force experienced by unit mass concentrated at the point x, y, z.[1])

Since any term $\frac{m_r}{r_r}$ possesses the same properties as the function $\frac{1}{r}$, § 118, we have for every term, for points where r is not equal to zero, $\varDelta\left(\frac{1}{r}\right) = 0$, and consequently

$$51)\qquad \varDelta V = m_1 \varDelta\left(\frac{1}{r_1}\right) + m_2 \varDelta\left(\frac{1}{r_2}\right) + \cdots + m_n \varDelta\left(\frac{1}{r_n}\right) = 0.$$

120. Potential of Continuous Distribution. Suppose now that the attracting masses, instead of being in discrete points, form a continuously extended body K.

Let the limit of the ratio of the mass to the volume of any infinitely small part be $\varrho = \lim_{\varDelta\tau = 0} \frac{\varDelta m}{\varDelta \tau}$, which is called the *density*. Let the coordinates of a point in the attracting body be a, b, c.

1) It is more usual among writers on attracting forces to write the force as the *positive* parameter of the potential. The convention above adopted in 49) amounts to defining the potential as the work necessary to remove the attracted particle of unit mass from the given poiut to infinity against the attracting forces, thus keeping the potential function positive, instead of negative as in § 28 (end). It is the usual practice to adopt such units that γ is equal to unity. In order to preserve consistency with the units previously employed and at the same time not to be obliged to introduce γ throughout all the equations of this chapter, we shall define potential as above 49) and introduce the factor γ into those equations which involve the relationship of the force to the potential. If the force is attractive, γ will be negative, and putting $\gamma = -1$, we get the usual formulae. Putting $\gamma = +1$, our notation agrees with that customary for electricity and magnetism, for example in the author's *Theory of Electricity and Magnetism*.

The potential at any point P, x, y, z, due to the mass dm at Q, a, b, c, is

$$dV = \frac{dm}{r},$$

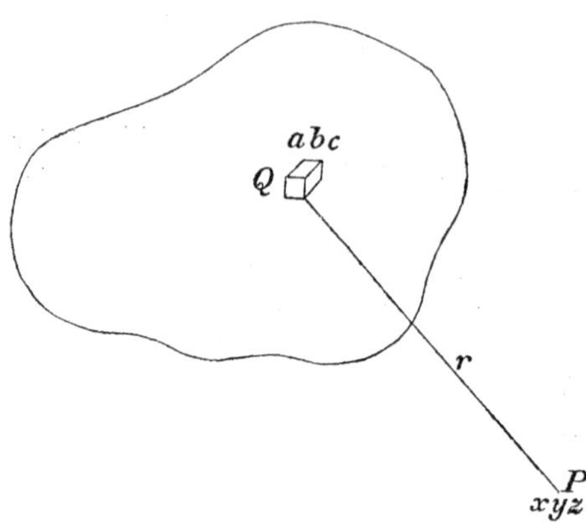

Fig. 126.

where r is the distance of the point x, y, z from the attracting point at a, b, c. The whole potential at x, y, z is the sum of that due to all parts of the attracting body, or the volume integral

$$52)\qquad V = \iiint\limits_K \frac{dm}{r}.$$

Now we have

$$dm = \varrho\, d\tau,$$

or in rectangular coordinates

$$d\tau = da\, db\, dc,$$

$$dm = \varrho\, da\, db\, dc.$$

If the body is not homogeneous, ϱ is different in different parts of the body K, and is a function of a, b, c, continous or discontinuous (e. g. a hole would cause a discontinuity). Since

$$r = \sqrt{(x-a)^2 + (y-b)^2 + (z-c)^2},$$

$$53)\qquad V = \iiint\limits_K \frac{dm}{r} = \iiint\limits_K \frac{\varrho\, da\, db\, dc}{\sqrt{(y-a)^2 + (y-b)^2 + (z-c)^2}}.$$

For every point x, y, z, V has a single, definite value. It is accordingly a uniform function of the point P, x, y, z.

It may be differentiated in any direction, we may find its level surfaces, its first differential parameter, whose negative multiplied by γ is equal to the whole action of K on a point of unit mass, and the *lines of force*, normal to the level, or *equipotential* surfaces.

If for any point x, y, z outside K, r_1 is the shortest distance to any point of K, and r_2 is the greatest distance, we have for any point in K

$$r_2 > r > r_1,$$

$$\frac{1}{r_2} < \frac{1}{r} < \frac{1}{r_1},$$

$$\frac{dm}{r_2} < \frac{dm}{r} < \frac{dm}{r_1};$$

$$\iiint\limits_K \frac{dm}{r_2} < \iiint\limits_K \frac{dm}{r} < \iiint\limits_K \frac{dm}{r_1}.$$

Since r_1 and r_2 are constant,

$$\frac{1}{r_2}\iiint\limits_K dm < \iiint\limits_K \frac{dm}{r} < \frac{1}{r_1}\iiint\limits_K dm.$$

Now since $\iiint\limits_K dm = M$, the whole mass of the body K, the above is

54) $$\frac{M}{r_2} < V < \frac{M}{r_1}.$$

Accordingly for an external point V is finite.

If R is the distance of x, y, z from some point in or at a finite distance from K,

55) $$\frac{RM}{r_2} < RV < \frac{RM}{r_1}.$$

If now we move off x, y, z to an infinite distance we have

$$\lim_{R=\infty} \frac{R}{r_2} = \lim_{R=\infty} \frac{R}{r_1} = 1,$$

and accordingly since RV lies between two quantities having the same limit,

56) $$\lim_{R=\infty} (RV) = M.$$

We say that V vanishes to the *first order* as R becomes infinite.

121. Derivatives. Consider the partial derivatives of V by x, y, z.

The element dm at a, b, c produces the potential

$$dV = \frac{dm}{r}$$

at x, y, z.

Differentiating by x, (dm and a, b, c being constant), we have

57) $$\frac{\partial}{\partial x}(dV) = dm\frac{\partial}{\partial x}\left(\frac{1}{r}\right) = -\frac{dm}{r^2}\frac{\partial r}{\partial x}.$$

By § 118, 45) $\frac{\partial r}{\partial x} = \frac{x-a}{r}$,

58) $$\frac{\partial}{\partial x}(dV) = -\frac{dm}{r^2}\frac{x-a}{r}.$$

Now

59) $$\frac{x-a}{r} = \cos(rx),$$

where the direction of r is taken from a, b, c to x, y, z. This being the derivative for that part of the potential due to dm, we have to take the sum of such expressions for all dm's in K, that is, perform a volume integration.

$$60)\qquad \frac{\partial V}{\partial x} = \frac{\partial}{\partial x}\iiint \frac{\varrho\, da\, db\, dc}{r} = \iiint \varrho \frac{\partial}{\partial x}\left(\frac{1}{r}\right) da\, db\, dc$$
$$= -\iiint \varrho \frac{x-a}{r^3} da\, db\, dc = -\iiint \frac{\varrho}{r^2} \cos(rx)\, da\, db\, dc.$$

Let the direction cosines of R be $\cos A$, $\cos B$, $\cos C$, and since

$$r_2 > r > r_1,$$
$$\frac{1}{r_2^2} < \frac{1}{r^2} < \frac{1}{r_1^2},$$
$$-\frac{1}{r_2^2} > -\frac{1}{r^2} > -\frac{1}{r_1^2},$$
$$-\frac{\varrho}{r_2^2}\cos(rx) > -\frac{\varrho}{r^2}\cos(rx) > -\frac{\varrho}{r_1^2}\cos(rx).$$

Multiplying and dividing the outside terms by $\cos A$ and integrating,

$$61)\qquad -\frac{\cos A}{r_2^2}\iiint \varrho \frac{\cos(rx)}{\cos A} d\tau > \frac{\partial V}{\partial x} > -\frac{\cos A}{r_1^2}\iiint \varrho \frac{\cos(rx)}{\cos A} d\tau.$$

Multiplying by R^2 and letting R increase without limit, since

$$\lim_{R=\infty} \frac{R^2}{r_1^2} = \lim_{R=\infty} \frac{R^2}{r_2^2} = \lim_{R=\infty} \frac{\cos(rx)}{\cos A} = 1,$$

$$62)\qquad \begin{aligned} \lim_{R=\infty}\left[R^2\frac{\partial V}{\partial x}\right] &= -M\cos A,\\ \lim_{R=\infty}\left[R^2\frac{\partial V}{\partial y}\right] &= -M\cos B,\\ \lim_{R=\infty}\left[R^2\frac{\partial V}{\partial z}\right] &= -M\cos C.\end{aligned}$$

Therefore the first derivatives of V, and hence the parameter, vanish at infinity to the *second* order.

In like manner for the second derivatives,

$$\frac{\partial^2 V}{\partial x^2} = \frac{\partial^2}{\partial x^2}\iiint \frac{\varrho\, d\tau}{r} = \iiint \varrho \frac{\partial^2}{\partial x^2}\left(\frac{1}{r}\right) d\tau$$
$$= \iiint \varrho \left[\frac{3(x-a)^2 - r^2}{r^5}\right] d\tau,$$
$$\frac{\partial^2 V}{\partial x \partial y} = \iiint \varrho \frac{3(x-a)(y-b)}{r^5} d\tau.$$

Every element in all the integrals discussed is finite, unless $r = 0$, hence all the integrals are finite. We might proceed in this manner, and should thus find that:

At points not in the attracting masses, V and all its derivatives are finite and (since their derivatives are finite) continuous, as well as uniform.

Also since

$$\frac{\partial^2 V}{\partial x^2} = \iiint \varrho \left[\frac{3(x-a)^2 - r^2}{r^5}\right] d\tau,$$

63)
$$\frac{\partial^2 V}{\partial y^2} = \iiint \varrho \left[\frac{3(y-b)^2 - r^2}{r^5}\right] d\tau$$

$$\frac{\partial^2 V}{\partial z^2} = \iiint \varrho \left[\frac{3(z-c)^2 - r^2}{r^5}\right] d\tau,$$

we have by addition

64)
$$\frac{\partial^2 V}{\partial x^2} + \frac{\partial^2 V}{\partial y^2} + \frac{\partial^2 V}{\partial z^2} = 0;$$

that is, V satisfies Laplace's equation.

This is also proved by applying Gauss's theorem [§ 118, 42)] to each element $\frac{dm}{r}$.

122. Points in the Attracting Mass. Let us now examine the potential and its derivatives at points in the substance of the attracting mass.

If P is within the mass, the element $\frac{dm}{r}$ at which the point Q, where dm is placed, coincides with P, becomes infinite. It does not however, therefore follow that the integral becomes infinite.

Let us separate from the mass K a small sphere of radius ε with the center at P. Call the part of the body within this sphere K' and the rest K''. Call the part of the integral due to K', V', and that due to K'', V''. Now since P is not in the mass K'', V'' and its derivatives are finite at P, and we have only to examine V' and its derivatives.

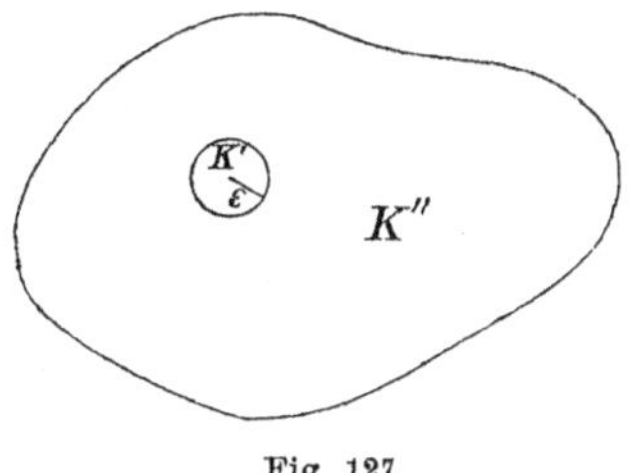

Fig. 127.

Let us insert polar coordinates

$$V' = \iiint\limits_{K'} \frac{\varrho\, d\tau}{r} = \int_0^{\varepsilon}\int_0^{\pi}\int_0^{2\pi} \frac{\varrho r^2 \sin\vartheta\, dr\, d\vartheta\, d\varphi}{r},$$

so that, integrating first with respect to φ and ϑ, since the absolute value of an integral is never greater than the integral of the absolute value of the integrand, and since $|\sin\theta| < 1$

65) $$| V' | < 2\pi^2 \varrho_m \int_0^\varepsilon r\,dr = 2\pi^2 \varrho_m \frac{\varepsilon^2}{2},$$

if ϱ_m is the greatest value of ϱ in K'.

As we make the radius ε diminish indefinitely this vanishes, hence the limit

$$\lim_{\varepsilon=0} (V' + V'')$$

is finite.

In like manner for the derivative

$$\frac{\partial V}{\partial x} = -\iiint \varrho \frac{x-a}{r^3} d\tau = -\iiint \frac{\varrho}{r^2} \cos(rx)\, d\tau.$$

Separate off K' from K''. The part of the integral from K'' is finite. In the other K' introduce polar coordinates, putting $\vartheta = (rx)$,

66) $$-\frac{\partial V'}{\partial x} = \int_0^\varepsilon \int_0^\pi \int_0^{2\pi} \frac{\varrho \cos\vartheta}{r^2} r^2 \sin\vartheta\, dr\, d\vartheta\, d\varphi,$$

$$\left|\frac{\partial V'}{\partial x}\right| < \varrho_m \int_0^\varepsilon dr \int_0^\pi \int_0^{2\pi} |\sin\vartheta \cos\vartheta|\, d\vartheta\, d\varphi,$$

$$< 2\pi^2 \varrho_m \varepsilon,$$

which also vanishes with ε. Therefore $\frac{\partial V}{\partial x}$ is everywhere finite, and in like manner $\frac{\partial V}{\partial y}, \frac{\partial V}{\partial z}$.

If we attempt this process for the second derivatives $\frac{\partial^2 V}{\partial x^2}, \ldots$ it fails on account of $\frac{dr}{r}$, which gives a logarithm becoming infinite in the limit.

We will give another proof of the finiteness of $\frac{\partial V}{\partial x}$. We have

67) $$\frac{\partial V}{\partial x} = \iiint \varrho \frac{a-x}{r^3}\, da\, db\, dc$$

$$= \iiint \varrho \frac{\partial}{\partial a}\left(-\frac{1}{r}\right) da\, db\, dc,$$

which by Green's theorem is equal to

$$\iint \frac{\varrho}{r} \cos(nx)\, dS + \iiint \frac{1}{r} \frac{\partial \varrho}{\partial a}\, d\tau.$$

This is however only to be applied in case the function $\frac{\varrho}{r}$ is everywhere finite and continuous. This ceases to be the case when P is in the attracting mass, hence we must exclude P by drawing a small sphere about it. Applying Green's theorem to the rest of the space K'', we have to add to the surface-integral the integral over the surface of the small sphere.

Since $\cos(nx) \leqq 1$, this is not greater than $\varrho_m \int\int \frac{dS}{r} = 4\pi\varepsilon\varrho_m$, which vanishes with ε. Hence the infinite element of the integrand contributes nothing to the integral.

In the same way that $\frac{\partial V}{\partial x}$ was proved finite, it may be proved continuous. Dividing it into two parts $\frac{\partial V'}{\partial x}$ and $\frac{\partial V''}{\partial x}$, of which the second is continuous, we may make, as shown, $\frac{\partial V'}{\partial x}$ as small as we please by making the sphere at P small enough. At a neighboring point P_1 draw a small sphere, and let the corresponding parts be $\frac{\partial V_1'}{\partial x}$ and $\frac{\partial V_1''}{\partial x}$. Then we can make $\frac{\partial V_1'}{\partial x}$ as small as we please, and hence also the difference $\frac{\partial V'}{\partial x} - \frac{\partial V_1'}{\partial x}$. Hence by taking P and P_1 near enough together, we can make the increment of $\frac{\partial V}{\partial x}$ as small as we please, or $\frac{\partial V}{\partial x}$ is continuous, and accordingly the second derivatives are finite.

123. Poisson's Equation. By Gauss's theorem [§ 118, 43)], we have

$$\int\int \frac{\cos(nr)\,dS}{r^2} = -4\pi,$$

when r is drawn from O, a point within S. Multiplying by m, a mass concentrated at O, and calling $V = \frac{m}{r}$,

$$68)\qquad \int\int \frac{m}{r^2}\cos(nr)\,dS = -\int\int \frac{\partial V}{\partial n}\,dS = -4\pi m.$$

The integral

$$-\gamma \int\int \frac{\partial V}{\partial n}\,dS = -\gamma \int\int P\cos(Pn)\,dS,$$

where n is the internal normal, is the surface integral of the *outward* normal component of the parameter γP, or the *inward* component of the force.

The surface integral of the normal component of force in the inward direction through S is called the *flux of force* into S, and we see that it is equal to $-4\pi\gamma$ times the element of mass within S. Masses without contribute nothing to the integral. Every mass dm

situated within S contributes $\frac{dm}{r}$ to the potential at any point and $-4\pi\gamma\, dm$ to the flux through the surface S. Therefore the whole mass, when the potential is $V = \iiint\limits_K \frac{dm}{r}$, contributes to the flux

$$-4\pi\gamma M = -4\pi\gamma \iiint\limits_K \varrho\, d\tau,$$

and

$$69)\qquad -\iint\limits_S \frac{\partial V}{\partial n}\, dS = -4\pi \iiint\limits_K \varrho\, d\tau.$$

Now the surface integral is, by the divergence theorem, equal to

$$70)\qquad \iiint\limits_\tau \Delta V\, d\tau = -4\pi \iiint\limits_K \varrho\, d\tau.$$

The surface S may be drawn inside the attracting mass, providing that we take for the potential only that due to matter in the space τ within S.

Accordingly for τ we may take any part whatever of the attracting mass, and

$$\iiint \Delta V\, d\tau = -4\pi \iiint\limits_\tau \varrho\, d\tau,$$

$$71)\qquad \iiint (\Delta V + 4\pi\varrho)\, d\tau = 0.$$

As the above theorem applies to any field of integration whatever, we must have everywhere

$$72)\qquad \Delta V + 4\pi\varrho = 0.$$

This is Poisson's extension of Laplace's equation, and says that at any point the second differential parameter of V is equal to -4π times the density at that point. Outside the attracting bodies, where $\varrho = 0$, this becomes Laplace's equation.

In our nomenclature, the *concentration* of the potential at any point is proportional to the density at that point.

A more elementary proof of the same theorem may be given as follows. At a point x, y, z construct a small rectangular parallelopiped whose faces have the coordinates

$$x,\ x+\xi,\ y,\ y+\eta,\ z,\ z+\zeta,$$

and find the flux of force through its six faces. At the face normal to the x-axis whose x coordinate is x let the mean value of the force be $-\frac{\partial V}{\partial x} = -P_x$.

The area of the face is $\eta\zeta$, so that this face contributes to the integral $-\iint P \cos(Pn)\, dS$ the amount $-\frac{\partial V}{\partial x}\eta\zeta$.

At the opposite face, since $\frac{\partial V}{\partial x}$ is continuous, we have for its value

$$\frac{\partial V}{\partial x} + \xi \frac{\partial}{\partial x}\left(\frac{\partial V}{\partial x}\right) + \text{terms of higher order in } \xi,$$

and therefore, the normal being directed the other way, this side contributes to the integral the amount

$$\left\{\frac{\partial V}{\partial x} + \xi \frac{\partial}{\partial x}\left(\frac{\partial V}{\partial x}\right) + \cdots\right\}\eta\zeta,$$

and the two together

$$\xi\eta\zeta \frac{\partial^2 V}{\partial x^2} + \text{terms of higher order.}$$

Similarly the faces perpendicular to Y-axis contribute $\xi\eta\zeta\frac{\partial^2 V}{\partial y^2}$, and the others $\xi\eta\zeta\frac{\partial^2 V}{\partial z^2}$.

Thus the surface integral is

$$\xi\eta\zeta\left\{\frac{\partial^2 V}{\partial x^2} + \frac{\partial^2 V}{\partial y^2} + \frac{\partial^2 V}{\partial z^2}\right\},$$

and by Gauss's theorem this is equal to

$$-4\pi m = -4\pi\rho\xi\eta\zeta,$$

where ρ is the mean density in the parallelopiped. Now making the parallelopiped infinitely small, and dividing by $\xi\eta\zeta$, we get

$$\Delta V = -4\pi\rho.$$

An important application of Poisson's equation has been made to the attraction of the earth. The acceleration g is made up of the resultant of the attraction of the earth and of the centrifugal acceleration. Since the latter has the components $\Omega^2 x$, $\Omega^2 y$ along axes perpendicular to the axis of rotation (§ 104), it has the potential function $\frac{\Omega^2}{2}(x^2 + y^2)$, so that if γ denote the *positive* value of the gravitation constant, and n the inward normal to an equipotential surface, we have, putting

73) $$U = \left\{\frac{\Omega^2}{2\gamma}(x^2 + y^2) + V\right\}, \quad g = \gamma\frac{\partial U}{\partial n},$$

where

$$V = \iiint \frac{\rho\, d\tau}{r}$$

is the potential of the earth's attraction. But by Poisson's equation,

$$\Delta V = -4\pi\varrho,$$

so that we have,

$$\text{74)} \qquad \Delta U = \Delta V + \frac{\Omega^2}{2\gamma}\Delta(x^2+y^2) = -4\pi\varrho + \frac{2\Omega^2}{\gamma}.$$

Now by the divergence theorem,

$$\iiint \Delta U d\tau = -\iint \frac{\partial U}{\partial n} dS,$$

so that

$$\text{75)} \qquad -\iint \frac{\partial U}{\partial n} dS = -4\pi \iiint \varrho d\tau + \frac{2\Omega^2}{\gamma}\iiint d\tau.$$

Now if the volume of the earth be v, its mean density ϱ_m, the volume integrals are respectively equal to $\varrho_m v$ and v, so that, multiplying by $\frac{\gamma}{4\pi v}$, this becomes

$$\text{76)} \qquad \gamma\varrho_m = \frac{\Omega^2}{2\pi} + \frac{1}{4\pi v}\iint \gamma \frac{\partial U}{\partial n} dS = \frac{\Omega^2}{2\pi} + \frac{1}{4\pi v}\iint g\, dS.$$

Thus if we know the value of g at every point on an equipotential surface, we obtain the value of the product $\gamma\varrho_m$ in terms of the angular velocity, and the surface integral of g. Using a formula given by Helmert representing the results of geodetic determinations of g, Woodward[1]) finds for the value of $\gamma\varrho_m$

$$\gamma\varrho_m = 3.6797 \times 10^{-7} \text{ sec}^{-2}.$$

Richarz and Krigar-Menzel[2]) obtain, in a similar manner,

$$\gamma\varrho_m = 3.680 \times 10^{-7} \text{ sec}^{-2}.$$

Combining this result with Boys's value of γ, p. 30, we obtain for the mean density of the earth the value

$$\varrho_m = 5.532 \frac{\text{gm}}{\text{cm}^3}.$$

124. Characteristics of Potential Function. We have now found the following properties of the potential function.

1st. It is everywhere holomorphic, that is, uniform, finite, continuous.

---

1) Woodward, *The Gravitational Constant and the Mean Density of the Earth.* Astronomical Journal, Jan. 1898.

2) F. Richarz und O. Krigar-Menzel, *Gravitationsconstante und mittlere Dichtigkeit der Erde, bestimmt durch Wägungen.* Ann. der Phys. u. Chem. 36, p. 177, 1898.

2nd. Its first partial derivatives are everywhere holomorphic.

3rd. Its second derivatives are finite.

4th. V vanishes at infinity to the first order,

$$\lim_{R=\infty} (RV) = M;$$

$\frac{\partial V}{\partial x}, \frac{\partial V}{\partial y}, \frac{\partial V}{\partial z}$ vanish to second order,

$$\lim_{R=\infty} \left(R^2 \frac{\partial V}{\partial x}\right) = -M \cos A.$$

5th. V satisfies everywhere Poisson's differential equation

$$\frac{\partial^2 V}{\partial x^2} + \frac{\partial^2 V}{\partial y^2} + \frac{\partial^2 V}{\partial z^2} = -4\pi\varrho,$$

and outside of attracting matter, Laplace's equation

$$\frac{\partial^2 V}{\partial x^2} + \frac{\partial^2 V}{\partial y^2} + \frac{\partial^2 V}{\partial z^2} = 0.$$

Any function having all these properties is a Newtonian potential function.

The field of force X, Y, Z is a solenoidal vector at all points outside of the attracting bodies, and hence if we construct tubes of force, the flux of force is constant through any cross-section of a given tube. A tube for which the flux is unity will be called a *unit tube.* The conception of lines of force and of the solenoidal property is due to Faraday.

Since V is a harmonic function outside of the attracting bodies, it has neither maximum nor minimum in free space, but its maximum and minimum must lie within the attracting bodies or at infinity.

In the attracting bodies the equation $-\Delta V = 4\pi\varrho$ says that the concentration of the potential at any point, or the divergence of the force from it is proportional to the density at that point, except where ϱ is discontinuous.

125. Examples. Potential of a homogeneous Sphere. Let the radius of the sphere be R, h the distance of P from its center,

$$V = \iiint \frac{\varrho\, d\tau}{r}.$$

Let us put s instead of r, using the latter symbol for the polar coordinate,

$$V = \iiint \frac{\varrho r^2}{s} \sin\vartheta\, d\vartheta\, d\varphi\, dr.$$

Now

$$s^2 = h^2 + r^2 - 2hr\cos\vartheta.$$

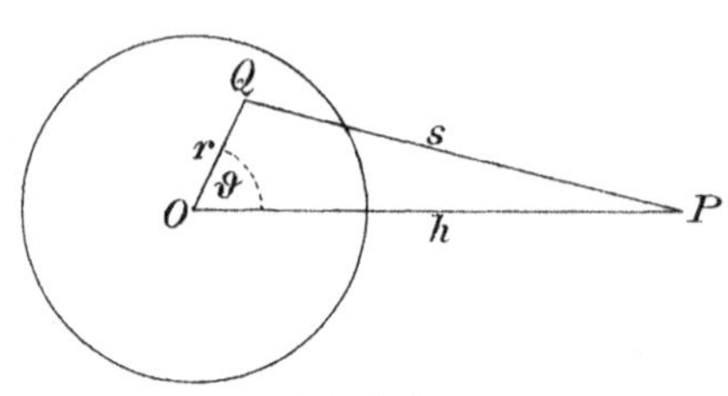

Fig. 128.

Differentiating, keeping r constant,

$$s\,ds = hr\sin\vartheta\,d\vartheta,$$

and introducing s as variable instead of ϑ,

$$V = \iiint \frac{\varrho r}{h}\,ds\,d\varphi\,dr.$$

If P is external we must integrate first with respect to s from $h - r$ to $h + r$.

77) $$V = \frac{2\pi\varrho}{h}\int_0^R\int_{h-r}^{h+r} r\,dr\,ds = \frac{4\pi\varrho}{h}\int_0^R r^2\,dr$$
$$= \frac{4\pi\varrho R^3}{3h} = \frac{M}{h}.$$

Hence the attraction of a sphere upon an external point is the same as if the whole mass were concentrated at the center.

A body having the property that the line of direction of its resultant attraction on a point passes always through a fixed point in the body is called *centrobaric.*

If instead of a whole sphere we consider a spherical shell of internal radius R_1 and outer R_2, the limits for r being R_1, R_2,

78) $$V = \frac{4\pi\varrho}{h}\int_{R_1}^{R_2} r^2\,dr = \frac{4\pi\varrho}{3h}(R_2^3 - R_1^3) = \frac{M}{h}.$$

We have

$$\frac{dV}{dh} = -\frac{M}{h^2},$$
$$\frac{d^2V}{dh^2} = \frac{2M}{h^3}.$$

If, on the other hand, P is in the spherical cavity, $h < R_1$, the limits for s are $r - h$, $r + h$

79) $$V = \frac{2\pi\varrho}{h}\int_{R_1}^{R_2}\int_{r-h}^{r+h} r\,dr\,ds = 4\pi\varrho\int_{R_1}^{R_2} r\,dr$$
$$= 2\pi\varrho(R_2^2 - R_1^2),$$

which is independent of h, that is, is constant in the whole cavity. Hence $\frac{\partial V}{\partial h} = 0$, and we get the theorem due to Newton that a homogeneous spherical shell exercises no force on a body within. (On account of symmetry the force can be only radial.)

If P is in the substance of the shell, we divide the shell into two by a concentric spherical surface passing through P, find the potential due to the part within P, and add it to that without, getting

80)
$$\begin{aligned} V &= \frac{4\pi\varrho}{3h}(h^3 - R_1^3) + 2\pi\varrho(R_2^2 - h^2) \\ &= 2\pi\varrho\left\{R_2^2 - \frac{h^2}{3}\right\} - \frac{4\pi\varrho R_1^3}{3h}, \\ \frac{dV}{dh} &= \frac{4\pi\varrho}{3}\left\{\frac{R_1^3}{h^2} - h\right\}, \\ \frac{d^2V}{dh^2} &= -\frac{4\pi\varrho}{3}\left\{\frac{2R_1^3}{h^3} + 1\right\}. \end{aligned}$$

Tabulating these results,

81)

	$h < R_1$	$R_1 < h < R_2$	$h > R_2$
	$2\pi\varrho(R_2^2 - R_1^2)$	$2\pi\varrho\left\{R_2^2 - \frac{h^2}{3}\right\} - \frac{4\pi\varrho R_1^3}{3h}$	$\frac{4\pi\varrho}{3h}(R_2^3 - R_1^3)$
$\frac{dV}{dh}$	0	$\frac{4\pi\varrho}{3}\left\{\frac{R_1^3}{h^2} - h\right\}$	$-\frac{4\pi\varrho}{3h^2}(R_2^2 - R_1^3)$
$\frac{d^2V}{dh^2}$	0	$-\frac{4\pi\varrho}{3}\left\{\frac{2R_1^3}{h^2} + 1\right\}$	$\frac{8\pi\varrho}{3h^2}(R_2^3 - R_1^3)$

Plotting the above results (Fig. 129) shows the continuity of V and its first derivative and the discontinuity of the second derivative at the surfaces of the attracting mass.

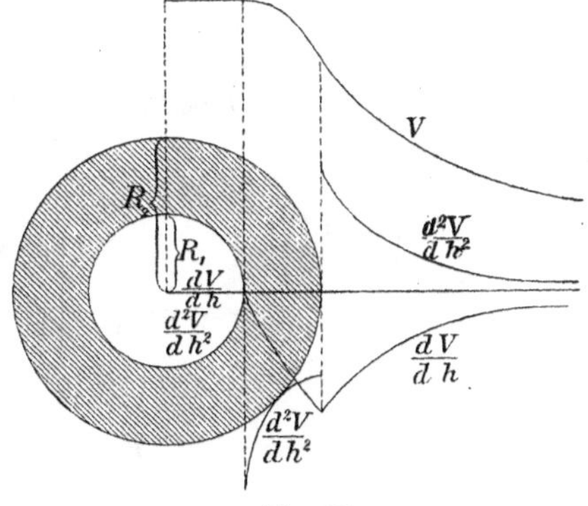

Fig. 129.

We see that the attraction of a solid sphere at a point within it is proportional to the distance from the center, for if $R_1 = 0$,

$$\frac{dV}{dh} = -\frac{4\pi\varrho h}{3},$$

and is independent of the radius of the sphere. Hence experiments on the decrease of the force of gravity in mines at known depths might give us the dimensions of the earth, if the earth were homogeneous. Experiment shows, however, that this is not the case.

126. Disc, Cylinder, Cone. Let us find the attraction of a circular disc of infinitesimal thickness at a point on a line normal to the disc at its center. Let the radius be R, thickness ε, distance of P from the center h.

82) $$V = \int_0^R \int_0^{2\pi} \varrho \frac{\varepsilon\, r\, dr\, d\varphi}{\sqrt{h^2 + r^2}}$$

$$= 2\pi\varepsilon\varrho \int_0^R \frac{r\, dr}{\sqrt{h^2 + r^2}} = 2\pi\varepsilon\varrho \left[\sqrt{h^2 + r^2}\right]_0^R$$

$$= 2\pi\varepsilon\varrho\, \{\sqrt{h^2 + R^2} - h\},$$

83) $$\frac{dV}{dh} = 2\pi\varepsilon\varrho \left\{\frac{h}{\sqrt{h^2 + R^2}} - 1\right\}.$$

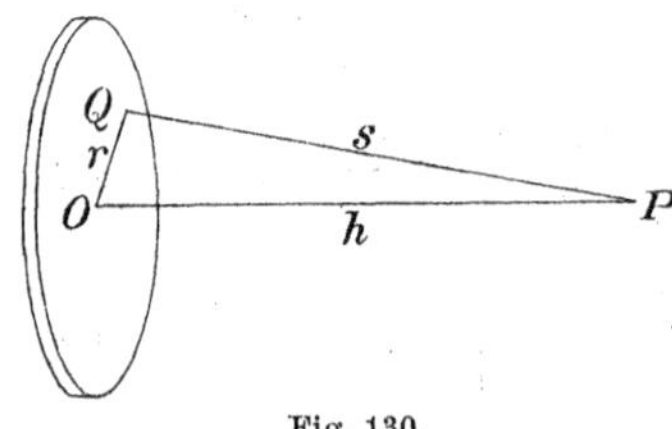

Fig. 130.

Attraction of circular cylinder on point in its axis. Let the length be l and let the point be external, at a distance h from the center.

By the above, a disc of thickness dx at a distance x from the center produces a potential at P

$$dV = 2\pi\varrho dx \{\sqrt{R^2 + (h - x)^2} - (h - x)\}.$$

Hence the whole is

84) $$V = 2\pi\varrho \int_{-\frac{l}{2}}^{\frac{l}{2}} \{\sqrt{R^2 + (h - x)^2} - (h - x)\}\, dx$$

$$= 2\pi\varrho \left\{\frac{x - h}{2}\sqrt{R^2 + (h - x)^2} + \frac{R^2}{2}\log\left(x - h + \sqrt{R^2 + (h - x)^2}\right)\right\}_{-\frac{l}{2}}^{\frac{l}{2}}$$

$$= \pi\varrho \left\{\left[\left(\frac{l}{2} - h\right)\sqrt{R^2 + \left(\frac{l}{2} - h\right)^2} + R^2 \log\left\{\frac{l}{2} - h + \sqrt{R^2 + \left(\frac{l}{2} - h\right)^2}\right\}\right]\right.$$

$$\left. - \left[\left(-\frac{l}{2} - h\right)\sqrt{R^2 + \left(-\frac{l}{2} - h\right)^2} + R^2 \log\left\{-\frac{l}{2} - h + \sqrt{R^2 + \left(-\frac{l}{2} - h\right)^2}\right\}\right]\right\}.$$

Circular cone on point in axis.

Let R be the radius of base, a the altitude, h the height of P above the vertex.

A disc at distance x below vertex and radius r causes potential at P,

$$dV = 2\pi\varrho\, dx\{\sqrt{(h+x)^2 + r^2} - (h+x)\};$$

and

$$\frac{r}{x} = \frac{R}{a}, \quad r = \frac{R}{a}x,$$

85) $$V = 2\pi\varrho \int_0^a dx \left\{\sqrt{(h+x)^2 + \frac{R^2}{a^2}x^2} - (h+x)\right\}.$$

If we have a conical mountain of uniform density on the earth, and determine the force of gravity at its summit and at the sea level, this gives us the ratio of the attraction of the sphere and cone to that of the sphere alone, and from this we get the ratio of the mass of the earth to the mass of the mountain. Such a determination was carried out by Mendenhall, on Fujiyama, Japan, in 1880, giving 5.77 for the earth's density.

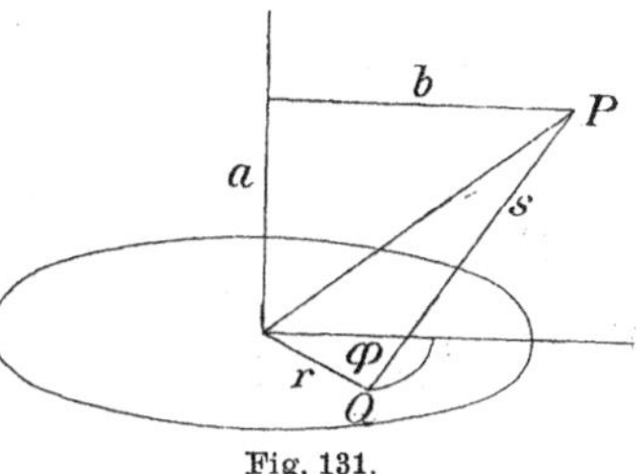

Fig. 131.

Circular disc on point not on axis. Let the coordinates of P with respect to the center be a, b, 0. Then

$$s^2 = a^2 + (b - r\cos\varphi)^2 + r^2\sin^2\varphi,$$

86) $$V = \int_0^R\int_0^{2\pi} \frac{\varepsilon r\, dr\, d\varphi}{\sqrt{a^2 + (b - r\cos\varphi)^2 + r^2\sin^2\varphi}},$$

an elliptic integral.

127. Surface Distributions. In the case of the circular disc of thickness ε, $\varepsilon\varrho$ is the amount of matter per unit of surface of the disc. It is often convenient to consider distributions of matter over surfaces, in such a manner that though ε be considered infinitesimal ϱ increases so that the product $\varepsilon\varrho$ remains finite. The product $\varepsilon\varrho = \sigma$ is called the surface density, and the distribution is called a surface distribution.

We have

87) $$dm = \sigma dS, \quad V = \iint \frac{\sigma dS}{r}.$$

In the case of the disc, we had

$$88)\qquad \frac{\partial V}{\partial h} = 2\pi\varepsilon\varrho\left\{\frac{h}{\sqrt{h^2+R^2}} - 1\right\}.$$

When $h = 0$ we have

$$89)\qquad \left(\frac{\partial V}{\partial h}\right)_{h=0} = -2\pi\sigma.$$

The attraction of a disc upon a particle in contact with it at its center is independent of the radius of the disc, and is equal to 2π times the surface density.

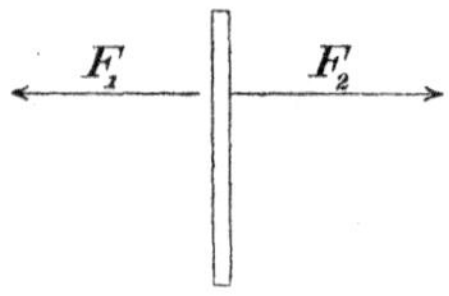

If the force on a particle in contact on the right be called F_2, positive if to the right, we have

$$90)\qquad F_2 = +2\pi\gamma\sigma.$$

Fig. 132.

By symmetry, the force on a particle at the left in contact with the disc is

$$F_1 = -2\pi\gamma\sigma,$$

$$91)\qquad F_2 - F_1 = 4\gamma\pi\sigma.$$

Now if x denote the direction of the normal to the right,

$$F_1 = -\gamma\left(\frac{\partial V}{\partial x}\right)_1,$$

$$F_2 = -\gamma\left(\frac{\partial V}{\partial x}\right)_2,$$

and we see that on passing through the surface there is a discontinuity in the value of $\frac{\partial V}{\partial x}$ of the magnitude $4\pi\sigma$.

Consider a thin spherical shell. We have for an external point

$$92)\quad V = \frac{4\pi\varrho}{3h}(R_2^3 - R_1^3 = \frac{4\pi\varrho}{3h}(R_2 - R_1)(R_2^2 + R_2R_1 + R_1^2),$$

and making $R_2 - R_1 = \varepsilon$, $\lim R_1 = \lim R_2 = R$,

$$V = \frac{4\pi\sigma}{3h}\cdot 3R^2,$$

$$\frac{dV}{dh} = -\frac{4\pi\sigma}{h^2}R^2,$$

and on the outside for $h = R$,

$$\frac{dV}{dh} = -4\pi\sigma.$$

Within we have everywhere

$$V = const.,\quad \frac{dV}{dh} = 0.$$

Thus there is in like manner a discontinuity in the first derivative of the potential in the direction of the normal, on passing through the attracting surface, of the amount $4\pi\sigma$.

Consider now any surface distribution of surface density σ. Apply Gauss's theorem to a small tube of force bounded by portions of two equipotential surfaces $d\Sigma_1$ and $d\Sigma_2$ on opposite sides of and near to the element of the attracting surface dS (Fig. 133). The flux out from the tubes is

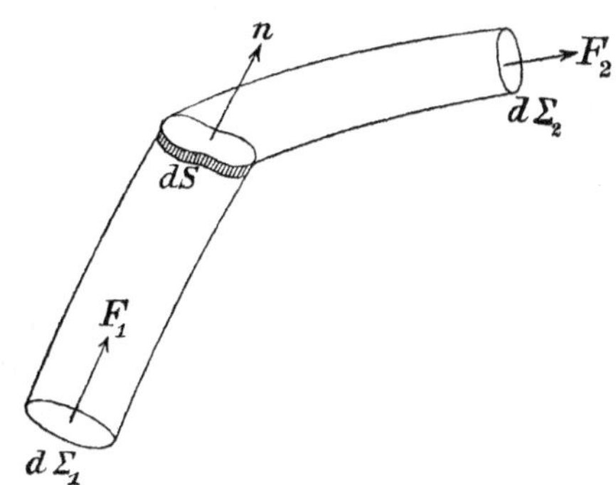

Fig. 133.

$$F_2 d\Sigma_2 - F_1 d\Sigma_1,$$

and this must be equal to $4\pi\gamma$ times the matter contained in the tube, which is σdS. Therefore

$$F_2 d\Sigma_2 - F_1 d\Sigma_1 = 4\pi\gamma\sigma dS.$$

But if the length and diameter of the tube are infinitesimal $d\Sigma_1$ and $d\Sigma_2$ are the projections of dS,

$$d\Sigma_1 = dS\cos(F_1\, n), \quad d\Sigma_2 = dS\cos(F_2\, n)$$

where n is the normal to the attracting surface. Accordingly

$$F_2\cos(F_2\, n)\, dS - F_1\cos(F_1\, n)\, dS = 4\pi\gamma\sigma\, dS,$$

and since

$$F_2\cos(F_2\, n) = -\gamma\left(\frac{\partial V}{\partial n}\right)_2,$$

$$F_1\cos(F_1\, n) = -\gamma\left(\frac{\partial V}{\partial n}\right)_2,$$

93) $$\left(\frac{\partial V}{dn}\right)_1 - \left(\frac{\partial V}{\partial n}\right)_2 = 4\pi\sigma.$$

The normal to S is here drawn toward the side 2. We find then that in general, on traversing a surface distribution, the normal force has a discontinuity equal to $4\gamma\pi\sigma$.

This is Poisson's equation for a surface distribution. If we draw the normal *away* from the surface on each side, we may write

94) $$\frac{\partial V}{\partial n_1} + \frac{\partial V}{\partial n_2} = -4\pi\sigma,$$

or

$$F_1\cos(F_1\, n_1) + F_2\cos(F_2\, n_2) = F_{1n_1} + F_{2n_2} = 4\pi\gamma\sigma.$$

128. Green's Formulae. Let us apply Green's theorem to two functions, of which one, V, is the potential function due to any distribution of matter, and the other, $U = \frac{1}{r}$, where r is the distance from a fixed point P, lying in the space τ over which we take the integral. Let the space τ concerned be that bounded by a closed surface S, a small sphere Σ of radius ε about P, and, if P is without S, a sphere of infinite radius with center P.

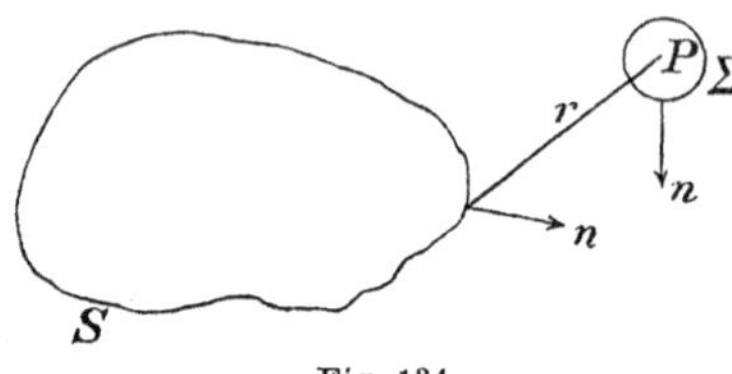

Fig. 134.

Now the theorem was stated in § 115, 22) for the normal drawn in toward τ, which means outward from S and Σ, and inward from the infinite sphere, as

$$1)\qquad \iint\left(\frac{\partial U}{\partial n} - U\frac{\partial V}{\partial n}\right) dS = \iiint (U\Delta V - V\Delta U)\, d\tau,$$

and since

$$U = \frac{1}{r}, \quad \Delta U = 0,$$

in the whole space τ, so that 1) becomes

$$2)\qquad \iint\left\{V\frac{\partial \frac{1}{r}}{\partial n} - \frac{1}{r}\frac{\partial V}{\partial n}\right\} dS = \iiint \frac{1}{r}\Delta V d\tau.$$

The surface integrals are to be taken over S, over the small sphere, and over the infinite sphere. For a sphere with center at P,

$$\frac{\partial \frac{1}{r}}{\partial n} = \pm \frac{\partial \frac{1}{r}}{\partial r} = \mp \frac{1}{r^2}, \quad dS = r^2 d\omega,$$

the upper or lower sign being taken according as the sphere is the inner or outer boundary of τ;

$$\iint V\frac{\partial \frac{1}{r}}{\partial n} dS = \mp \iint V d\omega,$$

and for

$$r = \infty$$

V vanishes, hence this integral vanishes. Also

$$3)\qquad -\iint \frac{1}{r}\frac{\partial V}{\partial n} dS = -\iint \frac{1}{r}\frac{\partial V}{\partial n} r^2 d\omega = -r\iint \frac{\partial V}{\partial n} d\omega.$$

Now at infinity, $\frac{\partial V}{\partial n}$ is of order $\frac{1}{r^2}$, and being multiplied by r, still vanishes. Accordingly the infinite sphere contributes nothing. For the small sphere the case is different. The first integral

$$-\iint V d\omega$$

becomes, as the radius ε of the sphere diminishes,

$$4) \qquad -V_P \iint d\omega = -4\pi V_P.$$

The second part

$$-\varepsilon \iint \frac{\partial V}{\partial n} d\omega,$$

however, since $\frac{\partial V}{\partial n}$ is finite on the sphere, vanishes with ε. Hence there remain on the left side of the equation only $-4\pi V_P$ and the integral over S. We obtain therefore

$$-4\pi V_P + \iint \left\{ V \frac{\partial \frac{1}{r}}{\partial n} - \frac{1}{r} \frac{\partial V}{\partial n} \right\} dS = \iiint \frac{\Delta V}{r} d\tau,$$

$$5) \; V_P = \frac{1}{4\pi} \iint \left\{ V \frac{\partial \frac{1}{r}}{\partial n} - \frac{1}{r} \frac{\partial V}{\partial n} \right\} dS - \frac{1}{4\pi} \iiint \frac{\Delta V}{r} d\tau \quad (P \text{ outside } S),$$

the normal being drawn *outward* from S. This formula is due to Green.

Therefore we see that any function which is uniform and continuous everywhere outside of a certain closed surface, which vanishes at infinity to the first order, and whose parameter vanishes at infinity to the second order, is determined at every point of space considered if we know at every point of that space the value of the second differential parameter, and in addition the values on the surface S of the function and its vector parameter resolved in the direction of the outer normal.

In particular, if V is *harmonic* in all the space considered, we have

$$6) \qquad V_P = \frac{1}{4\pi} \iint \left\{ V \frac{\partial \frac{1}{r}}{\partial n} - \frac{1}{r} \frac{\partial V}{\partial n} \right\} dS,$$

and a harmonic function is determined everywhere by its values and those of its normal component of parameter at all points of the surface S.

Since

$$\frac{\partial \frac{1}{r}}{\partial n} = -\frac{1}{r^2} \frac{\partial r}{\partial n}$$
$$= -\frac{1}{r^2} \left\{ \cos(nx) \frac{\partial r}{\partial x} + \cos(ny) \frac{\partial r}{\partial y} + \cos(nz) \frac{\partial r}{\partial z} \right\} = -\frac{\cos(nr)}{r^2},$$

we may write 6)

7) $$V_P = -\frac{1}{4\pi}\iint\left(\frac{V}{r^2}\cos(nr) + \frac{1}{r}\frac{\partial V}{\partial n}\right) dS.$$

Consequently, we may produce at all points outside of a closed surface S the same field of force as is produced by any distribution of masses lying *inside* of S, whose potential is V, if we distribute over the surface S a *surface* distribution of surface-density,

8) $$\sigma = -\frac{1}{4\pi}\left\{\frac{V\cos(nr)}{r} + \frac{\partial V}{\partial n}\right\}.$$

In the general expression, 5) the surface integral represents the potential due to the masses *within* S, while the volume integral

$$-\frac{1}{4\pi}\iiint\frac{\Delta V}{r}d\tau$$

since everywhere

$$-\frac{1}{4\pi}\Delta V = \varrho,$$

is equal to

$$\iiint\frac{\varrho\, d\tau}{r},$$

that is, the potential due to all the masses in the region τ, viz., outside S.

129. Equipotential Layers. As a still more particular case of 7), if the surface S is taken as one of the equipotential surfaces of the internal distribution, we have all over the surface $V = V_S = const.$, and the constant may be taken out from the first integral,

9) $$V_P = -\frac{V_S}{4\pi}\iint\frac{\cos(nr)}{r^2}dS - \frac{1}{4\pi}\iint\frac{1}{r}\frac{\partial V}{\partial n}dS.$$

Now by Gauss's theorem $\iint\frac{\cos(nr)}{r^2}dS = 0$; accordingly,

10) $$V_P = -\frac{1}{4\pi}\iint\frac{1}{r}\frac{\partial V}{\partial n}dS = \iint\frac{\sigma}{r}dS,$$

so that V_P is represented as the potential of a surface distribution of surface-density

$$\sigma = -\frac{1}{4\pi}\frac{\partial V}{\partial n} = \frac{1}{4\pi}\frac{F}{\gamma}\cos(Fn) = \frac{\pm 1}{4\pi}\frac{F}{\gamma}.$$

The whole mass of the equivalent surface distribution is

11) $$\iint\sigma\, dS = -\frac{1}{4\pi}\iint\frac{\partial V}{\partial n}dS$$
$$= \frac{1}{4\pi\gamma}\iint F\cos(Fn)\, dS,$$

which, being the flux of force outward from S, is by Gauss's theorem, § 128, 68), equal to M, the mass within S.

Accordingly we may enunciate the theorem, due to Chasles and Gauss[1]):

We may produce outside any equipotential surface of a distribution M the same effect as the distribution itself produces, by distributing over that surface a layer of surface-density equal to $\frac{1}{4\pi\gamma}$ times the outward force at every point of the surface. The mass of the whole layer will be precisely that of the original internal distribution. Such a layer is called an *equipotential layer.* (Definition — A superficial layer which coincides with one of its own equipotential surfaces.) Reversing the sign of this density will give us a layer which will, *outside,* neutralize the effect of the bodies within.

The above theorem has an important application in determining the attraction of the earth at outside points. Equation 10) shows that the potential and therefore the attraction is determined at all outside points if F, which is connected with g as in § 123, is known at all points of an equipotential surface. It will be shown later that the surface of the sea is an equipotential surface. Consequently if the value of g is known from pendulum observations at a sufficient number of stations distributed over the surface of the earth the attraction at all points outside the earth can be calculated.

Let us now suppose the point P is *within* S. In this case, we apply Green's theorem to the space within S, and we do not have the integrals over the infinite sphere. The normal in the above formulae is now drawn inward from S, or if we still wish to use the outward normal, we change the sign of the surface integral in 5),

$$12) \quad V_p = -\frac{1}{4\pi}\int\!\!\int\left\{V\frac{\partial\frac{1}{r}}{\partial n_e} - \frac{1}{r}\frac{\partial V}{\partial n_e}\right\}dS - \frac{1}{4\pi}\int\!\!\int\!\!\int\frac{\Delta V}{v}\,d\tau, \qquad (P \text{ inside } S).$$

Note that both formulae 5) are 12) are identical if the normal is drawn into the space in which P lies.

Hence within a closed surface a holomorphic function is determined at every point solely by its values and those of its normally resolved parameter at all points of the *surface,* and by the values of its *second* parameter at all points in the space within the surface.

A harmonic function may be represented by a potential function of a surface distribution.

---

1) Chasles, *Sur l'attraction d'une couche ellipsoidale infiniment mince,* Journ. Éc. Polytec., Cahier 25, p. 266, 1837; Gauss, *Allgemeine Lehrsätze,* § 36.

Now if the surface S is equipotential, the function V cannot be harmonic everywhere within unless it is constant. For since two equipotential surfaces cannot cut each other, the potential being a one-valued function, successive equipotential surfaces must surround each other, and the innermost one, which is reduced to a point, will be a point of maximum or minimum. But we have seen (§ 116) that this is impossible for a harmonic function. Accordingly a function constant on a closed surface and harmonic within must be a constant.

If however there be matter within and without S, the volume integral, as before, denotes the potential due to the matter in the space τ (within S), and the surface integral that due to the matter without. If the surface is equipotential, the surface integral

$$-\frac{V_S}{4\pi}\iint\frac{\partial\frac{1}{r}}{\partial n_e}+\frac{1}{4\pi}\iint\frac{1}{r}\frac{\partial V}{\partial n_e}dS$$
$$=\frac{V_S}{4\pi}\iint\frac{\cos(n_e r)}{r^2}dS+\frac{1}{4\pi}\iint\frac{1}{r}\frac{\partial V}{\partial n_e}dS.$$

The first integral is now equal to 4π, so that

$$13)\qquad V_P=V_S+\frac{1}{4\pi}\iint\frac{1}{r}\frac{\partial V}{\partial n_e}dS-\frac{1}{4\pi}\iiint\frac{\Delta V}{r}d\tau,$$

V_S being constant contributes nothing to the derivatives of V, so that the outside bodies may be replaced by a surface layer of density

$$14)\qquad \sigma=\frac{1}{2\pi}\frac{\partial V}{\partial n_e}=-\frac{1}{4\pi\gamma}F\cos(Fn_e)=\pm\frac{F}{4\pi\gamma}.$$

The mass of the surface distribution,

$$15)\qquad \iint\sigma\,dS=\frac{1}{4\pi}\iint\frac{\partial V}{\partial n_e}dS=-\frac{1}{4\pi\gamma}\iint F\cos(Fn_e)\,dS,$$

n_e being the outward normal, is the inward flux of force through S, which is equal to minus the mass of the *interior* matter, and is not, as in the former case, equal to the mass which it replaces.

130. Gauss's Mean Theorem. As an example of equation 6) let us make the surface S a sphere with center at P. Then in the first term of the integral we have

$$\frac{\partial\left(\frac{1}{r}\right)}{\partial n}=-\frac{\partial\left(\frac{1}{r}\right)}{\partial r}=\frac{1}{r^2},$$

which is constant and may be taken outside the integral. In the second term $\frac{1}{r}$ being similarly taken outside the integral, we have

$$\iint \frac{\partial V}{\partial n} dS = 0$$

since the function is harmonic in the sphere considered. Accordingly the formula reduces to the first term

$$V_P = \frac{1}{4\pi r^2} \iint V dS \tag{16}$$

The surface integral represents the mean value of V on the surface of the sphere multiplied by the area of the surface, $4\pi r^2$. Thus we have the theorem due to Gauss. The value of the potential at any point not situated in attracting matter is equal to the mean value of the potential at points on any sphere with center at the given point and not containing attracting matter. It at once follows from this theorem that a harmonic function cannot have a point of maximum or minimum, for making the sphere about such a point small enough the theorem would be violated.

131. Potential completely determined by its characteristic Properties. We have proved that the potential function due to any volume distribution has the following properties:

1. It is, together with its first differential parameter, uniform, finite, and continuous.

2. It vanishes to the first order at ∞, and its parameter to the second order.

3. It is harmonic outside the attracting bodies, and in them satisfies

$$\Delta V = -4\pi\varrho.$$

The preceding investigation shows that a function having these properties is a potential function, and is completely determined.

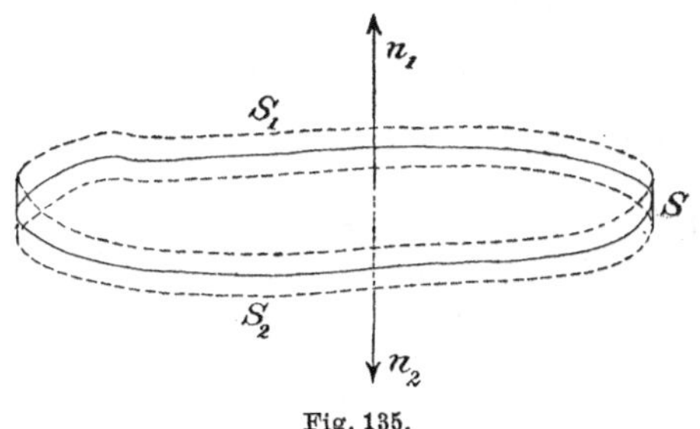

Fig. 135.

For we may apply the above formula 5) to all space, and then the only surface integral being that due to the infinite sphere, which vanishes, we have

$$V = -\frac{1}{4\pi} \iiint_\infty \frac{\Delta V}{r} d\tau = \iiint_\infty \frac{\varrho}{r} d\tau. \tag{17}$$

If however, the above conditions are fulfilled by a function V, except that a certain surfaces S its first parameter is discontinuous, let us draw on each side of the surface S surfaces at distances equal to ε from S, and exclude that portion of space lying between these, which we will call S_1 and S_2.

If the normals are drawn *into* τ we have

$$5)\quad V = \frac{1}{4\pi}\iint\left(V\frac{\partial}{\partial n}\left(\frac{1}{r}\right) - \frac{1}{r}\frac{\partial V}{\partial n}\right)dS - \frac{1}{4\pi}\iiint\frac{\Delta V}{r}\,d\tau.$$

The surface integrals are to be taken over both surfaces S_1 and S_2 and the volume integrals over all space except the thin layer between S_1 and S_2. This is the only region where there is discontinuity, hence in τ the theorem applies, and

$$18)\quad 4\pi V = \iint_{S_1} V\frac{\partial}{\partial n_1}\left(\frac{1}{r}\right)dS_1 + \iint_{S_2} V\frac{\partial}{\partial n_2}\left(\frac{1}{r}\right)dS_2$$
$$-\iint_{S_1}\frac{1}{r}\frac{\partial V}{\partial n_1}dS_1 - \iint_{S_2}\frac{1}{r}\frac{\partial V}{\partial n_2}dS_2 - \iiint_{\infty}\frac{\Delta V}{r}\,d\tau.$$

Now let us make ε infinitesimal, then the surfaces S_1, S_2 approach each other and S. V is continuous at S, that is, is the same on both sides, hence, since $\frac{\partial}{\partial n_1}\left(\frac{1}{r}\right) = -\frac{\partial}{\partial n_2}\left(\frac{1}{r}\right)$, in the limit the first two terms destroy each other. This is not so for the next two, for $\frac{\partial V}{\partial n_1}$ is *not* equal to $\frac{\partial V}{\partial n_2}$ because of the discontinuity.

In the limit, then

$$19)\quad V_P = -\frac{1}{4\pi}\iint\frac{1}{r}\left(\frac{\partial V}{\partial n_1} + \frac{\partial V}{\partial n_2}\right)dS - \frac{1}{4\pi}\iiint_{\infty}\frac{\Delta V}{r}\,d\tau.$$

The volume integral, as before, denotes the potential $\iiint\frac{\varrho}{r}\,d\tau$ due to the volume distribution, while the surface integral denotes the potential of a surface distribution $\iint\frac{\sigma\,dS}{r}$, where

$$20)\quad \sigma = -\frac{1}{4\pi}\left\{\frac{\partial V}{\partial n_1} + \frac{\partial V}{\partial n_2}\right\}.$$

Hence we get a new proof of Poisson's surface condition, § 127, 94)

132. Kelvin and Dirichlet's Principle. We shall now consider a question known on the continent of Europe as *Dirichlet's Problem.*

Given the values of a function at all points of a closed surface S — is it possible to find a function which, assuming these values on the surface is, with its parameters, uniform, finite, continuous, and is itself *harmonic* at all points within S?

This is the internal problem — the external may be stated in like manner, specifying the conditions as to vanishing at infinity.

Consider the integral

$$21)\qquad J(u)=\iiint\left\{\left(\frac{\partial u}{\partial x}\right)^2+\left(\frac{\partial u}{\partial y}\right)^2+\left(\frac{\partial u}{\partial z}\right)^2\right\}d\tau,$$

of a function u throughout the space τ within S.

J must be positive, for every element is a sum of squares. It cannot vanish, unless everywhere $\frac{\partial u}{\partial x}=\frac{\partial u}{\partial y}=\frac{\partial u}{\partial z}=0$, that is $u=const.$ But since u is continuous, unless it is constant on S, this will not be the case.

Accordingly
$$J(u)>0.$$

Now of the infinite variety of functions u there must be, according to Dirichlet, at least one which makes J less than for any of the others. Call this function v, and call the difference between this and any other hs, so that

$$u=v+hs,$$

h being constant.

The condition for a minimum is that

$$J(v)<J(v+hs),$$

for all values of h.

Now

$$\frac{\partial u}{\partial x}=\frac{\partial v}{\partial x}+h\frac{\partial s}{\partial x},\quad \text{etc.}$$

$$23)\quad \left(\frac{\partial u}{\partial x}\right)^2+\left(\frac{\partial u}{\partial y}\right)^2+\left(\frac{\partial u}{\partial z}\right)^2$$
$$=\left(\frac{\partial v}{\partial x}\right)^2+\left(\frac{\partial v}{\partial y}\right)^2+\left(\frac{\partial v}{\partial z}\right)^2+h^2\left\{\left(\frac{\partial s}{\partial x}\right)^2+\left(\frac{\partial s}{\partial y}\right)^2+\left(\frac{\partial s}{\partial z}\right)^2\right\}$$
$$+2h\left\{\frac{\partial v}{\partial y}\frac{\partial s}{\partial y}+\frac{\partial v}{\partial x}\frac{\partial s}{\partial x}+\frac{\partial v}{\partial z}\frac{\partial s}{\partial z}\right\}.$$

Integrating,

$$23)\quad J(u)=J(v)+h^2J(s)+2h\iiint\left(\frac{\partial v}{\partial x}\frac{\partial s}{\partial x}+\frac{\partial v}{\partial y}\frac{\partial s}{\partial y}+\frac{\partial v}{\partial z}\frac{\partial s}{\partial z}\right)d\tau.$$

Now in order that $J(v)$ may be a minimum, we must have

$$24)\qquad h^2J(s)+2h\iiint\left(\frac{\partial v}{\partial x}\frac{\partial s}{\partial x}+\frac{\partial v}{\partial y}\frac{\partial s}{\partial y}+\frac{\partial v}{\partial z}\frac{\partial s}{\partial z}\right)d\tau>0,$$

for *all* values h, positive or negative. But as s is as yet unlimited, we may take h so small that the absolute value of the term in h is greater than that of the term in h^2, and if we choose the sign of h opposite to that of the integral, making the product negative, the whole will be negative.

The only way to leave the sum always positive is to have the integral vanish. (It will be observed that the above is exactly the process of the calculus of variations. We might put δv instead of hs.)

The cond.tion for a minimum is then

$$25)\qquad \iiint\left\{\frac{\partial v}{\partial x}\frac{\partial s}{\partial x}+\frac{\partial v}{\partial y}\frac{\partial s}{\partial y}+\frac{\partial v}{\partial z}\frac{\partial s}{\partial z}\right\}d\tau=0.$$

But by Green's theorem, this is equal to

$$-\iint s\frac{\partial v}{\partial n}dS-\iiint s\Delta v\,d\tau.$$

Now at the surface the function is given, hence u and v must have the same values, and $s=0$.

Consequently the surface integral vanishes, and

$$26)\qquad \iiint s\Delta v\,d\tau=0.$$

But since s is arbitrary, the integral can vanish only if everywhere in τ, $\Delta v=0$, v is therefore the function which solves the problem.

The proof of the so-called *Existence-theorem*, namely, that there is such a function, depends on the assumption that there *is* a function which makes the integral J a minimum. This assumption has been declared by Weierstrass, Kronecker, and others, to be faulty. The principle of Lord Kelvin and Dirichlet, which declares that there is a function v, has been rigidly proved for a number of special cases, but the above general proof is longer admitted. It is given here on account of its historical interest.[1])

We can however prove that *if* there is a function v, satisfying the conditions, it is unique. For, if there is another, v', put

$$u=v-v'.$$

$$27)\qquad J(u)=\iiint\left\{\left(\frac{\partial u}{\partial x}\right)^2+\left(\frac{\partial u}{\partial y}\right)^2+\left(\frac{\partial u}{\partial z}\right)^2\right\}d\tau$$

$$=-\iint u\frac{\partial u}{\partial n}dS-\iiint u\Delta u\,d\tau.$$

1) Thomson, *Theorems with reference to the solution of certain Partial Differential Equations,* Cambridge and Dublin Math. Journ., Jan. 1848; Reprint of *Papers in Electrostatics and Magnetism,* XIII. The name *Dirichlets's Prinzip* was given by Riemann (*Werke*, p. 90). For a historical and critical discussion of this matter the student may consult Burkhardt, *Potentialtheorie* in the Encyklopädie der Mathematik, Bacharach, *Abriss der Geschichte der Potentialtheorie,* as well as Harkness and Morley, *Theory of Functions*, Chap. IX, Picard, *Traité d'Analyse,* Tom. II, p. 38. It has been quite recently shown by Hilbert that Riemann's proof given above can be so modified as to be made rigid. The modern subject of *integral equations* promises to be a powerful method for the establishment of such existence-theorems, but the matter is one in regard to which the physicist need have little concern.

On the surface, since $v = v'$, $u = 0$. In τ, since Δv and $\Delta v'$ are zero, $\Delta u = 0$. Accordingly $J(u) = 0$. But, as we have shown, this can only be if $u = const.$ But on S, $u = 0$, hence, throughout τ, $u = 0$ and $v = v'$.

133. Green's Theorem in Orthogonal Curvilinear Coordinates. We shall now consider Green's theorem in terms of any orthogonal coordinates, beginning with the special case forming the divergence theorem, § 117, 35).

$$28)\quad -\iint [X\cos(nx) + Y\cos(ny) + Z\cos(nz)]\,dS = \iiint \left\{\frac{\partial X}{\partial x} + \frac{\partial Y}{\partial y} + \frac{\partial Z}{\partial z}\right\} d\tau.$$

Instead of the components X, Y, Z, let us consider the projections P_1, P_2, P_3 of a vector P along the directions of the tangents to the coordinate lines q_1, q_2, q_3 at any point. Then projecting along the normal n to S, we have the integrand in the surface integral

$$29)\qquad P_1\cos(nn_1) + P_2\cos(nn_2) + P_3\cos(nn_3).$$

If we divide the volume T up into elementary curved prisms bounded by level surfaces of q_2 and q_3, as in the case of rectangular coordinates (Fig. 136), we have, at each case of cutting into or out of S respectively,

$$\pm\, dS\cos(nn_1) = dS_1,$$

where dS_1 is the area of the part cut by the prism from the level surface q_1.

Now by § 114,

$$dS_1 = \frac{dq_2}{h_2}\frac{dq_3}{h_3},$$

accordingly

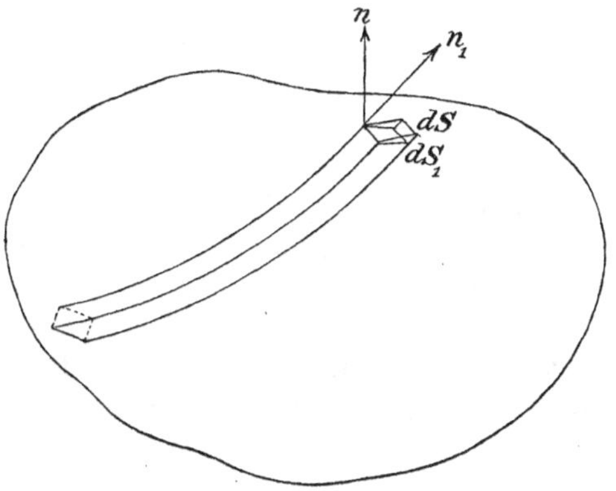

Fig. 136.

$$30)\quad -\iint P_1\cos(nn_1)\,dS = -\iint P_1\frac{dq_2}{h_2}\frac{dq_3}{h_3} = \iiint \frac{\partial}{\partial q_1}\left(\frac{P_1}{h_2 h_3}\right) dq_1\,dq_2\,dq_3,$$

the change from the double to the triple integral involving the same considerations as in the proof given for rectangular coordinates in § 115.

Transforming the other two integrals in like manner,

$$31)\quad -\iint\{P_1\cos(nn_1)+P_2\cos(nn_2)+P_3\cos(nn_3)\}dS$$

$$=\iiint\left\{\frac{\partial}{\partial q_1}\left(\frac{P_1}{h_2 h_3}\right)+\frac{\partial}{\partial q_2}\left(\frac{P_2}{h_3 h_1}\right)+\frac{\partial}{\partial q_3}\left(\frac{P_3}{h_1 h_2}\right)\right\}dq_1\,dq_2\,dq_3.$$

But this is equal to

$$\iiint \operatorname{div} P\cdot d\tau.$$

But since

$$d\tau=\frac{dq_1}{h_1}\frac{dq_2}{h_2}\frac{dq_3}{h_3},$$

multiplying and dividing the last integrand in 31) by $h_1 h_2 h_3$ we find that since the volume integrals are equal for any volume, the integrands must be equal, or

$$32)\quad \operatorname{div} P=h_1 h_2 h_3\left\{\frac{\partial}{\partial q_1}\left(\frac{P_1}{h_2 h_3}\right)+\frac{\partial}{\partial q_2}\left(\frac{P_2}{h_3 h_1}\right)+\frac{\partial}{\partial q_3}\left(\frac{P_3}{h_1 h_2}\right)\right\}.$$

If the vector is lamellar, its projections are the partial parameters according to q_1, q_2, q_3 of its potential V (§ 110),

$$P_1=h_1\frac{\partial V}{\partial q_1},\quad P_2=h_2\frac{\partial V}{\partial q_2},\quad P_3=h_3\frac{\partial V}{\partial q_3}.$$

Equation 32) then becomes

$$33)\quad \Delta V=h_1 h_2 h_3\left\{\frac{\partial}{\partial q_1}\left(\frac{h_1}{h_2 h_3}\frac{\partial V}{\partial q_1}\right)+\frac{\partial}{\partial q_2}\left(\frac{h_2}{h_3 h_1}\frac{\partial V}{\partial q_2}\right)+\frac{\partial}{\partial q_3}\left(\frac{h_3}{h_1 h_2}\frac{\partial V}{\partial q_3}\right)\right\}.$$

This result for the value of ΔV was given by Lamé, by means of a laborious direct transformation. The method here used is a modification of one given by Jacobi and Somoff.[1])

In order to prove Green's theorem in its general form, we remark that from the nature of the mixed parameter of the two functions U and V as a geometric product we have

$$34)\quad \Delta(U,V)=P_1^U P_1^V+P_2^U P_2^V+P_3^U P_3^V$$

$$=h_1^2\frac{\partial U}{\partial q_1}\frac{\partial V}{\partial q_1}+h_2^2\frac{\partial U}{\partial q_2}\frac{\partial V}{\partial q_2}+h_3^2\frac{\partial U}{\partial q_3}\frac{\partial V}{\partial q_3}.$$

Forming the volume integral, and integrating the first term partially according to q_1 we obtain

1) Lamé, *Journal de l'École Polytechnique,* Cahier 23, *p.* 215, 1833; *Leçons sur les Coordonnés curvilignes*, II. Jacobi, *Über eine partikuläre Lösung der partiellen Differentialgleichung* $\Delta V=0$, Crelle's Journal, Bd. 36, p. 113. Somoff, *Theoretische Mechanik*, II. §§ 51, 52.

$$35)\quad \iiint h_1^2 \frac{\partial U}{\partial q_1}\frac{\partial V}{\partial q_1} d\tau = \iiint h_1^2 \frac{\partial U}{\partial q_1}\frac{\partial V}{\partial q_1}\frac{dq_1}{h_1}\frac{dq_2}{h_2}\frac{dq_3}{h_3}$$
$$= \iint U h_1 \frac{\partial V}{\partial q_1}\frac{dq_2}{h_2}\frac{dq_3}{h_3}$$
$$-\iiint U \frac{\partial}{\partial q_1}\left(\frac{h_1}{h_2 h_3}\frac{\partial V}{\partial q_1}\right) dq_1\, dq_2\, dq_3,$$

which as above is equal to

$$-\iint U h_1 \frac{\partial V}{\partial q_1}\cos(n_1 n)\, dS - \iiint U \frac{\partial U}{\partial q_1}\left(\frac{h_1}{h_2 h_3}\frac{\partial V}{\partial q_1}\right) dq_1\, dq_2\, dq_3.$$

Integrating the other terms in like manner we obtain the general formula,

$$36)\quad \iiint \left\{ h_1^2 \frac{\partial U}{\partial q_1}\frac{\partial V}{\partial q_1} + h_2^2 \frac{\partial U}{\partial q_2}\frac{\partial V}{\partial q_2} + h_3^2 \frac{\partial U}{\partial q_3}\frac{\partial V}{\partial q_3} \right\} \frac{dq_1}{h_1}\frac{dq_2}{h_2}\frac{dq_3}{h_3}$$
$$= -\iint U \left\{ h_1 \frac{\partial V}{\partial q_1}\cos(n_1 n) + h_2 \frac{\partial V}{\partial q_2}\cos(n_2 n) h_3 \frac{\partial V}{\partial q_3}\cos(n_3 n) \right\} dS$$
$$-\iiint U \left\{ \frac{\partial}{\partial q_1}\left(\frac{h_1}{h_2 h_3}\frac{\partial V}{\partial q_1}\right) + \frac{\partial}{\partial q_2}\left(\frac{h_2}{h_3 h_1}\frac{\partial V}{\partial q_2}\right) \right.$$
$$\left. + \frac{\partial}{\partial q_3}\left(\frac{h_3}{h_1 h_2}\frac{\partial V}{\partial q_3}\right) \right\} dq_1\, dq_2\, dq_3,$$

in which each integrand is found to correspond to one of those in § 115, 20).

134. Stokes's Theorems in Orthogonal Curvilinear Coordinates. The proof of Stokes's theorem given in § 30 can be easily adapted to curvilinear coordinates.[1] Let P_1, P_2, P_3 be the projections of a vector P on the varying directions of the tangents to the coordinate lines at any point. Then, the projections of the arc ds being $ds_1\ ds_2\ ds_3$, we consider the line-integral

$$37)\quad I = \int_A^B P\cos(P, ds)\, ds = \int_A^B (P_1 ds_1 + P_2 ds_2 + P_3 ds_3)$$
$$= \int_A^B \left(P_1 \frac{dq_1}{h_1} + P_2 \frac{dq_2}{h_2} + P_3 \frac{dq_3}{h_3}\right)$$
$$= \int_A^B (R_1 dq_1 + R_2 dq_2 + R_3 dq_3),$$

where

$$R_s = \frac{P_s}{h_s}.$$

---

1) Webster. Note on Stokes's theorem in Curvilinear Coordinates. *Bull. Am. Math. Soc.*, 2nd Ser., Vol. IV., p. 438, 1898.

Let us now make an infinitesimal transformation of the curve as in § 30. Then the change in the integral is

$$38 \quad \delta I = \int_A^B (\delta R_1 dq_1 + \delta R_2 dq_2 + \delta R_3 dq_3 + R_1 d\delta q_1 + R_2 d\delta q_2 + R_3 d\delta q_3)$$

The last three terms can be integrated by parts, giving

$$39) \qquad \int_A^B R_s d\delta q_s = R_s \delta q_s \Big/_A^B - \int_A \delta q_s dRs. \qquad s=1, 2, 3,),$$

and, since the integrated terms vanish at the limits,

$$40) \quad \delta I = \int (\delta R_1 dq_1 + \delta R_2 dq_2 + \delta R_3 dq_3 - dR_1 \delta q_1 - dR_2 \delta q_2 - dR_3 \delta q_3),$$

Performing the operations denoted by δ and d, as on p. 85, six of the eighteen terms cancel, and there remain the terms,

$$41) \qquad \delta I = \int \Big[(\delta q_2 dq_3 - \delta q_3 dq_2) \left\{ \frac{\partial R_3}{\partial q_2} - \frac{\partial R_2}{\partial q_3} \right\} + (\delta q_3 dq_1 - \delta q_1 dq_3) \left\{ \frac{\partial R_1}{\partial q_3} - \frac{\partial R_3}{\partial q_1} \right\} + (\delta q_4 dq_2 - \delta q_2 dq_1) \left\{ \frac{\partial R_2}{\partial q_1} - \frac{\partial R_1}{\partial q_2} \right\} \Big].$$

Now the changes δq_2, dq_2, δq_3, dq_3, in the coordinates correspond to distances,

$$\frac{\delta q_2}{h_2}, \quad \frac{dq_2}{h_2}, \quad \frac{\delta q_3}{h_3}, \quad \frac{dq_3}{h_3},$$

measured along the coordinate lines, and the determinant of these distances,

$$\frac{1}{h_2 h_3} (\delta q_2 dq_3 - \delta q_3 dq_2)$$

is equal to the area of the projection on the surface q_1 of the infinitesimal parallelogram swept over by the arc ds during the transformation. Calling this area dS, and its normal n, we have

$$\frac{1}{h_2 h_3} (\delta q_2 dq_3 - \delta q_3 dq_2) = \cos(n n_1) dS.$$

If we now continually repeat the transformation, until the curve 1 joining AB is transformed into the curve 2, the total change in I is equal to the surface integral over the intervening surface,

$$42) \quad \Sigma \delta I = I - I_1$$

$$= \iint \Big[h_2 h_3 \left(\frac{\partial R_3}{\partial q_2} - \frac{\partial R_2}{\partial q_3} \right) \cos(nn_1) + h_3 h_1 \left(\frac{\partial R_1}{\partial q_3} - \frac{\partial R_3}{\partial q_2} \right) \cos(nn_2) + h_1 h_2 \left(\frac{\partial R_2}{\partial q_1} - \frac{\partial R_1}{\partial q_2} \right) \cos(n n_3) \Big] dS.$$

Accordingly the components of $\omega = \text{curl}\, P$ along the coordinate directions are, inserting the values of R_1, R_2, R_3.

$$\omega_1 = h_2 h_3 \left\{ \frac{\partial}{\partial q_2}\left(\frac{P_3}{h_3}\right) - \frac{\partial}{\partial q_3}\left(\frac{P_2}{h_2}\right) \right\},$$

43)
$$\omega_2 = h_3 h_1 \left\{ \frac{\partial}{\partial q_3}\left(\frac{P_1}{h_1}\right) - \frac{\partial}{\partial q_1}\left(\frac{P_3}{h_3}\right) \right\},$$

$$\omega_3 = h_1 h_2 \left\{ \frac{\partial}{\partial q_1}\left(\frac{P_2}{h_2}\right) - \frac{\partial}{\partial q_2}\left(\frac{P_1}{h_1}\right) \right\}.$$

If these vanish, the vector P is lamellar, and the above equations give the conditions that

$$\frac{P_1}{h_1} = \frac{\partial V}{\partial q_1}, \quad \frac{P_2}{h_2} = \frac{\partial V}{\partial q_2}, \quad \frac{P_3}{h_3} = \frac{\partial V}{\partial q_3},$$

so that

$$P_s = h_s \frac{\partial V}{\partial q_s}$$

as in the previous section.

135. Laplace's Equation in Spherical and Cylindrical Coordinates. Applying equation 33) to spherical coordinates

$$h_r = 1, \quad h_\vartheta = \frac{1}{r}, \quad h_\varphi = \frac{1}{r \sin\vartheta},$$

44)
$$\varDelta V = \frac{1}{r^2 \sin\vartheta}\left\{ \frac{\partial}{\partial r}\left(r^2 \sin\vartheta \frac{\partial V}{\partial r}\right) + \frac{\partial}{\partial \vartheta}\left(\sin\vartheta \frac{\partial V}{\partial \vartheta}\right) + \frac{\partial}{\partial \varphi}\left(\frac{1}{\sin\vartheta}\frac{\partial V}{\partial \varphi}\right) \right\}$$
$$= \frac{\partial^2 V}{\partial r^2} + \frac{2}{r}\frac{\partial V}{\partial r} + \frac{1}{r^2}\frac{\partial^2 V}{\partial \vartheta^2} + \frac{1}{r^2}\cot\vartheta \frac{\partial V}{\partial \vartheta} + \frac{1}{r^2 \sin^2\varphi}\frac{\partial^2 V}{\partial \varphi^2}.$$

We may apply this equation to determine the attraction of a sphere. For external points $\varDelta V = 0$, and since by symmetry V is independent of ϑ and φ,

45)
$$\frac{d^2 V}{dr^2} + \frac{2}{r}\frac{dV}{dr} = 0 \quad \text{or} \quad \frac{d}{dr}\left(r^2 \frac{dV}{dr}\right) = 0,$$

$$r^2 \frac{dV}{dr} = c, \qquad \frac{dV}{dr} = \frac{c}{r^2},$$

$$\mathrm{V} = -\frac{c}{r} + c'.$$

But since

$$\lim_{r=\infty} (r V) = M,$$

$$\lim_{r=\infty} [-c + c' r] = M,$$

we must have $c' = 0$, $-c = M$.

Apply the above transformation to cylindrical coordinates

$$h_z = 1, \quad h_\varrho = 1, \quad h_\omega = \frac{1}{\varrho},$$

46)
$$\begin{aligned}\Delta V &= \frac{1}{\varrho}\left\{\frac{\partial}{\partial z}\left(\varrho\frac{\partial V}{\partial z}\right) + \frac{\partial}{\partial \varrho}\left(\varrho\frac{\partial V}{\partial \varrho}\right) + \frac{\partial}{\partial \omega}\left(\frac{1}{\varrho}\frac{\partial V}{\partial \omega}\right)\right\}\\ &= \frac{\partial^2 V}{\partial z^2} + \frac{\partial^2 V}{\partial \varrho^2} + \frac{1}{\varrho}\frac{\partial V}{\partial \varrho} + \frac{1}{\varrho^2}\frac{\partial^2 V}{\partial \omega^2}.\end{aligned}$$

If we apply this to calculate the potential due to a cylindrical homogeneous body with generators parallel to the axis of z and of infinite length, the potential is independent of z and satisfies at external points,

47)
$$\begin{aligned}0 &= \frac{\partial^2 V}{\partial x^2} + \frac{\partial^2 V}{\partial y^2}\\ &= \frac{\partial^2 V}{\partial \varrho^2} + \frac{1}{\varrho}\frac{\partial V}{\partial \varrho} + \frac{1}{\varrho^2}\frac{\partial^2 V}{\partial \omega^2}.\end{aligned}$$

If the cylinder is circular, V is independent of ω, and we have the ordinary differential equation

$$\frac{d^2 V}{d\varrho^2} + \frac{1}{\varrho}\frac{dV}{d\varrho} = 0 \quad \text{or} \quad \frac{d}{d\varrho}\left(\varrho\frac{dV}{d\varrho}\right) = 0,$$

$$\varrho\frac{dV}{d\varrho} = C, \qquad \frac{dV}{d\varrho} = \frac{C}{\varrho},$$

48)
$$V = C\log\varrho + C'.$$

The force in the direction of ϱ is inversely proportional to the *first* power of ϱ.

We may verify this by direct calculation. Let us consider the cylinder as infinitely thin, with cross-section ϖ. We will find the component of *force* in the direction of ϱ. The action of dm at z on P at distance ϱ (Fig. 137) is

$$\frac{dm}{s^2} = \frac{dm}{\sqrt{\varrho^2 + z^2}}.$$

Fig. 137.

The component parallel to ϱ is

$$\frac{dm}{s^2}\cos(\varrho s) = \frac{\varrho\, dm}{s^3}.$$

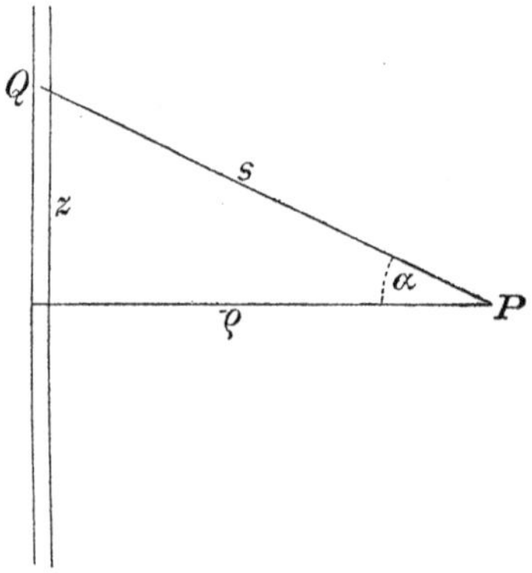

Now since, calling the density δ, $dm = \delta\varpi dz$, we have for the total force in direction ϱ

49)
$$F = 2\int_0^\infty \frac{\delta\varpi\varrho\, dz}{\sqrt{(\varrho^2 + z^2)^3}}.$$

Put

$$z = \varrho \tan \vartheta,$$
$$dz = \varrho \sec^2 \vartheta \, d\vartheta.$$

50) $$F = 2\,\varpi\,\delta \int_0^{\frac{\pi}{2}} \frac{\varrho^2 \sec^2 \vartheta \, d\vartheta}{\varrho^3 \sec^3 \vartheta} = \frac{2\,\varpi\,\delta}{\varrho} \Big[\sin \vartheta\Big]_0^{\frac{\pi}{2}}$$

$$= \frac{2\,\varpi\,\delta}{\varrho} = \frac{C}{\varrho}, \text{ as before.}$$

If we had attempted to verify the value 48) of V by direct calculation, we should have found a difficulty in the appearance of a logarithm which would have become infinite when the length of the cylinder became infinite. Nevertheless the *attraction* is finite, as just shown. It is to be noted that all the properties hitherto proved to hold have been for potentials of bodies of *finite* extent.

136. Logarithmic Potential. We may state the above result in terms of the following two-dimensional problem. Suppose that on a plane there be distributed a layer of mass in such a way that a point of mass m repels a point of unit mass in the plane with a force $\frac{m}{r}$ where r is their distance apart. The potential due to m is $V = -m \log r$ and it satisfies the differential equation

$$\frac{\partial^2 V}{\partial x^2} + \frac{\partial^2 V}{\partial y^2} = 0.$$

Similarly, in the case of any mass distributed in the plane, with surface-density μ, an element $dm = \mu dS$ produces the potential $-dm \log r$, and the whole potential

51) $$V = -\iint dm \log r = -\iint \mu \log r \, dS,$$

where r is the distance from the repelled point x, y to the repelling dm at a, b, so that

$$r^2 = (x-a)^2 + (y-b)^2.$$

We may verify by direct differentiation that, at external points, this V satisfies

$$\frac{\partial^2 V}{\partial x^2} + \frac{\partial^2 V}{\partial y^2} = 0,$$

$$\frac{\partial V}{\partial x} = -\frac{\partial}{\partial x}\iint \mu \log r \, da\, db = -\iint \mu \frac{\partial}{\partial x}(\log r)\, da\, db$$
$$= -\iint \frac{\mu}{r}\frac{\partial r}{\partial x}\, da\, db = -\iint \frac{\mu(x-a)}{r^2}\, da\, db,$$

$$\frac{\partial^2 V}{\partial x^2} = -\iint \mu \frac{\partial}{\partial x}\left\{\frac{x-a}{r^2}\right\} da\, db = -\iint \mu \left\{\frac{1}{r^2} - \frac{2(x-a)^2}{r^4}\right\} da\, db,$$

$$\frac{\partial^2 V}{\partial y^2} = -\iint \left\{\frac{1}{r^2} - \frac{2(y-b)^2}{r^4}\right\} da\, db,$$

$$\frac{\partial^2 V}{\partial x^2} + \frac{\partial^2 V}{\partial y^2} = -\iint \mu \left\{\frac{2}{r^2} - \frac{2[(x-a)^2 + (y-b)^2]}{r^4}\right\} da\, db = 0.$$

This potential is called the *logarithmic potential* and is of great importance in the theory of functions of a complex variable.

137. Green's Theorem for a Plane. In exactly the same manner that we proved Green's Theorem for three dimensions, we may prove it when the integral is the double integral in a plane

52) $$I = \iint_A \left\{\frac{\partial U}{\partial x}\frac{\partial V}{\partial x} + \frac{\partial U}{\partial y}\frac{\partial V}{\partial y}\right\} dx\, dy,$$

over an area A bounded by any closed contour C. Since we have for a continuous function W

53) $$\iint \frac{\partial W}{\partial x} dx\, dy = \int [W_2 - W_1 + \cdots + W_{2n} - W_{2n-1}]\, dy$$
$$= -\int_C W \cos(nx)\, ds,$$

where n is the inward normal, ds the element of arc of the contour.

Applying this to $W = U\frac{\partial V}{\partial x}$, we obtain

54) $$\iint \frac{\partial}{\partial x}\left\{U\frac{\partial V}{\partial x}\right\} = -\int U\frac{\partial V}{\partial x}\cos(nx)\, ds.$$

Treating the other term in like manner, we obtain

55) $$\iint_A \left\{\frac{\partial U}{\partial x}\frac{\partial V}{\partial x} + \frac{\partial U}{\partial y}\frac{\partial V}{\partial y}\right\} dx\, dy$$
$$= -\int_C U\frac{\partial V}{\partial n} ds - \iint_A U\left(\frac{\partial^2 V}{\partial x^2} + \frac{\partial^2 V}{\partial y^2}\right) dx\, dy.$$

Interchanging U and V we obtain the second form

56) $$\int_C \left(U\frac{\partial V}{\partial n} - V\frac{\partial U}{\partial n}\right) ds = \iint_A (V\Delta U - U\Delta V)\, dx\, dy,$$

where we write
$$\Delta V = \frac{\partial^2 V}{\partial x^2} + \frac{\partial^2 V}{\partial y^2}.$$

138. Application to Logarithmic Potential. If in 56) we put $U = 1$, we obtain

$$\int_C \frac{\partial V}{\partial n} ds = -\iint_A \Delta V \, dx \, dy, \tag{57}$$

which is the divergence theorem in two dimensions. If the function V is harmonic everywhere within the contour, we have

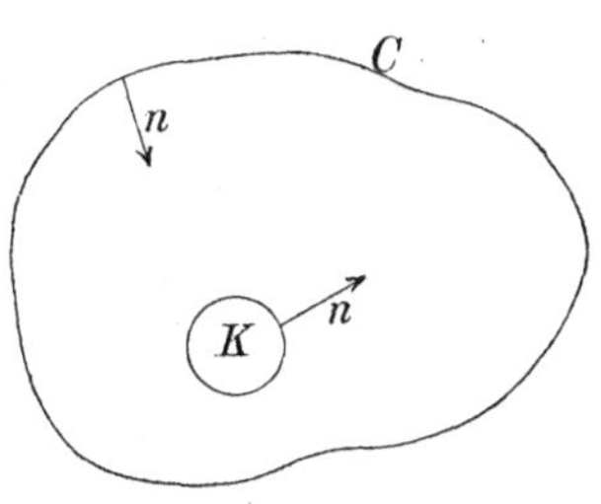

Fig. 138.

$$\int_C \frac{\partial V}{\partial n} ds = 0.$$

Applying this to the harmonic function $\log r$, where P, the fixed pole from which r is measured, is outside the contour,

$$\int_C \frac{\partial \log r}{\partial n} ds = \int_C \frac{1}{r} \frac{\partial r}{\partial n} ds = \int_C \frac{\cos (rn)}{r} ds = 0. \tag{58}$$

If the pole P is within the contour, we draw a circle K of any radius about the pole, and apply the theorem to the area outside of this circle and within the contour, obtaining the sum of the integrals around C and K equal to zero, or

$$\int_C \frac{\partial \log r}{\partial n} ds = -\int_K \frac{\cos (rn)}{r} ds = -\int_0^{2\pi} d\vartheta = -2\pi. \tag{59}$$

These two results are Gauss's theorem for two dimensions. They may of course be deduced geometrically in the same way as for three dimensions, § 118. We may now deduce Poisson's equation for the logarithmic potential as in § 123 for the Newtonian Potential. The logarithmic potential due to a mass dm being $-dm \log r$ gives rise to the flux of force $2\pi dm$ *outward* through any closed contour surrounding it, and a total mass m causes the flux

$$2\pi m = 2\pi \iint \mu \, dx \, dy.$$

Put in terms of the potential this is

$$\int_C \frac{\partial V}{\partial n_i} ds = -\iint_A \Delta V dx \, dy = 2\pi \iint_A \mu \, dx \, dy, \tag{60}$$

and since this is true for any area of the plane, we must have

$$\Delta V = -2\pi\mu. \tag{61}$$

This is Poisson's equation for the logarithmic potential.

139. Green's Formula for Logarithmic Potential. Applying Green's Theorem 56) to the functions $\log r$ and any harmonic function V, supposing the pole of P to be within the contour, and extending the integral to the area within the contour and without a circle K of radius ε about the pole,

$$62)\quad \int\limits_C \left(\log r \frac{\partial V}{\partial n} - V \frac{\partial \log r}{\partial n}\right) ds + \int\limits_K \left(\log r \frac{\partial V}{\partial n} - V \frac{\partial \log r}{\partial n}\right) ds = 0.$$

The third term is

$$\int\limits_K \log r \frac{\partial V}{\partial n} ds = \log \varepsilon \int \frac{\partial V}{\partial n} ds = 0,$$

(since V is harmonic in K) and the fourth,

$$-\int\limits_K V \frac{\partial \log r}{\partial n} ds = -\int\limits_K \frac{V}{r} r d\vartheta = -\int\limits_K V d\vartheta,$$

which, when we make ε decrease indefinitely, becomes

$$-2\pi V_P.$$

Accordingly we obtain the equation

$$63)\quad V_P = \frac{1}{2\pi} \int\limits_C \left(\log r \frac{\partial V}{\partial n} - V \frac{\partial \log r}{\partial n}\right) ds,$$

which is the analogue of equation 6), § 128. In a similar way we may find for nearly every theorem on the Newtonian Potential a corresponding theorem for the Logarithmic Potential. A comparison of the corresponding theorems will be found in C. Neumann's work, *Untersuchungen über das logarithmische und das Newtonsche Potential.*[1])

The Kelvin-Dirichlet Problem and Principle may be stated and demonstrated for the logarithmic potential precisely as in § 132.

140. Dirichlet's Problem for a Circle. Trigonometric Series. We shall call a homogeneous harmonic function of order n of the coordinates x, y of a point in a plane a Circular Harmonic, since it is equal to ϱ^n multiplied by a homogeneous function of $\cos\omega$ and $\sin\omega$, and consequently on the circumference of a circle about the origin is simply a trigonometric function of the angular coordinate ω. Any homogeneous function V_n of degree n satisfies the differential equation

$$64)\quad x\frac{\partial V_n}{\partial x} + y\frac{\partial V_n}{\partial y} = n V_n,$$

1) See also Harnack, *Die Grundlagen der Theorie des logarithmischen Potentiales;* Picard, *Traité d'Analyse,* tom. II; Poincaré, *Théorie du Potentiel Newtonien.*

so that a circular harmonic is a solution of this and Laplace's Equation simultaneously. The homogeneous function of degree n

$$a_n x^n + a_{n-1} x^{n-1} + \cdots a_1 x y^{n-1} + a \cdot y^n$$

contains $n+1$ terms, the sum of its second derivatives is a homogeneous function of degree $n-2$ containing $n-1$ terms, and if this is to vanish identically each of its $n-1$ coefficients must vanish, consequently there are $n-1$ relations between the $n+1$ coefficients of V_n, or only two are arbitrary. Accordingly all harmonics of degree n can be expressed in terms of two independent ones. The theory of functions of a complex variable[1]) tells us that the real functions $u(x, y)$, $v(x, y)$ in the complex variable $u + iv$, which is a function of the complex variable $x + iy$, are harmonic functions of x, y, and making use of Euler's fundamental formula,

65) $$x + iy = \varrho\{\cos\omega + i\sin\omega\} = \varrho e^{i\omega},$$

and raising to the nth power, we have

66) $$(x+iy)^n = \varrho^n e^{in\omega} = \varrho^n(\cos n\omega + i\sin n\omega).$$

Accordingly we have the two typical harmonic functions

67) $$u = \varrho^n \cos n\omega, \quad v = \varrho^n \sin n\omega.$$

It may be at once shown that these functions are harmonic by substitution in Laplace's equation in polar coordinates, equation 47). Accordingly the general harmonic of degree n is

68) $$V_n = \varrho^n\{A_n \cos n\omega + B_n \sin n\omega\} = \varrho^n T_n.$$

We may call the trigonometric factor T_n, which is the value of the harmonic on the circumference of a circle of radius unity, the peripheral harmonic of degree n.

If a function which is harmonic in a circular area can be developed in an infinite trigonometric series

69) $$V(x, y) = \sum_{n=0}^{n=\infty}\{A_n \cos n\omega + B_n \sin n\omega\} = \sum_0^\infty T_n$$

on the circumference of the circle of radius R, the solution of Dirichlet's Problem for the interior of the circle is given by the series

70) $$V = T_0 + \frac{\varrho}{R} T_1 + \frac{\varrho^2}{R^2} T_2 + \cdots.$$

For every term is harmonic, and therefore the series, if convergent, is harmonic. At the circumference $\varrho = R$, and the series takes the given values of V. The absolute value of every term is less than the absolute value of the corresponding term in the series 69), in

1) See § 197.

virtue of the factor $\frac{\varrho^n}{R^n}$, therefore if the series 69) converges, the series 70) does as well. Since the series fulfils all conditions, by Dirichlet's principle it is the only function satisfying them.

We may fulfil the outer problem by means of harmonics of negative degree. Taking n negative, the series

71) $$V = T_0 + \frac{R T_1}{\varrho} + \frac{R^2}{\varrho^2} T_2 + \cdots$$

is convergent, takes the required values on the circumference, and vanishes at infinity. For a ring-shaped area between two concentric circles, we may satisfy the conditions by a series in both positive and negative harmonics,

72) $$V = \sum_0^\infty \varrho^n \{A_n \cos n\omega + B_n \sin n\omega\} + \sum_1^\infty \varrho^{-n} \{A'_n \cos n\omega + B'_n \sin n\omega\}.$$

140a. Development in Circular Harmonics. We may use the formula 63), § 139, to obtain the development of a function in a trigonometric series on the circumference of a circle. Let the polar coordinates of a point on the circumference of the circle be R, ω and of a point P within the circumference ϱ, φ. Then we have for the distance between the two points

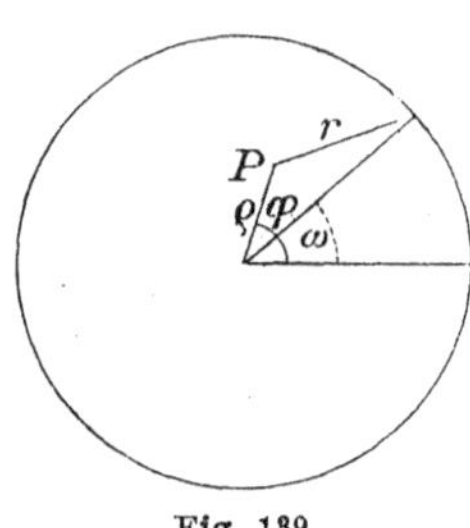

Fig. 139.

$$r = [R^2 + \varrho^2 - 2 R\varrho \cos(\omega - \varphi)]^{\frac{1}{2}}.$$

Removing the factor R^2, inserting for $\cos(\omega - \varrho)$ its value in exponentials, and separating into factors we obtain

73) $$r = R\left[1 + \frac{\varrho^2}{R^2} - \frac{\varrho}{R}\left(e^{i(\omega-\varphi)} + e^{-i(\omega-\varphi)}\right)\right]^{\frac{1}{2}}$$
$$= R\left[\left(1 - \frac{\varrho}{R} e^{i(\omega-\varphi)}\right)\left(1 - \frac{\varrho}{R} e^{-i(\omega-\varphi)}\right)\right]^{\frac{1}{2}}.$$

Taking the logarithm we may develop

$$\log\left(1 - \frac{\varrho}{R} e^{i(\omega-\varphi)}\right)$$

and

$$\log\left(1 - \frac{\varrho}{R} e^{-i(\omega-\varphi)}\right)$$

by Taylor's Theorem, obtaining

$$74)\qquad \log r = \log R - \frac{1}{2}\sum_{1}^{\infty}\frac{1}{n}\frac{\varrho^n}{R^n}\left(e^{ni(\omega-\varphi)} + e^{-ni(\omega-\varphi)}\right)$$

$$= \log R - \sum_{1}^{\infty}\frac{\varrho^n}{nR^n}\cos n(\omega-\varphi).$$

This series is convergent if $\varrho < R$, and also if $\varrho = R$, unless $\omega = \varphi$.

Inserting this value of $\log r$ in 63), differentiation with respect to the normal being according to $-R$, we have

$$75)\qquad \frac{\partial(\log r)}{\partial n} = -\left\{\frac{1}{R} + \sum_{1}^{\infty}\frac{\varrho^n}{R^{n+1}}\cos n(\omega-\varphi)\right\},$$

$$V_P = \frac{1}{2\pi}\int_{0}^{2\pi}\left\{\frac{\partial V}{\partial n}\left(\log R - \sum_{1}^{\infty}\frac{\varrho^n}{nR^n}\cos n(\omega-\varphi)\right) + V\left(\sum_{0}^{\infty}\frac{\varrho^n}{R^{n+1}}\cos n(\omega-\varphi)\right)\right\}R\,d\omega.$$

Expanding the cosines, we may take out from each term of the integral, except the first, a factor $\varrho^n \cos n\varphi$ or $\varrho^n \sin n\varphi$, so that V_P is developed as a function of its coordinates ϱ, φ, in an infinite series of circular harmonics, the coefficients of which are definite integrals around the circumference, involving the peripheral values of V and $\frac{\partial V}{\partial n}$. This does not establish the convergence of the series *on* the circumference. Admitting the possibility of the development, we may proceed to find it in a more convenient form. In order to do this let us apply the last equation to a function V_m, which is a circular harmonic of degree m. Then at the circumference we have

$$V_m = R^m T_m,\quad \frac{\partial V_m}{\partial n} = -\,mR^{m-1}T_m,$$

and

$$76)\qquad V_m(P) = \frac{R^m(1 - m\log R)}{2\pi}\int_{0}^{2\pi}T_m\,d\omega$$

$$+\frac{1}{2\pi}\sum_{n=1}^{n=\infty}\varrho^n R^{m-n}\left(\frac{m}{n}+1\right)\int_{0}^{2\pi}T_m\cos n(\omega-\varphi)\,d\omega.$$

The expression on the right is an infinite series in powers of ϱ, while $V_m(P)$ is simply $\varrho^m T_m$. As this equality must hold for all values of ϱ less than R, the coefficient of every power of ϱ except the mth must vanish, and we have the important equations

$$77)\qquad \int_0^{2\pi} T_m \cos n(\omega - \varphi)\, d\omega = 0, \qquad (m \neq n),$$

$$78)\qquad T_m(\varphi) = \frac{1}{\pi}\int_0^{2\pi} T_m(\omega) \cos m(\omega - \varphi)\, d\omega,$$

for all values of n, and for all values of m except 0. Since T_0 is a constant, we evidently have

$$79)\qquad T_0 = \frac{1}{2\pi}\int_0^{2\pi} T_0\, d\omega.$$

These two important results can be very simply deduced by direct integration, inserting the value of $T_m(\omega)$, but we have preferred to deduce them as a consequence af Green's formula 63), § 139, in order to show the analogy with Spherical Harmonics. Let us now suppose that the function $V(\omega)$ can be developed in the convergent infinite trigonometric series

$$80)\qquad V(\omega) = \sum_0^\infty (A_n \cos n\omega + B_n \sin n\omega) = \sum_0^\infty T_n(\omega).$$

Multiply both sides by $\cos m(\omega - \varphi)\, d\omega$ and integrate from 0 to 2π.

$$81)\qquad \int_0^{2\pi} V(\omega) \cos m(\omega - \varphi)\, d\omega = \sum_0^\infty \int_0^{2\pi} T_n(\omega) \cos m(\omega - \varphi)\, d\omega.$$

Every term on the right vanishes except the mth which is equal to $\pi T_m(\varphi)$. Accordingly we find for the circular harmonic T_m the definite integral

$$82)\qquad T_m(\varphi) = \frac{1}{\pi}\int_0^{2\pi} V(\omega) \cos m(\omega - \varphi)\, d\omega.$$

For $m = 0$, we must divide by 2.

Writing for $T_m(\varphi)$ its value

$$A_m \cos m\varphi + B_m \sin m\varphi,$$

expanding the cosine in the intregral, and writing the two terms separately, we obtain the coefficients

$$83)\qquad A_0 = \frac{1}{2\pi}\int_0^{2\pi} V(\omega)\, d\omega, \qquad A_m = \frac{1}{\pi}\int_0^{2\pi} V(\omega) \cos m\omega\, d\omega,$$

$$B_m = \frac{1}{\pi}\int_0^{2\pi} V(\omega) \sin m\omega\, d\omega$$

This form for the coefficients was given by Fourier[1]), who *assuming* that the development was possible, was able to determine the coefficients. The question of proving that the development thus found actually represents the function, and the determination of the conditions that the development shall be possible, formed one of the most important mathematical questions of the last century, which was first satisfactorily treated by Dirichlet.[2]) For the full and rigid treatment of this important subject, the student should consult Kronecker, *Theorie der einfachen und der vielfachen Integrale;* Picard, *Traité d'Analyse*, Tom. I, Chap. IX; Riemann-Weber, *Partielle Differentialgleichungen;* Poincaré, *Théorie du Potentiel Newtonien.*[3])

141. Spherical Harmonics. A Spherical Harmonic of degree n is defined as a homogeneous harmonic function of the coordinates x, y, z of a point in space, that is as a solution of the simultaneous equations

$$84)\qquad \frac{\partial^2 V}{\partial x^2}+\frac{\partial^2 V}{\partial y^2}+\frac{\partial^2 V}{\partial z^2}=0.$$

$$85)\qquad x\frac{\partial V}{\partial x}+y\frac{\partial V}{\partial y}+z\frac{\partial V}{\partial z}=nV.$$

The general homogeneous function of degree n

$$\begin{aligned} a_{n0}x^n + a_{n-1,0}\,x^{n-1}y + a_{n-2,0}\,x^{n-2}y^2\cdots + a_{0,0}\,y^n \\ + a_{n-1,1}\,x^{n-1}z + a_{n-2,1}\,x^{n-2}yz\cdots + a_{0,1}\,y^{n-1}z \\ + a_{n-2,2}\,x^{n-2}z^2\cdots + a_{0,2}\,y^{n-2}z^2 \\ \cdots\cdots\cdots\cdots \\ \cdots\cdots\cdots\cdots \\ + a_{0,n}z^n \end{aligned}$$

contains $1+2+3\cdots+n+1=\frac{(n+1)(n+2)}{2}$ terms. The sum of its second derivatives is a homogeneous function of degree $n-2$ and accordingly contains $\frac{(n-1)n}{2}$ terms. If the function is to vanish identically, these $\frac{(n-1)n}{2}$ coefficients must all vanish, so that there are $\frac{(n-1)n}{2}$ relations among the $\frac{(n+1)(n+2)}{2}$ coefficients of a harmonic of the nth degree, leaving $2n+1$ *arbitrary* coefficients. The general harmonic of degree n can accordingly be expressed as a linear function of $2n+1$ independent harmonics.

1) Fourier, *Théorie analytique de la Chaleur,* Chap. IX, 1822.

2) Dirichlet, "Sur la Convergence des Séries Trigonométriques", *Crelle's Journal,* Bd. 4, 1829.

3) A resumé of the literature is given by Sachse, *Bulletin des Sciences Mathématiques,* 1880.

Examples. Differentiating the arbitrary homogeneous function and determining the coefficients, we find for $n = 0, 1, 2, 3$, the following independent harmonics:

$n = 0$ a constant,

$n = 1$ $x,\quad y,\quad z,$

$n = 2$ $x^2 - y^2,\quad y^2 - z^2,\quad xy,\quad yz,\quad zx,$

$n = 3$ $3x^2y - y^3,\ 3x^2z - z^3,\ 3y^2x - x^3,\ 3y^2z - z^3,\ 3z^2x - x^3,\ 3z^2y - y^3,\ xyz.$

If we insert spherical coordinates r, ϑ, φ,

$$x = r \sin\vartheta \cos\varphi,$$
$$y = r \sin\vartheta \sin\varphi,$$
$$z = r \cos\vartheta$$

the harmonic V_n becomes

$$V_n = r^n Y_n(\vartheta, \varphi),$$

where Y_n is a homogeneous function of the trigonometric functions $\cos\vartheta$, $\sin\vartheta\cos\varphi$, and $\sin\vartheta\sin\varphi$. Y_n being the value of V_n on the surface of a sphere of unit radius, is called a surface harmonic. The equation $Y_n = 0$ represents a cone of order n, whose intersection with the sphere gives a geometrical representation of the harmonic V_n.

If u and v be any two continuous functions of x, y, z,

$$\frac{\partial^2 (uv)}{\partial x^2} = u\frac{\partial^2 v}{\partial x^2} + 2\frac{\partial u}{\partial x}\frac{\partial v}{\partial x} + v\frac{\partial^2 u}{\partial x^2}.$$

86) $$\Delta(uv) = u\Delta v + v\Delta u + 2\left(\frac{\partial u}{\partial x}\frac{\partial v}{\partial x} + \frac{\partial u}{\partial y}\frac{\partial v}{\partial y} + \frac{\partial u}{\partial z}\frac{\partial v}{\partial z}\right).$$

Put $u = r^m$, and since

$$\frac{\partial (r^m)}{\partial x} = m r^{m-1}\frac{\partial r}{\partial x} = m r^{m-2} x,$$

$$\frac{\partial^2 (r^m)}{\partial x^2} = m r^{m-2} + m(m-2) r^{m-4} x^2,$$

we get

87) $$\Delta(r^m) = 3 m r^{m-2} + m(m-2) r^{m-4}(x^2 + y^2 + z^2)$$
$$= m(m+1) r^{m-2}.$$

If V_n is a harmonic of degree n,

88) $$\Delta(r^m V_n) = r^m \Delta V_n + m(m+1) r^{m-2} V_n$$
$$+ 2 m r^{m-2}\left(x\frac{\partial V_n}{\partial x} + y\frac{\partial V_n}{\partial x} + z\frac{\partial V_n}{\partial z}\right)$$
$$= [m(m+1) + 2mn] r^{m-2} V_n,$$

by virtue of equations 84) and 85).

Consequently if $m = -(2n+1)$, the product $r^m V_n$ is a harmonic. Since V_n is of degree n, and r is of degree unity in the coordinates, $r^{-(2n+1)} V_n$ is of degree $-(n+1)$. Accordingly to any spherical harmonic $V_n = r^n Y_n$ of degree n there corresponds another,

$$V_{-(n+1)} = \frac{V_n}{r^{2n+1}} = \frac{Y_n}{r^{n+1}},$$

of degree $-(n+1)$. Compare this with the corresponding property of circular harmonics, where the degrees of the two corresponding harmonics are *equal* and opposite.

142. Dirichlet's Problem for Sphere. By means of these harmonics we may solve Dirichlet's problem for the sphere. If a function harmonic within a sphere of radius R can be developed at the surface in an infinite series of surface harmonics,

89) $$V = Y_0 + Y_1 + Y_2 \cdots,$$

the internal problem is solved by the series

90) $$V = Y_0 + \frac{r}{R} Y_1 + \frac{r^2}{R^2} Y_2 + \cdots.$$

For each term is harmonic, and therefore the series 90), if convergent, is harmonic. At the surface the series takes the given values of V. Every term of the series 90) is less than the corresponding term of the series 89) in virtue of the factor $\frac{r^n}{R^n}$, therefore if the series 89) converges, the series 90) does as well. Since the series fulfils all the conditions it is the only solution.

We may in like manner fulfil the outer problem by the series of harmonics of negative degree, which vanish at infinity.

91) $$V = \frac{R}{r} Y_0 + \frac{R^2}{r^2} Y_1 + \frac{R^3}{r^3} Y_2 + \cdots.$$

For the space bounded by two concentric spheres, we must use the series in harmonics of positive and negative degrees.

143. Forms of Spherical Harmonics. Before considering the question of development in spherical harmonics, we will briefly consider some convenient forms. Since if

$$\Delta V_n = 0,$$

we have

$$\frac{\partial}{\partial x} \Delta V_n = \Delta \frac{\partial V_n}{\partial x} = 0,$$

and any derivative of a harmonic is itself a harmonic, so that

$$\frac{\partial^\alpha}{\partial x^\alpha}\frac{\partial^\beta}{\partial y^\beta}\frac{\partial^\gamma}{\partial z^\gamma}V_n,$$

is a harmonic of degree $n-(\alpha+\beta+\gamma)$. Since to $V_0=c$ corresponds the harmonic $V_{-1}=\frac{c}{r}$, we have

$$92)\qquad \frac{\partial^\alpha}{\partial x^\alpha}\frac{\partial^\beta}{\partial y^\beta}\frac{\partial^\gamma}{\partial z^\gamma}\left(\frac{1}{r}\right)=V_{-(1+\alpha+\beta+\gamma)}.$$

If h_1 be any constant direction whose direction cosines are

$$\cos(h_1x)=l_1,\quad \cos(h_1y)=m_1,\quad \cos(h_1z)=n_1,$$

$$\frac{\partial}{\partial h_1}=l_1\frac{\partial}{\partial x}+m_1\frac{\partial}{\partial y}+n_1\frac{\partial}{\partial z},$$

and $\frac{\partial}{\partial h_1}\left(\frac{1}{r}\right)$ is a harmonic of degree -2, and to it corresponds the harmonic,

$$93)\qquad V_1=r^3\frac{\partial}{\partial h_1}\left(\frac{1}{r}\right),$$

which is of the first degree. Since $l_1^2+m_1^2+n_1^2=1$, the harmonic contains *two* arbitrary constants, and multiplying by a tird, A, we have the general harmonic of degree 1, in the form

$$94)\qquad V_1=Ar^3\frac{\partial}{\partial h_1}\left(\frac{1}{r}\right).$$

If in like manner $h_2, h_3, \dots h_n$, denote vectors with direction cosines $l_2, m_2, n_2, \dots l_n, m_n, n_n$.

$$\frac{\partial}{\partial h_1}\frac{\partial}{\partial h_2}\cdots\frac{\partial}{\partial h_n}\left(\frac{1}{r}\right)$$

is a spherical harmonic of degree $-(n+1)$ and to it corresponds

$$95)\qquad V_n=r^{2n+1}\frac{\partial}{\partial h_1}\frac{\partial}{\partial h_2}\cdots\frac{\partial}{\partial h_n}\left(\frac{1}{r}\right),$$

a harmonic of degree n, and since every h introduces two arbitrary constants, multiplying by another, A, gives us $2n+1$, and we have the general harmonic of degree n in the form,

$$96)\qquad V_n=A\frac{\partial}{\partial h_1}\frac{\partial}{\partial h_2}\cdots\frac{\partial}{\partial h_n}\left(\frac{1}{r}\right).$$

The directions $h_1, h_2, \dots h_n$ are called the *axes* of the harmonic. To illustrate the method of deriving the harmonics we shall find the first two.

$$V_1 = Ar^3 \frac{\partial}{\partial h}\left(\frac{1}{r}\right) = Ar^3\left(-\frac{lx}{r^3} - \frac{my}{r^3} - \frac{nz}{r^3}\right) = -A(lx + my + nz),$$

$$V_2 = Ar^5 \frac{\partial}{\partial h_1}\frac{\partial}{\partial h_2}\left(\frac{1}{r}\right)$$

$$= Ar^5\left(l_1\frac{\partial}{\partial x} + m_1\frac{\partial}{\partial y} + n_1\frac{\partial}{\partial z}\right)\left(l_2\frac{\partial}{\partial x} + m_2\frac{\partial}{\partial y} + n_2\frac{\partial}{\partial z}\right)\left(\frac{1}{r}\right)$$

$$= -Ar^5\left(l_1\frac{\partial}{\partial x} + m_1\frac{\partial}{\partial y} + n_1\frac{\partial}{\partial z}\right)\left(\frac{l_2 x + m_2 y + n_2 z}{r^3}\right)$$

$$= -Ar^5\left\{l_1\left(\frac{l_2}{r^5} - \frac{3l_2x^2}{r^5} - \frac{3m_2xy}{r^5} - \frac{3n_2xz}{r^5}\right)\right.$$

$$+ m_1\left(\frac{m_2}{r^3} - \frac{3m_2y^2}{r^5} - \frac{3l_2xy}{r^5} - \frac{3n_2yz}{r^5}\right)$$

$$\left. + n_1\left(\frac{n_2}{r^3} - \frac{3n_2z^2}{r^5} - \frac{3l_2xz}{r^5} - \frac{3m_2yz}{r^5}\right)\right\},$$

$$V_2 = A\{-(l_1l_2 + m_1m_2 + n_1n_2)(x^2 + y^2 + z^2)$$
$$+ 3(l_1l_2x^2 + m_1m_2y^2 + n_1n_2z^2) + (l_1m_2 + l_2m_1)xy$$
$$+ (m_1n_2 + m_2n_1)yz + (n_1l_2 + n_2l_1)zx)\}.$$

The coefficients are of course subject to the relations

$$l_1^2 + m_1^2 + n_1^2 = 1, \quad l_2^2 + m_2^2 + n_2^2 = 1.$$

144. Zonal Harmonics. If all the axes of the harmonic coincide, we may conveniently take the axis for one of the coordinate axes, and write

97) $$V_n = Ar^{2n+1}\frac{\partial^n}{\partial z^n}\left(\frac{1}{r}\right).$$

It is evident that this will contain only powers of z and r, so that the *surface* harmonic will be simply a polynomial in

$$\frac{z}{r} = \cos(rz).$$

The equation $Y_n\{\cos(rz)\} = 0$ may be shown to have n real roots lying between 1 and -1, and hence represents n circular cones of angles whose cosines are these roots, intersecting the surface of a sphere in n parallels of latitude which divide the surface into zones. The harmonics are therefore called Zonal Harmonics. The polynomial in $\cos(rz)$ which constitutes the zonal surface harmonic, when the value of the constant A is determined in the manner to be given in 106), is called a Legendre's Polynomial, and denoted by

$$P_n[\cos(rz)] = Ar^{n+1}\frac{\partial^n}{\partial z^n}\left\{\frac{1}{r}\right\}.$$

145. Harmonics in Spherical Coordinates. We have transformed Laplace's Operator into spherical coordinates in § 135, and $\Delta V = 0$ becomes

$$98)\quad \sin\vartheta \frac{\partial}{\partial r}\left(r^2 \frac{\partial V}{\partial r}\right) + \frac{\partial}{\partial \vartheta}\left(\sin\vartheta \frac{\partial V}{\partial \vartheta}\right) + \frac{\partial}{\partial \varphi}\left(\frac{1}{\sin\vartheta}\frac{\partial V}{\partial \varphi}\right) = 0.$$

If we put in this $V_n = r^n Y_n$ we obtain

$$99)\quad \sin\vartheta \cdot n(n+1) Y_n + \frac{\partial}{\partial \vartheta}\left\{\sin\vartheta \frac{\partial Y_n}{\partial \vartheta}\right\} + \frac{1}{\sin\vartheta}\frac{\partial^2 Y_n}{\partial \varphi^2} = 0,$$

as the differential equation satisfied by a surface harmonic $Y_n(\vartheta, \varphi)$. This is the form of Laplace's equation originally given by Laplace.[1]) If Y_n is the zonal harmonic P_n, which is independent of φ, we have

$$100)\quad n(n+1) P_n + \frac{1}{\sin\vartheta}\frac{d}{d\vartheta}\left\{\sin\vartheta \frac{dP_n}{d\vartheta}\right\} = 0,$$

or putting

$$\cos\vartheta = \mu,$$

$$101)\quad \frac{d}{d\mu}\left\{(1-\mu^2)\frac{dP_n}{d\mu}\right\} + n(n+1) P_n = 0.$$

This is known as Legendre's Differential Equation. We shall now, without considering more in detail the general surface harmonic, find the general expression for the zonal harmonic. It may be at once shown, by inserting for $P_n(\mu)$ a power-series in μ and determining the coefficients, that for integral values of n the differential equation is satisfied by a polynomial in μ. The form of these polynomials we shall find from one of their important properties.

146. Development of Reciprocal Distance. We know that $\frac{1}{r}$, the reciprocal of the distance of the point x, y, z from any fixed point P, is a harmonic function of the coordinates x, y, z, and although it is not a homogeneous function except when the fixed point is the origin, it may always be developed in a series of homogeneous functions, that is, in a series of spherical harmonics. We shall now use the letter d for the distance from any fixed point, reserving r for the distance from the origin. Let us for convenience take the axis of z as passing through the fixed point P, which lies at a distance r' from the origin, and put $\cos(rz) = \mu$. Then we have

$$102)\quad \frac{1}{d} = [r^2 + r'^2 - 2rr'\mu]^{-\frac{1}{2}} = [x^2 + y^2 + (z - r')^2]^{-\frac{1}{2}}.$$

1) Laplace, "Théorie des attractions des sphéroïdes et de la figure des planètes." *Mém. de l'Acad. de Paris.* Année 1782 (publ. 1785).

Considering this as a function of z let us develop by Taylor's Theorem,

103) $$\frac{1}{d}=f(z-r')=f(z)+(-r')\left(\frac{\partial f}{\partial z}\right)_{r'=0}+\frac{1}{2!}(-r')^2\left(\frac{\partial^2 f}{\partial z^2}\right)_{r'=0}+\cdots$$

and since for $r'=0$, $\frac{1}{d}=\frac{1}{r}$, $\frac{\partial^n f}{\partial z^n}=\frac{\partial^n}{\partial z^n}\left(\frac{1}{r}\right)$,

104) $$\frac{1}{d}=\frac{1}{r}+(-r')\frac{\partial}{\partial z}\left(\frac{1}{r}\right)+\frac{1}{2!}(-r')^2\frac{\partial^2}{\partial z^2}\left(\frac{1}{r}\right)+\cdots$$

Now multiplying and dividing each term by r^{n+1}, we find

105) $$\frac{1}{d}=\frac{1}{r}\left\{P_0+\frac{r'}{r}P_1+\frac{r'^2}{r^2}P_2+\cdots+\frac{r'^n}{r^n}P_n+\cdots\right\},$$

where

106) $$P_0=1,\quad P_n=\frac{(-1)^n}{n!}r^{n+1}\frac{\partial^n}{\partial z^n}\left(\frac{1}{r}\right).$$

This is the determination of the constant A, adopted by Legendre, for the reason that, since by the binomial theorem, for $r'<r$, and $\mu=1$,

$$\frac{1}{d}=\frac{1}{r}\left\{1-\frac{r'}{r}\right\}^{-1}=\frac{1}{r}\left\{1+\frac{r'}{r}+\frac{r'^2}{r^2}+\cdots+\frac{r'^n}{r^n}+\cdots\right\}$$

it makes for every n,

107) $$P_n(1)=1.$$

The term $\frac{P_n}{r^{n+1}}$ is a spherical harmonic of degree $-(n+1)$, and the series 105) is convergent for $r'<r$. In like manner if $r'>r$ we find

108) $$\frac{1}{d}=\frac{1}{r'}\left\{P_0+\frac{r}{r'}P_1+\frac{r^2}{r'^2}P_2+\cdots+\frac{r^n}{r'^n}P_n+\cdots\right\}.$$

In order to find P_n as a polynomial in μ we may write $\frac{r}{d}$ as

$$\frac{r}{d}=\left[1-2\frac{r'}{r}\left(\mu-\frac{r'}{2r}\right)\right]^{-\frac{1}{2}}$$

and develop by the binomial theorem.

109) $$\frac{r}{d}=\sum_{s=0}^{s=\infty}\frac{\frac{1}{2}\cdot\frac{3}{2}\cdot\frac{5}{2}\cdots\left(\frac{1}{2}+s-1\right)}{s!}\left(2\frac{r'}{r}\right)^s\left(\mu-\frac{r'}{2r}\right)^s.$$

Developing the last factor,

$$\left(\mu-\frac{r'}{2r}\right)^s=\sum_{t=0}^{t=s}(-1)^t\frac{s(s-1)\ldots(s-t+1)}{t!}\mu^{s-t}\left(\frac{r'}{2r}\right)^t.$$

$$110)\quad \frac{r}{d} = \sum_{s=0}^{s=\infty}\sum_{t=0}^{t=s}(-1)^t 2^{(s-t)} \frac{\frac{1}{2}\cdot\frac{3}{2}\cdot\frac{5}{2}\cdots\left(\frac{1}{2}+s-1\right)}{t!\,(s-t)!}\left(\frac{r'}{r}\right)^{s+t}\mu^{s-t}.$$

Picking out all the terms for which $s + t = n$ we get for the coefficient of $\left(\frac{r'}{r}\right)^n$

$$111)\quad P_n = \frac{1\cdot 3\cdot 5\cdots(2n-1)}{n!}\left[\mu^n - \frac{n(n-1)}{1\cdot 2\,(2n-1)}\mu^{n-2} + \frac{n(n-1)(n-2)(n-3)}{2\cdot 4\,(2n-1)(2n-3)}\mu^{n-4}\cdots\right].$$

The first polynomials have the values

$$P_0(\mu) = 1,$$
$$P_1(\mu) = \mu,$$
$$P_2(\mu) = \frac{1}{2}(3\mu^2 - 1),$$
$$P_3(\mu) = \frac{1}{2}(5\mu^3 - 3\mu),$$
$$P_4(\mu) = \frac{1}{8}(35\mu^4 - 30\mu^2 + 3),$$
$$P_5(\mu) = \frac{1}{8}(63\mu^5 - 70\mu^3 + 15\mu).$$

147. Development in Spherical Harmonics. We may use the formula 6), § 128, for an internal point, to obtain the development of a function of ϑ, φ, on the surface of a sphere in the same manner as in § 140a for the case of a circle. Since the polynomials in the development of the reciprocal distance involve only the cosine of the angle between the radii to the fixed and variable points, we have if $r' < r$,

$$105)\quad \frac{1}{d} = \frac{1}{r}\sum_0^\infty \left(\frac{r'}{r}\right)^s P_s(\mu) \qquad \mu = \cos(r'r),$$

and differentiating this with respect to $-r$, the internal normal,

$$117)\quad \frac{\partial\left(\frac{1}{d}\right)}{\partial n} = -\frac{\partial\left(\frac{1}{d}\right)}{\partial r} = \sum_0^\infty (s+1)\frac{r'^s}{r^{s+2}}P_s(\mu).$$

Inserting these values in 6), § 128, namely

$$6)\quad V_{(P)} = \frac{1}{4\pi}\iint\left\{V\frac{\partial\left(\frac{1}{d}\right)}{\partial n} - \frac{1}{d}\frac{\partial V}{\partial n}\right\}dS,$$

and applying it to the case that V is a spherical harmonic

$$V_m = r^m Y_m, \quad \frac{\partial V}{\partial n} = -mr^{m-1} Y_m,$$

we obtain, since

$$dS = r^2 \sin\vartheta\, d\vartheta\, d\varphi,$$

118) $$V_m(P) = \frac{1}{4\pi}\int_0^\pi\int_0^{2\pi}\Big\{r^m Y_m \sum_0^\infty (s+1)\frac{r'^s}{r^{s+2}} P_s(\mu) + mr^{m-2} Y_m \sum_0^\infty \left(\frac{r'}{r}\right)^s P_s(\mu)\Big\} r^2 \sin\vartheta\, d\vartheta\, d\varphi.$$

If the coordinates of P be r', ϑ', φ', we have,

$$V_m(P) = r'^m Y_m(\vartheta', \varphi'),$$

while on the right we have an infinite series in powers of r', with definite integrals as coefficients. Since the equality must hold for all values of r' less than r, we must have, collecting in terms in r'^s

119) $$\int_0^\pi\int_0^{2\pi} Y_m(\vartheta, \varphi)\, P_s(\mu) \sin\vartheta\, d\vartheta\, d\varphi = 0, \qquad s \gtrless m,$$

$$Y_m(\vartheta', \varphi') = \frac{m+s+1}{4\pi}\int_0^\pi\int_0^{2\pi} Y_m(\vartheta, \varphi) P_s(\mu) \sin\vartheta\, d\vartheta\, d\varphi, \quad s = m,$$

so that we have for the values of the integral

120) $$\int_0^\pi\int_0^{2\pi} Y_m(\vartheta, \varphi)\, P_m(\mu) \sin\vartheta\, d\vartheta\, d\varphi = \frac{4\pi}{2m+1} Y_m(\vartheta', \varphi').$$

In performing the integration, we must put for μ the value obtained by spherical trigonometry,

121) $$\mu = \cos(rr') = \cos\vartheta \cos\vartheta' + \sin\vartheta \sin\vartheta' \cos(\varphi - \varphi').$$

By means of the above integral expressions, 119) and 120) we may find the development of a function of ϑ, φ, *assuming that the development is possible.* Suppose we are to find the development

122) $$f(\vartheta, \varphi) = Y_0 + Y_1 + Y_2 + \cdots.$$

Multiply both sides by $P_n(\mu) \sin\vartheta\, d\vartheta\, d\varphi$, and integrate over the surface of the sphere and since every term vanishes except the nth we obtain

123) $$\int_0^\pi\int_0^{2\pi} f(\vartheta, \varphi)\, P_n(\mu) \sin\vartheta\, d\vartheta\, d\varphi = \frac{4\pi}{2n+1} Y_n(\vartheta', \varphi'),$$

124) $$Y_n(\vartheta', \varphi') = \frac{2n+1}{4\pi}\int_0^\pi\int_0^{2\pi} f(\vartheta, \varphi)\, P_n(\mu) \sin\vartheta\, d\vartheta\, d\varphi.$$

Accordingly to find the value of any term Y_n at any point P, (ϑ', φ') we find the zonal surface harmonic whose axis passes through the point P, multiply its value at every point of the sphere by the value of f for that point, by the element of area, and by $\frac{2n+1}{4\pi}$, and integrate the product over the surface. It remains to show that the development is possible, that is that the sum of the series

$$\frac{1}{4\pi}\sum_0^\infty (2n+1)\int_0^\pi\int_0^{2\pi} f(\vartheta, \varphi)\, P_n(\mu) \sin\vartheta\, d\vartheta\, d\varphi,$$

actually represents the function $f(\vartheta', \varphi')$. This theorem was demonstrated by Laplace, but without sufficient rigor, afterwards by Poisson, and finally in a rigorous manner by Dirichlet. A proof due to Darboux is given by Jordan, *Traité d'Analyse*, Tom. II, p. 249 (2me éd.).

148. Development of the Potential in Spherical Harmonics. In investigating the action of an attracting body at a distant point, and for many other purposes connected with geodesy and astronomy, it is convenient to develop the potential function in a series of spherical harmonics. If x, y, z denote the coordinates of the attracted point P, r its distance from the origin, a, b, c the coordinates of the attracting point Q, r' its distance from the origin, d the distance between them, $d\tau'$ the element of volume at Q, we have

$$V = \iiint \frac{\varrho}{d}\, d\tau',$$

and using the value of $\frac{1}{d}$ from 105), when $r > r'$,

$$125)\quad V = \iiint \frac{\varrho}{r}\, d\tau' + \iiint \frac{\varrho r' P_1(\mu)}{r^2}\, d\tau' + \iiint \frac{\varrho r'^2 P_2(\mu)}{r^3}\, d\tau' + \cdots$$

which, on removing the powers of r from under the integral signs, is the required development in spherical harmonics,

$$126)\quad V = \frac{Y_0}{r} + \frac{Y_1}{r^2} + \frac{Y_2}{r^3} + \cdots$$

where the surface harmonics Y_n are the volume integrals

$$127)\quad Y_n = \iiint \varrho r'^n P_n(\mu)\, d\tau'$$

taken over the space occupied by the attracting body. Since μ enters into the integrand, and, according to 121), it contains the angular coordinates ϑ, φ of P, the surface harmonics Y_n are functions of ϑ and φ.

If the body is homogeneous, and is symmetrical about an axis of revolution, since V is independeut of ϑ, φ, it is evident that all the harmonics are zonal, and we have

$$128)\qquad V = \frac{A_0 P_0}{r} + \frac{A_1 P_1}{r^2} + \frac{A_2 P_2}{r^3} + \cdots$$

where every P_n is the zonal harmonic in $\cos\vartheta$.

If we know the value of V for every point on the axis of revolution, so that we can develop it in powers of $\frac{1}{r}$ as

$$129)\qquad V_{\vartheta=0} = F(r) = \frac{B_0}{r} + \frac{B_1}{r^2} + \frac{B_2}{r^3} + \cdots$$

then putting $\cos\vartheta = 1$ in 128) and comparing with 129), we find $A_n = B_n$ so that V is completely determined as

$$130)\qquad V = \frac{B_0}{r} + \frac{B_1 P_1}{r^2} + \frac{B_2 P_2}{r^3} + \cdots.$$

If in addition the body has an equatorial plane of symmetry, so that $V(\cos\vartheta) = V(-\cos\vartheta)$, evidently the development will contain only harmonics of *even* order. As a case of this we shall develop the potential of a homogeneous ellipsoid of revolution in § 161.

Whether the body is homogeneous or not, we may easily obtain the physical significance of the first few terms in 126). For making use of the values in 111) since $\mu = \frac{ax+by+cz}{rr'}$ we have

$$131)\quad r' P_1(\mu) = \frac{(ax+by+cz)}{r},\quad r'^2 P_2(\mu) = \frac{1}{2}\frac{\{3(ax+by+cz)^2 - r'^2 r^2\}}{r^2},$$

$$132)\quad V = \frac{1}{r}\iiint \varrho\, da\, db\, dc + \frac{1}{r^3}\iiint \varrho(ax+by+cz)\, da\, db\, dc$$
$$+ \frac{1}{2r^5}\iiint \varrho\{3(ax+by+cz)^2 - r^2(a^2+b^2+c^2)\}\, da\, db\, dc.$$

There occur in the first three terms the volume integrals

$$\iiint \varrho\, da\, db\, dc = M,\qquad \iiint \varrho a\, da\, db\, dc = M\bar{a},$$
$$\iiint \varrho b\, da\, db\, dc = M\bar{b},\qquad \iiint \varrho c\, da\, db\, dc = M\bar{c},$$
$$\iiint \varrho a^2 da\, db\, dc = \frac{B+C-A}{2},\quad \iiint \varrho b^2 da\, db\, dc = \frac{C+A-B}{2},$$
$$\iiint \varrho c^2 da\, db\, dc = \frac{A+B-C}{2},$$
$$\iiint \varrho bc\, da\, db\, dc = D,\qquad \iiint \varrho ca\, da\, db\, dc = E,$$
$$\iiint \varrho ab\, da\, db\, dc = F,$$

where M is the mass, $\bar{a}$, $\bar{b}$, $\bar{c}$ the coordinates of the center of mass, A, B, C, D, E, F, the moments and products of inertia of the body at the origin. If we choose for origin the center of mass, and for axes the principal axes of inertia at that point, we have

$$\bar{a} = \bar{b} = \bar{c} = D = E = F = 0,$$

so that the second term of the development dissappears, and the third simplifies, so that we have

$$133) \quad V = \frac{M}{r} + \frac{1}{2}\frac{(B+C-2A)x^2 + (C+A-2B)y^2 + (A+B-2C)z^2}{r^5} + \cdots$$

In all these developments, it is to be borne in mind that r is greater than the greatest value of r' for any point Q in the body.

If the body is a homogeneous sphere, all terms dissappear except the first. If the attracted point is at a considerable distance compared with the dimensions of the attracting body, or if the body differs but slightly from a sphere, the terms decrease very rapidly in magnitude, so that the first is by far the most important. Thus under these circumstances bodies attract as if they were concentrated at their centers of mass, or were *centrobaric* (§ 125). The correction is in any case in which we are dealing with the actions of the planets, given with sufficient accuracy by the second term in 133), from which the moments causing precession were calculated in § 96. In § 161 we shall see how the terms depend upon the ellipticity of an ellipsoid of revolution.

149. Applications to Geodesy. Clairaut's Theorem. Although, as has been stated, the development 125) is not in general convergent inside of a sphere with center at the origin which just encloses the attracting body, on account of the divergence of the series 105) when $r' > r$, still it may occur that the performance of the integrations in 125) causes the latter series to converge even within this sphere. At any rate for a body having the properties of the earth, it has been shown by Clairaut[1]), Stokes[2]), and Helmert[3]), that the series 125) converges at all points on the surface of the body, and also that for the earth the two terms in 133) represent the attraction with quite sufficient approximation for applications to the figure of the earth. In order to exhibit the surface harmonics

1) Clairaut, *Théorie de la Figure de la Terre, tirée des Principes de l'Hydrostatique.* Paris, 1743.

2) Stokes, "On the Variation of Gravity at the Surface of the Earth". *Trans. Cambridge Phil. Soc.,* Vol. VIII, 1849.

3) Helmert, *Geodäsie.* 1884.

in terms of angular coordinates, let us introduce the geocentric latitude $\psi = \frac{\pi}{2} - \vartheta$ and longitude φ, in terms of which

$$x = r\cos\psi\cos\varphi, \quad y = r\cos\psi\sin\varphi, \quad z = r\sin\psi.$$

The second term of 133) thus becomes

$$\frac{1}{2r^3}\left[(B + C - 2A)\cos^2\psi\cos^2\varphi + (C + A - 2B)\cos^2\psi\sin^2\varphi + (A + B - 2C)\sin^2\psi\right]$$

which, on putting

$$\cos^2\varphi = \frac{1 + \cos 2\varphi}{2}, \quad \sin^2\varphi = \frac{1 - \cos 2\varphi}{2}, \quad \cos^2\psi = 1 - \sin^2\psi,$$

reduces to

134) $$\frac{1}{2r^3}\left[\left(C - \frac{A+B}{2}\right)(1 - 3\sin^2\psi) + \frac{3}{2}(B - A)\cos^2\psi\cos 2\varphi\right] = \frac{Y_2(\psi, \varphi)}{r^3}.$$

In order to deal with the apparent gravity g, we have to add to V, the potential of the attraction, that of the centrifugal force, as in § 123, 73), putting

135) $$\gamma V_c = \frac{1}{2}\omega^2(x^2 + y^2) = \frac{1}{2}\omega^2 r^2 \sin^2\vartheta.$$

It is to be noticed that by writing

$$\sin^2\vartheta = 1 - \mu^2 = \frac{2}{3}[1 - P_2(\mu)],$$

136) $$\gamma V_c = \frac{1}{3}\omega^2 r^2 - \frac{1}{3}\omega^2 r^2 P_2(\mu),$$

V_c is itself exhibited as $\frac{1}{3}\omega^2 r^2$ plus a spherical harmonic.

If we now write

$$K = \frac{C - \frac{A+B}{2}}{M},$$

we have the approximate expression for the potential of terrestrial gravity

137) $$U = \frac{M}{r}\left\{1 + \frac{K}{2r^2}(1 - 3\sin^2\psi) + \frac{3(B-A)}{4Mr^2}\cos^2\psi\cos 2\varphi + \frac{\omega^2 r^3}{2\gamma M}\cos^2\psi\right\},$$

with

138) $$g = -\gamma\frac{\partial U}{\partial n}.$$

If the surface of the earth is an ellipsoid whose radius vector differs at every point from that of a sphere by a small quantity of the first order, the angle between the normal and the radius vector will be small, its cosine will differ from unity by a small quantity of the second order, neglecting which we may put

$$139)\quad g = -\gamma \frac{\partial U}{\partial r} = \frac{\gamma M}{r^2}\left\{1 + \frac{K}{2r^2}(1 - 3\sin^2\psi) + \frac{9}{4}\frac{B-A}{4Mr^2}\cos^2\psi\cos 2\varphi - \frac{\omega^2 r^3}{\gamma M}\cos^2\psi\right\}.$$

Determinations with the pendulum show that g varies very slightly with the longitude, we may therefore put $B = A$, so that

$$140)\qquad U = \frac{M}{r}\left\{1 + \frac{K}{2r^2}(1 - 3\sin^2\psi) + \frac{\omega^2 r^3}{2\gamma M}\cos^2\psi\right\},$$

$$141)\qquad g = \frac{\gamma M}{r^2}\left\{1 + \frac{3K}{2r^2}(1 - 3\sin^2\psi) - \frac{\omega^2 r^3}{\gamma M}\cos^2\psi\right\}.$$

On a level surface, such as the surface of the ocean will be shown in § 179 to be,

$$U = const. = U_0.$$

For such a surface, equation 140) gives, putting $a = \frac{M}{U_0}$ and in the parenthesis substituting a for r, the equation of the surface,

$$142)\quad r = a\left\{1 + \frac{K}{2a^2}(1 - 3\sin^2\psi) + \frac{\omega^2 a^3}{2\gamma M}\cos^2\psi\right\} = a(1 + \eta).$$

The substitution of a for r is permitted in the higher powers because of the assumption that $\frac{r}{a}$ differs from unity only by a small quantity η, whose square is neglected, and thus $r^m = a^m(1 + m\eta)$. Inserting the value of r from 142) in 141) and approximating in like manner,

$$143)\quad g = \frac{\gamma M}{a^2}\left\{1 - \frac{K}{a^2}(1 - 3\sin^2\psi) - \frac{\omega^2 a^3}{\gamma M}\cos^2\psi\right\}\left\{1 + \frac{3K}{2a^2}(1 - 3\sin^2\psi) - \frac{\omega^2 a^3}{\gamma M}\cos^2\psi\right\}$$

or

$$144)\quad g = \frac{\gamma M}{a^2}\left\{1 + \frac{K}{2a^2} - \frac{2\omega^2 a^3}{\gamma M} + \left(\frac{2\omega^2 a^3}{\gamma M} - \frac{3K}{2a^2}\right)\sin^2\psi\right\},$$

$$145)\qquad g = g_0(1 + n\sin^2\psi),$$

$$146)\qquad n = \frac{2\omega^2 a^3}{\gamma M} - \frac{3K}{2a^2}.$$

The equation 142) is easily seen to be that of an ellipsoid of revolution, and putting $\psi = 0$, $\psi = \frac{\pi}{2}$ its semi-axes are found to be

$$147)\qquad r_e = a\left\{1 + \frac{K}{2a^2} + \frac{\omega^2 a^3}{2\gamma M}\right\},\quad r_p = a\left\{1 - \frac{K}{a^2}\right\}.$$

Accordingly the ellipticity, or flattening *(aplatissement, Abplattung)*, defined as the ratio of the difference of axes to the greater, is

148) $$e = \frac{r_e - r_p}{r_e} = \frac{3K}{2a^2} + \frac{\omega^2 a^3}{2\gamma M} = \frac{5}{2}\frac{\omega^2 a^3}{\gamma M} - n = \frac{5}{2}c - n.$$

The quantity $c = \frac{\omega^2 a^3}{\gamma M}$ is equal to the ratio of the centrifugal acceleration $\omega^2 a$ at the equator to the acceleration of gravity $\frac{\gamma M}{a^2}$ at the same place, while n is equal to the ratio of the excess of polar over equatorial gravity to the latter. Thus equation 148) gives us Clairaut's celebrated theorem,

149) $$e + n = \frac{5}{2}c.$$

$$\textit{Ellipticity of Sea-level} + \frac{\textit{Polar gravity} - \textit{equatorial gravity}}{\textit{Equatorial gravity}}$$
$$= \frac{5}{2}\frac{\textit{Centrifugal acceleration at equator}}{\textit{Gravity at equator}}.$$

The values of the constants in 145) adopted by Helmert as best representing the large number of pendulum observations that had been made up to 1884 are given by

150) $$g = 978.00(1 + 0.005310 \sin^2 \varphi),$$

agreeing closely with the formula given on p. 33. The value of the centrifugal acceleration is known from the length of the sidereal day, the time of the earth's rotation, giving

$$\omega = \frac{2\pi}{86{,}164.09 \text{ sec.}},$$

and the earth's equatorial radius, given by Bessel as 6,377,397 meters. From this is found

$$c = 0.0034672 = \frac{1}{288.41}$$

giving by 149)

$$e = \frac{5}{2} \times 0.0034672 - 0.005310 = 0.0033580 = \frac{1}{297.80}.$$

By a still closer approximation Helmert finds $e = \frac{1}{299.26}$. By a remarkable chance this coincides almost exactly with the value given by Bessel as the result of measurement of arcs of meridian. A third way of deducing the ellipticity is by means of the precession of the equinoxes, which, as has been shown in § 96, enables us to calculate the ratio

$$\frac{C - A}{C} = \frac{1}{297}$$

from which, though involving an assumption as to the distribution of density in the earth, the ellipticity may be derived. Finally, as the moon affects the motion of the earth, the earth disturbs the orbital motion of the moon, which gives a fourth method of obtaining the ellipticity, from which Helmert gives the value obtained from the moon's motion as $\frac{1}{297.8}$.

150. Potential of Tide-generating Forces. In order to study the theory of the tides, it is necessary to obtain the expression for the potential of the attraction of a distant body, such as the sun or moon, considered centrobaric, as a function of the geographical coordinates of a point on the surface of the earth. It is convenient to consider the earth's center of mass reduced to rest by the principle of § 102, according to which we impress upon every point of the earth an acceleration α, $(\alpha_x, \alpha_y, \alpha_z)$ equal and opposite to that impressed upon the earth by the distant body. But this acceleration which is the same for all points of the earth, is accordingly derivable from a potential

151) $$\alpha_x x + \alpha_y y + \alpha_z z = \alpha r \cos(\alpha r).$$

But if m is the mass of the distant body, D its distance from the earth's center we have, § 102,

$$\alpha = \frac{m}{D^2}.$$

Accordingly, if d is the distance of the distant body from the point P on the earth's surface (Fig. 140), Z the angle between the radii, or the zenith distance of the distant body at P, we have for the whole potential at P,

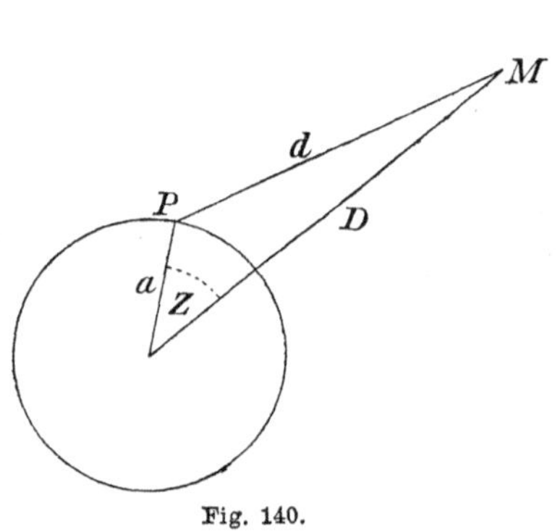

Fig. 140.

152) $$V = \frac{m}{d} - \frac{mr}{D^2}\cos Z.$$

Now developing $\frac{1}{d}$ by 105) and neglecting all but the first three terms,

153) $$V = \frac{m}{D}\left\{1 + \frac{rP_1}{D}(\cos Z) + \frac{r^2P_2}{D^2}(\cos Z)\right\} - \frac{mr}{D^2}\cos Z$$
$$= \frac{m}{D} + \frac{mr^2}{D^3}P_2(\cos Z).$$

The first term is the same for all points on the earth, and therefore may be neglected, so that the tide-generating potential is simply

154) $$\frac{mr^2}{D^3}P_2(\cos Z) = \frac{mr^2}{D^3}\frac{(3\cos^2 Z - 1)}{2}.$$

On account of the third power D^3 in the denominator, the effect of the moon is much larger than that of the sun, in spite of its comparatively microscopic mass.

151. Ellipsoidal Homoeoids. Newton's Theorem. If we transform Laplace's equation to elliptic coordinates and attempt to apply the methods of § 135 to the problem of finding the potential of a homogeneous ellipsoid, we are at once confronted with a difficulty. It is not evident, *nor is it true*, that the potential is independent of two of the coordinates, and that the equipotential surfaces are ellipsoids.

The following theorem was proved geometrically by Newton. A shell of homogeneous matter bounded by two similar and similarly placed ellipsoids exerts no force on a point placed anywhere within the cavity. Such a shell will be called an *ellipsoidal homœoid*.

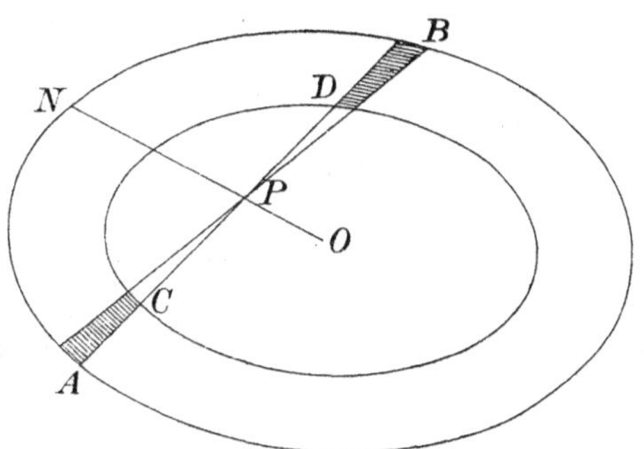

Fig. 141.

Let P, Fig. 141, be the attracted point inside. Since the attraction of a cone of solid angle $d\omega$ on a point of unit mass at its vertex is

$$\int_0^r \frac{dm}{r^2} = \int_0^r \frac{r^2\, d\omega\, dr}{r^2} = r\, d\omega,$$

we have for an element of the homœoid the attraction

$$d\omega\,(BP - DP),$$

in one direction, and

$$d\omega\,(AP - CP)$$

in the other, or in the direction PB,

$$d\omega\,(BD - AC).$$

Draw a plane through ABO, and let ON be the chord of the elliptical section conjugate to AB. Since the ellipsoids are similar and similarly placed, the same diameter is conjugate to the chord CD in both. But CD and AB being bisected in the same point,

$$AC = BD,$$

and the attraction of every part is counterbalanced by that of the opposite part.

152. Condition for Infinite Family of Equipotentials. Although the equipotentials of an ellipsoid are not in general ellipsoids, we may inquire whether there is any distribution of mass that will have ellipsoids as equipotential surfaces.

Let us examine, in general, whether any singly infinite system of surfaces depending upon a parameter q,

$$F(x, y, z, q) = 0$$

can be equipotential surfaces. If so, for any particular value of the parameter q, V must be constant, in other words V is a function of q alone, say $V = f(q)$. If x, y, z are given, q is found from $F(x, y, z, q) = 0$ and from that V from the preceding equation.

Now in free space, V satisfies the equation $\Delta V = 0$. But, since V is a function of q only,

$$\frac{\partial V}{\partial x} = \frac{dV}{dq}\frac{\partial q}{\partial x},$$

1)
$$\frac{\partial^2 V}{\partial x^2} = \frac{dV}{dq}\frac{\partial^2 q}{\partial x^2} + \frac{\partial q}{\partial x}\frac{\partial}{\partial x}\left(\frac{dV}{dq}\right)$$
$$= \frac{dV}{dq}\frac{\partial^2 q}{\partial x^2} + \left(\frac{\partial q}{\partial x}\right)^2 \frac{d^2 V}{dq^2}.$$

In like manner

$$\frac{\partial^2 V}{\partial y^2} = \frac{dV}{dq}\frac{\partial^2 q}{\partial y^2} + \left(\frac{\partial q}{\partial y}\right)^2 \frac{d^2 V}{dq^2},$$
$$\frac{\partial^2 V}{\partial z^2} = \frac{dV}{dq}\frac{\partial^2 q}{\partial z^2} + \left(\frac{\partial q}{\partial z}\right)^2 \frac{d^2 V}{dq^2},$$

2)
$$\Delta V = \frac{dV}{dq}\Delta q + \left\{\left(\frac{\partial q}{\partial x}\right)^2 + \left(\frac{\partial q}{\partial y}\right)^2 + \left(\frac{\partial q}{\partial z}\right)^2\right\}\frac{d^2 V}{dq^2}$$
$$= \frac{dV}{dq}\Delta q + \frac{d^2 V}{dq^2} h_q^2 = 0.$$

Accordingly

3)
$$\frac{\Delta p}{h_q^2} = -\frac{\dfrac{d^2 V}{dq^2}}{\dfrac{dV}{dq}} = -\frac{d}{dq}\left(\log\frac{dV}{dq}\right).$$

Now since V is a function of q only, the expression on the right must be a function of q only, say $\varphi(q)$. Consequently, that

$$F(x, y, z, q) = 0$$

may represent a set of equipotential surfaces, the parameter q must be such that the ratio of its second to the square of its first differential parameter is a function only of q,

$$\frac{\Delta q}{h_q^2} = \varphi(q).$$

This condition is due to Lamé.

If this is satisfied, we have

4) $$-\varphi(q) = \frac{d}{dq}\left(\log \frac{dV}{dq}\right),$$

$$\log \frac{dV}{dq} = -\int \varphi(q)\,dq + C,$$

$$\frac{dV}{dq} = A e^{-\int \varphi(q)\,dq},$$

5) $$V = A\int e^{-\int \varphi(q)\,dq}\,dq + B.$$

There must be one value q such that the level surface is a sphere of infinite radius, and for this V must vanish.

These conditions are satisfied by the polar coordinate r, for by § 141, 87)

$$\Delta r = \frac{2}{r},$$

$$h_r = 1,\quad \frac{\Delta r}{h_r^2} = \frac{2}{r} = \varphi(r),$$

$$V = A\int e^{-\int \frac{2}{r}dr}\,dr + B$$

$$= A\int \frac{dr}{r^2} + B$$

$$= -\frac{A}{r} + B.$$

For $r = \infty$, we must have $V = 0$, accordingly we must put $B = 0$.

We may get a convenient expression for $\frac{\Delta q}{h_q^2}$ by transforming Δq into terms of three orthogonal coordinates, of which it is itself one. Put $q = q_1$, and since it is independent of q_2 and q_3,

6) $$\Delta q_1 = h_1 h_2 h_3 \frac{\partial}{\partial q_1}\left\{\frac{h_1}{h_2 h_3}\frac{\partial q_1}{\partial q_1}\right\},$$

7) $$\frac{\Delta q_1}{h_1{}^2} = \frac{h_2 h_3}{h_1}\frac{\partial}{\partial q_1}\left\{\frac{h_1}{h_2 h_3}\right\}$$

$$= \frac{\partial}{\partial q_1}\log\left(\frac{h_1}{h_2 h_3}\right)$$

$$= \frac{1}{h_1}\frac{\partial h_1}{\partial q_1} - \frac{1}{h_2}\frac{\partial h_2}{\partial q_1} - \frac{1}{h_3}\frac{\partial h_3}{\partial q_1}.$$

153. Application to Elliptic Coordinates. Applying this to elliptic coordinates gives

8) $$\frac{\Delta\lambda}{h_\lambda^2} = \frac{\partial}{\partial\lambda}\left\{\log\frac{1}{2}\sqrt{\frac{(a^2+\lambda)(b^2+\lambda)(c^2+\lambda)(\mu-\nu)(\mu-\lambda)(\nu-\lambda)(\nu-\lambda)}{(\lambda-\mu)(\lambda-\nu)(a^2+\mu)(b^2+\mu)(c^2+\mu)(a^2+\nu)(b^2+\nu)(c^2+\nu)}}\right\}$$

$$= \frac{1}{2}\left\{\frac{1}{a^2+\lambda}+\frac{1}{b^2+\lambda}+\frac{1}{c^2+\lambda}\right\} = \varphi(\lambda),$$

which is independent of μ and ν, and therefore the system of ellipsoids λ *can* represent a family of equipotential surfaces. We have

9) $$\int\varphi(\lambda)\,d\lambda = \frac{1}{2}\int\left\{\frac{1}{a^2+\lambda}+\frac{1}{b^2+\lambda}+\frac{1}{c^2+\lambda}\right\}d\lambda$$

$$= \log\sqrt{(a^2+\lambda)(b^2+\lambda)(c^2+\lambda)},$$

$$e^{-\int\varphi(\lambda)\,d\lambda} = \frac{1}{\sqrt{(a^2+\lambda)(b^2+\lambda)(c^2+\lambda)}},$$

10) $$V = A\int\frac{d\lambda}{\sqrt{(a^2+\lambda)(b^2+\lambda)(c^2+\lambda)}} + B.$$

B must be such a constant that when $\lambda=\infty$, which gives the infinite sphere, $V=0$. This is obtained by taking the definite integral between λ and ∞,

11) $$V = A\int_\lambda^\infty\frac{ds}{\sqrt{(a^2+s)(b^2+s)(c^2+s)}},$$

λ being taken for the *lower* limit, so that A may be positive, making V decrease as λ increases. V is an elliptic integral in terms of λ, or λ is an elliptic function of V. For

12) $$\frac{dV}{d\lambda} = -\frac{A}{\sqrt{(a^2+\lambda)(b^2+\lambda)(c^2+\lambda)}},$$

13) $$A^2\left(\frac{d\lambda}{dV}\right)^2 = (a^2+\lambda)(b^2+\lambda)(c^2+\lambda),$$

a differential equation which is satisfied by an elliptic function.

We may determine the constant A by the property that

$$\lim_{x=\infty}(rV) = M,$$

or that

$$\lim_{r=\infty}\left(r^2\frac{\partial V}{\partial x}\right) = -M\cos(rx).$$

We have

$$\frac{\partial V}{\partial x} = \frac{dV}{d\lambda}\frac{\partial\lambda}{\partial x}$$

$$= -\frac{A}{\sqrt{(a^2+\lambda)(b^2+\lambda)(c^2+\lambda)}}\frac{2x\delta_\lambda^2}{(a^2+\lambda)}, \quad \text{[by § 73, 86)]}$$

$$r^2\frac{\partial V}{\partial x} = -\frac{2Ax\delta_\lambda^2 r^2}{(a^2+\lambda)\sqrt{(a^2+\lambda)(b^2+\lambda)(c^2+\lambda)}}.$$

From the geometrical definition of λ,

$$\lim_{r=\infty}\left(\frac{\lambda}{r^2}\right)=1.$$

Now consider, for simplicity, a point on the X-axis, where $\delta_\lambda = x = r$. The denominator becomes infinite in $\lambda^{\frac{5}{2}}$, that is, r^5, and so does the numerator. Hence

$$\lim\left\{r^2\frac{\partial V}{\partial x}\right\}=-2A=-M,$$

so that

14) $$V=\frac{M}{2}\int_{\lambda}^{\infty}\frac{ds}{\sqrt{(a^2+s)(b^2+s)(c^2+s)}}.$$

154. Chasles's Theorem. We have now found the potential due to a mass M of such nature that its equipotential surfaces are confocal ellipsoids, but it remains to determine the nature of the mass. This may be varied in an infinite number of ways; we will attempt to find an equipotential surface layer. By Green's theorem, § 129, 11), this will have the same mass as that of a body within it which would have the same potential outside.

If we find the required layer on an equipotential surface S, since the potential is constant on S, it must be constant at all points within, or the layer does not affect internal bodies.

The surface density must be given by 10), § 129,

$$\sigma=-\frac{1}{4\pi}\frac{\partial V}{\partial n_\lambda},\text{ where } n_\lambda \text{ is the } \textit{outward} \text{ normal to } \lambda,$$

and

$$\frac{\partial V}{\partial n_\lambda}=\frac{dV}{d\lambda}\frac{\partial\lambda}{\partial n_\lambda}=h_\lambda\frac{dV}{d\lambda}.$$

Now since

$$h_\lambda=2\delta_\lambda,$$

15) $$\sigma=-\frac{1}{2\pi}\delta_\lambda\frac{dV}{d\lambda}.$$

Since V is a function of λ alone, the same is true of $\frac{dV}{d\lambda}$, which for a constant value of λ is constant. Hence σ varies on the ellipsoid S as δ_λ. Therefore if we distribute on the given ellipsoid S a surface layer with surface density proportional at every point to the perpendicular from the origin on the tangent plane at the point, this layer is equipotential, and all its equipotential surfaces are ellipsoids confocal with it. Consequently if we distribute on *any* one of a set of confocal ellipsoids a layer of given mass whose surface density is proportional to δ the attraction of such various layers at given external points is the same, or if the masses differ, is proportional simply to the masses of the layers. For it depends only on λ, which depends only on the position of the point where we calculate the potential.

Since by the definition of a homœoid, the normal thickness of an infinitely thin homœoid is proportional at any point to the perpendicular on the tangent plane, we may replace the words *surface layer*, etc., above by the words *homogeneous infinitely thin homœoid.* The theorem was given in this form by Chasles.[1])

155. Maclaurin's Theorem. Consider two confocal ellipsoids, 1, Fig. 142, with semi-axes $\alpha_1, \beta_1, \gamma_1$, and 2, with semi-axes $\alpha_2, \beta_2, \gamma_2$. The condition of confocality is

$$16)\qquad \alpha_2^2 - \alpha_1^2 = \beta_2^2 - \beta_1^2 = \gamma_2^2 - \gamma_1^2 = s, \text{ say}.$$

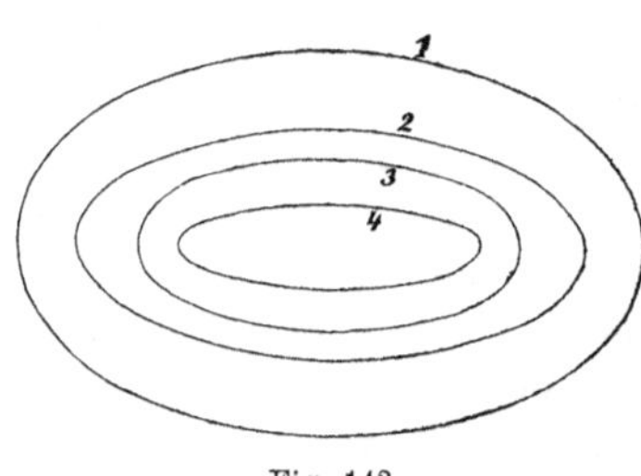

Fig. 142.

If we now construct two ellipsoids 3 and 4 *similar* respectively to 1 and 2, and whose axes are in the same ratio ϑ to those of 1 and 2, these two ellipsoids 3 and 4 are confocal (with each other, though not with 1 and 2). For the semi-axes of 3 are $\vartheta\alpha_1, \vartheta\beta_1, \vartheta\gamma_1$, and of 4 are $\vartheta\alpha_2, \vartheta\beta_2, \vartheta\gamma_2$, and hence the condition of confocality,

$$17)\qquad \vartheta^2\alpha_2^2 - \vartheta^2\alpha_1^2 = \vartheta^2\beta_2^2 - \vartheta^2\beta_1^2 = \vartheta^2\gamma_2^2 - \vartheta^2\gamma_1^2 = \vartheta^2 s$$

is satisfied. Now if on 3 we distribute one infinitely thin homœoidal layer between 3 and another ellipsoid for which ϑ is increased by $d\vartheta$, and on 4 a homœoidal layer given by the same values of ϑ and $d\vartheta$, and furthermore choose the densities such that these two homœoidal layers have the same mass, then (since these homœoids are confocal) their attractions at external points will be identical.

Now the volume of an ellipsoid with axes a, b, c, is $\frac{4}{3}\pi abc$, that of the inner ellipsoid of the shell 3 is accordingly

$$\frac{4}{3}\pi\vartheta^3\alpha_1\beta_1\gamma_1,$$

and that of the shell is the increment of this on increasing ϑ by $d\vartheta$, or

$$(\text{vol. } 3) = 4\pi\vartheta^2 d\vartheta\,\alpha_1\beta_1\gamma_1.$$

Similarly

$$(\text{vol. } 4) = 4\pi\vartheta^2 d\vartheta\,\alpha_2\beta_2\gamma_2.$$

1) Chasles, "Nouvelle solution du problème de l'attraction d'un ellipsoide hétérogène sur un point extérieur. *Journal de Liouville*, t. V. 1840.

Consequently, if we suppose the ellipsoids 1 and 2 filled with matter of uniform density ϱ_1 and ϱ_2, the condition of equal masses of the thin layers 3 and 4,

$$4\pi\varrho_1\vartheta^2 d\vartheta\,\alpha_1\beta_1\gamma_1 = 4\pi\varrho_2\vartheta^2 d\vartheta\,\alpha_2\beta_2\gamma_2,$$

is simply

18) $$\frac{4}{3}\pi\varrho_1\alpha_1\beta_1\gamma_1 = \frac{4}{3}\pi\varrho_2\alpha_2\beta_2\gamma_2,$$

that is, equality of masses of the two ellipsoids. And since for any two corresponding homœoids such as 3 and 4 (ϑ and $\vartheta + d\vartheta$) the attraction on outside points is the same, the attraction of the entire ellipsoids on external points is the same.

This is Maclaurin's celebrated theorem: Confocal homogeneous solid ellipsoids of equal masses attract external points identically, or the attractions of confocal homogeneous ellipsoids at external points are proportional to their masses.[1])

156. Potential of Ellipsoid. The potential due to any homœoidal layer of semi-axes α, β, γ is found to be from our preceding expression for V, 14),

14) $$V = \frac{M}{2}\int_\lambda^\infty \frac{ds}{\sqrt{(\alpha^2+s)(\beta^2+s)(\gamma^2+s)}},$$

where λ is the greatest root of

19) $$\frac{x^2}{\alpha^2+\lambda} + \frac{y^2}{\beta^2+\lambda} + \frac{z^2}{\gamma^2+\lambda} = 1.$$

Now if the semi-axes of the solid ellipsoid are a, b, c, those of the shell $\alpha = \vartheta a$, $\beta = \vartheta b$, $\gamma = \vartheta c$, we have $M = 4\pi\vartheta^2 d\vartheta\, abc$, if the density is unity, and

20) $$d_\vartheta V = 2\pi\vartheta^2 d\vartheta\, abc \int_\lambda^\infty \frac{ds}{\sqrt{(a^2\vartheta^2+s)(b^2\vartheta^2+s)(c^2\vartheta^2+s)}},$$

where λ is defined by

21) $$\frac{x^2}{a^2\vartheta^2+\lambda} + \frac{y^2}{b^2\vartheta^2+\lambda} + \frac{z^2}{c^2\vartheta^2+\lambda} = 1.$$

To get the potential of the whole ellipsoid, we must integrate for all the shells, and

22) $$V = 2\pi abc\int_0^1 \vartheta^2\, d\vartheta \int_\lambda^\infty \frac{ds}{\sqrt{(a^2\vartheta^2+s)(b^2\vartheta^2+s)(c^2\vartheta^2+s)}}.$$

---

1) Maclaurin, *A Treatise on Fluxions.* 1742.

For every value of ϑ there is *one* value of λ, given by the cubic 21).

Let us now change the variable s to t, where, ϑ being constant, $s = \vartheta^2 t$, $ds = \vartheta^2 dt$; and put $\lambda = \vartheta^2 u$.

Then

$$23)\qquad V = 2\pi abc \int_0^1 \vartheta\, d\vartheta \int_u^\infty \frac{dt}{\sqrt{(a^2+t)(b^2+t)(c^2+t)}},$$

where u is defined by

$$24)\qquad \frac{x^2}{a^2+u} + \frac{y^2}{b^2+u} + \frac{z^2}{c^2+u} = \vartheta^2.$$

Since ϑ^2 is thus given as a uniform function of u, we will now change the variable from ϑ to u.

Differentiating 24) by ϑ,

$$25)\qquad 2\vartheta\, d\vartheta = -\left\{\frac{x^2}{(a^2+u)^2} + \frac{y^2}{(b^2+u)^2} + \frac{z^2}{(c^2+u)^2}\right\} du.$$

When $\vartheta = 0$, $u = \infty$, and when $\vartheta = 1$, u has a value which we will call σ, defined by

$$26)\qquad \frac{x^2}{a^2+\sigma} + \frac{y^2}{b^2+\sigma} + \frac{z^2}{c^2+\sigma} = 1.$$

Accordingly, changing the variable,

$$27)\qquad V = \pi abc \int_\sigma^\infty \left\{\frac{x^2}{(a^2+u)^2} + \frac{y^2}{(b^2+u)^2} + \frac{z^2}{(c^2+u)^2}\right\} du \int_u^\infty \frac{dt}{\sqrt{(a^2+t)(b^2+t)(c^2+t)}}.$$

The three double integrals above are of the form

$$28)\qquad J = \int_\sigma^\infty \frac{du}{(C+u)^2} \int_u^\infty f(t)\, dt,$$

where

$$f(t) = \frac{1}{\sqrt{(a^2+t)(b^2+t)(c^2+t)}}.$$

This may be integrated by parts.
Call

$$\int_u^\infty f(t)\, dt = \varphi(u),$$

$$29)\qquad J = \int_\sigma^\infty \frac{\varphi(u)\, du}{(C+u)^2} = -\left[\frac{\varphi(u)}{C+u}\right]_\sigma^\infty + \int_\sigma^\infty \frac{\varphi'(u)\, du}{C+u}.$$

Now

$$\varphi(\infty)=\int_{\infty}^{\infty} f(t)\,dt=0, \quad [\text{since } f(\infty)=0],$$

$$\varphi(\sigma)=\int_{\sigma}^{\infty} f(t)\,dt,$$

$$\varphi'(u)=-f(u).$$

Inserting these values in 29),

$$30) \qquad J=\frac{1}{C+\sigma}\int_{\sigma}^{\infty} f(t)\,dt-\int_{\sigma}^{\infty}\frac{f(u)\,du}{C+u},$$

or the variable of integration being indifferent, we may put u for t in the first integral.

Applying this to our integral 27), by putting C successively equal to a^2, b^2, c^2, multiplying by x^2, y^2, z^2, and adding,

$$31) \quad V=\pi abc\int_{\sigma}^{\infty}\left\{\frac{x^2}{a^2+\sigma}+\frac{y^2}{b^2+\sigma}+\frac{z^2}{c^2+\sigma}-\frac{x^2}{a^2+u}-\frac{y^2}{b^2+u}-\frac{z^2}{c^2+u}\right\}f(u)\,du.$$

Now the first three terms of the integrand are, by definition, equal to 1, so that

$$32) \quad V=\pi abc\int_{\sigma}^{\infty}\left\{1-\frac{x^2}{a^2+u}-\frac{y^2}{b^2+u}-\frac{z^2}{c^2+u}\right\}\frac{du}{\sqrt{(a^2+u)(b^2+u)(c^2+u)}}.$$

This form was given by Dirichlet.[1])

If the point x, y, z lies on the surface of the ellipsoid,

$$\frac{x^2}{a^2}+\frac{y^2}{b^2}+\frac{z^2}{c^2}=1,$$

then $\sigma=0$ and

$$33) \quad V=\pi abc\int_{0}^{\infty}\left\{1-\frac{x^2}{a^2+u}-\frac{y^2}{b^2+u}-\frac{z^2}{c^2+u}\right\}\frac{du}{\sqrt{(a^2+u)(b^2+u)(c^2+u)}}.$$

We find for the derivatives of V,

$$\frac{\partial V}{\partial x}=-2\pi abcx\int_{\sigma}^{\infty}\frac{du}{(a^2+u)\sqrt{(a^2+u)(b^2+u)(c^2+u)}}$$

$$-\pi abc\frac{\partial\sigma}{\partial x}\left\{1-\frac{x^2}{a^2+\sigma}-\frac{y^2}{b^2+\sigma}-\frac{z^2}{c^2+\sigma}\right\}\frac{1}{\sqrt{(a^2+\sigma)(b^2+\sigma)(c^2+\sigma)}}.$$

1) Dirichlet, "Über eine neue Methode zur Bestimmung vielfacher Integrale". *Abh. der Berliner Akad.*, 1839. Translated in *Journ. de Liouville,* t. iv., 1839

By definition of σ, the parenthesis in the last term vanishes, and

$$\frac{\partial V}{\partial x} = -2\pi abcx \int_\sigma^\infty \frac{du}{(a^2+u)\sqrt{(a^2+u)(b^2+u)(c^2+u)}},$$

$$34) \qquad \frac{\partial V}{\partial y} = -2\pi abcy \int_\sigma^\infty \frac{du}{(b^2+u)\sqrt{(a^2+u)(b^2+u)(c^2+u)}},$$

$$\frac{\partial V}{\partial z} = -2\pi abcz \int_\sigma^\infty \frac{du}{(c^2+u)\sqrt{(a^2+u)(b^2+u)(c^2+u)}}.$$

157. Internal Point. In the case of an internal point, we pass through it an ellipsoid similar to the given ellipsoid, then by Newton's theorem it is unattracted by the homœoidal shell without, and we may use the above formulae for the attraction, putting for a, b, c, the values for the ellipsoid through x, y, z, say $\vartheta a, \vartheta b, \vartheta c$. Since the point is on the surface of this, $\sigma = 0$.

$$35) \qquad \frac{\partial V}{\partial x} = -2\pi\vartheta^3 abcx \int_0^\infty \frac{du}{(\vartheta^2 a^2+u)\sqrt{(\vartheta^2 a^2+u)(\vartheta^2 b^2+u)(\vartheta^2 c^2+u)}}.$$

Now let us insert a variable u' proportional to u, $u = \vartheta^2 u'$,

$$36) \qquad \frac{\partial V}{\partial x} = -2\pi\vartheta^3 abcx \int_0^\infty \frac{\vartheta^2 du'}{\vartheta^2(a^2+u')\vartheta^3\sqrt{(a^2+u')(b^2+u')(c^2+u')}}.$$

The ϑ divides out, and writing u for the variable of integration

$$\frac{\partial V}{\partial x} = -2\pi abcx \int_0^\infty \frac{du}{(a^2+u)\sqrt{(a^2+u)(b^2+u)(c^2+u)}}.$$

So that for *any* internal point, we put $\sigma = 0$ in the general formula. Integrating with respect to x, y, z, we have

$$37) \qquad V = \pi abc \int_0^\infty \left\{1 - \frac{x^2}{a^2+u} - \frac{y^2}{b^2+u} - \frac{z^2}{c^2+u}\right\} \frac{du}{\sqrt{(a^2+u)(b^2+u)(c^2+u)}}.$$

The constant term must be taken as above in order that at the surface V may be continuous.

In the case of an internal point the above four integrals may be made to depend upon the first. Calling

$$38)\qquad \Phi = \int_0^\infty \frac{du}{\sqrt{(a^2+u)(b^2+u)(c^2+u)}},$$

$$\frac{\partial \Phi}{\partial (a^2)} = \int_0^\infty -\frac{1}{2}\frac{du}{(a^2+u)\sqrt{(a^2+u)(b^2+u)(c^2+u)}},$$

and accordingly

$$39)\qquad V = \pi abc\left\{\Phi + 2\frac{\partial \Phi}{\partial (a^2)}x^2 + 2\frac{\partial \Phi}{\partial (b^2)}y^2 + 2\frac{\partial \Phi}{\partial (c^2)}z^2\right\}.$$

The integral Φ is an elliptic integral independent of x, y, z, and so are its derivatives with respect to a^2, b^2, c^2. Calling these respectively $\frac{L}{4}, \frac{M}{4}, \frac{N}{4}$, we have

$$40)\qquad V = \pi abc\left\{\Phi + \frac{1}{2}(Lx^2 + My^2 + Nz^2)\right\},$$

a symmetrical function of the second order, and since L, M, N are of the same sign, the equipotential surfaces are ellipsoids, similar to each other. Their relation to the given ellipsoid is however transcendental, their semi-axes being

$$41)\qquad \sqrt{\frac{\frac{V}{\pi abc}-\Phi}{2\frac{\partial\Phi}{\partial (a^2)}}},\quad \sqrt{\frac{\frac{V}{\pi abc}-\Phi}{2\frac{\partial\Phi}{\partial (b^2)}}},\quad \sqrt{\frac{\frac{V}{\pi abc}-\Phi}{2\frac{\partial\Phi}{\partial (c^2)}}}.$$

We have for the force,

$$42)\qquad \gamma\frac{\partial V}{\partial x} = X = \gamma Lx,\quad \gamma\frac{\partial V}{\partial y} = \gamma My,\quad \gamma\frac{\partial V}{\partial z} = \gamma Nz.$$

Therefore, since for two points on the same radius vector,

$$\frac{x_2}{x_1} = \frac{y_2}{y_1} = \frac{z_2}{z_1} = \frac{r_2}{r_1},\quad \text{we have}\quad \frac{X_2}{X_1} = \frac{Y_2}{Y_1} = \frac{Z_2}{Z_1} = \frac{r_2}{r_1}.$$

The forces are parallel and proportional to the distance from the center, though not directed toward the center.

158. Verification by Differentiation. For an outside point we have, differentiating 34)

$$43)\qquad \frac{\partial^2 V}{\partial x^2} = -2\pi abc\int_\sigma^\infty \frac{du}{(a^2+u)\sqrt{(a^2+u)(b^2+u)(c^2+u)}}$$

$$+ 2\pi abc\,x\frac{\partial\sigma}{\partial x}\left\{\frac{1}{(a^2+\sigma)\sqrt{(a^2+\sigma)(b^2+\sigma)(c^2+\sigma)}}\right\}.$$

Now by § 73, 86),

$$\frac{\partial\sigma}{\partial x} = \frac{2x}{a^2+\sigma} \Big/ \left\{\frac{x^2}{(a^2+\sigma)^2} + \frac{y^2}{(b^2+\sigma)^2} + \frac{z^2}{(c^2+\sigma)^2}\right\}.$$

Forming $\frac{\partial^2 V}{\partial y^2}$ and $\frac{\partial^2 V}{\partial z^2}$ and adding,

$$44)\quad \Delta V = -2\pi abc \int_\sigma^\infty \left\{\frac{1}{a^2+u} + \frac{1}{b^2+u} + \frac{1}{c^2+u}\right\} \frac{du}{\sqrt{(a^2+u)(b^2+u)(c^2+u)}}, + \frac{4\pi abc}{\sqrt{(a^2+\sigma)(b^2+\sigma)(c^2+\sigma)}}.$$

The integration may be at once effected.
Since

$$d(uvw) = uvw\left\{\frac{du}{u} + \frac{dv}{v} + \frac{dw}{w}\right\},$$

we have

$$45)\quad d\left\{\frac{1}{\sqrt{(a^2+u)(b^2+u)(c^2+u)}}\right\}$$

$$= \frac{du}{\sqrt{(a^2+u)(b^2+u)(c^2+u)}}\left\{-\frac{\sqrt{a^2+u}}{2\sqrt{(a^2+u)^3}} - \cdots\right\}$$

$$= -\frac{1}{2}\left\{\frac{1}{a^2+u} + \frac{1}{b^2+u} + \frac{1}{c^2+u}\right\}\frac{du}{\sqrt{(a^2+u)(b^2+u)(c^2+u)}}.$$

The integral becomes then

$$-\frac{4\pi abc}{\sqrt{(a^2+\sigma)(b^2+\sigma)(c^2+\sigma)}},$$

which cancels the second term, and $\Delta V = 0$.

For an internal point

$$46)\quad \frac{\partial^2 V}{\partial x^2} = -2\pi abc \int_0^\infty \frac{du}{(a^2+u)\sqrt{(a^2+u)(b^2+u)(c^2+u)}},$$

$$\Delta V = -4\pi.$$

At infinity $\sigma = \infty$, and V and its derivatives accordingly vanish.

Therefore the value of V found satisfies all the conditions.

159. Ivory's Theorem. If x, y, z is a point on the ellipsoid (1)

$$47)\quad \frac{x^2}{a_1^2} + \frac{y^2}{b_1^2} + \frac{z^2}{c_1^2} = 1,$$

the point

$$x\frac{a_2}{a_1},\quad y\frac{b_2}{b_1},\quad z\frac{c_2}{c_1}$$

lies on the ellipsoid (2)

$$48)\quad \frac{x^2}{a_2^2} + \frac{y^2}{b_2^2} + \frac{z^2}{c_2^2} = 1.$$

These will be called corresponding points. We shall now assume that these two ellipsoids are confocal, and (2) the smaller. Then

$$49)\qquad a_1^2 = a_2^2 + \lambda, \quad b_1^2 = b_2^2 + \lambda, \quad c_1^2 = c_2^2 + \lambda.$$

The action of (2) on the external point x, y, z is

$$50)\quad X_2 = -2\gamma\pi a_2 b_2 c_2 x \int_\sigma^\infty \frac{du}{(a_2^2+u)\sqrt{(a_2^2+u)(b_2^2+u)(c_2^2+u)}},$$

where

$$\frac{x^2}{a_2^2+\sigma} + \frac{y^2}{b_2^2+\sigma} + \frac{z^2}{c_2^2+\sigma} = 1,$$

and since

$$\frac{x^2}{a_1^2} + \frac{y^2}{b_1^2} + \frac{z^2}{c_1^2} = 1,$$

we must have $\sigma = \lambda$.

If now we substitute

$$u = u' + a_1^2 - a_2^2 = u' + \sigma,$$

$$51)\quad X_2 = -2\gamma\pi a_2 b_2 c_2 x \int_0^\infty \frac{du'}{(a_1^2+u')\sqrt{(a_1^2+u')(b_1^2+u')(c_1^2+u')}}.$$

Now the attraction of the ellipsoid (1) on the interior point $x\frac{a_2}{a_1}, y\frac{b_2}{b_1}, z\frac{c_2}{c_1}$ is

$$52)\quad X_1 = -2\gamma\pi a_1 b_1 c_1 x \frac{a_2}{a_1} \int_0^\infty \frac{du}{(a_1^2+u)\sqrt{(a_1^2+u)(b_1^2+u)(c_1^2+u)}}.$$

The definite integrals being the same in both cases, we have

$$53)\qquad \frac{X_2}{X_1} = \frac{b_2 c_2}{b_1 c_1}, \quad \frac{Y_2}{Y_1} = \frac{c_2 a_2}{c_1 a_1}, \quad \frac{Z_2}{Z_1} = \frac{a_2 b_2}{a_1 b_1}.$$

This is Ivory's theorem: Two confocal ellipsoids of equal density each act on corresponding points on the other with forces whose components are proportional to the areas of their principal sections normal to the components.[1])

160. Ellipsoids of Revolution. For an ellipsoid of revolution, the elliptic integrals reduce to inverse circular functions.

---

1) Ivory, "On the attractions of homogeneous Ellipsoids". *Phil. Trans.,* 1809.

Put $b = c$, a being the axis of revolution,

54) $$V = \pi a b^2 \int_\sigma^\infty \frac{du}{(b^2+u)\sqrt{a^2+u}} - \frac{1}{2}(Xx + Yy),$$

55) $$X = 2\pi a b^2 x \int_\sigma^\infty \frac{du}{(b^2+u)(a^2+u)^{\frac{3}{2}}},$$

56) $$Y = 2\pi a b^2 y \int_\sigma^\infty \frac{du}{(b^2+u)^2\sqrt{a^2+u}},$$

where

$$\frac{x^2}{a^2+\sigma} + \frac{y^2}{b^2+\sigma} = 1.$$

Put

$$b^2 + u = \frac{b^2 - a^2}{s^2},$$

$$a^2 + u = (b^2 - a^2)\left(\frac{1}{s^2} - 1\right) = (b^2 - a^2)\frac{(1 - s^2)}{s^2}$$

$$du = -\frac{2(b^2 - a^2)}{s^3} ds.$$

When $u = \infty$, $s = 0$; when $u = \sigma$, $s = \sqrt{\frac{b^2 - a^2}{b^2 + \sigma}}$, so that

57) $$V = \pi a b^2 \int_{u=\sigma}^{u=\infty} \frac{-2(b^2 - a^2)s^3\, ds}{s^3(b^2-a^2)\sqrt{(b^2 - a^2)(1 - s^2)}} - \frac{1}{2}(Xx + Yy)$$

$$= \frac{2\pi a b^2}{\sqrt{b^2 - a^2}} \int_0^{\sqrt{\frac{b^2-a^2}{b^2+\sigma}}} \frac{ds}{\sqrt{1-s^2}} - \frac{1}{2}(Xx + Yy),$$

58) $$V = \frac{2\pi a b^2}{\sqrt{b^2 - a^2}} \left\{\sin^{-1}\sqrt{\frac{b^2 - a^2}{b^2 + \sigma}}\right\} - \frac{1}{2}\{Xx + Yy\},$$

59) $$X = 2\pi a b^2 x \int - \frac{2(b^2 - a^2)s^2}{s^3(b^2 - a^2)(b^2 - a^2)^{\frac{3}{2}}} \frac{s^3\, ds}{(1-s^2)^{\frac{3}{2}}}$$

$$= \frac{4\pi a b^2 x}{(b^2 - a^2)^{\frac{3}{2}}} \int_0^{\sqrt{\frac{b^2-a^2}{b^2+\sigma}}} \frac{s^2\, ds}{(1 - s^2)^{\frac{3}{2}}}.$$

Now

$$\int \frac{s^2\, ds}{(1 - s^2)^{\frac{3}{2}}} = \frac{s}{\sqrt{1 - s^2}} - \int \frac{ds}{\sqrt{1 - s^2}},$$

so that

$$60)\qquad X = \frac{4\pi a b^2 x}{(b^2 - a^2)^{\frac{3}{2}}}\left\{\sqrt{\frac{b^2 - a^2}{a^2 + \sigma}} - \sin^{-1}\sqrt{\frac{b^2 - a^2}{b^2 + \sigma}}\right\},$$

$$61)\qquad Y = 2\pi a b^2 y \int \frac{-2(b^2 - a^2)}{s^3 (b^2 - a^2)^2}\,\frac{s^4 \cdot s\, ds}{\sqrt{(b^2 - a^2)(1 - s^2)}}$$

$$= \frac{4\pi a b^2 y}{(b^2 - a^2)^{\frac{3}{2}}}\int_0^{\sqrt{\frac{b^2 - a^2}{b^2 + \sigma}}} \frac{s^2\, ds}{\sqrt{1 - s^2}}.$$

Now

$$\int \frac{s^2\, ds}{\sqrt{1 - s^2}} = \frac{1}{2}\left\{\sin^{-1} s - s\sqrt{1 - s^2}\right\},$$

so that

$$62)\qquad Y = \frac{2\pi a b^2 y}{(b^2 - a^2)^{\frac{3}{2}}}\left\{\sin^{-1}\sqrt{\frac{b^2 - a^2}{b^2 + \sigma}} - \sqrt{\frac{(b^2 - a^2)(a^2 + \sigma)}{b^2 + \sigma}}\right\}.$$

For $\sin^{-1}\sqrt{\frac{b^2 - a^2}{b^2 + \sigma}}$ we may write $\tan^{-1}\sqrt{\frac{b^2 - a^2}{a^2 + \sigma}}$, for if

$$\sin \vartheta = \sqrt{\frac{b^2 - a^2}{b^2 + \sigma}},$$

then

$$\cos \vartheta = \sqrt{\frac{a^2 + \sigma}{b^2 + \sigma}},$$

$$\tan \vartheta = \sqrt{\frac{b^2 - a^2}{a^2 + \sigma}}.$$

These formulae all serve for an oblate spheroid, where $a < b$. For a prolate spheroid, $b > a$, they introduce imaginaries, from which they may be cleared as follows.

Call

$$\sin^{-1}(iu) = \vartheta,$$

then

$$iu = \sin \vartheta, \quad \sqrt{1 + u^2} = \cos \vartheta,$$

$$e^{-i\vartheta} = \cos \vartheta - i \sin \vartheta = \sqrt{1 + u^2} + u,$$

therefore

$$-i\vartheta = \log\left\{\sqrt{1 + u^2} + u\right\},$$

$$\sin^{-1}(iu) = \vartheta = i \log\left\{\sqrt{1 + u^2} +\right.$$

Put

$$u = \sqrt{\frac{a^2 - b^2}{b^2 + \sigma}},$$

$$\sin^{-1}\sqrt{\frac{b^2 - a^2}{b^2 + \sigma}} = i \log\left\{\frac{\sqrt{a^2 + \sigma} + \sqrt{a^2 - b^2}}{\sqrt{b^2 + \sigma}}\right\}.$$

Therefore

$$63)\quad V = \frac{2\pi ab^2}{\sqrt{a^2-b^2}} \log\left\{\frac{\sqrt{a^2-b^2}+\sqrt{a^2+\sigma}}{\sqrt{b^2+\sigma}}\right\} - \frac{1}{2}(Xx+Yy),$$

$$64)\quad X = \frac{4\pi ab^2 x}{(a^2-b^2)^{\frac{3}{2}}}\left\{\log\frac{\sqrt{a^2-b^2}+\sqrt{a^2+\sigma}}{\sqrt{b^2+\sigma}} - \sqrt{\frac{a^2-b^2}{a^2+\sigma}}\right\},$$

$$65)\quad Y = \frac{2\pi ab^2 y}{(a^2-b^2)^{\frac{3}{2}}}\left\{\frac{\sqrt{(a^2-b^2)(a^2+\sigma)}}{b^2+\sigma} - \log\frac{\sqrt{a^2-b^2}+\sqrt{a^2+\sigma}}{\sqrt{b^2+\sigma}}\right\}.$$

In all these formulae, σ is the larger root of the quadratic

$$66)\quad \frac{x^2}{a^2+\sigma}+\frac{y^2}{b^2+\sigma}=1,$$

for an outside point, and $\sigma=0$ for an inside point. In the latter case, we have functions only of the ratio $\frac{a}{b}$.[1])

161. Development of Potential of Ellipsoid of Revolution. We may develop the expression 58) for the potential of an oblate ellipsoid of revolution in a series of spherical harmonics. Considering first a point on the axis of revolution, let us put

$$x = r > a, \quad y = 0,$$

so that we have by 66), 61),

$$67)\quad \frac{x^2}{a^2+\sigma}=1, \quad Y=0$$

and using the $\tan^{-1}$ instead of $\sin^{-1}$ in 58) and 60), we obtain

$$68)\quad V_1 = \frac{2\pi ab^2}{\sqrt{b^2-a^2}}\left[\left\{1+\frac{r^2}{b^2-a^2}\right\}\tan^{-1}\frac{\sqrt{b^2-a^2}}{r} - \frac{r}{\sqrt{b^2-a^2}}\right].$$

Remarking that M the mass of the ellipsoid is $\frac{4}{3}\pi ab^2$, and developing the antitangent, we have, if $r > a > \sqrt{b^2-a^2}$, $b^2 < 2a^2$,

$$69)\quad V_1 = F(r) = \frac{3M}{2\sqrt{b^2-a^2}}\left[\left\{1+\frac{r^2}{b^2-a^2}\right\}\sum_0^\infty \frac{(-1)^n (b^2-a^2)^{\frac{2n+1}{2}}}{2n+1\quad r^{2n+1}} - \frac{r}{\sqrt{b^2-a^2}}\right]$$

$$= \frac{3M}{2}\left\{\sum_0^\infty \frac{(-1)^n}{2n+1}\frac{(b^2-a^2)^n}{r^{2n+1}} - \sum_0^\infty \frac{(-1)^n}{2n+3}\frac{(b^2-a^2)^n}{r^{2n+1}}\right\}$$

$$= \frac{M}{r} + 3M\sum_1^\infty \frac{(-1)^n (b^2-a^2)^n}{(2n+1)(2n+3)r^{2n+1}}$$

---

1) Thomson and Tait, *Natural Philosophy*, Part II, § 527.

from which, by § 148, 130), we obtain finally,

$$V = \frac{M}{r} + 3M \sum_{1}^{\infty} \frac{(-1)^n (b^2 - a^2)^n P_{2n}}{(2n+1)(2n+3) r^{2n+1}} \tag{70}$$

and this series is convergent for points on the surface of the ellipsoid itself, if $b^2 < 2a^2$. The series converges extremely rapidly if $\frac{a}{b}$ differs little from unity.

162. Energy of Distributions. Gauss's theorem. If a particle of unit mass be at P, (x, y, z) at a distance r from a particle of mass m_q, the work necessary to bring the unit particle from an infinite distance against the repulsion of the particle m_q will be

$$W = \gamma \frac{m_q}{r} = \gamma V(x, y, z) = \gamma V_p. \tag{71}$$

If, instead of a particle of unit mass, we have one of mass m_p the work necessary will be m_p times as great,

$$W_{pq} = \gamma \frac{m_q}{r} m_p = \gamma m_p V_p = \gamma m_q V_q, \tag{72}$$

where

$$V_q = \frac{m_p}{r}.$$

In other words, this is the amount of loss of the potential energy of the system on being allowed to disperse to an infinite distance from a distance apart r. Similarly, for any two systems of particles m_p, m_q,

$$W_{pq} = \gamma \sum_q \sum_p \frac{m_p m_q}{r_{pq}} = \gamma \sum_p m_p V_p' = \gamma \sum_q m_q V_q, \tag{73}$$

V_p' being the potential *at* any point p due to *all* the particles q and V_q being the potential *at* any point q due to *all* the particles p. This sum is called the *mutual potential energy* of the systems p and q. If however we consider all the particles to belong to one system, we must write

$$W = \frac{1}{2} \gamma \sum \sum \frac{m_p m_q}{r_{pq}} = \frac{1}{2} \gamma \sum m V, \tag{74}$$

where every particle appears both as p and q, the $\frac{1}{2}$ being put in because every *pair* would thus appear twice. This expression has been given in § 28, 36).

If the systems are continuously distributed over volumes τ, τ' we have

$$W_{pq} = \gamma \iiint\iiint \frac{\varrho_p \varrho_q}{r_{pq}} d\tau_p d\tau_q = \gamma \iiint_{\tau} \varrho_p V_p' d\tau_p \tag{75}$$

$$= \gamma \iiint_{\tau'} \varrho_q V_q d\tau_q.$$

The theorem expressed by the equality of the two integrals is known as Gauss's theorem on mutual energy, where V_p' represents the potential at p due to the whole mass M_q, V_q, that at q due to the whole mass M_p.[1])

The above equality may be also proved as follows. Since

$$\text{76)} \qquad \varrho_p = -\frac{1}{4\pi}\varDelta V_p,$$

and

$$\varrho_q = -\frac{1}{4\pi}\varDelta V_q',$$

the triple integrals in 75) become respectively,

$$\text{77)} \qquad -\frac{1}{4\pi}\iiint\limits_{\tau} V_p' \varDelta V_p \, d\tau_p,$$

and

$$-\frac{1}{4\pi}\iiint\limits_{\tau'} V_q \varDelta V_q' \, d\tau_q.$$

Now since outside of τ, $\varDelta V = 0$ and outside of τ', $\varDelta V' = 0$ the integrals may be extended to all space. But by Green's theorem, both these integrals are equal to

$$\frac{1}{4\pi}\iiint\limits_{\infty}\left\{\frac{\partial V}{\partial x}\frac{\partial V'}{\partial x} + \frac{\partial V}{\partial y}\frac{\partial V'}{\partial y} + \frac{\partial V}{\partial z}\frac{\partial V'}{\partial z}\right\} d\tau,$$

since the surface integrals

$$\iint V\frac{\partial V'}{\partial n}\,dS, \quad \iint V'\frac{\partial V}{\partial n}\,dS$$

vanish at infinity. Gauss's theorem accordingly follows from Green's theorem and Poisson's equation.

163. Energy in terms of Field. For the energy of any distribution consisting of both volume and surface distributions, the sum 74) becomes the integrals

$$\text{78)} \qquad W = \frac{\gamma}{2}\iint\limits_{S} V\sigma\,dS + \frac{\gamma}{2}\iiint\limits_{\tau} V\varrho\,d\tau.$$

Now at a surface distribution Poisson's equation is

$$\sigma = -\frac{1}{4\pi}\left\{\frac{\partial V}{\partial n_1} + \frac{\partial V}{\partial n_2}\right\}.$$

1) Gauss, "Allgemeine Lehrsätze in Beziehung auf die im verkehrten Verhältnisse der Entfernung wirkenden Anziehungs- und Abstossungskräfte." *Werke*, Bd. V, p. 197.

If, as in § 131, we draw surfaces close to the surface distributions, and exclude the space between them, we may, as above, extend the integrals to all other space, so that

$$79)\qquad W = -\frac{\gamma}{8\pi}\iint_S V\frac{\partial V}{\partial n}dS - \frac{\gamma}{8\pi}\iiint_\infty V\Delta V d\tau,$$

the normals being from the surfaces S toward the space τ. But by Green's theorem, as before, this is equal to the integral

$$80)\qquad W = \frac{\gamma}{8\pi}\iiint_\infty \left\{\left(\frac{\partial V}{\partial x}\right)^2 + \left(\frac{\partial V}{\partial y}\right)^2 + \left(\frac{\partial V}{\partial z}\right)^2\right\} d\tau.$$

Thus the energy is expressed in terms of the strength of the field

$$F^2 = \gamma^2\left[\left(\frac{\partial V}{\partial x}\right)^2 + \left(\frac{\partial V}{\partial y}\right)^2 + \left(\frac{\partial V}{\partial z}\right)^2\right]$$

at all points in space. This integral is of fundamental importance in the modern theory of electricity and magnetism.

It is at once seen that this always has the sign of γ, so that it is positive for electrical or magnetic, negative for gravitational distributions.

CHAPTER IX.

DYNAMICS OF DEFORMABLE BODIES.

164. Kinematics. Homogeneous Strain. We have now to consider the kinematics of a body that is not rigid, that is, one whose various points are capable of displacements relatively to each other. In the general displacement of such a body every point x, y, z moves to a new position x', y', z', so that x', y', z' are uniform functions of x, y, z. The functions must also be continuous, that is, two points infinitely near together remain infinitely near together, unless ruptures occur in the body.

The assemblage of relative displacements of all the points is called a *strain.* The simplest sort of strain is given when the functions are linear, that is,

$$1)\qquad \begin{aligned} x' &= a_1 x + a_2 y + a_3 z, \\ y' &= b_1 x + b_2 y + b_3 z, \\ z' &= c_1 x + c_2 y + c_3 z, \end{aligned}$$

where the a's, b's and c's are nine constants.

No constant terms are included because a displacement represented by $x' = a$, $y' = b$, $z' = c$, would denote a translation of the body as if rigid, which is unaccompanied by relative displacement or strain.

Let the solutions of the equations 1), which we shall term the direct substitution, be

$$
\begin{aligned}
x &= A_1 x' + B_1 y' + C_1 z', \\
y &= A_2 x' + B_2 y' + C_2 z', \qquad (2) \\
z &= A_3 x' + B_3 y' + C_3 z'
\end{aligned}
$$

where

$$A_1 = \begin{vmatrix} b_2, & b_3 \\ c_2, & c_3 \end{vmatrix} \div \Delta \qquad \text{etc.}$$

$$a_1 = \begin{vmatrix} B_2, & B_3 \\ C_2, & C_3 \end{vmatrix} \div \Delta', \qquad \text{etc.}$$

$$\Delta = \begin{vmatrix} a_1, & a_2, & a_3 \\ b_1, & b_2, & b_3 \\ c_1, & c_2, & c_3 \end{vmatrix}, \quad \Delta' = \begin{vmatrix} A_1, & B_1, & C_1 \\ A_2, & B_2, & C_2 \\ A_3, & B_3, & C_3 \end{vmatrix}.$$

A strain represented by the equations 1) is said to be *homogeneous.* If the accented letters denote initial positions, and the unaccented letters final positions, the strain represented by equations 2) is said to be inverse to the first strain.

In virtue of equations 1) or 2) a linear relation between x, y, z becomes a linear relation between x', y', z'. Accordingly in a homogeneous strain a plane remains a plane, and a straight line, being the intersection of two planes, remains a straight line. Finite points remain finite, since the coefficients are finite, accordingly parallel lines, intersecting at infinity remain parallel. Parallelograms remain parallelograms (their angles being in general changed), and therefore the changes of length experienced by equal parallel lines are equal, and for unequal parallel lines proportional to their lengths. Thus any portion of the body experiences the same change of size and shape as any equal and similarly placed portion at any other part of the body. This is the meaning of the term homogeneous, which signifies alike all over.

When two vectors OP of length r and OP' of length r' drawn from the same origin are so related that their respective components x, y, z, x', y', z' are connected by the equations 1) or their equivalents 2) either vector is said to be a *linear vector function* of the other. The properties of such linear functions are of great importance in mathematical physics, and will now be taken up before their application to strain.

Let us examine the conditions that the two vectors OP and OP' shall have the same direction. The condition for this is

$$3)\qquad \frac{x'}{x} = \frac{y'}{y} = \frac{z'}{z} = \lambda,$$

where λ is to be determined. Introducing the values $x' = \lambda x$, $y' = \lambda y$, $z' = \lambda z$ into equations 1) we obtain

$$4)\qquad \begin{aligned} (a_1 - \lambda)x + a_2 y + a_3 z &= 0, \\ b_1 x + (b_2 - \lambda) y + b_3 z &= 0, \\ c_1 x + c_2 y + (c_3 - \lambda) z &= 0, \end{aligned}$$

a set of linear equations to determine x, y, z. The condition that these shall be compatible is that the determinant of the coefficients vanishes.

$$5)\qquad \begin{vmatrix} a_1 - \lambda, & a_2 \;, & a_3 \\ b_1 \;, & b_2 - \lambda, & b_3 \\ c_1 \;, & c_2 \;, & c_3 - \lambda \end{vmatrix} = 0.$$

This is a cubic in λ. Let its roots be $\lambda_1, \lambda_2, \lambda_3$. Inserting any one of these in 4) we may find the ratios of x, y, z giving the direction of the vectors in question.

Supposing that $\lambda_1, \lambda_2, \lambda_3$ are real, let us find the condition that the three directions are mutually perpendicular.

Substituting first $\lambda = \lambda_1$ and then $\lambda = \lambda_2$ in 1) and 3), we have, denoting the values of x, y, z by corresponding subscripts,

$$6)\qquad \begin{aligned} a_1 x_1 + a_2 y_1 + a_3 z_1 &= \lambda_1 x_1, \\ b_1 x_1 + b_2 y_1 + b_3 z_1 &= \lambda_1 y_1, \\ c_1 x_1 + c_2 y_1 + c_3 z_1 &= \lambda_1 z_1, \\ a_1 x_2 + a_2 y_2 + a_3 z_2 &= \lambda_2 x_2, \\ b_1 x_2 + b_2 y_2 + b_3 z_2 &= \lambda_2 y_2, \\ c_1 x_2 + c_2 y_2 + c_3 z_2 &= \lambda_2 z_2. \end{aligned}$$

Multiplying the first three respectively by x_2, y_2, z_2 and adding, and subtracting the sum of the last three multiplied respectively by x_1, y_1, z_1, we obtain

$$7)\qquad \begin{aligned} (c_2 - b_3)(y_1 z_2 - z_1 y_2) + (a_3 - c_1)(z_1 x_2 - x_1 z_2) + (b_1 - a_2)(x_1 y_2 - y_1 x_2) \\ = (\lambda_1 - \lambda_2)(x_1 x_2 + y_1 y_2 + z_1 z_2). \end{aligned}$$

The condition for perpendicularity of r_1, r_2 is

$$x_1 x_2 + y_1 y_2 + z_1 z_2 = 0.$$

Accordingly the left-hand member of 7) must vanish. If r_3 is perpendicular to $r_1 r_2$ its coordinates must be proportional to their vector-product. Thus we may write 7) as

$$8)\qquad (c_2 - b_3)x_3 + (a_3 - c_1)y_3 + (b_1 - a_2)z_3 = 0.$$

Inasmuch as the order of suffixes 1, 2, 3 is indifferent, if the three vectors r_1, r_2, r_3 are to be mutually perpendicular, equation 8) must be satisfied by the components of all three. This can be true only if we have

$$c_2 = b_3, \quad a_3 = c_1, \quad b_1 = a_2,$$

that is, the determinants of the substitutions 1) and 2) are symmetrical. In this case the linear vector-function is said to be self-conjugate, and a strain represented by such a function is called a *pure strain.*

165. Self-conjugate Functions. Pure Strain. We will consider this important case in detail. Adopting a symmetrical notation, let us write

$$9)\qquad \begin{aligned} x' &= ax + hy + gz, \\ y' &= hx + by + fz, \\ z' &= gx + fy + cz. \end{aligned}$$

If by φ we denote the homogeneous quadratic function

$$10)\qquad \varphi \equiv ax^2 + by^2 + cz^2 + 2fyz + 2gzx + 2hxy,$$

equations 9) may be written

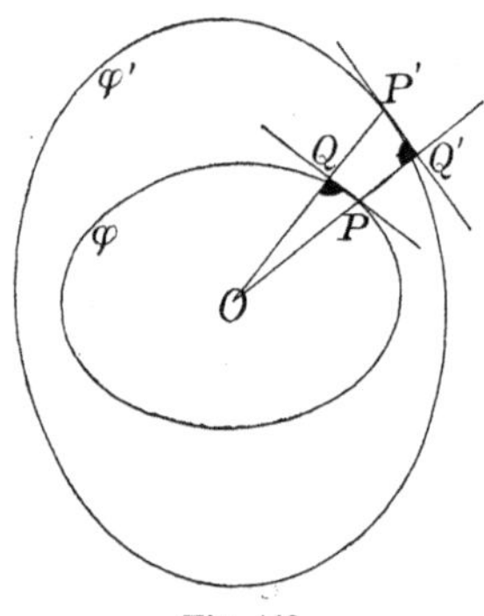

Fig. 143.

$$9)\quad x' = \frac{1}{2}\frac{\partial \varphi}{\partial x}, \quad y' = \frac{1}{2}\frac{\partial \varphi}{\partial y}, \quad z' = \frac{1}{2}\frac{\partial \varphi}{\partial z},$$

so that the vector OP' (Fig. 143) is parallel to the normal at the point P, whose coordinates are x, y, z, lying on the quadric $\varphi = \pm R^2$, where R is a constant introduced merely for the sake of homogeneity. In like manner calling

$$11)\quad \varphi' \equiv Ax'^2 + By'^2 + Cz'^2 + 2Fy'z' + 2Gx'z' + 2Hx'y',$$

equations 2) are

$$12)\qquad \begin{aligned} x &= Ax' + Hy' + Gz' = \frac{1}{2}\frac{\partial \varphi'}{\partial x'}, \\ y &= Hx' + By' + Fz' = \frac{1}{2}\frac{\partial \varphi'}{\partial y'}, \\ z &= Gx' + Fy' + Cz' = \frac{1}{2}\frac{\partial \varphi'}{\partial z'}, \end{aligned}$$

so that OP is parallel to the normal at P', whose coordinates are x', y', z', a point on the quadric $\varphi' = \pm R^2$.

By means of either of the quadrics φ and φ' we may thus obtain the mutual directional relations of every pair of vectors OP, OP' drawn from the origin, the relation being that of radius vector and normal at a given point, and reciprocal with respect to the two quadrics.

We see that in the application to strain any point P lying on the quadric $\varphi = \pm R^2$ is displaced to a point P' on the quadric $\varphi' = \pm R^2$.

If we call the feet of the perpendiculars from the origin on the tangent planes at P and P' respectively Q and Q' and write p and p' for the lengths OQ and OQ' we have

$$13)\qquad \begin{aligned} p &= r\cos(rr') = \frac{(xx'+yy'+zz')}{r'}, \\ p' &= r'\cos(rr') = \frac{(xx'+yy'+zz')}{r}, \end{aligned}$$

so that

$$14)\qquad \begin{aligned} pr' = p'r &= xx' + yy' + zz' \\ &= \frac{1}{2}\left(x\frac{\partial\varphi}{\partial x} + y\frac{\partial\varphi}{\partial y} + z\frac{\partial\varphi}{\partial z}\right) = \varphi \\ &= \frac{1}{2}\left(x'\frac{\partial\varphi'}{\partial x'} + y'\frac{\partial\varphi'}{\partial y'} + z'\frac{\partial\varphi'}{\partial z'}\right) = \varphi' = \pm R^2. \end{aligned}$$

The quadrics φ and φ' are then said to be *reciprocal* to each other with respect to a sphere of radius R. Since for the axes of the quadric $p=r$ and $p'=r'$, we have the relation between the axes,

$$15)\qquad rr' = \pm R^2,$$

so that the axes of reciprocal quadrics are inversely proportional.

Since all lines in the same direction are stretched in the same ratio a line OS of length $\varrho = S$ in the direction of OP is strained into a line OS' of length ϱ' in the direction OP', so that

$$16)\qquad \frac{\varrho'}{S} = \frac{r'}{r}.$$

Now if α, β, γ and α', β', γ' are the direction cosines of OP and OP' respectively, equations 9) and 12) may be written, when divided through by r and r' respectively,

$$17)\qquad \begin{aligned} \frac{r'}{r}\alpha' &= a\alpha + h\beta + g\gamma, \\ \frac{r'}{r}\beta' &= h\alpha + b\beta + f\gamma, \\ \frac{r'}{r}\gamma' &= g\alpha + f\beta + c\gamma, \end{aligned}$$

18)
$$\frac{r}{r'}\alpha = A\alpha' + H\beta' + G\gamma',$$
$$\frac{r}{r'}\beta = H\alpha' + B\beta' + F\gamma',$$
$$\frac{r}{r'}\gamma = G\alpha' + F\beta' + C\gamma'.$$

Squaring and adding 18) we have

19)
$$\left(\frac{r}{r'}\right)^2 = \frac{S^2}{\varrho'^2} = (A\alpha' + H\beta' + G\gamma')^2 + (H\alpha' + B\beta' + F\gamma')^2 + (G\alpha' + F\beta' + C\gamma')^2.$$

If now the coordinates of S and S' are ξ, η, ζ and ξ', η', ζ',

20)
$$\xi = \varrho\alpha, \quad \eta = \varrho\beta, \quad \zeta = \varrho\gamma,$$
$$\xi' = \varrho'\alpha', \quad \eta' = \varrho'\beta', \quad \zeta' = \varrho'\gamma',$$

and the equation 19) becomes on multiplication by ϱ'^2

21)
$$\psi' \equiv (A\xi' + H\eta' + G\zeta')^2 + (H\xi' + B\eta' + F\zeta')^2 + (G\xi' + F\eta' + C\zeta')^2 = S^2.$$

Consequently points originally situated on the sphere,

$$\varrho^2 = \xi^2 + \eta^2 + \zeta^2 = S^2,$$

lie after the displacement on the *strain-ellipsoid* $\psi' = S^2$. The quadric ψ' must be an ellipsoid because every point on it is finite.

In like manner if we find the locus of points on OP at such distances from the origin that *after* the strain they lie on a sphere, $\varrho'^2 = \xi'^2 + \eta'^2 + \zeta'^2 = S^2$, we have corresponding to 16),

22)
$$\frac{\varrho}{S} = \frac{r}{r'},$$

and from the equations 17) we find in like manner that the locus on which the points lay before the strain is the *inverse strain-ellipsoid*

23)
$$\psi \equiv (a\xi + h\eta + g\zeta)^2 + (h\xi + b\eta + f\zeta)^2 + (g\xi + f\eta + c\zeta)^2 = S^2$$

It is evident that the axes of the four quadrics φ, φ', ψ, ψ' coincide in direction.

Multiplying together 16) and 22) we obtain

24)
$$\varrho\varrho' = S^2,$$

and, since the directions of ϱ and ϱ' coincide for the axes of either ellipsoid, we see that the ellipsoids are reciprocal with respect to a sphere of radius S. Multiplying 15) by 16) and 22) respectively we get for the axial directions,

25)
$$\frac{\varrho'}{S} = \frac{r'^2}{R^2}; \quad \frac{\varrho}{S} = \frac{r^2}{R^2},$$

that is, the axes of the ellipsoids ψ, ψ' are proportional to the squares of the axes of the quadrics φ and φ'. By means of either pair of quadrics φ, ψ, or φ', ψ', we get a complete representation of the linear vector function, the directional relations being given by φ or φ' as above described, the quantitative relations by ψ or ψ' (Fig. 144).

Either quadric has three principal axes given by the three real roots of the determinantal cubic. These are called the *principal axes* of the strain. Conjugate diameters remain conjugate, since parallel lines remain parallel and bisected by corresponding lines. Accordingly the principal axes of the strain-ellipsoid must have been before the strain conjugate diameters of a sphere, and therefore mutually perpendicular. As the axes of the ellipsoid are the only set of mutually perpendicular conjugate diameters, there is no other set of mutually perpendicular lines which remain so after the strain.

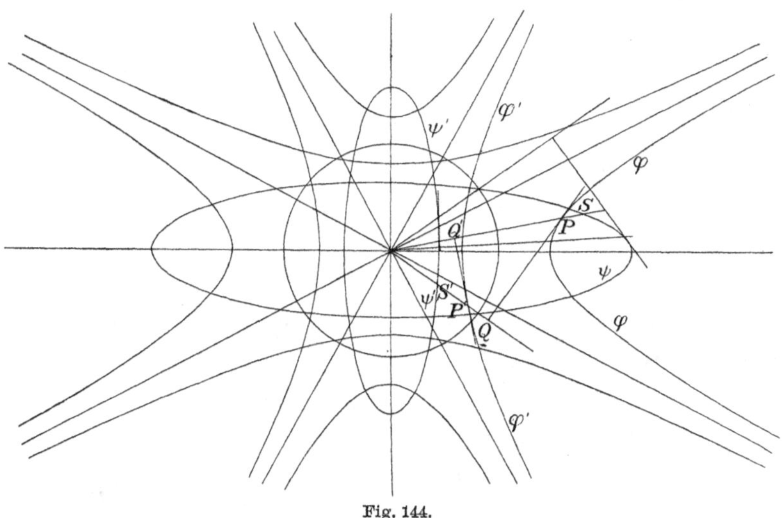

Fig. 144.

The equation 5) to determine the direction of lines which maintain their direction after the strain becomes for pure strain, represented by equations 9), the determinantal equation for the axes of the quadric φ. We have shown in Note IV that this equation always has three real roots and that the corresponding directions are mutually perpendicular. Thus the axes of a pure strain have their directions unaltered by the strain. A pure strain is therefore called *irrotational.* (It is to be noted that the axes are the only lines which are not rotated.)

If the quadrics are referred to their principal axes the equations assume the following simple forms from which the geometrical relatioes are easily seen,

10a) $$\varphi = ax^2 + by^2 + cz^2,$$

11a) $$\varphi' = \frac{x'^2}{a} + \frac{y'^2}{b} + \frac{z'^2}{c},$$

9a) $$x' = \frac{1}{2}\frac{\partial \varphi}{\partial x} = ax,\quad y' = \frac{1}{2}\frac{\partial \varphi}{\partial y} = by,\quad z' = \frac{1}{2}\frac{\partial \varphi}{\partial z} = cz,$$

12a) $$x = \frac{1}{2}\frac{\partial \varphi'}{\partial x} = \frac{x'}{a},\quad y = \frac{1}{2}\frac{\partial \varphi'}{\partial y} = \frac{y'}{b},\quad z = \frac{1}{2}\frac{\partial \varphi'}{\partial z} = \frac{z'}{c}.$$

23a) $$\psi = a^2\xi^2 + b^2\eta^2 + c^2\zeta^2$$

21a) $$\psi' = \frac{\xi'^2}{a^2} + \frac{\eta'^2}{b^2} + \frac{\zeta'^2}{c^2}.$$

In the cubic for the axes of φ,

26) $$\begin{vmatrix} a-\lambda, & h\;, & g \\ h\;, & b-\lambda, & f \\ g\;, & f\;, & c-\lambda \end{vmatrix} = 0,$$

the three roots are proportional to the squares of the reciprocals of the semi-axes (Note IV),

27) $$\lambda_1 = \frac{R^2}{r_1^2},\quad \lambda_2 = \frac{R^2}{r_2^2},\quad \lambda_3 = \frac{R^2}{r_3^2}.$$

Now by 25), and 24)

28) $$\frac{R^2}{r_1^2} = \frac{S}{\varrho_1} = \frac{\varrho_1'}{S},\quad \frac{R^2}{r_2^2} = \frac{S}{\varrho_2} = \frac{\varrho_2'}{S},\quad \frac{R^2}{r_3^2} = \frac{S}{\varrho_3} = \frac{\varrho_3'}{S},$$

and therefore

29) $$\lambda_1\lambda_2\lambda_3 = \frac{\varrho_1'\varrho_2'\varrho_3'}{S^3} = \frac{\textit{volume of strain-ellipsoid}}{\textit{volume of sphere}}.$$

But the product of the roots $\lambda_1\lambda_2\lambda_3$ is equal to the constant term in the cubic 26), that is,

30) $$\lambda_1\lambda_2\lambda_3 = \begin{vmatrix} a, & h, & g \\ h, & b, & f \\ g, & f, & c \end{vmatrix}.$$

The determinant of the coefficients of the substitution 9) accordingly represents the ratoi of expansion, and since parallel lines are stretched in the same ratio, the ratio of expansion of volume $\frac{V'}{V}$ is everywhere the same.

166. Rotation. Let us return to the case of the general homogeneous strain given by equations 1) and let us find the condition that all points situated before the strain on a sphere with center at the origin remain on the same sphere after the strain.

The condition

$$x'^2 + y'^2 + z'^2 = x^2 + y^2 + z^2$$

gives

$$31)\quad (a_1 x + a_2 y + a_3 z)^2 + (b_1 x + b_2 y + b_3 z)^2 + (c_1 x + c_2 y + c_3 z)^2 = x^2 + y^2 + z^2,$$

which being true for all values of x, y, z necessitates the equality of the coefficients of corresponding squares and products on both sides of the equation, that is,

$$32)\quad \begin{aligned} a_1^2 + b_1^2 + c_1^2 &= 1, \\ a_2^2 + b_2^2 + c_2^2 &= 1, \\ a_3^2 + b_3^2 + c_3^2 &= 1, \end{aligned}$$

$$33)\quad \begin{aligned} a_1 a_2 + b_1 b_2 + c_1 c_2 &= 0, \\ a_2 a_3 + b_2 b_3 + c_2 c_3 &= 0, \\ a_3 a_1 + b_3 b_1 + c_3 c_1 &= 0. \end{aligned}$$

Equations 32) show that a, b, c with the same suffix are direction-cosines of a line, equations 33) show that the three lines are mutually perpendicular, in other words, the equations of strain are merely those of transformation of coordinates, and the result of the strain is merely a rotation of the body as if rigid.

Let us obtain the analytical expression for an infinitesimal rotation about an axis. Let the direction-cosines of the axis be λ, μ, ν and the angle of rotation be $\delta\omega$. Since we have proved in § 57 that infinitesimal rotations may be resolved like vectors and treated like angular velocities, we have the components of rotation,

$$34)\quad \omega_x = \lambda\,\delta\omega, \quad \omega_y = \mu\,\delta\omega, \quad \omega_z = \nu\,\delta\omega,$$

from which by equations 119), § 76, we obtain the infinitesimal displacements,

$$35)\quad \begin{aligned} x' - x &= \delta x = z\,\omega_y - y\,\omega_z = (z\mu - y\nu)\,\delta\omega, \\ y' - y &= \delta y = x\,\omega_z - z\,\omega_x = (x\nu - z\lambda)\,\delta\omega, \\ z' - z &= \delta z = y\,\omega_x - x\,\omega_y = (y\lambda - x\mu)\,\delta\omega. \end{aligned}$$

From this we obtain the substitution for the rotation considered as a strain,

$$36)\quad \begin{aligned} x' &= 1 \cdot x - \nu\delta\omega \cdot y + \mu\delta\omega \cdot z, \\ y' &= \nu\delta\omega \cdot x + 1 \cdot y - \lambda\delta\omega \cdot z, \\ z' &= -\mu\delta\omega \cdot x + \lambda\delta\omega \cdot y + 1 \cdot z. \end{aligned}$$

The determinant of the substitution is skew symmetrical.

The ratio of expansion is

$$37)\qquad \frac{V'}{V} = \begin{vmatrix} 1\;, & -\nu\delta\omega, & \mu\delta\omega \\ \nu\delta\omega\;, & 1\;, & -\lambda\delta\omega \\ -\mu\delta\omega, & \lambda\delta\omega\;, & 1 \end{vmatrix},$$

which is equal to 1 plus terms involving $\delta\omega^2$ which are negligible, agreeing with the result that rotation as a rigid body is unaccompanied by change of volume.

167. General Small Strain. We shall now consider small strains in general, that is, strains in which the displacements of all points are small quantities whose squares and products may be neglected.

Let the components of the displacements or *shifts* be

$$38)\qquad x'-x=u,\quad y'-y=v,\quad z'-z=w,$$

so that if we now write the coefficients of the strain as

$$39)\qquad \begin{aligned} x' &= (1+a_1)x + a_2 y + a_3 z, \\ y' &= b_1 x + (1+b_2)y + b_3 z, \\ z' &= c_1 x + c_2 y + (1+c_3)z, \end{aligned}$$

we have the shifts given by the substitution

$$40)\qquad \begin{aligned} u &= a_1 x + a_2 y + a_3 z, \\ v &= b_1 x + b_2 y + b_3 z, \\ w &= c_1 x + c_2 y + c_3 z, \end{aligned}$$

and if the a's, b's and c's are small, u, v, w will be small quantities of the same order. The ratio of dilatation is by 30)

$$41)\qquad \frac{V'}{V} = \begin{vmatrix} 1+a, & a_2\;, & a_3 \\ b_1\;, & 1+b_2, & b_3 \\ c_1\;, & c_2\;, & 1+c_3 \end{vmatrix} = (1+a_1)(1+b_2)(1+c_3)$$

$$+ \text{ terms of higher order},$$

$$= 1 + a_1 + b_2 + c_3 + \text{ terms of higher order.}$$

The quantity

$$42)\qquad \frac{V'-V}{V} = a_1 + b_2 + c_3$$

is the increment of volume per unit volume, and will be referred to simply as the *dilatation* and denoted by σ.

Suppose two small strains take placee successively according to the equations 39) for the first, and

$$
\text{43)}\quad
\begin{aligned}
x'' &= (1+\alpha_1)x' + \alpha_2 y' + \alpha_3 z',\\
y'' &= \beta_1 x' + (1+\beta_2)y' + \beta_3 z',\\
z'' &= \gamma_1 x' + \gamma_2 y' + (1+\gamma_3)z',
\end{aligned}
$$

for the second.

Substituting the values of x', y', z' from 49) in 43) we obtain

$$
x'' = \left.\begin{matrix}(1+\alpha_1)(1+a_1)\\ +\alpha_2 b_1\\ +\alpha_3 c_1\end{matrix}\right| x + \left.\begin{matrix}(1+\alpha_1)a_2\\ +\alpha_2(1+b_2)\\ +\alpha_3 c_2\end{matrix}\right| y + \left.\begin{matrix}(1+\alpha_1)a_3\\ +\alpha_2 b_3\\ +\alpha_3(1+c_3)\end{matrix}\right| z
$$

. . . .

. . . .

Neglecting terms of the second order we obtain the equations of the resultant strain

$$
\text{44)}\quad
\begin{aligned}
x'' &= (1+a_1+\alpha_1)x + (a_2+\alpha_2)y + (a_3+\alpha_3)z,\\
y'' &= (b_1+\beta_1)x + (1+b_2+\beta_2)y + (b_3+\beta_3)z,\\
z'' &= (c_1+\gamma_1)x + (c_2+\gamma_2)y + (1+c_3+\gamma_3)z,
\end{aligned}
$$

and for the resultant shifts

$$
\text{45)}\quad
\begin{aligned}
u &= (a_1+\alpha_1)x + (a_2+\alpha_2)y + (a_3+\alpha_3)z,\\
v &= (b_1+\beta_1)x + (b_2+\beta_2)y + (b_3+\beta_3)z,\\
w &= (c_1+\gamma_1)x + (c_2+\gamma_2)y + (c_3+\gamma_3)z,
\end{aligned}
$$

that is, successive *small* strains are compounded by adding their shifts. This important proposition enables us conveniently to resolve small strains into types already studied. Every small strain represented by equations 40) can be written by addition and subtraction of equal terms

$$
\text{46)}\quad
\begin{aligned}
u &= a_1 x + \tfrac{1}{2}(a_2+b_1)y + \tfrac{1}{2}(a_3+c_1)z\\
&\quad + \tfrac{1}{2}(a_2-b_1)y + \tfrac{1}{2}(a_3-c_1)z,\\
v &= \tfrac{1}{2}(b_1+a_2)x + b_2 y + \tfrac{1}{2}(b_3+c_2)z\\
&\quad + \tfrac{1}{2}(b_1-a_2)x \qquad + \tfrac{1}{2}(b_3-c_2)z,\\
w &= \tfrac{1}{2}(c_1+a_3)x + \tfrac{1}{2}(c_2+b_3)y + c_3 z\\
&\quad + \tfrac{1}{2}(c_1-a_3)x + \tfrac{1}{2}(c_2-b_3)y.
\end{aligned}
$$

Accordingly we may write the strain as the resultant of two,

$$u = u_1 + u_2, \quad v = v_1 + v_2, \quad w = w_1 + w_2$$

where

47)
$$u_1 = a_1 x + \frac{1}{2}(a_2 + b_1) y + \frac{1}{2}(a_3 + c_1) z,$$
$$v_1 = \frac{1}{2}(a_2 + b_1) x + b_2 y + \frac{1}{2}(b_3 + c_2) z,$$
$$w_1 = \frac{1}{2}(a_3 + c_1) x + \frac{1}{2}(b_3 + c_2) y + c_3 z$$

denoting a pure strain, and

48)
$$u_2 = \frac{1}{2}(a_3 - c_1) z - \frac{1}{2}(b_1 - a_2) y,$$
$$v_2 = \frac{1}{2}(b_1 - a_2) x - \frac{1}{2}(c_2 - b_3) z,$$
$$w_2 = \frac{1}{2}(c_2 - b_3) y - \frac{1}{2}(a_3 - c_1) x,$$

denoting a rotation ω whose components are

49)
$$\omega_x = \frac{1}{2}(c_2 - b_3),$$
$$\omega_y = \frac{1}{2}(a_3 - c_1),$$
$$\omega_z = \frac{1}{2}(b_1 - a_2).$$

Thus every small strain may be resolved into a pure strain and a rotation.

In order to bring out the symmetry let us write the pure strain

50)
$$u_1 = s_x x + g_z y + g_y z,$$
$$v_1 = g_z x + s_y y + g_x z,$$
$$w_1 = g_y x + g_x y + s_z z,$$

where

51) $$g_x = \frac{1}{2}(c_2 + b_3), \quad g_y = \frac{1}{2}(a_3 + c_1), \quad g_z = \frac{1}{2}(a_2 + b_1),$$
$$s_x = a_1, \quad s_y = b_2, \quad s_z = c_3.$$

Thus the six quantities g and ω are respectlvely the half sums and half differences of shift-coefficients symmetrical about the main diagonal.[1])

1) In the usual notation the g's are defined as the above sums without the coefficient $\frac{1}{2}$, as stated by Todhunter and Pearson, *A History of Elasticity and Strength of Materials,* Vol. I, p. 882, "The advantage which would arise from

The general small strain is accordingly completely defined by the nine small coefficients,

$$s_x, \; s_y, \; s_z, \; g_x, \; g_y, \; g_z, \; \omega_x, \; \omega_y, \; \omega_z.$$

168. Simple Strains. Stretches and Shears. The pure strain 50) may be resolved into two parts

52)
$$\begin{array}{ll} \text{(a)} & \text{(b)} \\ u_1' = s_x x, & u_1'' = \;\; 0 \;\; + g_z y + g_y z, \\ v_1' = s_y y, & v_1'' = g_z x + \;\; 0 \;\; + g_x z, \\ w_1 = s_z z, & w_1'' = g_y x + g_x y + \;\; 0 \;. \end{array}$$

A strain whose equations contain but a single constant is called a *simple strain*. Thus we may resolve the strain (a) into three simple strains of which the first is given by

$$u = s_x x, \quad v = 0, \quad w = 0.$$

This represents a displacement in which each point is shifted parallel to the x-axis through a distance proportional to its x coordinate. Such a displacement is called a *stretch*. The constant s_x represents the distance moved by a plane at unit distance from the YZ-plane and measures the magnitude of the stretch or the linear expansion per unit length. If s is negative the stretch becomes a *squeeze*.

The strain (a) accordingly represents the resultant of three simple strains, namely stretches, of different amounts in the directions of the coordinate axes, which are evidently the axes of the strain. The semi-axes of the strain-ellipsoid are $1 + s_x$, $1 + s_y$, $1 + s_z$ and its equation

$$\frac{x^2}{(1+s_x)^2} + \frac{y^2}{(1+s_y)^2} + \frac{z^2}{(1+s\;)^2} = 1,$$

or neglecting squares of small quantities,

$$(1 - 2s_x)x^2 + (1 - 2s_y)y^2 + (1 - 2s_z)z^2 = 1.$$

The dilatation is by 42)

53)
$$\sigma = s_x + s_y + s_z.$$

Obviously we can have $\sigma = 0$ if at least one of the stretches is replaced by a squeeze. If the three s's are equal we have a simple

introducing the $\frac{1}{2}$ into the slides is thus obvious", and we have therefore so introduced it, although to them "it seemed too great an interference with the nearly general custom." We have also introduced a single suffix, g_x, instead of the more usual double suffix notation, g_{yz}, feeling that the brevity and analogy with ω_x thus gained justifies the change.

strain known as a uniform expansion for which the strain-ellipsoid is a sphere and the dilatation $\sigma = 3s$.

The part (b) of the pure strain represents a strain which, like a rotation, is unaccompanied by a dilatation, but which differs from a rotation in that it involves a change of form. We shall consider it, as we have the part (a) in three parts.

In the first,

$$u = g_z y, \quad v = g_z' x, \quad w = 0,$$

every point is shifted in the X direction through a distance proportional to its distance from the XZ-plane, while it is shifted in the Y direction through a distance proportional to its distance from the YZ-plane. Points at unit distance from the two named planes are shifted both ways by the same amount g_z, so that the new positions of the planes XZ, YZ make with the old angles whose tangents or sines are equal to g_z.

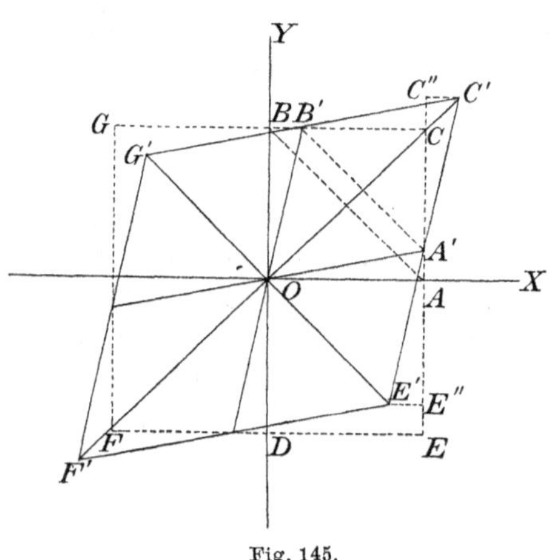

Fig. 145.

The square $OACB$ (Fig. 145) becomes the rhombus $OA'C'B'$, which is symmetrical about the diagonal OC bisecting the angle XOY. The diagonals AB and OC maintain their directions unchanged, and are accordingly two of the axes of the strain, the axis OZ being the third. The stretch-ratio along OC is

$$\frac{OC' - OC}{OC} = \frac{CC'}{OC} = \frac{C''C}{BC} = \frac{BB'}{OB} = g_z,$$

as may be seen by inspection of the figure. The stretch along the perpendicular axis OE is negative,

$$\frac{OE' - OE}{OE} = -\frac{EE'}{OE} = -\frac{EE''}{AE} = -g_z.$$

The stretch along the Z-axis is zero. Accordingly the sum of the three stretches along the axes of the strain is zero. Such a strain, involving a distortion but no expansion and depending upon a constant g, is called a *simple shear*. The plane of the shear is the plane parallel to which all points are displaced, in this case the XY-plane.

A shear may be defined as a stretch along one axis combined with a squeeze of equal magnitude along a perpendicular axis, and zero stretch along the axis perpendicular to both. The shear just considered is a *pure* shear, that is, without rotation. It is easily seen that the above shear might have been obtained if all planes parallel to XOZ had been moved parallel to themselves a distance in the X-direction equal to $2g_z y$, giving the rotational shear

$$u' = 2g_z y, \quad v' = 0, \quad w' = 0$$

and then rotating about the Z-axis through an angle $\omega_z = g_z$, according to the equations

$$u'' = -g_z y, \quad v'' = g_z x, \quad w'' = 0.$$

The lines OA', OB', which before the strain were perpendicular, have respectively the direction-cosines $1, g_z, 0$ and $g_z, 1, 0$ and the cosine of their included angle is accordingly $2g_z$. This change of cosine which, as we have just seen, is equal to the amount of sliding of the plane at unit distance from XOZ is commonly called the *amount of shear*, so that the stretch and squeeze of the axes are each one half of the amount of the shear.

We may now define the strain (b) as a combination of three simple shears of amounts $2g_x$, $2g_y$, $2g_z$, with planes mutually perpendicular and equivalent to stretches of amounts g_z, g_x and g_y along the bisectors of the angles XOY, YOZ, and ZOX respectively, which make angles of 60^0 with each other, together with squeezes of the same amounts along the bisectors of the other angles. We have thus the final positions of six points on these lines, or just sufficient to determine an ellipsoid whose center is given. This is the strain-ellipsoid.

The strain (b) will be called a *general* shear. The quadric φ is

$$54) \qquad \varphi \equiv x^2 + y^2 + z^2 + 2g_x yz + 2g_y xz + 2g_z xy = R^2,$$

and the shears are the coefficients of the product terms. If the equation of the quadric is transformed to its principal axes the product terms vanish. Accordingly we may always find three mutually perpendicular axes with respect to which the shear components vanish. These are the axes of the shear. (It may be remarked that the equations of the general rotational shear are obtained from 1) by putting $a_1 = b_2 = c_3 = 1$.)

In order to distinguish between the geometrical term shear and the dynamical shearing stress, to be presently considered, it will be convenient to characterize the coefficients g as the *slides* (corresponding to the French *glissement*, German *Gleitung*).

168a. Elongation and Compression Quadric. Since the equations 50) for the shifts, the components of the vector displacement q, as a function of r are of precisely the same form as equationss 9) for r' we have the complete geometrical representation here applicable. Of the four quadrics the first is the most important. The length r of the line OP is changed by the strain to r' which, when the strain is small, differs from it by a small quantity, so that the stretch

$$55)\qquad s_r = \frac{r'-r}{r}$$

is a small quantity of the first order. But, since the angle between r and r' is infinitesimal, we have to the first order, if q is the displacement PP',

$$56)\qquad r' = r + q\cos(qr),$$

$$57)\qquad s_r = \frac{q\cos(qr)}{r} = \frac{ux+vy+wz}{r^2}.$$

Now if u, v, w are given by equations 50) the numerator becomes

$$58)\qquad r^2 s_r = s_x x^2 + s_y y^2 + s_z z^2 + 2g_x yz + 2g_y zx + 2g_z xy \equiv \chi.$$

If we put this equal to unity we have

$$59)\qquad s_r = \frac{1}{r^2},$$

where r is the radius vector of the quadric

$$60)\qquad \chi = 1.$$

This is called the *elongation and compression quadric*, and it is to be noticed that the displacement of any of its points is in the direction of the normal, for

$$61)\qquad u = \frac{1}{2}\frac{\partial \chi}{\partial x},\quad v = \frac{1}{2}\frac{\partial \chi}{\partial y},\quad w = \frac{1}{2}\frac{\partial \chi}{\partial z}.$$

Since any one of the six coefficients may be positive or negative, the quadric may be an ellipsoid or an hyperboloid. In the latter case not all the lines drawn from the origin will meet the surface, and for those which do not r is imaginary and s_r is negative.

If we construct the conjugate hyperboloid, $\chi = -1$, those rays which do not meet the first hyperboloid meet this, and the magnitude of the compression is given by

$$62)\qquad s_r = -\frac{1}{r^2}.$$

Lines that meet both hyperboloids at infinity and therefore have a zero stretch or compression lie on the cone $\chi = 0$, asymptotic to the two hyperboloids, and known as the cone of no elongation.

All lines which are equally elongated with the stretch S, where

$$63)\qquad S = \frac{1}{r^2}\{s_x x^2 + s_y y^2 + s_z z^2 + 2g_x yz + 2g_y zx + 2g_z xy\},$$

lie on the cone

64) $(s_x - S)x^2 + (s_z - S_x)y^2 + (s_y - S)z^2 + 2g_x yz + 2g_y xz + 2g_z xy = 0,$

which may be called a cone of *equal* elongation S, of which the cone of no elongation is a particular case.

Let us form the elongation quadrics for expansions and shears. If the slides vanish we have

65) $$\chi \equiv s_x x^2 + s_y y^2 + s_z z^2 = 1,$$

and for a simple stretch in the X-direction

66) $$s_x x^2 = 1,$$

the elongation quadric breaks up into the two parallel planes,

$$\sqrt{s_x}\, x - 1 = 0 \quad \text{and} \quad \sqrt{s_x}\, x + 1 = 0,$$

at distances $\pm \frac{1}{\sqrt{s_x}}$ from the origin.

Since for any line making the angle ϑ with the X-axis we have

$$r = \frac{1}{\sqrt{s_x} \cos \vartheta},$$

the stretch is given by

67) $$s_r = \frac{1}{r^2} = s_x \cos^2 \vartheta.$$

The cone of no elongation is therefore the plane $\vartheta = \frac{\pi}{2}$ parallel to the above pair of planes. In equation 65) if s_x, s_y, s_z are of the same sign the quadric is an ellipsoid and the cone of no expansion is imaginary. If one s has a sign different from that of the others we have two hyperboloids and the cone of no expansion is real and separates the stretched from the squeezed lines.

In the general shear $s_x = s_y = s_z = 0$ we have

68) $$\chi \equiv 2(g_x yz + g_y zx + g_z xy) = 1,$$

and the cone of no elongation

69) $$g_x yz + g_y zx + g_z xy = 0,$$

contains the three coordinate-axes as generators. These are therefore unstretched. In a simple shear parallel to the XY-plane we have

70) $$\chi \equiv 2g_z xy = \pm 1$$

which represents equilateral hyperbolic cylinders with axes bisecting the angles between the x and y axes. The cone of no elongation, $xy = 0$, breaks up into two coordinate-planes, $x = 0$ and $y = 0$. These two planes are undistorted, and are the planes of circular section of the strain-ellipsoid.

A combination of two simple shears in planes at right angles obtained from 52b) by putting $g_z = 0$, has the elongation quadric

71) $$g_x yz + g_y xz = 0,$$

which breaks up into the two planes

72) $$z = 0 \quad \text{and} \quad g_x y + g_y x = 0,$$

at right angles to each other. It is to be noticed that the cone of the resultant of three simple shears in mutually perpendicular planes does not so break up.

We have seen that we require nine constants to specify the general homogeneous strain, of which three belong to the rotation, six to the pure strain. Let us consider the number of data required to specify a simple pure strain. To specify a uniform dilatation we require only the constant of dilatation σ; for a simple stretch, the direction of the axis, involving two data, and the magnitude of the stretch, making three in all; for a simple shear, four data, the magnitude of the shear, two to fix the plane of the shear and one additional for an axis. Consequently we may always represent a general strain as the resultant of three simple expansions, or of two simple shears and a uniform dilatation.

169. Heterogeneous Strain. If the displacements are not given by linear functions of the coordinates, the strain is said to be heterogeneous. In this case we may examine the relative displacements of two neighboring points. Let the coordinates of the first point P be before the strain x, y, z, and after it $x + u$, $y + v$, $z + w$, and those of the second, Q, be before $x + f$, $y + g$, $z + h$, and after $x + f + u'$, $y + g + v'$, $z + h + w'$. If the point Q be referred to P as an origin both before and after the strain, it has as relative coordinates before f, g, h, and after $f + u' - u$, $g + v' - v$, $h + w' - w$, so that the relative displacements are $u' - u$, $v' - v$, $w' - w$. Now u, v, w may be any functions of the coordinates x, y, z of P, but they must be continuous, otherwise the body would be split at surfaces of discontinuity. Accordingly u', v', w' being the values of u, v, w for $x + f, y + g, z + h$ may be developed by Taylor's theorem, so that, neglecting terms of order higher than the first in f, g, h

73) $$\begin{aligned} u' - u &= f\frac{\partial u}{\partial x} + g\frac{\partial u}{\partial y} + h\frac{\partial u}{\partial z}, \\ v' - v &= f\frac{\partial v}{\partial x} + g\frac{\partial v}{\partial y} + h\frac{\partial v}{\partial z}, \\ w' - w &= f\frac{\partial w}{\partial x} + g\frac{\partial w}{\partial y} + h\frac{\partial w}{\partial z}. \end{aligned}$$

Thus the relative displacements are given as linear functions of the relative coordinates f, g, h whose coefficients are the values of the nine first derivatives at the point P, that is to say, constants for all points Q in the neighborhood of P, consequently the *relative* strain of the portion of the body in the neighborhood of P is homogeneous. Thus we say that any continuous heterogeneous strain is homogeneous in its smallest parts.

Comparing with equations 49) and 51) we find the stretches, dilatation, slides and rotations at any point to be respectively

$$74)\qquad s_x = \frac{\partial u}{\partial y},\quad s_y = \frac{\partial v}{\partial y},\quad s_z = \frac{\partial w}{\partial z},$$

$$75)\qquad \sigma = \frac{\partial u}{\partial x} + \frac{\partial v}{\partial y} + \frac{\partial w}{\partial z},$$

$$76)\qquad g_x = \frac{1}{2}\left(\frac{\partial w}{\partial y} + \frac{\partial v}{\partial z}\right),\quad g_y = \frac{1}{2}\left(\frac{\partial u}{\partial z} + \frac{\partial w}{\partial x}\right),\quad g_z = \frac{1}{2}\left(\frac{\partial v}{\partial x} + \frac{\partial u}{\partial y}\right),$$

$$77)\qquad \omega_x = \frac{1}{2}\left(\frac{\partial w}{\partial y} - \frac{\partial v}{\partial z}\right),\quad \omega_y = \frac{1}{2}\left(\frac{\partial u}{\partial z} - \frac{\partial w}{\partial x}\right),\quad \omega_z = \frac{1}{2}\left(\frac{\partial v}{\partial x} - \frac{\partial u}{\partial y}\right).$$

Thus the volume dilatation is equal to the divergence of the displacement, while the rotation is equal to one half its curl.

We might have obtained the value of σ by the divergence theorem. Consider any closed surface S fixed in space so that portions of the deformable body flow through it during the strain, and let us find the volume of the matter which passes outward through S. Through an element dS at which the displacement is q there passes out a quantity filling a prism of slant-height q and base dS whose volume is therefore $q\cos(nq)dS$, where n is the *outward* normal to S. Through the whole surface there accordingly issues an amount whose volume is

$$78)\qquad \begin{aligned} Q &= \iint q\cos(nq)\,dS \\ &= \iint \{u\cos(nx) + v\cos(ny) + w\cos(nz)\}\,dS \\ &= \iiint \left(\frac{\partial u}{\partial x} + \frac{\partial v}{\partial y} + \frac{\partial w}{\partial z}\right) d\tau, \end{aligned}$$

by the divergence theorem. This is accordingly the increase in volume of the portion of substance originally included by the surface S. The ratio of this to the original volume is accordingly the mean value of the divergence in the volume in question, and making the volume infinitesimal, this becomes the dilatation σ.

In order that a strain shall be everywhere irrotational we must have the curl components of the displacement vanish everywhere.

But by § 31 this is the condition that the displacement is a lamellar vector and

79) $$u = \frac{\partial \varphi}{\partial x}, \quad v = \frac{\partial \varphi}{\partial y}, \quad w = \frac{\partial \varphi}{\partial z}.$$

Then φ is called the *strain-potential*. Only when the strain is irrotational can a strain-potential exist.

The line integral along any curve AB of the tangential component of the displacement

80) $$\int_A^B q \cos(q, ds)\, ds = \int_A^B (u\,dx + v\,dy + w\,dz)$$

is called the circulation along the path, and for irrotational strain is independent of the path, equal to $\varphi_B - \varphi_A$, and vanishes for a closed path.

Surfaces for which φ is constant are called equipotential surfaces, and the displacement-lines, or lines drawn so that their tangents are everywhere in the direction of the displacements, are normal to the equipotential surfaces.

The dilatation

81) $$\sigma = \frac{\partial u}{\partial x} + \frac{\partial v}{\partial y} + \frac{\partial w}{\partial z} = \Delta \varphi.$$

Accordingly in a non-dilatational, irrotational strain the displacement is a solenoidal vector and the displacement-potential a harmonic function. Since for a solenoidal vector the magnitude of the vector is inversely proportional to the cross-section of an infinitesimal tube, the displacement cannot vanish except at infinity. By the properties of harmonic functions φ cannot have a maximum or minimum unless σ is different from zero.

If φ be a homogeneous quadratic function of the coordinates, the strain is homogeneous throughout and not merely in its smallest parts. The equipotential surfaces are concentric quadric surfaces and since

82) $$\varphi = \frac{1}{2}\left(x\frac{\partial \varphi}{\partial x} + y\frac{\partial \varphi}{\partial y} + z\frac{\partial \varphi}{\partial z}\right)$$

the equipotential surfaces, $\varphi = const.$, are similar to the elongation quadric.

170. Stress. When a body is strained a system of forces is generally called into play tending to resist the strain. The system of forces is called a *stress*. In order to specify the stress at any point in the body, we draw a plane through the point separating the body into two parts. The parts on one side of this plane will exert certain forces on those on the other side, and the resultant of the forces which pass through an element dS of the plane will be a single force, proportional to the area dS, which we will write $F_n dS$. The stress at a point P is completely determined when we know the direction and magnitude of the force F_n for every possible direction of the normal to the element dS constructed at P. The *stress-vector* F_n, which is in general not normal to dS, may be resolved into its components, X_n, Y_n, Z_n, so that its direction-cosines are $\frac{X_n}{F_n}, \frac{Y_n}{F_n}, \frac{Z_n}{F_n}$. The normal component

$$83) \quad F_{nn} = F_n \cos(F_n n) = X_n \cos(nx) + Y_n \cos(ny) + Z_n \cos(nz).$$

If we draw the normal in either direction from the element dS, and if we understand by F_n the force exerted through dS by the portion of the body lying on the side toward which n is drawn on the portion lying on the other side, then if the normal component $F_{nn} = F_n \cos(F_n n)$ is positive it is called a *traction*, if negative, a *pressure*. In other words it is a *traction* if its effect is to cause the portions of the body to approach each other, a *pressure* if it is to make them recede.

The force upon any element dS can be expressed in terms of the forces upon three mutually perpendicular plane elements at the same point. Construct, enclosing the point P, an infinitesimal tetrahedron bounded by the element dS and three planes parallel to the coordinate planes (Fig. 146). Let the areas of the four triangular faces be dS, dS_x, dS_y, dS_z, the suffix in each case denoting the direction of the normal to the face. Further denote the stress-vector for any face by a sufffix giving the normal to that face, and let the stress-vectors be those for the portion of the body within the tetrahedron. Suppose that forces are applied to every portion of matter in proportion to its mass, such, for instance, as gravity, the components being X, Y, Z per unit mass. If $d\tau$ denote the volume of the tetrahedron the X-component of these external forces is accordingly $X \rho d\tau$.

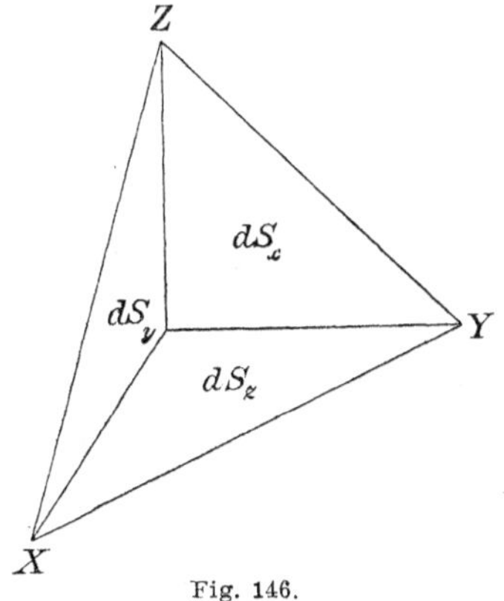

Fig. 146.

Let us now form the equations for equilibrium of the matter contained in the thetrahedron under the influce of the external forces and the stresses developed. The first of these is

$$84) \qquad X \rho d\tau + X_n dS - X_x dS_x - X_y dS_y - X_z dS_z = 0.$$

But since the three other sides are the projections of dS, we have

$$85)\quad dS_x = dS\cos (nx),\quad dS_y = dS\cos (ny),\quad dS_z = dS\cos (nz).$$

Inserting these in the equation 84), dividing through by dS, and taking the limit, as the edges of the tetrahedron become infinitely small the ratio of the volume to the surface disappears, so that we have finally

$$86)\qquad X_n = X_x \cos(nx) + X_y \cos(ny) + X_z \cos(nz),$$

and similarly

$$86)\qquad \begin{aligned} Y_n &= Y_x \cos(nx) + Y_y \cos(ny) + Y_z \cos(nz),\\ Z_n &= Z_x \cos(nx) + Z_y \cos(ny) + Z_z \cos(nz). \end{aligned}$$

Let us now consider the equilibrium of any portion of the body bounded by a closed surface S. Resolving in the X-direction, we have as the condition for equilibrium, considering both the stresses on the surface and the volume-forces,

$$87)\qquad \iint X_n dS + \iiint \varrho X d\tau = 0.$$

Making use of equations 86) for X_n,

$$88)\qquad \iint \{X_x \cos(nx) + X_y \cos(ny) + X_z \cos(nz)\} dS + \iiint \varrho X d\tau = 0,$$

and by the divergence theorem, n being the *outward* normal,

$$89)\qquad \iiint \left\{\frac{\partial X_x}{\partial x} + \frac{\partial X_y}{\partial y} + \frac{\partial X_z}{\partial z} + \varrho X\right\} d\tau = 0.$$

Since this must hold for every portion of the substance which is in equilibrium, the integrand must vanish, and we have consequently together with the result of resolving in the two other directions,

$$90)\qquad \begin{aligned} \varrho X + \frac{\partial X_x}{\partial x} + \frac{\partial X_y}{\partial y} + \frac{\partial X_z}{\partial z} &= 0,\\ \varrho Y + \frac{\partial Y_x}{\partial x} + \frac{\partial Y_y}{\partial y} + \frac{\partial Y_z}{\partial z} &= 0,\\ \varrho Z + \frac{\partial Z_x}{\partial x} + \frac{\partial Z_y}{\partial y} + \frac{\partial Z_z}{\partial z} &= 0. \end{aligned}$$

These are but three of the six equations for equilibrium. The other three are obtained by taking moments, the first being

$$91)\qquad \iint (yZ_n - zY_n)\, dS + \iiint \varrho (yZ - zY)\, d\tau = 0.$$

Introducing the values of ϱX, ϱY, ϱZ from equations 90) and of Y_n, Z_n from 86) this becomes

$$92)\quad \iint \{y[Z_x\cos(nx) + Z_y\cos(ny) + Z_z\cos(nz)]dS$$
$$- z[Y_x\cos(nx) + Y_y\cos(ny) + Y_z\cos(nz)]$$
$$+\iiint\left\{z\left(\frac{\partial Y_x}{\partial x} + \frac{\partial Y}{\partial y} + \frac{\partial Y_z}{\partial z}\right) - y\left(\frac{\partial Z_x}{\partial x} + \frac{\partial Z_y}{\partial y} + \frac{\partial Z_z}{\partial z}\right)\right\} d\tau = 0.$$

Writing the term

$$\iiint z\frac{\partial Y_z}{\partial z}\,d\tau = \iint zY_z\,dx\,dy - \iiint Y_z\,d\tau$$

and

$$\iiint y\frac{\partial Z_y}{\partial y}\,d\tau = \iint yZ_y\,dz\,dx - \iiint Z_y\,d\tau,$$

and applying the divergence theorem, all the surface integrals cancel each other and there remains only the volume integral

$$93)\quad \iiint (Z_y - Y_z)\,d\tau = 0.$$

As before, the field of integration being arbitrary, the integrand must vanish, and we obtain, after applying the same process to the remaining two equations,

$$94)\quad Y_z = Z_y,\quad Z_x = X_z,\quad X_y = Y_x.$$

We may also obtain these equations by considering the stresses on the faces of an infinitesimal cube (Fig. 147). We shall denote the tangential components or shearing stresses 94) by T_x, T_y, T_z, the normal components or tractions by P_x, P_y, P_z. The stress at any point is determined in terms of these six components, for we may find the stress-vector F_n, whose direction-cosines are α', β', γ' for any stress plane whose normal has the direction cosines α, β, γ by equations 86), which in our present notation become

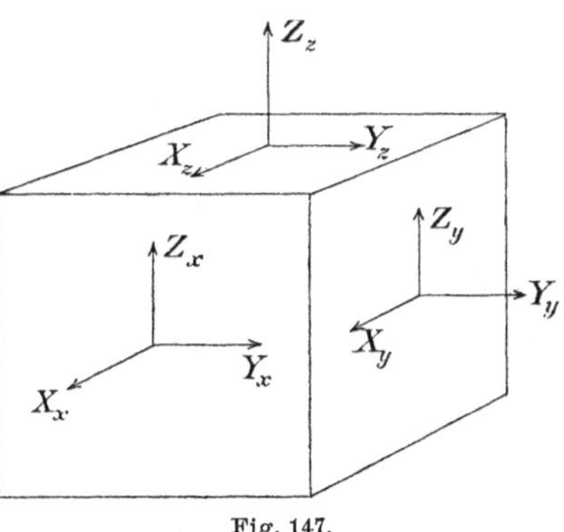

Fig. 147.

$$95)\quad \begin{aligned} X_n &= F_n\alpha' = P_x\alpha + T_z\beta + T_y\gamma,\\ Y_n &= F_n\beta' = T_z\alpha + P_y\beta + T_x\gamma\\ Z_n &= F_n\gamma' = T_y\alpha + T_x\beta + P_z\gamma.\end{aligned}$$

These are the exact analogues of equations 17). In other words, the stress-vector is a self-conjugate linear vector-function of the normal to the stress-plane. The stress-vector F_n occupies the place of $\frac{r'}{r}$ in 17). Accordingly the whole geometry of the linear vector function may be applied to the consideration of stress as follows.

171. Geometrical Representation of Stress. If we construct the quadric

$$96)\quad \varphi \equiv P_x x^2 + P_y y^2 + P_z z^2 + 2T_x yz + 2T_y zx + 2T_z xy = \pm R^2$$

any stress-vector F_n is perpendicular to the tangent plane drawn at the point where the normal to the stress-plane cuts the quadric φ (Fig. 148). This is known as Cauchy's stress-quadric. Let its equation, referred to its principal axes, which are known as the axes of the stress, be

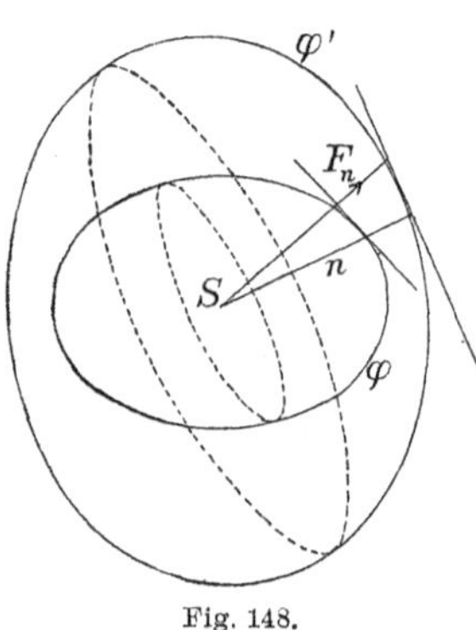

Fig. 148.

$$97)\quad \varphi \equiv P_1 x^2 + P_2 y^2 + P_3 z^2 = \pm R^2.$$

P_1, P_2, P_3 are called the principal tractions, being the normal stresses on the planes perpendicular to the axes, these planes being subject to no tangential stresses. Thus, as for any strain we may find three planes for which the slides vanish, so for stress we may find three planes for which the shearing stress vanishes.

In the reciprocal quadric,

$$98)\quad \varphi' \equiv \frac{x^2}{P_1} + \frac{y^2}{P_2} + \frac{z^2}{P_3} = \pm R^2,$$

the stress-vector is conjugate to its stress-plane, for the normal to the stress-plane is parallel to the normal to φ' where it is cut by the stress-vector. The quadric φ' is known as Lamé's stress-director quadric. In equations 17) and 14) putting F_n for $\frac{r'}{r}$ we obtain

$$99)\quad F_n = \pm \frac{R^2}{pr} = \pm \frac{R^2}{r^2 \cos(rr')}$$

or

$$100)\quad F_n \cos(rr') = F_{nn} = \pm \frac{R^2}{r^2} = \pm \frac{p'^2}{R^2}.$$

So that the traction or component of the stress-vector normal to its stress-plane is inversely proportional to the square of the radius-vector of the quadric φ in the direction of that normal, or is directly proportional to the square of the perpendicular upon the tangent plane to the quadric φ' parallel to the stress-plane.

If P_1, P_2, P_3 are all of the same sign the quadrics φ and φ' are ellipsoids. If they are positive we must take the positive sign with R^2, and the normal stress on every plane is a traction. If they are negative, we must take the negative sign, and the normal stress is always a pressure. If one of the P's has a different sign from the two others, we use both signs and have pairs of conjugate hyperboloids. In this case for directions parallel to the generators of the asymptotic cone $\varphi = 0$ to the stress quadric, we have r infinite and $F_{nn} = 0$. Accordingly for stress-planes perpendicular to these generators, the normal stress vanishes or the stress is a shearing stress. These planes envelop a cone called Lamé's shear-cone, which divides the directions for which the normal stress is a traction from those for which it is a pressure.

In the reciprocal quadric φ', when the radius vector is infinite, it lies in its conjugate plane, the stress-plane. But the radius vector to this quadric has the direction of the stress-vector, so that the shear-cone is the asymptotic cone to this quadric $\varphi' = 0$. If we construct the ellipsoid

101) $$\psi \equiv P_1^2 \xi^2 + P_2^2 \eta^2 + P_3^2 \zeta^2 = S^2,$$

we have by 22)

102) $$F_n = \frac{r'}{r} = \frac{S}{\varrho},$$

or the strain on a plane perpendicular to any radius vector is inversely proportional to that radius vector (it does not lie in the direction of the radius vector). This ellipsoid is called Cauchy's stress-ellipsoid and its axes are proportional to the squares of the stress quadric $\varphi = \pm R^2$. The reciprocal ellipsoid

103) $$\psi \equiv \frac{\xi'^2}{P_1^2} + \frac{\eta'^2}{P_2^2} + \frac{\zeta'^2}{P_3^2} = S^2,$$

has the property, since by 16) $F_n = \frac{\varrho'}{S}$, that the stress-vector for any plane is directly proportional to the radius vector in its own direction. This ellipsoid is called Lamé's stress-ellipsoid, or ellipsoid of elasticity.

171a. Simple Stresses. A simple stress is one that contains but a single constant in its specification. These are:

1°. Uniform traction or pressure.

104) $$P_1 = P_2 = P_3 = P,$$
$$T_x = T_y = T_z = 0.$$

All the quadrics are spheres and every stress is normal to its plane and of invariable amount P. Such a stress is physically realized by a body subjected to hydrostatic pressure.

2°. Simple traction,

105)
$$P_x = P, \quad P_y = P_z = 0,$$
$$T_x = T_y = T_z = 0.$$

The stress quadric is

106)
$$\varphi \equiv Px^2 = \pm 1,$$

a pair of planes perpendicular to the X-axis at a distance $\frac{1}{\sqrt{\pm P}}$ from the origin. The stress on any plane is parallel to the X-axis. The stress-director quadric and the shear-cone reduce to the axis of X, all planes tangent to which experience only shear.

Cauchy's ellipsoid,

107)
$$P^2x^2 + 0 \cdot y^2 + 0 \cdot z^2 = 1,$$

with axes, $\frac{1}{P}$, ∞, ∞, is a pair of planes perpendicular to the X-axis, and Lamé's ellipsoid with axes, P, 0, 0, becomes simply that part of the axis of X from $x = -P$ to $x = P$. From the property of this ellipsoid the stress-vector is proportional to the perpendicular on the tangent plane parallel to the stress plane. Since the tangent plane here always passes through one of the extremities we have

108)
$$F_n = P \cos(nx)$$

as is indeed evident from equations 95).

3°. Simple shearing stress.

109)
$$P_x = P_y = P_z = 0,$$
$$T_z = T, \quad T_x = T_y = 0.$$

Equations 95) become

110)
$$X_n = F_n \alpha' = T\beta,$$
$$Y_n = F_n \beta' = T\alpha,$$
$$Z_n = F_n \gamma' = 0.$$

The stress quadric is

111)
$$\varphi = 2Txy = \pm 1,$$

which represents a pair of rectangular hyperbolic cylinders with the semi-axes $\frac{1}{\sqrt{T}}$. The stress-director quadric is

112)
$$\frac{2xy}{T} = \pm 1.$$

The shear-cone $xy = 0$ represents the coordinate planes of XZ and YZ.

A shearing stress may also be written, referred to its principal axes,

113) $$P_x = -P_y = P, \quad P_z = 0,$$
$$T_x = T_y = T_z = 0,$$

when the stress quadric becomes

114) $$P(x^2 - y^2) = \pm 1,$$

the pair of hyperbolic cylinders referred to their axes and

115) $$\begin{aligned} X_n &= F_n\alpha' = \quad P\alpha, \\ Y_n &= F_n\beta' = -P\beta, \\ Z_n &= F_n\gamma' = \quad 0. \end{aligned}$$

We accordingly have
$$\alpha'\beta + \beta'\alpha = 0, \quad \text{or}$$
$$\frac{\alpha'}{\alpha} = -\frac{\beta'}{\beta}$$

that is, all stresses are parallel to the XY-plane, and the stress-vector and the projection on the XY-plane of the normal to the stress plane make equal angles with the X-axis on opposite sides (Fig. 149). Squaring equations 115) and adding,

116) $$F_n^2 = P^2(\alpha^2 + \beta^2) = P^2(1 - \gamma^2).$$

If $\gamma = 0$, that is, if the plane is tangent to the Z-axis
$$F_n = \pm P,$$
the normal stress being a traction if the normal to the stress-plane falls nearer to the X-axis, a pressure if nearer to the Y-axis.

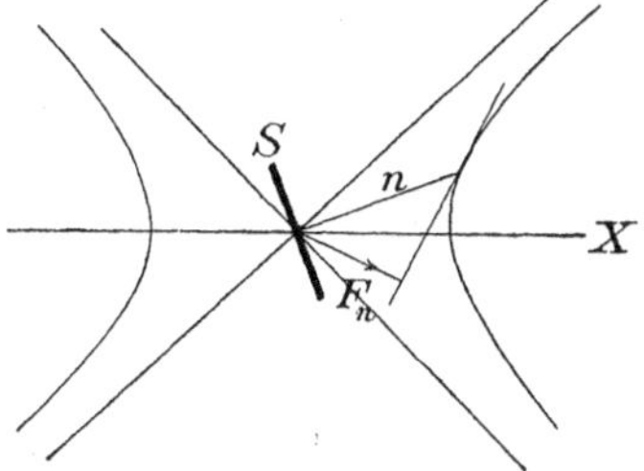

Fig. 149.

The shear cone $x^2 - y^2 = 0$ is composed of the two planes bisecting the dihedral angle between the XZ- and YZ-planes.

From this manner of representing the stress it is evident that a simple shearing stress is equivalent to an equal traction and pressure in two directions perpendicular to each other. Compare the representation of a shearing strain as an equal stretch and squeeze. For this case Lamé's ellipsoid $\frac{1}{P^2}(x^2 + y^2) + \frac{z}{0} = 1$ has the axes P, P, O and reduces to a circular disc normal to the Z-axis. Since all tangent planes pass through its edge,

117) $$F_n = P \sin(nz) = P\sqrt{1 - \gamma^2}$$

as above in 116).

172. Work of Stress in producing Strain. If every point in a body move a distance δq, whose components are δu, δv, δw, and if there act upon every unit of mass of the body the external forces X, Y, Z, and upon each unit of surface the forces X_n, Y_n, Z_n, the work done by all the forces in the displacement is

$$118)\qquad \delta W = \iint \{X_n \delta u + Y_n \delta v + Z_n \delta w\} dS + \iiint \varrho \{X \delta u + Y \delta v + Z \delta w\} d\tau,$$

which becomes by equations 86),

$$\begin{aligned}119)\qquad \delta W = \iint \{&[X_x \cos(nx) + X_y \cos(ny) + X_z \cos(nz)] \delta u \\ &+ [Y_x \cos(nx) + Y_y \cos(ny) + Y_z \cos(nz)] \delta v \\ &+ [Z_x \cos(nx) + Z_y \cos(ny) + Z_z \cos(nz)] \delta w\} dS \\ &+ \iiint \varrho (X \delta u + Y \delta v + Z \delta w) d\tau,\end{aligned}$$

and transforming surface integrals into volume integrals by differentiation in the manner of the divergence theorem and making use of equations 94),

$$\begin{aligned}\delta W = \iiint \Big\{ &\frac{\partial}{\partial x}(X_x \delta u + Y_x \delta v + Z_x \delta w) \\ &+ \frac{\partial}{\partial y}(X_y \delta u + Y_y \delta v + Z_y \delta w) \\ &+ \frac{\partial}{\partial z}(X_z \delta u + Y_z \delta v + Z_z \delta w) \\ &+ \varrho (X \delta u + Y \delta v + Z \delta w)\Big\} d\tau\end{aligned}$$

$$\begin{aligned}120)\qquad = \iiint \Big\{ &\left(\frac{\partial X_x}{\partial x} + \frac{\partial X_y}{\partial y} + \frac{\partial X_z}{\partial z} + \varrho X\right) \delta u \\ &+ \left(\frac{\partial Y_x}{\partial x} + \frac{\partial Y_y}{\partial y} + \frac{\partial Y_z}{\partial z} + \varrho Y\right) \delta v \\ &+ \left(\frac{\partial Z_x}{\partial x} + \frac{\partial Z_y}{\partial y} + \frac{\partial Z_z}{\partial z} + \varrho Z\right) \delta w \\ &+ X_x \frac{\partial \delta u}{\partial x} + Y_y \frac{\partial \delta v}{\partial y} + Z_z \frac{\partial \delta w}{\partial z} \\ &+ Y_z \left(\frac{\partial \delta w}{\partial y} + \frac{\partial \delta v}{\partial z}\right) + Z_x \left(\frac{\partial \delta u}{\partial z} + \frac{\partial \delta w}{\partial x}\right) \\ &+ X_y \left(\frac{\partial \delta v}{\partial x} + \frac{\partial \delta u}{\partial y}\right)\Big\} d\tau.\end{aligned}$$

By equations 90) the coefficients of δu, δv, δw vanish identically, so that, interchanging the order of differentiation and variation,

$$121)\quad \delta W = \iiint \left\{ X_x \delta \frac{\partial u}{\partial x} + Y_y \delta \frac{\partial v}{\partial y} + Z_z \delta \frac{\partial w}{\partial z} + Y_z \delta \left(\frac{\partial w}{\partial y} + \frac{\partial v}{\partial z} \right) + Z_x \delta \left(\frac{\partial u}{\partial z} + \frac{\partial w}{\partial x} \right) + X_y \delta \left(\frac{\partial v}{\partial x} + \frac{\partial u}{\partial y} \right) \right\} d\tau,$$

or in our later terminology,

$$122)\quad \delta W = \iiint \{ P_x \delta s_x + P_y \delta s_y + P_z \delta s_z + 2T_x \delta g_x + 2T_y \delta g_y + 2T_z \delta g_z \} d\tau.$$

Thus each of the six components of the stress does work on the corresponding component of the strain, and the work per unit volume in any infinitesimal strain is the sum of each stress component by the corresponding strain produced, except that with our terminology the shearing stresses are multiplied by *twice* the shearing strains or slides.

173. Relations between Stress and Strain. If a body is perfectly elastic the stresses at any point at any time depend simply upon the strain at the point at the time in question, so that if the elastic properties of the body are known at every point the stress components will be known functions of the strain components, which may differ from one point of the body to another. The stresses will be uniform and continuous functions of the strains and may be developed by Taylor's theorem. If then the strains are small, the terms of the lowest orders will be the most important. The strains dealt with by the ordinary theory of elasticity are so small that it is customary to neglect all terms above those of the first order. The results thus obtained are in good accordance with those obtained by experiment under the proper limitations. The law that for small strains the stresses are linear functions of the strains may be regarded as an extension of the law announced in 1676 by Hooke in the form of an anagram,

ceiiinosssttuu

Ut Tensio sic Vis.

The force varies as the stretch, or in our terminology the stress varies as the strain. Making this assumption we accordingly have

$$123)\quad \begin{aligned} X_x &= \varphi_{01} + \varphi_{11} s_x + \varphi_{21} s_y + \varphi_{31} s_z + \varphi_{41} g_x + \varphi_{51} g_y + \varphi_{61} g_z, \\ Y_y &= \varphi_{02} + \varphi_{12} s_x + \varphi_{22} s_y + \varphi_{32} s_z + \varphi_{42} g_x + \varphi_{52} g_y + \varphi_{62} g_z, \\ &\cdots\cdots\cdots\cdots\cdots\cdots \\ &\cdots\cdots\cdots\cdots\cdots\cdots \\ X_y &= \varphi_{06} + \varphi_{16} s_x + \varphi_{26} s_y + \varphi_{36} s_z + \varphi_{46} g_x + \varphi_{56} g_y + \varphi_{66} g_z. \end{aligned}$$

The φ's will in general be functions of the coordinates of the point, but if the body is homogeneous, that is alike at all points, they will be constants. We shall assume this to be true. If there be a natural state of the body or one in which the body is in equilibrium under the action of external forces, so that the stresses vanish for this state, it is convenient to measure the strains from the natural state. Then the stresses and strains vanish together, so that the terms $\varphi_{01}, \dots \varphi_{06}$ vanish. For such a body there are accordingly thirty-six constants φ, the so-called coefficients of elasticity. In the case of a gas there is no natural state, for a gas is never in equilibrium, unless kept so by an envelope, so that every portion of the gas always experiences pressure, consequently we cannot measure the strains from any natural state.

We have now the theory of elasticity as it was left by its founders, Navier and Cauchy. The idea is due to Green[1]) of supposing the elastic forces to be conservative and accordingly due to an energy function of the strains. If we call the function $\Phi(s_x, s_y, s_z, g_x, g_y, g_z)$ we have for the total potential energy due to any strain

124) $$W = \iiint \Phi \, d\tau.$$

The work done in changing the strain is then

125) $$\delta W = \iiint \delta\Phi \, d\tau = \iiint \left\{ \frac{\partial \Phi}{\partial s_x} \delta s_x + \frac{\partial \Phi}{\partial s_y} \delta s_y + \frac{\partial \Phi}{\partial s_z} \delta s_z + \frac{\partial \Phi}{\partial g_x} \delta g_x + \frac{\partial \Phi}{\partial g_y} \delta g_y + \frac{\partial \Phi}{\partial g_z} \delta g_z \right\} d\tau.$$

Comparing this with equations 121—122) we find

126) $$\begin{aligned} X_x &= \frac{\partial \Phi}{\partial s_x}, & Y_z = Z_y &= \frac{1}{2} \frac{\partial \Phi}{\partial g_x}, \\ X_y &= \frac{\partial \Phi}{\partial s_y}, & Z_x = X_z &= \frac{1}{2} \frac{\partial \Phi}{\partial g_y}, \\ X_z &= \frac{\partial \Phi}{\partial s_z}, & X_y = Y_x &= \frac{1}{2} \frac{\partial \Phi}{\partial g_z}. \end{aligned}$$

If then the stresses are to be linear functions of the strains, Φ must be a quadratic function, and, if we measure from the natural state, a homogeneous quadratic function. A homogeneous quadratic function of six variables contains twenty-one terms, so that instead of thirty-six elastic constants for the general homogeneous body we have only twenty-one, that is, the determinant of the φ's in equations 123) is symmetrical, fifteen coefficients on one side of the

1) Green, *Mathematical Papers*, p. 243.

principal diagonal being equal to the corresponding fifteen on the other side.

If the body besides being homogeneous is isotropic, that is, at any point its properties are the same with respect to *all directions*, there are many relations between the coefficients, so that the number of independent constants is much reduced. In an *anisotropic* or *eolotropic* body there are generally certain directions (the same for all parts of the body) with reference to which there is a certain symmetry, so that there are various relations involving a reduction in the number of constants. Such bodies are known as crystals. We shall deal here only with isotropic bodies.

174. Energy Function for Isotropic Bodies. In isotropic bodies the stresses developed depend only on the magnitude of the strains, not on their absolute directions with respect to the body. Accordingly if we change the axes of coordinates the expression for the energy must remain unchanged, or the energy function is an *invariant* for a change of axes. The cubic for the axes of the elongation quadric 58) belonging to the shift-equations 50) is the determinant

127)
$$\begin{vmatrix} s_x-\lambda, & g_z\ , & g_y \\ g_z\ , & s_y-\lambda, & g_x \\ g_y\ , & g_x\ , & s_z-\lambda \end{vmatrix}=0,$$

or expanding the determinant,

128)
$$\lambda^3-(s_x+s_y+s_z)\lambda^2+(s_ys_z+s_zs_x+s_xs_y-g_x^2-g_y^2-g_z^2)\lambda$$
$$+s_xg_x^2+s_yg_y^2+s_zg_z^2-s_xs_ys_z-2g_xg_yg_z=0.$$

If the roots are λ_1, λ_2, λ_3, the equation is

129)
$$\lambda^3-(\lambda_1+\lambda_2+\lambda_3)\lambda^2+(\lambda_1\lambda_2+\lambda_2\lambda_3+\lambda_3\lambda_1)\lambda-\lambda_1\lambda_2\lambda_3=0.$$

If we transform to another set of axes $X'Y'Z'$ with the same origin, so that the strain components are $s_{x'}$, $s_{y'}$, $s_{z'}$, $g_{x'}$, $g_{y'}$, $g_{z'}$, since the elongation quadric is a definite surface, the equation for its axes must have the same roots as before. Accordingly its coefficients are invariants. The roots λ_1, λ_2, λ_3 are the stretches for the directions of the principal axes of the strain. Therefore we have the three strain invariants, symmetrical functions of the roots,

130)
$$\begin{aligned} I_1&=\lambda_1+\lambda_2+\lambda_3=s_x+s_y+s_z,\\ I_2&=\lambda_1\lambda_2+\lambda_2\lambda_3+\lambda_3\lambda_1=s_ys_z+s_zs_x+s_xs_y-g_x^2-g_y^2-g_z^2.\\ I_3&=\lambda_1\lambda_2\lambda_3=2g_xg_yg_z+s_xs_ys_z-s_xg_x^2-s_yg_y^2-s_zg_z^2.\end{aligned}$$

The invariant I_1 represents the cubical dilatation σ, which by its geometrical definition is evidently independent of the choice of axes.

The energy function for an isotropic body, being unchanged when we change the axes, can contain the strains only in the combinations I_1, I_2, I_3, but these are of the first, second and third degrees respectively, and since Φ is of only the second degree it cannot contain I_3.

Since it is homogeneous (except for a gas) it can contain I_1 only through its square. We therefore have

131) $$\Phi = -PI_1 + AI_1^2 + BI_2,$$

where P, A, B are constants. P is zero, except for gases, and is then positive, for if the gas expands it loses energy. The constant A refers to a property common to all bodies, namely, resistance to compression, and is positive, for work must be done to compress a body. The constant B is peculiar to solids.

All symmetrical functions of the roots may be expressed in terms of the invariants, for example:

132) $$\begin{aligned}(\lambda_1-\lambda_2)^2+(\lambda_2-\lambda_3)^2+(\lambda_3-\lambda_1)^2 &= 2(\lambda_1^2+\lambda_2^2+\lambda_3^2)-2(\lambda_1\lambda_2+\lambda_2\lambda_3+\lambda_3\lambda_1)\\ &= 2(\lambda_1+\lambda_2+\lambda_3)^2-6(\lambda_1\lambda_2+\lambda_2\lambda_3+\lambda_3\lambda_1)\\ &= 2I_1^2-6I_2.\end{aligned}$$

Also

133) $$\begin{aligned}2(\lambda_1\lambda_2+\lambda_2\lambda_3+\lambda_3\lambda_1) &= (\lambda_1+\lambda_2+\lambda_3)^2-(\lambda_1^2+\lambda_2^2+\lambda_3^2), \quad \text{or}\\ 2I_2 &= I_1^2-(\lambda_1^2+\lambda_2^2+\lambda_3^2).\end{aligned}$$

We may accordingly write $AI_1^2+BI_2$ as a linear function, of I_1^2 and of either $(\lambda_1-\lambda_2)^2+(\lambda_2-\lambda_3)^2+(\lambda_3-\lambda_1)^2$, or of $\lambda_1^2+\lambda_2^2+\lambda_3^2$.

Suppose we write the quadratic terms

134) $$\begin{aligned}AI_1^2+BI_2 &= \frac{1}{2}H(\lambda_1+\lambda_2+\lambda_3)^2\\ &+\frac{1}{2}C\{(\lambda_1-\lambda_2)^2+(\lambda_2-\lambda_3)^2+(\lambda_3-\lambda_1)^2\},\end{aligned}$$

which is the form given by Helmholtz. The constant H, being multiplied by σ^2, refers to changes of volume without changes of form, representing in this case the whole energy, for if there is no change of form the stretches of the principal axes, $\lambda_1, \lambda_2, \lambda_3$ are equal. The term in C on the other hand refers to changes of form without change in volume, for it vanishes when $\lambda_1=\lambda_2=\lambda_3$, and represents the whole energy if $\sigma=0$. A *perfect fluid* is defined as a body in which changes of form produce no stress, so that for such bodies $C=0$.

We may also write

135) $$AI_1^2+BI_2=K(\lambda_1^2+\lambda_2^2+\lambda_3^2)+K\Theta(\lambda_1+\lambda_2+\lambda_3)^2,$$

which is the form used by Kirchhoff. We have then for the relations between the constants,

$$136)\qquad H - C = 2K\Theta \quad\text{and}\quad 3C = 2K$$

or

$$H = 2K\left(\Theta + \frac{1}{3}\right),\quad C = \frac{2}{3}K.$$

Accordingly for liquids in order to have $C = 0$ but H finite, we must put $K = 0$ and $\Theta = \infty$, so that $2K\Theta = H$.

Now since

$$\begin{aligned}137)\quad \lambda_1{}^2 + \lambda_2{}^2 + \lambda_3{}^2 &= I_1{}^2 - 2I_2\\ &= (s_x + s_y + s_z)^2 - 2(s_y s_z + s_z s_x + s_x s_y - g_x^2 - g_y^2 - g_z^2)\\ &= s_x^2 + s_y^2 + s_z^2 + 2(g_x^2 + g_y^2 + g_z^2),\end{aligned}$$

we have for solids, liquids and gases,

$$138)\quad \Phi = K\Theta\sigma^2 + K\{(s_x^2 + s_y^2 + s_z^2) + 2(g_x^2 + g_y^2 + g_z^2)\} - P(s_x + s_y + s_z).$$

We shall make use of the more common notation

$$139)\qquad 2K\vartheta = \lambda,\quad K = \mu.$$

(Thomson and Tait make use of the constants k for H and n for μ.)

We have accordingly

$$\begin{aligned}140)\quad \Phi = \frac{1}{2}\lambda(s_x + s_y + s_z)^2 &+ \mu\{s_x^2 + s_y^2 + s_z^2 + 2(g_x^2 + g_y^2 + g_z^2)\}\\ &- P(s_x + s_y + s_z).\end{aligned}$$

The constant μ like C refers to changes of form and vanishes for perfect fluids. In the present notation by equations 136) we have

$$141)\qquad H = \frac{2}{3}\mu + \lambda,$$

so that both λ and μ are involved in changes of volume. We thus see that isotropic bodies possess *two* elastic constants. By means of certain assumptions as to the nature of elastic stresses, making them depend upon actions between molecules, Cauchy and the earlier writers on elasticity reduced the energy function to a form depending on a single elastic constant, the same theory reducing the number of constants for an eolotropic body from twenty-one to fifteen. For this theory the reader may consult Neumann, *Theorie der Elastizität*, Todhunter and Pearson, *History of the Theory of Elasticity*. Experiments have not however confirmed this theory, and it is no longer generally held to be sound. Thomson and Tait inveigh against it with particular emphasis. We shall accordingly assume that an isotropic body has two independent constants of elasticity λ and μ.

175. Stresses in Isotropic Bodies. We may now calculate the stresses by means of equations 126), inserting the values of $s_x, s_y, s_z, g_x, g_y, g_z$.

142)
$$\begin{aligned} X_x &= \frac{\partial \Phi}{\partial s_x} = \lambda\sigma + 2\mu\frac{\partial u}{\partial x} - P, \\ Y_y &= \frac{\partial \Phi}{\partial s_y} = \lambda\sigma + 2\mu\frac{\partial v}{\partial y} - P, \\ Z_z &= \frac{\partial \Phi}{\partial s_z} = \lambda\sigma + 2\mu\frac{\partial w}{\partial z} - P, \\ Y_z = Z_y &= \frac{1}{2}\frac{\partial \Phi}{\partial g_x} = \mu\left(\frac{\partial w}{\partial y} + \frac{\partial v}{\partial z}\right), \\ Z_x = X_z &= \frac{1}{2}\frac{\partial \Phi}{\partial g_y} = \mu\left(\frac{\partial u}{\partial z} + \frac{\partial w}{\partial x}\right), \\ X_y = Y_x &= \frac{1}{2}\frac{\partial \Phi}{\partial g_z} = \mu\left(\frac{\partial v}{\partial x} + \frac{\partial u}{\partial y}\right). \end{aligned}$$

The first equation of equilibrium 90) becomes

143) $$\varrho X + \lambda\frac{\partial \sigma}{\partial x} + 2\mu\frac{\partial^2 u}{\partial x^2} + \mu\left(\frac{\partial^2 v}{\partial x\,\partial y} + \frac{\partial^2 u}{\partial y^2}\right) + \mu\left(\frac{\partial^2 u}{\partial z^2} + \frac{\partial^2 w}{\partial x\,\partial z}\right) = 0,$$

or, considering the value of σ, we may write the equation with its two companions,

144)
$$\begin{aligned} \varrho X + (\lambda + \mu)\frac{\partial \sigma}{\partial x} + \mu\Delta u &= 0, \\ \varrho Y + (\lambda + \mu)\frac{\partial \sigma}{\partial y} + \mu\Delta v &= 0, \\ \varrho Z + (\lambda + \mu)\frac{\partial \sigma}{\partial z} + \mu\Delta w &= 0. \end{aligned}$$

The equations at the surface of the body are by 86), using the above values of the stresses,

145)
$$\begin{aligned} X_n &= \left(\lambda\sigma + 2\mu\frac{\partial u}{\partial x} - P\right)\cos(nx) \\ &\quad + \mu\left(\frac{\partial v}{\partial x} + \frac{\partial u}{\partial y}\right)\cos(ny) + \mu\left(\frac{\partial u}{\partial z} + \frac{\partial w}{\partial x}\right)\cos(nz), \\ Y_n &= \mu\left(\frac{\partial v}{\partial x} + \frac{\partial u}{\partial y}\right)\cos(nx) + \left(\lambda\sigma + 2\mu\frac{\partial v}{\partial y} - P\right)\cos(ny) \\ &\quad + \mu\left(\frac{\partial w}{\partial y} + \frac{\partial v}{\partial z}\right)\cos(nz), \\ Z_n &= \mu\left(\frac{\partial u}{\partial z} + \frac{\partial w}{\partial x}\right)\cos(nx) + \mu\left(\frac{\partial w}{\partial y} + \frac{\partial v}{\partial z}\right)\cos(ny) \\ &\quad + \left(\lambda\sigma + 2\mu\frac{\partial w}{\partial z} - P\right)\cos(nz). \end{aligned}$$

176. Physical Meaning of the Constants. Let us consider a few simple cases of equilibrium with homogeneous strain under stress, there being no impressed bodily forces X, Y, Z, and putting $P = 0$.

1°. A simple dilatation.

$$u = ax, \quad v = ay, \quad w = az,$$

$$s_x = a, \quad s_y = a, \quad s_z = a, \quad g_x = g_y = g_z = 0,$$

$$\sigma = s_x + s_y + s_z = 3a,$$

146) $$X_x = Y_y = Z_z = (3\lambda + 2\mu)a = p,$$

$$Y_z = X_y = Z_x = 0.$$

The surface forces become simply

147) $$X_n = p\cos(nx), \quad Y_n = p\cos(ny), \quad Z_n = p\cos(nz),$$

or the surface force is normal to the surface, and

148) $$F_n = p = (3\lambda + 2\mu)a = \frac{3\lambda + 2\mu}{3} \cdot \sigma.$$

The ratio of normal traction to cubical dilatation, or of normal pressure to cubical compression,

149) $$\frac{p}{\sigma} = \frac{3\lambda + 2\mu}{3} = H$$

is called the *bulk-modulus* of elasticity, the term modulus being applied in general to the ratio of the stress to the strain thereby produced.

2°. A simple shear.

$$u = ay, \quad v = w = 0,$$

$$s_x = s_y = s_z = \sigma = g_x = g_y = 0,$$

150) $$g_z = \frac{a}{2}, \quad X_x = Y_y = Z_z = 0,$$

$$X_y = Y_x = \mu a = T,$$

$$Y_z = Z_y = Z_x = X_z = 0,$$

$$X_n = T\cos(ny), \quad Y_n = T\cos(nx), \quad Z_n = 0.$$

151) $$F_n = T\sqrt{\cos^2(nx) + \cos^2(ny)} = T\sin(nz), \quad \text{as in 117)}.$$

The ratio of the tangential stresses on the XZ and YZ planes to the amount of the shear produced,

$$152) \qquad \frac{T}{2g_z} = \mu,$$

is called the *shear modulus* of elasticity, or the simple rigidity.[1])

3^0. A stretch-squeeze.

$$153) \qquad \begin{gathered} u = ax, \quad v = -by, \quad w = -bz, \\ s_x = a, \quad s_y = s_z = -b, \quad \sigma = a - 2b, \\ X_x = \lambda(a - 2b) + 2\mu a, \\ Y_y = Z_z = \lambda(a - 2b) - 2\mu b, \end{gathered}$$

$$154) \qquad \begin{aligned} Y_z &= Z_y = X_z = 0, \\ X_n &= \{\lambda(a - 2b) + 2\mu a\}\cos(nx), \\ Y_n &= \{\lambda(a - 2b) - 2\mu b\}\cos(ny), \\ Z_n &= \{\lambda(a - 2b) - 2\mu b\}\cos(nz). \end{aligned}$$

If we choose a and b so that

$$155) \qquad b = \frac{\lambda}{2(\lambda + \mu)} \cdot a,$$

we have

$$156) \qquad \begin{aligned} Y_n &= Z_n = 0, \\ X_n &= 2\mu(a + b)\cos(nx) = \frac{\mu(3\lambda + 2\mu)}{\lambda + \mu} \cdot a\cos(nx). \end{aligned}$$

If the body is a cylinder, with generators parallel to the X-axis, bounded by perpendicular ends, experiencing a normal traction p, there is no force on the cylindrical surface, for which $\cos(nx) = 0$, and on the ends

$$157) \qquad p = \frac{\mu(3\lambda + 2\mu)a}{\lambda + \mu}.$$

The ratio of the tractive force to the stretch a,

$$158) \qquad E = \frac{p}{a} = \frac{\mu(3\lambda + 2\mu)}{\lambda + \mu} = \frac{9kn}{3k + n}$$

may be called the stretch or elongation modulus, and is generally known as *Young's modulus*.

The ratio of the lateral contraction to the longitudinal extension

$$159) \qquad \eta = \frac{b}{a} = \frac{\lambda}{2(\lambda + \mu)}$$

1) Thomson and Tait use the notation: bulk-modulus $= k = \lambda + \frac{2\mu}{3}$, simple rigidity $= n$, $m = \lambda + \mu = \frac{3k + n}{3}$.

is called *Poisson's ratio.* According to Poisson and the older writers $\lambda = \mu$, so that $\eta = \frac{1}{4}\cdot$ We must certainly have

$$-1 < \eta < \frac{1}{2},$$

for if $\eta > \frac{1}{2}$, $\mu < 0$, making the rigidity negative. If $\eta < -1$,

$$H = \left(\lambda + \frac{2}{3}\mu\right) < 0,$$

making the bulk-modulus negative. No known bodies have $\eta < 0$, and in experiments on isotropic bodies η has generally been found nearly equal to $\frac{1}{4}$, the value assumed by Poisson, the value being found to approach more nearly to Poisson's value the more pains were taken the secure isotropic specimens.

The bulk-, shear- and stretch-moduli and Poisson's ratio are the important elastic constants for an isotropic body, any two of which being known, all are known.

CHAPTER X.

STATICS OF DEFORMABLE BODIES.

177. Hydrostatics. Let us now consider the statics of a perfect fluid, that is, a body for which $\mu = 0$. If each element of the fluid is subjected to forces whose components are X, Y, Z per unit mass, equations 144), § 175 reduce to

$$\begin{aligned} \varrho X + \lambda \frac{\partial \sigma}{\partial x} &= 0, \\ \varrho Y + \lambda \frac{\partial \sigma}{\partial y} &= 0, \\ \varrho Z + \lambda \frac{\partial \sigma}{\partial z} &= 0, \end{aligned} \tag{1}$$

while the equations for the surface forces 145) become

$$\begin{aligned} X_n &= (\lambda\sigma - P)\cos(nx), \\ Y_n &= (\lambda\sigma - P)\cos(ny), \\ Z_n &= (\lambda\sigma - P)\cos(nz). \end{aligned} \tag{2}$$

The surface force is accordingly normal and equal to

$$\lambda\sigma - P = -p. \tag{3}$$

Thus we have the fundamental property of fluids, the force between any two elements of a perfect fluid is a pressure normal to the element of surface separating them and independent of its direction.

Differentiating equation 3) and replacing λ times the derivative of σ by the corresponding derivative of $(-p)$, our equations of equilibrium 1) are

$$4)\qquad \varrho X = \frac{\partial p}{\partial x}, \quad \varrho Y = \frac{\partial p}{\partial y}, \quad \varrho Z = \frac{\partial p}{\partial z}.$$

Thus the fluid can be in equilibrium only under the action of bodily forces of such a nature that ϱ times the resultant force per unit mass, that is to say, the force per unit volume, is a lamellar vector. If the pressure at any point depends only on the density, and conversely, and we put $\frac{1}{\varrho} = \frac{dP}{dp}$[1],

$$5)\qquad P = \int \frac{dp}{\varrho},$$

so that

$$6)\qquad \begin{aligned} \frac{\partial P}{\partial x} &= \frac{dP}{dp}\frac{\partial p}{\partial x} = \frac{1}{\varrho}\frac{\partial p}{\partial x}, \\ \frac{\partial P}{\partial y} &= \frac{dP}{dp}\frac{\partial p}{\partial y} = \frac{1}{\varrho}\frac{\partial p}{\partial y}, \\ \frac{\partial P}{\partial z} &= \frac{dP}{dp}\frac{\partial p}{\partial z} = \frac{1}{\varrho}\frac{\partial p}{\partial z}. \end{aligned}$$

Our equations 4) are

$$7)\qquad X = \frac{\partial P}{\partial x}, \quad Y = \frac{\partial P}{\partial y}, \quad Z = \frac{\partial P}{\partial z}.$$

Accordingly in this case the bodily forces per unit mass must be conservative. If V is their potential, multiplying equations 4) by dx, dy, dz respectively and adding, we have

$$8)\qquad \begin{aligned} \varrho(X\,dx + Y\,dy + Z\,dz) &= -\varrho\,dV \\ &= \frac{\partial p}{\partial x}dx + \frac{\partial p}{\partial y}dy + \frac{\partial p}{\partial z}dz = dp. \end{aligned}$$

If we have two fluids of different densities in contact we have at their common surface

$$9)\qquad -\varrho_1\,dV = dp, \quad -\varrho_2\,dV = dp,$$

so that

$$10)\qquad (\varrho_1 - \varrho_2)\,dV = 0,$$

therefore dV and dp are each equal to zero and the surface of separation is a surface of constant potential and constant pressure.

1) *Not* the *constant* P in 3).

Also, since, by 8) V differs from $-\dot{P}$ only by a constant, the surfaces of equal pressure are equipotential or level surfaces. If the fluid is incompressible ϱ is constant, so that we have

11) $$-V = const. + \frac{p}{\varrho}.$$

For gravity we have, if the axis of Z is measured vertically upward

$$X = Y = 0, \quad Z = -g,$$
$$V = gz,$$

so that

12) $$p = \varrho\,(C - gz).$$

If, neglecting the atmospheric pressure, we measure z from the level surface of no pressure

13) $$p = -g\varrho z,$$

which is the fundamental theorem for liquids, namely, that the pressure is proportional to the depth.

178. Height of the Atmosphere. If we consider a gas whose temperature is constant throughout, the relation between the pressure and volume is given by the law of Boyle and Mariotte

$$p = a\varrho,$$

accordingly

14) $$P = \int \frac{dp}{\varrho} = \int \frac{a\,d\varrho}{\varrho} = a \log \varrho + const.$$

and

15) $$V = gz = c - a \log \varrho,$$

16) $$\varrho = \varrho_0 e^{-\frac{gz}{a}},$$

where ϱ_0 is the density when $z = 0$.

Thus as we ascend to heights which are in arithmetical progression, the density decreases in geometrical progression, vanishing only for $z = \infty$. If on the other hand, we consider the relation between pressure and density to be that pertaining to adiabatic compression, that is compression in such a manner that the heat generated remains in the portion of the gas where it is generated, we shall obtain a law of equilibrium corresponding to what is known as convective equilibrium. The temperature then varies as we go upward in such a way that, if a portion of air is hotter than the stratum in which it lies, it will rise expanding and cooling at the same time until its temperature and density are the same as those of a higher layer. When there is no tendency for any portion of air to change its place convective equilibrium is established. The principles of thermodynamics give us the relation for adiabatic compression

$$17)\qquad p = b\varrho^{\varkappa},$$

where $\varkappa$ is the ratio of the specific beat at constant pressure to that at constant volume, whose numerical value is about 1.4. We then have

$$18)\qquad V = gz = -\int\frac{dp}{\varrho} = -\int b\varkappa\varrho^{\varkappa-2}d\varrho = c - \frac{b\varkappa\varrho^{\varkappa-1}}{\varkappa-1}.$$

Since $\varkappa > 1$, ϱ diminishes as z increases and is equal to zero when $gz = c$, so that on this hypothesis the atmosphere has an upper limit, which may be calculated when the value of ϱ for a single value of z is known.

It is obviously improper to consider the equilibrium of the atmosphere to an infinite distance without taking account of the variation of gravity as the distance above the surface of the earth increases. Considering the earth to be a sphere with a density that is a function only of the distance from the center, we have, with γ positive[1]), as in §§ 123, 149, instead of equation 8),

$$Xdx + Ydy + Zdz = \gamma d\left(\frac{M}{r}\right),$$

so that on the hypothesis of equal temperature

$$19)\qquad -\frac{\gamma M}{r} = const. - a\log\varrho,$$

$$20)\qquad \varrho = \varrho_0 e^{\frac{\gamma M}{ar}}.$$

On this hypothesis the density decreases as we leave the earth, but not so fast on account of the diminution of gravity, so that at infinity the density is not zero but equal to the constant ϱ_0.

In this example we have neglected the attraction on the gas of those layers lying below. From equation 20) the barometric formula is obtained.

Proceeding in the same manner for convective equilibrium, we have

$$21)\qquad -\frac{\gamma M}{r} = c - \frac{b\varkappa\varrho^{\varkappa-1}}{\varkappa-1}.$$

Here again ϱ decreases as r increases, giving an upper limit to the atmosphere for $\varrho = 0$ for a finite value of r.

1) It is to be observed that in equation 8) the forces are taken as the *negative* derivatives of the potential, but as in the following examples involving the earth's attraction they are obtained by multiplying the *positive* derivative by γ, we must in the integral equation change the sign of V and multiply by γ.

179. Rotating Mass of Fluid. If a mass of fluid rotates about an axis with a constant angular velocity ω, we may by the principle of § 104 treat the problem of motion like a statical problem, provided we apply to each particle a force equal to the centrifugal force. If we take the axis of rotation for the Z-axis the centrifugal force may be derived from the potential,

$$V_c = -\frac{\omega^2}{2}(x^2 + y^2). \tag{22}$$

If an incompressible liquid rotates about a vertical axis and is under the influence of gravity, we have by 11),

$$V = gz - \frac{\omega^2}{2}(x^2 + y^2), \tag{23}$$

$$p = \varrho\left(c - gz + \frac{\omega^2}{2}(x^2 + y^2)\right). \tag{24}$$

Consequently the surfaces of equal pressure are paraboloids of revolution. Measuring z from the vertex the equation of the free surface, for which $p = 0$, is

$$gz = \frac{\omega^2}{2}(x^2 + y^2). \tag{25}$$

The latus rectum is $\frac{2g}{\omega^2}$. On this principle centrifugal speed indicators are constructed.

An important case which we have already treated by this method in § 149 is the shape of the surface of the ocean. If we seek an approximation, assuming the earth to be centrobaric, the potential due to the attraction of the earth and centrifugal force will be, as we find either directly, or by putting $K = 0$ in § 149, 140),

$$U = \frac{M}{r}\left\{1 + \frac{\omega^2 r^3}{2\gamma M}\cos^2\psi\right\} \tag{26}$$

and the equation to the surface of the sea will be, $U = const.$, which may be written, writing $\frac{M}{a}$ for the constant,

$$\frac{1}{a} = \frac{1}{r}\left\{1 + \frac{\omega^2 r^3}{2\gamma M}\cos^2\psi\right\} \tag{27}$$

where ψ is the geocentric latitude, and a is the polar radius. In the case of the earth $\frac{\omega^2 r^2 a}{\gamma M} = \frac{1}{288.41}$ which is so small that the second term may be considered small with respect to the first, and its square neglected. Accordingly putting in this term $r = a$ the equation of the surface is

$$r = a\left\{1 + \frac{\omega^2 a^3}{2\gamma M}\cos^2\psi\right\} \tag{28}$$

which is the equation of an oblate ellipsoid of revolution. The ellipticity is

$$\frac{\omega^2 a^3}{2\gamma M} \bigg/ \left(1 + \frac{\omega^2 a^3}{2\gamma M}\right) = \frac{1}{577}.$$

The difference between this and the value $\frac{1}{299}$ given in § 149 is due to the fact that we have neglected the attraction of the water for itself and that the nucleus is not exactly centrobaric.

180. Gravitating, rotating Fluid. A problem of great importance in connection with the figure of the earth and other planets is the form of the bounding surface of a mass of homogeneous rotating liquid under the action of its own gravitation.

If V is the potential of the mass of fluid at any internal point, and we take the X-axis for the axis of rotation, we have

$$29)\qquad U = V + \frac{\omega^2}{2\gamma}(y^2 + z^2).$$

The form of the function V depends upon the shape of the bounding surface of the liquid, which is to be determined by the problem itself. The complete problem is thus one of very great difficulty and has been only partially solved.[1])

We will examine whether an ellipsoid is a possible figure of equilibrium.

We have found in § 157, 37) for the potential of a homogeneous ellipsoid

$$30)\qquad V = \pi\varrho abc \int_0^\infty \left\{1 - \frac{x^2}{a^2+u} - \frac{y^2}{b^2+u} - \frac{z^2}{c^2+u}\right\} \frac{du}{\sqrt{(a^2+u)(b^2+u)(c^2+u)}}$$

$$= const. - \frac{1}{2}\{Lx^2 + My^2 + Nz^2\},$$

where

$$L = 2\pi\varrho abc \int_0^\infty \frac{du}{(a^2+u)\sqrt{(a^2+u)(b^2+u)(c^2+u)}},$$

$$31)\qquad M = 2\pi\varrho abc \int_0^\infty \frac{du}{(b^2+u)\sqrt{(a^2+u)(b^2+u)(c^2+u)}},$$

$$N = 2\pi\varrho abc \int_0^\infty \frac{du}{(c^2+u)\sqrt{(a^2+u)(b^2+u)(c^2+u)}}.$$

---

1) Poincaré, "Sur l'equilibre d'une masse fluide animée d'un mouvement de rotation." *Acta math.*, t. VII, 1885. Also, *Figures d'equilibre d'une masse fluide.* Paris, 1903.

Inserting in the integral equation 11) (see footnote p. 466),

$$32)\qquad \frac{p}{\varrho} + \frac{\gamma}{2}\left\{Lx^2 + My^2 + Nz^2 - \frac{\omega^2}{\gamma}(y^2 + z^2)\right\} = const.,$$

the surfaces of equal pressure are similar to the ellipsoid

$$\frac{x^2}{a^2} + \frac{y^2}{b^2} + \frac{z^2}{c^2} = 1,$$

if

$$33)\qquad La^2 = \left(M - \frac{\omega^2}{\gamma}\right)b^2 = \left(N - \frac{\omega^2}{\gamma}\right)c^2,$$

from which we obtain

$$34)\qquad \frac{\omega^2}{\gamma} = \frac{Mb^2 - La^2}{b^2} = \frac{Nc^2 - La^2}{c^2} = \frac{Mb^2 - Nc^2}{b^2 - c^2}.$$

Equating the first and third values of $\frac{\omega^2}{\gamma}$

$$35)\qquad (b^2 - c^2)(Mb^2 - La^2) = b^2(Mb^2 - Nc^2),$$

or otherwise

$$36)\qquad a^2(c^2 - b^2)L = b^2c^2(M - N).$$

Since L, M and N are transcendental functions of a, b, c, this is a transcendental equation for the ratios of the axes. Since M is the same function of b that N is of c obviously the equation in satisfied if $b = c$ and $M = N$, giving an ellipsoid of revolution as a possible form of equilibrium. This is the celebrated solution given by Maclaurin in 1738.

If we put

$$\lambda = \frac{\sqrt{b^2 - a^2}}{a},\quad 1 + \lambda^2 = \frac{b^2}{a^2},$$

the formulae 60), 62), § 160 give

$$37)\qquad L = 4\pi\varrho\frac{1+\lambda^2}{\lambda^3}(\lambda - \tan^{-1}\lambda),$$

$$M = N = 2\pi\varrho\frac{1+\lambda^2}{\lambda^3}\left(\tan^{-1}\lambda - \frac{\lambda}{1+\lambda^2}\right).$$

Introducing these values the first value of $\frac{\omega^2}{\gamma}$ in equation 34) becomes

$$\frac{\omega^2}{\gamma} = M - \frac{L}{1+\lambda^2} = 2\pi\varrho\left\{\frac{1+\lambda^2}{\lambda^3}\left(\tan^{-1}\lambda - \frac{\lambda}{1+\lambda^2}\right) - \frac{2}{\lambda^3}(\lambda - \tan^{-1}\lambda)\right\},$$

$$38)\qquad \frac{\omega^2}{2\pi\gamma\varrho} = \frac{3+\lambda^2}{\lambda^3}\tan^{-1}\lambda - \frac{3}{\lambda^2} = \psi(\lambda).$$

By the development of $\tan^{-1}\lambda$ in a series we find

$$39)\qquad \psi(\lambda) = \sum_1^\infty (-1)^{n-1}\frac{4n}{(2n+1)(2n+3)}\lambda^{2n},$$

$$\psi(\lambda) = 0 \text{ for } \lambda = 0 \text{ and } \lambda = \infty,$$

and

$$\psi'(\lambda) = 0 \text{ for } \lambda = 2.5293,$$

for which value

$$\psi(\lambda) = 0.22467.$$

The course of the function $\psi(\lambda)$ is shown in Fig. 150, from which it is evident that if

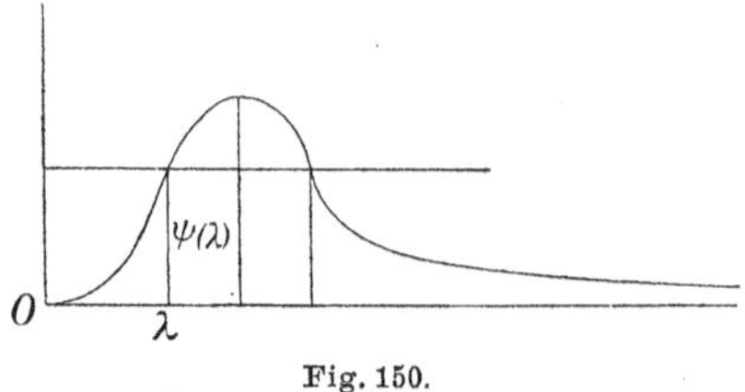

Fig. 150.

$$\frac{\omega^2}{2\pi\gamma\varrho} < 0.22467$$

there are two values of λ satisfying equation 38) and accordingly two possible ellipsoids of rotation. If on the contrary

$$\frac{\omega^2}{2\pi\gamma\varrho} > 0.22467$$

no possible ellipsoid of rotation is a figure of equilibrium.

When ω is very small one of the values of λ tends to zero and the other to infinity, that is, one of the ellipsoids is a sphere, the other a thin disc of infinite radius.

In the case of the earth using the value of $\gamma\varrho$ of § 123 and of ω of § 149, $\frac{\omega^2}{2\pi\gamma\varrho} = .00230$, and the smaller of the two values of λ coincides most nearly with the actual facts[1]), giving

$$\lambda^2 = 0.008688, \quad e = \frac{1}{231.7}.$$

The actual ellipticity being however $\frac{1}{299}$ we can only conclude that the earth when in its fluid state was not homogeneous.

The transcendental equation 36) written out is

$$40)\quad (b^2-c^2)\int\limits_0^\infty \left\{\frac{a^2}{a^2+u} + \frac{b^2c^2}{(b^2-c^2)(b^2+u)} - \frac{b^2c^2}{(b^2-c^2)(c^2+u)}\right\}\frac{du}{\sqrt{(a^2+u)(b^2+u)(c^2+u)}} = 0,$$

or otherwise

$$41)\quad (b^2-c^2)\int\limits_0^\infty \{a^2u^2 + [a^2(b^2+c^2) - b^2c^2]u\}\frac{du}{\sqrt{(a^2+u)^3(b^2+u)^3(c^2+u)^3}} = 0.$$

Besides the solution $b = c$ there is another given by putting the integral equal to zero. When $a = 0$, the integrand and consequently the definite integral is negative, when $a = \frac{bc}{\sqrt{b^2+c^2}}$ the integral is positive. There is accordingly a real value of a which satisfies the equation and there is an ellipsoid with three unequal axes which is a possible figure of equilibrium, if ω lies below a certain limit. This result was given by Jacobi in 1834. For further information on this subject the reader is referred to Thomson and Tait, Natural Philosophy, §§ 771—778.

1) Tisserand, *Traité de Mécanique Céleste,* Tom. II, p. 91.

181. Equilibrium of Floating Body. Let us apply the equations of equilibrium to a solid body immersed in a fluid under the action of any forces. Let us find the resultant force and moment of the pressure exerted by the liquid on the surface of the body. If we call the components of the resultant Ξ, H, Z, and of the moment L, M, N, we have

42)
$$\Xi = \iint p \cos(nx)\, dS,$$
$$H = \iint p \cos(ny)\, dS,$$
$$Z = \iint p \cos(nz)\, dS,$$

43)
$$L = \iint \{yp \cos(nz) - zp \cos(ny)\}\, dS,$$
$$M = \iint \{zp \cos(nx) - xp \cos(nz)\}\, dS,$$
$$N = \iint \{xp \cos(ny) - yp \cos(nx)\}\, dS.$$

If the body is in equilibrium it is evident that we may replace it by the fluid which it displaces, which would then be in equilibrium according to equations 4), and might then be solidified without disturbing the equilibrium.

If the body is only partly immersed we must apply the integration to the volume bounded by the wet surface and a horizontal plane forming a continuation of the free surface of the liquid and called the plane of flotation. Over this plane $p = 0$, consequently the surface integral is taken only over the wet surface, while the volume integral is as before taken over the volume of the fluid displaced. With this understanding we may convert the surface integrals into volume integrals taken throughout the space occupied by the displaced liquid, that is, within the surface of the solid body below the plane of flotation. We thus have

44)
$$\Xi = -\iiint \frac{\partial p}{\partial x}\, d\tau = -\iiint \varrho X\, d\tau,$$
$$H = -\iiint \frac{\partial p}{\partial y}\, d\tau = -\iiint \varrho Y\, d\tau,$$
$$Z = -\iiint \frac{\partial p}{\partial z}\, d\tau = -\iiint \varrho Z\, d\tau,$$

$$45)\qquad \begin{aligned} L &= -\iiint \left\{ y\frac{\partial p}{\partial z} - z\frac{\partial p}{\partial y} \right\} d\tau = -\iiint \varrho\{yZ - zY\} d\tau, \\ M &= -\iiint \left\{ z\frac{\partial p}{\partial x} - x\frac{\partial p}{\partial z} \right\} d\tau = -\iiint \varrho\{zX - xZ\} d\tau, \\ N &= -\iiint \left\{ x\frac{\partial p}{\partial y} - y\frac{\partial p}{\partial x} \right\} d\tau = -\iiint \varrho\{xY - yX\} d\tau. \end{aligned}$$

If the force acting on the liquid be gravity we have

$$X = Y = 0, \quad Z = -g,$$

accordingly

$$\Xi = H = 0, \quad Z = \quad g\iiint \varrho\, d\tau = mg,$$

where m is the mass of the fluid displaced by the body. This is the Principle of Archimedes: A body immersed in a liquid has its weight diminished by the weight of the displaced liquid.

For the moments we have

$$46)\qquad \begin{aligned} L &= \quad g\iiint \varrho y\, d\tau = \quad gm\bar{y}, \\ M &= -g\iiint \varrho x\, d\tau = -gm\bar{x}, \\ N &= \quad 0, \end{aligned}$$

where $\bar{x}$, $\bar{y}$ denote the coordinates of the center of mass of the displaced liquid. If the body is in equilibrium, by Archimedes' Principle the weight and therefore the mass of the body is equal to that of the displaced liquid. Consequently the resultant of the forces acting on the body is equivalent to a couple whose members are forces mg exerted downward at the center of mass of the body and upward at the center of the mass of the displaced liquid. If the couple is to vanish one of these must be vertically above the other. The center of mass of the displaced liquid is called the center of buoyancy of the body.

If the floating body is slightly displaced through a small angle $\delta\omega$ from the position of equilibrium by the application of a coulpe, the mass of the displaced fluid must remain unchanged, but the position of the center of buoyancy is slightly altered. Let us take the origin in the intersection of the old and new planes of flotation (Fig. 151). For the new position the figure is to be tilted in the direction of the arrow until the new position of the water line $W'L'$ is horizontal. The old center of buoyancy B is now no longer under the center of mass G and consequently, if the same portion were immersed, the body would be acted upon by a couple equal to mg times the horizontal distance between the verticals through G and B, and tending to increase the angular displacement. If b denote the length GB which now makes an angle $\delta\omega$ with the vertical, this horizontal distance is $b\delta\omega$, and the couple $mgb\delta\omega$. But this is not the only couple, for the immersed part is now different from that formerly immersed by the volume of the two wedges of small angle $\delta\omega$, the wedge of immersion EOE' and the wedge of emersion DOD'. The buoyancy added by the wedge of immersion and that lost by the wedge of emersion both produce moments in the direction tending to decrease the displacement. These moments must accordingly be subtracted from that previously found, to obtain the whole moment tending to overset the body.

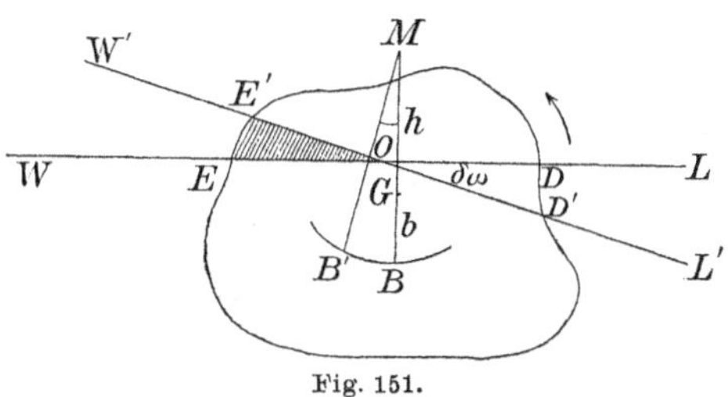

Fig. 151.

It is evident that if no vertical force is to be generated, the volumes of the wedges of immersion and emersion must be equal. Since the wedges are infinitely thin we may take for the element of volume

$$d\tau = z\,dx\,dy = y\,\delta\omega\,dx\,dy.$$

The condition for equal volumes is then

$$\text{47)} \qquad \iint z\,dx\,dy = \delta\omega \iint y\,dx\,dy = 0,$$

the integral being taken over the area of the plane of flotation. This will be the case if the axis taken through the center of mass of the area of flotation. The moment due to the wedges is

$$\text{48)} \qquad \begin{aligned} L' &= g\varrho \iiint y\,d\tau = g\varrho \iint zy\,dx\,dy \\ &= g\varrho\delta\omega \iint y^2 dx\,dy = g\varrho\delta\omega S\varkappa_x^2, \end{aligned}$$

where $\varkappa_x$ is the radius of gyration of the area of flotation about the X-axis.

In like manner

$$\text{49)} \qquad \begin{aligned} M' &= -g\varrho \iiint x\,d\tau = -g\varrho \iint xz\,dx\,dy \\ &= -g\varrho\delta\omega \iint xy\,dx\,dy. \end{aligned}$$

If we take for the axes of x and y the principal axes of the area of flotation the integral $\int\int xy\,dx\,dy$ vanishes. Accordingly a rotation about a principal axis through the center of mass of the plane of flotation developes a couple about that axis of magnitude $g\varrho\delta\omega S\varkappa_x^2$ tending to right the body. We have accordingly for the whole moment of the righting couple

$$L = g\delta\omega(\varrho S\varkappa_x^2 - mb). \tag{50}$$

On account of the change in the immersed part the center of buoyancy has moved from B to B'. If we draw a vertical in the new position through B', the point M in which it cuts the line BG is called the *metacenter* and the distance $MG = h$, the metacentric height. Since the couple acting on the body is composed of the two forces, mg acting downward at G and upward at B', it is evident that if the equilibrium is stable or the righting couple is positive M must be above G. The arm of the couple being the horizontal projection of MG is equal to $h_x \cdot \delta\omega$, and $L = mgh_x\delta\omega$. We accordingly have, inserting this value of L in equation 50) for the metacentric height

$$mh_x = (\varrho S\varkappa_x^2 - mb), \tag{51}$$

dividing by m and writing $\frac{m}{\varrho} = V$, the volume of the displaced liquid,

$$h_x = \frac{S\varkappa_x^2}{V} - b. \tag{52}$$

The equilibrium is stable or unstable according as this is positive or negative.

For the displacement about the Y-axis we have in like manner a couple proportional to the displacement, with a new metacentric height

$$h_y = \frac{S\varkappa_y^2}{V} - b, \tag{53}$$

where $\varkappa_y$ is the radius of gyration of the plane of flotation about the Y-axis. It is evident that the metacentric height is greater for the displacement about the shorter principal axis of the section. Thus it is easier to roll a ship than to tip it endwise.

Since the rotation about either axis is resisted by a couple proportional to the angular displacement, the body will perform small harmonic oscillations about the principal axes with the periodic times

$$T_x = \frac{2\pi K_x}{\sqrt{gh_x}} \quad \text{and} \quad T_y = \frac{2\pi K_y}{\sqrt{gh_y}},$$

where K_x and K_y are respectively the radii of gyration of the solid about the principal axes in the plane of flotation.

Introducing the values of h_x and h_y we may write

$$54)\qquad T_x = \frac{2\pi K_x}{\varkappa_x\sqrt{g\left(\frac{S}{V}-\frac{b}{\varkappa_x^2}\right)}},\qquad T_y = \frac{2\pi K_y}{\varkappa_y\sqrt{g\left(\frac{S}{V}-\frac{b}{\varkappa_y^2}\right)}}.$$

Since for a body of the shape of a ship K and $\varkappa$ increase together, we see that the larger $\varkappa$ corresponds to the shorter time. The pitching of a ship takes place more rapidly than the rolling.

The locus of the center of buoyancy for all possible displacements is called the surface of buoyancy, and the two metacenters are the centers of curvature of its principal sections. Evidently the body moves as if its surface of buoyancy rolled without friction on a horizontal plane, for it would then be acted on by the same couple as under the actual circumstances.

812. Solid hollow Sphere and Cylinder under Pressure. We have dealt in § 176 with a few cases of equilibrium of solid bodies under stress, where the strain produced was homogeneous. We shall now treat a number of cases in which the strain is not homogeneous. If there are no bodily forces the equations 144), § 175, become

$$55)\qquad \begin{aligned}(\lambda+\mu)\frac{\partial\sigma}{\partial x}+\mu\Delta u&=0,\\ (\lambda+\mu)\frac{\partial\sigma}{\partial y}+\mu\Delta v&=0,\\ (\lambda+\mu)\frac{\partial\sigma}{\partial z}+\mu\Delta w&=0.\end{aligned}$$

Forming the divergence, by differentiating respectively by x, y, z, and adding, and interchanging the order of the operations ∂ and Δ,

$$56)\qquad (\lambda+2\mu)\Delta\sigma=0,\quad \Delta\sigma=0,$$

so that σ is a harmonic function. Let us put $\sigma=a$. If the strain is pure, let φ be the strain potential, § 169, so that

$$57)\qquad u=\frac{\partial\varphi}{\partial x},\quad v=\frac{\partial\varphi}{\partial y},\quad w=\frac{\partial\varphi}{\partial z};\quad \sigma=\Delta\varphi=a,$$

and φ is equal to the potential of a mass of density $-\frac{a}{4\pi}$, coinciding with the body under investigation. If this be a sphere or spherical shell, we find, as in §§ 125, 135,

$$58)\qquad \Delta\varphi=\frac{1}{r^2}\frac{d}{dr}\left(r^2\frac{d\varphi}{dr}\right)=a,$$

$$59)\qquad \varphi=\frac{a}{6}r^2+\frac{b}{r},$$

where b is a second arbitrary constant, from which we have

60)
$$u = \frac{\partial \varphi}{\partial x} = \left(\frac{a}{3} r - \frac{b}{r^2}\right)\frac{x}{r},$$
$$v = \frac{\partial \varphi}{\partial y} = \left(\frac{a}{3} r - \frac{b}{r^2}\right)\frac{y}{r},$$
$$w = \frac{\partial \varphi}{\partial z} = \left(\frac{a}{3} r - \frac{b}{r^2}\right)\frac{z}{r},$$

61)
$$\frac{\partial u}{\partial x} = \frac{a}{3} - \frac{b}{r^3} + \frac{3bx^2}{r^5}, \quad \frac{\partial u}{\partial y} = \frac{3bxy}{r^5}, \quad \frac{\partial u}{\partial z} = \frac{3bxz}{r^5},$$
$$\frac{\partial v}{\partial x} = \frac{3bxy}{r^5}, \quad \frac{\partial v}{\partial y} = \frac{a}{3} - \frac{b}{r^3} + \frac{3by^2}{r^5}, \quad \frac{\partial v}{\partial z} = \frac{3byz}{r^5},$$
$$\frac{\partial w}{\partial x} = \frac{3bzx}{r^5}, \quad \frac{\partial w}{\partial y} = \frac{3bzy}{r^5}, \quad \frac{\partial w}{\partial z} = \frac{a}{3} - \frac{b}{r^3} + \frac{3bz^2}{r^5},$$

giving $\sigma = a$. The values of the surface forces are by 145), § 175,

62)
$$X_n = \left[\lambda a + 2\mu\left(\frac{a}{3} - \frac{b}{r^3} + \frac{3bx^2}{r^5}\right)\right]\cos(nx)$$
$$+ \frac{6\mu bx}{r^5}[y\cos(ny) + z\cos(nz)],$$
$$Y_n = \left[\lambda a + 2\mu\left(\frac{a}{3} - \frac{b}{r^3} + \frac{3by^2}{r^5}\right)\right]\cos(ny)$$
$$+ \frac{6\mu by}{r^5}[z\cos(nz) + x\cos(nx)],$$
$$Z_n = \left[\lambda a + 2\mu\left(\frac{a}{3} - \frac{b}{r^3} + \frac{3bz^2}{r^5}\right)\right]\cos(nz)$$
$$+ \frac{6\mu bz}{r^5}[x\cos(nx) + y\cos(ny)].$$

Collecting the terms
$$\frac{6\mu bx}{r^4}\left(\frac{x}{r}\cos(nx) + \frac{y}{r}\cos(ny) + \frac{z}{r}\cos(nz)\right) = \frac{6\mu bx}{r^4} = \frac{6\mu b}{r^3}\cos(nx),$$
and writing

63)
$$p = \lambda a + \frac{2}{3}\mu a + \frac{4\mu b}{r^3} = Ha + \frac{4\mu b}{r^3},$$
these become

64
$$X_n = p\cos(nx), \quad Y_n = p\cos(ny), \quad Z_n = p\cos(nz),$$
so that p is a normal traction or pressure. If R_1 and R_2 are the internal and external radii of the shell,

65)
$$p_1 = Ha + \frac{4\mu b}{R_1^3},$$
$$p_2 = Ha + \frac{4\mu b}{R_2^3},$$
which determine a and b.

66)
$$b = \frac{p_2 - p_1}{4\mu\left(\frac{1}{R_2^3} - \frac{1}{R_1^3}\right)},$$
$$a = \frac{p_2 R_2^3 - p_1 R_1^3}{H(R_2^3 - R_1^3)} = \sigma.$$

In Oersted's piezometer the internal and external pressures are equal, so that $b = 0$, and the sphere receives a homogeneous strain, which is of the same magnitude, $\sigma = \frac{p}{H}$, as if the sphere were solid. A second practical application is found in the correction of thermometers due to the pressure of the mercury causing the bulb to expand, the amount of expansion being found from 66).

In treating the case of a very long hollow cylinder, we proceed in precisely the same manner, except that the problem is a two dimensional one. We will number the equations in the same manner, with the addition of an accent. The formulae have an application in finding the pressure able to be borne by tubes and boilers.

57′) $$u = \frac{\partial \varphi}{\partial x}, \quad v = \frac{\partial \varphi}{\partial y}, \quad w = 0, \quad \sigma = \Delta \varphi = a,$$

58′) $$\Delta \varphi = \frac{1}{r}\frac{d}{dr}\left(r\frac{d\varphi}{dr}\right) = a,$$

59′) $$\varphi = \frac{a}{4}r^2 + b\log r,$$

60′) $$u = \frac{\partial \varphi}{\partial x} = \left(\frac{a}{2}r + \frac{b}{r}\right)\frac{x}{r}, \qquad v = \frac{\partial \varphi}{\partial y} = \left(\frac{a}{2}r + \frac{b}{r}\right)\frac{y}{r},$$

61′) $$\frac{\partial u}{\partial x} = \frac{a}{2} + \frac{b}{r^2} - \frac{2bx^2}{r^4}, \quad \frac{\partial u}{\partial y} = -\frac{2bxy}{r^4}, \qquad \frac{\partial v}{\partial x} = -\frac{2bxy}{r^4}, \quad \frac{\partial v}{\partial y} = \frac{a}{2} + \frac{b}{r^2} - \frac{2by^2}{r^4},$$

62′) $$X_n = \left[\lambda a + 2\mu\left(\frac{a}{2} + \frac{b}{r^2} - \frac{2bx^2}{r^4}\right)\right]\cos(nx) - \frac{4bxy}{r^4}\cos(ny),$$
$$Y_n = -\frac{4bxy}{r^4}\cos nx + \left[\lambda a + 2\mu\left(\frac{a}{2} + \frac{b}{r^2} - \frac{2bx^2}{r^4}\right)\right]\cos(ny),$$

63′) $$p = (\lambda + \mu)a - \frac{2\mu b}{r^2},$$

64′) $$X_n = p\cos(nx), \quad Y_n = p\cos(ny),$$

65′) $$p_1 = (\lambda + \mu)a - \frac{2\mu b}{R_1^{\,2}}, \qquad p_2 = (\lambda + \mu)a - \frac{2\mu b}{R_2^{\,2}},$$

66′) $$b = \frac{p_2 - p_1}{-2\mu\left(\frac{1}{R_2^{\,2}} - \frac{1}{R_1^{\,2}}\right)}, \qquad a = \frac{p_2 R_2^{\,2} - p_1 R_1^{\,2}}{(\lambda + \mu)(R_2^{\,2} - R_1^{\,2})}.$$

183. Problem of de Saint-Venant. We shall now treat a problem dealt with by the distinguished elastician Barré de Saint-Venant, in two celebrated memoirs on the torsion and flexion of prims, published in 1855 and 1856.[1]) For the full treatment of this subject the reader is referred to Clebsch, *Theorie der Elasticität,* or especially to the French translation of the same work, edited with notes by de Saint-Venant himself. The problem is thus defined by Clebsch, whose analysis is here followed.

What are the circumstances of equilibrium of a cylindrical body, on whose cylindrical surface no forces act, and wbose interior is not subjected to external forces, under which the longitudinal fibres composing the body experience no sidewise pressure?[2]) What forces must act on the free end surfaces, in order to bring about such circumstances?

We have already treated a special case of this problem in 3°, § 176. If the Z-axis be taken parallel to the generators of the cylindrical surface of the body, the conditions that adjacent fibres exercise no stress on each other perpendicular to their length are

$$67)\qquad X_x = Y_y = X_y = 0.$$

The conditions at the cylindrical surface are

$$68)\qquad \begin{aligned} X_n &= X_x \cos(nx) + X_y \cos(ny) + X_z \cos(nz) = 0,\\ Y_n &= Y_x \cos(nx) + Y_y \cos(ny) + Y_z \cos(nz) = 0,\\ Z_n &= Z_x \cos(nx) + Z_y \cos(ny) + Z_z \cos(nz) = 0,\end{aligned}$$

of which the first two are satisfied identically, since $\cos nz = 0$, and the last is simply,

$$69)\qquad Z_x \cos(nx) + Z_y \cos(ny) = 0.$$

In order to remove the possibility of a displacement of the cylinder like a rigid body involving six freedoms we will suppose fixed: a point, a line, and an infinitesimal element of surface of one orthogonal end or cross-section. Take the fixed point as origin, so that $u_0 = v_0 = w_0 = 0$. Take the fixed element of surface for the XY-plane, and the fixed line for the X-axis. For a point near the origin, the shifts are

---

1) Mémoire sur la torsion des prismes, avec des considerations sur leur flexion. *Mémoires des Savants étrangers,* 1855.

Mémoire sur la flexion des prismes. *Journ. de Math. de Liouville,* 2me Serie, T. I, 1856.

2) That is, the stress on a plane parallel to a generator is only tangential, and in the direction of the generator.

70)
$$u = u_0 + \left(\frac{\partial u}{\partial x}\right)_0 dx + \left(\frac{\partial u}{\partial y}\right)_0 dy + \left(\frac{\partial u}{\partial z}\right)_0 dz,$$
$$v = v_0 + \left(\frac{\partial v}{\partial x}\right)_0 dx + \left(\frac{\partial v}{\partial y}\right)_0 dy + \left(\frac{\partial v}{\partial z}\right)_0 dz,$$
$$w = w_0 + \left(\frac{\partial w}{\partial x}\right)_0 dx + \left(\frac{\partial w}{\partial y}\right)_0 dy + \left(\frac{\partial w}{\partial z}\right)_0 dz.$$

If the point is in the XY-plane, $dz = 0$, and as this plane is fixed, w must be 0, necessitating

$$\left(\frac{\partial w}{\partial x}\right)_0 = \left(\frac{\partial w}{\partial y}\right)_0 = 0.$$

If the point is on the X-axis, $dy = dz = 0$, and as it must remain on the X-axis, v and w must vanish, necessitating

$$\left(\frac{\partial v}{\partial x}\right)_0 = 0.$$

The six conditions at the origin are accordingly,

71)
$$u_0 = v_0 = w_0 = \left(\frac{\partial v}{\partial x}\right)_0 = \left(\frac{\partial w}{\partial x}\right)_0 = \left(\frac{\partial w}{\partial y}\right)_0 = 0,$$

while the conditions 67) everywhere satisfied are

72)
$$X_x = \lambda\sigma + 2\mu\frac{\partial u}{\partial x} = 0,$$
$$Y_y = \lambda\sigma + 2\mu\frac{\partial v}{\partial y} = 0,$$
$$X_y = \mu\left(\frac{\partial v}{\partial x} + \frac{\partial u}{\partial y}\right) = 0.$$

Of these the first two give

a)
$$\frac{\partial u}{\partial x} = \frac{\partial v}{\partial y} = -\frac{\lambda}{2\mu}\left(\frac{\partial u}{\partial x} + \frac{\partial v}{\partial y} + \frac{\partial w}{\partial z}\right) = -\frac{\lambda}{2\mu}\left(2\frac{\partial u}{\partial x} + \frac{\partial w}{\partial z}\right)$$
$$= -\frac{\lambda}{2(\lambda+\mu)}\frac{\partial w}{\partial z} = -\eta\frac{\partial w}{\partial z},$$

with the third

b)
$$\frac{\partial u}{\partial y} + \frac{\partial v}{\partial x} = 0.$$

The equations 144) of equilibrium are

73)
$$(\lambda + \mu)\frac{\partial \sigma}{\partial x} + \mu\Delta u = 0,$$
$$(\lambda + \mu)\frac{\partial \sigma}{\partial y} + \mu\Delta v = 0,$$
$$(\lambda + \mu)\frac{\partial \sigma}{\partial z} + \mu\Delta w = 0.$$

From the first of a) and b), by differentiation,

74) $$\Delta u = \frac{\partial^2 u}{\partial x^2} + \frac{\partial^2 u}{\partial y^2} + \frac{\partial^2 u}{\partial z^2} = \frac{\partial^2 u}{\partial z^2},$$
$$\Delta v = \frac{\partial^2 v}{\partial x^2} + \frac{\partial^2 v}{\partial y^2} + \frac{\partial^2 v}{\partial z^2} = \frac{\partial^2 v}{\partial z^2}.$$

From a),

75) $$\sigma = \frac{\partial u}{\partial x} + \frac{\partial v}{\partial y} + \frac{\partial w}{\partial z} = (1 - 2\eta)\frac{\partial w}{\partial z} = \frac{\mu}{\lambda + \mu}\frac{\partial w}{\partial z}.$$

The equations 73) become

c) $$\frac{\partial^2 w}{\partial x \partial z} + \frac{\partial^2 u}{\partial z^2} = 0,$$

d) $$\frac{\partial^2 w}{\partial y \partial z} + \frac{\partial^2 v}{\partial z^2} = 0,$$

e) $$\frac{\partial^2 w}{\partial x^2} + \frac{\partial^2 w}{\partial y^2} + 2\frac{\partial^2 w}{\partial z^2} = 0.$$

The equations a), b), c), d), e), together with the surface condition at the cylindrical surface

76) $$\left(\frac{\partial u}{\partial z} + \frac{\partial w}{\partial x}\right)\cos(nx) + \left(\frac{\partial v}{\partial z} + \frac{\partial w}{\partial y}\right)\cos(ny) = 0,$$

and the conditions 71) at the origin, constitute the mathematical statement of the problem.

Differentiating e) by z, and subtracting the derivative of c) by x and of d) by y,

77) $$2\frac{\partial^3 w}{\partial z^3} - \frac{\partial^2 u}{\partial z^2 \partial x} - \frac{\partial^2 v}{\partial z^2 \partial y} = 0.$$

Inserting the values of $\frac{\partial u}{\partial x}$ and $\frac{\partial v}{\partial y}$ in terms of $\frac{\partial w}{\partial z}$ from a) gives

78) $$\frac{\partial^3 w}{\partial z^3} = 0.$$

Adding the derivative of c) by y and of d) by x,

79) $$\frac{\partial^2}{\partial z^2}\left\{\frac{\partial u}{\partial y} + \frac{\partial v}{\partial x}\right\} + 2\frac{\partial^3 w}{\partial x \partial y \partial z} = 0,$$

which, by use of b), gives

80) $$\frac{\partial^3 w}{\partial x \partial y \partial z} = 0.$$

Differentiating e) by z, and using 78),

81) $$\frac{\partial^3 w}{\partial x^2 \partial z} + \frac{\partial^3 w}{\partial y^2 \partial z} = 0.$$

Differentiating c) by x, d) by y, and comparing with a),

82) $$\frac{\partial^3 w}{\partial x^2 \partial z} = \frac{\partial^3 w}{\partial y^2 \partial z}.$$

Comparing with 81), we find that both of these derivatives must be zero. Accordingly we have

83) $$\frac{\partial^2}{\partial x^2}\left(\frac{\partial w}{\partial z}\right) = \frac{\partial^2}{\partial y^2}\left(\frac{\partial w}{\partial z}\right) = \frac{\partial^2}{\partial z^2}\left(\frac{\partial w}{\partial z}\right) = \frac{\partial^2}{\partial x\,\partial y}\left(\frac{\partial w}{\partial z}\right) = 0.$$

Hence $\frac{\partial w}{\partial z}$ cannot contain any power of x, y or z above the first, nor xy, so that we may put

84) $$\frac{\partial w}{\partial z} = a + a_1 x + a_2 y + z(b + b_1 x + b_2 y).$$

Hence from a),

85) $$\frac{\partial u}{\partial x} = \frac{\partial v}{\partial y} = -\eta\,\{a + a_1 x + a_2 y + z(b + b_1 x + b_2 y)\}.$$

Differentiating 84), equations c) and d) become

86) $$\frac{\partial^2 u}{\partial z^2} = -(a_1 + b_1 z), \quad \frac{\partial^2 v}{\partial z^2} = -(a_2 + b_2 z).$$

Accordingly u and v contain powers of z not above z^3, u contains powers of x not above x^2, v contains powers of y not above y^2 and w, powers of z not above z^2.

Integrating 85) and 86),

87) $$\begin{aligned} u &= -\eta\left\{ax + \frac{a_1 x^2}{2} + a_2 xy + z\left(bx + \frac{b_1 x^2}{2} + b_2 xy\right)\right\} \\ &\quad - \left(\frac{a_1 z^2}{2} + \frac{b_1 z^3}{6}\right) + z\psi(y) + \chi(y), \\ v &= -\eta\left\{ay + a_1 xy + \frac{a_2 y^2}{2} + z\left(by + b_1 xy + \frac{b_2 y^2}{2}\right)\right\} \\ &\quad - \left(\frac{a_2 z^2}{2} + \frac{b_2 z^3}{6}\right) + z\varphi(x) + f(x). \end{aligned}$$

Putting these in b),

88) $$\begin{aligned} &-\eta(a_2 x + b_2 xz) + z\psi'(y) + \chi'(y), \\ &-\eta(a_1 y + b_1 yz) + z\varphi'(x) + f'(x) = 0, \end{aligned}$$

so that ψ, χ, φ, f are of the second degree, in the variables indicated, say,

89) $$\begin{aligned} \chi(y) &= a' + a_1' y + a_2' y^2, \\ \psi(y) &= b' + b_1' y + b_2' y^2, \\ f(x) &= a'' + a_1'' x + a_2'' x^2, \\ \varphi(x) &= b'' + b_1'' x + b_2'' x^2, \end{aligned}$$

in which

90) $$\begin{aligned} a_2'' &= \frac{1}{2}\eta a_2, \quad a_2' = \frac{1}{2}\eta a_1, \quad a_1' + a_1'' = 0, \\ b_2'' &= \frac{1}{2}\eta b_2, \quad b_2' = \frac{1}{2}\eta b_1, \quad b_1' + b_1'' = 0. \end{aligned}$$

Put

$$a_1' = - a_1'' = \alpha,$$
$$b_1' = - b_1'' = \beta,$$

and we have

91)
$$u = - \eta\left[ax + a_1 \frac{x^2 - y^2}{2} + a_2 xy + z\left(bx + b_1 \frac{x^2 - y^2}{2} + b_2 xy\right)\right] - \frac{a_1 z^2}{2} - \frac{b_1 z^3}{6} + (b' + \beta y) z + a' + \alpha y,$$
$$v = - \eta\left[ay + a_1 xy + a_2 \frac{y^2 - x^2}{2} + z\left(by + b_1 xy + b_2 \frac{y^2 - x^2}{2}\right)\right] - \frac{a_2 z^2}{2} - \frac{b_2 z^3}{6} + (b'' - \beta x) z + a'' - \alpha x.$$

Integrating 84),

92) $$w = (a + a_1 x + a_2 y) z + \frac{z^2}{2}(b + b_1 x + b_2 y) + F(x, y),$$

which inserted in e) gives

93) $$\frac{\partial^2 F}{\partial x^2} + \frac{\partial^2 F}{\partial y^2} + 2(b + b_1 x + b_2 y) = 0.$$

If we put

94) $$F = \Omega - \left(b \frac{x^2 + y^2}{2} + b_1 xy^2 + b_2 x^2 y\right) - b'x - b''y$$

we get finally

95) $$\Delta \Omega = \frac{\partial^2 \Omega}{\partial x^2} + \frac{\partial^2 \Omega}{\partial y^2} = 0,$$

or Ω is a harmonic function of the variables x, y. The conditions 71) at the origin give

96)
$$a' = a'' = \Omega_0 = 0,$$
$$b' = \left(\frac{\partial \Omega}{\partial x}\right)_0, \quad b'' = \left(\frac{\partial \Omega}{\partial y}\right)_0, \quad \alpha = 0.$$

We have now for the traction,

97)
$$Z_z = \lambda \sigma + 2\mu \frac{\partial w}{\partial z} = E \frac{\partial w}{\partial z}$$
$$= E\{a + a_1 x + a_2 y + z(b + b_1 x + b_2 y)\}.$$

If we take for the origin the center of mass of the cross-section, so that

$$\iint x\,dS = \iint y\,dS = 0,$$

we have, integrating over the cross-section, $z = const.$,

98) $$\iint Z_z dS = E \iint (a + bz)\, dS = E(a + bz) S.$$

Now since the beam is in equilibrium, this integral Z-traction must be the same for all values of z, hence $b = 0$. We have then finally,

99)
$$
\begin{aligned}
u &= -\eta\left[ax + a_1\frac{x^2-y^2}{2} + a_2 xy + z\left(b_1\frac{x^2-y^2}{2} + b_2 xy\right)\right] \\
&\quad - \frac{a_1 z^2}{2} - \frac{b_1 z^3}{6} + z\left(\frac{\partial \Omega}{\partial x}\right)_0 + \beta y z, \\
v &= -\eta\left[ax + a_2\frac{y^2-x^2}{2} + a_1 xy + z\left(b_2\frac{y^2-x^2}{3} + b_1 xy\right)\right] \\
&\quad - \frac{a_2 z^2}{2} - \frac{b_2 z^2}{6} + z\left(\frac{\partial \Omega}{\partial y}\right)_0 - \beta x z, \\
w &= (a + a_1 x + a_2 y) z + (b_1 x + b_2 y)\frac{z^2}{2} \\
&\quad - (b_1 xy^2 + b_2 x^2 y) + \Omega - x\left(\frac{\partial \Omega}{\partial x}\right)_0 - y\left(\frac{\partial \Omega}{\partial y}\right)_0.
\end{aligned}
$$

From these we deduce the non-vanishing stress-components,

100)
$$
\begin{aligned}
Z_z &= E\frac{\partial w}{\partial z} = E\left[a + a_1 x + a_2 y + (b_1 x + b_2 y) z\right], \\
Z_x &= \mu\left(\frac{\partial w}{\partial x} + \frac{\partial u}{\partial z}\right) \\
&= \mu\left[\beta y - b_2(2+\eta)xy - \frac{\eta b_1 x^2}{2} + b_1\left(\frac{\eta}{2} - 1\right)y^2 - \frac{\partial \Omega}{\partial x}\right],
\end{aligned}
$$

and changing x to y, a_1, a_2 to b_1, b_2, β to $-\beta$,

100) $$Z_y = \mu\left[-\beta x - b_1(2+\eta)xy - \frac{\eta b_2 y^2}{2} + b_2\left(\frac{\eta}{2} - 1\right)x^2 + \frac{\partial \Omega}{\partial y}\right].$$

The condition 69) at the cylindrical surface gives

101)
$$
\begin{aligned}
&\frac{\partial \Omega}{\partial x}\cos(nx) + \frac{\partial \Omega}{\partial y}\cos(ny) = \frac{\partial \Omega}{\partial n} \\
&= \left[-\beta y + b_2(2+\eta)xy + \frac{\eta b_1 x^2}{2} + b_1\left(1 - \frac{\eta}{2}\right)y^2\right]\cos(nx) \\
&+ \quad \left[\beta x + b_1(2+\eta)xy + \frac{\eta b_2 y^2}{2} + b_2\left(1 - \frac{\eta}{2}\right) \quad\right]\cos(ny) \equiv f(x, y).
\end{aligned}
$$

184. Determination of Function for Particular Cases. Torsion. We may easily show that a function harmonic within a certain plane contour, and on the contour satisfying the condition,

101) $$\frac{\partial \Omega}{\partial n} = f(x, y)$$

is uniquely determined. For if there be two functions Ω_1 and Ω_2 satisfying the conditions, the difference $V = \Omega_1 - \Omega_2$ is harmonic, and on the contour

$$\frac{\partial V}{\partial n} = \frac{\partial \Omega_1}{\partial n} - \frac{\partial \Omega_2}{\partial n} = 0.$$

Now by Green's theorem, § 137, 55),

$$102)\qquad \int\!\!\int\left\{\left(\frac{\partial V}{\partial x}\right)^2+\left(\frac{\partial V}{\partial y}\right)^2\right\}dx\,dy$$

$$=-\int\!\!\int V\frac{\partial V}{\partial n}\,ds-\int\!\!\int\!\!\int V\varDelta V\,dx\,dy,$$

and both integrals on the right vanish. Accordingly, as in the demonstration of Dirichlet's Principle,

$$\frac{\partial V}{\partial x}=\frac{\partial V}{\partial y}=0,\quad V=\Omega_1-\Omega_2=const.$$

But we have the condition $\Omega_0=0$, so that the difference V is zero, and the solution is unique.

If the contour is a circle, the solution is immediate, by the aid of circular harmonics. For the function Ω being developed in a series of such,

$$103)\qquad \Omega=\sum_0^\infty(A_n r^n\cos n\varphi+B_n r^n\sin n\varphi)$$

we have

$$104)\qquad \frac{\partial\Omega}{\partial n}=\frac{\partial\Omega}{\partial r}=\sum_0^\infty n\,R^{n-1}(A_n\cos n\varphi+B_n\sin n\varphi)$$

so that if $f(x, y)$ is given developed as a trigonometric series,

$$105)\qquad f(x, y)=\sum_0^0(C_n\cos n\varphi+D_n\sin\varphi)$$

we must have,

$$106)\qquad A_n=\frac{C_n}{nR^{n-1}},\quad B_n=\frac{D_n}{nR^{n-1}}$$

which determines Ω.

It may also be proved by the principles of the conformal representation of plane areas, that the problem can be solved for any contour whose area can be conformally represented upon a circle.

Let us now put

$$107)\qquad \Omega=\beta V+b_1V_1+b_2V_2,$$

so that if in the whole cross-section,

$$108)\qquad \varDelta V^1=\varDelta V_2=\varDelta V_3=0$$

and on the contour,

$$\frac{\partial V}{\partial n}=-y\cos(nx)+x\cos(ny),$$

$$109)\qquad \frac{\partial V_1}{\partial n}=\left[\frac{\eta}{2}x^2+\left(1-\frac{\eta}{2}\right)y^2\right]\cos(nx)+(2+\eta)xy\cos(ny),$$

$$\frac{\partial V_2}{\partial n}=(2+\eta)xy\cos(nx)+\left[\frac{\eta}{2}y^2+\left(1-\frac{\eta}{2}\right)x^2\right]\cos(ny),$$

V, V_1, V_2 are uniquely determined, each vanishing at the origin. The origin is any point on the line of centers of mass of the cross-sections.

The most general solution of de Saint-Venant's problem now contains six constants, $a, a_1, a_2, \beta, b_1, b_2$. We may examine separately the corresponding simple strains, by putting in each case all the constants but one equal to zero.

I. $a \neq 0$.

$$\text{110)} \quad \begin{aligned} u &= -\eta a x, \quad v = -\eta a y, \quad w = a z, \\ Z_z &= E a, \quad Z_x = Z_y = 0. \end{aligned}$$

This is a stretch-squeeze of ratio η, which has been already treated in § 176.

II. $\beta \neq 0$.

$$\text{111)} \quad u = \beta y z, \quad v = -\beta x z, \quad w = \beta V.$$

These equations represent a torsion, whose rotation

$$\omega_z = \frac{1}{2}\left(\frac{\partial v}{\partial x} - \frac{\partial u}{\partial y}\right) = -\beta z,$$

is proportional to the distance from the origin, or fixed section.

We have $u = v = w = 0$ for $x = y = 0$ the line of centers. For the stresses we find

$$\text{112)} \quad Z_z = E\frac{\partial w}{\partial z} = 0, \quad Z_x = \mu\beta\left(y + \frac{\partial V}{\partial x}\right), \quad Z_y = \mu\beta\left(-x + \frac{\partial V}{\partial y}\right),$$

so that the stress on any cross-section is completely tangential.

If the contour of the bar is circular, we have

$$\cos(nx) : \cos(ny) = x : y,$$

so that $\frac{\partial V}{\partial n} = 0$ and $V = w = 0$, and accordingly we see that the plane cross-sections remain planes. For any other form of cross-section than a circle, V does not vanish, so that the cross-sections *buckle* into non-plane surfaces. This buckling was neglected in the old Bernoulli-Euler theory of beams, and constitutes, as was shown by de Saint-Venant, a serious defect in that theory, which is still largely used by engineers.

In order to produce this torsion, we should apply at the end cross-section the stresses

$$\text{113)} \quad Z_x = X_z = \mu\beta y, \quad Z_y = Y_z = -\mu\beta x,$$

that is at every point a tangential stress perpendicular to the radius vector of the point in the plane and proportional to its distance from the line of centers. As in practice it would be impossible to apply stresses varying in this manner, we make use, in this and the other cases, of de Saint-Venant's principle of *equipollent loads*, viz:

If the cross-section is small with respect to the length, stresses applied to a body near the ends produce approximately the same effect if their statical resultants and moments are the same.

Consequently we may apply to the end faces or to the convex surface near them the forces and couples,

$$X = \iint X_z dS = \mu\beta \iint y dS = 0,$$

$$Y = \iint Y_z dS = -\mu\beta \iint x dS = 0,$$

$$Z = \iint Z_z dS = 0,$$

114)
$$L = \iint (y Z_z - z Y_z) dS = \mu\beta \iint xz dS = 0,$$

$$M = \iint (z X_z - x Z_z) dS = \mu\beta \iint yz dS = 0,$$

$$N = \iint (x Y_z - y X_z) dS = -\mu\beta \iint (x^2 + y^2) dS$$

$$= -\mu\beta \int_0^R \int_0^{2\pi} r^3 dr\, d\varphi = -\frac{\mu\beta\pi R^4}{2}.$$

The twisting couple divided by the rotation per unit of length,

115)
$$\frac{Nz}{\omega_z} = \frac{\mu\pi R^4}{2},$$

is proportional to the fourth power of the radius. This is the law announced by Coulomb, in his work on the torsion-balance. This factor of the applied couple multiplying the twist per unit length is called the torsional rigidity of the prism. Thus the shear-modulus μ may be determined by experiments on torsion.

For other contours than the circle it is convenient to introduce into the problem instead of V its so-called conjugate function Ψ, defined by the equations,

116)
$$\frac{\partial V}{\partial x} = \frac{\partial \Psi}{\partial y}, \quad \frac{\partial V}{\partial y} = -\frac{\partial \Psi}{\partial x}, \quad \Delta \Psi = 0.$$

If we now form the line-integral around the contour of the cross-section, of $\frac{\partial V}{\partial n}$ from 109),

$$117) \qquad \int \frac{\partial V}{\partial n} ds = \int \left(\frac{\partial V}{\partial x}\cos(nx) + \frac{\partial V}{\partial y}\cos(ny)\right) ds$$

$$= \int [-y\cos(nx) + x\cos(ny)]\, ds$$

$$= \int \left(\frac{\partial \Psi}{\partial y}\cos(nx) - \frac{\partial \Psi}{\partial x}\cos(ny)\right) ds,$$

since if we circulate anti-clockwise (Fig. 152),

$$ds\cos(nx) = dy, \quad ds\cos(ny) = -dx$$

the integrals become

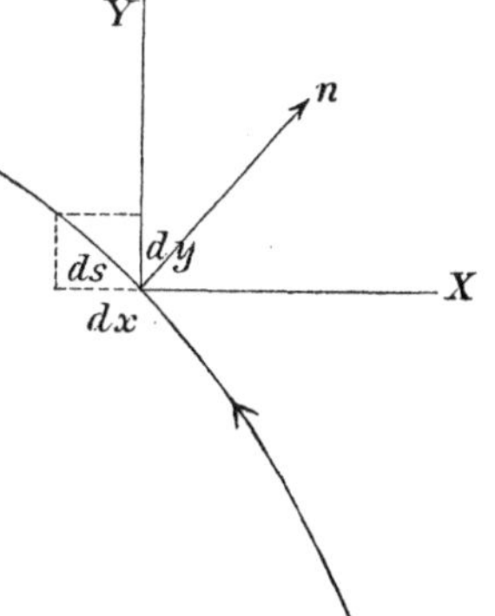

Fig. 152.

$$118) \quad I = \int \left(\frac{\partial \Psi}{\partial x} dx + \frac{\partial \Psi}{\partial y} dy\right)$$

$$= -\int (y\,dy + x\,dx),$$

and since both differentials are perfect, integrating,

$$119) \qquad \Psi = C - \frac{1}{2}(x^2 + y^2).$$

Accordingly if we can determine a harmonic function Ψ which *on the contour* shall have the value 119), the problem is solved. For example take the functions,

$$\Psi = A(x^2 - y^2), \quad V = -2Axy,$$

giving

$$\Psi + \frac{1}{2}(x^2 + y^2) = \left(A + \frac{1}{2}\right)x^2 + \left(\frac{1}{2} - A\right)y^2$$

equal to a constant on the ellipse

$$\frac{x^2}{a^2} + \frac{y^2}{b^2} = 1,$$

if we take

$$\left(A + \frac{1}{2}\right)a^2 = \left(\frac{1}{2} - A\right)b^2,$$

giving

$$A = \frac{1}{2}\frac{b^2 - a^2}{b^2 + a^2}, \quad V = \frac{a^2 - b^2}{a^2 + b^2}xy.$$

We thus obtain

$$u = \beta y z, \quad v = -\beta x z, \quad w = \frac{\beta(a^2 - b^2)}{a^2 + b^2} xy.$$

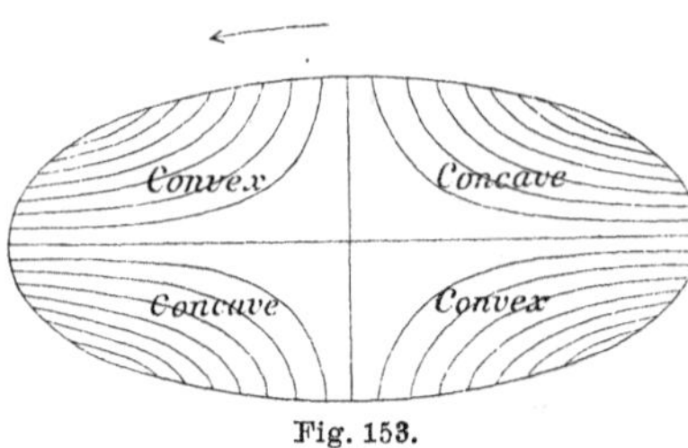

Fig. 153.

The curves of equal longitudinal displacement, or contour-lines of the buckled sections, are equilateral hyperbolas (Fig. 153). The stresses are, by 112),

$$X_z = \frac{2\mu\beta a^2}{a^2 + b^2} y, \quad Y_z = -\frac{2\mu\beta b^2}{a^2 + b^2} x.$$

The twisting couple is

120) $$N = \iint (x Y_z - y X_z)\, dS = -\frac{2\mu\beta}{a^2 + b^2} \iint (a^2 y^2 + b^2 x^2)\, dx\, dy,$$
$$= -\pi \mu \beta \frac{a^3 b^3}{a^2 + b^2}.$$

In a similar we may deal with an equilateral triangular prism. If a be the radius of the inscribed circle, the equation of the boundary may be written,

$$(x - a)(x - y\sqrt{3} + 2a)(x + y\sqrt{3} + 2a) = 0,$$

or

$$x^3 - 3xy^2 + 3a(x^2 + y^2) - 4a^3 = 0.$$

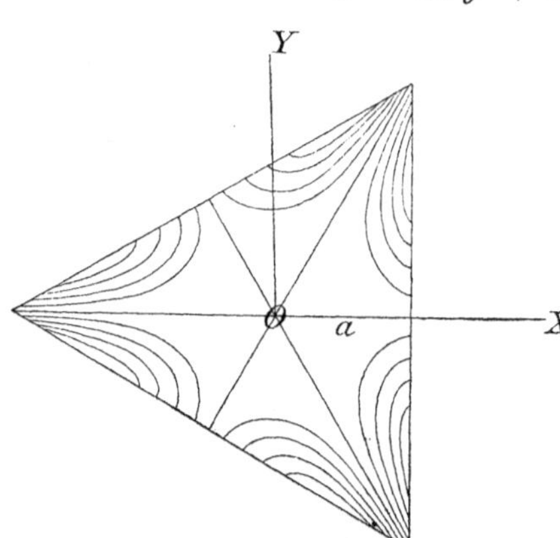

Fig. 154.

The function $A(x^3 - 3xy^2)$ is a circular harmonic, and

$$A(x^3 - 3xy^2) + \frac{1}{2}(x^2 + y^2)$$

is constant on the boundary if we take

$$A = \frac{1}{6a}, \quad \Psi = \frac{1}{6a}(x^3 - 3xy^2),$$
$$V = \frac{1}{6a}(y^3 - 3x^2 y).$$

The curves of equal distortion are shown in Fig. 154.

For any contour we have for the couple, from 112)

121) $$N = \iint (x Y_z - y X_z)\, dS = -\mu\beta \iint (x^2 + y^2)\, dS$$
$$+ \mu\beta \iint \left(x \frac{\partial V}{\partial y} - y \frac{\partial V}{\partial x}\right) dS$$
$$= -\mu\beta I + \mu\beta \iint \left\{x \frac{\partial V}{\partial y} - y \frac{\partial V}{\partial x}\right\} dS,$$

where I is the moment of inertia of the section about the Z-axis. If we call

122) $$\iint\left(x\frac{\partial V}{\partial y}-y\frac{\partial V}{\partial x}\right)dS=(1-q)\iint(x^2+y^2)\,dS,$$

then

123) $$N=-q\mu\beta I$$

and the moment per unit twist per unit length $q\mu I$ is called the torsional rigidity of the bar, and q is de Saint-Venant's torsion-factor. For the circle $q=1$, for the ellipse $q=\frac{4a^2b^2}{(a^2+b^2)^2}$, since $I=\frac{\pi}{4}ab(a^2+b^2)$. If $S=\pi ab$ is the area of the ellipse, the rigidity may be written,

$$q\mu I=\frac{\mu}{4\pi^2}\frac{S^4}{I}$$

and by a generalization of this formula, de Saint-Venant writes

$$qI=\varkappa\frac{S^4}{I}.$$

Having solved the problem for a great variety of sections, he found that, when the section is not very elongated, and has no reentrant angles, $\varkappa$ varies only between .0228 and .026, its value for the ellipse being .02533. We may thus put in practice

$$qI=\frac{1}{40}\frac{S^4}{I},$$

obtaining a most valuable engineering formula. Considering the dimensions of S and I, we see that for similar cross-sections, the rigidity varies as the square of the area of the section, as stated by Coulomb, but for different sections the results differ much from those of the old theory, in which q was supposed to be unity.

185. Flexion. For the third case we put

III. $a_1 \neq 0$,

124) $$u=-\frac{a_1}{2}\{z^2+\eta(x^2-y^2)\},\quad v=-\eta a_1xy,\quad w=a_1xz.$$

125) $$Z_z=E\frac{\partial w}{\partial z}=Ea_1x,\quad Z_x=Z_y=0.$$

The force on any section,

$$Z=\iint Z_z dS=Ea_1\iint x\,dx\,dy=0$$

since the origin is the center of mass of the section. In this case the couples L and N vanish, while

$$126)\quad M = \iint (zX_z - xZ_z)\,dS = -Ea_1 \iint x^2\,dx\,dy = -Ea_1 I_y$$

where I_y is the moment of inertia of the cross-section about the Y-axis.

The line of centers of mass $x = y = 0$ becomes strained into $x' = -\frac{a_1 z^2}{2}$, a parabola, or, since the displacement is supposed small, a circle of radius $\frac{1}{a}$. This displacement is called *uniform* flexure, for the curvature of the central line is constant. It is produced by the action of no forces, but of a couple applied at the ends. The couple is the same for all cross-sections, and is equal to the product of Young's modulus by I_y and the curvature of the central line

$$127)\quad M = EI_y \frac{\partial^2 x'}{\partial z^2}.$$

This is the theorem of the *bending moment*. Such a strain is produced in a bar when we take it in our hands and bend it by turning them outwards. If the bar has a length l from the fixed section, the deflection of the end is

$$128)\quad u = -\frac{1}{2} a l^2 = \frac{l^2 M}{2EI_y}$$

and the flexural rigidity, or moment per unit displacement per unit of length is $2EI_y$.

For a rectangular section of breadth b and height h,

$$I_y = \int_{-\frac{h}{2}}^{\frac{h}{2}} \int_{-\frac{b}{2}}^{\frac{b}{2}} x^2\,dx\,dy = \frac{bh^3}{12}.$$

For a circular section of radius R,

$$I_y = \iint x^2\,dx\,dy = \frac{1}{2} \iint (x^2 + y^2)\,dx\,dy = \frac{\pi R^4}{4}.$$

For a circular and rectangular beam of equal cross-sections, since $\pi R^2 = bh$ the ratio of stiffnesses is

$$\frac{I_{rectangle}}{I_{circle}} = \frac{bh^3}{12} \frac{4}{\pi R^4} = \frac{h^2}{3R^2}$$

and if $b = h$,

$$\frac{I_{square}}{I_{circle}} = \frac{\pi}{3} = 1.0472.$$

Since $w = a_1 xz$, a plane cross-section at a distance z from the origin remains a plane which is rotated through the angle $\frac{w}{x} = a_1 z$, cutting the XY-plane at a distance $\frac{1}{a_1}$ below the origin. Such a plane remains normal to the line of centers, and the flexure is circular (Fig. 155). If the section is rectangular, the sides $y = \pm \frac{b}{2}$ become, since $v = -\eta a_1 xy$,

$$y' = y + v = \pm \frac{b}{2}(1 - a_1 \eta x)$$

while the sides $x = \pm \frac{h}{2}$ become, since

$$u = -\frac{a_1}{2}\{z^2 + \eta(x^2 - y^2)\},$$

$$x' = \pm \frac{h}{2} - \frac{a_1}{2}\left\{z^2 + \eta\left(\frac{h^2}{4} - y^2\right)\right\},$$

or since z is constant, circles, so that the cross-section becomes like Fig. 156.

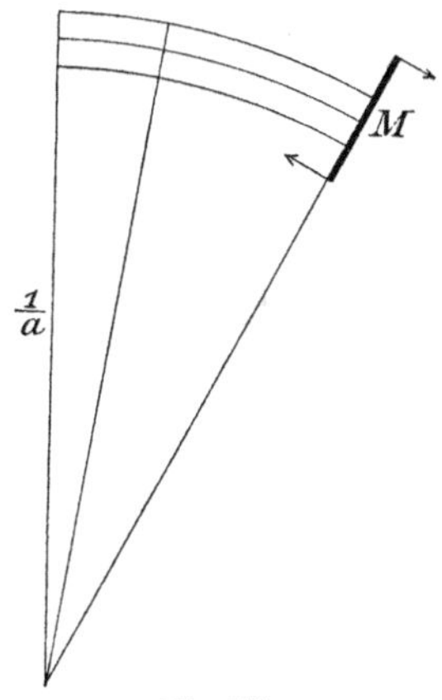

Fig. 155.

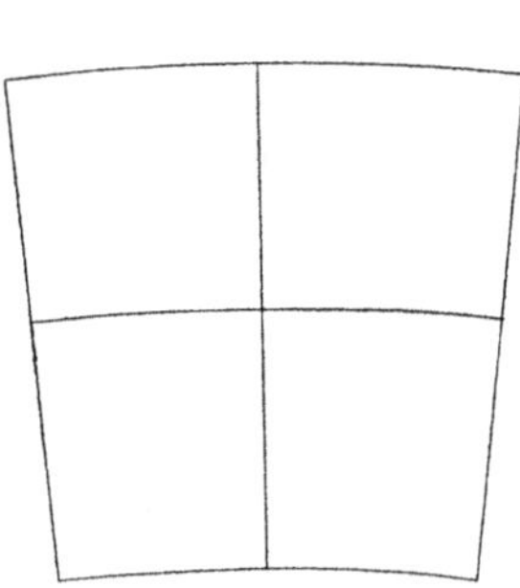

Fig. 156.

In an experiment by Cornu a bar of glass was thus bent, and points of equal vertical displacement were observed by Newton's fringes produced by a parallel plate of glass placed over the strained plate. The curves $u = const.$ are

$$z^2 - \eta y^2 = const.$$

a set of hyperbolas whose asymptotes make an angle α with the Z-axis given by $\operatorname{ctn}^2 \alpha = \eta$. By photographing the fringes and measuring the angle Cornu obtained a value of η very near to $\frac{1}{4}$, the value given by Poisson.

The case $a_2 \neq 0$ is precisely like that just treated, except that the rôles of the X and Y-axes are interchanged. We pass therefore to:

IV. $b_1 \neq 0$.

$$u = -\frac{\eta b_1}{2} z(x^2 - y^2) - \frac{b_1}{6} z^3 + b_1 z \left(\frac{\partial V_1}{\partial x}\right)_0,$$

129) $$v = -\eta b_1 xyz + b_1 z \left(\frac{\partial V_1}{\partial y}\right)_0,$$

$$w = \frac{b_1}{2} xz^2 - b_1 xy^2 + b_1 V_1 - b_1 x \left(\frac{\partial V_1}{\partial x}\right)_0 - b_1 y \left(\frac{\partial V_1}{\partial y}\right)_0,$$

where V_1 is defined by

130) $$\frac{\partial V_1}{\partial n} = \left[\frac{\eta}{2} x^2 + \left(1 - \frac{\eta}{2}\right) y^2\right] \cos(nx) + (2 + \eta) xy \cos(ny).$$

If the cross-section is symmetrical with respect to both axes of X and Y, evidently the boundary condition is satisfied by a function V_1 *even* in y, and *odd* in x, consequently $\frac{\partial V_1}{\partial y}$ is odd, and vanishes at the origin.

The line of centers is deformed into a plane curve (since v vanishes with x and y), having the equation

$$x' = -\frac{b_1}{6} z^3 + b_1 z \left(\frac{\partial V_1}{\partial x}\right)_0,$$

a cubical parabola. The strain is called a non-uniform flexure. The stresses are

$$Z_z = E\frac{\partial w}{\partial z} = Eb_1 zx,$$

131) $$X_z = \mu b_1 \left[-\frac{\eta}{2} x^2 + \left(\frac{\eta}{2} - 1\right) y^2 + \frac{\partial V_1}{\partial x}\right],$$

$$Y_z = \mu b_1 \left[-(2 + \eta) xy + \frac{\partial V_1}{\partial y}\right],$$

$$Z = \iint Z_z dS = Eb_1 z \iint x\,dx\,dy = 0,$$

$$Y = \iint Y_z dS = \mu b_1 \left[-(2 + \eta) \iint xy\,dx\,dy + \iint \frac{\partial V_1}{\partial y} dy\,dx\right] = 0,$$

the latter integral vanishing because $V(y) = V(-y)$, and the section is symmetrical.

132) $$X = \iint X_z dS = \mu b_1 \iint \left\{-\frac{\eta}{2} x^2 + \left(\frac{\eta}{2} - 1\right) y^2 + \frac{\partial V_1}{\partial x}\right\} dx\,dy$$

$$= \mu b_1 \left[-\frac{\eta}{2} I_y + \left(\frac{\eta}{2} - 1\right) I_x + 2\int V_1(x, y)\,dy\right]$$

the last integral taking the above form since $V_1(x) = -V_1(-x)$.

The moments are

$$L = \iint (yZ_z - zY_z)\,dS = Eb_1 z \iint xy\,dx\,dy - zY = 0,$$

133) $$M = \iint (zX_z - xZ_z)\,dS = z(X - Eb_1 I_y),$$

$$N = \iint (xY_z - yX_z)\,dS = \mu b_1 \iint \Big\{ -(2+\eta)x^2 y + x\frac{\partial V_1}{\partial y} + \frac{\eta}{2}x^2 y - \Big(\frac{\eta}{2} - 1\Big)y^3 - y\frac{\partial V_1}{\partial x}\Big\}\,dx\,dy = 0,$$

every integral in N vanishing on account of the symmetry.

We have as a result that to produce this strain we require a transverse force X applied at the free end, together with a couple M about the Y-axis. The transverse sections are buckled, contrary to the old theory and to the case of uniform flexure, while as there the fibres for which x is positive are in longitudinal tension, those for which it is negative in compression.

It was shown by Clebsch that the integral $\iint \frac{\partial V_1}{\partial x}\,dx\,dy$, which occurs in X and M, could be calculated without determining the function V_1 itself.

Putting in Green's formula, § 137, 55), $U = x$, we have, since

$$\frac{\partial x}{\partial x} = 1, \quad \frac{\partial x}{\partial y} = 0, \quad \Delta V_1 = 0,$$

134) $$\iint \frac{\partial V_1}{\partial x}\,dx\,dy = \int x\frac{\partial V_1}{\partial n}\,ds$$

and for this problem, taking the value of $\frac{\partial V_1}{\partial n}$ from 130)

135) $$\iint \frac{\partial V_1}{\partial x}\,dx\,dy = \int \{\mathfrak{X}\cos(nx) + \mathfrak{Y}\cos(ny)\}\,ds$$

where

$$\mathfrak{X} = \frac{\eta}{2}x^3 + \Big(1 - \frac{\eta}{2}\Big)xy^2, \quad \mathfrak{Y} = (2+\eta)x^2 y$$

Converting the line-integral into a surface-integral by the divergence theorem,

136) $$\iint \frac{\partial V_1}{\partial x}\,dx\,dy = \iint \Big(\frac{\partial \mathfrak{X}}{\partial x} + \frac{\partial \mathfrak{Y}}{\partial y}\Big)\,dx\,dy = \Big(\frac{5\eta}{2} + 2\Big)I_y + \Big(1 - \frac{\eta}{2}\Big)I_x.$$

Inserting this value in 132),

137) $$X = \mu b_1 (2\eta + 2) I_y = Eb_1 I_y$$

so that 133) gives, $M = 0$. Thus the moment of the applied force X and of the couple about the *origin* is zero, while the moment of the *couple* alone due to the forces Z_z symmetrically applied is $M' = -Eb_1 I_y z$. It is to be noted that while the rod is bent down by the couple M' (Fig. 157), the tendency of the force X is to pull the end up. We see also that $\frac{\partial x'}{\partial z}$ for the line of centers does not vanish in general at the origin, being equal to $b_1 \left(\frac{\partial V_1}{\partial x}\right)_0$.[1]

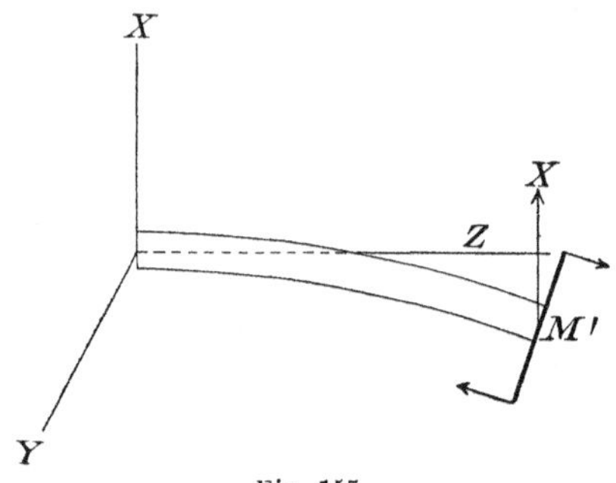

Fig. 157.

The solution of the practical problem of the deflection of a beam by a transverse force applied at one end is obtained by a combination of the results for uniform and non-uniform flexure, as just investigated. Since the former requires a couple which is the same for all sections, the latter one which is proportional to z, by a suitable combination of the two with opposite signs, we may make the couple upon the end section equal to zero, so that we have to apply only a force. Since the couple due to the uniform flexure is $-Ea_1 I_y$, and that due to the non-uniform flexure $-Eb_1 I_y z$, if we put $a_1 + b_1 l = 0$, where l is the length of the beam, there will be no couple to be applied to the end of the beam, but only the force X. Determining b_1 from 137) and adding eqnations 124) and 129), we obtain the shifts,

$$u = \frac{X}{2EI_y}\left\{\eta(l-z)(x^2-y^2) + \left(l - \frac{z}{3}\right)z^2 + 2z\left(\frac{\partial V_1}{\partial x}\right)_0\right\},$$

$$138)\quad v = \frac{X}{EI_y}\left\{\eta(l-z)xy + z\left(\frac{\partial V_1}{\partial y}\right)_0\right\},$$

$$w = \frac{X}{EI_y}\left\{\left(\frac{z}{2} - l\right)xz - xy^2 + V_1 - x\left(\frac{\partial V_1}{\partial x}\right)_0 - y\left(\frac{\partial V_1}{\partial y}\right)_0\right\}.$$

The equation of the central line is

$$139)\qquad x' = \frac{X}{2EI_y}\left\{\left(l - \frac{z}{3}\right)z^2 + 2z\left(\frac{\partial V_1}{\partial x}\right)_0\right\}$$

and the deflection of the end of the beam

$$\frac{X}{EI_y}\left\{\frac{l^3}{3} + l\left(\frac{\partial V_1}{\partial x}\right)_0\right\}.$$

1) The terms in $\left(\frac{\partial V_1}{\partial x}\right)_0$ denote merely a rigid body rotation.

The stresses to be applied are the force

$$140)\qquad X = -EI_y \frac{d^3 x'}{dz^3},$$

and the couple

$$141)\qquad M = X(l - z) = EI_y \frac{d^2 x'}{dz^2}.$$

Thus we still have the theorem of the bending moment. The meaning of equation 141) is seen by considering the stresses across a section of the beam at any distance z from the origin, and noticing that if the beam were cut at this section in order to hold it in equilibrium we should have to apply to the two cut ends couples as shown in Fig. 158, together with equal and opposite forces X, the latter being independent of the position of the cut, the former proportional to the distance from the free end of the beam, which is also proportional to the curvature of the central line.

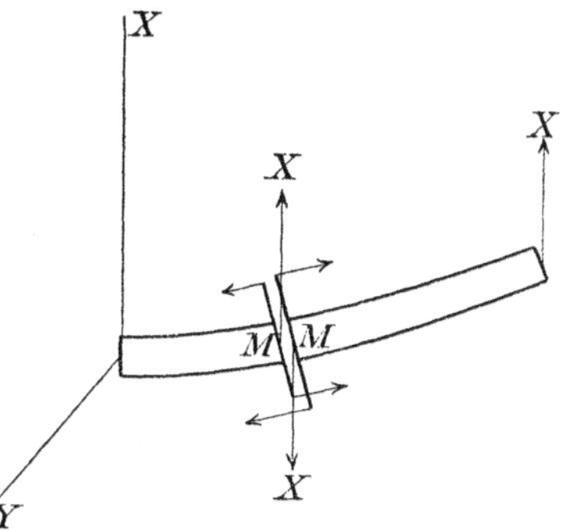

Fig. 158.

We shall close this subject by the determination of the function V_1 for the case of an elliptic cross-section. If the equation of the ellipse is

$$\frac{x^2}{a^2} + \frac{y^2}{b^2} = 1,$$

we have

$$\cos(nx) : \cos(ny) = \frac{x}{a^2} : \frac{y}{b^2}$$

and the equation 130) becomes

$$b^2 x \frac{\partial V_1}{\partial x} + a^2 y \frac{\partial V_1}{\partial y} = \left[\frac{\eta}{2} x^2 + \left(1 - \frac{\eta}{2}\right) y^2\right] b^2 x + (2 + \eta) a^2 x y^2.$$

As in the case of torsion, a solution is given by a circular harmonic of the third degree,

$$V_1 = Cx + D(x^3 - 3xy^2)$$

if we have

$$b^2 x\{C + 3D(x^2 - y^2)\} - 6Da^2 xy^2$$
$$= b^2\left\{\frac{\eta}{2} x^3 + \left(1 - \frac{\eta}{2}\right) xy^2\right\} + (2 + \eta) a^2 xy^2.$$

Dividing through by x, transposing and putting the coefficients of x^2, y^2 and 1 respectively proportional to $\frac{1}{a^2}$, $\frac{1}{b^2}$, -1, we have

$$\left(\frac{\eta}{2} - 3D\right) a^2 = \left\{3Db^2 + 6Da^2 + \left(1 - \frac{\eta}{2}\right) b^2 + (2 + \eta) a^2\right\} = C$$

which are satisfied by

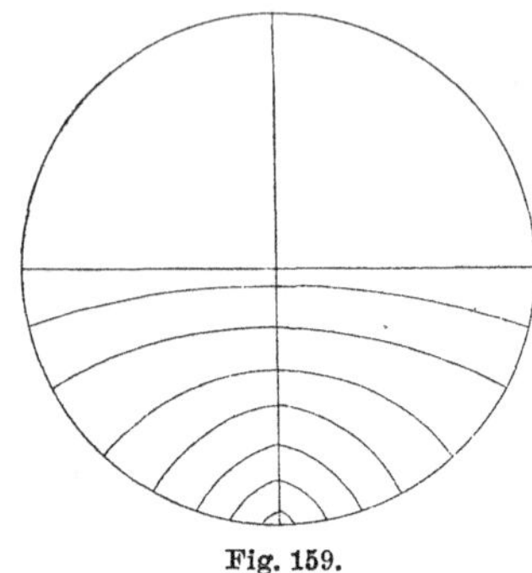

Fig. 159.

$$D = \frac{\left(\frac{\eta}{2} - 1\right) b^2 - \left(2 + \frac{\eta}{2}\right) a^2}{9a^2 + 3b^2},$$

$$C = \frac{\eta}{2} a^2 - 3Da^2.$$

We find the buckling of the sections by 129),

$$w = \frac{X}{EI_y}\left[\frac{xz^2}{2} - xy^2 + D(x^3 - 3xy^2)\right].$$

In the case of a circular beam, we have $D = -\frac{1}{4}$, and the curves of equal w are

$$x\left\{\frac{z^2}{2} - \frac{1}{4}(x^2 + y^2)\right\} = const.,$$

the contour lines of which are shown in Fig. 159.

CHAPTER XI.

HYDRODYNAMICS.

186. Equations of Motion. The equations of hydrostatics 4) § 177 being

$$
1)\qquad
\begin{aligned}
\varrho X - \frac{\partial p}{\partial x} &= 0,\\
\varrho Y - \frac{\partial p}{\partial y} &= 0,\\
\varrho Z - \frac{\partial p}{\partial z} &= 0,
\end{aligned}
$$

where X, Y, Z are the components of the applied forces per unit mass, we may obtain the equations of motion by d'Alembert's Principle.

Suppose the velocity at any point in a perfect fluid of density ϱ is a vector q whose components u, v, w are uniform, continuous differentiable functions of the point x, y, z and the time t. (The notation is now changed from that of Chapter IX where u, v, w denoted *displacements.*) Then if we consider the motion of the fluid contained in an element of volume $d\tau$ of mass $dm = \varrho d\tau$, we have the effective forces

$$
\begin{aligned}
&\varrho d\tau \frac{d^2x}{dt^2} = \varrho d\tau \frac{du}{dt},\\
2)\qquad &\varrho d\tau \frac{d^2y}{dt^2} = \varrho d\tau \frac{dv}{dt},\\
&\varrho d\tau \frac{d^2z}{dt^2} = \varrho d\tau \frac{dw}{dt},
\end{aligned}
$$

and these are to be subtracted from the applied forces

$$\varrho X d\tau, \quad \varrho Y d\tau, \quad \varrho Z d\tau,$$

and introduced in 1). Consequently we have the equations of motion

$$
\begin{aligned}
&\varrho \left\{ X - \frac{du}{dt} \right\} = \frac{\partial p}{\partial x},\\
3)\qquad &\varrho \left\{ Y - \frac{dv}{dt} \right\} = \frac{\partial p}{\partial y},\\
&\varrho \left\{ Z - \frac{dw}{dt} \right\} = \frac{\partial p}{\partial z}.
\end{aligned}
$$

Now by the ordinary derivative $\frac{du}{dt}$ is meant the rate of change of velocity of a *particular particle* as it moves about. If we have any function F pertaining to a particular particle we may write its derivative

$$4)\qquad \frac{dF}{dt} = \frac{\partial F}{\partial t} + \frac{\partial F}{\partial x}\frac{dx}{dt} + \frac{\partial F}{\partial y}\frac{dy}{dt} + \frac{\partial F}{\partial z}\frac{dz}{dt},$$

where $\frac{\partial F}{\partial t}$ would be the rate of change of F if the particle were at rest. The derivatives $\frac{dx}{dt}, \frac{dy}{dt}, \frac{dz}{dt}$ are the velocity components of the particle, u, v, w. Accordingly we have

$$5)\qquad \frac{dF}{dt} = \frac{\partial F}{\partial t} + u\frac{\partial F}{\partial x} + v\frac{\partial F}{\partial y} + w\frac{\partial F}{\partial z}.$$

We shall call this mode of differentiation particle differentiation.[1])

Introducing this terminology, dividing by ϱ and transposing, our equations of motion 3) become

$$
\begin{aligned}
&\frac{\partial u}{\partial t} + u\frac{\partial u}{\partial x} + v\frac{\partial u}{\partial y} + w\frac{\partial u}{\partial z} = X - \frac{1}{\varrho}\frac{\partial p}{\partial x},\\
6)\qquad &\frac{\partial v}{\partial t} + u\frac{\partial v}{\partial x} + v\frac{\partial v}{\partial y} + w\frac{\partial v}{z} = Y - \frac{1}{\varrho}\frac{\partial p}{\partial y},\\
&\frac{\partial w}{\partial t} + u\frac{\partial w}{\partial x} + v\frac{\partial w}{\partial y} + w\frac{\partial w}{\partial z} = Z - \frac{1}{\varrho}\frac{\partial p}{\partial z}.
\end{aligned}
$$

1) In most English books the symbol $\frac{DF}{Dt}$, used by Stokes, is used for particle differentiation because of the very objectionable practice of making no distinction in the symbol for ordinary and partial differentiation.

If we consider any closed surface fixed in space, expressing the fact that the increase of mass of the fluid contained therein is represented by the mass of fluid which flows into the surface, we shall obtain an additional equation. The velocity being q, the volume of fluid entering through an area dS in unit time is, as in § 169, equation 78), $q\cos(q, n)\,dS$, and the mass, $\varrho q\cos(q, n)\,dS$. We have therefore for the total amount entering in unit time

$$7)\quad \iint \varrho q \cos(q, n)\,dS = \iint \varrho\{u\cos(nx) + v\cos(ny) + w\cos(nz)\,dS$$
$$= -\iiint\left\{\frac{\partial(\varrho u)}{\partial x} + \frac{\partial(\varrho v)}{\partial y} + \frac{\partial(\varrho w)}{\partial z}\right\}d\tau.$$

But this is equal to the increase of mass per unit of time,

$$8)\quad \frac{\partial m}{\partial t} = \frac{\partial}{\partial t}\iiint \varrho\,d\tau = \iiint \frac{\partial \varrho}{\partial t}\,d\tau,$$

for the volume of integration is fixed, that is, independent of the time, consequently we may differentiate under the integral sign. Writing this equal to the volume integral in 7) and transposing,

$$9)\quad \iiint\left\{\frac{\partial \varrho}{\partial t} + \frac{\partial(\varrho u)}{\partial x} + \frac{\partial(\varrho v)}{\partial y} + \frac{\partial(\varrho w)}{\partial z}\right\}d\tau = 0.$$

Since this holds true for any volume whatever the integrand must vanish, so that we have

$$10)\quad \frac{\partial \varrho}{\partial t} + \frac{\partial(\varrho u)}{\partial x} + \frac{\partial(\varrho v)}{\partial y} + \frac{\partial(\varrho w)}{\partial z} = 0,$$

which is known as the Equation of Continuity.

Performing the differentiations we have

$$11)\quad \frac{\partial \varrho}{\partial t} + u\frac{\partial \varrho}{\partial x} + v\frac{\partial \varrho}{\partial y} + w\frac{\partial \varrho}{\partial z} + \varrho\left\{\frac{\partial u}{\partial x} + \frac{\partial v}{\partial y} + \frac{\partial w}{\partial z}\right\} = 0,$$

or in the notation of particle differentiation,

$$12)\quad \frac{1}{\varrho}\frac{d\varrho}{dt} + \frac{\partial u}{\partial x} + \frac{\partial v}{\partial y} + \frac{\partial w}{\partial z} = 0.$$

If now we fix our attention upon a small portion of the fluid of volume V *as it moves,* its mass will be constant, say, $m = \varrho V = const.$ By logarithmic differentiation,

$$13)\quad \frac{1}{\varrho}\frac{d\varrho}{dt} + \frac{1}{V}\frac{dV}{dt} = 0,$$

so that the expression

$$14)\quad \frac{\partial u}{\partial x} + \frac{\partial v}{\partial y} + \frac{\partial w}{\partial z} = -\frac{1}{\varrho}\frac{d\varrho}{dt} = \frac{1}{V}\frac{dV}{dt},$$

that is, the divergence of the velocity is the time rate of increase of volume per unit volume. This corresponds with the expression for the dilatation found in equation 75) § 169, the divergence, which we shall still call σ, being now the *time rate* of dilatation. Accordingly the equation of continuity is purely kinematical in character and expresses the conservation of mass of every part of the fluid.

If the fluid is incompressible, ϱ is constant, and consequently

15) $$\sigma = \frac{\partial u}{\partial x} + \frac{\partial v}{\partial y} + \frac{\partial w}{\partial z} = 0,$$

that is, the velocity of an incompressible fluid is a solenoidal vector. This is the property that give the name to such vectors, and we see, as in § 117, that the flux across every cross-section of a tube of flow is constant.

Besides the three dynamical and one kinematical equation there will be a physical equation involving the nature of the fluid, giving the relation connecting the density with the pressure,

$$\varrho = f(p),$$

making five equations to determine the five functions u, v, w, p, ϱ of the four variables x, y, z, t.

We have here made use of two distinct methods. In one we fix our attention on a definite point in space and consider what takes place there as different particles of fluid pass through it. This is called the statistical method, for by the statistics of all points we get a complete statement of the motion. This method is commonly associated with the name of Euler and the equations 6) are called the Eulerian equations of motion. The second method consists in fixing our attention upon a given particle and following it in its travels. In this we use the notation of ordinary derivatives. This is called the historical or Lagrangian method. Obviously if we know the history of all particles we also have a complete representation of the motion. Both methods are due to Euler. We shall not here make use of the Lagrangian equations and shall therefore not write them down. The student will find them in the usual treatises on Hydrodynamics of which Lamb's and Basset's Hydrodynamics, Kirchhoff's Dynamik and Wien's Hydrodynamik may be especially commended.

187. Hamilton's Principle. We shall now deduce the equations by means of Hamilton's Principle.

The kinetic energy of the fluid contained in an element $d\tau$ being $\frac{1}{2}\varrho d\tau$ times the square of its velocity, we have for the kinetic energy of the fluid contained in a given fixed volume,

16) $$T = \frac{1}{2}\iiint \varrho \left\{\left(\frac{dx}{dt}\right)^2 + \left(\frac{dy}{dt}\right)^2 + \left(\frac{dz}{dt}\right)^2\right\} d\tau.$$

For the potential energy, besides that due to the external forces, for which

$$17) \qquad \delta W' = -\iiint \varrho \{X\delta x + Y\delta y + Z\delta z\} d\tau,$$

we have the energy stored up by the pressure doing work in compressing the fluid, if it be not incompressible. This potential energy is the δW of equation 122) § 172. If we put for the displacements of the particle δx, δy, δz, we have

$$\delta s_x = \frac{\partial \delta x}{\partial x}, \quad \delta s_y = \frac{\partial \delta y}{\partial y}, \quad \delta s_z = \frac{\partial \delta z}{\partial z}.$$

Putting

$$P_x = P_y = P_z = -p,$$

$$T_x = T_y = T_z = 0,$$

we have for this part of the energy, sometimes called *intrinsic* energy,

$$18) \qquad \delta W'' = -\iiint p \left\{\frac{\partial \delta x}{\partial x} + \frac{\partial \delta y}{\partial y} + \frac{\partial \delta z}{\partial z}\right\} d\tau.$$

We accordingly have as the equations of Hamilton's Principle,

$$\int_{t_0}^{t_1} (\delta T - \delta W' - \delta W'')\, dt = 0,$$

$$19) \qquad \int_{t_0}^{t_1} dt \Big[\delta \iiint \frac{\varrho}{2}\left\{\left(\frac{dx}{dt}\right)^2 + \left(\frac{dy}{dt}\right)^2 + \left(\frac{dz}{dt}\right)^2\right\} d\tau$$

$$+ \iiint \varrho \{X\delta x + Y\delta y + Z\delta z\} d\tau$$

$$+ \iiint p \left\{\frac{\partial \delta x}{\partial x} + \frac{\partial \delta y}{\partial y} + \frac{\partial \delta z}{\partial z}\right\} d\tau\Big] = 0.$$

Performing the variations,

$$20) \qquad \int_{t_0}^{t_1} dt \Big[\iiint \Big\{\varrho \Big(\frac{dx}{dt}\frac{d\delta x}{dt} + \frac{dy}{dt}\frac{d\delta y}{dt} + \frac{dz}{dt}\frac{d\delta z}{dt}$$

$$+ X\delta x + Y\delta y + Z\delta z\Big)$$

$$+ p\left(\frac{\partial \delta x}{\partial x} + \frac{\partial \delta y}{\partial y} + \frac{\partial \delta z}{\partial z}\right)\Big\} d\tau\Big] = 0.$$

Integrating the first three terms by parts with respect to t we have,

$$21)\quad \iiint\left[\varrho\left(\frac{dx}{dt}\delta x+\frac{dy}{dt}\delta y+\frac{dz}{dt}\delta z\right)\Bigg/_{t_0}^{t_1}\right.$$

$$-\int_{t_0}^{t_1}dt\left\{\varrho\left(\frac{d^2x}{dt^2}\delta x+\frac{d^2y}{dt^2}\delta y+\frac{d^2z}{dt^2}\delta z-\{X\delta x+Y\delta y+Z\delta z\}\right)\right.$$

$$\left.\left.-p\left(\frac{\partial\delta x}{\partial x}+\frac{\partial\delta y}{\partial y}+\frac{\partial\delta z}{\partial z}\right)\right\}\right]d\tau=0.$$

The integrated terms vanish as usual at the limits for the time. We may now integrate the last three terms with respect to the space variables, obtaining

$$22)\quad -\iiint p\left(\frac{\partial\delta x}{\partial x}+\frac{\partial\delta y}{\partial y}+\frac{\partial\delta z}{\partial z}\right)d\tau$$

$$=\iint p\{\delta x\cos(nx)+\delta y\cos(ny)+\delta z\cos(nz)\}\,dS$$

$$+\iiint\left(\frac{\partial p}{\partial x}\delta x+\frac{\partial p}{\partial y}\delta y+\frac{\partial p}{\partial z}\right)d\tau.$$

If we assume that δx, δy, δz vanish for the particles of the fluid at the bounding surface, the surface integral vanishes. We therefore have, collecting the terms accordiug to δx, δy, δz,

$$23)\quad \int_{t_0}^{t_1}dt\left[\iiint\left(\varrho\frac{d^2x}{dt^2}-\varrho X+\frac{\partial p}{\partial x}\right)\delta x+\left(\varrho\frac{d^2y}{dt^2}-\varrho Y+\frac{\partial p}{\partial y}\right)\delta y\right.$$

$$\left.+\left(\varrho\frac{d^2z}{dt^2}-\varrho Z+\frac{\partial p}{\partial z}\right)\delta z\right]d\tau=0.$$

By the usual reasoning the coefficients of δx, δy δz must vanish, giving as the equations of motion 6).

188. Equation of Activity. Subtracting from both sides of the first equation 6) the quantity

$$24)\quad \frac{\partial}{\partial x}\left(\frac{1}{2}q^2\right)=\frac{1}{2}\frac{\partial}{\partial x}(u^2+v^2+w^2)=u\frac{\partial u}{\partial x}+v\frac{\partial v}{\partial x}+w\frac{\partial w}{\partial x},$$

we obtain

$$25)\quad \frac{\partial u}{\partial t}+w\left(\frac{\partial u}{\partial z}-\frac{\partial w}{\partial x}\right)-y\left(\frac{\partial v}{\partial x}-\frac{\partial u}{\partial y}\right)=X-\frac{1}{\varrho}\frac{\partial p}{\partial\varrho}-\frac{\partial\left(\frac{1}{2}q^2\right)}{\partial x}.$$

If the applied forces are conservative and derived from a potential V, the right-hand member is the derivative,

$$-\frac{\partial}{\partial x}\left(V+P+\frac{1}{2}q^2\right),$$

where P has the value of § 177, 5). In the left-hand member occur in this and the companion equations the expressions

$$
\begin{aligned}
&\frac{1}{2}\left(\frac{\partial w}{\partial y}-\frac{\partial v}{\partial z}\right)=\xi,\\
26)\qquad &\frac{1}{2}\left(\frac{\partial u}{\partial z}-\frac{\partial w}{\partial x}\right)=\eta,\\
&\frac{1}{2}\left(\frac{\partial v}{\partial x}-\frac{\partial u}{\partial y}\right)=\zeta.
\end{aligned}
$$

which represent the components of a vector ω, which by comparison with the expressions for ω_x, ω_y, ω_z in § 169, 77) is seen to be the angular velocity of rotation of a particle or the *vorticity*. Let us accordingly write, putting

$$V+\frac{1}{2}q^2=-U,$$

$$
\begin{aligned}
&\frac{\partial u}{\partial t}+2(w\eta-v\zeta)=\frac{\partial U}{\partial x}-\frac{1}{\varrho}\frac{\partial p}{\partial x},\\
27)\qquad &\frac{\partial v}{\partial t}+2(u\zeta-w\xi)=\frac{\partial U}{\partial y}-\frac{1}{\varrho}\frac{\partial p}{\partial y},\\
&\frac{\partial w}{\partial t}+2(v\xi-u\eta)=\frac{\partial U}{\partial z}-\frac{1}{\varrho}\frac{\partial p}{\partial z}.
\end{aligned}
$$

Multiplying respectively by ϱu, ϱv, ϱw and adding, the terms in ξ, η, ζ disappear and integrating over any volume, we have

$$
\begin{aligned}
28)\qquad &\iiint\varrho\left\{u\frac{\partial u}{\partial t}+v\frac{\partial v}{\partial t}+w\frac{\partial w}{\partial t}\right\}d\tau\\
&=\iiint\left[\varrho\left\{u\frac{\partial U}{\partial x}+v\frac{\partial U}{\partial y}+w\frac{\partial U}{\partial z}\right\}-\left\{u\frac{\partial p}{\partial x}+v\frac{\partial p}{\partial y}+w\frac{\partial p}{\partial z}\right\}\right]d\tau\\
&=-\iint[u\cos(nx)+v\cos(ny)+w\cos(nz)](U\varrho-p)\,dS\\
&\quad-\iiint\left[U\left(\frac{\partial(\varrho u)}{\partial u}+\frac{\partial(\varrho v)}{\partial y}+\frac{\partial(\varrho w)}{\partial z}\right)-p\left(\frac{\partial u}{\partial x}+\frac{\partial v}{\partial y}+\frac{\partial w}{\partial z}\right)\right]d\tau.
\end{aligned}
$$

Introducing the value of U and of its multiplier in the last integrand, which by equation 10) is equal to $-\frac{\partial\varrho}{\partial t}$, and transposing, we obtain,

$$
\begin{aligned}
29)\qquad &\iiint\left\{\varrho\frac{\partial}{\partial t}\left(\frac{q^2}{2}\right)+\left(V+\frac{q^2}{2}\right)\frac{\partial\varrho}{\partial t}\right\}d\tau\\
&=\iint q\cos(qn)\left\{\varrho\left(V+\frac{q^2}{2}\right)+p\right\}dS\\
&\quad+\iiint p\left\{\frac{\partial u}{\partial x}+\frac{\partial v}{\partial y}+\frac{\partial w}{\partial z}\right\}d\tau.
\end{aligned}
$$

If the applied forces are independent of the time, $\frac{\partial V}{\partial t}=0$, and we may write the left-hand member,

$$30)\quad \iiint \frac{\partial}{\partial t}\left(\frac{1}{2}\varrho q^2 + \varrho V\right) d\tau = \frac{\partial}{\partial t}\iiint \left(\frac{1}{2}\varrho q^2 + \varrho V\right) d\tau,$$

since the volume of integration is fixed. The first term of the integral represents the kinetic energy and the second term, the potential energy due to the applied forces. The term on the left in 29) is accordingly the rate of increase of the energy, kinetic and potential.

Of the terms on the right, the amount of matter $\varrho q \cos(qn)\, dS$ flowing through dS in unit time brings with it the energy

$$\varrho q \cos(qn)\, dS\left(V + \frac{1}{2}q^2\right),$$

so that the first part of the surface integral represents the total inflow of energy. The remaining surface integral and volume integral containing p represent the work done by the pressure, for at the surface the velocity q and the force $p\,dS$ give the activity

$$pq \cos(qn)\, dS,$$

so that the surface integral represents the activity of the pressure at the surface.

If we consider a small element of volume V, the work done in compressing it by an amount dV is as above $-p\,dV$, and the activity

$$31)\quad -p\frac{dV}{dt} = -pV\left\{\frac{\partial u}{\partial x} + \frac{\partial v}{\partial y} + \frac{\partial w}{\partial z}\right\}.$$

Putting $V = d\tau$ and integrating, we find that

$$-\iiint p\left\{\frac{\partial u}{\partial x} + \frac{\partial v}{\partial y} + \frac{\partial w}{\partial z}\right\} d\tau$$

is the activity of the pressure in producing changes of density in the whole mass. Transposing this term we find that equation 29) expresses the following: The rate of increase of energy of the fluid, both kinetic and potential, due to the external forces plus the activity of compression (production of intrinsic energy) is equal to the rate of inflow of energy plus the activity of pressure at the surface. Equation 29) is therefore the equation of activity or conservation of energy.

189. Steady Motion. Steady motion is defined as a motion which is the same at all times. Assuming that not only X, Y, Z, V but u, v, w, p, ϱ are independent of t, equations 27) for steady motion become

$$\begin{aligned} 2(w\eta - v\zeta) &= -\frac{\partial}{\partial x}\left(V + P + \frac{1}{2}q^2\right),\\ 32)\quad 2(u\zeta - w\xi) &= -\frac{\partial}{\partial y}\left(V + P + \frac{1}{2}q^2\right),\\ 2(v\xi - u\eta) &= -\frac{\partial}{\partial z}\left(V + P + \frac{1}{2}q^2\right). \end{aligned}$$

If the motion is non-vortical, the left-hand members vanish, and we immediately obtain the integral

33) $$V + P + \frac{1}{2} q^2 = const.,$$

for the expression on the left has been assumed independent of t, and by the equations is shown to be independent of x, y, z.

It is to be noticed that if we multiply equations 32) respectively by u, v, w, or by ξ, η, ζ and add, the left-hand member vanishes identically. But the operator $u\frac{\partial}{\partial x} + v\frac{\partial}{\partial y} + w\frac{\partial}{\partial z}$ denotes differentiation in the direction of the line of the vector-velocity q, or *stream-line* (see p. 333), and $\xi\frac{\partial}{\partial x} + \eta\frac{\partial}{\partial y} + \zeta\frac{\partial}{\partial z}$ differentiation in the direction of the line of the vector ω, or vortex-line. Consequently even though there is vortical motion, along a stream-line or a vortex-line the sum $V + P + \frac{1}{2} q^2$ is constant in steady motion, though its value changes as we go from one line to another.

If the fluid is incompressible $P = \frac{p}{\varrho}$, if there are no applied forces $V = 0$, and equation 33) becomes

34) $$\frac{p}{\varrho} = const. - \frac{1}{2} q^2,$$

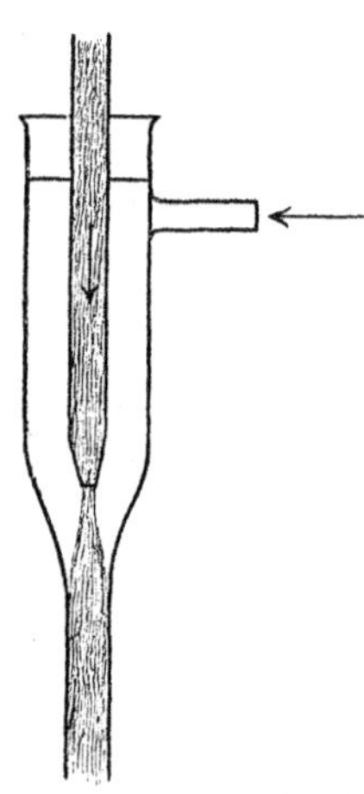

Fig. 160.

so that where the velocity is small the pressure is great and vice versa. By constricting a tube the velocity is made large and the pressure accordingly is smaller than at other parts of the tube. This is the principle of jet exhaust pumps, like that of Bunsen (Fig. 160), the air being sucked in at the narrow portion of the jet. The same principle is made use of in the Venturi water-meter. The main being reduced in diameter at a certain portion and the difference of pressure at that point and in the main being measured, the velocity is computed. If the pressure at two cross-sections S_1 and S_2 are p_1 and p_2 we have

35) $$p_1 + \frac{1}{2}\varrho q_1^2 = p_2 + \frac{1}{2}\varrho q_2^2,$$

or

36) $$p_1 - p_2 = \frac{\varrho}{2}(q_2^2 - q_1^2).$$

But by the equation of continuity, the velocity being solenoidal,

37) $$S_1 q_1 = S_2 q_2.$$

Combining this with equation 36),

$$38) \qquad p_1 - p_2 = \frac{1}{2}\varrho q_1^2\left[\left(\frac{S_1}{S_2}\right)^2 - 1\right],$$

which determines q_1 in terms of the difference of pressures. The flux in unit time is then

$$\varrho q_1 S.$$

The theorem expressed by equation 33) is known as Daniel Bernoulli's theorem.

For gases expanding isothermally

$$P = a \log \varrho = a \log p + const.$$

Consequently equation 33) becomes

$$39) \qquad a \log p + \frac{1}{2} q^2 = const.$$

This formula may be used to calculate the velocity of efflux through an orifice from a vessel containing gas under pressure. If the pressure in the vessel at a point so remote from the orifice that the air may be considered at rest is p and if the pressure of the atmosphere at the orifice where the velocity is q is p_0, we have

$$a \log p = a \log p_0 + \frac{1}{2} q^2,$$

$$40) \qquad q^2 = 2a \log \frac{p}{p_0}.$$

If the efflux is adiabatic, as in practice it nearly is, by § 178, 18)

$$P = \frac{b\varkappa\varrho^{\varkappa-1}}{\varkappa-1} = \frac{b\varkappa}{\varkappa-1}\left(\frac{p}{b}\right)^{\frac{\varkappa-1}{\varkappa}}.$$

Accordingly

$$\frac{b\varkappa}{\varkappa-1}\left(\frac{p}{b}\right)^{\frac{\varkappa-1}{\varkappa}} = \frac{b\varkappa}{\varkappa-1}\left(\frac{p_0}{b}\right)^{\frac{\varkappa-1}{\varkappa}} + \frac{1}{2}q^2,$$

$$41) \qquad q^2 = 2b^{\frac{1}{\varkappa}}\frac{\varkappa}{\varkappa-1}\cdot\left\{p^{\frac{\varkappa-1}{\varkappa}} - p_0^{\frac{\varkappa-1}{\varkappa}}\right\},$$

which is the usual formula for the efflux of gases.

If the external force is gravity $V = gz$, so that equation 33) becomes for an incompressible fluid,

$$42) \qquad \frac{p}{\varrho} + gz + \frac{1}{2}q^2 = const.$$

If we consider efflux from a reservoir whose upper free surface is so large that q is negligible, the pressure being that of the atmosphere, the z-coordinate z_1, the velocity of efflux q at a point where $z = z_2$ is given by

$$\frac{p_0}{\varrho} + g z_1 = \frac{p_0}{\varrho} + g z_2 + \frac{1}{2} q^2,$$

$$q^2 = 2g(z_1 - z_2), \tag{43}$$

or the velocity of efflux is equal to that acquired by a body falling freely from a height equal to that of the free surface to the orifice. This is Torricelli's theorem.

190. Circulation. We define the circulation along any path as the line integral of the resolved tangential velocity,

$$\varphi_{AB} = \int_A^B q \cos(q, ds)\, ds = \int_A^B (u dx + v dy + w dz), \tag{45}$$

corresponding to the circulation for displacement in § 169. By Stokes's theorem this is converted for a closed path into the surface integral

$$\iint \Big[\Big\{\frac{\partial w}{\partial y} - \frac{\partial u}{\partial z}\Big\} \cos(nx) + \Big\{\frac{\partial u}{\partial z} - \frac{\partial w}{\partial x}\Big\} \cos(ny) + \Big\{\frac{\partial v}{\partial x} - \frac{\partial u}{\partial y}\Big\} \cos(nz)\Big] dS \tag{45}$$

over any surface bounded by the path. But this is by 26),

$$2\iint \{\xi \cos(nx) + \eta \cos(ny) + \zeta \cos(nz)\}\, dS = 2\iint \omega \cos(n\omega)\, dS, \tag{46}$$

that is, the circulation around any closed contour is equal to twice the surface integral, over a cap bounded by the contour, of the resolved normal vorticity. By the definition of ξ, η, ζ, 26)

$$\frac{\partial \xi}{\partial x} + \frac{\partial \eta}{\partial y} + \frac{\partial \zeta}{\partial z} = 0 \tag{47}$$

identically, or the vorticity is a solenoidal vector. Accordingly by applying the divergence theorem to a vortex-tube, or tube whose generators are vortex-lines (lines whose tangents have the direction of ω), we find that the integral over any section of the vortex-tube is

$$\iint \omega \cos(n\omega)\, dS = \varkappa$$

a constant for the tube.

The fluid within any vortex-tube constitutes a *vortex*. If the vortex is contained in a tube of infinitesimal cross-section S, the constant $\varkappa = \omega S$ is called the *strength* of the filament. The strength of any vortex is the sum or integral of the strengths of all its filaments. If ω is finite, S cannot be zero, so that a vortex-filament cannot end except at the free surface of the fluid. We see a case of this in the vortics produced by an oar or paddle in rowing and by a spoon drawn across the surface of a cup of coffee. We see by equation 46) that the strength of a vortex is equal to one-half the circulation around any closed path drawn embracing the vortex on its surface, which is independent of the path. In particular, in any non-vortical region the circulation around any closed path is zero, or the circulation along an open path φ_{AB} is independent of the path, depending only on A and B, or the velocity is a lamellar vector. We then have

$$48)\qquad u = \frac{\partial \varphi}{\partial x}, \quad v = \frac{\partial \varphi}{\partial y}, \quad w = \frac{\partial \varphi}{\partial z},$$

and φ is called the *velocity potential*, a term introduced by Lagrange. (When there is vorticity there is no velocity potential.)

Before 1858 only cases of motion had been treated in which a velocity potential existed. In that year appeared the remarkable paper by Helmholtz[1]) on Vortex kotion.

Let us now find the change of circulation along a path moving with the fluid, that is, composed of the same particles, the forces being conservative.

Our equations of motion 3) may be written, putting $U' = -(V + P)$

$$49)\qquad \frac{du}{dt} = \frac{\partial U'}{\partial x}, \quad \frac{dv}{dt} = \frac{\partial U'}{\partial y}, \quad \frac{dw}{dt} = \frac{\partial U'}{\partial z}.$$

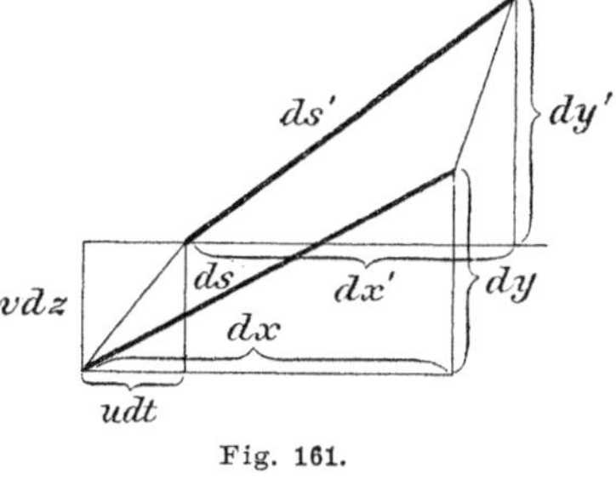

Fig. 161.

The change of circulation along the path AB is

$$50)\qquad \frac{d\varphi_{AB}}{dt} = \frac{d}{dt}\int_A^B (u\,dx + v\,dy + w\,dz),$$

in which dx, dy, dz vary with the time, being the projections of an arc ds composed of parts which move about. If after a time dt the arc ds assumes a length ds' whose components are dx', dy', dz' we have (Fig. 161).

1) *Über Integrale der hydrodynamischen Gleichungen, welche den Wirbelbewegungen entsprechen.* Wissenschaftliche Abhandlungen I, p. 101.

$$dx' = x + dx + dt\left\{u + \frac{\partial u}{\partial x}dx + \frac{\partial u}{\partial y}dy + \frac{\partial u}{\partial z}dz\right\} - (x + udt)$$

$$dx' - dx = dt\left\{\frac{\partial u}{\partial x}dx + \frac{\partial u}{\partial y}du + \frac{\partial u}{\partial z}dz\right\}$$

and therefore the change per unit time in the projections are

$$\begin{aligned} &\frac{d}{dt}(dx) = \frac{\partial u}{\partial x}dx + \frac{\partial u}{\partial y}dy + \frac{\partial u}{\partial z}dz, \\ \text{51)} \quad &\frac{d}{dt}(dy) = \frac{\partial v}{\partial x}dx + \frac{\partial v}{\partial y}dy + \frac{\partial v}{\partial z}dz, \\ &\frac{d}{dt}(dz) = \frac{\partial w}{\partial x}dx + \frac{\partial w}{\partial y}dy + \frac{\partial w}{\partial z}dz. \end{aligned}$$

Thus we have

$$\text{52)} \quad \frac{d\varphi_{AB}}{dt} = \int_A^B \left\{\frac{du}{dt}dx + \frac{dv}{dt}dy + \frac{dw}{dt}dz + u\frac{d}{dt}(dx) + v\frac{d}{dt}(dy + w\frac{d}{dt}(dz)\right\}$$

and substituting from equations 49) and 51)

$$\begin{aligned} \text{53)} \quad \frac{d\varphi_{AB}}{dt} &= \int_A^B \Big[\frac{\partial U'}{\partial x}dx + \frac{\partial U'}{\partial y}dy + \frac{\partial U'}{\partial z}dz \\ &\quad + \left(u\frac{\partial u}{\partial x} + v\frac{\partial v}{\partial x} + w\frac{\partial w}{\partial x}\right)dx \\ &\quad + \left(u\frac{\partial u}{\partial y} + v\frac{\partial v}{\partial y} + w\frac{\partial w}{\partial y}\right)dy \\ &\quad + \left(u\frac{\partial u}{\partial z} + v\frac{\partial v}{\partial z} + w\frac{\partial w}{\partial z}\right)dz\Big] \\ &= \int_A^B \left\{\frac{\partial}{\partial x}\left(U' + \frac{1}{2}q^2\right)dx + \frac{\partial}{\partial y}\left(U' + \frac{1}{2}q^2\right)dy \right. \\ &\quad \left. + \frac{\partial}{\partial z}\left(U' + \frac{1}{2}q^2\right)dz\right\} = \left(U' + \frac{1}{2}q^2\right)\Big/_A^B, \end{aligned}$$

which vanishes for a closed curve.

Therefore if the forces are conservative, the circulation around any closed path moving with the fluid is independent of the time. Thus if the circulation around any closed path is zero at one time, it is always zero, or in other words if a velocity potential once exists, it always exists. This theorem is due to Lagrange.

191. Vortex Motion. We will now consider the case in which no velocity potential exists, that is, the case of vortex-motion, according to the methods of Helmholtz.

From the equations 27), whose right-hand members are the derivatives of $-\left(V+P+\frac{1}{2}q^2\right)$, this quantity may be eliminated by differentiation. Differentiating the last equation by y, the second by z, and subtracting, we obtain

54) $$\frac{1}{2}\frac{\partial}{\partial t}\left(\frac{\partial w}{\partial y}-\frac{\partial v}{\partial z}\right)+\xi\frac{\partial v}{\partial y}+v\frac{\partial \xi}{\partial y}-\eta\frac{\partial u}{\partial y}-u\frac{\partial \eta}{\partial y}$$
$$-\zeta\frac{\partial u}{\partial z}-u\frac{\partial \zeta}{\partial z}+\xi\frac{\partial w}{\partial z}+w\frac{\partial \xi}{\partial z}=0,$$

or otherwise

55) $$\frac{\partial \xi}{\partial t}+u\frac{\partial \xi}{\partial x}+v\frac{\partial \xi}{\partial y}+w\frac{\partial \xi}{\partial z}=u\left\{\frac{\partial \xi}{\partial x}+\frac{\partial \eta}{\partial y}+\frac{\partial \zeta}{\partial z}\right\}$$
$$-\xi\left\{\frac{\partial u}{\partial x}+\frac{\partial v}{\partial y}+\frac{\partial w}{\partial z}\right\}+\xi\frac{\partial u}{\partial x}+\eta\frac{\partial u}{\partial y}+\eta\frac{\partial u}{\partial z}.$$

On the right the coefficient of u vanishes identically by 47), and that of ξ is by the equation of continuity 12) equal to $-\frac{1}{\varrho}\frac{d\varrho}{dt}$, thus equation 55) becomes

56) $$\frac{d\xi}{dt}=\frac{\xi}{\varrho}\frac{d\varrho}{dt}+\xi\frac{\partial u}{\partial x}+\eta\frac{\partial u}{\partial y}+\zeta\frac{\partial u}{\partial z}.$$

Now we have

$$\varrho\frac{d}{dt}\left(\frac{\xi}{\varrho}\right)=\frac{d\xi}{dt}-\frac{\xi}{\varrho}\frac{d\varrho}{dt},$$

and accordingly we may write our equation 56) and its two companions

57) $$\begin{aligned}\frac{d}{dt}\left(\frac{\xi}{\varrho}\right)&=\frac{\xi}{\varrho}\frac{\partial u}{\partial x}+\frac{\eta}{\varrho}\frac{\partial u}{\partial y}+\frac{\zeta}{\varrho}\frac{\partial u}{\partial z},\\ \frac{d}{dt}\left(\frac{\eta}{\varrho}\right)&=\frac{\xi}{\varrho}\frac{\partial v}{\partial x}+\frac{\eta}{\varrho}\frac{\partial v}{\partial y}+\frac{\zeta}{\varrho}\frac{\partial v}{\partial z},\\ \frac{d}{dt}\left(\frac{\zeta}{\varrho}\right)&=\frac{\xi}{\varrho}\frac{\partial w}{\partial x}+\frac{\eta}{\varrho}\frac{\partial w}{\partial y}+\frac{\zeta}{\varrho}\frac{\partial w}{\partial z}.\end{aligned}$$

Thus the time derivatives of $\frac{\xi}{\varrho}$, $\frac{\eta}{\varrho}$, $\frac{\zeta}{\varrho}$ for a given particle are homogeneous linear functions of these quantities. By continued differentiation with respect to t and substitution of the derivatives from these equations, we see that all the time derivatives are homogeneous linear functions of the three quantities themselves. Consequently if at a certain instant a particle does not rotate, it never acquires a rotation. This we find by developing $\frac{\xi}{\varrho}$, $\frac{\eta}{\varrho}$, $\frac{\zeta}{\varrho}$ as functions of t by Taylor's theorem, for if the derivatives of every order vanish for a certain instant, the function always vanishes. Stokes[1]) objects to this method of proof as not rigorous inasmuch as it is not evident that the functions ξ, η, ζ can be developed by Taylor's theorem, and replaces it by the following demonstration.

Let L be a superior limit to the numerical values of the coefficients of $\frac{\xi}{\varrho}$, $\frac{\eta}{\varrho}$, $\frac{\zeta}{\varrho}$ in the second member of equations 57). Then evidently ξ, η, ζ cannot increase faster than if their numerical or absolute values satisfied the equations

$$(58)\qquad \begin{aligned} \frac{d}{dt}\left|\frac{\xi}{\varrho}\right| &= L\left(\left|\frac{\xi}{\varrho}\right| + \left|\frac{\eta}{\varrho}\right| + \left|\frac{\zeta}{\varrho}\right|\right), \\ \frac{d}{dt}\left|\frac{\eta}{\varrho}\right| &= L\left(\left|\frac{\xi}{\varrho}\right| + \left|\frac{\eta}{\varrho}\right| + \left|\frac{\zeta}{\varrho}\right|\right), \\ \frac{d}{dt}\left|\frac{\zeta}{\varrho}\right| &= L\left(\left|\frac{\xi}{\varrho}\right| + \left|\frac{\eta}{\varrho}\right| + \left|\frac{\zeta}{\varrho}\right|\right), \end{aligned}$$

instead of 57), ξ, η, ζ vanishing in this case also when $t = 0$. Adding these three equations and writing

$$\left|\frac{\xi}{\varrho}\right| + \left|\frac{\eta}{\varrho}\right| + \left|\frac{\zeta}{\varrho}\right| = \Omega,$$

we obtain

$$59)\qquad \frac{d\Omega}{dt} = L\Omega.$$

The integral of this equation is

$$\Omega = ce^{Lt},$$

and since $\Omega = 0$ when $t = 0$, c must be zero, and Ω is always zero. Since the sum of the absolute values cannot vanish unless the separate values vanish, the theorem is proved.

Let us now consider two points A and B lying on the same vortex line at a distance apart $ds = \varepsilon\frac{\omega}{\varrho}$, where ε is a small constant. Since the particles lie on a vortex-line we have

$$60)\qquad \frac{dx}{\xi} = \frac{dy}{\eta} = \frac{dz}{\zeta} = \frac{ds}{\omega} = \frac{\varepsilon}{\varrho}.$$

We have for the difference of velocity at A and B

$$61)\qquad \begin{aligned} u_B - u_A &= \frac{\partial u}{\partial x}dx + \frac{\partial u}{\partial y}dy + \frac{\partial u}{\partial z}dz \\ &= \varepsilon\left\{\frac{\xi}{\varrho}\frac{\partial u}{\partial x} + \frac{\eta}{\varrho}\frac{\partial u}{\partial y} + \frac{\zeta}{\varrho}\frac{\partial u}{\partial z}\right\}, \end{aligned}$$

or by equations 57),

$$62)\qquad u_B - u_A = \varepsilon\frac{d}{dt}\left(\frac{\xi}{\varrho}\right).$$

---

1) Stokes, *Math. and Phys. Papers*, Vol. II, p. 36.

Now at an instant later by dt, when the particles are at A' and B', we have

$$\begin{aligned} dx' &= dx + (u_B - u_A)\,dt = \varepsilon\left[\frac{\xi}{\varrho} + \frac{d}{dt}\left(\frac{\xi}{\varrho}\right)dt\right],\\ dy' &= dy + (v_B - v_A)\,dt = \varepsilon\left[\frac{\eta}{\varrho} + \frac{d}{dt}\left(\frac{\eta}{\varrho}\right)dt\right],\\ dz' &= dz + (w_B - w_A)\,dt = \varepsilon\left[\frac{\zeta}{\varrho} + \frac{d}{dt}\left(\frac{\zeta}{\varrho}\right)dt\right]. \end{aligned} \tag{63}$$

Therefore the projections of the arc ds' in the new position are proportional to the new values of $\frac{\xi}{\varrho}$, $\frac{\eta}{\varrho}$, $\frac{\zeta}{\varrho}$, as they originally were, so that the particles still lie on a vortex-line. Accordingly a vortex-line is always composed of the same particles of fluid. Also since the components of ds have increased or have changed so as to be always proportional to the components of $\frac{\omega}{\varrho}$, if the liquid is incompressible the rotation is proportional to the distance between the particles. And whether ϱ vary or not, if S be the area of the cross-section of a vortex-filament, $\varrho S ds$, the mass of a length ds remaining constant, so does $S\omega$, the strength of the filament.

It is easy to see that this is equivalent to a statement of the conservation of angular momentum for each portion of the fluid. Evidently in a perfect fluid no moment can be exerted on any portion, since the tangential forces vanish.

Accordingly the strength of a vortex-filament is constant, not only at all points in the filament but at all times, consequently a vortex existing in a perfect fluid is indestructible, howeyer it may move. It is from this remarkable property of vortices discovered by Helmholtz that Lord Kelvin was lead to imagine atoms as consisting of vortices in a perfect fluid.

192. Vector Potential. Helmholtz's Theorem. We have seen that any curl is a solenoidal vector. We may naturally ask whether conversely any solenoidal vector can be replaced by the curl of another vector. It was shown by Helmholtz that any uniform continuous vector point-function vanishing at infinity can be expressed as the sum of a lamellar and a solenoidal part, and the solenoidal part may be expressed as the curl of a vector point-function. A vector point-function is completely determined if its divergence and curl are everywhere given. Let q be the given vector, which in our case is the velocity of the fluid. Let us suppose that it is possible to express it as the sum of the vector-parameter of a scalar function φ and the curl of a vector function Q, whose components are U, V, W. Accordingly let us put

$$
64)\qquad
\begin{aligned}
u &= \frac{\partial \varphi}{\partial x} + \frac{\partial W}{\partial y} - \frac{\partial V}{\partial z},\\
v &= \frac{\partial \varphi}{\partial y} + \frac{\partial U}{\partial z} - \frac{\partial W}{\partial x},\\
w &= \frac{\partial \varphi}{\partial z} + \frac{\partial V}{\partial x} - \frac{\partial U}{\partial y}.
\end{aligned}
$$

Finding first the divergence of q we have

$$
65)\qquad \text{div. } q = \frac{\partial u}{\partial x} + \frac{\partial v}{\partial y} + \frac{\partial w}{\partial z} = \varDelta \varphi,
$$

the divergence of the curl part vanishing. But by § 128, 5) we know that if φ and its first derivatives are everywhere finite and continuous, we have

$$
66)\qquad \varphi = -\frac{1}{4\pi}\iiint_\infty \frac{\varDelta \varphi}{r}\, d\tau = -\frac{1}{4\pi}\iiint_\infty \frac{\text{div. } q}{r}\, d\tau.
$$

Since q is continuous by hypothesis, div. q is finite. Consequently the lamellar part of q is determined by its divergence.

Secondly finding the curl of q,

$$
\begin{aligned}
67)\qquad 2\xi = \frac{\partial w}{\partial y} - \frac{\partial v}{\partial z} &= \frac{\partial}{\partial y}\left(\frac{\partial \varphi}{\partial z} + \frac{\partial V}{\partial x} - \frac{\partial U}{\partial y}\right) - \frac{\partial}{\partial z}\left(\frac{\partial \varphi}{\partial y} + \frac{\partial U}{\partial z} - \frac{\partial W}{\partial x}\right)\\
&= -\varDelta U + \frac{\partial}{\partial x}\left(\frac{\partial U}{\partial x} + \frac{\partial V}{\partial y} + \frac{\partial W}{\partial z}\right).
\end{aligned}
$$

Since the vector Q is as yet undetermined except as to its partial derivatives by equations 64), let us assume that it is solenoidal, or

$$
68)\qquad \frac{\partial U}{\partial x} + \frac{\partial V}{\partial y} + \frac{\partial W}{\partial z} = 0,
$$

then we have

$$
69)\qquad
\begin{aligned}
2\xi &= -\varDelta U,\\
2\eta &= -\varDelta V,\\
2\zeta &= -\varDelta W,
\end{aligned}
$$

the integrals of which are as above,

$$
70)\qquad
\begin{aligned}
U &= \frac{1}{2\pi}\iiint \frac{\xi}{r}\, d\tau,\\
V &= \frac{1}{2\pi}\iiint \frac{\eta}{r}\, d\tau,\\
W &= \frac{1}{2\pi}\iiint \frac{\zeta}{r}\, d\tau.
\end{aligned}
$$

Equations 66) and 70) were given by Stokes, in his paper, On the Dynamical Theory of Diffraction, 1849.

Where the vorticity vanishes the space contributes nothing to the integrals, so that the latter may be taken over all the vortices. Thus we see that Q, which determines the solenoidal part of q, is determined by curl q. Consequently both parts of q are completely determined and the theorem is proved. If q is solenoidal div. q vanishes, $\varphi = 0$, and $q = \text{curl } Q$. Accordingly every solenoidal vector may be represented as a curl. If q is irrotational curl $q = 0$ and $Q = 0$, so that every irrotational vector is lamellar, as we saw in § 31.

The vector Q, whose *components* are formed as potential functions for densities $\frac{\xi}{2\pi}, \frac{\eta}{2\pi}, \frac{\zeta}{2\pi}$ respectively, is called the *vector potential* of the vector $\frac{\omega}{2\pi}$. We may thus abbreviate our results in the vector equations,

71) $$\overline{q} = \text{vector parameter } \varphi + \text{curl } \overline{Q},$$

72) $$\varphi = -\frac{1}{4\pi}\iiint_{\infty} \frac{\text{div. } q}{r}\, d\tau,$$

73) $$\overline{Q} = \frac{1}{4\pi}\iiint_{\infty} \frac{\overline{\text{curl} q}}{r}\, d\tau.$$

Let us verify that $\overline{Q}$ as determined is solenoidal. We shall distinguish the point of integration by accents, so that

70) $$\begin{aligned} U &= \frac{1}{2\pi}\iiint \frac{\xi'}{r}\, d\tau', \\ V &= \frac{1}{2\pi}\iiint \frac{\eta'}{r}\, d\tau', \\ W &= \frac{1}{2\pi}\iiint \frac{\zeta'}{r}\, d\tau', \end{aligned}$$

$$r^2 = (x - x')^2 + (y - y')^2 + (z - z')^2.$$

Differentiating we have

74) $$\begin{aligned} \frac{\partial U}{\partial x} &= \frac{1}{2\pi}\iiint \xi' \frac{\partial}{\partial x}\left(\frac{1}{r}\right) d\tau' = -\frac{1}{2\pi}\iiint \xi' \frac{\partial}{\partial x'}\left(\frac{1}{r}\right) d\tau' \\ &= \frac{1}{2\pi}\left[\iint \frac{\xi'}{r} \cos(nx)\, dS + \iiint \frac{1}{r}\frac{\partial \xi'}{\partial x'}\, d\tau'\right]. \end{aligned}$$

In like manner

75) $$\begin{aligned} &\frac{\partial U}{\partial x} + \frac{\partial V}{\partial y} + \frac{\partial W}{\partial z} \\ &= \frac{1}{2\pi}\left[\iint \left(\xi' \cos(nx) + \eta' \cos(ny) + \zeta' \cos(nz)\right) \frac{dS}{r} \right. \\ &\left. + \iiint \frac{1}{r}\left(\frac{\partial \xi'}{\partial x'} + \frac{\partial \eta'}{\partial y'} + \frac{\partial \zeta'}{\partial z'}\right) d\tau'\right]. \end{aligned}$$

But since ω is solenoidal the volume integral vanishes, and, since at the surface of the vortices, which are composed of vortex-lines, ω is tangential, the surface integral vanishes and div. $Q = 0$.

193. Velocity due to Vortex. Let us now consider in an incompressible fluid the velocity at any point due to vortical motion. We have

$$76)\quad u = \frac{\partial W}{\partial y} - \frac{\partial V}{\partial z} = \frac{1}{2\pi}\left[\frac{\partial}{\partial y}\left(\iiint \frac{\zeta'}{r} d\tau'\right) - \frac{\partial}{\partial z}\left(\iiint \frac{\eta'}{r} d\tau'\right)\right]$$
$$= \frac{1}{2\pi}\iiint\left\{\zeta' \frac{\partial}{\partial y}\left(\frac{1}{r}\right) - \eta' \frac{\partial}{\partial z}\left(\frac{1}{r}\right)\right\} d\tau'.$$

Thus the portions of the velocity contributed by an element $d\tau'$ of the vortex are:

$$du = \frac{1}{2\pi r^2}\left\{\zeta'\left(\frac{y'-y}{r}\right) - \eta'\left(\frac{z'-z}{r}\right)\right\} d\tau',$$
$$77)\quad dv = \frac{1}{4\pi r^2}\left\{\xi'\left(\frac{z'-z}{r}\right) - \zeta'\left(\frac{x'-x}{r}\right)\right\} d\tau',$$
$$dw = \frac{1}{2\pi r^2}\left\{\eta'\left(\frac{x'-x}{r}\right) - \xi'\left(\frac{y'-y}{r}\right)\right\} d\tau',$$

or the velocity at the point x, y, z due to the element $d\tau'$ is $\frac{1}{2\pi r^3}$ multiplied by the vector-product of the vorticity and the vector r drawn from the element $d\tau'$ of the vortex to the point x, y, z. If dq be the magnitude of the resultant of du, dv, dw we thus have

$$73)\quad dq = \frac{\omega \sin(\omega, r)}{2\pi r^2} d\tau'.$$

Let us take for the element $d\tau'$ a length ds of a vortex filament of cross-section S. Then $d\tau' = S ds$ and since $S\omega = \varkappa$, the strength of the filament,

$$79)\quad dq = \frac{\varkappa\, ds \sin(\omega, r)}{2\pi r^2}.$$

The velocity is connected with the vorticity in the same may that the magnetic field is connected with the electric current density producing it, and equation 79) gives us the magnetic field produced by a linear current element of length ds and strength $\frac{\varkappa}{2\pi}$.[1])

---

1) See the author's treatise on *The Theory of Electricity and Magnetism*, §§ 222—226.

194. Kinetic Energy of Vortex. The kinetic energy of the incompressible liquid moving vortically is

$$80)\quad T = \frac{1}{2}\varrho\iiint(u^2 + v^2 + w^2)\,d\tau$$
$$= \frac{\varrho}{2}\iiint\left[u\left(\frac{\partial W}{\partial y} - \frac{\partial V}{\partial z}\right) + v\left(\frac{\partial U}{\partial z} - \frac{\partial W}{\partial x}\right) + w\left(\frac{\partial V}{\partial x} - \frac{\partial U}{\partial y}\right)\right]d\tau$$
$$= \frac{\varrho}{2}\iint[(vW - wV)\cos(nx) + (wU - uW)\cos(ny)$$
$$+ (uV - vU)\cos(nz)]\,dS$$
$$+ \frac{\varrho}{2}\iiint\left[U\left(\frac{\partial w}{\partial y} - \frac{\partial v}{\partial z}\right) + V\left(\frac{\partial u}{\partial z} - \frac{\partial w}{\partial x}\right) + W\left(\frac{\partial v}{\partial x} - \frac{\partial u}{\partial y}\right)\right]d\tau.$$

If the integral be taken over all space, since the motion is supposed to vanish at infinity the surface integrals vanish, and

$$81)\qquad T = \varrho\iiint[U\xi + V\eta + W\zeta]\,d\tau,$$

or inserting the values of U, V, W from 70)

$$82)\qquad T = \frac{\varrho}{2\pi}\iiiint\!\!\iint\frac{\xi\xi' + \eta\eta' + \zeta\zeta'}{r}\,d\tau\,d\tau',$$

and the integration may now be restricted to the vortices.

If again we integrate by filaments, we find

$$83)\qquad T = \frac{\varrho}{2\pi}\iiiint\!\!\iint\frac{\varkappa\varkappa'\cos(ds, ds')}{r}\,ds\,ds',$$

where the integration is expressed as over the length of each of the double infinity of vortex-filaments constituting the vortices. This is the form obtained for the energy of two electric currents by Franz Neumann.

195. Straight parallel Vortices. Let us now consider the case in which the vorticity is everywhere parallel to a single direction, that of the axis of z. Let the motion be uniplanar, that is parallel to a single plane, the XY-plane, and the same in all planes parallel to it. All quantities are therefore independent of z. The vortices are columnar and either of infinite length or end at the free surface of the liquid. Such vortices may be produced standing vertically in a tank with a horizontal bottom. Under the conditions imposed we have

$$84)\qquad 0 = w = \frac{\partial u}{\partial z} = \frac{\partial v}{\partial z} = \frac{\partial w}{\partial z} = \xi = \eta = U = V,$$

and

$$85)\qquad u = \frac{\partial W}{\partial y},\quad v = -\frac{\partial W}{\partial x},$$

86) $$2\zeta = \frac{\partial v}{\partial x} - \frac{\partial u}{\partial y},$$

and by equations 57)

$$\frac{d\zeta}{dt} = 0,$$

so that ζ is independent of the time for any given vortex-filament.

The function W is not a velocity potential, but is said to be *conjugate* to a velocity potential φ for which

87) $$u = \frac{\partial \varphi}{\partial x}, \quad v = \frac{\partial \varphi}{\partial y}.$$

The function W has a simple physical meaning. If we find the amount of liquid which flows across a cylindrical surface with generators parallel to the z-axis of height unity in unit of time, we have

88) $$\psi = \int_A^B q \cos(qn)\, ds = \int_A^B [u \cos(nx) + v \cos(ny)]\, ds,$$

the line integral being taken around any orthogonal section of the cylinder. Now we have

$$ds \cos(nx) = \quad dy,$$
$$ds \cos(ny) = -\, dx,$$

so that

89) $$\psi = \int (u\, dy - v\, dx) = \int_A^B \left(\frac{\partial W}{\partial x} dx + \frac{\partial W}{\partial y} dy\right) = W_B - W_A.$$

A function, the difference of whose values at two points A and B gives the quantity flowing in unit time across a cylinder of unit height drawn on any curve with ends at A and B, is called a flux or current function. The quantity crossing is independent of the curve because the fluid is incompressible. In the present case the vector potential W is a current function. The stream lines being lines across which no current flows are given by the equation $\psi = const.$ Substituting the values of u and v from 85) in 86), we have

90) $$-2\zeta = \frac{\partial^2 W}{\partial x^2} + \frac{\partial^2 W}{\partial y^2}.$$

But this is the equation for a logarithmic potential with density $\frac{\zeta}{\pi}$, § 138, 61), so that we have as the integral

91) $$W = -\frac{1}{\pi}\iint \zeta' \log r\, dS',$$

as may also be found from equation 70) by integrating over the infinite cylinder as in § 135, subject to the difficulty mentioned on p. 385. The value of W given in equation 91) satisfies the equation

92) $$\frac{\partial^2 W}{\partial x^2} + \frac{\partial^2 W}{\partial y^2} = 0$$

outside of the vortices and equation 90) at points within them, as shown in § 138. If we have a single vortex filament of cross-section dS and strength $\varkappa = \zeta' dS$,

93) $$W = -\frac{\zeta'}{\pi} \log r \, dS = -\frac{\varkappa}{\pi} \log r,$$

and the lines of flow are circles, $r = const.$ Then

94) $$\begin{aligned} u &= \frac{\partial W}{\partial y} = \frac{\varkappa}{\pi r} \cdot \frac{y' - y}{r}, \\ v &= -\frac{\partial W}{\partial x} = \frac{\varkappa}{\pi r} \cdot \frac{x - x'}{r}, \end{aligned}$$

95) $$q^2 = u^2 + v^2 = \frac{\varkappa^2}{\pi^2 r^2},$$

the velocity is perpendicular to the radius joining the point x, y with the vortex and inversely proportional to its length. It is to be observed that although the motion is whirling, every point describing a circle about the center, the motion is irrotational except at the center, where the vortex-filament is situated, each particle describing its path without turning about itself, like a body of soldiers obliquing or changing direction while each man faces in the same unchanging direction. The motion in the vortex on the contrary is similar to that of a body of soldiers *wheeling* or changing direction like a rigid body rotating.

If we have a number of vortices of strengths $\varkappa_1, \varkappa_2, \ldots \varkappa_n$, and form the linear functions of the velocities of each,

96) $$\begin{aligned} U &= \Sigma \varkappa_s u_s, \text{[1]} \\ V &= \Sigma \varkappa_s v_s, \end{aligned}$$

where u_s, v_s, is the velocity at the vortex s both vanish. For any pair of vortices r and s we have

$$u_r = \varkappa_s \cdot \frac{x_s - x_r}{\pi r^2},$$

where u_r is the part of the velocity at x_r, y_r due to the vortex of strength $\varkappa_s$ situated at $x_s y_s$. Thus

$$\varkappa_r u_r = \varkappa_s \varkappa_r \cdot \frac{x_s - x_r}{\pi r^2},$$

while similarly

$$\varkappa_s u_s = \varkappa_s \varkappa_r \frac{x_r - x_s}{\pi r^2},$$

so that the therms of the sum destroy each other in pairs.

1) U and V have nothing here to do with the components of the vector-potential.

Similarly for vortices continuously distributed, the strength of any elementary filament being ζdS,

97) $$U = \iint u\zeta dS = \iiiint \zeta\zeta' \cdot \frac{x'-x}{r} \cdot dS\, dS',$$

which again vanishes, since every point is covered by both dS and dS'.

If we define the center of the vortex as $x_0 y_0$ where

98) $$x_0 \iint \zeta dS = \iint x\zeta dS,$$
$$y_0 \iint \zeta dS = \iint y\zeta dS,$$

then since ζ does not depend upon t, if we follow the particle differentiating,

98) $$\frac{dx_0}{dt} \iint \zeta dS = \iint u\zeta dS = 0,$$

the integrals being taken over areas moving with the liquid. Therefore

100) $$\frac{dx_0}{dt} = \frac{dy_0}{dt} = 0,$$

or the center of all columnar vortices present remains at rest.

If we have a single vortex filament of infinitesimal cross section S, for which

101) $$\iint \zeta dS = \varkappa,$$

the velocity depends on the current function $W = -\frac{\varkappa}{\pi}\log r$. In the vortex and close to it, if $\varkappa$ is finite, ζ, W, u, v are infinite. But at the center $u = v = 0$, the vortex stands still and the fluid moves about it in circles with velocity $\frac{\varkappa}{\pi r}$. The angular velocity and the area of the cross-section remain constant, although the shape of the latter may vary. If we have two such vortex-filaments each urges the other in a direction perpendicular to the line joining them, they accordingly revolve about their center, maintaining a constant distance from each other. If they are whirling in the same direction the center is between them (Fig. 162), but if in opposite directions, it is outside, and if they are equal it lies at infinity. Such a pair of vortices may be called a vortex-couple or doublet, and they advance at a constant velocity, keeping symmetrical with respect to the plane bisecting perpendicularly the line joining them. This plane is a stream-plane and may accordingly be taken as a boundary of the

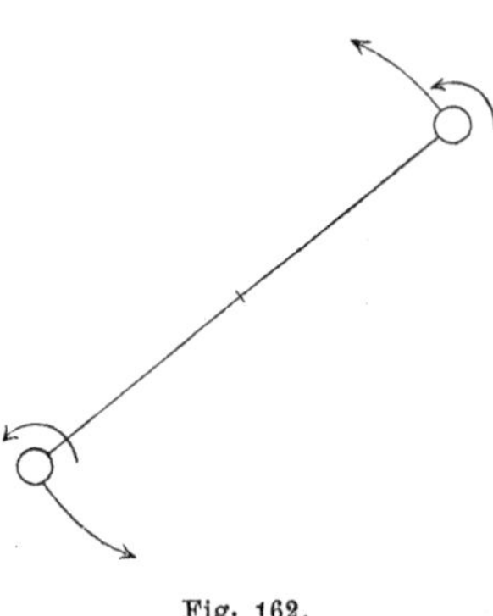

Fig. 162.

fluid. Since either vortex moves with a velocity $\frac{\varkappa}{\pi r}$ and half way between them the velocity being due to both is $\frac{2\varkappa}{\frac{\pi r}{2}} = \frac{4\varkappa}{\pi r}$, we find that a single vortex near a plane wall moves parallel to it with a velocity one fourth that of the water at the wall. This is an illustration of the method of images, of frequent application in hydrodynamics.

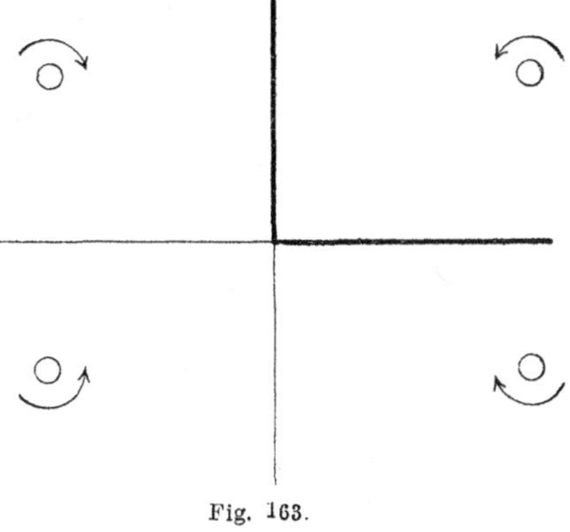
Fig. 163.

As another illustration consider the motion of a single vortex-filament in a square corner inclosed by two infinite walls. The motion is evidently the same as if we had a pair of vortex-couples formed by the given vortex and its images in the two walls, turning as shown in Fig. 163 and forming what may be called a vortex kaleidoscope. From the symmetry it is evident that the planes of the walls are stream-planes, so that we may consider the motion in one corner alone. If x and y be the coordinates of the vortex considered, we have as due to the others,

102)
$$u = \frac{\varkappa}{2\pi y} - \frac{\varkappa}{2\pi} \cdot \frac{y}{x^2+y^2} = \frac{\varkappa}{2\pi} \frac{x^2}{y(x^2+y^2)},$$
$$v = -\frac{\varkappa}{2\pi x} + \frac{\varkappa}{2\pi} \frac{x}{x^2+y^2} = -\frac{\varkappa}{2\pi} \frac{y^2}{x(x^2+y^2)}.$$

Since u and v are the velocities $\frac{dx}{dt}, \frac{dy}{dt}$ of the vortex, we have for the equation of its path

103)
$$\frac{\frac{dx}{dt}}{\frac{dy}{dt}} = \frac{u}{v} = -\frac{x^3}{y^3} = \frac{dx}{dy}, \text{ or}$$
$$\frac{dx}{x^3} = -\frac{dy}{y^3},$$

whose integral is

104)
$$\frac{1}{x^2} + \frac{1}{y^2} = \frac{1}{a^2}, \quad x^2 + y^2 = \frac{x^2 y^2}{a^2},$$

and in polar coordinates,

$$r^2 = \frac{r^4}{a^2} \sin^2\vartheta \cos^2\vartheta,$$

105)
$$r \sin 2\vartheta = \pm 2a,$$

the equation of a Cotes's spiral, having one of the axes as an asymptote.

The same problem gives us the motion of two equal vortex-couples approaching each other head on, or a single vortex-couple approaching a plane boundary, showing how as they are stopped they spread out. The behavior of vortex-couples will serve to illustrate that of circular vortex rings, for the theory of which the reader is referred to Helmholtz's original paper. The two opposite parts of a circular vortex appear to be rotating in opposite directions if viewed on their intersection by a diametral plane normal to the circle, thus resembling a vortex-couple. It is found that the circular vortex advances with a constant velocity in the direction of the fluid in the center, maintaining its diameter, but that when approaching a wall head on it spreads out like the vortex-couple. Two circular vortices approaching each other do the same thing, but if moving in the same direction the forward one spreads out, the following one contracts and is sucked through the foremost vortex, when it in turn spreads out and the one which is now behind passes through it, and so on in turn, as may also be shown for two colummar vortex-couples traveling in the same direction.

Most of these properties of circular vortices may be realized with smoke rings made by causing smoke to puff out through a circular hole in a box, or mouth of a smoker, or smoke-stack of a locomotive. The friction at the edge of the hole holds the outside of the smoke back, while the inside goes forward, establishing thereby the vortical rotation. As previously stated no vortex could be formed if there were no friction. It is to be noticed that the direction of the fluid on the *inside* of the vortex gives the direction of advance.

196. Irrotational Motion. We shall now consider the non-vortical motion of an incompressible fluid. We then have a velocity potential φ and

106) $$u=\frac{\partial \varphi}{\partial x}, \quad v=\frac{\partial \varphi}{\partial y}, \quad w=\frac{\partial \varphi}{\partial z}.$$

The equation of continuity becomes

107) $$\Delta \varphi = 0,$$

and the potential is harmanic at all points except where liquid is being created (*sources*) or withdrawn (*sinks*). The volume of flow per unit time outward from any closed surface S is

108) $$-\iint [u \cos(nx) + v \cos(ny) + w \cos(nz)]\, dS$$
$$= -\iint \frac{\partial \varphi}{\partial n}\, dS = \iiint \Delta \varphi\, d\tau,$$

so that if this is not equal to zero, it is equal to the quantity created in the space considered in unit time,

109) $$\frac{dQ}{dt} = \iiint \Delta\varphi\, d\tau,$$

so that if we put $\Delta\varphi = \sigma$, σ is the amount of liquid produced per unit volume per unit of time. The total amount

$$\frac{dQ}{dt} = \iiint \sigma\, d\tau$$

is called the *strength of the source.* If σ is given as a function of the point we have

110) $$\varphi = -\frac{1}{4\pi}\iiint \frac{\Delta\varphi}{r}\, d\tau = -\frac{1}{4\pi}\iiint \frac{\sigma}{r}\, d\tau.$$

Accordingly the velocity potential has the properties of a force potential, the density of attracting matter being represented by $\frac{1}{4\pi}$ times the strength of source per unit volume. The negative sign occours here from the different convention employed, it being customary to define the force as the negative parameter, the velocity as the positive parameter of its potential. In particular a point source of strength m produces a radial velocity of magnitude $\frac{1}{4\pi}\frac{m}{r^2}$. This system is called by Clifford a *squirt.*

197. Uniplanar Motion. A simple and interesting case is that of uniplanar flow as defined above. We then have all quantities independent of z, so that Laplace's equation reduces to

111) $$\frac{\partial^2\varphi}{\partial x^2} + \frac{\partial^2\varphi}{\partial y^2} = 0.$$

A powerful method of treatment of such problems is furnished by the method of functions of a complex variable. The complex number $a + ib$, where a and b are real numbers and i is a unit defined by the equation

$$i^2 = -1,$$

(the same root being always taken) is subject to all the laws of algebra, and vanishes only when a and b both vanish separately. Any function of the complex number obtained by algebraic operations, after substituting for every factor i^2 its value -1, becomes the sum of a real number plus a pure imaginary, that is a real number multiplied by i. Any equation between complex numbers is equivalent to two equations between real numbers, being satisfied only when the real parts in both numbers are equal as well as the real coefficients of i in both members. If z denote the complex variable $x + iy$, any function of z may be written

$$w = f(z) = u + iv,$$

where u and v are real functions of the two real variables x and y. For instance

112) $$z^2 = (x+iy)^2 = x^2 - y^2 + 2ixy, \quad u = x^2 + y^2, \quad v = 2xy,$$

113) $$\frac{1}{z} = \frac{1}{x+iy} = \frac{x-iy}{x^2+y^2} = \frac{x}{x^2+y^2} - \frac{iy}{x^2+y^2}, \quad u = \frac{x}{x^2+y^2}, \qquad v = -\frac{y}{x^2+y^2}.$$

Let us examine the relation between an infinitesimal change in z and the corresponding change in $f(z)$. We have, x and y being real variables capable of independent variation,

114) $$dz = dx + i\,dy,$$

115) $$df(z) = dw = du + idv = \frac{\partial u}{\partial x}dx + \frac{\partial u}{\partial y}dy + i\left(\frac{\partial v}{\partial x}dx + \frac{\partial v}{\partial y}dy\right)\cdot$$

Consequently by division,

116) $$\frac{dw}{dz} = \frac{du+idv}{dx+idy} = \frac{\frac{\partial u}{\partial x}dx + \frac{\partial u}{\partial y}dy + i\left(\frac{\partial v}{\partial x}dx + \frac{\partial v}{\partial y}dy\right)}{dx+idy} = \frac{\frac{\partial u}{\partial x} + i\frac{\partial v}{\partial x} + \left(\frac{\partial u}{\partial y} + i\frac{\partial v}{\partial y}\right)\frac{dy}{dx}}{1+i\frac{dy}{dx}}.$$

The ratio of the differentials of w and z accordingly depends in general on the ratio of dy to dx, that is, if x and y represent the coordinates of a point in a plane, on the direction of leaving the point. If the ratio of dw to dz is to be independent of this direction and to depend only on the position of the point x, y, the numerator must be a multiple of the denominator, so that the expression containing $\frac{dy}{dx}$ divides out. In order that this may be true we must have

$$\left(\frac{\partial u}{\partial x} + i\frac{\partial v}{\partial x}\right) : 1 = \left(\frac{\partial u}{\partial y} + i\frac{\partial v}{\partial y}\right) : i,$$

that is

117) $$i\left(\frac{\partial u}{\partial x} + i\frac{\partial v}{\partial x}\right) = \frac{\partial u}{\partial y} + i\frac{\partial v}{\partial y}\cdot$$

Putting real and imaginary parts on both sides equal,

118) $$\frac{\partial u}{\partial x} = \frac{\partial v}{\partial y}, \quad \frac{\partial v}{\partial x} = -\frac{\partial u}{\partial y},$$

and

119) $$\frac{dw}{dz} = \frac{\partial u}{\partial x} + i\frac{\partial v}{\partial x} = \frac{\partial v}{\partial y} - i\frac{\partial u}{\partial y}\cdot$$

In this case the function w is said to have a definite derivative defined by

$$f'(z) = \lim_{\substack{dx=0\\dy=0}} \frac{dw}{dz},$$

and it is only when the functions u and v satisfy these conditions 118) that $u + iv$ is said to be an *analytic function* of z. This is Riemann's definition of a function of a complex variable.[1]) The real functions u and v are said to be *conjugate* functions of the real variables x, y.

It is obvious that if w is given as an analytic expression involving z, $w = f(z)$, then w always satisfies this condition. For

$$\frac{\partial w}{\partial x} = \frac{df(z)}{dz}\frac{\partial z}{\partial x} = f'(z), \quad \frac{\partial w}{\partial y} = \frac{df(z)}{dz}\frac{\partial z}{\partial y} = if'(z).$$

Accordingly

$$i\frac{\partial w}{\partial x} = i\left(\frac{\partial u}{\partial x} + i\frac{\partial v}{\partial x}\right) = \frac{\partial w}{\partial y} = \frac{\partial u}{\partial y} + i\frac{\partial v}{\partial y},$$

$$\frac{\partial u}{\partial x} = \frac{\partial v}{\partial y}, \quad -\frac{\partial v}{\partial x} = \frac{\partial u}{\partial y}.$$

If we differentiate the equations 118), the first by x, and the second by y and add, since

$$\frac{\partial^2 v}{\partial x \partial y} = \frac{\partial^2 v}{\partial y \partial x},$$

we obtain

120) $$\frac{\partial^2 u}{\partial x^2} + \frac{\partial^2 u}{\partial y^2} = 0.$$

Differentiating the second by x and the first by y and subtracting, we find that v satisfies the same equation

121) $$\frac{\partial^2 v}{\partial x^2} + \frac{\partial^2 v}{\partial y^2} = 0.$$

Thus every function of a complex variable gives a pair of solutions of Laplace's equation, either one of which may be taken for the velocity potential, representing two different states of flow.

It is to be noticed that the question here dealt with is simply one of kinematics, since Laplace's equation is simply the equation of continuity and there is no reference to the dynamical equations.

The question arises whether any two solutions of Laplace's equation will conversely give us the function of a complex variable. It obviously will not answer to take *any* two harmonic functions, for they must be related so as to satisfy the equations 118) or be mutually conjugate. In order to avoid confusion with the velocity

1) Riemann, *Mathematische Werke*, p. 5.

components u and v, let us call the two conjugate functions φ and ψ, satisfying the equations

122) $$\frac{\partial \varphi}{\partial x} = \frac{\partial \psi}{\partial y}, \quad \frac{\partial \varphi}{\partial y} = - \frac{\partial \psi}{\partial x}.$$

It is evident that φ and ψ have the relation of the velocity potential and stream function defined in § 195. If one function is given we con find the conjugate, for we must have

$$d\psi = \frac{\partial \psi}{\partial x} dx + \frac{\partial \psi}{\partial y} dy,$$

which by equations 122) is

$$d\psi = - \frac{\partial \varphi}{\partial y} dx + \frac{\partial \varphi}{\partial x} dy.$$

Now if we call this $X dx + Y dy$ it satisfies the condition for a perfect differential

$$\frac{\partial X}{\partial y} = \frac{\partial Y}{\partial x}$$

that is, in this case,

$$- \frac{\partial^2 \varphi}{\partial y^2} = \frac{\partial^2 \varphi}{\partial x^2}.$$

Consequently the line integral

$$\int \left\{ - \frac{\partial \varphi}{\partial y} dx + \frac{\partial \varphi}{\partial x} dy \right\}$$

from a given point x_0, y_0 to a variable point x, y, is a function only of its upper limit and represents ψ. Similarly if ψ is given

123) $$\varphi = \int \left\{ \frac{\partial \varphi}{\partial x} dx + \frac{\partial \varphi}{\partial y} dy \right\} = \int \left\{ \frac{\partial \psi}{\partial y} dx - \frac{\partial \psi}{\partial x} dy \right\}.$$

Furthermore the first of the equations 122) is the condition that $\psi\, dx + \varphi\, dy$ is a perfect differential and the second that $\varphi\, dx - \psi\, dy$ is such. Accordingly the line integrals

124) $$\Phi = \int \{ \psi\, dx + \varphi\, dy \},$$
$$\Psi = \int \{ \varphi\, dx + \psi\, dy \},$$

give two new point functions Φ, Ψ which in virtue of the equations

125) $$\psi = \frac{\partial \Phi}{\partial x} = - \frac{\partial \Psi}{\partial y},$$
$$\varphi = \frac{\partial \Phi}{\partial y} = \frac{\partial \Psi}{\partial x},$$

are conjugate to each other and give a new analytic function of z,

$$(\Psi + i\Phi),$$

whose derivative is $\varphi + i\psi$. From these by new integrations we may obtain any number. The method of the complex variable accordingly gives us the solution of an unlimited number of uniplanar problems.

The equations 122) are geometrically the condition that the lines $\varphi = const.$, $\psi = const.$, intersect each other everywhere at right angles. If ψ is the stream function the lines $\psi = const.$ are the lines of flow, which we know intersect the equipotential surfaces at right angles. As examples consider the cases worked above,

$$w = z^2, \quad \varphi = x^2 - y^2, \quad \psi = 2xy.$$

The equipotential lines are sets of equilateral hyperbolas, intersected at right angles by the system of equilateral hyperbolas forming the stream lines (Fig. 164). The stream line $\psi = 0$ consists of the X and Y axes, which may accordingly be a boundary, so that one quarter

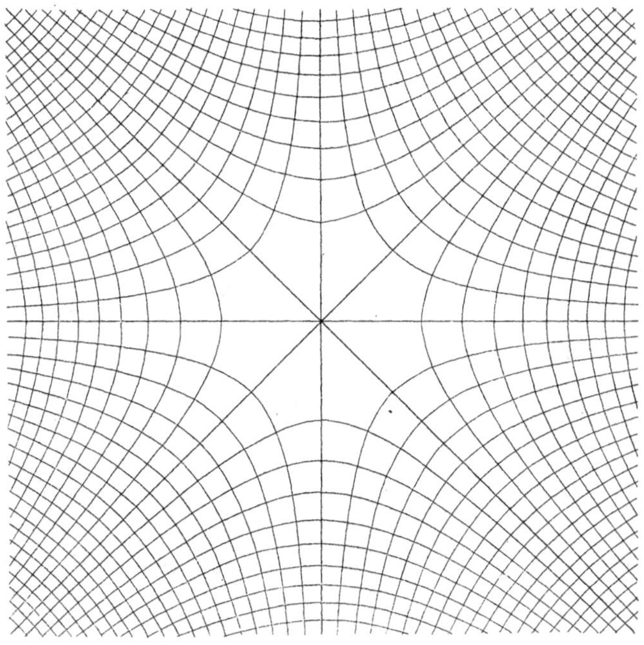

Fig. 164.

of the figure represents the flow in a square corner of a stream of infinite extent.

The function $w = \frac{1}{z}$ gives

$$\varphi = \frac{x}{x^2 + y^2}, \quad \psi = -\frac{y}{x^2 + y^2}.$$

The equipotential lines give a set of circles all tangent to the Y-axis at the origin, while the lines of flow are a similar set all tangent to the X-axis (Fig. 165). The water flows in on one side of the

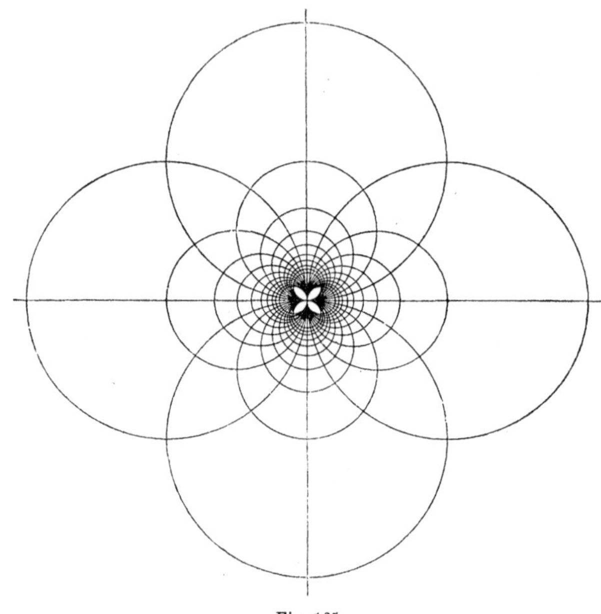

Fig. 165.

origin and out at the other as if there were a source on one side and an equal sink on the other close together.

The function z^n, of which the two examples just treated are particular cases, gives an interesting case which is most simply worked out by the introduction of polar coordinates.

$$x = r\cos\omega, \quad y = r\sin\omega,$$
$$z = x + iy = r(\cos\omega + i\sin\omega) = re^{i\omega},$$
$$z^n = r^n e^{in\omega} = r^n(\cos n\omega + i\sin n\omega),$$

from which we obtain the two conjugate functions

126) $$u = r^n\cos n\omega, \quad v = r^n\sin n\omega.$$

If we multiply these two harmonic functions by constants and add, the sum

127) $$r^n[A_n\cos(n\omega) + B_n\sin n\omega]$$

is the circular harmonic function treated in § 140. We may accordingly develop the velocity potential in a series of circular harmonics,

$$128) \qquad \varphi = \sum_n r^n \{A_n \cos n\omega + B_n \sin n\omega\},$$

and if we know the values of φ on the circumference of a circle with center at the origin, we may find the coefficients by the method of Fourier as in § 140a, 83).

Let us examine the motion in a segment between two walls making an angle 2α at the origin and reaching to infinity. If we use the value of φ given by equation 128), the coefficients and the values of n admissible are to be determined by the condition

$$\frac{\partial \varphi}{\partial n} = 0$$

along each wall. But since $dn = r d\omega$, we have

$$129) \qquad \frac{1}{r}\frac{\partial \varphi}{\partial \omega} = \sum_n n r^{n-1} \{B_n \cos n\omega - A_n \sin n\omega\}$$

which must vanish for $\omega = \pm \alpha$. If

$$n\alpha = \frac{2\varkappa + 1}{2}\pi, \quad \cos n\alpha = 0,$$

and if

$$n\alpha = \varkappa\pi, \quad \sin(\pm n\alpha) = 0,$$

$\varkappa$ being any integer. Therefore if we put when n is an odd multiple of $\frac{\pi}{2\alpha}$, $A_n = 0$ and $B_n = C_{2\varkappa+1}$ and for even multiples, $B_n = 0$ and $A_n = C_{2\varkappa}$, we shall have as a solution of the problem

$$120) \quad \varphi = \sum_\varkappa \left[C_{2\varkappa+1} r^{\frac{2\varkappa+1}{2}\frac{\pi}{\alpha}} \sin\left(\frac{2\varkappa+1}{2} \cdot \frac{\pi}{\alpha} \cdot \omega\right) + C_{2\varkappa} r^{\frac{\varkappa\pi}{\alpha}} \cos\left(\frac{\varkappa\pi}{\alpha}\omega\right) \right].$$

The tangential velocity at the wall is given by

$$131) \qquad \left(\frac{\partial \varphi}{\partial r}\right)_{\omega=\alpha} = \left[C_1 \frac{\pi}{2\alpha} r^{\frac{\pi}{2\alpha} - 1} \sin\frac{\pi}{2} + \cdots \right].$$

The exponent of the lowest power of r is $\frac{\pi}{2\alpha} - 1$. If this is negative, that is if $\alpha > \frac{\pi}{2}$, the velocity is infinite for $r = 0$, that is at the corner, unless $C_1 = 0$.

The pressure is given by the equation

$$p = const. - \frac{\varrho}{2} q^2,$$

so that at a sharp projecting edge around which the water flows there would be an infinite negative pressure. This being impossible, around such an edge the motion is discontinuous, so that instead of flowing as in a) Fig. 166, the water flows as in b), the flow being discontinuous at the dotted line. In actual fluids such surfaces of discontinuity give rise to vortex motion, so that we see eddies formed at projecting corners.

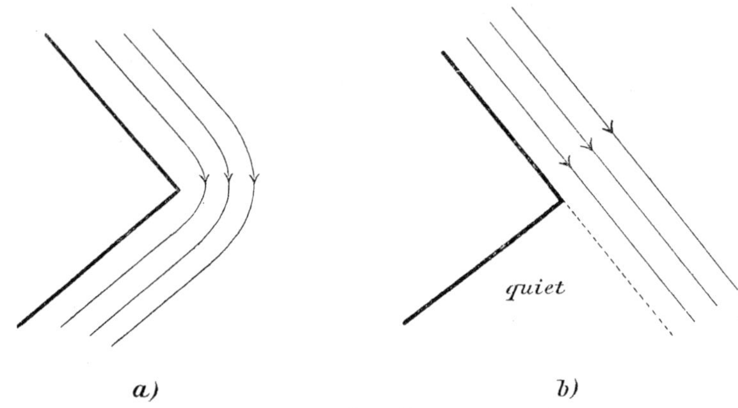

Fig. 166.

The function

$$\log z = \log r + i\omega$$

with

$$\varphi = \log r, \quad \psi = \omega,$$

gives us radial stream lines forming a uniplanar squirt, while

$$\varphi = \omega, \ \psi = \log r$$

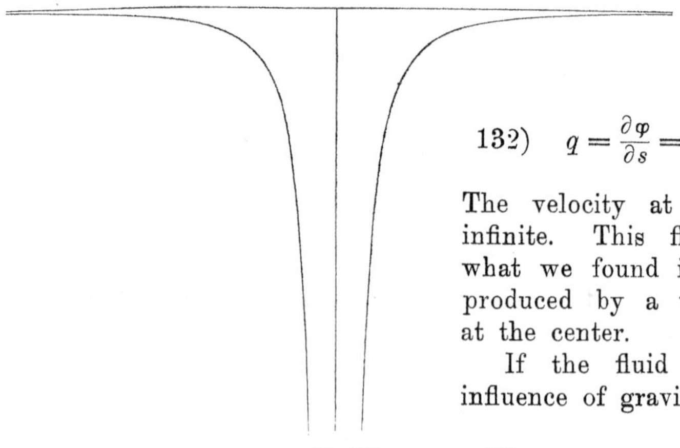

gives us flow in circles with a velocity

$$132) \quad q = \frac{\partial \varphi}{\partial s} = \frac{1}{r}\frac{\partial \varphi}{\partial \omega} = \frac{1}{r}.$$

The velocity at the center is infinite. This flow is exactly what we found in § 195 to be produced by a vortex filament at the center.

If the fluid is under the influence of gravity, we have

$$42) \quad p = const. - g\varrho z - \frac{1}{2}\varrho q^2$$

Fig. 167.

and for a free surface, $p = 0$,

$$133) \qquad C - g\varrho z - \frac{1}{2}\frac{\varrho}{r^2} = 0.$$

If z is zero when $r=\infty$, $C=0$, and the equation of the surface is

$$-2gz=\frac{1}{r^2}.$$

The form of the surface is shown in Fig. 167. This is approximately the form taken by the water running out of a circular orifice in the bottom of a tank, although the above investigation takes no account of the vertical motion.

198. Wave Motion. The case of uniplanar water waves may be dealt with by the method of the preceding section.[1] Let us take the XY-plane vertical, the Y-axis pointing vertically upward and the motion as before independent of the z coordinate, so that we may use z to denote the complex variable. We shall find that the waves travel with a constant velocity and it will therefore simplify the problem if we impress upon the whole mass of liquid an equal and opposite velocity so that the waves stand still and the motion is *steady*. Such still waves are actually seen on the surface of a running stream.

Let us first consider waves in very deep water. At a great depth the vertical motion will disappear and we shall have only the constant horizontal velocity that we have impressed, so that

$$u=-a,\quad v=0,$$

from which

$$\varphi=-ax.$$

The function

$$f(z)=-az+Ae^{-ikz}=-a(x+iy)+Ae^{-ik(x+iy)}$$

gives

$$\varphi+i\psi=-a(x+iy)+Ae^{ky}(\cos kx-i\sin kx),$$

134)
$$\varphi=-ax+Ae^{ky}\cos kx,$$
$$\psi=-ay-Ae^{ky}\sin kx.$$

When $y=-\infty$ this makes $\varphi=-ax$, as required. The free surface of the water being composed of stream lines is represented by one of the lines $\psi=const.$ and if we take the origin in the surface its equation is consequently

135)
$$ay+Ae^{ky}\sin kx=0,$$

which shows that y is a periodic function of x with the wave-length $\lambda=\frac{2\pi}{k}$. The longer the wave-length, that is the smaller k, the more

1) Rayleigh, *On Waves*. Phil. Mag. I, pp. 257—279, 1876. Scientific Papers, Vol. I, p. 261.

nearly does the exponential reduce to unity and the more nearly is the profile a curve of sines. The velocity is given by

136) $$q^2 = u^2 + v^2,$$

where

137) $$u = \frac{\partial \varphi}{\partial x} = \frac{\partial \psi}{\partial y} = -a - Ake^{ky}\sin kx,$$
$$v = \frac{\partial \varphi}{\partial y} = -\frac{\partial \psi}{\partial x} = Ake^{ky}\cos kx,$$

138) $$q^2 = a^2 + A^2k^2e^{2ky} + 2Aake^{ky}\sin kx.$$

So far all our work has been kinematical. The relation to dynamics is given by introducing the equation 33) for steady motion,

139) $$\frac{p}{\varrho} + gy + \frac{1}{2}q^2 = C,$$

and at the surface putting $p = 0$, and making use of the equation 135),

140) $$gy + \frac{1}{2}\{a^2 + A^2k^2e^{2ky} - 2a^2ky\} = C.$$

Since the surface passes through the origin, putting $y = 0$ we obtain

$$C = \frac{1}{2}\{a^2 + A^2k^2\},$$

inserting which gives

141) $$(g - a^2k)y + \frac{1}{2}A^2k^2(e^{2ky} - 1) = 0.$$

This equation can be only approximately fulfilled, but if the height of the waves is small compared with the wave-length, so that $2ky$ is small, developing the exponential and neglecting terms of higher order than the first in ky we have

$$(g - a^2k + A^2k^3)y = 0,$$

giving the equation connecting the velocity and wave-length

142) $$g - a^2k + A^2k^3 = 0.$$

If ky is small the equation of the surface 135) is approximately

143) $$y = -\frac{A}{a}\sin kx$$

so that the maximum height of the waves above the origin is $B = \frac{A}{a}$. Inserting the values of the height and wave-length in equation 142) it becomes

144) $$a^2\left\{\frac{2\pi}{\lambda}\left(1 - \frac{4\pi^2B^2}{\lambda^2}\right)\right\} = g,$$

an equation connecting the wave-length, height and velocity. For waves long enough in comparison with their height to neglect $\frac{4\pi^2 B^2}{\lambda^2}$, we have

$$145) \qquad a^2 = \frac{g\lambda}{2\pi}.$$

If s is the height from which a body must fall to acquire a velocity equal to the wave-velocity, since $a^2 = 2gs$, the equation becomes

$$146) \qquad \lambda = 4\pi s,$$

accordingly the velocity of propagation of long waves in deep water is equal to the velocity acquired by a body falling freely from a height equal to one-half the radius of a circle whose circumference is the wave-length.

In order to study the motions of individual particles of water let us now impress upon the motion given by 137) a uniform velocity a in the X-direction. Equations 137) now give the motion with respect to moving axes travelling with the waves, so that in order to obtain the motion with respect to fixed axes we have to add a to the u of 137) and replace x by $x - at$, obtaining

$$147) \qquad \begin{aligned} u &= -Ake^{ky}\sin k(x-at), \\ v &= \quad Ake^{ky}\cos k(x-at), \end{aligned}$$

for the equations of the unsteady motion of the actual wave-propagation. For the velocity of a particle we have

$$148) \qquad q = \sqrt{u^2+v^2} = Ake^{ky}$$

showing that the velocity decreases rapidly as we go below the surface, so that for every increase of depth of one wave-length it is reduced in the ratio $e^{-2\pi} = .001\,867$. If the displacement of a particle which when at rest was at x, y is ξ, η we have

$$149) \qquad \begin{aligned} \frac{d\xi}{dt} &= -Ake^{ky}\sin k(x-at), \\ \frac{dy}{dt} &= \quad Ake^{ky}\cos k(x-at), \end{aligned}$$

if we neglect the small change of velocity from x, y to $x+\xi, y+\eta$, so that we obtain by integration

$$150) \qquad \begin{aligned} \xi &= -Be^{ky}\cos k(x-at), \\ \eta &= -Be^{ky}\sin k(x-at). \end{aligned}$$

Thus each particle performs a uniform revolution in a circle of radius Be^{ky} in the periodic time $\frac{2\pi}{ka} = \frac{\lambda}{a}$. We thus see how the motion is confined to the surface layers. The direction of the motion in the orbit is such that particles at the crest of the wave move in the direction of the wave-propagation, those at a trough in the opposite direction.

Let us now discuss the form of the wave-profile 135) when the restriction that the height of the waves is small in comparison with the wave-length is removed. The equation of the surface is

151) $$y + Be^{ky} \sin kx = 0.$$

This may be conveniently done by means of a graphical construction, Fig. 168. Let us construct two curves, with the running coordinates X, Y, the first the logarithmic curve

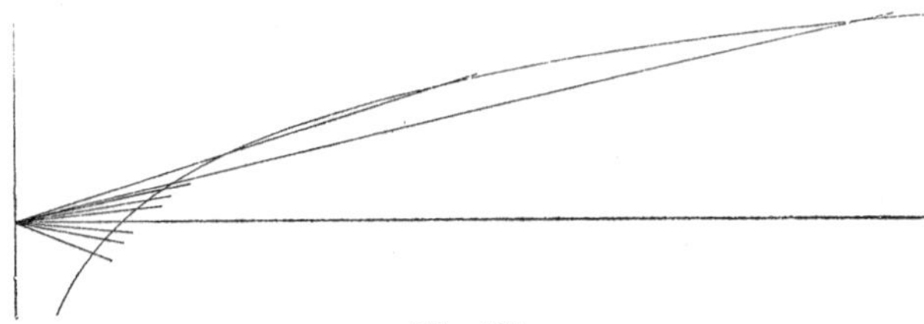

Fig. 168.

$$X = e^{kY}$$

and the second the straight line $X = -\frac{Y}{B \sin kx}$, which must be separately constructed for each value of x. At the intersection of the line and curve, we have

$$Y + Be^{kY} \sin kx = 0,$$

so that the value of Y thus obtained may be taken for the y coordinate of the wave-profile with the abscissa x. As x varies, the line swings back and forth about the X-axis, and we see that when $\sin kx$ is positive there is one intersection of the line and curve, while if $\sin kx$ is negative there are two, giving two values of y, bot positive. Any positive y is greater in absolute value than the corresponding negative for the symmetrical position of the line. Thus the unsymmetrical nature of trough and crest is made evident. Beginning with $x = 0$, the two values of y are one zero, the other

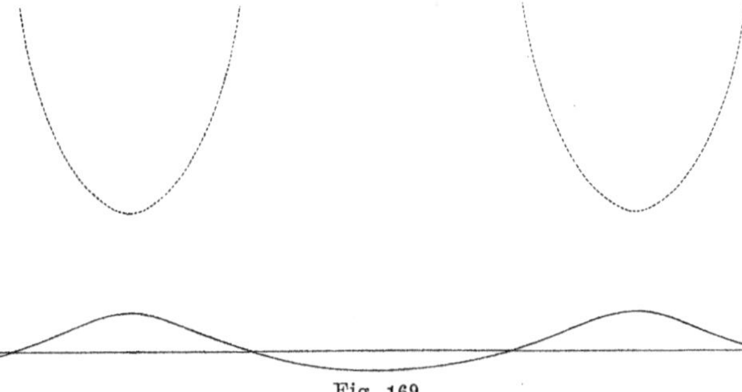

Fig. 169.

infinity, and as x increases, y has a single negative value. When $x = \frac{\pi}{k} = \frac{\lambda}{2}$, y is again zero and infinity, and as x increases the two values of y, both positive approach each other until $y = \frac{3}{4}\lambda$, then recede until $y = \lambda$. The form of the curve as constructed in this manner is shown in Fig. 169, the lower branch representing the wave-profile. If B is greater than a certain quantity the values of y between certain limits are imaginary. This limiting value of B is that which makes the highest position of the straight line, for which $\sin kx = -1$, tangent to the exponential curve. We then have

$$\frac{dX}{dY} = kX \text{ for the curve, equal to } \frac{X}{Y} = \frac{1}{B} \text{ for the line,}$$

from which

$$kY = 1 = ky, \quad \frac{1}{B} = ke.$$

The upper and lower branches of the curve 151) then come together, and the wave-profile has an angle. Waves cannot be higher than this without breaking. By differentiation of 151) we find for the summit, $\frac{dy}{dx} = \pm 1$, so that the angle between the two sides of the wave is a right angle (Fig. 170). As a matter of fact, before the waves are as high as this, the equation 141) is no longer satisfied with sufficient approximation for the waves to have the form in question. By an elaborate system of approximation, Michell[1]) has shown that the highest waves have a height $.142\lambda$, while the equation 151) gives $.203\lambda$. It was shown by Stokes[2]) that at the crest the angle was not 90^0, but 120^0, as follows.

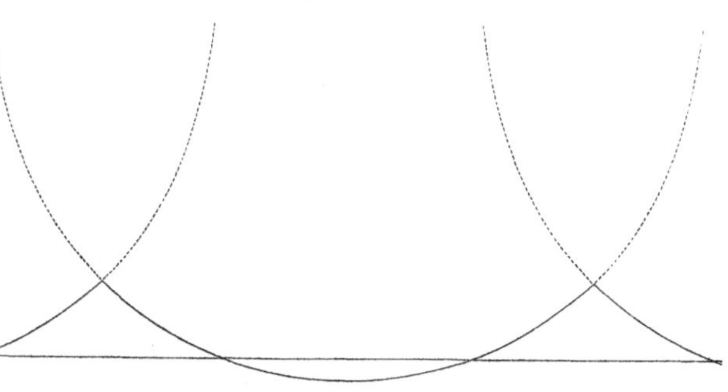

Fig. 170.

In the stationary wave, in order to have an edge, the velocities u and v for a particle at the surface must both vanish together, for if v alone vanishes, there will be a horizontal tangent. Consequently, if we place the origin at the crest, equation 139) becomes

$$gy + \frac{1}{2}q^2 = 0.$$

But if we represent the surface by a deyelopment of the form of equation 128), on account of symmetry there will be only sine terms, and if in the neigborhood of the origin we retain only the most important term, we may put

1) Michell, *The highest Waves in Water.* Phil. Mag. 36, p. 430, 1893.

2) Stokes, *On the Theory of Oscillatory Waves.* Trans. Cambridge Philosophical Society, Vol. VIII, p. 441, 1847. Math. and Phys. Papers, Vol. I, p. 227.

152) $$\varphi = Ar^n \sin n\omega, \quad \psi = Ar^n \cos n\omega,$$

ω being the angle measured from the vertical. We have for the radial velocity

$$q_r = \frac{\partial \varphi}{\partial r} = Anr^{n-1} \sin n\omega,$$

and if α is the inclination of the surface to the vertical at the crest $q = Anr^{n-1} \sin n\alpha$. But we have $q^2 = -2gy = 2gr \cos\alpha$ and accordingly $2(n-1) = 1$, $n = \frac{3}{2}$. Also as in 120), $\cos n\alpha = 0$. Thus

$$\frac{3}{2}\alpha = \frac{\pi}{2}, \quad 2\alpha = \frac{2\pi}{3} = 120^0.$$

The problem of waves in water of finite depth may be treated in a similar manner, by putting instead of 134),

$$\varphi + i\psi = -az + Ae^{-ikz} + Be^{ikz},$$

153)
$$\begin{aligned} \varphi &= -ax + (Ae^{ky} + Be^{-ky}) \cos kx, \\ \psi &= -ay - (Ae^{ky} - Be^{-ky}) \sin kx, \\ u &= -a - k(Ae^{ky} + Be^{-ky}) \sin kx, \\ v &= k(Ae^{ky} - Be^{-ky}) \cos kx. \end{aligned}$$

If the depth is h, we must have $v = 0$ for $y = -h$, giving

$$Ae^{-kh} = Be^{kh}.$$

Calling this value C, we have

154) $$\psi = -ay - C(e^{k(h+y)} - e^{-k(h+y)}) \sin kx = 0,$$

as the equation for the wave-profile. For the first approximation, for waves whose height is small compared to their length, replacing e^{ky}, e^{-ky} by unity, we have

155) $$ay = -C(e^{kh} - e^{-kh}) \cdot \sin kx,$$

and neglecting $(Ck)^2$,

156) $$u^2 + v^2 = a^2 + 2Cak(e^{kh} + e^{-kh}) \sin kx.$$

Thus the surface equation 139) becomes

157) $$const. = \frac{p}{\varrho} - \frac{Cg}{a}(e^{kh} - e^{-kh}) \sin kx + \frac{1}{2}\left[a^2 + 2Cak(e^{kh} + e^{-kh}) \sin kx\right]$$

which is satisfied by

158) $$ak(e^{kh} + e^{-kh}) - \frac{g}{a}(e^{kh} - e^{-kh}) = 0,$$

giving the velocity

159) $$a^2 = \frac{g}{k} \frac{e^{kh} - e^{-kh}}{e^{kh} + e^{-kh}}.$$

If h is infinite this reduces to 145), while if the depth is very small with respect to the wave-length, it reduces to $a^2 = gh$. Accordingly long waves in shallow water travel with a velocity independent of their length, being the velocity acquired by a body falling through a distance equal to one-half the depth of the water. Consequently the resultant of such waves of different wave-lengths is propagated without change, contrary to what is the case in deep water.

Changing to fixed axes, we have for the running wave

$$160)\quad \begin{aligned} \varphi &= \quad C\left(e^{k(h+y)} + e^{-k(h+y)}\right) \cos k(x - at), \\ \psi &= -C\left(e^{k(h+y)} - e^{-k(h+y)}\right) \sin k(x - at), \end{aligned}$$

and by comparison with 147), 150), we find that the particles describe ellipses with semi-axes equal to

$$161)\quad \frac{C}{a}\left(e^{k(h+y)} + e^{-k(h+y)}\right), \quad \frac{C}{a}\left(e^{k(h+y)} - e^{-k(h+y)}\right).$$

If we consider the resultant of two equal wave-trains running in opposite directions, we have

$$162)\quad \begin{aligned} \varphi &= \quad C\left(e^{k(h+y)} + e^{-k(h+y)}\right) \left[\cos k(x - at) + \cos k(x + at)\right] \\ &= \quad 2C\left(e^{k(h+y)} + e^{-k(h+y)}\right) \cos kx \cos kat, \\ \psi &= - \quad C\left(e^{k(h+y)} - e^{-k(h+y)}\right) \left[\sin k(x - at) + \sin k(x + at)\right] \\ &= -2C\left(e^{k(h+y)} - e^{-k(h+y)}\right) \sin kx \cos kat. \end{aligned}$$

The equation of the profile is now of the form, y is equal to a function of x multiplied by a function of t, so that the profile is always of the same shape, with a varying vertical scale. Such waves are called *standing* waves, and we see them in a chop sea. The difference between them and the stationary wave in a running stream, with which we began, is very marked, as here every point on the surface oscillates up and down, while there the water-profile was invariable both as to time and place.

199. Equilibrium Theory of the Tides. We shall now briefly consider some aspects of the phenomena of the tides, the general theory of which is far too complicated to be dealt with here. The earliest theory historically is that proposed by Newton, which supposes that the water covering the earth assumes, under the attraction of a disturbing body, the form that it would have if *at rest* under the action of the forces in question. This so-called equilibrium theory, which neglects the inertia of the water, belongs logically to the subject of hydrostatics, but will be now treated. If U denote the potential of gravity, including the centrifugal force, as in § 149, we have, as there, for the undisturbed surface of the ocean,

163) $$U(r_0, \psi, \varphi) = const.$$

r_0, ψ, φ, denoting the radius, latitude, and longitude. If V denote the potential of the disturbing body, we have according to the equilibrium theory, for the disturbed surface,

164) $$U(r, \psi, \varphi) + V = const.$$

and subtracting equation 163) from this, we have

165) $$U(r, \psi, \varphi) - U(r_0, \psi, \varphi) + V = const. = C.$$

But if we put $h = r - r_0$, h is the height of the tide, and being small with respect to the radius, we may put

166) $$U(r, \psi, \varphi) - U(r_0, \psi, \varphi) = h\frac{\partial U}{\partial r},$$

giving

167) $$V - C = -h\frac{\partial U}{\partial r}.$$

But $g = -\gamma\frac{\partial U}{\partial r}$, as in § 149, so that we obtain for the height of the tide

168) $$h = \gamma\frac{V-C}{g}.$$

We may determine the constant in 168) by the consideration that the total volume of the water is constant. If dS is the area of an element of the earth's surface, the total volume of the tide above the surface of equilibrium must vanish, giving

169) $$0 = \iint h\,dS, \quad \iint V dS = C\iint dS, \quad \overline{V} = C,$$

where $\overline{V}$ is the mean value of the disturbing potential over the earth's surface. Now we have found in § 150, equation 154), the value of the potential of the tide-generating forces,

170) $$V = \frac{mr^2}{2D^3}(3\cos^2 Z - 1),$$

where Z is the zenith-distance of the heavenly body at the point in question. If we refer other points on the earth's surface to polar coordinates with respect to this point and any plane through it, with coordinates Z, Φ, we have

$$dS = r^2 \sin Z\,dZ\,d\Phi,$$

$$\iint V dS = \frac{mr^4}{2D^3}\int_0^{2\pi} d\Phi \int_0^{\pi}(3\cos^2 Z - 1)\sin Z\,dZ = 0,$$

so that the mean of V vanishes. Accordingly we have

171) $$h = \frac{\gamma m r^2}{2gD^3}(3\cos^2 Z - 1).$$

This equation shows that the tidal surface is a prolate ellipsoid of revolution, with its axis pointing at the disturbing body.

Let us now express $\cos Z$ in terms of the latitude ψ of the point of observation and of the declination δ and hour-angle H of the disturbing body, which for brevity we shall call the moon. If we take axes in the earth as usual, with the XZ-plane passing through the point of observation and measure H from this plane, we have for its coordinates and those of the moon respectively

$$\begin{array}{ll} r\cos\psi, & D\cos\delta\cos H, \\ 0\quad, & D\cos\delta\sin H, \\ r\sin\psi, & D\sin\delta\quad, \end{array}$$

from which we obtain the cosine of the angle included by their radii

$$\cos Z = \cos\psi\cos\delta\cos H + \sin\psi\sin\delta.$$

Squaring this, replacing $\cos^2 H$ by $\frac{1}{2}(1+\cos 2H)$, $\cos^2\psi\cos^2\delta$ by $(1-\sin^2\psi)(1-\sin^2\delta)$, we easily obtain

$$3\cos^2 Z - 1 = \frac{3}{2}\left[\cos^2\delta\cos^2\psi\cos 2H + \sin 2\delta\sin 2\psi\cos H + \frac{(1-3\sin^2\delta)(1-3\sin^2\psi)}{3}\right].$$

Inserting this in 171), replacing g by its approximate value $\frac{\gamma M}{r^2}$ and, as we have already done, neglecting the attraction of the disturbed water, we have the equation for the tide,

$$172)\quad \frac{h}{r} = \frac{3mr^3}{4MD^3}\left[\cos^2\delta\cos^2\psi\cos 2H + \sin 2\delta\sin 2\psi\cos H + \frac{(1-3\sin^2\delta)(1-3\sin^2\psi)}{3}\right].$$

The first term in the brackets, containing the factor $\cos 2H$, where H is the moon's hour-angle at the point of the earth in question, is periodic in one-half a lunar day, consequently this term has a maximum when the moon is on the meridian, both above and below, low water when the moon is rising or setting. The effect of this term is the semi-diurnal tide, which is the most familiar, with two high and two low waters each day. This tide is a maximum for points on the equator, where $\cos^2\psi = 1$, and for those times of the month when $\cos^2\delta = 1$, that is when the moon is crossing the equator. These are the so-called equinoctial tides.

The second term, containing the factor $\cos H$, is periodic in a lunar day, and gives the diurnal tide. This gives high water under the moon, and low water on the opposite side of the earth. On the side toward the moon, these two tides are therefore added, while on the opposite side we have their difference. Consequently, at any point, the difference of two consecutive high waters is twice the diurnal tide. This difference is generally small, showing that the latter tide is small. It vanishes for points on the equator, and at the times of the equinoctial tides.

The third term, which vanishes for latitude 35° 16′, does not depend on the moon's hour-angle, but only on its declination. This declinational tide, depending on the square of $\sin\delta$, has a period of one-half a lunar month.

Beside the tides due to the moon, we must add those due to the sun, for which the factor outside the brachets in 172) is somewhat less than one-half that due to the moon. The highest tides therefore occur at those times in the month when the sun and the moon are on the meridian together, namely at new and full moon. These are known as spring-tides. The lowest occur when the moon is in quadrature with the sun, and the lunar and solar tides are in opposition. These are known as neap-tides, and according to this theory would be only one-third the height of the spring-tides. The greatest spring-tides would be those in which the moon was on the equator, or the equinoctial spring-tides. Now it is found that, instead of this, the high tides come about a day and a half later. Consequently, although the equilibrium theory indicates to us the general nature of the different tides to be expected, it does not give us an accurate expression for their values. Roughly speaking we may say that the tides act as if they were produced as described by the action of the sun and moon, but that the time of arrival of the effects produced was delayed.

A correction was introduced into the equilibrium theory by Lord Kelvin, to take account of the effect of the continents For if the height of the tide were given by the equation 171), removing the various volumes of water in the space actually occupied by land would subtract an amount of water now positive, now negative, so that the condition of constant volume would not be fulfilled. In order that it still may do so, the integral 169) is to be taken only over those parts of the earth's surface covered by the sea. The value of $\overline{V}$ is then not zero. The effect of this is to introduce at each point on the earth's surface change of time of the arrival of each tide, varying from point to point. The practical effect of this correction is not large.

200. Tidal Waves in Canals. In the dynamical theory of the tides, taking account of the inertia of the water, we have the problem of the forced oscillations of the sea under periodic forces. As a simple example illustrating this method we shall consider waves in straight canals. Let the motion be in the plane of XY, as in § 198, and let h, the depth of the canal, be small in comparison with the wave-length. We shall suppose the displacements of all the particles, with their velocities and their space-derivatives, to be small quantities whose squares and products may be neglected. We shall also neglect the vertical acceleration, so that the equation for y is that of hydrostatics, giving the pressure proportional to the distance below the surface. If the ordinate of the free surface is $h + \eta$, this gives

173) $$p = g\varrho\,(h + \eta - y),$$

174) $$\frac{\partial p}{\partial x} = g\varrho\frac{\partial \eta}{\partial x},$$

while the equation of motion, the first of equations 6), is

175) $$\frac{\partial u}{\partial t} = X - \frac{1}{\varrho}\frac{\partial p}{\partial x}.$$

Combining these two equations, we have

176) $$\frac{\partial u}{\partial t} = X - g\frac{\partial \eta}{\partial x},$$

and if X is independent of y, since $\frac{\partial \eta}{\partial x}$ is also, this shows that u depends only on x and t, or vertical planes perpendicular to the XY plane remain such during the motion.

Integrating the equation of continuity

$$\frac{\partial u}{\partial x} + \frac{\partial v}{\partial y} = 0,$$

with respect to y from the bottom to the surface,

177) $$v = -\int\frac{\partial u}{\partial x}dy = -(h + \eta)\frac{\partial u}{\partial x},$$

or approximately, at the surface,

178) $$v = \frac{\partial \eta}{\partial t} = -h\frac{\partial u}{\partial x}.$$

Now putting $u = \frac{\partial \xi}{\partial t}$, the equation of continuity 178) becomes,

179) $$\frac{\partial \eta}{\partial t} = -h\frac{\partial^2 \xi}{\partial t\,\partial x}$$

and on integration with respect to the time,

180) $$\eta = -h\frac{\partial \xi}{\partial x}.$$

Substituting in 176) we have for the horizontal displacement

181) $$\frac{\partial^2 \xi}{\partial t^2} = X + gh\frac{\partial^2 \xi}{\partial x^2}.$$

If there is no disturbing force, $X = 0$, and we have the equation

$$182)\qquad \frac{\partial^2 \xi}{\partial t^2} = gh\frac{\partial^2 \xi}{\partial x^2}$$

for the propagation of free waves, which we might have used in order to obtain the results of § 198, for instance it is satisfied by equations 150) if we put $a^2 = gh$. This is the same equation as we had in § 46, equation 109), for the motion of a stretched string, and the standing waves of 162), § 198, putting $y = 0$, are the normal vibrations of equation 115), § 46. The general solution of equation 182) is obtained in the next section, for the present it is sufficient to consider the wave already obtained which advances unchanged in form with the velocity a. We have then, in the case of an endless canal encircling the earth, the curvature of which we may neglect, the case of a free wave, running around and around, without change, so that at any point, the motion is periodic in the time $\frac{l}{a}$, where l is the length of the endless canal. We thus have a system with free periods, and when we consider the action upon it of periodic disturbing forces, we may expect the phenomena of resonance, as described in Chapter V.

Let us now suppose the canal coincides with a parallel of latitude, and that x is measured to the westward from a certain meridian. We then have for the horizontal component of the disturbing force

$$X = -\gamma\frac{\partial V}{\partial x}$$

where V is given by 170), and H, the hour-angle of the moon at the point x, is

$$183)\qquad H = \omega t - \frac{x}{r\cos\psi}, \quad \text{so that } -\frac{\partial V}{\partial x} = \frac{\partial V}{\partial H}\frac{1}{r\cos\psi},$$

ω being the angular velocity of the moon with respect to the earth. We accordingly find X to be composed of two terms each of the form

$$-A\sin(mt - kx),$$

where for the semi-diurnal part

$$184)\qquad A = \frac{3}{2}\frac{\gamma m r}{D^3}\cos^2\delta\cos\psi, \quad m = 2\omega, \quad k = \frac{2}{r\cos\psi} = \frac{4\pi}{l}.$$

Introducing this into the equation 181),

$$185)\qquad \frac{\partial^2 \xi}{\partial t^2} = a^2\frac{\partial^2 \xi}{\partial x^2} - A\sin(mt - kx),$$

we may find a solution

186) $$\xi = B \sin (mt - kx),$$

where by insertion in 185) we find

187) $$B = \frac{A}{m^2 - a^2 k^2}.$$

From 180) we obtain

188) $$\eta = \frac{Akh}{a^2 k^2 - m^2} \cos (mt - kx).$$

The coefficient of the cosine is positive or negative according as ak is greater or less than m, so that we have, according to circumstances, high or low water under the moon. In the former case, the tides are said to be direct, as in the equilibrium theory, in the latter they are inverted. But $\frac{ka}{m}$ is the ratio of the time period of the force, or half a lunar day, to the time required for a free wave to travel half around the earth, and the tide is direct or inverted according as this is greater or less than unity. Equation 188) is the analogue of equation 50), § 44. Inserting the values of the constants in 188) we find that the canal theory gives the height of the tide as given by the equilibrium theory in 172) (which we also obtain by putting $m = 0$), multiplied by the factor

$$\frac{1}{1 - \left(\frac{m}{ak}\right)^2},$$

exactly as described for the system with one degree of freedom on page 155. If we introduced into our equations a term giving the effect of friction we should obtain a change of phase, as in § 44, of amount other than a half-period, or inversion.

In order to determine the directness or inversion of the tides, let us insert the values of m, k from 184) in 188), by which we find that the tides are direct or inverted according as we have the upper or lower sign in the inequality

189) $$gh \gtrless r^2 \omega^2 \cos^2 \psi.$$

Supposing the lunar day to be 24 hours, 50 minutes, the earth's circumference forty million meters, we find at the equator the critical depth, determining the inversion, to be 20.46 kilometers, or 12.7 miles. As the depth is less than this, the tides are inverted. For any depth less than the critical depth, there will be a latitude beyond which the tides will be direct. Accordingly we see that even if we consider the ocean to be composed of parallel canals separated by partitions, the tides will be very different in different latitudes, so that if the partitions be removed, water will flow north and south. We thus obtain an idea of the complication of the actual motion of the tides.

By introducing the complete expressions for the accelerations with respect to revolving axes, given in § 104, and applying the principles of forced oscillations, we obtain the more complete theory given by Laplace.

201. Sound-Waves. Let us now consider the motion of a compressible fluid which takes place in the propagation of sound. In the production of all ordinary sounds, except those violent ones produced by explosions, the motion of each particle of air is extremely minute. We shall therefore suppose that the velocity components u, v, w and their space derivatives are so small that their squares and products may be neglected. Let us put

$$190)\qquad \varrho = \varrho_0 (1 + s),$$

where ϱ_0 is a constant and s is a small quantity, of the same order as the velocities, called the compression. From the equation of continuity we have

$$191)\qquad \sigma = \frac{\partial u}{\partial x} + \frac{\partial v}{\partial y} + \frac{\partial w}{\partial z} = -\frac{1}{\varrho}\frac{\partial \varrho}{\partial t} = -\frac{1}{1+s}\frac{\partial s}{\partial t},$$

or neglecting the product of s and its derivative,

$$192)\qquad \sigma = -\frac{\partial s}{\partial t}.$$

In order to calculate P, we have, since the changes in ϱ are small

$$193)\qquad dp = a^2\, d\varrho = a^2 \varrho_0\, ds,$$

where a^2 is a constant representing the value of the derivative $\frac{dp}{d\varrho}$ for $\varrho = \varrho_0$, the density of the air at atmospheric pressure. We therefore have

$$194)\qquad P = \int \frac{dp}{\varrho} = \int \frac{a^2 ds}{1+s} = a^2 \log (1+s) = a^2 s,$$

to the same degree of approximation.

Neglecting small quantities the equations of hydrodynamics 6) become, when there are no applied forces,

$$195)\qquad \begin{aligned} \frac{\partial u}{\partial t} &= -a^2 \frac{\partial s}{\partial x}, \\ \frac{\partial v}{\partial t} &= -a^2 \frac{\partial s}{\partial y}, \\ \frac{\partial w}{\partial t} &= -a^2 \frac{\partial s}{\partial z}, \end{aligned}$$

with

$$192)\qquad \sigma = -\frac{\partial s}{\partial t}.$$

Differentiating the equations respectively by x, y, z, adding and observing the definition of σ, we obtain

196) $$\frac{\partial \sigma}{\partial t} = - a^2 \Delta s,$$

and differentiating 192) by t and combining with this

197) $$\frac{\partial^2 s}{\partial t^2} = a^2 \Delta s.$$

Since the motion is assumed to be irrotational, introducing the velocity potential into equations 195) they become the derivatives by x, y, z respectively of the equation

198) $$\frac{\partial \varphi}{\partial t} = - a^2 s.$$

Differentiating by t, making use of equation 192),

199) $$\frac{\partial^2 \varphi}{\partial t^2} = - a^2 \frac{\partial s}{\partial t} = a^2 \sigma = a^2 \Delta \varphi.$$

Thus both the velocity potential and the compression satisfy the differential equation

200) $$\frac{\partial^2 \varphi}{\partial t^2} = a^2 \Delta \varphi.$$

This is known as the differential equation of wave-motion and is the basis of the theories of sound and light.

202. Plane Waves. Let us suppose first that the motion is the same at all points in each plane perpendicular to a given direction which we will take for that of the X-axis. Thus all the quantities concerned become independent of y and z and equation 200) reduces to

201) $$\frac{\partial^2 \varphi}{\partial t^2} = a^2 \frac{\partial^2 \varphi}{\partial x^2},$$

which is equation 109) § 46, the equation for the motion of a continous string, or equation 182) § 200, for the propagation of long waves in shallow water.

The general solution of this equation is found by introducing the two new independent variables $p = x - at$, $q = x + at$.

We have then

202) $$\begin{aligned} \frac{\partial \varphi}{\partial t} &= \frac{\partial \varphi}{\partial p}\frac{\partial p}{\partial t} + \frac{\partial \varphi}{\partial q}\frac{\partial q}{\partial t} = - a \frac{\partial \varphi}{\partial p} + a \frac{\partial \varphi}{\partial q}, \\ \frac{\partial \varphi}{\partial x} &= \frac{\partial \varphi}{\partial p}\frac{\partial p}{\partial x} + \frac{\partial \varphi}{\partial q}\frac{\partial q}{\partial x} = \frac{\partial \varphi}{\partial p} + \frac{\partial \varphi}{\partial q}, \\ \frac{\partial^2 \varphi}{\partial t^2} &= a^2 \left\{ \frac{\partial^2 \varphi}{\partial p^2} + \frac{\partial^2 \varphi}{\partial q^2} - 2 \frac{\partial^2 \varphi}{\partial p \partial q} \right\}, \\ \frac{\partial^2 \varphi}{\partial x^2} &= \frac{\partial^2 \varphi}{\partial p^2} + \frac{\partial^2 \varphi}{\partial q^2} + 2 \frac{\partial^2 \varphi}{\partial p \partial q}. \end{aligned}$$

Inserting these values in 201), we have

203) $$\frac{\partial^2 \varphi}{\partial p \partial q} = 0.$$

Integrating with respect to q, we find that $\frac{\partial \varphi}{\partial p}$ is independent of q, but may be an arbitrary function of p, say

$$\frac{\partial \varphi}{\partial p} = F_1'(p).$$

Integrating again we find

$$\varphi = F_1(p) + F_2(q), \tag{204}$$

where F_2 is an arbitrary function of q. Whatever the functions F_1 and F_2, this value will satisfy the equation 203). Replacing p and q by their values, we have the general solution

$$\varphi = F_1(x - at) + F_2(x + at). \tag{205}$$

Let us first assume $F_2 = 0$ and consider the solution

$$\varphi = F_1(p) = F_1(x - at). \tag{206}$$

The value of φ depending only on p is unchanged when x has increased by the amount at, that is to say, if φ be represented graphically as a function of x at the time $t = 0$, it will be represented at the time t by the same curve moved to the right a distance at. Such a motion is termed a *wave* moving with the velocity a ln the positive X-direction.

Similarly the solution

$$\varphi = F_2(q) = F_2(x + at) \tag{207}$$

represents a wave moving in the negative X-direction with the same velocity.

If the function $F(p)$ is zero except for a certain small range of values p_0, p_1, the motion is sometimes called a pulse. A pulse is none the less a wave.

Thus the general solution of equation 201) represents two plane waves propagated in opposite directions with the same velocity a.

The velocity of sound $a = \sqrt{\frac{dp}{d\varrho}}$ depends upon the elasticity of the air and was calculated by Newton, assuming that the process was isothermal, using Boyle's law. As this was found to give results not agreeing with experiment Laplace suggested that the compression was adiabatic, the vibrations being so rapid that the heat generated did not have time to flow from the heated to the cooled parts. Thus the constant $\varkappa$, equation 17) § 178 representing the ratio of the two specific heats of the air is introduced. The velocity of sound gives one of the most accurate ways of determining this ratio $\varkappa$.

The velocity of the particle of air is obtained by

$$u = \frac{\partial \varphi}{\partial x} = F_1'(x - at) \tag{208}$$

in the wave going to the right. The compression by

209) $$s = -\frac{1}{a^2}\frac{\partial \varphi}{\partial t} = \frac{1}{a} F_1'(x - at),$$

so that

$$u = as$$

Thus the velocity of the particle is in the same direction as that of the wave where there is condensation, or s is positive, in the opposite direction where there is rarefaction, or s is negative.

203. Echo. Organ-pipes. Suppose there is a rigid wall whose equation is $x = l$. The velocity of a particle normal to the wall must be zero, so that

210 $$u = F_1'(x - at) + F_2'(x + at) = 0$$

when $x = l$, or

211) $$F_1'(l - at) + F_2'(l + at) = 0$$

for all positive values of t. Thus one of the functions is determined by the other. Put

$$l + at = y,$$

so that our equation 211) is

212) $$F_2'(y) = - F_1'(2l - y),$$

a differential equation connecting F_1 and F_2, the integral of which is

213) $$F_2(y) = F_1(2l - y) + C$$

for $y \geqq l$. Since the velocity depends only on the derivative of φ, the value of C is immaterial, and we will put it equal to zero.

The equation $F_2(y) = F_1(2l - y)$ indicates that the curve representing F_2 is the geometrical reflection in the wall of that representing F_1, in other words the function F_2 represents a wave travelling to the left, which after $x - at$ is greater than l represents the motion on the left of the wall, the values of φ at points a certain distance to the left of the wall being the same as they would have been at the same distance to the right of the wall had the direct wave gone on unchanged. Since the values of u depend on the derivative of φ according to x, the velocity changes sign in the reflection. This must be the case for the condition producing reflection is that $u = 0$ at $x = l$, so that the wave coming to the left must have a velocity equal and opposite to that of the wave going to the right. If there is a wall at $x = 0$ as well as at $x = l$, the wave is reflected in that also, so that the motion consists in the continual to and fro motion of the original disturbance. The motion at any point is periodic in the time $\frac{2l}{a}$. We may accordingly develop the motion in a series of normal vibrations as in § 46. If we take the particular solution of equation 109) given in equation 115) of that section, and write

213) $$\varphi = A \cos kx \cos nt,$$

where $n = ka$, we have

214) $$u = \frac{\partial \varphi}{\partial x} = -Ak \sin kx \cos nt,$$
$$s = -\frac{1}{a^2}\frac{\partial \varphi}{\partial t} = A\frac{n}{a^2}\cos kx \sin nt.$$

Every particle oscillates with a simple harmonic motion with an amplitude $\frac{A}{a}\sin kx$, and we have a *pure tone.* The compression also varies harmonically with an amplitude $\frac{Ak}{a}\cos kx$. Thus the maximum pressure occurs at points of no motion, such points being called nodes. These occur where $kx = r\pi$, where r is any integer, or $x = \frac{\pi}{k}, \frac{2\pi}{k}, \frac{3\pi}{k}, \dots$ The wave-length being $\lambda = \frac{2\pi}{k}$, the nodes are separated by distances $\frac{\lambda}{2}$. The condensation s follows a similar law, but vanishes half way between the nodes where the motion is a maximum. The regions between the nodes are called loops. The maximum changes of s are at the nodes.

As there is no motion at the nodes, but only changes of pressure, we may place reflecting walls there and apply the theory to the case of a stopped organ-pipe, whose length is accordingly any number of half wave-lengths. If the ends of the pipe are at $x = 0$ and $x = l$ we have

$$l = r\frac{\lambda}{2}, \quad kl = r\pi, \quad k = \frac{r\pi}{l},$$

and the frequency is determined by

$$\frac{n}{2\pi} = \frac{ak}{2\pi} = \frac{ar}{2l}.$$

Consequently the possible frequencies for a simple harmonic vibration of a stopped pipe are in the ratio of the integers 1, 2, 3, etc.

For a pipe open at the end the condition is that the pressure is that of the external air, that is, there is a loop. Thus a pipe open at both ends has its length equal to an integral number of half wave-lengths, and has the same hormonics as a closed pipe. Opening one of the holes in a flute produces a loop, so that the tones of a flute are produced by the column of air between the mouth-piece and the first open hole.

For a pipe open at one end and stopped at the other, the length is equal to an odd number of quarter wave-lengths, so that the frequencies are proportional to the odd integers 1, 3, 5, etc.

204. Spherical Waves. If the velocity potential depends only on the distance from a fixed point, using the expression of § 135, equation 44), the wave-equation 200) becomes

$$215)\qquad \frac{\partial^2\varphi}{\partial t^2} = a^2\left\{\frac{\partial^2\varphi}{\partial r^2} + \frac{2}{r}\frac{\partial\varphi}{\partial r}\right\} = \frac{a^2}{r}\frac{\partial^2(r\varphi)}{\partial r^2}.$$

Multiplying by r, this is

$$216)\qquad \frac{\partial^2(r\varphi)}{\partial t^2} = a^2\frac{\partial^2(r\varphi)}{\partial r^2},$$

so that the product $r\varphi$ satisfies an equation like 201), of which the solution is

$$r\varphi = F_1(r-at) + F_2(r+at).$$

Accordingly we have

$$217)\qquad \varphi = \frac{1}{r}\{F_1(r-at) + F_2(r+at)\},$$

of which the first term represents a wave proceeding outwards, the second one proceeding inwards, the magnitude however varying according to the factor $\frac{1}{r}$.

For a periodic solution representing a simple tone proceeding from a single point-source we may take

$$218)\qquad \varphi = -\frac{A}{4\pi r}\cos k(at-r).$$

The physical meaning of the constant A is obtained as follows. Let us find the volume of air flowing in unit of time through the surface of a sphere with center at the source. We will call this the total current,

$$219)\qquad I = \iint q\cos(qn)\,dS = \iint \frac{\partial\varphi}{\partial r}dS = 4\pi r^2\frac{\partial\varphi}{\partial r}$$
$$= A\{\cos k(at-r) - kr\sin k(at-r)\}.$$

Accordingly when $r=0$ we have $I = A\cos kat$ and A, the maximum rate of emission of air per unit of time, is called the *strength* of the source, agreeing with the definition of § 196.

In order to obtain the activity of the source, that is the rate of emission of energy per unit of time, we may find the rate of working of the pressure at the surface of a sphere, as explained in § 188,

$$220)\qquad P = \iint pq\,dS = 4\pi r^2 p\frac{\partial\varphi}{\partial r}.$$

In order to find p, we have, if p_0 is the undisturbed atmospheric pressure, by integration of 193), and by 198),

221) $$p - p_0 = a^2 \varrho_0 s = - \varrho_0 \frac{\partial \varphi}{\partial t}$$

from which we obtain

222) $$P = A\left\{\cos k(at - r) - kr \sin k(at - r)\right\}\left\{p_0 - \frac{A \varrho_0 k a}{4 \pi r} \sin k(at - r)\right\}.$$

This contains a part which is alternately positive and negative, and also one which is always positive. If we seek the mean value of P throughout the period, that is

$$\overline{P} = \frac{1}{T}\int_0^T P\, dt, \quad T = \frac{2\pi}{ka},$$

we easily find, since the mean of $\cos\vartheta$, $\sin\vartheta$, $\cos\vartheta \sin\vartheta$, is zero, while the mean of $\sin^2\vartheta$ is $\frac{1}{2}$,

223) $$\overline{P} = \frac{A^2 \varrho_0 k^2 a}{8\pi}$$

which is independent of the radius, as it should be. The mean energy-flow per unit of time and per unit of area of the sphere is

224) $$\frac{\overline{P}}{S} = \frac{A \varrho_0 k^2 a}{32 \pi^2 r^2}$$

which is a measure of the intensity of the sound. This decreases as the inverse square of the distance. In order to give an idea of the extremely small dynamical magnitudes involved in musical sounds, it may be stated that measurements made by the author[1]) showed that the energy emitted by a cornet, playing with an average loudness, was 770 ergs per second, or about one ten-millionth of a horse-power, while a steam-whistle that could under favorable circumstances be heard twenty miles away emitted but $\frac{1.25 \times 10^2 \text{ ergs}}{\text{sec.}}$, or one-sixtieth of a horse-power (see note, p. 153).

205. Waves in a Solid. The equations of motion for an elastic solid are obtained from the equations of equilibrium 144, § 175 by the application of d'Alembert's principle in the same manner as the equations of hydrodynamics were deduced from those of hydrostatics. It will be convenient here to revert to the notation of Chapter IX where u, v, w and σ refer to displacements rather than to velocities. Applying d'Alembert's principle we thus obtain

225) $$\varrho\left(X - \frac{\partial^2 u}{\partial t^2}\right) + (\lambda + \mu)\frac{\partial \sigma}{\partial x} + \mu \Delta u = 0, \text{ etc.}$$

1) Webster, *On the Mechanical Efficiency of the Production of Sound.* Boltzmann-Festschrift, p. 866, 1904.

If there are no bodily forces we have the equations of motion

$$\begin{aligned} \varrho \frac{\partial^2 u}{\partial t^2} &= (\lambda + \mu)\frac{\partial \sigma}{\partial x} + \mu \Delta u, \\ \varrho \frac{\partial^2 v}{\partial t^2} &= (\lambda + \mu)\frac{\partial \sigma}{\partial y} + \mu \Delta v, \\ \varrho \frac{\partial^2 w}{\partial t^2} &= (\lambda + \mu)\frac{\partial \sigma}{\partial z} + \mu \Delta w. \end{aligned} \tag{226}$$

Differentiating respectively by x, y, z and adding we obtain

$$\varrho \frac{\partial^2 \sigma}{\partial t^2} = (\lambda + 2\mu)\Delta\sigma, \tag{227}$$

which is the equation for the propagation of wave-motion, the dilatation being propagated with a velocity $b = \sqrt{\frac{\lambda + 2\mu}{\varrho}}$. Taking the curl of equations 226) we have

$$\begin{aligned} \varrho \frac{\partial^2 \xi}{\partial t^2} &= \mu \Delta \xi, \\ \varrho \frac{\partial^2 \eta}{\partial t^2} &= \mu \Delta \eta, \\ \varrho \frac{\partial^2 \zeta}{\partial t^2} &= \mu \Delta \zeta. \end{aligned} \tag{228}$$

Thus the components of the curl are propagated independently, each with a velocity $a = \sqrt{\frac{\mu}{\varrho}}$. The velocity of the compressional wave which is unaccompanied by rotation depends upon the bulk modulus and the modulus of shear. The velocity of the torsional wave which is unaccompanied by change of density depends only upon the modulus ot shear. The general motion of an elastic body is a combination of waves of compression and of torsion. The wave of torsion is that upon which the dynamical theory of light is founded. Inasmuch as μ vanishes for a perfect fluid no wave of torsion is propagated, so that the luminiferous ether must have the properties of a solid and not those of a fluid.

206. Viscous Fluids. We have now to consider a class of bodies intermediate in their properties between solids and perfect fluids, namely the viscous fluids. By definition a perfect fluid is one in which no tangential stresses exist. We have then

$$X_x = Y_y = Z_z = -p, \qquad X_y = Y_z = Z_x = 0. \tag{229}$$

In a fluid which is not perfect no tangential stresses can exist in a state of rest, but during motion such stresses can exist. While in a solid the stresses depend on the change of size and shape of the small portions of the solid, in the case of a viscous fluid the stresses depend on the time-rates of change, that is on the velocities of the shears, stretches, and dilatations. The simplest assumption that we can make is that the stress-components are linear functions of the strain-velocities. The fluid being isotropic, considerations regarding invariance bring us to precisely similar conclusions to those we reached in § 175, so that to the stresses of equations 229) for a perfect fluid are added stresses given by equations 142), § 175, λ and μ being constants for the fluid, and u, v, w, σ now denoting velocities, instead of displacements, returning to the notation of this chapter. (We put $P = 0$, since these additional terms vanish with the velocities.) We thus obtain

230)
$$\begin{aligned}
X_x &= -p + \lambda\sigma + 2\mu\frac{\partial u}{\partial x},\\
Y_y &= -p + \lambda\sigma + 2\mu\frac{\partial v}{\partial y},\\
Z_z &= -p + \lambda\sigma + 2\mu\frac{\partial w}{\partial z},\\
Y_z &= Z_y = \mu\left(\frac{\partial w}{\partial y} + \frac{\partial v}{\partial z}\right),\\
Z_x &= X_z = \mu\left(\frac{\partial u}{\partial z} + \frac{\partial w}{\partial x}\right),\\
X_y &= Y_x = \mu\left(\frac{\partial v}{\partial x} + \frac{\partial u}{\partial y}\right),
\end{aligned}$$

which are of the same form, with a different meaning, as 142), § 175. If the fluid is incompressible we find, putting $\sigma = 0$,

$$X_x + Y_y + Z_z = -3p,$$

and assuming that this holds also for compressible fluids we must have

231) $$3\lambda + 2\mu = 0.$$

Replacing λ by its value $-\frac{2}{3}\mu$, we find for the forces, as in § 175, 144),

232)
$$\begin{aligned}
\varrho X - \frac{\partial p}{\partial x} + \frac{1}{3}\mu\frac{\partial \sigma}{\partial x} + \mu \Delta u,\\
\varrho Y - \frac{\partial p}{\partial y} + \frac{1}{3}\mu\frac{\partial \sigma}{\partial y} + \mu \Delta v,\\
\varrho Z - \frac{\partial p}{\partial z} + \frac{1}{3}\mu\frac{\partial \sigma}{\partial z} + \mu \Delta w,
\end{aligned}$$

which are to be introduced into the equations of hydrodynamics 6). Thus we obtain the general equations, putting $\frac{\mu}{\varrho} = \nu$,

233)
$$\begin{aligned}
\frac{\partial u}{\partial t} + u\frac{\partial u}{\partial x} + v\frac{\partial u}{\partial y} + w\frac{\partial u}{\partial z} - \frac{\nu}{3}\frac{\partial \sigma}{\partial x} - \nu\Delta u &= X - \frac{1}{\varrho}\frac{\partial p}{\partial x},\\
\frac{\partial v}{\partial t} + u\frac{\partial v}{\partial x} + v\frac{\partial v}{\partial y} + w\frac{\partial v}{\partial z} - \frac{\nu}{3}\frac{\partial \sigma}{\partial y} - \nu\Delta v &= Y - \frac{1}{\varrho}\frac{\partial p}{\partial y},\\
\frac{\partial w}{\partial t} + u\frac{\partial w}{\partial x} + v\frac{\partial w}{\partial y} + w\frac{\partial w}{\partial z} - \frac{\nu}{3}\frac{\partial \sigma}{\partial z} = \nu\Delta w &= Z - \frac{1}{\varrho}\frac{\partial p}{\partial z},
\end{aligned}$$

which reduce to 6) when $\mu = 0$. The coefficient μ is called the viscosity of the fluid, and its quotient by the density, ν, is called by Maxwell the kinematical coefficient of viscosity.

The equations 233) are too complicated to be used in all their generality. We shall here consider only the case of incompressible fluids, for which the terms in σ vanish. If we form the equation of activity as in § 188, we obtain beside the terms in the first integral of 29) the additional terms

$$-\mu\iiint(u\Delta u + v\Delta v + w\Delta w)\,d\tau,$$

which by Green's theorem may be converted into

$$\iint\left(u\frac{\partial u}{\partial n} + v\frac{\partial v}{\partial n} + w\frac{\partial w}{\partial n}\right)dS$$
$$+\iiint\left\{\left(\frac{\partial u}{\partial x}\right)^2 + \left(\frac{\partial u}{\partial y}\right)^2 + \left(\frac{\partial v}{\partial z}\right)^2 + \left(\frac{\partial v}{\partial x}\right)^2 + \cdots\right\}d\tau.$$

If the integration be extended to a region where the liquid is at rest, say the surface of a containing solid, where the liquid does not slip, the surface integrals vanish, and the volume integrals give a *positive* addition. That is to say, the applied forces have to do an amount of work over and above that going into kinetic and potential energy, and this work is *dissipated* into heat. If there are no applied forces, the energy of the fluid is dissipated, and it will eventually come to rest.

In order to find simple solutions of our equations, we may deal either with steady motion, or with motions so slow that we may neglect the terms of the second order in u, v, w and their derivatives. Let us first consider steady motion. The simplest case is uniplanar flow parallel to a single direction, or as we may call it, *laminar* flow. If we take

234) $$0 = v = w = \frac{\partial u}{\partial z} = \frac{\partial p}{\partial z},$$

the equation of continuity gives

235) $$\frac{\partial u}{\partial x} = 0.$$

If there are no applied forces, equations 233) reduce to

236) $$\mu\frac{\partial^2 u}{\partial y^2} = \frac{\partial p}{\partial x}, \quad \frac{\partial p}{\partial y} = 0.$$

Since u depends only on y and p only on x, this equation cannot hold unless each side is constant. Accordingly

237) $$\mu\frac{d^2 u}{dy^2} = a = \frac{dp}{dx}, \quad u = c + by + \frac{1}{2}\frac{a}{\mu}y^2, \quad p = d + ax,$$

where a, b, c, d are constants. If we determine them so that the velocity vanishes for planes at a distance $\pm h$ from the X-axis, we have

$$238)\qquad u = \frac{a}{2\mu}(y^2 - h^2).$$

The amount of liquid that flows through such a laminar tube per unit of breadth parallel to the Z-axis is accordingly

$$239)\qquad Q = \int_{-h}^{h} u\,dy = -\frac{2}{3}\frac{ah^3}{\mu}$$

and for a length of tube l the difference of pressures at the ends is

$$240)\qquad p_2 - p_1 = al,\quad Q = \frac{2}{3}\frac{h^3}{\mu}\frac{(p_1 - p_2)}{l},$$

so that the flow is proportional to the difference of pressures at the ends and inversely to the viscosity.

For the practical determination of viscosity, we may take the almost equally simple case of cylindrical flow, where the velocity has everywhere the same direction, and depends upon the distance r from the axis of a circular tube, at the surface of which it is at rest.

If we put $u = v = 0$ we have the equations of motion and of continuity

$$241)\qquad \mu\Delta w = \frac{\partial p}{\partial z},\quad \frac{\partial p}{\partial x} = \frac{\partial p}{\partial y} = 0,\quad \frac{\partial w}{\partial z} = 0,$$

and since w depends only on r the first becomes

$$242)\qquad \mu\left\{\frac{d^2 w}{dr^2} + \frac{1}{r}\frac{dw}{dz}\right\} = \frac{dp}{dz} = a,$$

where a is a constant as before. This equation is integrated as in § 182, 58'),

$$284)\qquad w = \frac{a}{4\mu}r^2 + b\log r + c.$$

Since R is finite when $r = 0$ we must have $b = 0$ and if w vanishes for $r = w$ we obtain

$$244)\qquad w = \frac{a}{4\mu}(r^2 - R^2).$$

For the flow we find

$$245)\qquad Q = \int_0^R 2\pi w r\,dr = \frac{a\pi R^4}{8\mu} = \frac{\pi R^4}{8\mu}\frac{(p_1 - p_2)}{l}.$$

This method was invented by Poiseuille[1]) for the measurement of

1) Poiseuille, *Recherches expérimentales sur le mouvement des liquides dans les tubes de très petits diamètres.* Comptes Rendus, 1840—41; Mém. des Savants Étrangers, t. 9, 1846.

the viscosity of fluids. His verification of the proportionality of the flow to the fourth power of the radius of the tube has been taken as a proof that the liquid does not slide when in contact with a solid.

As another example of steady flow let us cosider uniplanar cylindrical flow, in which each particle moves in a circle with velocity depending only on the distance from the axis, as in the case of the lubricant between a journal and its bearing. Each cylindrical stratum then revolves like a rigid body, which requires

246) $$u = -\omega y, \quad v = \omega x,$$

where ω depends only on $r = \sqrt{x^2 + y^2}$. We then find

247) $$\begin{aligned} \frac{\partial u}{\partial x} &= -\frac{xy}{r}\frac{d\omega}{dr}, & \frac{\partial u}{\partial y} &= -\omega - \frac{y^2}{r}\frac{d\omega}{dr}, \\ \frac{\partial v}{\partial x} &= \omega + \frac{x^2}{r}\frac{d\omega}{dr}, & \frac{\partial v}{\partial y} &= \frac{xy}{r}\frac{d\omega}{dr}, \end{aligned}$$

and; most easily by the application of equation 86), § 141, and by the expression of $\Delta\omega$ in terms of r,

248) $$\begin{aligned} \Delta u &= -y\left(\frac{d^2\omega}{dr^2} + \frac{1}{r}\frac{d\omega}{dr}\right) - 2\frac{y}{r}\frac{d\omega}{dr}, \\ \Delta v &= \quad x\left(\frac{d^2\omega}{dr^2} + \frac{1}{r}\frac{d\omega}{dr}\right) + 2\frac{x}{r}\frac{d\omega}{dr}. \end{aligned}$$

Thus the first two of equations 233) become

249) $$\begin{aligned} -\omega^2 x + \nu y\left(\frac{d^2\omega}{dr^2} + \frac{3}{r}\frac{d\omega}{dr}\right) &= -\frac{1}{\varrho}\frac{x}{r}\frac{dp}{dr}, \\ -\omega^2 y - \nu x\left(\frac{d^2\omega}{dr^2} + \frac{3}{r}\frac{d\omega}{dr}\right) &= -\frac{1}{\varrho}\frac{y}{r}\frac{dp}{dr}. \end{aligned}$$

Multiplying the first by y, the second by x and subtracting,

250) $$\frac{d^2\omega}{dr^2} + \frac{3}{r}\frac{d\omega}{dr} = 0,$$

a differential equation whose solution is

251) $$\omega = \frac{a}{r^2} + b.$$

Determining the constants so that $\omega = 0$ for $r = R_1$ and $\omega = \Omega$ for $r = R_2$,

252) $$\omega = \frac{\Omega R_2^{\,2}}{R_1^{\,2} - R_2^{\,2}}\left\{\frac{R_1^{\,2}}{r^2} - 1\right\}.$$

Multiplying equations 249) by x and y respectively, and adding, we have to determine p,

253) $$\varrho\,\omega^2 r = \frac{dp}{dr}.$$

For the stresses we obtain, using equations 230),

$$X_n = \left(-p - 2\mu\frac{xy}{r}\frac{d\omega}{dr}\right)\cos(nx) + \mu\left(\frac{x^2 - y^2}{r}\frac{d\omega}{dr}\right)\cos(ny),$$

$$254)\quad Y_n = \mu\left(\frac{x^2-y^2}{r}\frac{d\omega}{dr}\right)\cos(nx) + \left(-p + 2\mu\frac{xy}{r}\frac{d\omega}{dr}\right)\cos(ny),$$

$$Z_n = -p\cos(nz) = 0.$$

This shows that there is a normal pressure p, together with a tangential stress which we obtain by resolving along the tangent,

$$255)\quad T = Y_n\cos(nx) - X_n\cos(ny)$$

$$= \mu\frac{d\omega}{dr}\left\{\frac{x^2-y^2}{r}\cos^2 nx - \cos^2 ny) + \frac{4xy}{r}\cos(nx)\cos(ny)\right\}$$

and since $\cos(nx) = \frac{x}{r}$, $\cos(ny) = \frac{y}{r}$,

$$256)\quad T = \mu r\frac{d\omega}{dr} = -\frac{2\mu\Omega R_1{}^2 R_2{}^2}{R_1{}^2 - R_2{}^2}\frac{1}{r^2}.$$

The moment of the tangential stress on the cylinder of radius r and unit length, is accordingly

$$257)\quad 2\pi r T\cdot r = -\frac{4\pi\mu\Omega R_1{}^2 R_2{}^2}{(R_1{}^2 - R_2{}^2)}.$$

We may accordingly use this method to determine the viscosity, as is in fact done in apparatus for the testing of lubricants. We see that if the linear dimensions are multiplied in a certain ratio, the moment is increased in the square of that ratio. We also see that the moment of the force required to twist the cylinder is independent of the pressure p, which contains an arbitrary constant, not given by the equation 253), but depending on the hydrostatic pressure applied at the ends.

Let us now consider some simple cases where the flow is not steady, limiting ourselves to the case of small velocities, so that the terms in 233) involving the first space derivatives, being of the second order, are negligible. Let us once more consider laminar flow, defined by equations 234), 235). Let us also put $p = const.$ Instead of 236) we now have for the first of equations 233),

$$258)\quad \frac{\partial u}{\partial t} = \nu\frac{\partial^2 u}{\partial y^2}.$$

This equation is the same as that which represents the conduction of heat in one direction. Let us first consider a solution periodic in the time, such as may be realized physically by the harmonic small oscillation in its own plane of a material lamina constituting the plane $y = 0$, along which the liquid does not slip. We may take as a particular solution

$$e^{my+nt},$$

which inserted in 258) gives

$$n = \nu m^2.$$

If this is to be periodic in t, n must be pure imaginary, say $n = ip$.

Then we have

$$m = \sqrt{\frac{ip}{\nu}} = \pm (1+i)\sqrt{\frac{p}{2\nu}},$$

and

$$e^{\pm\sqrt{\frac{p}{2\nu}}\, y + i\left(pt \pm \frac{p}{2\nu} y\right)},$$

of which complex quantity both the real and the imaginary parts must separately satisfy the equation 258), when multiplied by arbitrary constants. Let us accordingly take

$$250)\quad u = \left\{A \cos\left(pt - \sqrt{\frac{p}{2\nu}} \cdot y\right) + B \sin\ pt - \sqrt{\frac{p}{2\nu}} y)\right\} e^{-\sqrt{\frac{p}{2\nu}} \cdot y}.$$

This represents a wave of frequency $\frac{p}{2\pi}$ and wave-length $2\pi\sqrt{\frac{2 \cdot \nu}{p}}$ travelling with velocity $\sqrt{2\nu p}$, which as we see varies as the square root of the frequency. Unlike our waves in perfect fluids however it falls off in amplitude, being rapidly damped as we go into the fluid, being reduced in the ratio $e^{-2\pi} = \frac{1}{535}$ in each wave-length. Thus such motions are propagated but a short distance into a fluid. In a similar manner the absorption of light by non-transparent media is explained, the ether there having the properties of a viscous *solid*.

If we treat the equations 233) in the same way as we did 27) in obtaining equation 57), § 191, we obtain instead the following,

$$260)\quad \begin{aligned} \frac{d}{dt}\left(\frac{\xi}{\varrho}\right) &= \frac{\xi}{\varrho}\frac{\partial u}{\partial x} + \frac{\eta}{\varrho}\frac{\partial u}{\partial y} + \frac{\zeta}{\varrho}\frac{\partial u}{\partial z} + \nu \varDelta \xi, \\ \frac{d}{dt}\left(\frac{\eta}{\varrho}\right) &= \frac{\xi}{\varrho}\frac{\partial v}{\partial x} + \frac{\eta}{\varrho}\frac{\partial v}{\partial y} + \frac{\zeta}{\varrho}\frac{\partial v}{\partial z} + \nu \varDelta \eta, \\ \frac{d}{dt}\left(\frac{\zeta}{\varrho}\right) &= \frac{\xi}{\varrho}\frac{\partial w}{\partial x} + \frac{\eta}{\varrho}\frac{\partial w}{\partial y} + \frac{\zeta}{\varrho}\frac{\partial w}{\partial z} + \nu \varDelta \zeta. \end{aligned}$$

Under the circumstances of slow motions these also reduce to

$$261)\quad \begin{aligned} \frac{\partial \xi}{\partial t} &= \mu \varDelta \xi, \\ \frac{\partial \eta}{\partial t} &= \mu \varDelta \eta, \\ \frac{\partial \zeta}{\partial t} &= \mu \varDelta \zeta. \end{aligned}$$

Thus we see that the three components of the vorticity are propagated independently, each according to the equation for the conduction of heat. The example just treated is an example of this, for we find at once

$$262)\qquad \xi = \eta = 0, \quad \xi = -\frac{1}{2}\frac{\partial u}{\partial y},$$

and the vorticity is propagated like the velocity.

As a final example, let us consider a case of laminar motion in which u, as a function of y, has a discontinuity, this having an important application to the theory of thin plane jets and flames, including sensitive flames.[1]) We will suppose that at the time $t = 0$ for $y < 0$ u has a certain constant value, and that for $y > 0$ it has a different constant value. It is easy to see that this is equivalent io supposing that there is no vorticity except in an infinitely thin lamina at $y = 0$. For we have

$$263)\qquad \int_{-\varepsilon}^{\varepsilon} \xi\, dy = -\frac{1}{2}\int_{-\varepsilon}^{\varepsilon} \frac{\partial u}{\partial y}\, dy = \frac{1}{2}(u_1 - u_2)$$

where u_1 is the velocity on one side u_2 that an the other of the layer of thickness 2ε. Now if the thickness decrease without limit, while ξ increases without limit, the integral may still be finite, as we shall suppose.

We have then to find two solutions u and ξ of equation 258), so related that $\xi = -\frac{1}{2}\frac{\partial u}{\partial y}$. Let us put $s = \frac{y}{\sqrt{t}}$, and try to find a particular solution that is a function of s alone. We have.

$$264)\qquad \begin{aligned} \frac{\partial u}{\partial t} &= \frac{du}{ds}\frac{\partial s}{\partial t} = -\frac{1}{2}\frac{du}{ds}\frac{y}{\sqrt{t^3}}, \\ \frac{\partial u}{\partial y} &= \frac{du}{us}\frac{\partial s}{\partial y} = \frac{du}{ds}\frac{1}{\sqrt{t}}, \\ \frac{\partial^2 u}{\partial y^2} &= \frac{1}{\sqrt{t}}\frac{d^2u}{ds^2}\frac{\partial s}{\partial y} = \frac{1}{t}\frac{d^2u}{ds^2}, \end{aligned}$$

so that our equation becomes the ordinary differential equation,

$$265)\qquad -\frac{1}{2}s\frac{du}{ds} = \mu\frac{d^2u}{ds^2},$$

or

$$266)\qquad \frac{d}{ds}\left(\log\frac{du}{ds}\right) = -\frac{s}{2\mu}.$$

1) Rayleigh, *On the Stability, or Instability, of certain Fluid Motions* Proc. London Math. Soc., xi., pp. 57—70, 1880. Scientific Papers, Vol. I, p. 474.

The integral of this equation is given by

$$267)\quad \log\frac{du}{ds} = -\frac{s^2}{4\mu} + const.,\quad \frac{du}{ds} = ce^{-\frac{s^2}{4\mu}},\quad u = c\int e^{-\frac{s^2}{4\mu}} ds.$$

The last indicated quadrature cannot be effected except by development of the integrand in series, but if we take for the lower limit the value zero, we may express u in terms of the so-called error-function, occurring in the theory of probability,

$$268)\quad Erf(x) = \int_0^x e^{-x^2} dx.$$

Tables of the values of $Erf(x)$ have been calculated, and are found in treatises on probability. (Lord Kelvin reprints one such on p. 434 of Vol. 3 of his collected papers.) Since the integrand is an even function of x, it is evident that $Erf(x)$ is an odd function of its upper limit x. It may be easily shown that the definite integral between zero and infinity has the value $\frac{\sqrt{\pi}}{2}$, so that putting $x^2 = \frac{s^2}{4\mu}$ and adding a constant, we have

$$269)\quad u = \frac{u_2 + u_1}{2} + \frac{u_2 - u_1}{\sqrt{\pi}} \int^{\frac{y}{2\sqrt{\mu t}}} e^{-x^2} dx.$$

This determination of the constants makes, for *all* positive values of y and for $t = 0$, $u = u_1$ (the upper limit being $+\infty$), and for *all* negative values $u = u_2$, thus giving the discontinuity required at $y = 0$. For all other values of t however, no matter how small, the values from the negative side run smoothly into those on the positive, showing how the discontinuity is instantly lost. This is shown in Fig. 171, in which successive curves show values of u at times equal to 1, 2, 3, 4, 5, 6 times $\frac{1}{\mu}$.

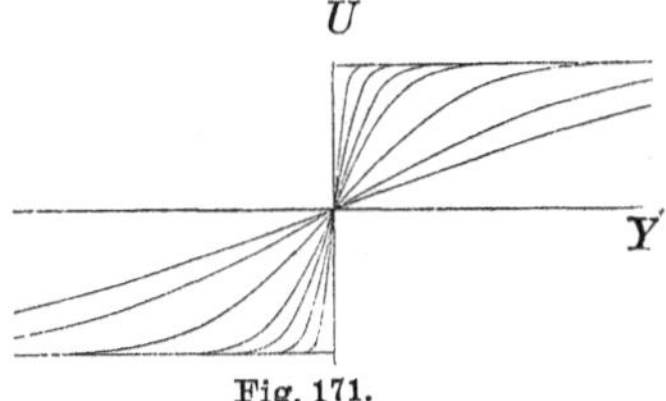

Fig. 171.

Differentiating by the limit, we find

$$270)\quad \zeta = -\frac{1}{2}\frac{\partial u}{\partial y} = \frac{u_1 - u_2}{4\sqrt{\pi\mu t}} e^{-\frac{y^2}{4\mu t}},$$

which is infinite when $t = 0$, $\frac{y}{\sqrt{t}} = 0$, as we supposed, but which immediately drops to a finite value, and, no matter what the value of y, immediately acquires values different from zero. Thus the

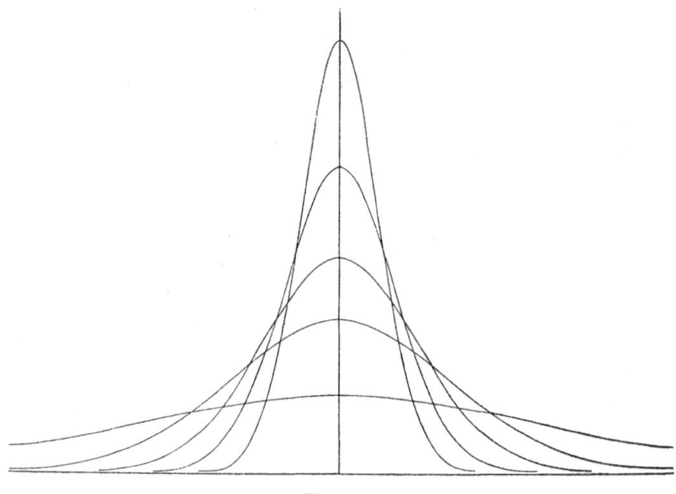

Fig. 172.

vorticity, originally confined within the infinitely thin sheet of discontinuity, is instantaneously distributed throughout the liquid, as shown in Fig. 172, for the times $\frac{1}{8}$, $\frac{1}{4}$, $\frac{1}{2}$, 1, 4 times $\frac{1}{\mu}\cdot$ Thus we see how discontinuities of the sort shown in Fig. 166 are impossible in nature, being replaced by the formation of eddies.

NOTES.

NOTE I.

DIFFERENTIAL EQUATIONS.

The differential equations of mechanics are of the type known as *ordinary*, as opposed to *partial*, that is they involve a number of functions of a single variable, the time, and the derivatives of the functions with respect to that variable. Suppose for simplicity that we have three functions x, y, z of the variable t, and that instead of being given explicitly, they are defined by the equations

$$1) \quad F_1(x, y, z, t) = 0, \quad F_2(x, y, z, t) = 0, \quad F_3(x, y, z, t) = 0.$$

If we now differentiate these equations, bearing in mind that x, y, z, are dependent on t, we obtain

$$2) \quad \begin{aligned} \frac{\partial F_1}{\partial x}\frac{dx}{dt} + \frac{\partial F_1}{\partial y}\frac{dy}{dt} + \frac{\partial F_1}{\partial z}\frac{dz}{dt} + \frac{\partial F_1}{\partial t} &= 0, \\ \frac{\partial F_2}{\partial x}\frac{dx}{dt} + \frac{\partial F_2}{\partial y}\frac{dy}{dt} + \frac{\partial F_2}{\partial z}\frac{dz}{dt} + \frac{\partial F_2}{\partial t} &= 0, \\ \frac{\partial F_3}{\partial x}\frac{dx}{dt} + \frac{\partial F_3}{\partial y}\frac{dy}{dt} + \frac{\partial F_3}{\partial z}\frac{dz}{dt} + \frac{\partial F_3}{\partial t} &= 0. \end{aligned}$$

Suppose now that the functions F contained, besides the variables indicated, certain constants, $c_1, c_2 \ldots$ Each time that we obtain an equation by differentiation, we may utilize it in order to eliminate from the equations 1) one of the constants c. Thus we obtain (since the *partial* derivatives are given functions of x, y, z, t), instead of the equations 1), the following,

$$3) \quad G_1\left(x, y, z, t, \frac{dx}{dt}, \frac{dy}{dt}, \frac{dz}{dt}\right) = 0, \quad G_2\left(x, \ldots \frac{dz}{dt}\right) = 0, \quad G_3\left(x, \ldots \frac{dz}{dt}\right) = 0,$$

which, since they contain the derivatives $\frac{dx}{dt}, \frac{dy}{dt}, \frac{dz}{dt}$, ere *differential* equations, of which equations 1) are said to be *integrals.*

If we again differentiate equations 2), we obtain

$$4) \quad \frac{\partial F_1}{\partial x}\frac{d^2x}{dt^2} + \frac{\partial^2 F_1}{\partial x^2}\left(\frac{dx}{dt}\right)^2 + 2\frac{\partial x\,\partial y}{\partial^2 F_1}\frac{dx}{dt}\frac{dy}{dt} + \cdots = 0, \text{ etc.}$$

which we may again use to eliminate constants c from 3), so that instead of 1) or 3) we now have the system

$$H_1\left(x, y, z, t, \frac{dx}{dt}, \frac{dy}{dt}, \frac{dz}{dt}, \frac{d^2x}{dt^2}, \frac{d^2y}{dt^2}, \frac{d^2z}{dt^2}\right) = 0,$$

5) $$H_2\left(x, y, z, t, \ldots \ldots \frac{d^2z}{dt}\right) = 0,$$

$$H_3\left(x, y, z, t, \ldots \ldots \frac{d^2z}{ct^2}\right) = 0.$$

These differential equations, since the order of the derivatives of the highest order contained in them is the second, are said to be of the second order. In like manner we may continue, and successively eliminate all the constants $c_1, c_2 \ldots$, obtaining differential equations of successively higher orders. Reversing the process, each set of a given order is said to be the *integral* of the set of order next higher.

Any of the sets of differential equations represents the functions x, y, z, but with the following distinction. If the equations 1) contain constants, to which different values may be assigned,

6) $$F_1(x, y, z, t, c_1, c_2 \ldots c_n) = 0, \quad F_2(x, y, z, t, c_1, c_2 \ldots c_n) = 0,$$
$$F_3(x, y, z, t, c_1, c_2 \ldots c_n) = 0,$$

for every set of values that may be assigned to the constants, a different set of functions is represented, so that we have an infinity of different functions, the order of the infinity being the number of constants contained in the equations. Now the differential equations obtained by eliminating the arbitrary constants represent all the functions obtained by giving the constants any set of values whatever. Thus the information contained in the differential equations is in a sense more general than that contained in the equations 6), in which we give the constants any particular values.

If we reverse the process which we have here followed to form the differential equations, we see that every time that we succeed, by integration, in making derivatives of a certain order disappear, we introduce at the same time a number of arbitrary constants equal to the number of derivatives which disappear. Thus the integral equations of a set of differential equations of any order will contain a number of arbitrary constants equal to the order of the differential equations multiplied by the number of dependent variables. As an example consider the very simple case of equations 38), § 13.

38) $$\frac{d^2x}{dt^2} = 0, \quad \frac{d^2y}{dt^2} = 0, \quad \frac{d^2z}{dt^2} = 0.$$

Integrating these we obtain

39) $$x = c_1t + d_1, \quad y = c_2t + d_2, \quad z = c_3t + d_3,$$

containing the six arbitrary constants $c_1, c_2, c_3, d_1, d_2, d_3$. The meaning of these integral equations is that the point x, y, z describes a straight line with a constant velocity. But the differential equations 38) represent the motion of a point describing any line in space with any velocity. Now there are a four-fold infinity of lines in space, and a single infinity of velocities. We therefore see the very general nature of the information contained in the differential equations. So in the example of § 13 the statement that all the planets experience an acceleration toward the sun which is proportional to the inverse square of the distance expresses a very general and simple truth, in the form of a set of differential equations, while the integral states that the planets describe some conic section in some plane through the sun, in some periodic time, all the particulars of which statement are arbitrary.

The characteristic property of the differential equations of mechanics, for the phenomena furnished us by Nature, is apparently that they are of the second order. This, although leaving possibilities of great generality, suffices to limit natural phenomena to a certain class, in contrast to what would be conceivable. For the consequences of the removal of this limitation, the student is refered to the very interesting work by Königsberger, *Die Principien der Mechanik.*

In order to determine the particular values of the arbitrary constants applicaple to any particular problem, some data must be given in addition to the differential equations. It is customary to furnish these by stating for a particular instant of time, the values of the coordinates of each point of the system, and of their first time-derivatives, which amounts to specifying for each point its position and its vector velocity for the particular instant in question. This furnishes six data for each independent point, which is just sufficient to determine the constants. Thus if we are dealing with a system of n points free to move in any manner, under the action of any forces, the statement of the problem will consist in the giving of the differential equations

$$F_1\left(x_1, y_1, z_1, x_2, y_2, \ldots z_n, t, \frac{dx}{dt}, \ldots \frac{dz_n}{dt}, \frac{d^2x}{dt^2}, \ldots \frac{d^2z_n}{dt^2}\right),$$

$$F_2\left(x_1, \ldots \ldots \ldots \ldots \ldots \ldots \ldots \ldots \frac{d^2z_n}{dt^2}\right),$$

$$\ldots\ldots\ldots\ldots\ldots\ldots\ldots\ldots\ldots$$

$$\ldots\ldots\ldots\ldots\ldots\ldots\ldots\ldots\ldots$$

$$F_{3n}\left(x_1, \ldots \ldots \ldots \ldots \ldots \ldots \ldots \ldots \frac{d^2z_n}{dt^2}\right),$$

together with the so-called initial conditions, that for $t = t_0$,

$$x_1 = x_1^0, \quad y_1 = y_1^0 \ldots \ldots z_n = z_n^0$$

$$\frac{dx_1}{dt} = [x'_1]^0, \frac{dy_1}{dt} = [y'_1]^0, \ldots \frac{dz_n}{dt} = [z'_n]^0.$$

From these it is required to find the integrals

$$x_1 = f_1(t), \; y_1 = f_2(t), \; z_1 = f_3(t), \; \ldots \; z_n = f_{3n}(t).$$

Cases involving the motion of points whose freedom of motion is limited are dealt with in subsequent chapters.

NOTE II.

ALGEBRA OF INDETERMINATE MULTIPLIERS.

On page 61 we have an example of the use of indeterminate multipliers in elimination. It may be somewhat more clear if we examine in just what the process involved consists. Equation 12) is a linear equation involving the $3n$ quantities $\delta x_1, \ldots \delta z_n$, each multiplied by a coefficient which is independent of the δ's. Besides this equation the quantities δ satisfy the equation 14), which are of the same form, that is, linear in all the δ's, with coefficients independent of them. Aside from this the δ's may have any values whatever. It is for the purposes of this discussion quite immaterial that the δ's are small quantities, we are concerned simply with a question of elimination. Let us accordingly represent them by the letters $x_1, x_2, \ldots x_m$, between which we havn the linear equation

$$1) \qquad A_1 x_1 + A_2 x_2 + \cdots + A_m x_m = 0.$$

The x's are however not independent, but must in addition satisfy the equations

$$2) \qquad \begin{aligned} B_{11} x_1 + B_{12} x_2 \cdots + B_{1m} x_m &= 0, \\ B_{21} x_1 + B_{22} x_2 \cdots + B_{2m} x_m &= 0, \\ \cdots\cdots\cdots\cdots\cdots\cdots& \\ B_{k1} x_1 + B_{k2} x_2 \cdots + B_{km} x_m &= 0. \end{aligned}$$

The number of these equations, k, is less than m, the number of the x's. The question is now, what relations are involved among the A's and B's when the x's have any values whatever compatible with the equations 2).

We may evidently proceed as follows. Transposing $m - k$ terms in 2), say the last, we may solve the equations for the quantities $x_1, x_2 \ldots x_k$, as linear functions of the remaining $x_{k+1}, \ldots x_m$. These $m - k$ quantities are now perfectly arbitrary. Inserting the values of $x_1 \ldots x_k$ in equation 1), this becomes linear in the $m - k$ quantities $x_{k+1}, \ldots x_m$, which being purely arbitrary, in order for equation 1) to hold for all values of the x's, the coefficient of each must vanish, giving us the required $m - k$ relations between the A's and B's.

Instead of proceeding in the manner described, the method of Lagrange is to multiply the equations 2) respectively by multipliers $\lambda_1, \lambda_2, \ldots \lambda_k$, to which any convenient value may be given, and then to add them to equation 1). We thus obtain

$$3) \qquad \begin{aligned} &(A_1 + \lambda_1 B_{11} + \lambda_2 B_{21} \cdots + \lambda_k B_{k1}) x_1 \\ &+ (A_2 + \lambda_1 B_{12} + \lambda_2 B_{22} \cdots + \lambda_k B_{k2}) x_2 \\ &\cdots\cdots\cdots\cdots\cdots\cdots \\ &+ (A_m + \lambda_1 B_{1m} + \lambda_2 B_{2m} \cdots + \lambda_k B_{km}) x_m = 0. \end{aligned}$$

In this equation the x's are not all arbitrary, but as before k may be determined as linear functions of the remainder, say $x_1, \ldots x_k$ in terms of $x_{k+1}, \ldots x_m$, which are arbitrary. But the multipliers λ are as yet arbitrary. Let us determine them so that they satisfy the equations

$$4)\qquad \begin{aligned} A_1 + \lambda_1 B_{11} + \lambda_2 B_{21} \cdots + \lambda_k B_{k1} &= 0,\\ A_2 + \lambda_1 B_{12} + \lambda_2 B_{22} \cdots + \lambda_k B_{k2} &= 0,\\ \cdots\cdots\cdots\cdots\cdots\cdots\cdots\cdots\\ A_k + \lambda_1 B_{1k} + \lambda_2 B_{2k} \cdots + \lambda_k B_{kk} &= 0, \end{aligned}$$

which are just sufficient to determine them. We thus have

$$5)\qquad \begin{aligned} &(A_{k+1} + \lambda_1 B_{1,\,k+1} + \cdots + \lambda_k B_{k,\,k+1}) x_{k+1}\\ &\cdots\cdots\cdots\cdots\cdots\cdots\cdots\cdots\\ +\,&(A_m + \lambda_1 B_{1m} + \cdots + \lambda_k B_{km}) x_m = 0, \end{aligned}$$

in which the x's are all arbitrary, so that the $m-k$ coefficients must vanish, giving the $m-k$ equations.

$$6)\qquad \begin{aligned} A_{k+1} + \lambda_1 B_{1,\,k+1} + \cdots + \lambda_k B_{k,\,k+1} &= 0\\ \cdots\cdots\cdots\cdots\cdots\cdots\cdots\cdots\\ A_m + \lambda_1 B_{1m} + \cdots + \lambda_k B_{km} &= 0. \end{aligned}$$

Inserting in these the values of the λ's already found, we have the $m-k$ required relations between the A's and B's. Obviously the result of the elimination may be expressed in the form obtained by writing equal to zero each of the determinants of order $k+1$ obtained from the array of A's and B's in equations 4) and 5) by omitting $m-k-1$ rows, only $m-k$ of the determinants thus obtained being independent.

NOTE III.

QUADRATIC DIFFERENTIAL FORMS. GENERALIZED VECTORS.

The method of transforming the equations of motion used in § 37 and the application of hyperspace there occurring render a somewhat more detailed treatment of the question desirable. In order to elucidate matters, we will begin with the very simple case of a space which is included in ordinary space, namely the space of two dimensions forming the surface characterized by two coordinates q_1, q_2, as on page 110. We have seen that this space is completely characterized by the expression for the arc as the quadratic differential form]

$$1)\qquad ds^2 = E\,dq_1{}^2 + 2F\,dq_1\,dq_2 + G\,dq_2{}^2.$$

A point lying on this surface may be displaced in any manner, in or out of the surface. If it is displaced in the surface, its displacements is a vector belonging to the two-dimensional space considered. We will call the changes dq_1, dq_2 the *coordinates* of the displacement. We have found that when the displacement is made so as to change only one of the coordinates of the point q_1 or q_2, the arcs are respectively $ds_1 = \sqrt{E}\,dq_1$, $ds_2 = \sqrt{G}\,dq_2$, and that the angle included by them is given by

$$\cos\vartheta = \frac{F}{\sqrt{EG}}.$$

If now we have any displacement ds, whose coordinates are dq_1, dq_2, and project it orthogonally upon the directions of ds_1, ds_2, we easily see (Fig. 26) that the projections $d\sigma_1$, $d\sigma_2$ are

$$2)\quad \begin{aligned} d\sigma_1 &= ds_1 + ds_2\cos\vartheta = \sqrt{F}\,dq_1 + \frac{F\sqrt{G}\,dq_2}{\sqrt{EG}} = \frac{E\,dq_1 + F\,dq_2}{\sqrt{E}},\\ d\sigma_2 &= ds_2 + ds_1\cos\vartheta = \sqrt{G}\,dq_2 + \frac{F\sqrt{E}\,dq_1}{\sqrt{EG}} = \frac{F\,dq_1 + G\,dq_1}{\sqrt{G}}. \end{aligned}$$

We shall now, following Hertz, introduce the *reduced component* of the displacement along either coordinate-line, defined as the orthogoneal projection divided by the rate of change of the coordinate with respect to the distance traveled in its own direction. These reduced components we shall denote by a bar, so that

$$3)\quad \begin{aligned} d\bar{q}_1 &= \frac{d\sigma_1}{\frac{dq_1}{ds_1}} = E\,dq_1 + F\,dq_2,\\ d\bar{q}_2 &= \frac{d\sigma_2}{\frac{dq_2}{ds_2}} = F\,dq_1 + G\,dq_2, \end{aligned}$$

The fundamental property of these reduced components is found in the equation giving the magnitude of the displacement

$$4)\quad ds^2 = dq_1\,d\bar{q}_1 + dq_2\,d\bar{q}_2,$$

that is the square of an infinitesimal displacement is the sum of products of each coordinate of the displacement multiplied by the respective reduced component.

In like manner the geometric product of two different displacements ds, ds', whose coordinates are dq_1, dq_2, dq_1', dq_2' is found to be

$$5)\quad \begin{aligned} ds\,ds'\cos(ds, ds') &= dx\,dx' + dy\,dy' + dz\,dz'\\ &= \left(\frac{\partial x}{\partial q_1}dq_1 + \frac{\partial x}{\partial q_2}dq_2\right)\left(\frac{\partial x}{\partial q_1}dq_1' + \frac{\partial x}{\partial q_2}dq_2'\right)\\ &+ \left(\frac{\partial y}{\partial q_1}dq_1 + \frac{\partial y}{\partial q_2}dq_2\right)\left(\frac{\partial y}{\partial q_1}dq_1' + \frac{\partial y}{\partial q_2}dq_2'\right)\\ &+ \left(\frac{\partial z}{\partial q_1}dq_1 + \frac{\partial z}{\partial q_2}dq_2\right)\left(\frac{\partial z}{\partial q_1}dq_1' + \frac{\partial z}{\partial q_2}dq_2'\right)\\ &= E\,dq_1\,dq_1' + F(dq_1\,dq_2' + dq_2\,dq_1') + G\,dq_2\,dq_2'\\ &= dq_1\,d\bar{q}_1' + dq_2\,d\bar{q}_2' = dq_1'\,d\bar{q}_1 + dq_2'\,d\bar{q}_2. \end{aligned}$$

The geometric product of two displacements is equal to the sum of products of the coordinates of either vector by the reduced components of the other. Thus the geometric product is defined by means of the quadratic differential form 1) defining the space in question.

Solving the equations 3) for dq_1, dq_2, we obtain

$$6) \quad \begin{aligned} dq_1 &= \frac{G d\bar{q}_1 - F d\bar{q}_2}{EG - F^2} = R_{11} d\bar{q}_1 + R_{12} d\bar{q}_2 \\ dq_2 &= -\frac{F d\bar{q}_1 - E d\bar{q}_2}{EG - F_2} = R_{12} d\bar{q}_1 + R_{22} d\bar{q}_2, \end{aligned}$$

from which we obtain

$$7) \quad ds^2 = R_{11} d\bar{q}_1^{\,2} + 2 R_{12} d\bar{q}_1 d\bar{q}_2 + R_{22} d\bar{q}_2^{\,2}.$$

The expression 7) is called the reciprocal form to 1). Corresponding to it we obtain the form of the geometric product

$$8) \quad ds\,ds' \cos(ds, ds') = R_{11} d\bar{q}_1 d\bar{q}_1' + R_{12}(d\bar{q}_1 d\bar{q}_2' + d\bar{q}_2 d\bar{q}_1') + R_{22} d\bar{q}_2 d\bar{q}_2'.$$

We may now define any vector belonging to the space considered, as one whose components have the same properties as those possessed by those of an infinitesimal displacement. Suppose that X, Y, Z are the rectangular components of a vector R, it does not belong to the space 1) unless it is tangent to the surface in question. If so, we have a displacement such that

$$9) \quad \frac{R}{ds} = \frac{X}{dx} = \frac{Y}{dy} = \frac{Z}{dz} = \frac{Q_1}{dq_1} = \frac{Q_2}{dq_2} = \frac{1}{\varepsilon}.$$

Then Q_1, Q_2 are the coordinates of the vector in the system q_1, q_2', and the magnitude of the vector is given by the equations

$$10) \quad \begin{aligned} R^2 = X^2 + Y^2 + Z^2 &= \frac{1}{\varepsilon^2}(dx^2 + dy^2 + dz^2) \\ &= \frac{1}{\varepsilon^2}(E dq_1^{\,2} + 2F dq_1 dq_2 + G dq_2^{\,2}) \\ &= E Q_1^{\,2} + 2 F Q_1 R_2 + G Q_2^{\,2} \\ &= Q_1 \bar{Q}_1 + Q_2 \bar{Q}_2, \end{aligned}$$

where

$$11) \quad \begin{aligned} \bar{Q}_1 &= E Q_1 + E Q_2 \\ \bar{Q}_2 &= F Q_1 + G Q_2 \end{aligned}$$

are its reduced components belonging to its coordinates Q_1, Q_2. The geometric product of two vectors R, R' is

$$12) \quad Q_1 \bar{Q}_1' + Q_2 \bar{Q}_2' = Q_1' \bar{Q}_1 + Q_2' \bar{Q}_3.$$

If now one of the vectors is finite, the other an infinitesimal displacement, we have the geometric product

$$13) \quad \begin{aligned} R\,ds \cos(R, ds) &= X dx + Y dy + Z dz = \bar{Q}_1 dq_1 + \bar{Q}_2 dq_2 \\ &= \left(X \frac{\partial x}{\partial q_1} + X \frac{\partial y}{\partial q_1} + Z \frac{\partial z}{\partial q_1}\right) dq_1 + \left(X \frac{\partial x}{\partial q_2} + Y \frac{\partial y}{\partial q_2} + Z \frac{\partial z}{\partial q_2}\right) dq_2. \end{aligned}$$

Thus the reduced components $\bar{Q}_1$, $\bar{Q}_2$ are given by the definition which we have adopted on page 116, equation 42).

Having now illustrated the subject by a space of two dimensions we can easily extend our notions to space of any number of dimensions m, defined by the form

14) $$ds^2 = \sum_{r=1}^{r=m} \sum_{r=1}^{r=m} Q_{rs} dq_r dq_s.$$

For any one of the coordinate directions we have

15) $$ds_r^2 = Q_{rr} dq_r^2,$$

and for the geometric product of two displacements,

16) $$ds\,ds' \cos(ds,\ ds') = \sum^r \sum^s Q_{rs} dq_r dq_s'.$$

If one is in the direction ds_r, all dq's being 0 except dq_r,

17) $$ds\,ds_r \cos(ds,\ ds_r) = \sum^s Q_{rs} dq_r dq_s,$$

and dividing by ds_r we obtain the orthogonal projection of ds on ds_r

18) $$d\sigma_r = ds \cos(ds,\ ds_r) = \frac{\Sigma^s Q_{rs}\, dq_s}{\sqrt{Q_{rr}}\, dq_r},$$

from which we obtain by the definition of the reduced component

19) $$d\bar{q}_r = \frac{d\sigma_r}{\frac{dq_r}{ds_r}} = \sum^s Q_{rs} dq_s.$$

We have as before

20) $$ds^2 = \sum^r dq_r d\bar{q}_r,$$

21) $$ds\,ds' \cos(ds,\ ds') = \sum^r \sum^s dq_r d\bar{q}_s,$$

and if the solution of 19) is

22) $$dq_r = \sum^s R_{rs} d\bar{q}_s,$$

we have the reciprocal form

23) $$ds^2 = \sum^r \sum^s R_{rs} d\bar{q}_r d\bar{q}_s.$$

Again we may define a vector belonging to the hyperspace considered, and now the rectangular components may be of any number, the limitation of the vector to the space in question reducing the number of generalized components to accord with the number of dimensions of the hyperspace. The geometric product of the vector with an infinitesimal displacement defines the generalized coordinates of the vector, so that

$$\sum_{r=1}^{r=n}(X_r dx_r + Y_r dy_r + Z_r dz_r)$$

$$=\sum_{r=1}^{r=n}\sum_{s=1}^{s=n}\left(X_r\frac{\partial x_r}{\partial q_s} + Y_r\frac{\partial y_r}{\partial q_s} + Z_r\frac{\partial z_r}{\partial q_s}\right)dq_s = \sum_{s=1}^{s=m} P_s dq_s,$$

and we find that the *reduced component* of the vector is what is defined by the formula 42) of page 116).

In our application to mechanics the differential form in question is $2Tdt^2$, where T is the kinetic energy of the system. It is immaterial wheter we speak of vectors in a hyperspace, as we have here done, or, as Hertz does, speak of vectors with respect to our mechanical system. The meaning in either case is plain. On dividing the above formulae by dt^2, we find that the generalized velocities and momenta have the relation to each other of coordinates and reduced components of the same vector in the hyperspuce. The two reciprocal forms 14) and 23) have the relation of the Lagrangian and Hamiltonian forms of the kinetic energy. The equations of motion of the system say that, no matter how the forces are applied, or how parts of them are equilibrated by the constraints, the *reduced components* of the applied and the effective forces are equal for every coordinate.

NOTE IV.

AXES OF CENTRAL QUADRIC.

The principal axes of a central quadric surface,

1) $$F(x, y, z) \equiv Ax^2 + By^2 + Cz^2 + 2Dyz + 2Ezx + 2Fxy = 1,$$

are defined as the radii vectores in the directions for which the radius vector is a maximum or minimum. If we put

$$x = r\alpha, \qquad y = r\beta, \qquad z = r\gamma,$$

we have

2) $$\frac{1}{r^2} = F(\alpha, \beta, \gamma),$$

and the maxima and minima of r occurring for the same directions as the minima and maxima of $1/r^2$, are obtained by finding the maxima and minima of $F(\alpha, \beta, \gamma)$ subject to the condition

3) $$\varphi(\alpha, \beta, \gamma) \equiv \alpha^2 + \beta^2 + \gamma^2 - 1 = 0.$$

If we multiply this equation by an arbitrary constant $-\lambda$ and add it to $F(\alpha, \beta, \gamma)$, we obtain the condition by writing the derivatives of $F - \lambda\varphi$ equal to zero. Thus we obtain

$$4)\qquad \begin{aligned} \frac{\partial F}{\partial \alpha} - 2\lambda\alpha &= 0, \\ \frac{\partial F}{\partial \beta} - 2\lambda\beta &= 0, \\ \frac{\partial F}{\partial \gamma} - 2\lambda\gamma &= 0. \end{aligned}$$

Now the direction cosines of the normal to the quadric at a point x, y, z are proportional to

$$\frac{\partial F}{\partial x}, \quad \frac{\partial F}{\partial y}, \quad \frac{\partial F}{\partial z}.$$

At points where the normal is in the direction of the radius vector we have

$$\frac{\dfrac{\partial F}{\partial x}}{x} = \frac{\dfrac{\partial F}{\partial y}}{y} = \frac{\dfrac{\partial F}{\partial z}}{z}.$$

But

$$\frac{\dfrac{\partial F(x, y, z)}{\partial x}}{x} = \frac{\dfrac{\partial F(\alpha, \beta, \gamma)}{\partial \alpha}}{\alpha} \quad \text{etc.},$$

so that the equations 4) show that at the ends of the principal axes the tangent plane is perpendicular to the radius vector.

Effecting the differentiations the equations 4) become

$$5)\qquad \begin{aligned} (A-\lambda)\alpha + F\beta \qquad + E\gamma \qquad &= 0, \\ F\alpha + (B-\lambda)\beta + D\gamma \qquad &= 0, \\ E\alpha + D\beta \qquad + (C-\lambda)\gamma &= 0. \end{aligned}$$

The condition that these equations, linear in α, β, γ, shall be compatible for values of α, β, γ, other than zero is that the determinant of the coefficients shall vanish.

$$6)\qquad \begin{vmatrix} A-\lambda, & F\;, & E \\ F\;, & B-\lambda, & D \\ E\;, & D\;, & C-\lambda \end{vmatrix} = 0.$$

This is a cubic in λ, which being expanded is

$$7)\qquad \begin{aligned} f(\lambda) \equiv (A-\lambda)(B-\lambda)(C-\lambda) - D^2(A-\lambda) - E^2(B-\lambda) \\ - F^2(C-\lambda) + 2DEF = 0. \end{aligned}$$

We shall show that this always has three real roots. Put

$$8)\qquad \begin{aligned} A-\lambda &= u+q,\\ B-\lambda &= v+r,\\ C-\lambda &= w+s, \end{aligned}$$

where q, r, s are to be determined later. Then

$$8)\qquad \begin{aligned} (u+q)(p+r)(w+s) - D^2(u+q) - E^2(v+r)\\ + F^2(w+s) + 2DEF = 0 \end{aligned}$$

or arranged according to powers of u, v, w,

$$9)\qquad \begin{aligned} uvw + qvw + rwu + suv\\ + u(rs-D^2) + v(sq-E^2) + w(qr-F^2)\\ + qrs + 2DEF - D^2q - E^2r - F^2s = 0. \end{aligned}$$

Let us now determine q, r, s, so as to make the terms of first order in u, v, w vanish.

$$rs = D^2, \quad sq = E^2, \quad qr = F^2,$$

from which by multiplication and division

$$10)\qquad qrs = DEF, \quad q = \frac{EF}{D}, \quad r = \frac{FD}{E}, \quad s = \frac{DE}{F}.$$

Thus there remains

$$11)\qquad uvw + \frac{EF}{D}vw + \frac{FD}{E}wu + \frac{DE}{F}uv = 0.$$

Now from 8)

$$12)\qquad \begin{aligned} u &= A-\lambda-q = A - EF/D - \lambda = a-\lambda,\\ v &= B-\lambda-r = B - FD/E - \lambda = b-\lambda,\\ w &= C-\lambda-s = C - DE/F - \lambda = c-\lambda, \end{aligned}$$

if we write

$$A - EF/D = a, \quad B - FD/E = b, \quad C - DE/F = c.$$

Also since from 10) q, r, s are all of the same sign, let us call them $\pm\, l^2$, m^2, n^2, so that we have from 11)

$$13)\qquad \begin{aligned} f(\lambda) \equiv (a-\lambda)(b-\lambda)(c-\lambda) \pm [l^2(b-\lambda)(c-\lambda)\\ + m^2(c-\lambda)(a-\lambda) + n^2(a-\lambda)(b-\lambda)]. \end{aligned}$$

Substituting for λ in turn the values $-\infty$, c, b, a, $+\infty$, we obtain

$$14)\qquad \begin{aligned} f(-\infty) &= \infty,\\ f(c) &= \pm\, n^2(a-c)(b-c)\\ f(b) &= \pm\, m^2(c-b)(a-b)\\ f(a) &= \pm\, l^2(b-a)(c-a)\\ f(\infty) &= -\infty. \end{aligned}$$

Let us suppose $a > b > c$, and take the upper sign in 14). Then for

$\lambda = -\infty$	$f(\lambda)$ is $+$
c	$+$
b	$-$
a	$+$
∞	$-$

and the function $f(\lambda)$ behaves as shown in Fig. 13. As there are three changes of sign, there are three real roots. It is to be noticed that the reality of all the roots depends on q, r, s being of the same sign. Let us call the roots λ_1, λ_2, λ_3. Either one of these being inserted in the equations 5), the equations become compatible, and suffice to determine the ratios of the direction cosines. There are therefore always three principal axes to a central quadric surface. If we call the cosines belonging to the roots λ_1, α_1, β_1, γ_1, those belonging λ_2, α_2, β_2, γ_2, equations 5) become

$$15)\qquad \begin{aligned} \lambda_1\alpha_1 &= A\alpha_1 + F\beta_1 + E\gamma_1, \\ \lambda_1\beta_1 &= F\alpha_1 + B\beta_1 + D\gamma_1, \\ \lambda_1\gamma_1 &= E\alpha_1 + D\beta_1 + C\gamma_1, \\ \lambda_2\alpha_2 &= A\alpha_2 + F\beta_2 + E\gamma_2, \\ \lambda_2\beta_2 &= F\alpha_2 + B\beta_2 + D\gamma_2, \\ \lambda_2\gamma_2 &= E\alpha_2 + D\beta_2 + C\gamma_2, \end{aligned}$$

Multiplying the first three respectively by α_2, β_2, γ_2 and adding

$$16)\qquad \begin{aligned} \lambda_1(\alpha_1\alpha_2 + \beta_1\beta_2 + \gamma_1\gamma_2) &= A\alpha_1\alpha_2 + B\beta_1\beta_2 + C\gamma_1\gamma_2 \\ &+ D(\beta_2\gamma_1 + \beta_1\gamma_2) + E(\gamma_2\alpha_1 + \gamma_1\alpha_2) + F(\alpha_2\beta_1 + \alpha_1\beta_2). \end{aligned}$$

If we multiply the second three equations respectively by α_1, β_1, γ_1 and add we obtain for

$$\lambda_2(\alpha_1\alpha_2 + \beta_1\beta_2 + \gamma_1\gamma_2)$$

the same expression. Accordingly we have

$$17)\qquad (\lambda_1 - \lambda_2)(\alpha_1\alpha_2 + \beta_1\beta_2 + \gamma_1\gamma_2) = 0.$$

so that if the roots λ_1, λ_2 are unequal the corresponding axes are perpendicular. In like manner if the determinantal cubic has three unequal roots, the quadric has three mutually perpendicular principal axes.

If two roots are equal the position of the corresponding axes becomes indeterminate, and it may be shown that all radii perpendicular to the direction given by the third root are principal axes of the same length. The surface is then one of revolution about the determinate axis. If all three roots are equal, the surface is a sphere, and any axis is a principal axis.

We will now transform the equation of the quadric 1) to a new set of axes coinciding in direction with its principal axes. Let the new coordinates be x', y', z', and let the direction cosines of the angles made by the new with the old axes be given in the table below.

	x	y	z
x'	α_1	β_1	γ_1
y'	α_2	β_2	γ_2
z'	α_3	β_3	γ_3

The equations of transformation of coordinates are then

18)
$$\begin{aligned} x' &= \alpha_1 x + \beta_1 y + \gamma_1 z, \\ y' &= \alpha_2 x + \beta_2 y + \gamma_2 z, \\ z' &= \alpha_3 x + \beta_3 y + \gamma_3 z, \end{aligned}$$

19)
$$\begin{aligned} x &= \alpha_1 x' + \alpha_2 y' + \alpha_3 z', \\ y &= \beta_1 x' + \beta_2 y' + \beta_3 z', \\ z &= \gamma_1 x' + \gamma_2 y' + \gamma_3 z'. \end{aligned}$$

Now using equations 19), we obtain

$$\begin{aligned} Ax + Fy + Ez = x'(A\alpha_1 + F\beta_1 + E\gamma_1) \\ + y'(A\alpha_2 + F\beta_2 + E\gamma_2) \\ + z'(A\alpha_3 + F\beta_3 + E\gamma_3), \end{aligned}$$

which in virtue of equations 15) is equal to

$$\lambda_1 \alpha_1 x' + \lambda_2 \alpha_2 y' + \lambda_3 \alpha_3 z'.$$

In like manner

$$\begin{aligned} Fx + By + Dz &= \lambda_1 \beta_1 x' + \lambda_2 \beta_2 y' + \lambda_3 \beta_3 z', \\ Ex + Dy + Cz &= \lambda_1 \gamma_1 x' + \lambda_2 \gamma_2 y' + \lambda_3 \gamma_3 z'. \end{aligned}$$

Multiplying respectively by x, y, z and adding, we obtain

$$\begin{aligned} &Ax^2 + By^2 + Cz^2 + 2Dyz + 2Ezx + 2Fxy \\ &= \lambda_1 x'(\alpha_1 x + \beta_1 y + \gamma_1 z) + \lambda_2 y'(\alpha_2 x + \beta_2 y + \gamma_2 z) + \lambda_3 z'(\alpha_3 x + \beta_3 y + \gamma_3 z) \\ &= \lambda_1 x'^2 + \lambda_2 y'^2 + \lambda_3 z'^2. \end{aligned}$$

Consequently the equation of the quadric referred to its principal axes is

20) $$\lambda_1 x'^2 + \lambda_2 y'^2 + \lambda_3 z'^2 = 1,$$

and the three roots of the cubic are equal to the squares of the reciprocals of the lengths of the semi-axes. Accordingly in order to find the equation referred to the axes it is not necessary to solve the linear equations 5), but only to solve the cubic 6).

NOTE V.

TRANSFORMATION OF QUADRATIC FORMS.

The last two notes have dealt with quadratic forms, and in Note IV we have by a linear transformation of the variables 19) transformed the form F into a form 20) in which no product terms appear, and we find that the coefficients of the squares are the roots of the determinant 6). In this note we shall consider similar transformations of forms of any number of variables, and shall incidentally obtain a proof of the reality of the roots of Lagrange's determinant, 65), page 159, for the case of no dissipation.

We shall require a number of elementary properties of both linear and quadratic forms, which we shall now set forth. Suppose we have n linear forms

$$1)\qquad \begin{aligned} u_1 &= a_{11}x_1 + a_{12}x_2 + \cdots + a_{1n}x_n, \\ u_2 &= a_{21}x_1 + a_{22}x_2 + \cdots + a_{2n}x_n, \\ &\cdots\cdots\cdots\cdots\cdots\cdots \\ &\cdots\cdots\cdots\cdots\cdots\cdots \\ u_n &= a_{n1}x_1 + a_{n2}x_2 + \cdots + a_{nn}x_n, \end{aligned}$$

and let us call R the determinant

$$2)\qquad R = \begin{vmatrix} a_{11}, & a_{12}, & \ldots & a_{1n} \\ a_{21}, & a_{22}, & \ldots & a_{2n} \\ \cdots & \cdots & \cdots & \cdots \\ a_{n1}, & a_{n2}, & \ldots & a_{nn} \end{vmatrix}.$$

If we multiply the kth column of R by x_k, and then add to this column the first, second, etc., multiplied respectively by $x_1, x_2, \ldots,$ we obtains.

$$3)\qquad Rx_k = \begin{vmatrix} a_{11}, & \ldots & a_{1,k-1}, & u_1, & a_{1,k+1}, & \ldots & a_{1n} \\ a_{21}, & \ldots & a_{2,k-1}, & u_2, & a_{2,k+1}, & \ldots & a_{2n} \\ \cdots & \cdots & \cdots & \cdots & \cdots & \cdots & \cdots \\ a_{n1}, & \ldots & a_{n,k-1}, & u_n, & a_{n,k+1}, & \ldots & a_{nn} \end{vmatrix}.$$

If now the determinant R is zero, the determinant on the right vanishes, expanding which we obtain

$$4)\qquad c_1u_1 + c_2u_2 + \cdots + c_nu_n = 0,$$

where the c's are the minors of the elements of the kth column of R. Thus if the determinant of the forms vanishes, the forms are not independent, but satisfy indentically the linear relation 4).

Consider now the quadratic form

$$5)\qquad f(x_1, x_2 \ldots x_n) = \sum_{r=1}^{r=n}\sum_{s=1}^{s=n} a_{rs}x_rx_s,$$

for which $a_{rs} = a_{sr}$, and

$$6) \qquad u_r = \frac{1}{2}\frac{\partial f}{\partial x_r} = \frac{1}{2} f_r' = \sum_{s=1}^{s=n} a_{rs} x_s,$$

$$7) \qquad f = \frac{1}{2}\sum_{r=1}^{r=n} f_r' x_r = \sum_{r=1}^{r=n} u_r x_r.$$

If $y_1, y_2, \ldots y_n$ are another set of variables, and we put for each x_r the value $x_r + \lambda y_r$ we have

$$f(x_1 + \lambda y_1, \ldots x_n + \lambda y_n) = \sum_{r=1}^{r=n}\sum_{s=1}^{s=n} a_{rs}(x_r + \lambda y_r)(x_s + \lambda y_s)$$

$$8) \qquad = \sum_{r=1}^{r=n}\sum_{s=1}^{s=n} a_{rs} x_r x_s + 2\lambda \sum_{r=1}^{r=n}\sum_{s=1}^{s=n} a_{rs} x_r y_s + \lambda^2 \sum_{r=1}^{r=n}\sum_{s=1}^{s=n} a_{rs} y_r y_s$$

$$= f(x_1, \ldots x_n) + \lambda \sum_{r=1}^{r=n} f_r'(x) y_r + \lambda^2 f(y_1, \ldots y_n).$$

If now R, the *determinant of the form* f vanishes, we have a relation

$$9) \qquad c_1 u_1 + c_2 u_2 + \cdots + c_n u_n \equiv \frac{1}{2}\sum_{r=1}^{r=n} c_r f_r'(x_1, \ldots x_n) = 0$$

for *all* values of $x_1, \ldots x_n$.

Let us now put for the y's of equation 8) the values of c of equation 9). We then have by 7) and 9)

$$f(c_1, \ldots c_n) = \frac{1}{2}\sum_{r=1}^{r=n} c_r f_r'(c_1, \ldots c_n) = 0,$$

so that becomes

$$10) \qquad \begin{aligned} f(x_1 + \lambda c_1, \ldots x_n + \lambda c_n) &= f(x_1, \ldots x_n) \\ &+ \lambda \sum_{r=1}^{r=n} c_r f_r'(x_1, \ldots x_n) = f(x_1, \ldots x_n). \end{aligned}$$

We thus find that in this case f is independent of λ and of $y_1, \ldots y_n$. Accordingly if c_n is not zero, let us put $\lambda = -\frac{x_n}{c_n}$, so that we obtain for all values of $x_1, \ldots x_n$,

$$11) \quad f(x_1, \ldots x_n) = f(x_1 - \frac{c_1}{c_n} x_n,\ x_2 - \frac{c_2}{c_n} x_n,\ \ldots\ x_{n-1} - \frac{c_{n-1}}{c_n} x_n,\ 0).$$

Thus we obtain the theorem: every quadratic form in n variables whose determinant vanishes, may be expressed as a quadratic form of *less* than n other variables y, which are linear combinations of the original variables. Such forms are called *singular*, as opposed to *ordinary* forms, whose determinant does not vanish. As an example the form

$$x_1^2 + x_2^2 - x_3^2 + 2x_1x_2,$$

whose determinant

$$\begin{vmatrix} 1 & 1 & 0 \\ 1 & 1 & 0 \\ 0 & 0 & -1 \end{vmatrix}$$

vanishes, may be written

$$(x_1 + x_2 + x_3)(x_1 + x_2 - x_3).$$

If f is an ordinary form and a_{rr} is not zero, we may write

$$f = a_{rr}x_r^2 + 2x_rp + q, \tag{12}$$

where p is a linear form containing all the other variables except x_r and q is a quadratic form in the same variables. Completing the square we may then write

$$a_{rr}f = (a_{rr}x_r + p)^2 + a_{rr}q - p^2. \tag{13}$$

If on the other hand every coefficient of a square a_{rr} is zero, we may write

$$f = 2a_{rs}x_rx_s + 2x_rp + 2x_sq + r, \tag{14}$$

where p, q are linear, r a quadratic form not containing either x_r or x_s, we may then write

$$\begin{aligned} 2a_{rs}f &= 4(a_{rs}x_r + q)(a_{rs}x_s + p) + 2a_{rs}r - 4pq \\ &= \{a_{rs}(x_r + x_s) + p + q\}^2 - \{a_{rs}(x_r - x_s) - p + q\}^2 + 2a_{rs}r - 4pq. \end{aligned} \tag{15}$$

In the former case we have exhibited f as the sum of a square and a form in $n-1$ variables, in the latter as two terms in squares and a form in $n-2$ variables. In either case the remaining form may be treated in a similar manner, and so on, so that the form is eventually reduced to a sum of terms in squares. If the coefficients A of all the squares in

$$f = A_1y_1^2 + A_2y_2^2 \cdots + A_ny_n^2, \tag{16}$$

when the y's are linear forms in $x_1, \ldots x_n$, have the same sign, the form is said to be *definite*, for it can not change sign however the values of the variables be altered. If the coefficients A are not all of the same sign the form is indefinite.

If we transform the linear forms 1) by means of the linear substitution

$$\begin{aligned} x_1 &= \beta_{11}y_1 + \beta_{12}y_2 + \cdots + \beta_{1n}y_n, \\ x_2 &= \beta_{21}y_1 + \beta_{22}y_2 + \cdots + \beta_{2n}y_n, \\ &\cdots\cdots\cdots\cdots\cdots\cdots \\ x_n &= \beta_{n1}y_1 + \beta_{n2}y_2 + \cdots + \beta_{nn}y_n, \end{aligned} \tag{17}$$

we obtain linear forms in the new variables $y_1, \ldots y_n$, so that if we write

$$18)\qquad \begin{aligned} u_1 &= c_{11}y_1 + c_{12}y_2 + \cdots + c_{1n}y_n, \\ u_2 &= c_{21}y_1 + c_{22}y_2 + \cdots + c_{2n}y_n, \\ &\ldots\ldots\ldots\ldots\ldots\ldots \\ u_n &= c_{n1}y_1 + c_{n2}y_2 + \cdots + c_{nn}y_n, \end{aligned}$$

we find by carrying out the transformation,

$$19)\qquad c_{rs} = \sum_{t=1}^{t=n} a_{rt}\beta_{ts}.$$

But this is, according to the rule for multiplication of two determinants the condition that the determinant of the forms u in y,

$$20)\qquad \begin{vmatrix} c_{11}, \ldots c_{1n} \\ \ldots\ldots \\ \ldots\ldots \\ c_{n1}, \ldots c_{nn} \end{vmatrix} = \begin{vmatrix} a_{11}, \ldots a_{1n} \\ \ldots\ldots \\ \ldots\ldots \\ a_{n1}, \ldots a_{nn} \end{vmatrix} \begin{vmatrix} \beta_{11}, \ldots \beta_{1n} \\ \ldots\ldots \\ \ldots\ldots \\ \beta_{n1}, \ldots \beta_{nn} \end{vmatrix}$$

is the product of the determinant of the forms in x by the determinant of the substitution 17).

The determinant in which the element in the rth row and s^{th} column is a derivative $\dfrac{\partial u_r}{\partial x_s}$ is called the Jacobian of the functions $u_1, \ldots u_n$ with respect to the variables $x_1, \ldots x_n$, and is often denoted by

$$\frac{\partial(u_1, \ldots u_n)}{\partial(x_1, \ldots x_n)}.$$

In this notation 20) becomes

$$21)\qquad \frac{\partial(u_1, \ldots u_n)}{\partial(y_1, \ldots y_n)} = \frac{\partial(u_1, \ldots u_n)}{\partial(x_1, \ldots x_n)} \begin{vmatrix} \beta_{11}, \ldots \beta_{1n} \\ \ldots\ldots \\ \beta_{n1}, \ldots \beta_{nn} \end{vmatrix}.$$

If the functions $u_1, \ldots u_n$ in a Jacobian are the partial derivatives of the same function f, $u_r = \dfrac{\partial f}{\partial x_r}$, so that the element in the rth row and s^{th} column is $\dfrac{\partial^2 f}{\partial x_r \partial x_s}$ the determinant is called the Hessian of the function. Thus the determinant R of 2) is the Hessian of f, and will be denoted by $H_x f$.

If now we transform the quadratic form 5) by the substitution 17), so that

$$22)\qquad f = \sum_{r=1}^{r=n}\sum_{s=1}^{s=n} a_{rs}x_r x_s = \sum_{r=1}^{r=n}\sum_{s=1}^{s=n} A_{rs}y_r y_s,$$

we may find a relation between the Hessians of f with respect to the x's and that with respect to the y's. Using the notation for Jacobians, by 21),

$$23)\quad H_y f = \frac{\partial\left(\frac{\partial f}{\partial y_1}, \dots \frac{\partial f}{\partial y_n}\right)}{\partial(y_1, \dots y_n)} = \frac{\partial\left(\frac{\partial f}{\partial y_1}, \dots \frac{\partial f}{\partial y_n}\right)}{\partial(x_1, \dots x_n)} \begin{vmatrix} \beta_{11}, \dots \beta_{1n} \\ \cdots\cdots \\ \cdots\cdots \\ \beta_{n1}, \dots \beta_{nn} \end{vmatrix}.$$

But in every derivative

$$\frac{\partial^2 f}{\partial x_r \partial y_s} = \frac{\partial^2 f}{\partial y_s \partial x_r} = \frac{\partial u_r}{\partial y_s}.$$

Consequently the Jacobian on the right of 23) is the same as that on the *left* of 21).

Thus we find

$$24)\qquad H_y f = \frac{\partial(u_1 \dots u_n)}{\partial(x_1 \dots x_n)} \begin{vmatrix} \beta_{11} \dots \beta_{1n} \\ \cdots\cdots \\ \beta_{n1} \dots \beta_{nn} \end{vmatrix}^2 = H_x f \begin{vmatrix} \beta_{11} \dots \beta_{1n} \\ \cdots\cdots \\ \beta_{n1} \dots \beta_{nn} \end{vmatrix}^2$$

or the determinant of the transformed form is equal to that of the original form multiplied by the square of the determinant of transformation.

If an ordinary form f vanishes for a particular set of values

$$x_1 = c_1, \quad x_2 = c_2 \ \dots \ x_n = c_n,$$

where the c's are not all zero, we can show that the form is indefinite. For if we substitute for the x's excepting x_n, the values

$$25)\quad x_1 = c_1 \frac{x_n}{c_n} + y_1,\ x_2 = c_2 \frac{x_n}{c_n} + y_2, \ \dots\ x_{n-1} = c_{n-1} \frac{x_n}{c_n} + y_{n-1},$$

we have

$$f = \sum_{r=1}^{r=n} \sum_{s=1}^{s=n} a_{rs} x_r x_s = \sum_{r=1}^{r=n-1} \sum_{s=1}^{s=n-1} a_{rs}\left(c_r \frac{x_n}{c_n} + y_r\right)\left(c_s \frac{x_r}{c_n} + y_s\right)$$

$$26)\quad + x_n \sum_{s=1}^{s=n-1} a_{sn}\left(c_s \frac{x_n}{c_n} + y_s\right) + a_{nn} x_n^2$$

$$= \frac{x_n^2}{c_n^2} \sum_{r=1}^{r=n} \sum_{s=1}^{s=n} a_{rs} c_r c_s + \frac{2x_n}{c_n} \sum_{r=1}^{r=n} \sum_{s=1}^{s=n-1} a_{rs} c_r y_s + \sum_{r=1}^{r=n-1} \sum_{s=1}^{s=n-1} a_{rs} y_r y_s.$$

The first term, containing $f(c_1, \dots c_n)$ as a factor, vanishes by hypothesis. The sums in the other terms contain only $n - 1$ variables. If then there are any values of $y_1, \dots y_{n-1}$, for which $\sum_{r=1}^{r=n} \sum_{s=1}^{s=n-1} a_{rs} c_r y_s$ does not vanish, since x_n is independent of these variables, we may by suitably choosing the value of x_n make the form have either sign, it is therefore indefinite.

(If the sum $\sum_{r=1}^{r=n}\sum_{s=1}^{s=n-1} a_{rs}c_r y_s$ is zero for all values of $y_1, \dots y_{n-1}$, we must have

$$\sum_{r=1}^{r=n} a_{rs}c_r = 0$$

for $s = 1, 2, , \dots n-1$, but since $f(c_1, \dots c_n)$ is zero we must have also $\sum_{r=1}^{r=n} a_{rn}c_r$, and these n equations require the determinant of the form to vanish, and the form is singular.)

As a result of this theorem we see that if a form is to be definite, no coefficient a_{rr} of a square x_r^2 must be absent, and all must have the same sign. For if $a_{rr} = 0$, putting all the variables equal to zero except x_r would make the form vanish, and if a_{rr} is not zero, the same assumption would make the form have the sign of a_{rr}. Consequently all these coefficients must be of the same sign.

Let us now consiedr two ordinary quadratic forms of the same variables, with real coefficients[1])

$$27)\qquad \varphi = \sum_{r=1}^{r=n}\sum_{s=1}^{s=n} a_{rs}x_r x_s, \quad \psi = \sum_{r=1}^{r=n}\sum_{s=1}^{s=n} c_{rn}x_r x_s,$$

from which with an arbitrary multiplier λ we construct the from

$$28)\qquad \lambda\varphi + \psi.$$

As we give λ an infinite set of real values, we obtain an infinite *sheaf* of forms. Let us examine wheter they are definite or not.

The determinant of the form $\lambda\varphi + \psi$

$$28)\qquad \begin{vmatrix} \lambda a_{11} + c_{11}, & \lambda a_{12} + c_{12}, & \dots \lambda a_{1n} + e_{1n} \\ \lambda a_{21} + c_{21}, & \lambda a_{22} + c_{22}, & \dots \lambda a_{2n} + c_{2n} \\ \dots & \dots & \dots \\ \dots & \dots & \dots \\ \lambda a_{n1} + c_{n1}, & \lambda a_{n2} + c_{n2}, & \dots \lambda a_{nn} + c_{nn} \end{vmatrix} = f(\lambda)$$

is identical with Lagrange's determinant, page 159, when the $\varkappa$'s are zero. (We here have written λ for the λ^2 on p. 159) We shall now prove that if the equation $f(\lambda) = 0$ has a complex root, all the forms of the sheaf $\lambda\varphi + \psi$ are indefinite.

Let $\lambda = \alpha + i\beta$ be a complex root of the determinantal equation $f(\lambda) = 0$. Then since the form $(\alpha + i\beta)\varphi + \psi$ is singular, it may be represented as a sum of less than n squares, and since it is complex, these may be squares of complex variables, so that we have

$$29)\qquad (\alpha + i\beta)\varphi + \psi = (y_1 + iz_1)^2 + (y_2 + iz_2)^2 + \cdots + (y_{n-1} + iz_{n-1})^2,$$

1) Kronecker, *Über Schaaren quadratischer Formen.* Monatsber. der Königl. Preuß. Akad. d. Wiss. zu Berlin, 1868. pp. 339—346. Werke, Bd. I, p. 165.

the y's and z's being real linear forms in the x's, any of which, but not all, may be zero. Separating the real and the imaginary terms, we obtain

30) $$\alpha\varphi + \psi = \sum_{r=1}^{r=n-1}(y_r^2 - z_r^2), \quad \beta\varphi = 2\sum_{r=1}^{r=n-1} y_r z_r.$$

From the two forms

31) $$\sum_{r=1}^{r=n-1}(y_r^2 - z_r^2), \quad \sum_{r=1}^{r=n-1} y_r z_r$$

the whole sheaf may be obtained. Solving 30) for φ and ψ,

32) $$\lambda\varphi + \psi = \sum_{r=1}^{r=n-1}(y_r^2 - z_r^2) + \frac{2(\lambda - \alpha)}{\beta}\sum_{r=1}^{r=n-1} y_r z_r,$$

which we may write

33) $$\lambda\varphi + \psi = \sum_{r=1}^{r=n-1}(y_r - \mu z_r)\left(y_r + \frac{1}{\mu} z_r\right),$$

where

$$\frac{1}{\mu} - \mu = \frac{2(\lambda - \alpha)}{\beta};$$

a quadratic in μ, giving for every real value of λ a real value of μ. Now each of the forms 33) vanishes for values of $x_1, \ldots x_n$ other than zero, which satisfy the $n - 1$ linear equations

34) $$0 = y_1 - \mu z_1 = y_2 - \mu z_2 = \cdots = y_{n-1} - \mu z_{n-1},$$

and is accordinlgy indefinite. Conversely if there is in the sheaf a single definite form, the roots of $f(\lambda) = 0$ are all real. Now in the mechanical application, the form φ, which is proportional to that given by the value $\lambda = \infty$, is the kinetic energy, a definite positive form, consequently the reality of the roots is proved.

If $\lambda_\varkappa$ is one of the real roots of the equation $f(\lambda) = 0$ the form $\lambda_\varkappa\varphi + \psi$ being singular, can be expressed in terms of less than n linear functions of $x_1, \ldots x_n$, say $y_1, \ldots y_{n-1}$. Let y_n be any other linear function of the x's, such that the determinant of the functions $y_1, \ldots y_n$ is not zero, then we can express the function φ in terms of these n variables y, and if it is definite and positive, the coefficient of every square will have a positive sign, accordingly, as in 12), 13), we may separate off from φ a square $z_\varkappa^2$, where z contains y_n, and accordingly we have

35) $$\begin{aligned}\lambda\varphi + \psi &= (\lambda - \lambda_\varkappa)\varphi + \lambda_\varkappa\varphi + \psi \\ &= (\lambda - \lambda_\varkappa) z_\varkappa^2 + \lambda\varphi' + \psi',\end{aligned}$$

where $\lambda\varphi' + \psi'$ contains only the $n - 1$ variables $y_1, \ldots y_{n-1}$. Now a definite form in n variables remains a definite form in $n - 1$ variables if we put any variable equal to zero, consequently φ' is, like φ, a definite positive form.

A linear divisor such as $\lambda - \lambda_r$ of the determinant of the form $\lambda\varphi' + \psi'$ is also a divisor of the determinant of $\lambda\varphi + \psi$, for on writing out the determinant of the form 35) in terms of $y_1, \ldots y_{n-1} z_\varkappa$, we find

$$36)\quad H_{z_\varkappa, y_1, \ldots y_{n-1}}(\lambda\varphi + \psi) = (\lambda - \lambda_\varkappa) H_{y_1, \ldots y_{n-1}}(\lambda\varphi' + \psi'),$$

so that the vanishing of the determinant of order $n - 1$ on the right makes the determinant on the left vanish. But this equal to the determinant of $\lambda\varphi + \psi$ in the variables $x_1, \ldots x_n$ multiplied by a constant

We may now treat the form $\lambda\varphi' + \psi'$ in the same manner, and so on, so that finally we obtain

$$37)\quad \lambda\varphi + \psi = (\lambda - \lambda_1)z_1^2 + (\lambda - \lambda_2)z_2^2 + \cdots + (\lambda - \lambda_n)z_n^2,$$

where $\lambda_1, \ldots \lambda_n$ are the roots of the determinantal equation $f(\lambda) = 0$. Since this is true for all values of λ we have

$$38)\quad \begin{aligned} \varphi &= z_1^2 + z_2^2 + \cdots + z_n^2, \\ -\psi &= \lambda_1 z_1^2 + \lambda_2 z_2^2 + \cdots + \lambda_n z_n^2, \end{aligned}$$

which is the simultaneous transformation of two quadratic forms required in the treatment of principal coordinates. It is obvious according to this method that it makes no difference whether the determinant has equal roots or not.

INDEX.

(The numbers refer to pages.)

ERRATUM.

„205, line 7 from bottom, for AB and PQ read AP and BQ.

Printed by B. G. Teubner, Dresden.

Abraham, Dr. **M.**, Professor am R. Instituto Tecnico Superiore, Mailand, Theorie der Elektrizität. In 2 Bänden. gr. 8. In Leinw. geb.

I. Band. Einführung in die Maxwellsche Theorie der Elektrizität. Mit einem einleitenden Abschnitte über das Rechnen mit Vektorgrößen in der Physik. Von A. Föppl. 3., völlig umgearbeitete Auflage von M. Abraham. Mit 11 Figuren. [XVIII u. 460 S.] 1907. ℳ 12.—

II. — Elektromagnetische Theorie der Strahlung. Von M. Abraham. 2. Auflage. Mit 6 Figuren. [XII u. 404 S.] 1908. ℳ 10.—

Brion, Dr. **G.**, Privatdozent an der Technischen Hochschule zu Dresden, Leitfaden zum elektrotechnischen Praktikum. Mit 380 Figuren. [XIV u. 404 S.] gr. 8. 1910. Geh. ℳ 10.—, in Leinw. geb. ℳ 11.—

Ferraris, **G.**, weil. Professor an der Universität Turin, wissenschaftliche Grundlagen der Elektrotechnik. Nach den Vorlesungen über Elektrotechnik, deutsch herausgegeben von Dr. L. Finzi, Professor an der Technischen Hochschule zu Aachen. Mit 161 Figuren. [XII u 358 S.] gr. 8. 1901. In Leinw. geb. ℳ 12.—. 2. Auflage. 1912. [Unter der Presse.]

Fleming, Dr. **J. A.**, Professor am University College zu London, elektrische Wellen-Telegraphie. 4 Vorlesungen. Autorisierte deutsche Ausgabe von weil. Professor Dr. E. Aschkinaß. Mit 53 Abbildungen. [IV u. 185 S.] gr. 8. 1906. In Leinw. geb. ℳ 5.—

Föppl, Dr. **A.**, Professor an der Kgl. Technischen Hochschule zu München, Vorlesungen über technische Mechanik. 6 Bände. gr. 8. In Leinw. geb.

I. Band. Einführung in die Mechanik. 4. Auflage. Mit 104 Figuren. [XV u. 424 S.] 1911. ℳ 10.—
II. — Graphische Statik. 3. Aufl. Mit 209 Figuren. [XII u. 419 S.] 1912. ℳ 8.—
III. — Festigkeitslehre. 4. Aufl. Mit 86 Figuren. [XVI u. 426 S.] 1909. ℳ 10.—
IV. — Dynamik. 3., stark veränd. Aufl. Mit 71 Figuren. [VIII u. 422 S.] 1909. ℳ 10.—
V. — Die wichtigsten Lehren der höheren Elastizitätstheorie. Mit 44 Figuren. [XII u. 391 S.] 1907. ℳ 10.—
VI. — Die wichtigsten Lehren der höheren Dynamik. Mit 30 Fig. [XII u 490 S.] 1910. ℳ 12.—

Heun, Dr. **K.**, Professor an der Technischen Hochschule zu Karlsruhe, die kinetischen Probleme der wissenschaftlichen Technik. Mit 18 Figuren. [VI u. 123 S.] gr. 8. 1900. Geh. ℳ 4.—

Janet, **P.**, Professor an der École superieure d'Electricité und Direktor des Laboratoire central zu Paris, Vorlesungen über allgemeine Elektrotechnik. Deutsche Ausgabe nach der 2. franz. Aufl. von Fr. Süchting, Direktor des Elektrizitätwerkes zu Bremen. 3 Bände (Band II und III bearbeitet von E. Rieke.) gr. 8. Geb. [Band I unter der Presse.]

Lanchester, **F. W.**, Aërodynamik. Ein Gesamtwerk über das Fliegen. 2 Bände. Aus dem Englischen übersetzt von C. und A. Runge. In Leinw. geb.

I. Band: Mit Anhängen über die Geschwindigkeit und den Impuls von Schallwellen, über die Theorie des Segelfluges usw. Mit 162 Figuren und 1 Tafel. [XIV u. 360 S.] gr. 8. 1909. ℳ 12.—

II. — Aërodonetik. Mit Anhängen über die Theorie und Anwendung des Gyroskops, über den Flug der Geschosse usw. Mit 208 Figuren und 1 Titelbild. [XVI u. 327 S.] gr. 8. 1911. ℳ 12.—

Lorenz, **H.**, Professor an der Technischen Hochschule zu Danzig, Dynamik der Kurbelgetriebe mit besonderer Berücksichtigung der Schiffsmaschinen. Mit 66 Figuren. [V u. 156 S.] gr. 8. 1901. Geh. ℳ 5.—

Meyer, **G. W.**, Elektroingenieur, Maschinen und Apparate der Starkstromtechnik, ihre Wirkungsweise und Konstruktion. Ein Lehrbuch für den Gebrauch an technischen Lehranstalten, zum Selbststudium und für den in der Praxis stehenden Ingenieur. I. Teil: Gleichstrom. II. Teil: Wechselstrom. gr. 8. ca. 480 S. [Unter der Presse.]

v. Mises, Dr. **R.** in Straßburg, Theorie der Wasserräder. Mit 24 Figuren. [120 S.] gr. 8. 1909. Geh. ℳ 3.60.

Müller, **S.**, Professor an der Technischen Hochschule zu Charlottenburg, technische Hochschulen in Nordamerika. Mit zahlreichen Abbildungen, 1 Karte und 1 Lageplan. [VI u. 103 S.] gr. 8. 1907. Geh. ℳ 1.—, in Leinw. geb. ℳ 1.25.

Ostenfeld, Dr. **A.**, Professor an der Technischen Hochschule zu Kopenhagen, technische Statik. Vorlesungen über die Theorie der Tragkonstruktionen. Deutsch von D. Skouge. Mit 336 Figuren auf 33 Tafeln. [VIII u. 456 S.] gr. 8. 1904. In Leinw. geb. ℳ 12.—

Perry, Dr. **J.**, F. R. S., Professor am Royal College of Science in London, höhere Analysis für Ingenieure. Autorisierte deutsche Bearbeitung von Dr. R. Fricke, Professor an der Technischen Hochschule zu Braunschweig, und F. Süchting, Direktor des Elektrizitätswerkes in Bremen. 2. verbesserte und vermehrte Aufl. Mit 106 Figuren. [XI u. 464 S.] gr. 8 1910. In Leinwand geb. ℳ 13.—

——— die Dampfmaschine, einschließlich der Dampfturbine und Gas- und Ölmaschinen. Autor., erweiterte deutsche Bearbeitung von Dr.-Ing. H. Meuth. Mit 350 Figuren und 1 Wärmetafel. [XII u. 708 S.] gr. 8. 1909. In Leinw. geb. ℳ 22.—

——— angewandte Mechanik. Ein Lehrbuch für Studierende, die Versuche anstellen und numerische und graphische Beispiele durcharbeiten wollen. Deutsche Ausgabe von R. Schick. Mit 371 Figuren [VIII u. 666 S.] gr. 8. 1908. In Leinwand geb. ℳ 18.—

Petit-Bois, G., Bergingenieur in José bei Herve, Tafeln unbestimmter Integrale. [XII u. 154 S.] 4. 1906. Geh. ℳ 8.—

Rinkel, R., Ingenieur und Professor an der Handelshochschule zu Köln, Einführung in die Elektrotechnik, Physikalische Grundlagen und technische Ausführung. Mit 445 Abbildungen. [VI u. 464 S.] gr. 8. 1908. In Leinwand geb. ℳ 12.—

Schaefer, Dr. **C.**, Privatdozent an der Universität Breslau, Einführung in die Maxwellsche Theorie der Elektrizität und des Magnetismus. Mit Bildnis Maxwells und 32 Figuren. [VIII u. 174 S.] gr. 8. 1908. Kart. ℳ 3.40, geb. ℳ 3.80.

Schwaiger, Dr.-Ing. **A.**, Diplom-Ingenieur in Karlsruhe i. B., das Regulierproblem in der Elektrotechnik. Mit 28 Abbildungen. [VI u. 102 S.] gr. 8. 1909. Geh. ℳ 2.80, in Leinwand geb. ℳ 3.60.

Starke, Dr. **H.**, Professor an der Universität Greifswald, experimentelle Elektrizitätslehre, verbunden mit einer Einführung in die Maxwellsche und die Elektronentheorie der Elektrizität und des Lichts. 2., auf Grund der Fortschritte der Wissenschaft umgearbeitete und erweiterte Auflage. Mit 334 Abbildungen. [XVI u. 662 S.] gr. 8. 1910. In Leinw. geb. ℳ 12.—

Wagner, Dr. **K. W.**, Kaiserl. Telegraphen-Ingenieur in Berlin, elektro-magnetische Ausgleichsvorgänge in Freileitungen u. Kabeln. Mit 23 Figuren. [IV u. 109 S.] 8. 1908. Geh. ℳ 2.40, in Leinw. geb. ℳ 2.80.

Weber, Professor Dr. **R.**, in Neuchâtel (Schweiz), Beispiele und Übungen aus Elektrizität und Magnetismus. Nach dem Manuskript der 5. französischen Auflage. Mit 74 Figuren. [VIII u. 330 S.] 8. 1910. Geh. ℳ 4.80, in Leinw. geb. ℳ 5.25.

Weber, Dr. **H.**, und Dr. **J. Wellstein**, Professoren an der Universität Straßburg i. E., Encyklopädie der Elementar-Mathematik. Ein Handbuch für Lehrer und Studierende. In 3 Bänden. gr. 8. In Leinwand geb.

I. Band: Elementare Algebra und Analysis. Bearbeitet von H. Weber. 3. Auflage. Mit 40 Figuren. [XVIII u. 532 S.] 1910. ℳ 10.—

II — Elemente der Geometrie. Bearbeitet von H. Weber, J. Wellstein und W. Jacobsthal. 2. Auflage. Mit 251 Figuren. [XII u. 596 S.] 1907. ℳ 12.—

III — Angewandte Elementar-Mathematik. In 2 Teilen. 2. Auflage. I. Teil: Mathematische Physik. Mit einem Buche über Maxima und Minima von H. Weber und J. Wellstein. Bearbeitet von R. H. Weber, Professor in Rostock. Mit 254 Figuren. [XII u. 536 S.] 1910. ℳ 12.—. II. Teil. Angewandte Mathematik und Astronomie. [Unter der Presse.]